T0140338

Advances in Intelligent Systems and Computing

Volume 921

Series editor

Janusz Kacprzyk, Systems Research Institute, Polish Academy of Sciences, Warsaw, Poland

The series "Advances in Intelligent Systems and Computing" contains publications on theory, applications, and design methods of Intelligent Systems and Intelligent Computing. Virtually all disciplines such as engineering, natural sciences, computer and information science, ICT, economics, business, e-commerce, environment, healthcare, life science are covered. The list of topics spans all the areas of modern intelligent systems and computing such as: computational intelligence, soft computing including neural networks, fuzzy systems, evolutionary computing and the fusion of these paradigms, social intelligence, ambient intelligence, computational neuroscience, artificial life, virtual worlds and society, cognitive science and systems, Perception and Vision, DNA and immune based systems, self-organizing and adaptive systems, e-Learning and teaching, human-centered and human-centric computing, recommender systems, intelligent control, robotics and mechatronics including human-machine teaming, knowledge-based paradigms, learning paradigms, machine ethics, intelligent data analysis, knowledge management, intelligent agents, intelligent decision making and support, intelligent network security, trust management, interactive entertainment, Web intelligence and multimedia.

The publications within "Advances in Intelligent Systems and Computing" are primarily proceedings of important conferences, symposia and congresses. They cover significant recent developments in the field, both of a foundational and applicable character. An important characteristic feature of the series is the short publication time and world-wide distribution. This permits a rapid and broad dissemination of research results.

** Indexing: The books of this series are submitted to ISI Proceedings, EI-Compendex, DBLP, SCOPUS, Google Scholar and Springerlink **

Advisory Editors

Nikhil R. Pal, Indian Statistical Institute, Kolkata, India

Rafael Bello Perez, Faculty of Mathematics, Physics and Computing, Universidad Central de Las Villas, Santa Clara, Cuba

Emilio S. Corchado, University of Salamanca, Salamanca, Spain

Hani Hagras, Electronic Engineering, University of Essex, Colchester, UK

László T. Kóczy, Department of Automation, Széchenyi István University, Gyor, Hungary

Vladik Kreinovich, Department of Computer Science, University of Texas at El Paso, EL PASO, TX, USA

Chin-Teng Lin, Department of Electrical Engineering, National Chiao Tung University, Hsinchu, Taiwan

Jie Lu, Faculty of Engineering and Information Technology, University of Technology Sydney, Sydney, NSW, Australia

Patricia Melin, Graduate Program of Computer Science, Tijuana Institute of Technology, Tijuana, Mexico

Nadia Nedjah, Department of Electronics Engineering, University of Rio de Janeiro, Rio de Janeiro, Brazil

Ngoc Thanh Nguyen, Faculty of Computer Science and Management, Wrocław University of Technology, Wrocław, Poland

Jun Wang, Department of Mechanical and Automation Engineering, The Chinese University of Hong Kong, Shatin, Hong Kong

More information about this series at http://www.springer.com/series/11156

Aboul Ella Hassanien · Ahmad Taher Azar ·
Tarek Gaber · Roheet Bhatnagar ·
Mohamed F. Tolba
Editors

The International Conference on Advanced Machine Learning Technologies and Applications (AMLTA2019)

 Springer

Editors
Aboul Ella Hassanien
Faculty of Computers and Information
Cairo University
Giza, Egypt

Ahmad Taher Azar
Faculty of Computers and Information
Benha University
Benha, Egypt

Tarek Gaber
School of Computing, Science and
Engineering
University of Salford
Salford, Greater Manchester, UK

Faculty of Computers and Informatics
Suez Canal University
Ismailia, Egypt

Roheet Bhatnagar
Department of Computer Science
and Engineering
School of Computing and IT
Faculty of Engineering
Manipal University Jaipur
Jaipur, Rajasthan, India

Mohamed F. Tolba
Faculty of Computer and Information
Science
Ain Shams University
Cairo, Egypt

ISSN 2194-5357   ISSN 2194-5365   (electronic)
Advances in Intelligent Systems and Computing
ISBN 978-3-030-14117-2   ISBN 978-3-030-14118-9   (eBook)
https://doi.org/10.1007/978-3-030-14118-9

Library of Congress Control Number: 2019933214

This Springer imprint is published by the registered company Springer Nature Switzerland AG
The registered company address is: Gewerbestrasse 11, 6330 Cham, Switzerland

To the scientific Research Group in Egypt members and my wife Nazaha Hassan El Saman

Preface

This volume constitutes the refereed proceedings of the fourth International Conference on Advanced Machine Learning Technologies and Applications, AMLTA2019, will be held in Cairo, Egypt, in March 28–30, 2019. In response to the call for papers for AMLTA2019, 199 papers were submitted for presentation and inclusion in the proceedings of the conference. After a careful blind refereeing process, 93 papers were selected for inclusion in the conference proceedings. The papers were evaluated and ranked on the basis of their significance, novelty, and technical quality by at least two reviewers per paper. The papers cover current research in machine learning, data sciences, bioinformatics, complex network, renewable energy, swarm intelligence, biomedical engineering, complex control systems, cyber security, data mining, and deep learning. Also, five special sessions and one workshop will be organized: Special Session (1) Recent Trends in Data Science and Analytics by Prof Mrinal Kanti Ghose, and Prof Roheet Bhatnagar, India; Special Session (2) Advances and Applications of Modern Control Systems by Professor Ahmad Taher Azar, IEEE Senior Member; Special Session (3) Soft Computing Applications in Environmental and Earth Sciences by Prof. Richa N. K. Sharma and Prof. Shamama Anwar; Special Session (4) Machine Learning in Bioinformatics: Current Trends and Challenges by Dr. Mohamed Hamed and Dr. Mohamed Tahoon; Special Session (5) Image Processing: Classical and Modern Techniques by Dr. Mohamed Wagieh and Dr. Fayez Al Fayez. The workshop titled Machine Learning for Control and Power Energy will be organized by Professor Ashraf Darwish, Egypt.

In addition to these papers, the program will include two keynote talks: the first talk by Professor Swagatam Das (Indian Institute of Technology) titled Multi-modal and Uncertain Optimization with Differential Evolution - Some Recent Approaches and Future Challenges, and the second talk by Professor Ivan Zelinka (Department of Computer Science, Faculty of Electrical Engineering and Computer Science, VŠB-TUO, 17. listopadu 15, 708 33 Ostrava-Poruba, Czech Republic) titled Modern Algorithms in Control and Design of Complex Systems. In addition, one tutola will be given by Professor Ibrahim A. Hameed (Norwegian University of Science and Technology, Noway) titled Hands-on to TensorFlow. We express our sincere thanks to the plenary and tutorial speakers, workshop/special session chairs, and international program committee members for helping us to formulate a rich technical program. We would like to extend our sincere appreciation for the outstanding work contributed over many months by the Organizing Committee: Local Organization Chair, and Publicity Chair. We also wish to express our appreciation to the SRGE members for their assistance. We would like to emphasize that the success of AMLTA2019 would not have been possible without the support of many committed volunteers who generously contributed their time, expertise, and resources toward making the conference an unqualified success. Finally, thanks to Springer team for their support in all stages of the production of the proceedings. We hope that you will enjoy the conference program.

Aboul Ella Hassanien
Mohamed F. Tolba
Ahmad Taher Azar
Roheet Bhatnagar
Tarek Gaber

Organization

Honorary Chair

M. F. Tolba, Egypt

General Chair

Aboul Ella Hassanien

International Advisory Board

Nagwa Badr	Egypt
Howaida Shedeed	Egypt
Yudong Zhang	UK
Alok Kole	India
Ibrahim A. Hameed	NTNU, Ålesund, Norway
Siddhartha Bhattacharyya	India
Subarna Shakya	Nepal
Fatos Xhafa	Spain
Kazumi Nakamatsu	Japan
P. K. Mahanti	Canada
Xiaohui Yuan	University of North Texas, Denton, Texas, USA
Kumkum Garg	India
Ahmed Sharaf Eldin	Egypt
Thomas Loruenser	Austrian Institute of Technology, Austria
Feihu Xu	Cambridge, UK
Vaclav Snasel	Czech Republic
Janusz Kacprzyk	Poland
Tai-hoon Kim	Korea

| M. K. Ghose | India |
| Ahmed Abdel Rehiem | Egypt |

Program Chairs

Ahmad Taher Azar, Egypt
Roheet Bhatnagar, India
Tarek Gaber, UK
Mohamed Elhoseny, Egypt

Publicity Chairs

Manik Sharma, India
Kais Haddar, Tunisia
Nour Mahmoud, Egypt
Mohamed Hamed, Egypt
Dariusz Jacek Jak贸bczak, Poland

Technical Program Committee

Alok Kole	India
Hala Mousher	Egypt
Safwat Hamad	Egypt
Kelton Augusto Pontara da Costa	Brazil
Auzuir Ripardo de Alexandria	Brazil
Rodrigo C. Bortoletto	S. Paulo Federal Institute of Education, Brazil
Khalid Mohamed Hosny	Egypt
Sami Ghonam	Egypt
Sayed Hassan Ahmed	University of Central Florida, Orlando, USA
Mahmoud Zaher	Prince Sattam Bin Abdulaziz University, KSA
Amit Kumar Singh	National Institute of Technology Patna, India
Noura Metawa	University of New Orleans, USA
Sonali Vyas	AIIT, Amity University, Rajasthan, India
Hosny Abbas	Assiut University, Egypt
Mohamed Elsharkwaey	Suez Canal University, Egypt
Vinod Kumar Verma	Sant Longowal Institute of Engineering and Technology, India
Robert John Collins	Heriot-Watt University, UK
Ujjwal Sen	Harish-Chandra Research Institute, Allahabad, India

Hani M. K. Mahdi	Egypt
Vandana Bhattacharjee	India
Anandhavalli Gauthaman	Saudi Arabia
Karanjeet Singh Kahlon	India
Binod Singh	India
Moheet Bhatnagar	USA
Neeraj Bhargava	India
Mohammed Abdel-Megeed	Egypt
Abdelhameed Ibrahim	Egypt
Valentina Emilia Balas	Romania
Camelia Pintea	Romania
Marius M. Balas	Romania
Shikha Singh	India
Muaz A. Niazi	Pakistan
Jothi Ganesan	India
Kemal Polat	Turkey
Nizar Banu	India
Hannah Inbarani	India
Shifei Ding	China
A. V. Senthil Kumar	India
Anjali Awasthi	Canada
Rabie A. Ramadan	Saudi Arabia
Abdelkrim Haqiq	Morocco
Hajar Mousannif	Morocco
Pavel Kromer	Czech Republic
Jan Platos	Czech Republic
Ivan Zelinka	Czech Republic
Sebastian Tiscordio	Czech Republic
Natalia Spyropoulou	Hellenic Open University, Greece
Dimitris Sedaris	Hellenic Open University, Greece
Vassiliki Pliogou	Metropolitan College, Greece
Pilios Stavrou	Metropolitan College, Greece
Eleni Seralidou	University of Piraeus, Greece
Stelios Kavalaris	Metropolitan College, Greece
Litsa Charitaki	University of Athens, Greece
Elena Amaricai	University of Timisoara, Greece
Qing Tan	Athabasca University, Greece
Pascal Roubides	Broward College, Greece
Alaa Tharwat	Germany
Amira S. Ashour	KSA
Gurjot Singh Gaba	India
Thamer Ba Dhafari	University of Leeds, UK
Eman Nashnush	University of Salford, UK
Tooska Dargahi	University of Salford, UK

Sana Belguith	University of Salford, UK
Santosh More	University of Salford, UK
Julian Bass	University of Salford, UK

Local Arrangement Chairs

Mohamed Abd Elfattah (Chair), Egypt
Mourad Raft, Egypt
Hassan Aboul Ella, Egypt
Taha Aboul Ella, Egypt
Heba Aboul Ella, Egypt

Keynote Speakers and Tutorial

Multi-modal and Uncertain Optimization with Differential Evolution - Some Recent Approaches and Future Challenges

Swagatam Das

Indian Institute of Technology
https://www.isical.ac.in/~swagatam.das/

Abstract. Differential Evolution (DE) is arguably one of the most powerful stochastic real-parameter optimization algorithms of current interest. DE operates through similar computational steps as employed by a standard Evolutionary Algorithm (EA). However, unlike traditional EAs, the DE variants perturb the current-generation population members with the scaled differences of distinct population members. Therefore, no separate probability distribution has to be used for generating the offspring. Since its inception in 1995, DE has drawn the attention of many researchers all over the world resulting in a lot of variants of the basic algorithm with improved performance. This talk will begin with a brief but comprehensive overview of the basic concepts related to DE, its algorithmic components and control parameters. It will subsequently discuss some of the significant algorithmic variants of DE for bound-constrained single-objective optimization for high-dimensional search spaces. The talk will then focus on some interesting DE variants with additional mechanisms like a distance-based selection, a clustering procedure and aging mechanisms for optimization of objective functions corrupted with additive noise from various sources (with various probability distributions) and also optimization over dynamic/noisy fitness landscapes where the optima can shift with time. The talk will finally discuss a few interesting applications of DE and highlight a few open research problems.

Bio: Swagatam Das is currently serving as an associate professor at the Electronics and Communication Sciences Unit of the Indian Statistical Institute, Kolkata, India. His research interests include evolutionary computing, pattern recognition, multi-agent systems, and wireless communication. Dr. Das has published one research monograph, one edited volume, and more than 200 research articles in peer-reviewed journals and international conferences. He is the founding co-editor-in-chief of Swarm and Evolutionary Computation, an international journal from Elsevier.

He has also served as or is serving as the associate editors of Pattern Recognition (Elsevier), IEEE Trans. on Systems, Man, and Cybernetics: Systems, IEEE Computational Intelligence Magazine, IEEE Access, Neurocomputing (Elsevier), Engineering Applications of Artificial Intelligence (Elsevier), and Information Sciences (Elsevier). He is an editorial board member of Progress in Artificial Intelligence (Springer), PeerJ Computer Science, International Journal of Artificial Intelligence and Soft Computing, and International Journal of Adaptive and Autonomous Communication Systems. Dr. Das has 14000+ Google Scholar citations and an H-index of 56 till date. He has been associated with the international program committees and organizing committees of several regular international conferences including IEEE CEC, IEEE SSCI, SEAL, GECCO, and SEMCCO. He has acted as guest editors for special issues in journals like IEEE Transactions on Evolutionary Computation and IEEE Transactions on SMC, Part C. He is the recipient of the 2012 Young Engineer Award from the Indian National Academy of Engineering (INAE). He is also the recipient of the 2015 Thomson Reuters Research Excellence India Citation Award as the highest cited researcher from India in Engineering and Computer Science category between 2010 to 2014.

Modern Algorithms in Control and Design of Complex Systems

Ivan Zelinka

Department of Computer Science, Faculty of Electrical Engineering
and Computer Science VŠB-TUO, 17. listopadu 15, 708 33 Ostrava-Poruba,
Czech Republic
ivan.zelinka@vsb.cz
www.ivanzelinka.eu

Abstract. This keynote discusses the mutual intersection of exciting fields of research: bio-inspired algorithms, deterministic chaos and complex systems. The first one will discuss main principles of bio-inspired methods, its historical background and its use on various examples including real-world ones. Examples include plasma reactor control, optimal signal routing in the network of portable meteorological stations, complex system design as antenna design, nonlinear system and controllers design and more. Also, its use in deterministic chaos control with focusing on simple chaotic systems (logistic, Hennon,...), as well as CML systems exhibiting spatiotemporal chaos, will be mentioned and explained. The second part will discuss the use of deterministic chaos instead of pseudo-random number generators inside evolutionary algorithms with the application on well known evolutionary algorithms (differential evolution, PSO, SOMA, genetic algorithms,..) and test functions. A mutual comparison will be presented, based on our research. Also will be discussed the question whether evolutionary dynamics need pseudo-random numbers or no. At the end will be mentioned a novel approach joining evolutionary dynamics, complex networks and CML systems exhibiting chaotic behaviour. Reported methodology and results are based on the actual state of the art (that is a part of this keynote) as well as on our research. The keynote is designed as an introduction; no advanced or expert knowledge from complex networks, chaos and control is expected.

Bio: Ivan Zelinka is currently working at the Technical University of Ostrava (VSB-TU), Faculty of Electrical Engineering and Computer Science. He graduated consequently at Technical University in Brno (1995 – MSc.), UTB in Zlin (2001 – PhD) and again at Technical University in Brno (2004 – assoc. prof.) and VSB-TU (2010 - professor). Before academic career, he was an employed like TELECOM technician, computer specialist (HW+SW) and Commercial Bank (computer and LAN supervisor). During his career at UTB, he proposed and opened 7 different lectures. He also has been invited for lectures at numerous universities in different EU countries plus the role of the *keynote speaker* at the Global Conference on Power, Control and Optimization in Bali, Indonesia (2009), Interdisciplinary Symposium on Complex Systems (2011), Halkidiki, Greece and IWCFTA 2012, Dalian China. The field of his expertise if mainly on unconventional algorithms and cybersecurity. He is and was responsible supervisor of 3 grant of fundamental research of Czech grant agency GAČR, co-supervisor of grant FRVŠ - Laboratory of parallel computing. He was also working on numerous grants and two EU project like a member of the team (FP5 - RESTORM) and supervisor (FP7 - PROMOEVO) of the Czech team and supervisor of international research (founded by TACR agency) focused on the security of mobile devices (Czech - Vietnam). Currently, he is a professor at the Department of Computer Science and in total, he has been the supervisor of more than 40 MSc. and 25 Bc. diploma thesis. Ivan Zelinka is also supervisor of doctoral students including students from the abroad. He was awarded by Siemens Award for his PhD thesis, as well as by journal Software news for his book about artificial intelligence. Ivan Zelinka is a member of British Computer Society, Editor in chief of Springer book series: Emergence, Complexity and Computation (http://www.springer.com/series/10624), Editorial board of Saint Petersburg State University Studies in Mathematics, a few international program committees of various conferences and international journals. He is the author of journal articles as well as of books in Czech and English language and one of three founders of TC IEEE on big data http://ieeesmc.org/about-smcs/history/2014-archives/44-about-smcs/history/2014/technical-committees/204-big-data-computing/. He is also head of research group NAVY http://navy.cs.vsb.cz.

Hands-on to TensorFlow
(Tutorial on)

Ibrahim A. Hameed

Norwegian University of Science and Technology, Noway

Abstract. TensorFlow is an open source software library for numerical computation using data and flow graphs. In this tutorial, you will be introduced to TensorFlow and how to use for training deep learning models using CPU and GPU. You will learn how to use colab notebooks with real-world examples.

Speaker: Ibrahim A. Hameed has a Ph.D. in artificial intelligence from South Korea. A Ph.D. in field robotics from Denmark. He is working as an Associate Professor at the department of ICT at the Norwegian University of Science and Technology, Noway. He is a deputy head of research and innovation. He is the head of the international master program in Simulation and Innovation.

Contents

Swarm Intelligence and Applications

Machine Learning in Biomedical

Control and Chaotic Systems

Machine Learning in Image and Signal Processing

Renewable Energy

Complex Networks and Intelligent Systems

Deep Learning Applications

A Deep Learning Approach
for Tongue Diagnosis

Meng Xiao[1], Guozheng Liu[1], Yu Xia[1], and Hao Xu[1,2,3](✉)

[1] College of Computer Science and Technology,
Jilin University, Changchun, China
xuhao@jlu.edu.cn
[2] School of Management, Jilin University, Changchun, China
[3] Department of Computer Science and Technology,
Zhuhai College of Jilin University, Zhuhai, China

Abstract. With the improvement of living standards, people are paying more attention to healthcare, but there is still a long way to go to improve healthcare. A usable, intelligent aided diagnosis measure can be helpful for people to achieve daily health management. Several studies suggested that tongue features can directly reflect a person's physical state. In this paper, we apply tongue diagnosis to daily health management. To this end, this paper proposes and implements a classification model of tongue image syndromes based on convolutional neural network and carries out an experiment to verify the feasibility and stability of the model. Finally, a tongue diagnosis platform that can be used for daily health management is implemented. In the two-class experiment, our model has achieved a good result. In addition, our model performs better on classifying the tongue image syndrome compared with traditional machine learning methods.

Keywords: Healthcare · Health management · Tongue image diagnosis · Convolutional neural network

1 Introduction

In recent years, with the improvement of living standards, people are paying more attention to healthcare. Healthcare is a universally used service that can significantly affect the quality of daily living [1]. Healthcare has become a hot topic in the research community [2]. The development of science and technology has provided more opportunities for improving healthcare [3]. Several image recognition algorithms and machine learning methods have been applied to the medical field, such as medical image recognition [4–6] and the classification and recognition of the tongue image [7], to achieve the effect of auxiliary diagnosis. At the same time, health management tools have emerged to help people participate in health management, such as Keep [8] and Lifesum [9]. But it is still meaningful to provide people with a way of daily available intelligent aided diagnosis to have a general idea of their own bodies.

A. E. Hassanien et al. (Eds.): AMLTA 2019, AISC 921, pp. 3–12, 2020.
https://doi.org/10.1007/978-3-030-14118-9_1

Traditional Chinese medicine plays an important role in the Chinese civilization. In traditional Chinese medicine, the human body is defined as an organic whole, in which every part is related. External changes may indicate that there are diseases or lesions in the corresponding positions. In this process, the tongue image, just like a mirror, can directly reflect the physical state and help the doctor understand the patient's condition [10]. The so-called tongue diagnosis is a traditional Chinese medicine diagnosis method that aids in judging the condition by observing the patient's tongue, according to the colour and morphological changes of the appearance.

With the popularization of traditional Chinese medicine, people have a deeper and more comprehensive understanding of tongue diagnosis. Many researchers have devoted themselves to the field of tongue image analysis and have proposed many mature and stable methods and solutions. These include the study of the texture features [11], colour features [12–15], shape features [16, 17] and segmentation of tongue images. As a result, a set of universal tongue features, including shapes, contours, specific colour channels and tongue moss lines, has been discovered and integrated [18]. The method of tongue shape segmentation proposed by Kanawong et al. [17] has obtained good results that provide a basis for the pre-processing of tongue images in this paper.

Deep learning has been widely used in medical image processing. It helps provide technical support for auxiliary diagnosis and treatment. The deep learning algorithm applied to tongue image processing is mainly based on convolutional neural network (CNN). As a widely used neural network, CNN can weaken the influence of geometric transformation, deformation and illumination on the classification effect of this algorithm [19]. In traditional tongue diagnosis, light source, weather and even the season could have a slight impact on the result. Therefore, CNN can be a good way to reduce the impact of these factors. Many studies have been conducted on tongue image using CNN. Hou et al. used CNN to study the details and characteristics of tongues [20]. Based on the traditional CaffeNet model, they explored a new approach to classify tongue colour from tongue images. Hu et al. used CNN based on the Softmax classifier to extract relevant features from tongue images and provide relevant Chinese herbal prescriptions [21], introducing a double convolutional channel frame for higher accuracy. Meng et al. proposed a CNN-based deep tongue image feature analysis model that can learn high-level features and increase the accuracy to 91.49% [22].

This paper is organized as follows: in Sect. 2, we propose a tongue image syndrome classification model based on CNN. We verify the accuracy of the model under the two-class experiment and three-class experiment in Sect. 3. The experimental results are shown in Sect. 4 and a tongue diagnosis platform is presented in Sect. 5. Finally, our work is discussed and the conclusion is presented in Sect. 6.

2   Tongue Image Syndrome Classification Model Based on CNN

To obtain a convolutional neural network model that can be applied to the diagnosis of tongue images, the process is divided into two parts, the data pre-processing stage and the deep learning stage. The data pre-processing stage of tongue images mainly includes tongue image collection, diagnosis and labelling and tongue image segmentation.

The deep learning stage mainly includes convolutional neural network selection, parameter adjustment, model training and cross-validation. Finally, a CNN-based tongue image syndrome classification model was obtained. Figure 1 shows the specific process.

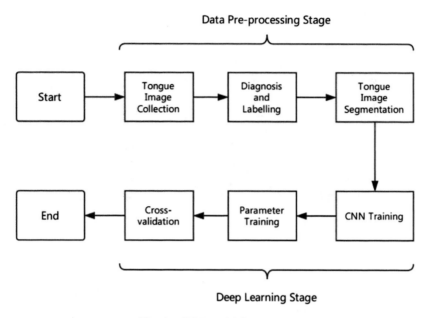

Fig. 1. CNN model flow chart.

2.1 Data Pre-processing Stage

Data pre-processing is an important part of the image recognition process. It mainly includes collecting accurate tongue image data, screening and kicking noise data and segmenting tongue images to eliminate redundant information. All steps are helpful for the fast convergence in the training phase and achieving good training results. The proportion of each type in the sample data occupies an important position in the processing of image classification. Usually, the proportion of each type in the training data should be balanced as much as possible to avoid the model being prominent only in a certain type. Therefore, to achieve a balanced effect, some data amplification work should be carried out or some excess samples should be eliminated.

Tongue Image Collection. The collection of tongue image is different from other images because external factors, such as the angle of the light source, have some influence on the classification of tongue image syndromes, and some details may lead to unsatisfactory experimental results. Therefore, in terms of data scale, tongue images often do not have access to large quantities of standard data. In this paper, we collaborated with Shanghai University of Traditional Chinese Medicine to collect 398 samples of standard tongue images using tongue image collection instruments.

Diagnosis and Labelling. In our model, the syndromes of the tongue are divided into three types, cold syndrome, hot syndrome, and normal (Fig. 2) [23]. Before collecting the tongue image, the professional TCM doctor needs to diagnose and label the person and correlate with respective tongue image information to ensure the accuracy of the data. After sorting and classification, the tongue image data used in this paper included 79 cases in normal group, 149 cases in hot syndrome group, and 170 cases in cold syndrome group.

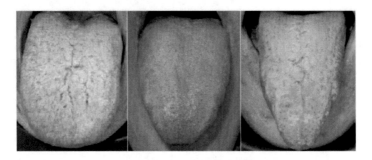

Fig. 2. Three types of syndromes (cold, normal, hot from left to right).

Tongue Image Segmentation. The collected tongue image inevitably includes other information about the non-target tongue image, such as the lips, face and tongue imager. Therefore, to ensure that the tongue image can effectively represent the body's syndrome, factors outside the tongue itself need to be excluded. In combination with the tongue image segmentation method [17] in the previous work, the single tongue image itself is retained and is regarded as input sample data for subsequent training and testing.

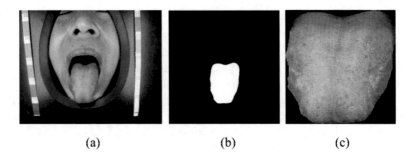

(a)   (b)   (c)

Fig. 3. Tongue image segmentation.

According to the segmentation algorithm, this paper performs the edge check to obtain the outline image of the tongue image in the centre, as shown in Fig. 3a. By matching it with the original image (Fig. 3b), the tongue image required for this paper is obtained, as shown in Fig. 3c.

2.2   Deep Learning Stage

At this stage, we mainly select the appropriate convolutional neural network structure and design the convolutional neural network model. Based on the model's performance in automatically learning and extracting the features corresponding to the sample tags, better model parameters need to be selected.

The convolutional neural network continuously performs the operation of convolutions to extract relevant features in the image. Then the extracted features can be assigned corresponding weights. Finally, the results of the classification are obtained. This paper designs a four-layer convolutional neural network based on LeNet network structure for the recognition of tongue image syndrome (hereinafter Conv4). The network structure includes four convolution layers, four pooling layers and two fully connected layers.

In this paper, the sample data are input in the form of 256*256 pixels. If the original data sample is too large, the time increases and more training resources are wasted. However, if the data sample is too small, the feature information can be lost. Finally, it is determined that the form of 256*256 pixels is more suitable. Each convolution layer contains sixteen 3*3 three-channel convolution kernels. The pooling layer uses the max-pooling method to get the feature information to the greatest extent. The filter in the pooling layer has a size of 3*3 and a step size of 2. Finally, a two-layer fully connected layer and a Softmax function are included for classifying the extracted feature data.

After each layer of convolution operation, an activation function can be used. The activation function can add the nonlinear factors for the learned features and enhance the description capabilities of the neural network. In the choice of activation function, this paper chose ReLU function to replace the traditional Sigmoid function because when the upper input is too large or too small, the gradient of the neuron is close to zero. This causes the weight of the neurons to be hardly updated during the back propagation. Furthermore, the output of the Sigmoid function is non-zero, making it easy to perform only in the same direction during the weight updating process. The ReLU function solves these problems well. There is no saturation problem in the gradient, which greatly reduces the amount of calculation and accelerates the convergence of the training model and can also solve the problem of over-fitting. Therefore, in this experiment, the ReLU function was chosen as the activation function of Conv4.

Some important parameters in the convolutional neural network need to be manually judged and preset, such as learning rate and batch. A suitable value can improve the efficiency of model fitting on the premise of ensuring the accuracy of the model. Firstly, the learning rate has an optimization effect in the process of gradient descent of the neural network. Excessive learning rate can lead to an increase in convergence speed, but it is easy to make the loss value into the local minimum, thus missing the optimal solution; a too small value can seriously affect the speed of the operation and cause the convergence to be calculated too slowly. After exploration and experiments, the learning rate is set to 0.0001. The batch value is the number of sample data entering the convolutional neural network at the same time. A larger batch value can cause the model to learn more features at the same time, but the calculation amount can increase correspondingly. In this paper, a larger batch value of 256 is selected.

3   Experiment

3.1   Two-Class Experiment

From the perspective of the balance of data, this paper collects 398 cases, of which the proportion of the normal, hot syndrome and cold syndrome is 79:149:170. In the experiment, this paper selects the normal group and the hot syndrome group, the normal group and the cold syndrome group, and the hot syndrome group and the cold syndrome group for comparative experiments. The Conv4 designed in this paper is trained on each group, and the respective accuracy rate curve on the test set is obtained. It aims to verify the feasibility and stability of the convolutional neural network in the tongue image classification model.

At the same time, the effect of our model on the two-class experiment is compared with traditional machine learning methods such as Support Vector Machine (SVM), Multilayer Perceptron Network (MLP) and Random Forest (RF).

Normal Group and Hot Syndrome Group. There were 79 cases in the normal group and 149 cases in the hot syndrome group. The accuracy rate curve on the test set was in a relatively steady trend of slow fitting. At the initial stage, the accuracy of the test set was about 60%, but after a few rounds of training, the accuracy of the model fluctuated between 75% and 78%. At 522 rounds, the best accuracy of the model was 77.8%. After 90% off cross validation, the accuracy of the model was 82.8%. The accuracy of the three traditional machine learning methods, SVM, MLP and RF, is 84.4%, 83.9% and 80.6%, respectively [24].

Normal Group and Cold Syndrome Group. There were 79 cases in the normal group and 170 cases in the cold syndrome group. The accuracy rate curve on the test set was in a relatively steady trend of slow fitting. In the initial stage, the accuracy rate of the test set was about 63%, but after 50 rounds, the downward trend appeared and then gradually fitted. The best accuracy rate of the model was 72.8% in the 122nd round. After 90% off cross validation, the accuracy of the model was 82.8%. The accuracy of the three traditional machine learning methods, SVM, MLP and RF, is 81.9%, 80.2% and 80.2%, respectively [24].

Hot Syndrome Group and Cold Syndrome Group. There were 149 cases in the hot group and 170 cases in the cold syndrome group. The accuracy rate curve on the test set was in a relatively steady trend of slow fitting. In the initial stage, the accuracy rate of the test set was around 48%. Then the model was fitted at a fast speed. The accuracy rate of the model was continuously increased to above 70%, and the best accuracy rate of the model was 84% in the 522th round. After 90% off cross validation, the accuracy of the model was 82.8%. The accuracy of the three traditional machine learning methods, SVM, MLP and RF, is 70.0%, 74.0% and 73.5%, respectively [24].

Normal Group and Abnormal Group. There were 79 cases in the normal group and 170 cases in the abnormal group. Overall, the accuracy rate on the test set tended to decline. In the first 50 rounds of training, the accuracy rate on the test set remained at around 80%, but as the training progressed, the accuracy rate on the test set dropped to between 0.6 and 0.7. After 90% off cross validation, the accuracy of the model was 82.8%. The accuracy of the three traditional machine learning methods, SVM, MLP and RF, is 89.2%, 91.1% and 89.5%, respectively [24].

3.2   Three-Class Experiment

The traditional machine learning method often does not perform well on the three-class problem. The main method of the multi-classification research is to carry out multiple two-class experiments. In this paper, we try to realize the three-class model of tongue image by training convolutional neural network. This experiment includes 349 total samples, including 79 cases in normal group, 149 in hot syndrome group and 170 in cold syndrome group. The result of the three-class experiment is as follows.

The accuracy rate varied with the number of training round. After entering the 180th round, the test set began to converge. But compared to the test set, the training samples continued to fit until the difference of the accuracy rate reached nearly 50%, which showed that there was over-fitting in the training situation of this model. After 90% off cross validation, the accuracy of the model was 53.6%.

4   Result

To verify the effectiveness of the model, we conducted two-class and three-class experiments separately.

In the two-class experiment, our model achieved a good result. For normal group and hot syndrome group, the accuracy is 82.2%, which is close to that of traditional machine learning methods. For normal group and cold syndrome group, and hot syndrome group and cold syndrome group, the accuracy is higher than one of traditional machine learning methods, especially hot syndrome group and cold syndrome group. For normal group and abnormal group, the effect of our model is not as good as that of traditional machine learning methods. But in general, the effect of the model on tongue image syndrome classification was slightly better than that of traditional machine learning methods.

The effect of the model on the three-class experiment does not reach the expected result. The main reason could be that the network structure is relatively shallow, and it is impossible to extract the exact features or over-fitting due to the imbalance of the data.

5   Platform

Based on the model, a mobile health management platform is implemented for tongue diagnosis. People can take photos of their tongues and upload the images to the platform. The platform provides feedback related to physical state based on the tongue image. People can more easily understand their physical condition and make appropriate adjustments, such as changing their work schedule, diet, etc. We hope this can help people have a holistic idea of their own health, improve their health awareness and enable them to participate in daily health management. Figure 4 shows the main page of the platform.

Fig. 4. Pages of the platform.

6   Discussion and Conclusion

By studying and analysing the process of TCM tongue diagnosis and combined with the previous experience, this paper explores the feasibility of convolutional neural network applied to the classification of TCM tongue images. Therefore, we propose and implement a CNN-based tongue image syndrome classification model, which can help people understand their physical condition more easily. We also analyse the performance of the model in two-class and three-class experiments. In the two-class experiment, compared with the traditional machine learning methods, our model gets a good effect. A mobile health management platform is implemented for tongue diagnosis. The effect of the model on the three-class experiment does not reach the expected result, but this paper shows a context of deep learning combined with healthcare. This model still has a positive effect on improving healthcare.

However, there are some deficiencies in this paper. Firstly, the accuracy of the model is not high enough. We need to improve our model for higher accuracy in the future. Secondly, through the analysis and design of the convolutional neural network, we know that it has a good performance in dealing with the classification of TCM tongue images. We should continue to expand the problem to the multi-classification problem. The cold syndrome and hot syndrome are only the most basic two types in TCM syndromes. The human body can be divided into more detailed classifications in TCM. Studying the problem of multi-classification can better aid tongue diagnosis and treatment. Finally, in the previous work, we learned that Chinese medicine plays a key role in the diagnosis of Western medicine. In the future, the relationship between the tongue image and the specific Western medical condition can be established. The deep learning method is used to explain the auxiliary effect of TCM tongue diagnosis on the treatment of the disease. From the perspective of data science, the relevance between the TCM theory and the Western medical condition should be explained. We hope that our work can further contribute to the development of healthcare.

Acknowledgements. This research was funded by the [Development Project of Jilin Province of China] grant number [20160414009GH, 20170101006JC, 20160204022GX], the [National Natural Science Foundation of China] grant number [61472159, 71620107001, 71232011], the [Jilin Provincial Key Laboratory of Big Date Intelligent Computing] grant number [20180622002JC]. The Premier-Discipline Enhancement Scheme was supported by Zhuhai Government and Premier Key-Discipline Enhancement Scheme was supported by Guangdong Government Funds.

References

1. Berry, L., Bendapudi, N.: Health care: a fertile field for service research. J. Serv. Res. **10**(2), 111–122 (2007)
2. Miotto, R., Wang, F., Wang, S., Jiang, X., Dudley, J.: Deep learning for healthcare: review, opportunities and challenges. Brief. Bioinform. (2017)
3. Compston, H.: King Trends and the Future of Public Policy. Palgrave Macmillan, Basingstoke (2006)
4. Yezzi, A., Kichenassamy, S., Kumar, A., Olver, P., Tannenbaum, A.: A geometric snake model for segmentation of medical imagery. IEEE Trans. Med. Imaging **16**(2), 199–209 (1997)
5. Tomczyk, A., Szczepaniak, P., Pryczek, M.: Cognitive hierarchical active partitions in distributed analysis of medical images. J. Amb. Intel. Hum. Comp. **4**(3), 357–367 (2012)
6. Quist, M.J.: A Method of Image Registration and Medical Image Data Processing Apparatus (2017)

7. Zhi, L., Zhang, D., Yan, J., Li, Q., Tang, Q.: Classification of hyperspectral medical tongue images for tongue diagnosis. Comput. Med. Imag. Grap. **31**(8), 672–678 (2007)
8. Gotokeep.com. Keep (2018). https://www.gotokeep.com. Accessed 21 Aug 2018
9. Lifesum.com. Lifesum (2018). https://lifesum.com. Accessed 21 Aug 2018
10. Zhang, B., Wang, X., You, J., Zhang, D.: Tongue color analysis for medical application. Evid. Based Complementary Altern. Med. **2013**, 1–11 (2013)
11. Chiu, C.C., Lin, H.S., Lin, S.L.: A structural texture recognition approach for medical diagnosis through tongue. Biomed. Eng. Appl. Basis Commun. **7**(2), 143–148 (1995)
12. Wang, Y., Yang, J., Zhou, Y., Wang, Y.: Region partition and feature matching based color recognition of tongue image. Pattern Recogn. Lett. **28**(1), 11–19 (2007)
13. Li, C., Yuen, P.: Tongue image matching using color content. Pattern Recogn. **35**(2), 407–419 (2002)
14. Pang, B., Zhang, D., Wang, K.: The bi-elliptical deformable contour and its application to automated tongue segmentation in Chinese medicine. IEEE Trans. Med. Imaging **24**(8), 946–956 (2005)
15. Liu, Z., Yan, J., Zhang, D., Li, Q.: Automated tongue segmentation in hyperspectral images for medicine. Appl. Opt. **46**(34), 8328 (2007)
16. Zhang, D., Liu, Z., Yan, J.: Dynamic tongueprint: a novel biometric identifier. Pattern Recogn. **43**(3), 1071–1082 (2010)
17. Obafemiajayi, T., Kanawong, R., Xu, D., Duan, Y.: Features for automated tongue image shape classification. In: IEEE International Conference on Bioinformatics and Biomedicine Workshops, pp. 273–279 (2013)
18. Chiu, C.: A novel approach based on computerized image analysis for traditional Chinese medical diagnosis of the tongue. Comput. Methods Programs Biomed. **61**(2), 77–89 (2000)
19. Ma, C., Sun, C., Song, D., Li, X., Xu, H.: A deep learning approach for online learning emotion recognition. In: 13th International Conference on Computer Science & Education, pp. 1–5 (2018)
20. Hou, J., Su, H.Y., Yan, B., Zheng, H., Sun, Z.L., Cai, X.C.: Classification of tongue color based on CNN. In: IEEE International Conference on Big Data Analysis, pp. 725–729 (2017)
21. Hu, Y., Wen, G., Liao, H., Wang, C., Dai, D., Yu, Z., Zhang, J.: Automatic construction of Chinese herbal prescription from tongue image via CNNs and auxiliary latent therapy topics (2018)
22. Meng, D., Cao, G., Duan, Y., Zhu, M., Tu, L., Xu, J., Xu, D.: A deep tongue image features analysis model for medical application. In: IEEE International Conference on Bioinformatics and Biomedicine, pp. 1918–1922 (2017)
23. Kanawong, R., Obafemi-Ajayi, T., Ma, T., Xu, D., Li, S., Duan, Y.: Automated tongue feature extraction for ZHENG classification in traditional Chinese medicine. Evid. Based Complementary Altern. Med. **2012**, 1–14 (2012)
24. Obafemi-Ajayi, T., Xu, D., Yu, J., Duan, Y., Kanawong, R., Li, S.: ZHENG classification in Traditional Chinese Medicine based on modified specular-free tongue images. In: IEEE International Conference on Bioinformatics and Biomedicine Workshops, pp. 288–294 (2013)

Reduced 3-D Deep Learning Framework for Hyperspectral Image Classification

Noureldin Laban[1](✉), Bassam Abdellatif[1], Hala M. Ebeid[2],
Howida A. Shedeed[2], and Mohamed F. Tolba[2]

[1] Data Reception and Analysis Division,
National Authority for Remote Sensing and Space Science, Cairo, Egypt
{nourlaban,bassam.abdellatif}@narss.sci.eg
[2] Faculty of Computer and Information Sciences, Ain Shams University, Cairo, Egypt
{halam,Hoveyda.Saber,fahmytolba}@cis.asu.edu.eg

Abstract. In recent years, machine learning has achieved a break-through step due to re-branding of Convolutional Neural Networks (CNNs). Advancement in machine learning algorithms makes it easier to process big and information-rich images such as hyper-spectral images. Hyperspectral imaging (HSI) technology has also shown obvious increase in number of satellites and increased number of bands which lead to a huge amount of data generated every day. In this paper, we propose a reduced version for 3-dimensional convolutional neural network (3D-CNN) as a deep learning framework for hyperspectral image classification. The latest proposed CNNs models, especially 3D ones, have achieved near 100% of accuracy with benchmark hyperspectral data sets. Our proposed framework explores the effect of dimensions reduction on the performance with respect to total classification accuracy. In our experiments, two benchmarks HSIs are used to evaluate performance of reduced framework with different number of bands. The experimental results demonstrate that the reduced 3D-CNN framework has significantly reduced the time of training of CNN with more than 60% compared to the full bands training almost without affecting the accuracy of classification.

Keywords: Deep learning · Hyperspectral image classification ·
Principal Component Analysis (PCA) ·
3D-Convolutional Neural Network

1 Introduction

Hyperspectral images are captured with both spectral and spatial features using hyperspectral sensors [1]. These hyperspectral sensors collect electromagnetic spectrum information in hundreds of narrow and contiguous bands. They are able to discriminate different ground land cover according to their unique spectral signatures [2]. The hyperspectral sensors technology records the spatial and spectral data simultaneously [3]. Hyperspectral images (HSIs) are used in many

© Springer Nature Switzerland AG 2020
A. E. Hassanien et al. (Eds.): AMLTA 2019, AISC 921, pp. 13–22, 2020.
https://doi.org/10.1007/978-3-030-14118-9_2

applications as urban mapping, environmental management, crop analysis, and mineral detection [4]. It is possible now to analyze small spatial structures in images due to the fine spatial resolution of sensors. The increased number of bands of the images leads to many theoretical and practical problems [5].

Classification of hyperspectral data aims at predicting the labels of each pixel in a hyperspectral image [3]. Several machine learning methods have been proposed in the last few years. They are competing on speeding up the classification process and improving classification accuracy [6]. There are many methods used in hyperspectral image classification including Support Vector Machines (SVMs), Multinomial Logistic Regression (MLR) and Markov Random Fields Models, Genetic Algorithms (GA), and Artificial Neural Networks [1,6–8]. Among these methods, the most commonly used classifier is SVMs for HSI classification as they perform robustly with high-dimensional data [9].

Convolutional Neural Networks (CNNs) and their extensions have obtained unprecedented advances in machine learning tasks [9]. Typical deep neural network models include Stacked Autoencoders (SAEs), Deep Belief Networks (DBNs) and CNNs based on which a series of methods have been proposed, aiming at the HSI classification task [10]. CNN is an end-to-end machine learning as the model is learning directly from raw data without intermediate feature extraction [11]. Different CNNs frameworks have been developed to extract high-level semantic features from the massive amounts of data collected recently [12].

Fully 3D CNNs have a massive number of parameters with significant memory and processing requirements [13]. For hyperspectral image classification, the third dimension refers to the spectral bands in the framework of 3D-CNN models [14]. The main difference between the 3D-CNN model and 2D-CNN model is that the 3D-CNN model has an extra process of reordering. In this process the hyperspectral bands are rearranged in an ascending order so the spectral bands preserve their correlations [14]. Also, the 3D-CNN model has a 3D convolution operation instead of 2D convolution operation.

The rest of this paper is organized as follows. Section 2 introduces the proposed framework with its two components. Section 3 reports the experimental results including used data sets, framework setting and experiments results comparisons. Finally, we give our conclusion and the future research work.

2   Proposed Framework

Our proposed framework depends upon two basic components; the hyperspectral dimension reduction and 3D Convolutional Neural Network. The first hyperspectral dimension reduction component aims at reducing the number of hyperspectral bands with the most informative bands, and hence, reducing the computations needed during training. The second component which represents the 3D Convolutional Neural Network is responsible for the feature extraction and classification processes. Figure 1 illustrates the overall framework architecture.

2.1 Hyperspectral Dimension Reduction

There are many approaches used to reduce high dimensional space into a low dimensional one [10]. Typical approaches include Principal Component Analysis (PCA)[15], Independent Component Analysis (ICA) [16], Fisher's Linear Discriminant Analysis(FLDA) [17], and local Linear Embedding [18]. Principal Component Analysis (PCA) is the most widely used technique for dimension reduction in remote sensing applications, specifically in hyperspectral images [19]. The PCA uses the statistical properties of hyperspectral image bands to explore band dependency and correlation. It is applied to the original image and return only the first principle components. To compute the PCA we use mean (Eq. 1) and co-variance matrix (Eqs. 2 and 3).

$$m = \frac{1}{M} \sum_{i=1}^{M} [x_1 x_2 ..... x_N]_i^T \tag{1}$$

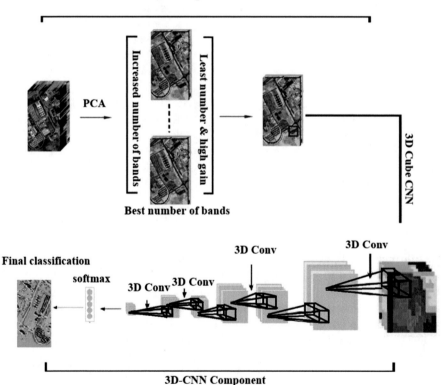

Fig. 1. Proposed architecture

$$C_x = \frac{1}{M} \sum_{i=1}^{M} (x_i - m)(x_i - m)^T \tag{2}$$

$$C_x = ADA^T \tag{3}$$

where N is the number of bands, m is the mean vector, C is the co-variance matrix, D is the diagonal matrix composed of the eigenvalues of the co-variance matrix C_x and A is the orthogonal matrix Eigen vectors.

Using the information gain of the PCA returned bands, we can decide how many number of bands we use for later classification process.

2.2 3D Convolutional Neural Network

Deep learning frameworks consist of multiple successive layers of nonlinear neurons that can learn hierarchical representations using a large number of labeled images [20]. Convolution operations enable CNNs to learn more discriminative features directly from raw data with sparsity constraint [21]. Recent CNNS architectures have very huge structure arrange from 10 to 20 layers, hundreds of millions of interconnection weights [21].

In 3D-CNN model, the convolution operation is applied on the hyperspectral cube to extract deep features other than 2D operation [22]. 3D-CNN can extract the spatial and spectral information simultaneously using 3-D convolution. 3D-CNN can learn both spatial and spectral features of hyperspectral image using 1D-CNN to extract spectral features and 2-D CNN to extract the local spatial features [22].

A hyperspectral image is captured using very close spectral bands, so they may have correlations. Different hyperspectral images may have same bands due to these correlations [14]. 3D-CNN model processes are similar to those of the 2D-CNN model [13]. A 3-D convolution operator is applied instead of 2-D convolution operator to the 3-DCNN image. Equation 4 explains the 3D convolution operation [14] as follows:

$$v_{ij}^{xyz} = F\left(b_{ij} + \sum_{m} \sum_{p=0}^{N_i-1} \sum_{q=0}^{M_1-1} \sum_{r=0}^{D_i-1} w_{ijm}^{pqr} v_{(i-1)m}^{(x+p)(y+q)(z+r)} \right) \tag{4}$$

where i indicates the current layer, j is the number of feature maps of the layer i, v_{ij}^{xyz} stands for the output at position (x, y, z) of the jth feature map at the ith layer, b_{ij} refers to the bias term, $F(.)$ denotes the activation function of the layer, and m indexes over the set of feature maps of the $(i-1)^{th}$ layer, which are the inputs to the ith layer. D_i is the size of the 3D kernel along the spectral dimension and j is the number of kernels of the i layer; w_{ijm}^{pqr} is the value at the $(p, q, r)^{th}$ position of the kernel connected to the mth feature map (a cube) of the preceding layer.

Several Rectified Linear Units (ReLUs) and dropouts are used to achieve powerful sparse-based regularization for the deep network and prevent the overfitting of classification. Batch Normalization [23] is conducted also at every convolutional layer to make the training processing more efficient.

3   Experimental Results

In this section, we report the experimental data sets, settings and discuss the results of the reduced 3D-CNN framework on the data set classification.

3.1   Experimental Datasets

We have used two classical hyperspectral data sets in our experiments, including Pavia University and Indian Pines, which are used to evaluate accuracy and performance of the reduced framework.

Pavia University data set [9,10] acquired by the Reflective Optics Systems Imaging Spectrometer sensor in Northern Italy in 2001. This dataset consists of an image with 103 spectral bands that has 610×340 pixels with spectral coverage from 430 to 860 nm and a spatial resolution of 1.3 m. Figure 2 shows the false-color composition of the Pavia University image and the ground truth map. Table 1 shows the training, validation and test sets.

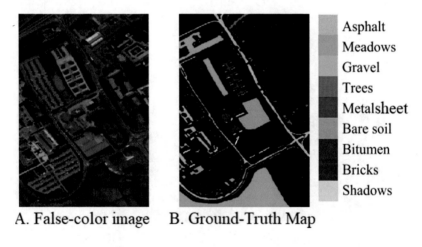

A. False-color image   B. Ground-Truth Map

Fig. 2. The Pavia University data set.

Indian Pines dataset [9,10] is gathered by Airborne Visible/Infrared Imaging Spectrometer (AVIRIS) over the Indian Pines in Northwestern Indiana. The dataset consists of an image that has 200 spectral bands with 145x145 pixels with a spatial resolution of 20 m by pixel. Figure 3 shows the false-color composition of the Pavia University image and the ground truth map. Table 2 shows the training, validation and test set.

3.2   Experimental Design

CNN setting can extract high level spatial features. It can learn automatically the spatial parameters. Training process is configured and the parameters of 3-D filter banks is updated through the back-propagation of the gradients. Three factors that control the training process and classification performance are explored. The three factors are the number of bands, the spatial size of the input 3D cubes and the batch size. Through the training process for each setting, models with the highest classification accuracy and the least processing time in validation groups were preserved and reported for results.

Table 1. The Pavia University data set details

No.	Class	Train.	Val. .	Test	No.	Class	Train.	Val. .	Test
1	Asphalt	664	670	5297	6	Bare Soil	503	500	4026
2	Meadows	1865	1810	14974	7	Bitumen	133	133	1046
3	Gravel	210	241	1648	8	Bricks	369	363	2950
4	Trees	307	333	2424	9	Shadows	95	97	755
5	Metal Sheets	135	134	1076					
						TOTAL	4281	4281	34214

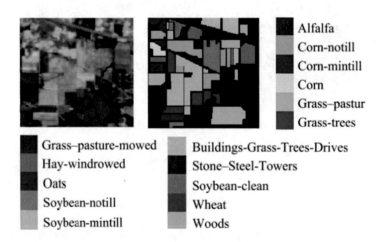

Fig. 3. The Indian Pines dataset data set.

Experiments were carried out to determine the effect of the reduced model on classification performance on the Dell Precision Tower 5810 Workstation with Quad-Core Intel Xeon E5-1620V3/3.5 GHz and 32 GB RAM. Training was performed on a single NVIDIA quadro 600m GPU.

3.3   Classification Results

Results for Pavia University Data Set: Pavia university image has 103 spectral bands. using the PCA reduction algorithm, we get 3, 8, 16 and 59 most informative spectral bands as shown in Table 3. Figure 4 presents the experimental results of selected numbers of spectral bands of the Pavia University image data set. As shown in Table 3, although we reduce the number of bands from 103 to 8, we found that these 8 bands have 99.6% of information content. Results show that the reduced framework achieves 99.6% of classification accuracy in about 1703 s using only 16 spectral bands instead of 103 bands. The improvement in training time achieved by the reduced framework using 16 bands is 17% with almost the same overall accuracy.

Table 2. The Indian Pines data set details

No.	Class	Train.	Val.	Test.	No.	Class	Train.	Val.	Test.
1	Alfalfa	10	1	35	9	Oats	4	4	12
2	Com-noli]l	286	131	1011	10	Soybean-nolill	195	94	683
3	Corn-mintill	166	83	581	11	Soybean-mimill	491	264	1700
4	Corn	48	22	167	12	Soybean-clean	119	56	418
5	Grass-pasture	97	42	344	13	Wheat	41	26	138
6	Gmss-tree	146	69	515	14	Wood,	253	136	876
7	Grass-pasture-mowed	6	3	19	15	Buildings-Grass-Trees	78	34	274
8	Hay-windrowed	96	55	327	16	Stone-Stcel-Towers	19	5	69
						TOTAL	2055	1025	7169

Results for the Indian Pines Data Set. Indian Pines data set shows almost same results as Pavia university data set. Indian Pines image has 200 spectral band. We get 3, 8, 25 and 69 most informative spectral bands as shown in Table 3. Figure 5 presents the experimental results of selected numbers of spectral bands of the Indian Pine image data set. As shown in Table 4, although we reduced the number of bands from 200 to 25, we still found that these 25 bands have 99% of information content. Results show that the reduced framework achieves 99% of classification accuracy in about 750 s using only 25 spectral bands from 200 bands. The improvement in training time achieved by the reduced framework using 25 bands is 39% with almost the same overall accuracy.

Table 3. Number of bands versus information gain for Pavia University dataset

Number of bands	Information gain
3	98%
8	99.6%
16	99.9%
59	99.99%
103	100%

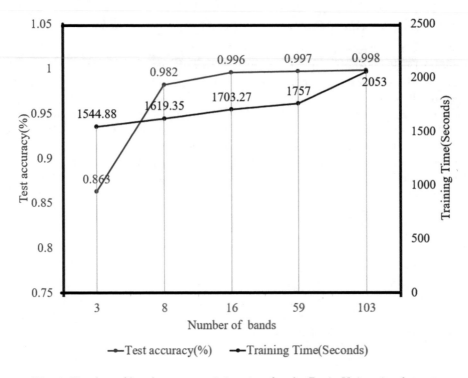

Fig. 4. Number of bands versus training time for the Pavia University dataset

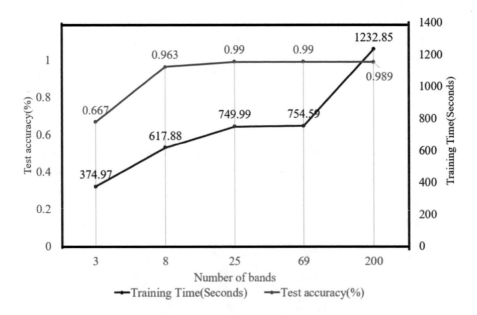

Fig. 5. Number of bands versus training time for THE Indian Pines dataset

Table 4. Number of bands versus information gain for Indian Pines data set

Number of bands	Information gain
3	93%
8	96%
25	99%
69	99.9%
200	100%

4 Conclusion

In this paper, we have presented a reduced 3D deep learning framework for hyperspectral image classification. We explored the effect of reduction of the number of hyperspectral bands on hyperspectral image classification performance. We used the PCA as a dimension reduction algorithm. Also, we get the information content of the reduced image. Experiments results show that using only the most informative bands of hyperspectral image is enough to obtain state-of-the-art classification performance. The reduced 3D-CNN framework decreased training time with about 17% and 40% for Pavia University data sets and Indian Pines data sets respectively in preserving a classification accuracy of about 99%. The experimental results demonstrate that the reduced 3D-CNN framework has significantly reduced the time of training. In the future, further experiments will be conducted to more hyperspectral data sets with larger size and harder classification problems.

References

1. Yue, J., Mao, S., Li, M.: A deep learning framework for hyperspectral image classification using spatial pyramid pooling. Remote Sens. Lett. **7**(9), 875–884 (2016)
2. Mei, S., Ji, J., Hou, J., Li, X., Du, Q.: Learning sensor-specific spatial-spectral features of hyperspectral images via convolutional neural networks. IEEE Trans. Geosci. Remote Sens. **55**(8), 4520–4533 (2017)
3. Zhou, X., Li, S., Member, F.T., Qin, K., Hu, S., Liu, S.: Deep learning with grouped features for spatial spectral classification of hyperspectral images. IEEE Geosci. Remote Sens. Lett. **14**(1), 1–5 (2017)
4. Zhao, W., Du, S.: Spectral-spatial feature extraction for hyperspectral image classification: a dimension reduction and deep learning approach. IEEE Trans. Geosci. Remote Sens. **54**(8), 4544–4554 (2016)
5. Chen, Y., Jiang, H., Li, C., Jia, X., Ghamisi, P.: Deep feature extraction and classification of hyperspectral images based on convolutional neural networks. IEEE Trans. Geosci. Remote Sens. **54**(10), 6232–6251 (2016)
6. Jain, D.K., Dubey, S.B., Choubey, R.K., Sinhal, A., Arjaria, S.K., Jain, A., Wang, H.: An approach for hyperspectral image classification by optimizing SVM using self organizing map. J. Comput. Sci. (2017)

7. Yue, J., Zhao, W., Mao, S., Liu, H.: Spectral-spatial classification of hyperspectral images using deep convolutional neural networks. Remote Sens. Lett. **6**(6), 468–477 (2015)
8. Li, W., Wu, G., Zhang, F., Du, Q.: Hyperspectral image classification using deep pixel-pair features. IEEE Trans. Geosci. Remote Sens. **55**(2), 844–853 (2017)
9. Zhong, Z., Li, J., Luo, Z., Chapman, M.: Spectral-spatial residual network for hyperspectral image classification: a 3-D deep learning framework. IEEE Trans. Geosci. Remote Sens. **56**(2), 847–858 (2018)
10. Xu, Y., Zhang, L., Du, B., Zhang, F.: Spectral-spatial unified networks for hyperspectral image classification. IEEE Trans. Geosci. Remote Sens. 1–17 (2018)
11. Basaeed, E., Bhaskar, H., Al-Mualla, M.: Supervised remote sensing image segmentation using boosted convolutional neural networks. Knowl.-Based Syst. **99**, 19–27 (2016)
12. Liu, W., Wang, Z., Liu, X., Zeng, N., Liu, Y., Alsaadi, F.E.: A survey of deep neural network architectures and their applications. Neurocomputing **234**(October 2016), 11–26 (2016)
13. Kamnitsas, K., Ledig, C., Newcombe, V.F., Simpson, J.P., Kane, A.D., Menon, D.K., Rueckert, D., Glocker, B.: Efficient multi-scale 3D CNN with fully connected CRF for accurate brain lesion segmentation. Med. Image Anal. **36**, 61–78 (2017)
14. Yang, X., Ye, Y., Li, X., Lau, R.Y.K., Zhang, X., Huang, X.: Hyperspectral image classification with deep learning models. IEEE Trans. Geosci. Remote Sens. **56**(9), 1–16 (2018)
15. Craig, R., Shan, J.: Principal component analysis for hyperspectral image classification. Surveying Land Inf. Sci. **62**(2), 115 (2002)
16. Wang, J., Chang, C.I.: Independent component analysis-based dimensionality reduction with applications in hyperspectral image analysis. IEEE Trans. Geosci. Remote Sens. **44**(6), 1586–1600 (2006)
17. Li, W., Prasad, S., Fowler, J.E., Bruce, L.M.: Locality-preserving dimensionality reduction and classification for hyperspectral image analysis. IEEE Trans. Geosci. Remote Sens. **50**(4), 1185–1198 (2012)
18. Roweis, S.T., Saul, L.K.: Nonlinear dimensionality reduction by locally linear embedding. Science **290**(5500), 2323–2326 (2000)
19. Lazcano, R., Madroñal, D., Salvador, R., Desnos, K., Pelcat, M., Guerra, R., Fabelo, H., Ortega, S., Lopez, S., Callico, G.M., Juarez, E., Sanz, C.: Porting a PCA-based hyperspectral image dimensionality reduction algorithm for brain cancer detection on a manycore architecture. J. Syst. Architect. **77**, 101–111 (2017)
20. Krizhevsky, A., Hinton, G.E.: ImageNet classification with deep convolutional neural networks. In: NIPS 2012 Proceedings of the 25th International Conference, vol. 1, pp. 1–9 (2012)
21. Lecun, Y., Bengio, Y., Hinton, G.: Deep learning. Nature **521**(7553), 436–444 (2015)
22. Chen, Y., Jiang, H., Li, C., Jia, X., Member, S.: Deep feature extraction and classification of hyperspectral images based on convolutional neural networks. IEEE Trans. Geosci. Remote Sens. **54**(10), 1–20 (2016)
23. Ioffe, S., Szegedy, C.: Batch normalization: accelerating deep network training by reducing internal covariate shift. In: 32nd International Conference on Machine Learning, pp. 448–456 (2015)

An Efficient Deep Convolutional Neural Network for Visual Image Classification

Basma Abd El-Rahiem[1], Muhammad Atta Othman Ahmed[2(✉)] (iD),
Omar Reyad[3], Hani Abd El-Rahaman[1], Mohamed Amin[1], and Fathi Abd
El-Samie[4]

[1] Faculty of Science, Menoufia University, Shibin El Kom, Egypt
basma.rahiem@gmail.com
[2] Faculty of Computers and Information, South Valley University, Luxor, Egypt
muahann@gmail.com
[3] Faculty of Science, Sohag University, Sohag, Egypt
ormak4@yahoo.com
[4] Faculty of Electronic Engineering, Menoufia University, Shibin El Kom, Egypt

Abstract. Such a hot open issue in the area of computer vision is the classification of visual images especially in Internet of Things (IoT) and remote mid-band and high-band based connections. In this paper, we propose a robust and efficient taxonomy framework. The proposed model utilizes the well-known convolutional neural network composites to construct a robust Visual Image Classification Network (VICNet). The VICNet consists of three convolutional layers, four Relu/Leaky Relu activation layers, three max-pooling layers and only two fully connected layers for extracting expected input image features. To make the training process faster, we used non-saturating neurons with a very efficient Graphics Processing Unit (GPU) implementation for the convolution operation. To minimize over-fitting issue in the fully-connected layers, we utilized a recently-developed regularization approach "dropout" with a dropping probability of 50%. The proposed VICNet framework has a high potential capability in the recognition of test images. The experimental and simulations results proven the efficacy of the proposed model.

Keywords: Deep learning · Visual image classification ·
Convolutional neural network · Deep feature extraction ·
Transfer learning

1 Introduction

In artificial intelligence, machine learning and deep learning are attracted the attention of researchers from various scholars. Artificial intelligence improvement over years yielded the well known as machine learning area as a consequence of evolution in needed technology. Indeed, machine learning allow the systems, without being explicitly programmed, to automatically classify and detect patterns in data for further purposes [1–5]. It has various implementations and

© Springer Nature Switzerland AG 2020
A. E. Hassanien et al. (Eds.): AMLTA 2019, AISC 921, pp. 23–31, 2020.
https://doi.org/10.1007/978-3-030-14118-9_3

applications in different fields from security issues to the detection of diseases in the area of medicine and computer vision community [6–12]. Deep learning, as a new subset from machine learning, refers to deeply linear combinations of several simple layers or functions parameterized by variables. The architecture, particularly the composition of layers or functions, defines a parametric function with dozens of parameters to be optimized in order to minimize the 'loss' or an objective function by some form of gradient descent based on artificial neural networks (ANNs) [13–15]. Generally, Deep learning is the multilayered hierarchical data representation in the form of a ANNs with more than two layers. It has the ability to generate new features from limited set of features located in the training data without a human intervention [16,17]. ANNs have been applied to classification, clustering, approximation and recognition problems in many applications. The progress in ANNs has been caused by the innovation of deep learning methods [1,2,9,18].

2   Related Work, Contribution, and Organization of this Paper

Convolutional neural network (CNN) is a successful deep learning approach based on artificial neural networks and attracted the attention of several scholars due to its similarity to the biological systems [19–22]. Deep convolutional neural network (DCNN) is more efficient technique than CNN and shown promising performance in visual image classification. It comprise several interconnected components and requires large datasets to train networks of a certain depth from scratch. The common processing layers of deep learning approach is as follows: data separation (training, testing and validation sets), randomized sampling during training, loading and sampling of image data, data augmentation, DCNN, fast computational structure for optimization and inference, and evaluation metrics performance during training and inference.

In the literature, there are many DCNN models have been emerged such as GoogleNet [23], AlexNet [24], ResNet [25], VGGNet [26], and others. However, most of them, if not all, are computationally overheads due to the extensive number of layers in the training and updating of the weights in the network. In addition, they have difficulties to train the first few layers in the transfer learning efficiently. This paper proposed a new Visual Image Classification Network, namely VICNet.

The main contributions of this work are as follows: a very efficient method for visual image classification based on DCNN is introduced. It adapts the well-known DCNN concepts to construct a robust VICNet. The architecture of the proposed VICNet comprises of three convolutional layers, four Relu activation layers, three max-pooling layers and only two fully connected layers for extracting candidate input image features. In order to make the training phase more faster, we utilized non-saturating neurons with a very efficient graphics processing unit (GPU) implementation for the convolution operation. The resulted over-fitting issue in the fully-connected layers is reduced by "dropout" with a dropping probability of 50%. The experimental modelling simulations proven that the proposed model has high classification accuracy. The proposed VICNet framework has a high potential capability in the recognition of test images.

The rest of the paper is organized as follows: In Sect. 3, the proposed deep learning approach for images is proposed. In Sect. 4, we discussed the proposed schemes related results while conclusions are given in Sect. 5.

3   The Proposed Deep Model (VICNet)

Here, we present an efficient deep learning approach for the recognition of visual images. The proposed topology architecture contains Convolution, ReLU, Cross Channel Normalization, Pooling, Dropout in various blend connections as in Table 1. VICNet consists of 18 layers of convolutional, some of which are followed by max-pooling layers, and fully-connected layers with a final two megabits softmax. In order to accelerate the training phase, we utilized the non-saturating neurons and a very efficient GPU implementation of the convolution operation. To eliminate over-fitting in the fully-connected layers, we adapted the recently-developed regularization method called "dropout" that proved to be very effective. We use Using Stratified Sampling Division (SSDiv [22]) to split the data into Design/Test separated subsets. The detailed steps of the proposed framework are declared in Algorithm 3:

Algorithm 1. Describes the pseudo steps of proposed VICNet image classification framework.

Require: An image data-store $\mathcal{I} = \{I_1, I_2, \ldots, I_n\}$.

Ensure: A label for visual images $\mathcal{I}' \notin \mathcal{I}$.

Training phase:

1. Using SSDiv split $\mathcal{I}$ into $\text{Design}_{Set}/\text{Test}_{Set}$; $\mathcal{I} = (D_{Set} \bigcup T_{Set})$.
2. Using SSDiv split D_{Set} into Training/Validation; $D_{Set} = (Tr_{Set} \bigcup Val_{Set})$.
3. Train(VICNet, Tr_{Set}, Val_{Set}).
4. Stop training when predefined stopping criteria is satisfied.
5. **Return:** Trained VICNet.

Test phase:

1. Evaluate VICNet($\mathcal{I}'$) metrics listed in Table 2.
2. Extract the 2-megabits of Test_{Set} features $Test_{\mathcal{F}}$ of $\mathcal{I}'$ at FC$_2$ layer.
3. Train SVM, KNN, NB and DT classifiers Using $test\mathcal{F}$.
4. Compute T_{Set} classification evaluation measures listed in Table 2.
5. **Stop.**

---

4   Experimental Evaluations

4.1   ImageNet Dogs

The database used in this article is ImageNet Dogs Dataset [27, 28]. It contains images of 120 dog breeds, 150 images per class, and total images are 20,580. Animal samples are shown in Fig. 1.

Fig. 1. Samples of ImageNet dogs dataset.

4.2   Features Classification

As an alternation for softmax classification accuracy computed at layer 17, we extracted the train/test features of length 2-megabits at the fully connected layer FC_2. The extract features are employed to train a set of base classifiers such as Support Vector Machine (SVM), K-Nearest Neighbour (K-NN), Naive Bayesian (NB) and Decision Tree (DT). We used the default Matlab implementation [29] for those classifiers.

4.3   Experimental Evaluation Metrics

To evaluate the proposed VICNet on the ImageNet dogs [27] dataset, the overall classification model is evaluated by common measures such as Sensitivity, Accuracy, Specificity, Precision, Recall, F-Measure and G-mean [30]. Table 2 shows the set of evaluation metrics for overall performance of the proposed model. The training stage recorded accuracy besides the validation check accuracy are

Table 1. Hierarchical layers topology and features of the proposed VICNet

Layer index	Layer name	Layer type	Description
1	Data	Image input	$300 \times 300 \times 3$ images with *zero center* normalization
2	$Conv_1$	Convolution	100 $10 \times 10 \times 3$ convolutions
3	$Relu_1$	ReLU	ReLU
4	$Norm_1$	Cross channel normalization	Cross channel 7 channels/element
5	$Pool_1$	Max pooling	3×3 max pooling with stride [2 2]
6	$Conv_2$	Convolution	250 $5 \times 5 \times 45$ convolutions with padding [2 2 2 2]
7	$Relu_2$	Leaky ReLU	Leaky ReLU
8	$Norm_2$	Cross channel normalization	Cross channel 5 channels/element
9	$Pool_2$	Max pooling	3×3 max pooling with stride [2 2]
10	$Conv_3$	Convolution	380 $3 \times 3 \times 256$ convolutions
11	$Relu_3$	ReLU	ReLU
12	$Pool_3$	Max pooling	3×3 max pooling with stride [2 2]
13	FC_1	Fully connected	4096 fully connected layer
14	$Drop_1$	Dropout	50% dropout
15	$Relu_4$	Leaky ReLU	Leaky ReLU
16	FC_2	Fully connected	2048 fully connected layer
17	softmax	Softmax	Softmax
18	Output	Classification label	Cross entropy with classes

Table 2. Set of evaluation metrics for overall performance of the proposed model

Classifier	Soft-Max	SVM	K-NN, $(k = 1)$	Naive Bayesian	Decision Tree
Accuracy	0.9814	**0.9949**	0.9786	0.9772	0.9246
Sensitivity	1	1	1	1	1
Specificity	**0.9809**	0.8784	0.9568	0.8700	0.8955
Precision	**0.9870**	0.9243	0.9012	0.9788	0.9489
Recall	1	1	1	1	1
F_1-Measure	0.9538	**0.9744**	0.9234	0.9222	0.9533
G-Mean	0.9352	0.9501	0.9234	**0.9788**	0.9340

satisfying. deep dream image visualization of top channels features learned by VICNet of Blenheim dog breed at different pyramidal levels are shown in Figs. 2 and 3. The trained VICNet can be found at [31].

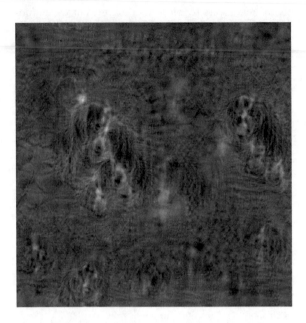

Fig. 2. Visualize deep dream synthetic features of blenheim spaniel dog breeds that learned by VICNet, pyramidal level = 1.

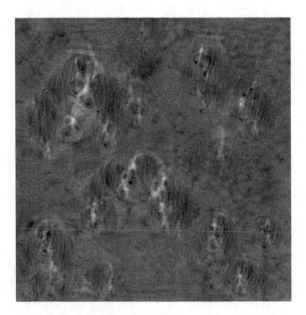

Fig. 3. Visualize deep dream synthetic features of blenheim spaniel dog breeds that learned by VICNet, pyramidal level = 4.

5   Results and Discussion

Deep learning is the newborn generation of machine learning. In this work we proposed an efficient visual image classification network of category convolutional network. The proposed approach adapted Convolution, Cross Channel Normalization, Max Pooling, Fully Connected, and Softmax as components of CNN procedures to classify the ImageNet dog breeds images. The trained VIC-Net performance is evaluated via a subset of evaluation metrics (see Sect. 4.3). Applying network deep activation to extract the features of the applied at input layer images. Using the 2-megabits extracted activation features, we train a set of features classifiers (see Sect. 4.2), using the training set then the classifiers were evaluated using the test set. Results reported in Table 2 shows an evidence for the proposed model robustness and reliability as an empirical DNN composite layered model. Our future extended work on VICNet is to present a full demonstration on the utilization of the proposed model on full ImageNet database.

References

1. Lee, J.-G., Jun, S., Cho, Y.-W., Lee, H., Kim, G.B., Seo, J.B., Kim, N.: Deep learning in medical imaging general overview. Korean J. Radiol. **18**(4), 570–584 (2017)
2. Ahmed, M.A.O., Didaci, L., Lavi, B., Fumera, G.: Using diversity for classifier ensemble pruning: an empirical investigation. Theor. Appl. Inform. **29**(1&2), 25–39 (2018)
3. Cai, J., Luo, J., Wang, S., Yang, S.: Feature selection in machine learning: a new perspective. Neurocomputing **300**, 70–79 (2018)
4. Zhang, T., El-Latif, A.A.A., Wang, N., Li, Q., Niu, X.: A new image segmentation method via fusing NCut eigenvectors maps. In: Proceedings of SPIE 8334, Fourth International Conference on Digital Image Processing (ICDIP 2012), p. 833430 (2012)
5. Bai, X., Zhang, T., Wang, C., El-Latif, A.A.A., Niu, X.: A fully automatic player detection method based on one-class SVM. IEICE Trans. Inf. Syst. **96**(2), 387–391 (2013)
6. Shi, Z., Yu, L., El-Latif, A.A.A., Niu, X.: Skeleton modulated topological perception map for rapid viewpoint selection. IEICE Trans. Inf. Syst. **95**(10), 2585–2588 (2012)
7. Khfagy, M., AbdelSatar, Y., Reyad, O., Omran, N.: An integrated smoothing method for fingerprint recognition enhancement. In: International Conference on Advanced Intelligent Systems and Informatics, pp. 407–416. Springer, Cham (2016)
8. Ahmed, M.A.O., Reyad, O., AbdelSatar, Y., Omran, N.F.: Multi-filter score-level fusion for fingerprint verification. In: International Conference on Advanced Machine Learning Technologies and Applications, pp. 624–633. Springer, Cham (2018)
9. El-Sayed, M.A., Khafagy, M.A.: An identification system using eye detection based on wavelets and neural networks. arXiv preprint arXiv:1401.5108 (2014)
10. Nife, F., Kotulski, Z., Reyad, O.: New SDN-oriented distributed network security system. Appl. Math. Inf. Sci. **12**(4), 673–683 (2018)

11. Gad, R., Talha, M., El-Latif, A.A.A., Zorkany, M., El-Sayed, A., El-Fishawy, N., Muhammad, G.: Iris recognition using multi-algorithmic approaches for cognitive internet of things (CIoT) framework. Future Gener. Comput. Syst. **89**, 178–191 (2018)
12. Peng, J., El-Latif, A.A.A., Belazi, A., Kotulski, Z.: Efficient chaotic nonlinear component for secure cryptosystems. In: Ninth International Conference on Ubiquitous and Future Networks (ICUFN), pp. 989–993. IEEE (2017)
13. Shiddieqy, H.A., Hariadi, F.I., Adiono, T.: Implementation of deep-learning based image classification on single board computer. In: 2017 International Symposium on Electronics and Smart Devices (ISESD), pp. 133–137. IEEE (2017)
14. Liu, R., Yang, B., Zio, E., Chen, X.: Artificial intelligence for fault diagnosis of rotating machinery: a review. Mech. Syst. Sig. Process. **108**, 33–47 (2018)
15. Gibson, E., Li, W., Sudre, C., Fidon, L., Shakir, D.I., Wang, G., Rosen, Z.E., Gray, R., Doel, T., Hu, Y., Whyntie, T., Nachevc, P., Modat, M., Barratta, D.C., Ourselin, S., Cardoso, M.J., Vercauteren, T.: NiftyNet: a deep-learning platform for medical imaging. Comput. Methods Programs Biomed. **158**, 113–122 (2018)
16. Hu, W., Huang, Y., Wei, L., Zhang, F., Li, H.: Deep convolutional neural networks for hyperspectral image classification. J. Sens. **2015**, 12 (2015)
17. El-Sayed, M.A., Estaitia, Y.A., Khafagy, M.A.: Automated edge detection using convolutional neural network. Int. J. Adv. Comput. Sci. Appl. **4**(10), 10–20 (2013)
18. Lavi, B., Ahmed, M.A.O.: Interactive fuzzy cellular automata for fast person re-identification. In: The International Conference on Advanced Machine Learning Technologies and Applications (AMLTA2018), pp. 147–157. Springer, Cham (2018)
19. LeCun, Y., Bengio, Y., Hinton, G.: Deep learning. Nature **521**(7553), 436 (2015)
20. Li, J., Zhang, B., Lu, G., Zhang, D.: Generative multi-view and multi-feature learning for classification. Inf. Fusion **45**, 215–226 (2019)
21. Russakovsky, O., Deng, J., Su, H., Krause, J., Satheesh, S., Ma, S., Huang, Z., Karpathy, A., Khosla, A., Bernstein, M., et al.: ImageNet large scale visual recognition challenge. Int. J. Comput. Vis. **115**(3), 211–252 (2015)
22. Ahmed, M.A.O.: Trained neural networks ensembles weight connections analysis. In: International Conference on Advanced Machine Learning Technologies and Applications, pp. 242–251. Springer, Cham (2018)
23. Szegedy, C., Liu, W., Jia, Y., et al.: Going deeper with convolutions. In: IEEE Conference on Computer Vision and Pattern Recognition, Boston, USA, pp. 1–9, (2015). IEEE
24. Krizhevsky, A., Sutskever, I., Hinton, G.E.: Imagenet classification with deep convolutional neural networks. In: Advances in Neural Information Processing Systems, California, USA, pp. 1097–105 (2012)
25. He, K., Zhang, X., Ren, S., et al.: Deep residual learning for image recognition. In: IEEE Conference on Computer Vision and Pattern Recognition, Las Vegas, USA, pp. 770–778, (2016)
26. Simonyan, K., Zisserman, A.: Very deep convolutional networks for large-scale image recognition (2015, 2017). https://arxiv.org/pdf/1409.1556v6.pdf
27. Khosla, A., Jayadevaprakash, N., Yao, B., Li, F.-F.: Stanford dogs dataset (2011). http://vision.stanford.edu/aditya86/ImageNetDogs
28. Khosla, A., Jayadevaprakash, N., Yao, B., Li, F.-F.: Novel dataset for fine-grained image categorization. In: First Workshop on Fine-Grained Visual Categorization, IEEE Conference on Computer Vision and Pattern Recognition, Colorado Springs, CO, June 2011
29. MATLAB: Statistics and Machine Learning Toolbox. The mathworks (2018)

30. Sokolova, M., Lapalme, G.: A systematic analysis of performance measures for classification tasks. Inf. Process. Manag. **45**(4), 427–437 (2009)
31. Khfagy, M.A.O.A.: Visual image classification convolutional network (VICNET) (2019). https://github.com/mkhfagy/VICNet

Deep Learning for Predictive Analytics in Healthcare

Anandhavalli Muniasamy[1(✉)], Sehrish Tabassam[1],
Mohammad A. Hussain[1], Habeeba Sultana[1], Vasanthi Muniasamy[1],
and Roheet Bhatnagar[2]

[1] College of Computer Science, King Khalid University,
Abha, Kingdom of Saudi Arabia
anandhavalli.dr@gmail.com
[2] Department of Computer Science and Engineering,
Manipal University Jaipur, Rajasthan, India

Abstract. Despite a recent wealth of data and information, the healthcare sector is lacking in actionable knowledge. The healthcare industry faces challenges in essential areas like electronic record management, data integration, and computer-aided diagnoses and disease predictions. It is necessary to reduce healthcare costs and the movement towards personalized healthcare. The rapidly expanding fields of deep learning and predictive analytics has started to play a pivotal role in the evolution of large volume of healthcare data practices and research. Deep learning offers a wide range of tools, techniques, and frameworks to address these challenges. Health data predictive analytics is emerging as a transformative tool that can enable more proactive and preventative treatment options. In a nutshell, this paper focus on the framework for deep learning data analysis to clinical decision making depicts the study on various deep learning techniques and tools in practice as well as the applications of deep learning in healthcare.

Keywords: Healthcare data · Electronic medical records ·
Deep learning (DL) · Predictive analytics

1 Introduction

Healthcare is a domain in which prediction is perhaps more important than explanation, considering the formidable cost of delay in diagnosis and treatment [1]. The value of predictive analytics in healthcare repeatedly highlighted in prior information systems (IS) research. The digitization of healthcare is resulting in the creation of massive new data sets in the healthcare industry. Potential sources of clinical information include computerized physician order entries, physicians' notes, and imaging devices, just to name a few. These datasets are particularly complex and fragmented compared to other industries [36], which involves huge problems in diagnosis, treatment, and prevention, and their improvement represents incalculable value.

Predictive analytics supports health care life sciences and providers and applies many techniques from data mining, statistics, modeling, machine learning, and artificial

© Springer Nature Switzerland AG 2020
A. E. Hassanien et al. (Eds.): AMLTA 2019, AISC 921, pp. 32–42, 2020.
https://doi.org/10.1007/978-3-030-14118-9_4

intelligence to investigate current findings to make predictions about the future. It helps healthcare organizations to prepare for health care by optimizing the cost [5], diagnosing the diseases accurately, enhancement of patient care, resource optimization and improves clinical outcomes.

The concept of deep learning is to dig large volume of data to automatically identify patterns and extract features from complex unsupervised data without the involvement of human, which makes it an important tool in big data analysis [19]. Deep learning plays an important role in diagnostic applications [15]. Deep learning techniques can reveal clinically relevant information hidden in the large data with a guidance of relevant clinical questions to assist clinical decision-making and in turn provides the physicians the analysis of any disease accurately for better treatment, thus resulting in better medical decisions.

This paper summarizes the status of deep learning for predictive analysis in the health sector, as well as discuss its future. We briefly review four relevant aspects from medical investigators' perspectives:

- Motivations of applying deep learning in healthcare
- Deep Learning Framework for Healthcare predictions
- Disease types that the deep learning communities are currently tackling
- Tools that enable deep learning systems to generate clinical meaningful results.

2 Deep Learning Predictive Analytics Survey in Health Care

Healthcare predictive analytics aims to predict future health-related outcomes or events based on clinical and/or nonclinical patterns in the data. In pharmaceutical research, applications of deep learning have emerged in recent years, have shown promise in addressing diverse problems in drug discovery [8] by analyzing the patient's medical history, and provides the best treatment for the patients by gaining insights from their symptoms and tests. The outcomes of the studies such as medical complications [38], hospital readmissions [3], treatment responses [31], and patient mortality [39], are often of great practical importance in healthcare predictive analytics. The current trend for deep learning is given in (see Fig. 1) shows its prominent role in the healthcare data analysis.

Development of healthcare predictive model involves in two ways: firstly, the collection of patient data in clinical trials with a set of predefined protocols. For example, Lung cancer risk prediction model [40]; Framingham Heart Study [13]; UK Prospective Diabetes Study (UKPDS) [38]. Secondly, the use of existing patient data collected in clinical practice, such as EHRs, insurance claims, and clinical registries. For instance, inpatient mortality predictive model [39]; readmission model [3].

While healthcare predictive analytics can support clinical decisions, the actual use of predictive models in clinical practice remains limited [33]. The barriers in the use of predictive models in healthcare eased with the implementation of EHR systems [33, 42], such as inadequate integration with existing clinical workflow, requiring variables that are expensive to obtain or not immediately accessible, and the need to adapt the models from the study population to the local population.

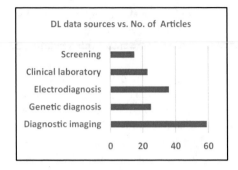
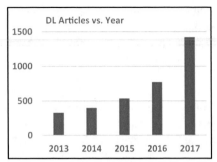
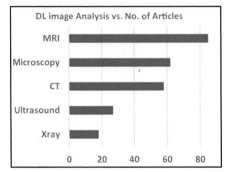
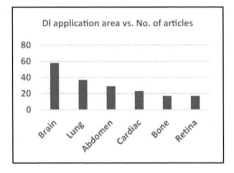

Fig. 1. Current trend for deep learning. The data are generated through searching the deep learning in healthcare and specific keywords on PubMed.

Predictive models capture the characteristics of the specific event. The UKPDS risk engine can predict coronary heart disease and stroke in patients with type 2 diabetes [28, 38]. 37 developed a two-part model to predict lymph node ratio and survival in pancreatic cancer patients. [3] investigated the readmissions of patients with congestive heart failure, and developed a model to answer whether, when, and how often the patients would have been readmitted.

Deep learning is widely used for medical imaging analysis in several different application domains, which an evident from the survey conducted by [30] includes over 300 papers, most of them recent. Medical imaging techniques such as MRI scans, CT scans, ECG, used to diagnose dreadful diseases such as heart disease, cancer, brain tumor. Hence, with the help of deep learning, the doctors can analyze the disease better and provide patients with the best treatment. In addition, the deep learning used to analyze the medical insurance fraud claims. However, along with predictive analytics, it can predict fraud claims that are likely to happen in the future. Moreover, deep learning helps the insurance industry to send out discounts and offers to their target patients. Deep learning technique used to detect Alzheimer's disease [32] at an early stage in which the medical industry faces the challenges currently. Deep learning technique used to understand a genome [45] and help patients get an idea about the disease that might affect them, which has a promising future also.

3 Deep Learning Framework for Healthcare Predictions

Deep learning combines advances in computing power and neural networks with many layers (See Fig. 2) to learn complicated patterns in large amounts of data. It is an extension of classical neural network and uses more hidden layers so that the algorithms can handle complex data with various structures [19].

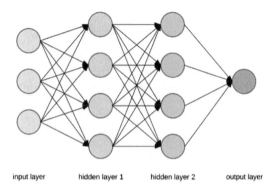

input layer   hidden layer 1   hidden layer 2   output layer

Fig. 2. An illustration of deep learning with two hidden layers.

In the Gartner-selected top ten technology trends of 2018, deep learning represented artificial intelligence technologies were ranked at the top position [43]. In recent days, deep learning has been applied in many computer vision applications [23] like machine vision, robotic vision, natural language processing, neurosciences, text recognition, natural language processing, image recognition and classification, audio and speech recognition, etc. Deep learning collects a large volume of data, including patients' records, medical reports, and insurance records, and applies its neural networks to provide the best outcomes. Therefore, it is important to involve deep learning role to resolve healthcare issues due to its representational and recognition supremacy that assists healthcare personals to determine, predict, analyze, and practice its theories for the delivery of healthcare.

3.1 Deep Learning Models

The main difference between traditional machine learning and deep learning algorithms is in the feature engineering which requires domain expertise and a time-consuming process. Deep learning algorithms involve automatic feature engineering, whereas we need to handcraft the features in traditional machine learning algorithms (see Fig. 3).

In the medical applications, the commonly used deep learning algorithms include

- Convolution neural network (CNN)
- Recurrent neural network (RNN)
- Deep belief network (DBN)
- Deep neural network (DNN)
- Generative Adversarial Network (GAN)

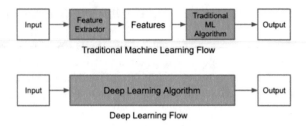

Fig. 3. Deep learning Vs. Machine learning flow

The following Table 1 gives a summary of the recent applications in healthcare, technical advantages and limitations of each deep learning models.

Table 1. Deep learning model summary

Model	Recent applications in healthcare	Advantages	Limitations
CNN	Abnormal Heart Sound Detection [22], Myocardial Infarction Detection [26], Radiology [44]	Provide very good performance for 2D data. Model learning is fast	Need lots of labeled data for classification
RNN	Detection of heart failure onset [11], Classification of lung abnormalities [25]	Learn sequential events and model time dependencies. Provide good accuracy in speech & character recognition and NLP related tasks	Need big datasets. Has many issues due to gradient vanishing
DBN	Predict Drug combination [9], Detection of type 1 diabetes [35]	Supports both supervised and unsupervised learning model	Initialization process makes the training process computationally expensive.
DNN	Heart Sound Recognition [10], Phonocardiography [17]	Widely used with great accuracy	Training process is not trivial as the error propagated back to the previous layers and becomes very small. Learning process is also too much slow
GAN	Generating synthetic brain CTs [16], Reconstructing natural images from brain activity [24], Medical imaging platform [18]	Good method for training classifiers in a semi-supervised way	Hard to learn to generate discrete data, like text and very hard to train.

Convolution neural network (CNN): The CNN was first proposed and applied method for the high-dimensional image analysis by [29]. It consists of convolutional filters, which transform 2D into 3D.

Recurrent neural network (RNN): It is a neural net architecture with recurrent connections between hidden states and has the capability of learning sequences and model time dependencies also. The recurrent connections are used to detect relationships not only between inputs, but also over time. Therefore, it is well suited to health problems that often involve modeling clinical data changes over time [46].

Deep belief network (DBN): This model has a unidirectional connection at two layers on the top of layers. The hidden layers of each sub-network serve as a visible layer for the next layer.

Deep neural network (DNN): It has more than two layers, which allows the complex non-linear relationship.

Generative Adversarial Network (GAN): GAN architecture consists of a two networks, generator and discriminator, both of which compete in the training phase. GAN has been widely used to generate realistic images.

3.2 Framework of Deep Learning in Healthcare

When building an analytical environment, it is necessary to have an infrastructure layer, data layer, the analytics layer, and application layer. The infrastructure layer supports various frameworks and applications. Data layer covers multiple data streams such as electronic medical records, images, genetic and eligible provider (EP) data. The analytics layer should support different classification, clustering, regression etc. algorithms for predictive analytics. Since healthcare data is unstructured, data analytic tools like natural language processing can be incorporated to analyze specific data sources. Application layer supports the visualization of predictive model results. The proposed framework (see Fig. 4) starts with the clinical data generation and ends with clinical decision making; motivated by clinical problems and be applied to assist clinical practice in the end with the help of deep learning techniques. The proposed framework involves the following phases:

- Clinical data generation
- Natural language processing for unstructured data enhancement
- Deep learning data analysis to find patterns of interest
- Clinical decision making

Developing a medical database involves the collection of data from an electronic medical record (EMR) and clinical activities. The clinical activities involve diagnosis, screening, and treatment data of the patients. The image, eligible provider (EP) and genetic data are machine-understandable so that the DL algorithms can be directly performed. Clinical information in the form of narrative text, such as physical examination, clinical laboratory reports, etc., are unstructured. For unstructured data processing, natural language processing (NLP) is useful for extracting useful information from the narrative text to assist clinical decision making. An NLP pipeline covers two main components such as text processing and classification. NLP identifies a series of disease-relevant keywords in the clinical notes based on the historical databases during

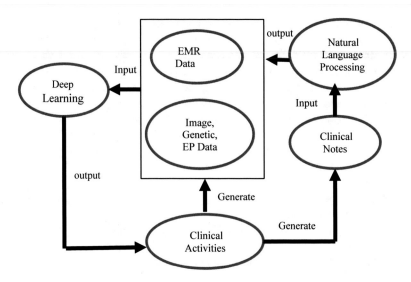

Fig. 4. Deep learning data analysis to clinical decision making

text processing. Then a subset of the keywords is selected through examining their effects on the classification of the normal and abnormal cases. The NLP pipeline will be developed to assist clinical decision making. The authors are working to realize the models as discussed above.

4   Deep Learning Tools

Deep learning offers a collection of tools that use computers to convert data into actionable information. Tools are a huge part of machine learning and selecting the right tool can be as important as working with the best algorithms. Excellent tools shown in Table 2 can automate each step in the applied deep learning process by reducing the time.

Table 2. Deep learning tools.

Tools	Features
Tensorflow [41]	An open source artificial intelligence library, using data flow graphs to build models and allows developers to create large-scale neural networks with many layers
MXNet [34]	An open source software library for numerical computation using data flow graphs and supports DL architectures CNN and RNN
Caffe [6]	A cross-platform support C++, Matlab, and Python programming interfaces
Theano	Provides capabilities like symbolic API supports looping control (scan), which makes implementing RNNs easy and efficient

(continued)

Table 2. (*continued*)

Tools	Features
Keras	Theano based deep learning library
ConvNet [12]	Matlab based convolutional neural network toolbox
Deeplearning4j [14]	An open-source, Apache 2.0-licensed with distributed neural net library written in Java and Scala
Apache Singa [2]	Open source library for deep learning
H20.ai [21]	Open source used by leading healthcare companies to deliver AI solutions that are changing the industry

5 Future Trends of Deep Learning in Healthcare Predictions

Researchers and vendors in medical sector have confidence in that deep learning applications will take over human and not only the diagnosis process will be performed by intelligent machines but will also help to prescribe medicine, predict disease, and guide in treatment in future. Deep learning techniques applied to medical imaging is evident since the advent of digital imaging. Google DeepMind Health [20] is working with the National Health Service of the United Kingdom to process the medical data of more patients across the five hospitals. IBM Watson [27] recently enhanced itself through billion-dollar entry into the imaging arena by the acquisition of Merge's medical management platform, which covers radiology, cardiology, orthopedics, eye care, and other healthcare specialties. [4] says that they use deep learning technique to distill actionable insights from billions of clinical cases and makes doctors faster and more accurate. [7] uses deep learning technique and helps parents to monitor the health of their children through a smart device in real time, thus minimizing frequent visits to the doctor. Machine learning will be used to combine visual data and motor patterns within devices such as the da Vinci to allow machines to master surgeries in the future. Deep learning future in medical imaging data analysis is an emerging trend but still, there are challenges involved in this field. The idea of applying deep learning algorithms to medical imaging data is a fascinating and growing research area; the challenges are the unavailability of the dataset, dedicated medical experts, nonstandard data machine learning algorithms, privacy, and legal issues. With the help of deep learning, personalized medicine involves health recommendations and disease treatments based on their medical history, genetic lineage, past conditions, diet, stress levels, and more.

6 Conclusion

In this paper, we have provided a brief overview of deep learning research as it pertains to healthcare data predictive analysis with the motivation of using deep learning in healthcare. We presented the various healthcare area that DL has analyzed and surveyed the major disease types that DL has been deployed. The major intention of this study is to propose a framework for monitoring healthcare data using DL with predictive analysis. So, the proposed model can be practically implemented to verify the empirical results to show the benefits of the proposed model. Looking at current trends in the deep learning, we identify a key area which can be highly relevant for medical imaging and is receiving interest in unsupervised learning. We also foresee deep learning approaches will be used for the variable domain of the health industry. Deep learning will thus have a great impact on healthcare data analysis.

References

1. Agarwal, R., Dhar, V.: Big data, data science, and analytics: the opportunity and challenge for IS research. Inf. Syst. Res. **25**(3), 443–448 (2014)
2. Apache Singa: http://singa.apache.org/en/index.html. Accessed 15 Nov 2018
3. Bardhan, I., Oh, J., Zheng, Z., Kirksey, K.: Predictive analytics for readmission of patients with congestive heart failure. Inf. Systems. Res. **26**(1), 19–39 (2014)
4. Enlitic: https://www.enlitic.com/. Accessed 15 Nov 2018
5. Nithya, B.: Study on predictive analytics practices in health care system. IJETTCS **5**, 98–102 (2016)
6. Caffe: http://caffe.berkeleyvision.org/. Accessed 15 Nov 2018
7. Cellscope: https://www.cellscope.com/. Accessed 15 Nov 2018
8. Hongming, C., Engkvist, O., Wang, Y., Olivecrona, M., Blaschke, T.: The rise of deep learning in drug discovery. Drug Discov. Today **23**(6), 1241–1250 (2018)
9. Chen, G., Tsoi, A., Xu, H., Zheng, W.J.: Predict effective drug combination by deep belief network and ontology fingerprints. J. Biomed. Inform. **85**, 149–154 (2018)
10. Chen, T.E., et al.: S1 and S2 heart sound recognition using deep neural networks. IEEE Trans. Biomed. Eng. **64**(2), 372–380 (2017)
11. Choi, E., Schuetz, A., Stewart, W.F., Sun, J.: Using recurrent neural network models for early detection of heart failure onset. J. Am. Med. Inform. Assoc. **24**(2), 361–370 (2017)
12. ConvNet: https://github.com/sdemyanov/ConvNet. Accessed 15 Nov 2018
13. D'Agostino, R.B., Vasan, R.S., Pencina, M.J., Wolf, P.A., Cobain, M., Massaro, J.M., Kannel, W.B.: General cardiovascular risk profile for use in primary care: the framingham heart study. Circulation **117**(6), 743–753 (2008)
14. Deeplearning4j: https://deeplearning4j.org/. Accessed 15 Nov 2018
15. Deep learning in Oncology: https://www.techemergence.com/deep-learning-in-oncology/. Accessed 15 Nov 2018
16. Emami, H., Dong, M., Nejad-Davarani, S.P., Glide-Hurst, C.K.: Generating synthetic CTs from magnetic resonance images using generative adversarial networks. Med. Phys. **45**, 3627–3636 (2018)
17. Gharehbaghi, A., Babic, A.: Structural risk evaluation of a deep neural network and a Markov model in extracting medical information from phonocardiography. Stud. Health Technol. Inform **251**, 157–160 (2018)
18. Gibson, E., et al.: NiftyNet: a deep-learning platform for medical imaging. Comput. Methods Programs Biomed. **158**, 113–122 (2018)
19. Goodfellow, I., Bengio, Y., Courville, A.: Deep Learning, 1st edn. The MIT Press, Cambridge (2016)
20. Google DeepMind Health: https://deepmind.com/applied/deepmind-health/. Accessed 15 Nov 2018
21. H20.ai: https://www.h2o.ai/. Accessed 15 Nov 2018

22. Humayun, A.I., Ghaffarzadegan, S., Feng, Z., Hasan, T.: Learning front-end filter-bank parameters using convolutional neural networks for abnormal heart sound detection. In: Conference on Proceedings of IEEE Engineering in Medicine Biology Society (2018)
23. Howard, J.: The business impact of deep learning. In: Proceedings of the 19th ACM SIGKDD International Conference on Knowledge Discovery and Data Mining, p. 1135 (2013)
24. Seeliger, K., Güçlü, U., Ambrogioni, L., Güçlütürk, Y., van Gerven, M.A.J.: Generative adversarial networks for reconstructing natural images from brain activity. NeuroImage **181**, 775–785 (2018)
25. Khodabakhshi, M.B., Moradi, M.H.: The attractor recurrent neural network based on fuzzy functions: an effective model for the classification of lung abnormalities. Comput. Biol. Med. **1**(84), 124–136 (2017)
26. Liu, W., et al.: Real-time multilead convolutional neural network for myocardial infarction detection. IEEE J. Biomed. Health Inform. **22**(5), 1434–1444 (2018)
27. Lohr, S.: IBM is counting on its bet on Watson, and Paying Big Money for It: https://www.nytimes.com/2016/10/17/technology/ibm-is-counting-on-its-bet-on-watson-and-paying-big-money-for-it.html. Accessed 15 Nov 2018
28. Kothari, V., Stevens, R.J., Adler, A.I., Stratton, I.M., Manley, S.E., Neil, H.A., Holman, R. R.: UKPDS 60: risk of stroke in type 2 diabetes estimated by the uk prospective diabetes study risk engine. Stroke **33**(7), 1776–1781 (2002)
29. Lecun, Y., Bottou, L., Bengio, Y., et al.: Gradient-based learning applied to document recognition. Proc. IEEE Inst. Electr. Electron. Eng. **86**, 2278–2324 (1998)
30. Litjens, G., et al.: A survey on deep learning in medical image analysis. Med. Image Anal. **42**, 60–88 (2017)
31. Meyer, G., Adomavicius, G., Johnson, P.E., Elidrisi, M., Rush, W.A., Sperl-Hillen, J.M., O'Connor, P.J.: A machine learning approach to improving dynamic decision making. Inf. Syst. Res. **25**(2), 239–263 (2014)
32. Mirzaei, G., Adeli, A., Adeli, H.: Imaging and machine learning techniques for diagnosis of Alzheimer's disease. Rev. Neurosci. **27**(8), 857–870 (2018)
33. Moons, K.G., Royston, P., Vergouwe, Y., Grobbee, D.E., Altman, D.G.: Prognosis and prognostic research: what, why, and how? BMJ **338**, 1317–1320 (2009)
34. MXNet: http://mxnet.incubator.apache.org/. Accessed 15 Nov 2018
35. San, P.P., Ling, S.H., Nguyen, H.T.: Deep learning framework for detection of hypoglycemic episodes in children with type 1 diabetes. In: Conference on Proceedings of IEEE Engineering in Medicine Biology Society (2016)
36. Raghupathi, W., Raghupathi, V.: Big data analytics in healthcare: promise and potential. Health Inf. Sci. Syst. **2**(1), 3 (2014)
37. Smith, B.J., Mezhir, J.J.: An interactive Bayesian model for prediction of lymph node ratio and survival in pancreatic cancer patients. J. Am. Med. Inform. Assoc. **21**(2), 203–211 (2016)
38. Stevens, R.J., Kothari, V., Adler, A.I., Stratton, I.M.: The UKPDS risk engine: a model for the risk of coronary heart disease in type II diabetes (UKPDS 56). Clin. Sci. **101**(6), 671–679 (2001)
39. Tabak, Y.P., Sun, X., Nunez, C.M., Johannes, R.S.: Using electronic health record data to develop inpatient mortality predictive model: acute laboratory risk of mortality score (ALaRMS). J. Am. Med. Inform. Assoc. **21**(3), 455–463 (2014)
40. Tammemägi, M.C., Katki, H.A., Hocking, W.G., Church, T.R., Caporaso, N., Kvale, P.A., Chaturvedi, A.K., Silvestri, G.A., Riley, T.L., Commins, J., Berg, C.D.: Selection criteria for lung-cancer screening. New Engl. J. Med. **368**(8), 728–736 (2013)
41. Tensorflow: https://www.tensorflow.org/. Accessed 15 Nov 2018

42. Toll, D.B., Janssen, K.J.M., Vergouwe, Y., Moons, K.G.M.: Validation, updating and impact of clinical prediction rules: a review. J. Clin. Epidemiol. **61**(11), 1085–1094 (2008)
43. Top Strategic Technology Trends for 2018: http://www.gartner.com/technology/research/top-10-technology-trends/. Accessed 15 Nov 2018
44. Yamashita, R., Nishio, M., Do, R.K.G., Togashi, K.: Convolutional neural networks: an overview and application in radiology. Insights Imaging **9**, 611–629 (2018)
45. Yue, T., Wang, H.: Deep Learning for Genomics: A Concise Overview (2018)
46. Lipton, Z.C., Kale, D.C., Elkan, C., Wetzel, R.: Learning to diagnose with LSTM recurrent neural networks. In: International Conference on Learning Representations (ICLR) (2016)

Deep Layer CNN Architecture for Breast Cancer Histopathology Image Detection

Zanariah Zainudin[✉], Siti Mariyam Shamsuddin,
and Shafaatunnur Hasan

School of Computing, Faculty of Engineering, Universiti Teknologi Malaysia,
81310 Skudai, Johor, Malaysia
zanariah86@gmail.com, sitimariyams@gmail.com,
shafaatunnur@gmail.com, shafaatunnur@utm.my

Abstract. In recent years, there are various improvements in computational image processing methods to assist pathologists in detecting cancer cells. Consequently, deep learning algorithm known as Convolutional Neural Network (CNN) has now become a popular method in the application image detection and analysis using histopathology image (images of tissues and cells). This study presents the histopathology image related to breast cancer cells detection (mitosis and non-mitosis). Mitosis is an important parameter for the prognosis/diagnosis of breast cancer. However, mitosis detection in histopathology image is a challenging problem that needs a deeper investigation. This is because mitosis consists of small objects with a variety of shapes, and is easily confused with some other objects or artefacts present in the image. Hence, this study proposed three types of deep layer CNN architecture which are called 6-layer CNN, 13-layer CNN and 17-layer CNN, respectively in detecting breast cancer cells using histopathology image. The aim of this study is to detect the breast cancer cell which is called mitosis from histopathology image using suitable layer in deep layer CNN with the highest accuracy and True Positive Rate (TPR), and the lowest False Positive Rate (FPR) and loss performances. The result shows a promising performance for deep layer CNN architecture of 17-layer CNN is suitable for this dataset with the highest average accuracy, 84.49% and True Positive Rate (TPR), 80.55%; while the least False Positive Rate (FNR), 11.66% and loss 15.50%.

Keywords: Breast cancer image detection · Deep Learning ·
Histopathology image · Convolutional Neural Network (CNN)

1 Introduction

Cancer has become the most popular health problem that can be dangerous to human life especially for women. This is supported by the statistics from Europe and United State health reports that 30% women from a total of 852,630 was estimated to be suffering from breast cancer. Another supporting statistics is based on The National Cancer Institute in 2014, which reported on breast cancer based on the population statistics in the US population. It shows that the number of cancer patients kept increasing year by year. Based on this report, the study conducted by the National Cancer Registry (NCR), US women are more exposed to breast cancer where in 1999

© Springer Nature Switzerland AG 2020
A. E. Hassanien et al. (Eds.): AMLTA 2019, AISC 921, pp. 43–51, 2020.
https://doi.org/10.1007/978-3-030-14118-9_5

to 2013, some 231,840 new estimated cases related to breast cancer were reported to NCR in 2014 and the number kept increasing from year to year. The number of new cases of female breast cancer increased to 852,630 in 2017 [1, 2, 25]. Moreover, it is also stated that the age-adjustment for female breast cancer amongst women was lower compared to several age gaps such as stated and illustrated below (Fig. 2) 20 years (0.00%), 20 to 34 year (1.8%), 35 to 44 years (9.1%), 45 to 54 years (21.6%), 55 to 64 year (25.6%), 65 to 74 years (21.9%), 75 to 84 years (14.2%) and 84 year and above (5.7%). It can be summarized that, Fig. 1 provides comparisons for age–adjustment for female breast cancer per 100,000 women from 2008 until 2012 for all women in United State (US) [1, 2].

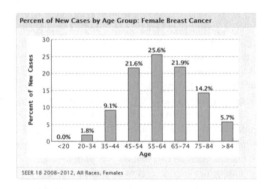

Fig. 1. Age-adjustment incidences of breast cancer per 100,000 population

This situation can be reduced if there are improvements in early detection and treatment [3, 4]. In order to detect breast cancer, various clinical tests have been used in hospitals such as ultrasound scan, mammogram, and histopathology. Normally, a lot of patients use mammogram to detect possibility of having cancerous cells. However, these tests have several disadvantages where younger women may get non accurate results as their breasts give clearer picture compared to older women. As an alternative test, most younger women are advised to use histopathology image for more accurate result. Usually biopsy test is a supportive test after the first test is conducted. Using histopathology image is the most efficient test to detect cancerous cells [4, 5]. Although histopathology image has been proven to be the best way to classify the mitosis and nuclei for breast cancer, getting True Positive Rate (TPR) result from the image however requires the skills and experience of a pathology expert [6, 7]. Based on previous researcher as stated by Khuzi *et al.* and Singh *et al.,* the misclassification of the mitosis and non-mitosis are based on many criteria such as image quality is not very good, physician eye, poor segmentation problem in image and many other causes where it can lead to incorrect misclassification, which is also called False Positive Rate (FPR) [6, 7]. The main contribution of this studies is to find the suitable layer for Deep Learning algorithm for histopathology image dataset.

Previous research shows that the False Positive Rate (FPR) results on diagnosis of mitosis and non-mitosis using the normal deep learning algorithm were quite high [8–12]. Ciresan *et al.* [13], in their paper entitled "Mitosis detection in breast cancer histology images with deep neural networks" using Deep Neural Network and MITOS dataset stated that their result of using Deep Neural Network is that the precision is 0.88, re-call is 0.70 and F1-Score is 0.782 which better than the algorithm. Meanwhile, Su *et al.* [14], said that using deep learning algorithm are improving the result on detection of the breast cancer cell. They used image of size 1000 × 1000 for the experimental result and they enable to shorten the computational time. Most of the traditional algorithm are not good in term of computational time and the performance of the breast cancer detection. Spanhol *et al.* [17], 2016 also discussed on DeCAF features where they compare three type of layer Deep Neural Network and reusing the CAFFE architecture and parameters of a pre-trained neural network. Wahlstr [15], mentioned that using deep learning algorithm successfully decreased the error for his testing model for detecting Sensor Network Data. The result for the joint training error is 0.0011 and separate training error is 0.0011. His also said that, for the future work that he will used the sampling-based approaches to get better result. Feng *et al.* [16] stated that Deep Learning model is the best algorithm compare other machine learning classifiers.

2   The Proposed Deep Layer CNN Architecture for Breast Cancer Image Detection

Recently, using Deep Learning algorithm is better compared to segmentation due to the misclassification of the mitosis with non-mitosis [13, 18, 19]. This is caused by the complicated structure of the histopathology image and also the difficulty in the detection of the mitosis part due to its similarity [3, 13]. Figure 2 shows the Deep Layer CNN framework for breast cancer image detection.

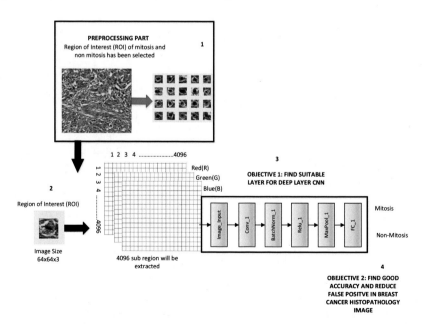

Fig. 2. Deep Layer CNN framework for breast cancer image detection.

Various applications of classifiers especially in medical image has great accuracy using Deep Learning classifier. The Deep Learning can be trained with many training images inside the algorithm. The advantage of the Deep Learning is it can divide each image into sub regions. For an example, 4096 × 3 sub regions can be extracted from a Region of Interest (ROI) with an image size of 64 × 64 × 3. As a result, the training of Deep Learning is done on sub regions basis; not on the whole region basis as shown in Fig. 2. Deep Learning with huge number of samples from the training sample dataset can allow some enrichment process which will help the training to avoid over fitting problem of an Artificial Neural Network (ANN) [20–22]. Deep learning algorithm

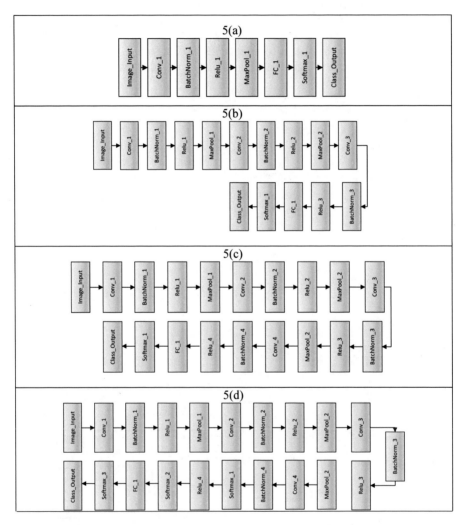

Fig. 3. (a) The proposed Deep Layer CNN Architecture of 6-layer CNN, (b) The proposed Deep Layer CNN Architecture of 13-layer CNN, (c), (d) The proposed Deep Layer CNN Architecture of 17-layer CNN

which is called Convolutional Neural Network (CNN) is used in this studies and we proposed suitable layer for CNN for this histopathology image. The deep learning framework for breast cancer image detection (mitosis and non-mitosis) is shown in Fig. 2 from pre-processing part until classification part. For data preparation, the sample of mitosis images were collected from opened source database which is called MITOS ATYPHIA (See Fig. 2). The dataset was divided into two datasets; 70% for training dataset and 30% for testing dataset. This was to ensure the learning generalization had been achieved by feeding the classifier to a new datasets.

In this study, three types of deep layer CNN architecture have been proposed for breast cancer image detection, namely, 6-layer CNN, 13-layer CNN 17-layer CNN and 19-layer CNN. The proposed deep layer of CNN architecture is shown in Fig. 3(a)–(d).

3 Performance Measure for Breast Cancer Image Detection

The performance measures for previous researchers on image cancer detection were typically measured using the Receiver Operating Characteristic (ROC) curve analysis. It had four types of performance measures, and the tests were called true positive rate (TPR), false positive rate (FPR), true negative rate (TNR) and false negative rate (FNR). In this case, the sensitivity and specificity tests had also been used to get the accuracy from the histopathology image with mitosis being the True Positive Rate (TPR) class, and non-mitosis samples being the True Negative Rate (TNR) class [23, 24]. Hence, the performance measure for the breast cancer image detection is described in Table 1.

Table 1. Performance measure test (TP, FP, TN and FN) are represented for this study

Performance measurement					
Actual Vs Predicted	Predicted Yes	Parameter	Predicted No	Parameter	Total parameter
Actual Yes	True Positive (TP)	TP	False-Negative (FN)	FN	TP+FN
Actual No	False Positive (FP)	FP	True Negative (TN)	TN	FP+TN
Total parameter		TP+FP		FN+TN	TP+FP +FN+TN

Where the elaboration of all tests for the performance measurement are listed below:

- **True Positive Rate/Recall/Sensitivity (TP/(TP+FN))** is the number of correct predictions that an instance is positive,
- **False Positive Rate (FP/(TP+FP))** is the number of incorrect predictions that an instance is positive,

- **Accuracy ((TP+TN)/(TP+FN+FP+TN))** is the number of correct classifications, and
- **Misclassification Rate/Error Rate/Loss ((FP+FN)/(TP+FN+FP+TN)** is the number of wrong classifications.

The experimental results will be discussed in the next section.

4 Experimental Results

The experimental results for breast cancer image detection using the proposed deep Convolutional Neural Network (CNN) layer are reported in Tables 2, 3 and 4, accordingly. The Graphical Processing Unit (GPU) device NVDIA GeForce 940MX has been implemented in this study for better speed up.

Table 2. Experimental result using 6-layer CNN

Experiments	Accuracy	Loss	True positive rate	False positive rate
1	68.06%	31.94%	75.00%	38.89%
2	77.78%	22.22%	80.56%	25.00%
3	62.50%	37.50%	66.67%	41.97%
4	50.00%	50.00%	100%	100%
5	66.67%	33.33%	58.33%	25.00%
6	61.11%	38.89%	63.89%	41.67%
7	59.72%	40.68%	66.67%	47.22%
8	65.28%	34.72%	58.33%	27.78%
9	73.61%	26.39%	66.67%	19.44%
10	55.56%	44.44%	55.56%	44.44%
Average	**64.03%**	**35.97%**	**69.17%**	**41.44%**

Table 3. Experimental result using 13-layer CNN

Experiments	Accuracy	Loss	True positive rate	False positive rate
1	75.00%	25.00%	69.44%	19.44%
2	76.39%	23.61%	72.22%	19.44%
3	72.22%	27.78%	69.44%	25.00%
4	76.39%	23.61%	75.00%	22.22%
5	68.06%	31.94%	66.67%	30.56%
6	68.06%	31.94%	66.67%	30.56%
7	76.39%	23.61%	80.56%	27.78%
8	72.22%	27.78%	80.56%	36.11%
9	65.28%	34.72%	58.33%	27.78%
10	73.61%	26.39%	75.00%	27.78%
Average	**72.36%**	**27.64%**	**71.39%**	**26.67%**

Table 4. Experimental result using 17-layer CNN

Experiments	Accuracy	Loss	True positive rate	False positive rate
1	79.17%	20.83%	86.11%	27.78%
2	83.33%	16.67%	86.11%	19.44%
3	86.61%	13.39%	83.33%	11.11%
4	81.94%	18.06%	83.33%	19.44%
5	84.72%	15.28%	83.33%	13.89%
6	86.11%	13.89%	80.56%	8.33%
7	81.94%	18.06%	86.11%	22.22%
8	81.94%	18.06%	77.78%	13.89%
9	83.33%	16.67%	83.33%	16.67%
10	80.56%	19.44%	83.33%	22.22%
Average	**82.96%**	**17.04%**	**83.33%**	**17.50%**

Table 5. Experimental result using 19-layer CNN

Experiments	Accuracy	Loss	True positive rate	False positive rate
1	81.94%	18.06%	77.78%	13.89%
2	86.11%	13.89%	80.56%	8.33%
3	86.61%	13.39%	83.33%	11.11%
4	86.11%	13.89%	80.56%	8.33%
5	81.94%	18.06%	77.78%	13.89%
6	86.11%	13.89%	80.56%	8.33%
7	81.94%	18.06%	86.11%	22.22%
8	81.94%	18.06%	77.78%	13.89%
9	86.11%	13.89%	80.56%	8.33%
10	86.11%	13.89%	80.56%	8.33%
Average	**84.49%**	**15.50%**	**80.55%**	**11.66%**

In overall performance, the proposed deep layer CNN architecture with 17-layer CNN gave better result compared to 6-layer CNN and 13-layer CNN. The results from Table 2, 3, 4 and 5 shows that deeper layer of CNN architecture produced better result than the least layer. For example, the highest accuracy of 84.49% and True Positive Rate (TPR) of 80.55%; while the least loss of 15.50% and False Positive Rate (FPR) of 11.66% performances were produced by 19-layer CNN compared to 17-layer CNN, 13-layer CNN and 6-layer CNN.

Based on Table 2, 3, 4 and 5, the proposed 19-layer CNN and 17-layer CNN produced the best accuracy compared to 6-layer CNN and 13-layer CNN with the most of the accuracy results from 10 experiments were maintained above 80%. In conclusion, the aim of this study has been achieved where deep layer CNN architecture was capable on detecting the breast cancer (which is known as mitosis and non-mitosis) with feasible results of accuracy, loss, TPR and FPR.

5   Conclusion

In this study, the proposed deep layer CNN architecture which is called 6-layer CNN, 13-layer CNN, 17-layer CNN and 19-layer CNN has been successfully implemented in detecting the mitosis and non-mitosis breast cancer histopathology images. Hence, we can conclude that, the deeper the CNN layer, the better performance result for breast cancer image detection. This is because more training layers had been fed into the proposed model and the learning multiple level of abstraction yielded better generalization of high level features.

Acknowledgement. This work is supported by Ministry of Education (MOE), Malaysia, Universiti Teknologi Malaysia (UTM), Malaysia and ASEAN-Indian Research Grant. This paper is financially supported by MYBRAIN, Grant No. 17H62, 03G91, and 04G48. The authors would like to express their deepest gratitude to the Bram van Ginneken, SjoerdKerkstra, and James Meakin for their support in providing the MITOS-ATYPHIA datasets to ensure the success of this research.

References

1. Howlader, C.K., Noone, N., Krapcho, M., Garshell, J., Miller, D., Altekruse, S.F., Kosary, C.L., Yu, M., Ruhl, J., Tatalovich, Z., Mariotto, A., Lewis, D.R., Chen, H.S., Feuer, E.J., Cancer statistics review 1975-2012: introduction, pp. 1–101 (2015)
2. Siegel, R., Naishadham, D., Jemal, A., Ma, J., Zou, Z., Jemal, A.: Cancer statistics, 2014. CA Cancer J. Clin. **64**(1), 9–29 (2014)
3. Veta, M., et al.: Assessment of algorithms for mitosis detection in breast cancer histopathology images. Med. Image Anal. **20**(1), 237–248 (2015)
4. Zhang, S., Grave, E., Sklar, E., Elhadad, N.: Longitudinal analysis of discussion topics in an online breast cancer community using convolutional neural networks. J. Biomed. Inform. **69**, 1–9 (2017)
5. Dalle, J.-R., Leow, W.K., Racoceanu, D., Tutac, A.E., Putti, T.C.: Automatic breast cancer grading of histopathological images. In: The 30th Annual International Conference of the IEEE Engineering in Medicine and Biology Society, 20–25 August 2008, Vancouver, BC, Canada (2008)
6. Mohd, A., Khuzi, R., Besar, W.M.D., Zaki, W., Ahmad, N.N.: Identification of masses in digital mammogram using gray level co-occurrence matrices. Biomed. Imaging Interv. J. **5**, 1–13 (2009)
7. Singh, S., Gupta, P., Sharma, M.: Breast cancer detection and classification of histopathological images. Int. J. Eng. **3**(5), 4228–4232 (2010)
8. Lim, G.C.C., Halimah, Y.: Cancer incidence in Peninsular Malaysia 2003-2005. National Cancer Registry (2008)
9. Khan, A.M., Rajpoot, N., Treanor, D., Magee, D.: A nonlinear mapping approach to stain normalization in digital histopathology images using image-specific color deconvolution. IEEE Trans. Biomed. Eng. **61**(6), 1729–1738 (2014)
10. Kothari, S., Phan, J., Wang, M.: Eliminating tissue-fold artifacts in histopathological whole-slide images for improved image-based prediction of cancer grade. J. Pathol. Inform. **4**, 22 (2013)

11. Veta, M., van Diest, P.J., Kornegoor, R., Huisman, A., Viergever, M.A., Pluim, J.P.W.: Automatic nuclei segmentation in H&E stained breast cancer histopathology images. PLoS ONE **8**, e70221 (2013)
12. Shen, W., et al.: Multi-crop convolutional neural networks for lung nodule malignancy suspiciousness classification. Pattern Recogn. **61**, 663–673 (2017)
13. Ciresan, D.C., Giusti, A., Gambardella, L.M., Schmidhuber, J.: Mitosis detection in breast cancer histology images with deep neural networks. In: Lecture Notes Computer Science (including Subseries Lecture Notes Artificial Intelligent Lecture Notes Bioinformatics), vol. 8150, LNCS, PART 2, pp. 411–418 (2013)
14. Su, H., Liu, F., Xie, Y., Xing, F., Meyyappan, S., Yang, L.: Region segmentation in histopathological breast cancer images using deep convolutional neural network. In: 2015 IEEE 12th International Symposium on Biomedical Imaging, pp. 55–58 (2015)
15. Wahlstr, N.: Learning deep dynamical models from image pixels (2016)
16. Feng, Y., Zhang, L., Yi, Z.: Breast cancer cell nuclei classification in histopathology images using deep neural networks. Int. J. Comput. Assist. Radiol. Surg. **13**(2), 179–191 (2018)
17. Spanhol, F.A., Oliveira, L.S., Petitjean, C., Heutte, L.: Breast cancer histopathological image classification using convolutional neural networks. In: 2016 International Joint Conference on Neural Networks, pp. 2560–2567 (2016)
18. Roux, L., et al.: Mitosis detection in breast cancer histological images an ICPR 2012 contest. J. Pathol. Inform. **4**(1), 8 (2013)
19. Veta, M., Pluim, J.P.W., van Diest, P.J., Viergever, M.A.: Breast cancer histopathology image analysis: a review **61**, 1400–1411 (2014)
20. Kotzias, D.: From Group to Individual Labels using Deep Features (2015)
21. Wu, S., Zhong, S., Liu, Y.: Deep residual learning for image steganalysis. Multimed. Tools Appl. **77**, 1–17 (2017)
22. Wahab, N., Khan, A., Lee, Y.S.: Two-phase deep convolutional neural network for reducing class skewness in histopathological images based breast cancer detection. Comput. Biol. Med. **85**(April), 86–97 (2017)
23. Demir, C., Yener, B.: Automated cancer diagnosis based on histopathological images: a systematic survey. Technical Report vol. TR-05-09, pp. 1–16, Department of Computer Science, Rensselaer Polytechnic Institute, Troy, NY, USA (2005)
24. Bhattacharjee, S., Mukherjee, J., Nag, S., Maitra, I.K., Bandyopadhyay, S.K.: Review on histopathological slide analysis using digital microscopy. Int. J. Adv. Sci. Technol. **62**, 65–96 (2014)
25. Hassanien, A.E., Ali, J.M., Nobuhara, H.: Detection of spiculated masses in Mammograms based on fuzzy image processing, In: Rutkowski, L., Siekmann, J.H., Tadeusiewicz, R., Zadeh, L.A. (eds.) Artificial Intelligence and Soft Computing - ICAISC 2004. Lecture Notes in Computer Science, vol. 3070, pp. 102–107. Springer, Berlin, Heidelberg (2004)

Regression Task on Big Data with Convolutional Neural Network

Chang Liu[1], Ziheng Wang[2], Su Wu[1], Shaozhi Wu[1(✉)], and Kai Xiao[3]

[1] University of Electronic Science and Technology of China, Chengdu, China
wszfrank@uestc.edu.cn
[2] Tongji University, Shanghai, China
[3] Shanghai Jiao Tong University, Shanghai, China

Abstract. As one of most widely utilized methods in deep learning, convolutional neural network (CNN) has been proven effective in many machine learning applications, especially in the areas of image understanding and computer vision. However, CNN is mainly used for applications with the approach of classification, while its usage for regression is not well-studied. In this work, we propose a strategy based on CNN with Visual Geometry Group Network (VGG) for image regression task. We have applied this method on images of MNIST processed with labels of continuous number. In our study, the original discrete classes of handwriting numbers are converted into float numbers with respect to normal distribution, thereby the traditional classification task in MNIST becomes a regression one. In our study, different loss functions such as Mean Absolute Error (MAE) and Log-cosh have been applied and validated. Final results generated by model trained with CNN with VGG with 10-fold cross-validation can be obtained, where MAE is less than 0.25, compared to the much higher error of around 3 with the use of other loss functions and convolutional layers. The significantly reduced error suggests the applicability of our proposed method.

Keywords: Regression · CNN · VGG · Loss function · Normal distribution

1 Introduction

Convolutional Neural Network (CNN) is a deep learning architecture inspired by biological natural visual cognition mechanisms. In 1959, Hubel and Wiesel found that animal visual cortical cells are responsible for detecting optical signals [1]. Enlightened by this, Kunihiko Fukushima proposed the predecessor of CNN – neocognitron in 1980 [2]. In the 1990s, LeCun et al. established the modern structure of CNN and later improved it [3]. They designed a multi-layered artificial neural network called LeNet-5, which could be trained using the back-propagation [4] algorithm. Originally invented for computer vision, CNN has achieved remarkable results in recent years. For instance, in 2012, AlexNet [5],

© Springer Nature Switzerland AG 2020
A. E. Hassanien et al. (Eds.): AMLTA 2019, AISC 921, pp. 52–58, 2020.
https://doi.org/10.1007/978-3-030-14118-9_6

the winner of the ILSVRC (ImageNet Large Scale Visual Recognition Competition) achieved great advance in image classification [6]. In 2017, Paoletti presented a CNN architecture for the classification of multi-spectral and hyperspectral images [7]. In the classification task of MNIST handwritten character recognition dataset [8], CNN architectures have reached an error rate less than 1% [9]. However, CNNs' usage for regression is not well-studied. In this paper, we propose a method based on CNN for image regression task. We build our convolutional neural network based on the VGG Network [10]. In the Sect. 2 we will introduce our network structure in detail, and the dataset we use. In the experiment, we adjusted the parameters and functions of our neural network. In addition, we changed the continuous labels' standard deviation, which is shown in Sect. 3. The results are displayed in Sect. 4. We demonstrate that the suggested strategy is very effective in dealing with image regression problems.

2   Methods and Data

Machine learning is a subset of artificial intelligence in the field of computer science, which typically uses statistical techniques to enable computers to "learn" from data [11]. The method of regression is widely utilized to analyze the connection between labels of continuous number and data. Applying machine learning to regression allows us to make predictions from data by learning the relationship between features of our data and labels. The trained model will preform the task of predicting a continuous quantity output for an unknown input.

Convolutional neural network is a kind of feed-forward artificial neural networks, which is often used in the areas of image understanding and computer vision. It has been proven effective in many machine learning applications. The overall architecture of our CNN is depicted in the left part of Fig. 1. As shown in the figure, the net contains four convolutional blocks, followed by four fully connected layers, each with a dropout layer in front of it. Fully connected layers connect every neuron in one layer to every neuron in another layer. The structure of each convolutional block is as shown in the right part of Fig. 1. In these convolutional blocks, we use a VGG-style convolutional layers. The first convolutional layer filters the 42×42 input image with kernels of size 3×3. The second convolutional layer utilizes 1×1 convolution filters, which can be regraded as a linear transformation of the input channels. Both convolutional layers are followed by batch normalization [12] and the Exponential Linear Unit(ELU) activation function [13]

$$f(x) = \begin{cases} x, & x \geq 0 \\ \alpha(e^x - 1), & x < 0 \end{cases} \tag{1}$$

where x is the input and α is an adjustable parameter, which controls when the negative part of ELU saturates. The ELU is followed by max poolings performed over a 2×2 pixel window. The model was first trained with log-cosh loss function

$$L(x, y) = \frac{1}{n} \sum_{i=1}^{n} log[cosh(y_i - x_i)] \tag{2}$$

using adam optimizer [14], where y_i is the label and x_i is the corresponding prediction.

MNIST is a handwritten digits dataset containing a training set of 60,000 examples, and a test set of 10,000 examples. It is a subset of a larger set available from NIST. The handwritten digits in it have been size-normalized and centered in a fixed-size image. It is a good database to validate machine learning models in image understanding for we don't need spending too much efforts on preprocessing and formatting images. Original MNIST dataset is classification task. In our experiment, we transform the classification task to a regression one. First, we compress the original 28*28 pixels MNIST images into 14*14 pixels. Then, we put seven images into one. Figure 2 shows an example of the digit 1,2,3,...,9. It consists of three rows and three columns while we vacate the position of the first column in the first row and the third column in the third row. The size of the image after the combination is 42*42. By attaching labels of continuous numbers to the images we made, the classification task is transformed into a regression one. Our model is used to predict the images' labels.

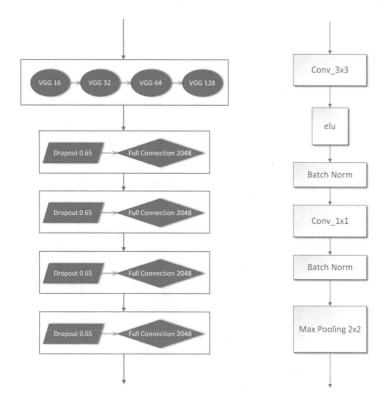

Fig. 1. The architecture of our network.

Fig. 2. An example of the handwritten digits image we made.

3 Experiments

We adopted a 10-fold cross-validation method to conduct the experiment. Cross-validation [15] is a computer intensive technique, which utilizes all available examples as training and test examples. It mimics the use of training and test sets by training the algorithm ten times with 1/10 of training examples left out for testing purposes. This method helps us test the model's ability to predict new data that was not used in estimating it, in order to flag issues such as overfitting and to give an insight on how the model will generalize to an unknown dataset in the real problem.

To minimize overfitting, we added dropout [16] layers after each fully connected layers. Dropout is a technique for improving neural networks by reducing overfitting [17]. The dropout layers set to zero the output of each hidden neuron with probability of 0.5. The neurons which are dropped out in this way would not contribute to the forward pass and participate in backpropagation. Without dropout, our network exhibited considerable overfitting. In the experiment, different activation functions were tested, for instance, softmax, Rectified Linear Unit (ReLU) [18] and Exponential Linear Unit (ELU). By comparison, ELU significantly outperformed other activation functions on the dataset. ELU has a clear saturation plateau in its negative regime, allowing them to learn a more robust and stable representation.

To test the stability of our model, we changed the standard deviation of the continuous labels. The distribution of labels is shown in Fig. 3.

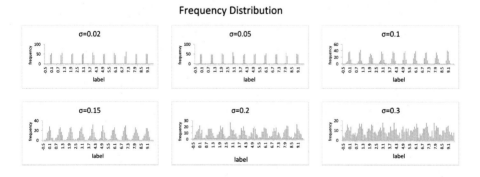

Fig. 3. The histogram of label distribution.

4   Result

In order to improve the accuracy, we adjusted the numbers of neurons in the fully connected layer. Table 1 is an overview of the results. The concrete outcomes are displayed in Fig. 4. We use mean absolute error (MAE)

$$MAE = \frac{1}{n} \sum_{i=1}^{n} |y_i - x_i| \qquad (3)$$

to measure the accuracy of predictions, which is an average of the absolute error, where x_i is the prediction and y_i is the true value. MAE is the most natural measure of average error magnitude [19]. As shown in Fig. 4, the MAE of prediction reaches 0.23, when the standard deviation of the label is 0.02. When the standard deviation is 0.05, MAE stands at 0.47. As the σ increases, MAE grows but eventually stabilizes at around 0.5. As can be seen from the line charts below, our prediction curve fits well with the actual curve. The prediction accuracy of our model is positive.

Table 1. The standard deviation (σ) of labels' distribution and the corresponding prediction MAE.

σ	0.02	0.05	0.1	0.15	0.2	0.3
MAE	0.23	0.47	0.46	0.46	0.47	0.54

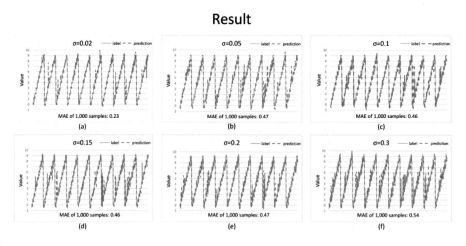

Fig. 4. The prediction result is shown in the figure. In (a), MAE reaches minimum at 0.23 when σ is 0.02. In (b), MAE is 0.47 when σ is 0.05. In (c), (d) and (e), MAE stabilizes at 0.47. As the σ increases to 0.3 in (f), MAE eventually reaches around 0.5.

5   Conclusion and Future Work

In this paper, we have proposed a method based on CNN with VGG for image regression task. Through applying this method on processed images of MNIST, we have verified our strategy is effective for regression tasks. The explanation of such good result may be hidden in the choice of appropriate activation function and loss function, as well as the struction of fully connected layers. In addition, we changed the standard deviation of the continuous labels to add difficulty to the task to check out the stability of the model. However, we are only experimenting with a specific dataset. In future work, we will evaluate our model on other different dataset. Besides, our neural network consumes considerable computing resources due to the inclusion of hidden layers. Next, we will also try to use transfer learning, so that we can make small changes to the already trained models according to our demand.

References

1. Hubel, D.H., Wiesel, T.N.: Early exploration of the visual cortex. Neuron **20**(3), 401–412 (1998)
2. Fukushima, K., Miyake, S.: Neocognitron: a self-organizing neural network model for a mechanism of visual pattern recognition. In: Amari, S., Arbib, M.A. (eds.) Competition and Cooperation in Neural Nets, pp. 267–285. Springer, Berlin, Heidelberg (1982)
3. LeCun, Y., et al.: Gradient-based learning applied to document recognition. In: Proceedings of the IEEE 86.11, pp. 2278–2324 (1998)
4. Hecht-Nielsen, R.: Theory of the backpropagation neural network. In: Neural networks for perception, pp. 65–93 (1992)
5. Krizhevsky, A., Sutskever, I., Hinton, G.E.: Imagenet classification with deep convolutional neural networks. In: Advances in Neural Information Processing Systems (2012)
6. Deng, J., et al.: Imagenet large scale visual recognition competition 2012 (ILSVRC2012). See net.org/challenges/LSVRC (2012)
7. Paoletti, M.E., et al.: A new deep convolutional neural network for fast hyperspectral image classification. ISPRS J. Photogramm. Remote Sens. **145**, 120–147 (2018)
8. LeCun, Y.: The MNIST database of handwritten digits (1998). http://yann.lecun.com/exdb/mnist/
9. Simard, P.Y., Steinkraus, D., Platt, J.C.: Best practices for convolutional neural networks applied to visual document analysis. IEEE (2003)
10. Simonyan, K., Zisserman, A.: Very deep convolutional networks for large-scale image recognition. arXiv preprint arXiv:1409.1556 (2014)
11. Samuel, A.L.: Some studies in machine learning using the game of checkers. IBM J. Res. Dev. **3**(3), 210–229 (1959)
12. Ioffe, S., Szegedy, C.: Batch normalization: Accelerating deep network training by reducing internal covariate shift. arXiv preprint arXiv:1502.03167 (2015)
13. Clevert, D.-A., Unterthiner, T., Hochreiter, S.: Fast and accurate deep network learning by exponential linear units (elus). arXiv preprint arXiv:1511.07289 (2015)

14. Kingma, D.P., Ba, J.: Adam: a method for stochastic optimization. arXiv preprint arXiv:1412.6980 (2014)
15. Kohavi, R.: A study of cross-validation and bootstrap for accuracy estimation and model selection. In: Ijcai, vol. 14., no. 2 (1995)
16. Hinton, G.E., et al.: Improving neural networks by preventing co-adaptation of feature detectors. arXiv preprint arXiv:1207.0580 (2012)
17. Srivastava, N., et al.: Dropout: a simple way to prevent neural networks from overfitting. J. Mach. Learn. Res. **15**(1), 1929–1958 (2014)
18. Nair, V., Hinton, G.E.: Rectified linear units improve restricted boltzmann machines. In: Proceedings of the 27th International Conference on Machine Learning (ICML-10) (2010)
19. Willmott, C.J., Matsuura, K.: Advantages of the mean absolute error (MAE) over the root mean square error (RMSE) in assessing average model performance. Clim. Res. **30**(1), 79–82 (2005)

The Regression of MNIST Dataset Based on Convolutional Neural Network

Ziheng Wang[1], Su Wu[2], Chang Liu[2], Shaozhi Wu[3], and Kai Xiao[4(✉)]

[1] School of Aerospace Engineering and Applied Mechanics, Tongji University,
No. 1239 Siping Road, Yangpu District, Shanghai, China
wangzh1364@hotmail.com
[2] School of Information and Software Engineering,
University of Electronic Science and Technology of China, Chengdu, China
m18781078969@163.com, luucas799@gmail.com
[3] School of Computer Science and Engineering,
University of Electronic Science and Technology of China, Chengdu, China
wszfrank@uestc.edu.cn
[4] School of Electronic Information and Electrical Engineering,
Shanghai Jiao Tong University, 800 Dongchuan Road, Shanghai, China
showkey@sjtu.edu.cn

Abstract. The MNIST dataset of handwritten digits has been widely used for validating the effectiveness and efficiency of machine learning methods. Although this dataset was primarily used for classification and results of very high accuracy (99.3%+) on it have been obtained, its important application of regression is not directly applicable, thus substantially deteriorates its usefulness and the development of regression methods for such types of data. In this paper, to allow MNIST to be usable for regression, we firstly apply its class/label with normal distribution thereby convert the original discrete class numbers into float ones. Modified Convolutional Neural Networks (CNN) is then applied to generate a regression model. Multiple experiments have been conducted in order to select optimal parameters and layer settings for this application. Experimental results suggest that, optimal outcome of mean-absolute-error (MAE) value can be obtained when ReLu function is adopted for the first layer with other layers activated by the softplus functions. In the proposed approach, two indicators of MAE and Log-Cosh loss have been applied to optimize the parameters and score the predictions. Experiments on 10-fold cross-validation demonstrate that, desired low values of MAE and Log-Cosh error respectively at 0.202 and 0.079 can be achieved. Furthermore, multiple values of standard deviation of the normal distribution have been applied to verify the applicability when data of label number at varied distributions is used. The experimental results suggest that a positive correlation exists between the adopted standard deviation and the loss value, that is, the higher concentration degree of data will contribute to the lower MAE value.

Keywords: MNIST dataset · Convolutional neural network · Regression

© Springer Nature Switzerland AG 2020
A. E. Hassanien et al. (Eds.): AMLTA 2019, AISC 921, pp. 59–68, 2020.
https://doi.org/10.1007/978-3-030-14118-9_7

1   Introduction

The MNIST dataset, which was first introduced in 1998 by LeCun et al. [1], is one of the most widely known datasets in the computer vision and neural networks community. It is a subset of a much larger dataset known as the NIST Special Database 19 [2] which contains both handwritten numerals and letters. The accessibility of the MNIST dataset has almost certainly contributed to its widespread use. The entire dataset is relatively small (by comparison to more recent benchmarking dataset), free to access and use, and is encoded and stored in an entirely straightforward manner. The encoding does not make use of complex storage structures, compression, or proprietary data formats. For this reason, it is easy to access and include the dataset from any platform or through any programming language.

Meanwhile, the MNIST dataset, which consists of 60000 handwritten digits and the following labels, has provided a means to test and validate a classification system. And there is an excellent applicability in classification experiment whose accuracy can achieve a rate of over 99%. Despite this, it has never been applied for regression. Nevertheless, it can be seen that labels of the MNIST dataset naturally follow a distribution of continuous integer which ranges from 1 to 10 so that the dataset can be changed into a regression task with original discrete class numbers converted into float ones.

The CNN is a class of deep, feed-forward artificial neural networks, which has been widely applied in image processing and analysis. It was inspired by biological processes [3] in that the connectivity pattern between neurons resembles the organization of the animal visual cortex. Individual cortical neurons respond to stimuli only in a restricted region of the visual field known as the receptive field. The receptive fields of different neurons partially overlap such that they cover the entire visual field. That only part of the neurons are connected between the convolutional layers makes a CNN occupy less memory than traditional neural network and a higher speed when training images. CNNs are often used in image recognition systems. In 2012 an error rate of 0.23% on the MNIST database was reported [4]. When applied to facial recognition, CNNs achieved a large decrease in error rate [5]. Another paper reported a 97.6% recognition rate on "5,600 still images of more than 10 subjects" [3]. CNNs were used to assess video quality in an objective way after manual training; the resulting system had a very low root mean square error [5]. In addition, CNNs have applications in video analysis, recommender systems [6] and natural language processing [8] as well. For health risk assessment and biomarkers of aging discovery, there is a simple CNN combined with Cox-Gompertz proportional hazards model and used to produce a proof-of-concept example of digital biomarkers of aging in the form of all-causes-mortality predictor [9]. End-to-end training and prediction are common practice in computer vision. However, human interpretable explanations are still required. With recent advances in visual salience, spatial and temporal attention, the most critical spatial regions/temporal instants could be visualized to justify the CNN predictions [10]. Although the CNN has grown to be a mature machine learning method, its application in intuitive regression problems can still be very few.

Therefore, in this paper, we want to study the regression task of MNIST dataset based on the Convolutional Neural Network. To make the MNIST dataset usable for regression tasks, we firstly apply the class/label of the dataset with normal distribution thereby convert the original discrete class numbers into float continuous ones. After experiments, the optimal neural architecture is established by reasonably building blocks, choosing hyperparameters and setting the other extra relevant parameters that achieves the minimum values of loss functions [11].

2   Method

2.1   Established Relationship Between Regression and MNIST Dataset

The MNIST database of handwritten digits from zero to nine that have been size-normalized and centered in a fixed-size image has a training set of 60,000 examples, and a test set of 10,000 examples (see Fig. 1). Both training set and test set are composed of half patterns collected among Census Bureau employees, while the other half was from high-school students. Meanwhile, the label/class is just the digit of each image. Regression analysis is a way to find the relationship between data features and continuous target values. At the same time, regression analysis has been not only applied in pure mathematics, but also has good practical significance.

Fig. 1. The MNIST database of handwritten digits

As we all know, the MNIST database has been used extensively for classification in recent years and obtained a very high accuracy of 99.3%. In order to figure out the feasibility of the dataset used in the regression, we have processed the class/label from the discrete class numbers to float ones by normal distribution. Note that the processed labels tend to be more clustered around each value of the handwritten digit. That is to say, we got these continuous target values with a small standard deviation. We do this just because we suspect that the distribution of new labels closer to the original classification label is more likely to achieve a low error. In this way, the dataset can be used for regression.

2.2   Description of the CNN in Deep Learning

Deep learning is a class of machine learning algorithms that [12] use a cascade of multiple layers of nonlinear processing units for feature extraction and transformation. Each successive layer uses the output from the previous layer as input. The CNN is just a kind of deep, feed-forward artificial neural network which uses a variation of multilayer perceptrons designed to require minimal preprocessing [13]. It is also known as shift invariant or space invariant artificial neural networks (SIANN), based on their shared-weights architecture and translation invariance characteristics [14,15]. More specifically, sharing the convolution kernel with neurons of the same layer and extracting image features automatically can be realized in the CNN to attain a high efficiency compared with the traditional neural network. In other words, as long as we set the relevant sound parameters, CNN can automatically extract the features and train the data to fit the targets. The target is the expected value which will then be compared with the predicted value to get an error.

To be specific, the CNN consists of several basic layers. First, the convolutional layer. It is the core building block of a CNN. The layer's parameters are comprised of a set of learnable filters which are also called convolution kernels. During the forward pass, each filter is convolved across the width and height of the input volume, computing the dot product between the entries of the filter and the input and producing a 2-dimensional activation map. Meanwhile, the filters achieve a parameter sharing scheme used in convolutional layers to control the number of free parameters. Second, the pooling layer. Pooling is another important concept of the CNN in which the max pooling can be most widely used. The max pooling partitions the input image into a set of non-overlapping rectangles and outputs the maximum of each sub-region. The pooling layer serves to progressively reduce the spatial size of the representation, to reduce the number of parameters and amount of computation in the network, and hence to also avoid overfitting. Third, the flatten layer and the fully connected layer. After pooling, the flatten layer is applied to flatten these 2-dimensional feature maps into a 1-dimension map so that it can be put into fully-connected layers. Through some fully connected layers, the output size of the layer is one. In addition, the output of each convolutional layer and fully connected layer also entitled the dense layer in the Keras library should be processed by an activation function before entering the next layer. The reason why we have adopted the CNN is because of its accessibility and effectiveness when learning images. Besides, it is because the CNN has been rarely applied to regression that we'd liked to experiment it on a basic data set to explore its effect on regression tasks.

2.3   Predicting the Target of MNIST by the CNN

At first, we put all the training images into the CNN. Then filters of convolutional layer will extract the features automatically and all the layers we set will find the relationship between the features and the successive data, which is often called the target or the label. The above process is called training a model. Finally, we use this trained model on the test data set to get the predicted values. This process is called prediction.

The process of training the model is actually finding the values of parameters in this neural network. The process of finding parameters requires mathematics and some optimization programs that are already well developed. Machine learning also includes some hyperparameters, which are parameters that are set before the training starts. Optimizing these hyperparameters can improve the efficiency of model training and the accuracy of the model. The process of optimizing hyperparameters, known as tuning, requires some inspiration and trying different values. Specifically, we have set the initial loss function [11] and the optimizer before training, which can adjust the weight when the convolutional neural network propagates backward to gain the more accurate result.

It is noted that we use 10-fold cross-validation [16], to make full use of the data and provide a more objective result. In k-fold cross-validation, the original sample is randomly partitioned into 10 equal sized subsamples. Of the 10 subsamples, a single subsample is retained as the validation data for testing the model, and the remaining 9 subsamples are used as training data. The cross-validation process is then repeated 10 times, with each of the 10 subsamples used exactly once as the validation data. The 10 results can then be averaged to produce a single estimation. The advantage of this method over repeated random sub-sampling is that all observations are used for both training and validation, and each observation is used for validation exactly once.

Our experiment is divided into two parts: get a good regression effect based on MNIST dataset by changing the arrangement of layers, training times and various parameters in the CNN. Furthermore, greater values of standard deviation of the normal distribution which make the label a more dispersed distribution will be applied to verify if the result will become worse.

3   Experiments and Results

We have decided to apply the CNN to realize the regression of MNIST database. This deep learning method has been chosen mainly because of its effectiveness in image learning and probable good accuracy in dealing with regression problems that we were looking for. On the other hand, we have adopted 10-fold cross-validation [16] method to make full use of the data and acquire a more objective outcome. Specifically, for a training set of 60000 handwritten digits divided into ten parts, we apply nine parts to train the model each time and then the trained model is used to predict the target value of the remaining one part. By performing these steps 10 times, the 10 results can then be averaged to produce a single estimation as a credible result.

3.1   Establishing the Optimal Network Architecture

Our model is a CNN based on Keras with a regression output. Keras is an open source neural network library written in Python. It is capable of running on top of TensorFlow, Microsoft Cognitive Toolkit, or Theano [17]. It takes $28 \times 28 \times 1$ images as the input of the first layer with 128 such pictures a batch. And the network is trained by minimizing two indicators of Mean Absolute Error (MAE) [18] and Log-Cosh respectively with Adam optimizer [19]. We begin training with the learning rate 10^{-3} and then progressively increase it to 10^{-2} which brings about a better outcome. The values of these two indicators can be obtained by Eq. (1).

$$MAE = \frac{1}{n} \sum_{i=1}^{n} |p_i - e_i|$$

$$L(p_i, e_i) = \sum_{i=1}^{n} log(cosh(p_i - e_i)) \tag{1}$$

In Eq. (1), p_i represents the predicted value, e_i represents the expected value, and n is the total number of sample points. MAE [18] and Log-Cosh are often used to optimize the parameters and score the predictions, and the smaller their values are, the higher the accuracy of the predictions is.

In order to select optimal hyperparameters and layer settings for this application, multiple experiments have been conducted. Firstly, we applied four convolutional layers with 16, 32, 64, 128 filters, three max-pooling layers, four fully-connected layers of 2048, 500, 100, 20 neurons and a single output. However, the result isn't ideal that the MAE value [18] is 2.160. Thereafter, we have considered if the arrangement of convolutional layers and fully-connected layers is not reasonable enough so that we try to adjust the amount of each sort of layer. A great reduction of loss appears though multiple experiments that the MAE value [18] jumps to 0.493. After this, we continue to look for ways to make the prediction more precise. Finally, we find that optimal outcome of MAE value [18] can be obtained when ReLu function [20] is adopted for the first layer with other layers activated by the softplus function [21]. Before this, ReLu function [20] has been the activation function of all the convolutional layers and fully-connected layers.

The overall network architecture for MNIST shown in Table 1 that acquires the best result represents a stack of three convolutional layers with 16, 32, 64 filters followed by two max-pooling layers and three fully-connected layers of 500, 100, 20 neurons respectively and a single output. For better generalization, we have applied dropout layers after the first two fully-connected layers. For regression targets, we have processed the class/label from the discrete class numbers to float ones by normal distribution. It should be stated that parameters not described in the figure are consistent with the default values in the convolutional neural network framework of Keras. After training with 11 epochs of each fold, the result shows that the values of MAE [18] and Log-Cosh value are equal to 0.202 and 0.079 respectively. Compared with the outcome that the MAE value [18] is 0.493 and the Log-Cosh value is 0.471 of the experiment which ReLu function [20] is applied in all the convolutional layers and fully- connected layers, the loss value has been reduced. The CNN has also been used in regression experiments for bone age assessment [22] in which the best accuracy with MAE [18] equal to about 6. In comparison, our model has achieved a very high accuracy. Objectively speaking, the fewer image interference, normalization of the MNIST dataset and centrally distributed labels contribute to it in a way.

Table 1. Optimal CNN configuration for the regression of MNIST

Name	Filters/units(FC)	Activation function
Conv2D-1	$3 \times 3 \times 16$	ReLu
Conv2D-2	$3 \times 3 \times 32$	softplus
MP2D-2	2×2	
Conv2D-3	$3 \times 3 \times 64$	softplus
MP2D-3	2×2	
Flatten		
Dense-1	500	softplus
Dropout-1(0.5)		
Dense-2	100	softplus
Dropout-2(0.25)		
Dense-3	20	softplus
Dense-4	1	

3.2 Results of Regression with Labels of Different Standard Deviation

We supposed that the reason why the regression of MNIST dataset could gain a good result is because the continuous targets with the concentrated distribution simulated by normal distribution were very similar to the original classified targets. Thus, we inferred that the difference between targets and predictions would be greater as the targets become more scattered. To verify it, we gradually increase the value of the standard deviation so that the targets will be more evenly distributed. Figure 2 below shows this more visually. As we can see, when the value goes from 0.02 to 0.5, we will get more mean labels. New labels are adopted in the optimal network structure for next training and the result is represented in the Table 2. And you can get more intuition from Fig. 3. The data shows that as the standard deviation value of the normal distribution increases, both the MAE value [18] and the Log-Cosh value progressively increase. In other words, there is a positive correlation between the adopted standard deviation and the loss value. Compared with the results of the bone age analysis [22] in that paper, we can find that each age can be compared with each of our numerical values, and the label distribution is more average. Therefore, the result of that paper should be about 0.5 in our experiment, which is almost equivalent to the result when the standard deviation is 0.5, which also supports our results to some extent.

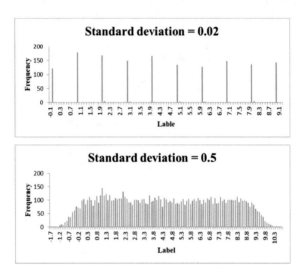

Fig. 2. Frequency profile of labels under two standard deviations. Each subgraph takes 10,000 samples as data.

Table 2. MAE, Log-Cosh values of experimental results with different values of standard deviation

Standard deviation	0.020	0.100	0.200	0.300	0.500
MAE	0.202	0.257	0.308	0.392	0.530
Log-Cosh	0.079	0.083	0.096	0.129	0.226

The values in the table are rounded and kept to three digits after the decimal point.

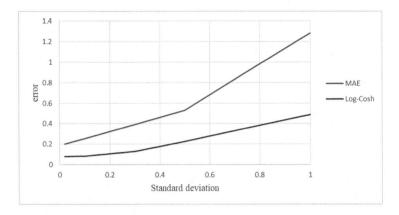

Fig. 3. MAE and Log-Cosh values of experimental results with different data distribution

4   Conclusion

In this paper, we have applied the CNN to generate a regression model of MNIST dataset. Through experiments, the optimal outcome of both MAE value [18] and Log-Cosh value has been attained by optimizing the network architecture progressively. In addition, we have tried to use different values of standard deviation of the normal distribution to verify the applicability when data of label number at varied distributions is used. The result shows that a positive correlation exists between the adopted standard deviation and the loss value so that the more concentrated the data distribution is, the lower MAE value [18] will be. Overall, the regression experiment of MNIST based on the CNN has yielded an excellent result. However, we have only experimented the regression task with a certain dataset (MNIST dataset). In future work, we will try to study a more common regression task based on the CNN and other machine learning methods.

References

1. LeCun, Y., Bottou, L., Bengio, Y., Haffner, P.: Gradient-based learning applied to document recognition. Proc. IEEE **86**(11), 2278–2324 (1998)
2. Grother, P.J.: NIST special database 19. Handprinted forms and characters database, National Institute of Standards and Technology (1995)
3. Matsugu, M., Mori, K., Mitari, Y., Kaneda, Y.: Subject independent facial expression recognition with robust face detection using a convolutional neural network. Neural Netw. **16**(5–6), 555–559 (2003)
4. Cireşan, D., Meier, U., Schmidhuber, J.: Multi-column deep neural networks for image classification (2012). arXiv preprint arXiv:1202.2745
5. Lawrence, S., Giles, C.L., Tsoi, A.C., Back, A.D.: Face recognition: a convolutional neural-network approach. IEEE Trans. Neural Netw. **8**(1), 98–113 (1997)

6. Le Callet, P., Viard-Gaudin, C., Barba, D.: A convolutional neural network approach for objective video quality assessment. IEEE Trans. Neural Netw. **17**(5), 1316–1327 (2006)
7. van den Oord, A., Dieleman, S., Schrauwenvan, B.: Deep content-based music recommendation. Curran Associates, Inc., pp. 2643–2651 (2013)
8. Collobert, R., Weston, J.: A unified architecture for natural language processing: deep neural networks with multitask learning. In: Proceedings of the 25th International Conference on Machine learning, pp. 160–167. ACM (2008)
9. Pyrkov, T.V., Slipensky, K., Barg, M., Kondrashin, A., Zhurov, B., Zenin, A., Fedichev, P.O.: Extracting biological age from biomedical data via deep learning: too much of a good thing? Sci. Reports **8**(1), 5210 (2018)
10. Zang, J., Wang, L., Liu, Z., Zhang, Q., Hua, G., Zheng, N.: Attention-based temporal weighted convolutional neural network for action recognition. In: IFIP International Conference on Artificial Intelligence Applications and Innovations, pp. 97–108. Springer, Cham (2018)
11. Wald, A.: Statistical decision functions (1950)
12. Deng, L., Yu, D.: Deep learning: methods and applications. Found. Trends® in Signal Process. **7**(3–4), 197–387 (2014)
13. LeCun, Y.: LeNet-5, convolutional neural networks (2015). http://yann.lecun.com/exdb/lenet, 20
14. Zhang, W.: Shift-invariant pattern recognition neural network and its optical architecture. In: Proceedings of Annual Conference of the Japan Society of Applied Physics (1988)
15. Zhang, W., Itoh, K., Tanida, J., Ichioka, Y.: Parallel distributed processing model with local space-invariant interconnections and its optical architecture. Appl. Opt. **29**(32), 4790–4797 (1990)
16. McLachlan, G., Do, K.A., Ambroise, C.: Analyzing Microarray Gene Expression Data, vol. 422. Wiley, London (2005)
17. Keras backends. keras.io. Accessed 23 Feb 2018
18. Willmott, C.J., Matsuura, K.: Advantages of the mean absolute error (MAE) over the root mean square error (RMSE) in assessing average model performance. Clim. Res. **30**(1), 79–82 (2005)
19. Kingma, D.P., Ba, J.: Adam: a method for stochastic optimization (2014). arXiv preprint arXiv:1412.6980
20. Nair, V., Hinton, G.E.: Rectified linear units improve restricted boltzmann machines. In: Proceedings of the 27th International Conference on Machine Learning (ICML-10), pp. 807–814 (2010)
21. Dugas, C., Bengio, Y., Bélisle, F., Nadeau, C., Garcia, R.: Incorporating second-order functional knowledge for better option pricing. In: Advances in Neural Information Processing Systems, pp. 472–478 (2001)
22. Iglovikov, V.I., Rakhlin, A., Kalinin, A.A., Shvets, A.A.: Paediatric Bone age assessment using deep convolutional neural networks. In: Deep Learning in Medical Image Analysis and Multimodal Learning for Clinical Decision Support, pp. 300–308. Springer, Cham (2018)

Swarm Intelligence and Applications

Intelligent Hybrid Approach
for Feature Selection

Ahmed M. Anter[1], Ahmad Taher Azar[2,3($\boxtimes$)], and Khaled M. Fouad[2]

[1] Faculty of Computers and Information,
Beni-Suef University, Beni Suef, Egypt
[2] Faculty of Computer and Information, Benha University, Benha, Egypt
ahmad_t_azar@ieee.org, ahmad.azar@fci.bu.edu.eg
[3] School of Engineering and Applied Sciences, Nile University,
6th of October City, Giza, Egypt

Abstract. The issues of multitude of noisy, irrelevant, misleading features, and the capability to tackle inaccurate and inconsistent data in real world topics are the justification to turn into one of the most significant needs for feature selection. This paper proposes an intelligent hybrid approach using Rough Set Theory (RST), Chaos Theory and Binary Grey Wolf Optimization Algorithm (CBGWO) for feature selection problems. Ten different chaotic maps are used to estimate and tune GWO parameters. Experiments are applied on complex medical datasets with various uncertainty features and missing values. The performance of the proposed approach is extensively examined and compared with that of existent feature selection algorithms; such as ant lion optimization (ALO), chaotic ant lion optimization (CALO), bat optimization (BAT), whale optimization algorithm (WOA), chaotic whale optimization algorithm (CWOA), binary crow search algorithm (BCSA), and chaotic binary crow search algorithm (BCCSA) algorithms. The achievement of the proposed approach is analyzed using different evaluation criteria. The overall result indicates that the proposed approach delivers better performance, lower error, higher speed and shorter execution time.

Keywords: Grey Wolf Optimization Algorithm · Bio-inspired ·
Rough set theory · Optimization · Feature selection · Classification

1 Introduction

The achievement of the classification [3] is degraded by evolving information technology quickly, the complexity and scalability of real-world datasets. There are irrelevant attributes in the dataset, therefore, an efficient and effective mechanism is required to determine and detach relevant attributes from irrelevant attributes in a dataset [1, 2]. Attribute selection or feature selection is nominating optimal features that deliver the most accurate prediction outcome. The main issue is that not every feature is significant. Many features in a dataset may be dispensable, irrelevant or noisy. Feature selection is nominating a subset of mightily relevant features, which is an input of the required analysis [6, 19]. A tool is required to minimize the data dimensionality using the rough set (RS) theory [9] that is exploited to tackle vagueness and uncertainty in a

© Springer Nature Switzerland AG 2020
A. E. Hassanien et al. (Eds.): AMLTA 2019, AISC 921, pp. 71–79, 2020.
https://doi.org/10.1007/978-3-030-14118-9_8

dataset [12]. The classification issues are applied by many rough sets systems [18]. The reduction of attributes is based on data dependencies [17]. The RS theory splits a dataset into some equivalent (indiscernibility) classes, and estimates uncertain and vague concepts based on the parts [12]. On the other hand, the more recently Grey Wolf optimization algorithm (GWO) has been proposed by Seyedali and Andrew [9]. It is a meta-heuristic optimization technique, population-based and inspired by the hunting strategy and behavior of the grey wolves. Several engineering optimization problems have been solved by the GWO such as [7, 9, 10]. This paper will tune GWO parameters and improve its performance in terms of local minima avoidance and convergence speed using chaos theory.

This paper presents an intelligent hybrid approach based on rough set optimized with Chaos Theory [15] and Binary Grey Wolf Optimization algorithm (RSCBGWOFS) to enhance the processes of feature selection problems applied in medical domain. The standard GWO algorithm is worked in binary version and integrated with chaos maps to estimate random parameters. These parameters are replaced by using the ten chaos maps to exploit the better statistical and the best dynamical properties of these maps. Five standard datasets are taken from UCI Machine learning repository [8]. Moreover, the data should be preprocessed to tackle missing values [5, 11], transform categorical values to numerical values and tackle data inconsistency. Then, the proposed approach is applied to perform the processes of feature selection. The overall results show that the proposed approach provides the optimal estimated parameters with higher convergence speed, with shorter execution time and with better accuracy. The performance of the proposed approach is compared with different met-heuristic optimization techniques [16].

This paper is structured to many sections. Section 2 shows a Grey Wolf Optimization Algorithm, Sect. 3 presents the Proposed Hybrid RSCBGWOFS Approach. Section 4 illustrates results and discussion. Section 5 presents the conclusion the future work.

2   Grey Wolf Optimization Algorithm (GWO)

The Grey Wolf Optimizer algorithm (GWO) is a meta-heuristic technique, which was presented in 2014 by Mirjalili [9]. GWO is devised from the hunting task and disposal of the grey wolves. Grey wolves are structured in a group from five to eleven members. They regulate the set into four kinds of wolves' categories, through hunting process, to remain the discipline in a group (alpha (α), beta (β), omega (ω), delta (δ)). Hunters are accountable to hunt the prey. When the wolves are encircling around the prey through hunting, this strategy is computationally modelled [7, 18] by showing the following equations.

$$X_{t+1} = X_{p,t} - Ad \tag{1}$$

$$d = |cX_{p,t} - X_t| \tag{2}$$

$$A = 2br_1 - b \tag{3}$$

$$c = 2r_2 \tag{4}$$

where X_{t+1} is the position of the wolf at $(t + 1)^{\text{th}}$ iteration, $X_{p,t}$ is the position of the prey at t^{th} iteration, d is a variation vector, r_1 and r_2 are uniformly distributed random vectors generated between [0,1]. A and c are coefficient vectors, and b is a linearly reducing vector from 2 to 0 over iterations, indicated in Eq. (5).

$$b = 2 - 2\left(\frac{t}{max\ no\ of\ iterations}\right) \tag{5}$$

In the hunting process, grey wolves are computationally modeled by approximating the prey position by knowing α, β and δ solutions (wolves). Furthermore, by presenting this approximation, each wolf can vary their positions by the following equations.

$$X_1' = X_\alpha - A_\alpha d_\alpha \tag{6}$$

$$X_2' = X_\beta - A_\beta d_\beta \tag{7}$$

$$X_3' = X_\delta - A_\delta d_\delta \tag{8}$$

$$X_{t+1} = \frac{\left(X_1' + X_2' + X_3'\right)}{3} \tag{9}$$

X_α, X_β, X_δ are approximated places by α, β and δ solutions. The optimization framework of GWO is shown in the Fig. 1.

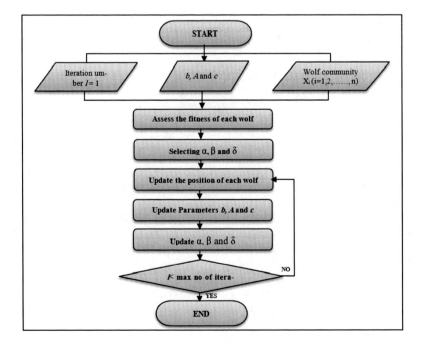

Fig. 1. The optimization framework of GWO.

3 Proposed Hybrid RSCBGWOFS Approach

In this section, the integrated rough set, chaos theory and improved population-based algorithm grey wolf optimization algorithm (GWO), namely; (RSCBGWOFS) for feature selection problems in complex medical datasets are presented. Rough set (RS) theory is used to decrease the dimensionality of the dataset. Using the concept of discernibility matrix and attribute dependency of RS reduct is computed from the dataset, the association of each attribute is computed, and the optimal candidate is nominated. The main challenging task in GWO is the balancing between exploration and exploitation with randomly distribution behavior. In non-linear system, exploration and exploitation are in conflict and there is no clear boundary between themes. For these reasons, GWO suffers from low convergence, fall in local minima, and not guarantee for global solution. Therefore, ten chaotic sequence maps are embedded in GWO to tune GWO position instead of the random behavior. The schematic view of the proposed approach RS-CBGWO-FS is shown in Fig. 2.

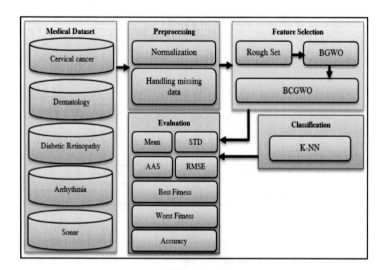

Fig. 2. General architecture of the proposed model

Firstly, preprocessing of data is used in two steps, first step, k-Nearest Neighbors (k-NN) algorithm is utilized to estimate missing. In second step, attributes of medical datasets are normalized between [0, 1] using Min-max method for the purpose to increase the efficiency by reducing the fluctuation among features. After normalization, chaotic maps are selected and merged with the algorithm in the initialization process. The population of wolves are initialized chaotically. Then, the parameters of GWO which are (A and C) are tuned by chaotic vector map to dominate the exploration and exploitation rate. In the next phase, fitness of all wolves are initiated in the search space using cost function based on k-NN method. The fitness values are assigned to each wolf in the search space. These wolves are evaluated in each iteration and the best wolf fitness is selected as the best solution. To obtain the stability and high accuracy for GWO, the fitness function is minimized by the optimization process. Wolves move in the search space chaotically to alter their positions in the space, this is named the continuous space. The continuous space should be transformed to corresponding binary space and then, persuades to produce a recent version of the BGWO based chaos theory.

4   Results and Discussion

In this section, the experimental scenario of the proposed RSCBGWOFS approach on the medical diagnosis datasets are presented.

4.1   GWO Static Parameters Setting

Static parameters of GWO are summarized in Table 1.

Table 1. GWO initial parameter setting

Parameters	Initial values	Description
Agents	5	No. of grey wolfs
Iter	10	No. of iterations
U	0	Upper value
L	1	Lower value
Dim	Based on the data set problem	
K	5	K-NN neighbors
Tr	10	Number of run

4.2   GWO Dynamic Parameters Tuning

In the standard GWO, two essential parameters that are named *(A, C)* controlling the behavior of the algorithm to balance between the two main factor intensification and diversification rate. Chaotic maps are utilized to tune these parameters as in the following:

$$b_{\text{Choas Vec}} = \left\{ (a_i - t) \left(\frac{a_f - a_i}{Max_{iter}} \right).Choas Vec(t) \right. \tag{10}$$

$$A_{\text{Choas Vec}} = 2b.Choas Vec(t) - b \tag{11}$$

$$C_{\text{Choas Vec}} = 2.Choas Vec(t) \tag{12}$$

a_i, a_f are the elementary and latest values adjusted as 2, 0 respectively. t is the current iteration and Max_{iter} are the maximal number of iterations.

4.3 Numerical Results

In this work, an improved binary grey wolf optimization (BGWO) algorithm integrated with rough set theory for feature subset selection is presented. The proposed model is executed using MATLAB 2016a in a PC Intel 2.5 GHz processor with 8 GB RAM. Several benchmark datasets are used for evaluation phase. Five standard datasets are taken from UCI Machine learning repository, which are Cervical cancer (DS1), Dermatology (DS2), Diabetic Retinopathy (DS3), Arrhythmia (DS4), Sonar (DS5), with different instances, attributes and different classes [10]. The experimental results are evaluated for different chaotic maps using seven different measures to evaluate the proposed approach: Average Accuracy (AC), Root Mean Square Error (RMSE), Mean (μ_f), Standard derivation (STD), Best fitness score (BS), Worst fitness score (WS), and Average Attributes Selected (AAS) [14]. BGWOFS algorithm works stochastically in binary space instead of continues search space. Each position is an attribute subset. BGWOFS owns a sturdy robustness and has the faster convergence velocity to locate the minimal reduct subset. To balance between intensification and diversification rates, chaos theory is used to improve the performance of WOA in terms of convergence speed and local minima avoidance. The chaos maps used to tune GWO random parameters are Chebyshev map (C1), Circle map (C2), Gauss/mouse map (C3), Iterative map (C4), Logistic map (C5), Piecewise map (C6), Sine map (C7), Singer map (C8), Sinusoidal map (C9), Tent map (C10). The performance of the rough set method is improved by using the integration between rough set and GWO algorithm is proposed. The running time of CBGWO is influenced less by the problem dimension. The different evaluation criteria are presented in Table 2 for five medical datasets with ten chaos maps using the proposed approach. We can see that the proposed approach attained the highest classification accuracy ($\approx$0.97) and *RMSE* ($\approx$0.42) using Piecewise map and Singer map for DS1, followed by DS2 with accuracy ($\approx$0.96) and *RMSE* ($\approx$0.2) using Chebyshev map. Further details of other maps can be seen in Table 2.

4.4 Comparison with Other Meta-Heuristics

In this subsection, the performance of the proposed approach integrated with piecewise and Singer chaotic maps are compared with other optimization algorithms. These algorithms are WOA [14], ALO [4], CALO integrated with three chaotic versions (CALO + Tent; CALO + Sinusoidal; CALO + Singer) [4, 16], BAT [13], and BCCSA integrated with four chaotic versions (BCCSA + Gauss; BCCSA + Sine; BCCSA + Sinusoidal; BCCSA + Tent) [14]. The initial parameters are the same setting for these optimization algorithms; number of populations (Agents = 5), number of iterations (Iter = 50), upper and lower limit (u = 1 and l = 0), and problem dimension (Dim) which is based on number of features in the whole dataset.

    Figure 3 shows the best accuracy of the proposed approach vs. WOA, ALO, CALO with three chaotic versions, BAT, BCCSA with four chaotic versions algorithms. For Cervical cancer (DS1), the highest value of accuracy attained by ALO with different chaotic maps ($\approx$0.99) followed by CBGWOFS with two chaotic (Piecewise, Singer maps) ($\approx$0.97), BCCSA with Sinusoidal map ($\approx$0.947), and WOA ($\approx$0.9444) respectively. Further these algorithms applied on Dermatology data sets DS2, the highest

Table 2. Results of RSCBGWOFS approach using different evaluation criteria

	Eval.	C1	C2	C3	C4	C5	C6	C7	C8	C9	C10
DS1	AC	0.845	0.718	0.634	0.817	0.747	0.648	0.704	0.775	0.662	0.775
	RMSE	0.831	1.295	1.369	0.949	0.949	1.25	1.3	1.404	1.492	0.986
	WS	0.048	0.037	0.032	0.032	0.055	0.044	0.045	0.034	0.04	0.034
	BS	0.048	0.037	0.032	0.032	0.055	0.044	0.045	0.034	0.04	0.034
	Mean	0.052	0.045	0.05	0.038	0.058	0.051	0.06	0.042	0.04	0.047
	Std	0.003	0.009	0.019	0.017	0.007	0.01	0.017	0.014	0	0.027
	No.At.	13	31	32	33	19	19	22	21	24	21
	Time	1.13	0.74	0.85	0.73	0.73	0.79	0.85	0.74	0.98	0.88
DS2	AC	0.947	0.936	0.936	0.936	0.936	0.936	0.936	0.936	0.86	0.936
	RMSE	0.229	0.254	0.254	0.254	0.254	0.254	0.254	0.254	0.375	0.254
	WS	0.082	0.058	0.078	0.074	0.055	0.07	0.075	0.064	0.081	0.066
	BS	0.082	0.058	0.078	0.074	0.055	0.07	0.075	0.064	0.081	0.066
	Mean	0.055	0.058	0.078	0.074	0.083	0.071	0.076	0.067	0.081	0.072
	Std	0	0	0	0	0	0.003	0.002	0.004	0	0.007
	No.At.	14	18	13	15	21	18	12	5	18	14
	Time	1.43	1.12	0.9	1.24	1.1	1.32	0.96	1.1	1.07	0.96
DS3	AC	0.592	0.59	0.648	0.653	0.62	0.568	0.639	0.634	0.599	0.59
	RMSE	0.639	0.64	0.594	0.589	0.617	0.658	0.601	0.605	0.633	0.64
	WS	0.324	0.333	0.32	0.346	0.323	0.327	0.317	0.336	0.324	0.306
	BS	0.324	0.333	0.32	0.346	0.323	0.327	0.317	0.336	0.324	0.306
	Mean	0.328	0.333	0.327	0.348	0.325	0.331	0.324	0.343	0.325	0.32
	Std	0.006	0	0.002	0.004	0.005	0.007	0.003	0.006	0.002	0.015
	No.At.	9	12	13	17	12	11	12	12	12	7
	Time	1.65	1.26	1.2	1.32	1.31	1.02	0.811	1.34	1.26	0.92
DS4	AC	0.59	0.59	0.59	0.59	0.59	0.59	0.59	0.59	0.59	0.59
	RMSE	4.332	4.332	4.332	4.332	4.332	4.332	4.332	4.332	4.332	4.332
	WS	0.421	0.42	0.423	0.424	0.411	0.392	0.411	0.389	0.421	0.401
	BS	0.421	0.42	0.423	0.424	0.411	0.392	0.411	0.389	0.421	0.401
	Mean	0.425	0.42	0.43	0.43	0.411	0.408	0.422	0.409	0.421	0.405
	Std	0.007	0	0.003	0.002	0	0.016	0.01	0.018	0	0.005
	No.At.	138	123	198	218	136	165	145	91	134	129
	Time	1.77	0.88	0.87	0.79	0.82	0.76	0.79	0.74	0.7	0.74
DS5	AC	0.732	0.756	0.732	0.829	0.805	0.707	0.707	0.732	0.805	0.756
	RMSE	0.518	0.494	0.518	0.413	0.442	0.541	0.541	0.518	0.442	0.494
	WS	0.195	0.168	0.205	0.214	0.167	0.178	0.137	0.158	0.197	0.137
	BS	0.195	0.168	0.205	0.214	0.167	0.178	0.137	0.158	0.197	0.137
	Mean	0.196	0.168	0.205	0.214	0.167	0.19	0.152	0.178	0.198	0.171
	Std	0.003	0	0	0	0	0.017	0.024	0.023	0.003	0.034
	No.At.	29	35	30	28	33	38	24	33	38	23
	Time	1.1	0.6	0.62	0.57	0.71	0.73	0.63	0.56	0.63	0.79

accuracy value obtained by CBGWOFS for two chaotic (Piecewise, Singer maps) (≈0.95) in comparison with CALO and its chaotic (≈0.9447), BCCSA with Sine map (≈0.9395), WOA (≈0.93), Bat (≈0.93) and BALO (≈0.930). It is illustrated that the integration of rough set, chaos theory and GWO algorithm improved the accuracy performance of the proposed RSCBGWOFS approach. The proposed approach is superior and outperformed in the complex and non-linear medical data set in relation to other optimization algorithms.

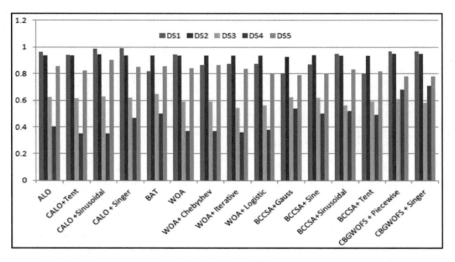

Fig. 3. The accuracy results obtained from different meta-heuristic algorithms

5   Conclusions and Future Directions

Feature selection is nontrivial and substantial in high-scale data; furthermore, it causes an efficient classification process. In this research, a recent approach for feature selection is presented through medical domain based on rough set hybridized with Chaos Theory and Binary Grey Wolf Optimization Algorithm (RSCBGWOFS). The proposed approach is performed to be applied to five variant complex datasets that are retrieved from UCI repository. Ten chaotic maps are used to tune GWO algorithm, to improve performance and to increase convergence speed. The experimental results show that the proposed approach provides superior classification accuracy through comparing it with different optimization techniques ALO, CALO, BAT, WOA, CWOA, BCSA, and BCCSA. In a future work, the proposed approach is assessed with large-scale data or high dimensional data. The process of pruning a rule can be performed to enhance the accuracy of the classification.

Acknowledgement. The study is supported and funded under the auspices of the Benha University, Egypt research project titled "Rough Set Hybridization with Metaheuristic Optimization Techniques for Dimensionality Reduction of Big-Data". We would like to show our gratitude to the Benha University, Egypt for funding the research project.

References

1. Anter, A.M., Hassenian, A.E.: Normalized multiple features fusion based on PCA and multiple classifiers voting in CT liver tumor recognition. In: Advances in Soft Computing and Machine Learning in Image Processing, pp. 113–129. Springer, Cham (2018)
2. Azar, A.T., Hassanien, A.E.: Dimensionality reduction of medical big data using neural-fuzzy classifier. Soft. Comput. **19**(4), 1115–1127 (2014). https://doi.org/10.1007/s00500-014-1327-4
3. Elbably, D.L., Fouad, K.M.: A hybrid approach for improving data classification-based on PCA and enhanced ELM. Int. J. Adv. Intell. Paradigms (2018). Forthcoming articles
4. Emary, E., Zawbaa, H.M., Hassanien, A.E.: Binary ant lion approaches for feature selection. Neurocomputing **213**, 54–65 (2016)
5. Fouad, K.: A hybrid approach of missing data imputation for upper gastrointestinal diagnosis. Int. J. Adv. Intell. Paradigms (2018). Forthcoming articles
6. Inbarani, H.H., Banu, P.K.N., Azar, A.T.: Feature selection using swarm-based relative reduct technique for fetal heart rate. Neural Comput. Appl. **25**(3–4), 793–806 (2014). https://doi.org/10.1007/s00521-014-1552
7. Kohli, M., Arora, S.: Chaotic grey wolf optimization algorithm for constrained optimization problems. J. Comput. Des. Eng. (2017)
8. Lichman, M.: UCI Machine Learning Repository. University of California, School of Information and Computer Science, Irvine, CA (2013). http://archive.ics.uci.edu/ml
9. Mirjalili, S., Mirjalili, S., Lewis, A.: Grey wolf optimizer. Adv. Eng. Softw. **69**, 46–61 (2014)
10. Mirjalili, S.: How effective is the Grey Wolf optimizer in training multi-layer perceptrons. Appl. Intell. **43**(1), 150–161 (2015)
11. ElSoud, M.A., Anter, A.M.: Computational intelligence optimization algorithm based on meta-heuristic social-spider: case study on CT liver tumor diagnosis. Comput. Intell. 7(4) (2016)
12. Pawlak, Z., Skowron, A.: Rough sets and Boolean reasoning. Inf. Sci. **177**(1), 41–73 (2007)
13. Rodrigues, D., Pereira, L.A., Nakamura, R.Y., Costa, K.A., Yang, X.S., Souza, A.N., Papa, J.P.: A wrapper approach for feature selection based on bat algorithm and optimum-path forest. Expert Syst. Appl. **41**(5), 2250–2258 (2014)
14. Sayed, G.I., Hassanien, A.E., Azar, A.T.: Feature selection via a novel chaotic crow search algorithm. Neural Comput. Appl. 1–18 (2017)
15. Strogatz, S.H.: Nonlinear Dynamics and Chaos: With Applications to Physics, Biology, Chemistry, and Engineering. CRC Press, Boca Raton (2018)
16. Hassanien, A.E., Alamry, E.: Swarm Intelligence: Principles, Advances, and Applications. CRC, Taylor & Francis Group, Boca Raton (2015). ISBN 9781498741064 - CAT# K26721
17. Swiniarski, R.W., Skowron, A.: Rough set method in feature selection and recognition. Pattern Recogn. Lett. **24**(6), 833–849 (2003)
18. Hassanien, A.E., Suraj, Z., Slezak, D., Lingras, P.: Rough Computing: Theories, Technologies, and Applications. Idea Group Inc. (2008)
19. Anter, A.M., Hassanien, A.E., ElSoud, M.A., Kim, T.H.: Feature selection approach based on social spider algorithm: case study on abdominal CT liver tumor. In: 2015 Seventh International Conference on Advanced Communication and Networking (ACN), pp. 89–94. IEEE, July 2015

Parameter Estimation for Chaotic Systems Using the Fruit Fly Optimization Algorithm

Saad M. Darwish[1], Amr Elmasry[2], and Asmaa H. Ibrahim[1(✉)]

[1] Department of Information Technology, Institute of Graduate Studies
and Research, Alexandria University, Alexandria, Egypt
saad.darwish@alexu-igsr.edu.eg,
igsr.asmaahassan@alexu.edu.eg
[2] Department of Computer Engineering and Systems, Faculty of Engineering,
Alexandria University, Alexandria, Egypt
elmasry@alexu.edu.eg

Abstract. A chaotic system is a one of encryption technique used in wireless communication. The chaotic system is a nonlinear dynamic system, which provides a mechanism for signal design and generation to cover the transmitting information and data. The chaotic system is high security, fast and difficult to predict because of its parameters sensitive to initial conditions. Parameter estimation of chaotic systems is a big challenge and an active area of study. This paper is concerned with the parameter estimation of a chaotic system can be formulated as a multidimensional optimization problem. Optimization algorithms can be solving this problem because of its good computational performance and robustness. Fruit Fly Optimization Algorithm (FOA) is used to estimate the parameters of chaotic systems which belong to an optimization algorithm. Simulation results which applied on the Lorenz system show that the FOA can identify the parameters of the chaotic systems more accurately, more rapidly, and more stable.

Keywords: Chaotic system · Fruit fly optimization algorithm (FOA) ·
Parameter estimation

1 Introduction

For the last several years, chaotic nonlinear dynamic systems have gained wide attention in the applications of biological systems, economics, and communication [1]. As a matter of fact, the most important characteristic of the chaotic system lies in the dynamical system which follows no pattern, does not repeat itself, require a precision that is unable to be computed and the initial condition sensitivity [2]. Furthermore, the advances in wireless communication devices used in military, mobile communication, and private data encryption have led to the development of highly secure and fast data encryption techniques. In these applications, real-time, fast, secure, and reliable monitoring are considered the most essential requirements [3]. Actually, many encryption techniques can be used such as the traditional encryption algorithms and chaotic-based encryption systems. However, some of these traditional encryption

© Springer Nature Switzerland AG 2020
A. E. Hassanien et al. (Eds.): AMLTA 2019, AISC 921, pp. 80–90, 2020.
https://doi.org/10.1007/978-3-030-14118-9_9

algorithms are hard to understand, complex, not suitable for real-time applications, and slow for encryption. On the other hand, the main advantage of using chaos lies in the observation that a chaotic signal looks like noise for the unauthorized users. Therefore, chaotic-based encryption systems have highly secure, fast, and easy implemented encryption systems for secure transmission networks [4].

In addition, the control of chaotic electronic circuits, suitable for secure communications, depends on two main factors. The First is the way a non-linear circuit begins to operate in chaotic mode (route to chaos), and the second factor is achieving transmitter-receiver synchronization [5]. A chaotic system, called the master, generates a signal which is sent over a channel to a slaver, which uses this signal to synchronize itself with the master. To use a chaotic signal in communications the receiver must have a duplicate of the transmitter's chaotic signal; synchronized with the transmitter [6]. However, the two identical chaotic systems which have started with the same initial conditions will be uncorrelated if any small change in their initial conditions has occurred. In other words, the synchronization cannot effectively perform if there is an unknown parameter in the system [7]. Due to the complexity of chaotic systems, it is difficult to determine whether the parameters are not known in case of secure communication. Accordingly, estimating unknown system parameters of chaotic systems to control or synchronize chaotic systems represents a big challenge [8].

Furthermore, many methods have been applied to estimate the parameters of chaotic systems, including the active-passive-decomposition method, the minimization strategy, the iterative method, statistical methods, and optimization algorithms [9]. Optimization algorithms are a group of mathematical algorithms utilized in machine learning to find the best available alternative under the given constraints, which try to find the best solution of the problem by minimizing or maximizing the objective function [10]. In fact, optimization algorithms require fewer experiments to achieve an optimum formulation, and they can trace and rectify "problem" in a remarkably easier manner. Hence, optimization algorithms are far better than other methods. Furthermore, optimization algorithms include many algorithms like Swarm Intelligence (SI), Genetic Algorithm (GA), Tabu Search (TS) and Simulated Annealing (SA) [11].

Recently, swarm intelligence has been employed for the estimation of unknown model parameters in chaotic systems. Basically, SI can be defined as the searching behavior of intelligent swarms. The main properties of SI are self-organization and division of labor [12]. SI includes bee algorithms (BA), artificial bee colony algorithm (ABC), particle swarm optimization (PSO), ant colony optimization (ACO), and fruit fly optimization algorithm (FOA) [13, 14]. Compared with other swarm-based optimization algorithms, FOA has the advantages of being easy to understand and a simple computational process. What's more, FOA can quickly search the feature space for an optimal or near-optimal feature subset minimizing a given fitness function [15]. Thus, this paper proposes a fruit fly optimization algorithm for the estimation of a chaotic system parameter to achieve synchronization in communication wireless network.

The rest of the paper is organized as follows. Depicted the related works in Sect. 2. The detailed description of the proposed system has been made in Sect. 3. The experimental result and discussion are written in Sect. 4. Finally, we conclude the paper in Sect. 5.

2 Literature Survey

In the last years, researchers have begun to explore how the understanding of nature can be utilized for the purpose of estimating the chaotic system parameters in communication networks. There are a number of biologically inspired algorithms concepts that have inspired significantly improved performance over traditional approaches. Actually, Li et al. [16] suggested a hybrid between differential evolution algorithm and artificial bee colony algorithm to identify the chaotic system parameters. Moreover, differential evolution is used to find the region of the global minimum. On the other hand, artificial bee colony algorithm is utilized to get the optimal solution quickly. However, while this method has a quick convergence rate, it has a high computational time, and is complex.

Additionally, Li et al. [17] adopted fireworks algorithm (FWA) to estimate the parameters of the chaotic system. The sparks of firework will search the whole solution space in a fine structure and focus on a small place to find the optimal solution, which renders this algorithm effective in finding the global optimal value. Despite the fact that FWA has a good computational performance and robustness, it has a large time consuming, and is weak on shifted functions and lack of mathematical foundation.

The authors in [18] presented the chaotic firefly optimization algorithm (FA) for the identification of chaotic system parameter. Actually, this method has improved both the optimization accuracy and the convergence speed. On the other hand, this algorithm requires more parameters to be set, which would subsequently make the system more complex.

Another approach for estimating the parameters of Lorenz and Chen chaotic system is introduced by Xiang-Tao et al. [19]. This approach combined the stochastic exploration of the cuckoo search (CS) and the exploitation capability of the orthogonal learning strategy to find the chaotic system parameters. Likewise, this approach is good at exploring the optimal solution; nevertheless, the convergence rate is relatively slow.

Moreover, He et al. [20] adopted particle swarm optimization (PSO) algorithm to solve the problem of parameters estimation of the chaotic system. The particle swarm optimization has a quick convergence and high performance. However, the results are poor when the population size is too small because the solution space will not be explored sufficiently.

Similarly, the authors in [21] presented the chaotic ant swarm algorithm (CAS) to solve the problem of estimating the parameters of the chaotic system. Hence, the proposed method merges the effects of chaotic dynamics with the swarm-based search to search for the optimal solution. Remarkably, CAS has a fast convergence rate and effective. Nevertheless, this method has some disadvantages such as a low solution precision and long computational time.

Another work, related to chaotic system parameter estimation, is introduced in [22]. The author has suggested an artificial bee colony algorithm combined with cuckoo search strategy for solving the chaotic system problem. In fact, the artificial bee colony (ABC) algorithm is motivated by the intelligent behavior of honey bees which has good global optimization ability when seeking for an optimal solution in the search space. The local search mechanism of cuckoo search optimization (CS) and the chaotic search mechanism are introduced into the onlooker bee phase to balance the bee colony global search and local search abilities. The numerical results illustrate that this algorithm is a powerful tool for parameter estimation with low deviations even when using limited input data. However, the optimal solution will be given up if it is repeated, which affects the accuracy of the system.

Following this further, the work in this paper proposes to use FOA to address the problem of the chaotic system by estimating its parameters. Therefore, FOA is used to optimize the solution accuracy and system performance. Compared with other swarm intelligent algorithms, FOA is simple and only requires two parameters to be set, namely, population size and maximum generation number. Conversely, the other traditional intelligent algorithms require at least three parameters [23]. In addition, the improper parameter will seriously affect the performance of the algorithm, and cause difficulty in estimating the parameters of the system. Thus, the FOA system has the ability to find the optimal solution with high accuracy.

3 Proposed Methodology

3.1 Problem Formulation

The parameters estimation of chaotic systems is formulated as a continuous nonlinear multidimensional complexity problem in the following equation [22]:

$$\dot{X} = F(X, X_0, \theta) \tag{1}$$

where $X = (x_1, x_2, \ldots x_N)^T \in R^N$ indicates n-dimensional state variable of the original chaotic system, X_0 indicates system's initial state, and $\theta = (\theta_1, \theta_2, \ldots \theta_d)^T$ is the set of original unknown parameters. Assume knowing the system (1) structure, the estimated system can be expressed as:

$$\dot{\widetilde{X}} = F\left(\widetilde{X}, X_0, \widetilde{\theta}\right) \tag{2}$$

where $\widetilde{X} = (\widetilde{X}_1, \widetilde{X}_2, \ldots \widetilde{X}_N)^T \in R^N$ indicates n-dimensional state variable of estimation chaotic system state and $\widetilde{\theta} = (\widetilde{\theta}_1, \widetilde{\theta}_2, \ldots \widetilde{\theta}_m)^T$ is the set of estimation parameters.

The heuristic algorithms depend only on the objective or fitness function to guide the search; the three-dimensional objective function can be defined as follows [19]:

$$f\left(\widetilde{\theta}_i^n\right) = \sum_{t=0}^{W} \left[\left(x_1(t) - \widetilde{x}_{i,1}^n(t)\right)^2 + \left(x_2(t) - \widetilde{x}_{i,2}^n(t)\right)^2 + \left(x_3(t) - \widetilde{x}_{i,3}^n(t)\right)^2 \right] \tag{3}$$

where $t = 0, 1, \ldots W$, i denotes the i-th state vector, and $x(t)$ and $\widetilde{x}(t)$ are real and estimated values at time t, respectively. The goal of estimating the parameters of the chaotic system (2) is to find out the suitable value of $\widetilde{\theta}_i^n$ such that fitness function (3) is globally minimized. The block diagram of parameter estimation for the chaotic system (see Fig. 1). The initial state is given to both the real system and the estimated model.

Then, outputs from the real system and its estimated model are input to the optimization algorithm, where the objective function $f\left(\tilde{\theta}_i^n\right)$ will be calculated. The parameters may be difficult to determine due to the dynamic and unstable behavior of chaotic systems. Moreover, there are multiple variables in the problem and multiple local optima in the landscape of objective function. Hence, the FOA algorithm will be adopted to overcome trapped in local optima and achieve the global optimal parameters. In this paper, FOA is applied to estimate the parameters for Lorenz chaotic system and compared with other typical algorithms.

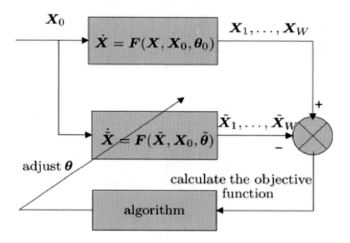

Fig. 1. The principle of parameter estimation for the chaotic system [19].

3.2 Fruit Fly Optimization Algorithm

The fruit fly optimization algorithm is based on the feeding behavior of fruit fly. The fruit fly itself is superior to other species in sensing and perception, especially in smell and vision. The food finding process of fruit fly has two steps [24]:

(a) It smells the food source by using osphresis organ and flies towards that direction.
(b) After it gets close to the food location, it can also use its sensitive vision to find food and fruit flies' flocking location and flies towards that direction.

According to the food searching characteristics of a fruit fly swarm, the FOA can be divided into several steps as follows [25]:

Step 1: Set the main parameters population size (n), the maximum number of iterations (m), and random initial location of fruit fly group (X, Y).
Step 2: Give the random location (X_i, Y_i) and distance for the search for food using the sense of smell of a fruit fly individual.

$$X_i = X + Random\ Value\ 1 \tag{4}$$

$$Y_i = Y + Random\ Value\ 2 \tag{5}$$

Step 3: The smell concentration judgment value for each fly is represented as the objective function value, which calculated according to (3).

Step 4: Sort the fruit flies and find out the fruit fly with minimal objective function among the fruit fly swarm.

Step 5: Keep the best objective function value and X, Y coordinate (best objective function location), and at this moment, the fruit fly swarm will use vision to fly toward that location.

Step 6: repeat iterations until the maximum iteration number is reached. The proposed model (see Fig. 2).

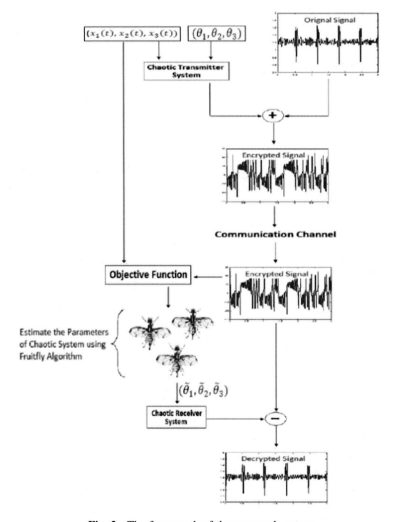

Fig. 2. The framework of the proposed system

4 Simulation Results

This section shows the performance of the Fruit fly algorithm on extracting of unknown parameters of Lorenz chaotic system. The FOA parameters are assigned as discussed in [19] the number of fireflies is as follows: the number of fruit flies = 30 and the maximum generation number is 100, and the total number of the run is chosen as 30 times for each system. Sampling intervals (h) and total sampling points (W) for each chaotic system are set to $h = 0.01$ and $W = 30$ respectively. Fourth order adaptive Runge-Kutta method is applied for solving the system of differential equations which are selected for testing the effectiveness of the algorithm. The experiments were implemented in MATLAB on a computer with Intel® Core Core™i5 – 2.40 GHz 2.40 GHz and 4 GB memory.

The mathematical description of the Lorenz system is as follows:

$$\begin{cases} \dot{x}_1 = \theta_1(x_2 - x_1) \\ \dot{x}_2 = (\theta_2 - x_3)x_1 - x_2 \\ \dot{x}_3 = x_1 x_2 - \theta_3 x_3 \end{cases} \tag{6}$$

where x_1, x_2, and x_3 are the state variables; θ_1, θ_2, θ_3 are the unknown chaotic system parameters. The searching ranges are set as follows: $9 \leq \theta_1 \leq 11$, $20 \leq \theta_2 \leq 30$, and $2 \leq \theta_3 \leq 3$. In the Lorenz system (6), three-dimensional parameters are unknown and need to be estimated. Table 1 lists the statistical results of the mean of best fitness values, the standard deviations, and the identified parameters of the Lorenz system, which obtained from FOA, are compared with the solutions acquired by [19]. From Table 1, it can be seen that all of the estimated parameters obtained by FOA are very close to the true values. Therefore, it is concluded that FOA is more effective and robust. The value of f decreases very fast to zero, which implies that FOA can converge to the global optimum very quickly (see Fig. 3). Moreover, the parameters θ_1, θ_2, and θ_3 converge to the true values rapidly, which demonstrates the great efficiency of FOA to achieve global optimization (see Figs. 4, 5 and 6). The time response of the model with estimated parameters (see Figs. 7, 8 and 9). The synchronization of the model response with that of the actual system is clearly obvious. From the experimental result can note that the FOA able to achieve the synchronization between the transmitter and receiver chaotic systems to transfer data correctly by estimating the unknown parameter of the receiver chaotic system. Though the FOA algorithm does not guarantee to find

Table 1. Statistical results for the Lorenz system

Algorithm	Means of the best fitness	Std. dev. of the best fitness	θ_1	θ_2	θ_3
FOA	1.01×10^{-8}	6.20×10^{-9}	10.0000	28.000	2.6667
CS	1.81×10^{-4}	1.66×10^{-4}	10.0000	28.000	2.6665
PSO	0.11788	0.268094	9.9999	27.9999	2.6666
GA	1.331401	2.783125	10.0274	28.0058	2.6692

the actual parameters values, this algorithm will find an acceptable solution in relatively a smaller number of iterations and able to perform the synchronization effectively between two identical transmitter and receiver chaotic systems.

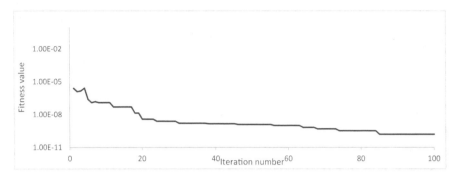

Fig. 3. The convergence process of the fitness value

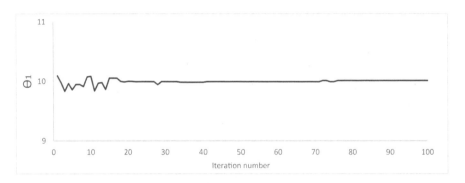

Fig. 4. The convergence process of θ_1.

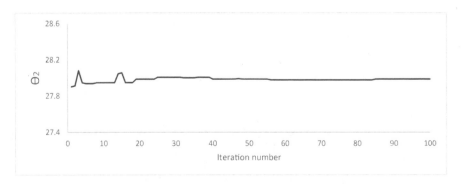

Fig. 5. The convergence process of θ_2

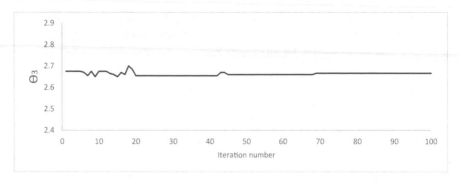

Fig. 6. The convergence process of θ_3

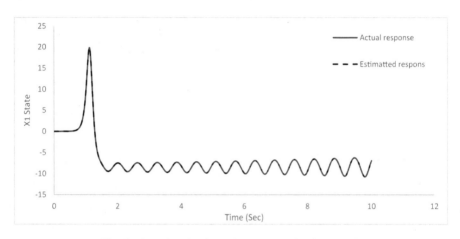

Fig. 7. Actual and estimated responses of x_1 in Eq. (6)

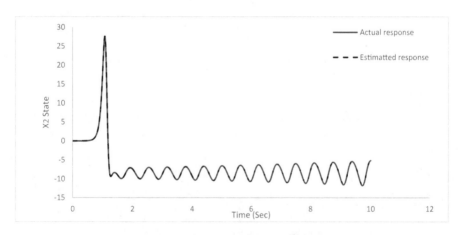

Fig. 8. Actual and estimated responses of x_2 in Eq. (6)

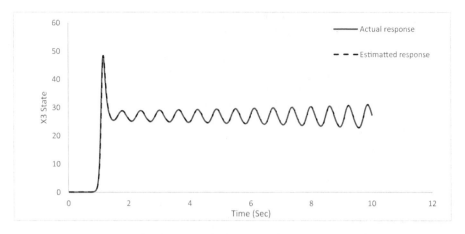

Fig. 9. Actual and estimated responses of x_3 in Eq. (6)

5 Conclusions

In this paper, the fruit fly optimization algorithm was applied to estimate the parameters for a chaotic system, which is formulated as a multi-dimensional optimization problem. The results those are obtained from FOA are compared with the solutions acquired by cuckoo search, particle swarm optimization, and genetic algorithm. The FOA show a better performance in estimating the parameters of the chaotic dynamical system than other algorithms. As the experimental results of the chaotic Lorenz system show that FOA can identify the parameters of the Lorenz system more accurately, more rapidly, and more stable.

References

1. Bakhache, B., Ahmad, K., Assad, S.: A new chaotic encryption algorithm to enhance the security of Zigbee and wi-fi networks. Int. J. Intell. Comput. Res. **2**(4), 219–227 (2011)
2. Banerjee, S., Kurths, J.: Chaos and cryptography: a new dimension in secure communications. Eur. Phys. J. Spec. Top. **223**(8), 1441–1445 (2014)
3. Shukla, P., Khare, A., Rizvi, M., Stalin, S., Kumar, S.: Applied cryptography using chaos function for fast digital logic-based systems in ubiquitous computing. J. Entropy **17**(3), 1387–1410 (2015)
4. Shaerbaf, S., Seyedin, S.: Nonlinear multiuser receiver for optimized chaos-based DSCDMA systems. Iran. J. Electr. Electron. Eng. **7**(3), 149–160 (2011)
5. Stavrinides, S., Anagnostopoulos, A., Miliou, A., Valaristos, L., Kosmatopulos, K., Papaioannou, S.: Digital chaotic synchronized communication system. J. Eng. Sci. Technol. Rev. **2**(1), 82–86 (2009)
6. Homer, M., Hogan, S., Bernardo, M., Williams, C.: The importance of choosing attractors for optimizing chaotic communications. IEEE Trans. Circuits Syst. **51**(10), 511–516 (2004)
7. He, J., Cai, J.: Parameter modulation for secure communication via the synchronization of chen hyperchaotic systems. Syst. Sci. and Control Eng. **2**(1), 718–726 (2014)

8. Wang, J., Zhou, B., Zhou, S.: An improved cuckoo search optimization algorithm for the problem of chaotic systems parameter estimation. Comput. Intell. Neurosci. **2016**(8), 1–8 (2016)
9. Chen, M., Kurths, J.: Chaos synchronization and parameter estimation from a scalar output signal. Phys. Rev. E **76**(2), 1–4 (2007)
10. George, G., Raimond, K.: A survey on optimization algorithms for optimizing the numerical functions. Int. J. Comput. Appl. **61**(6), 41–46 (2013)
11. Binitha, S., Sathya, S.: A survey of bio-inspired optimization algorithms. Int. J. Soft Comput. Eng. **2**(2), 137–151 (2012)
12. Wahab, M., Meziani, S., Atyabi, A.: A comprehensive review of swarm optimization algorithms. PLoS ONE **10**(5), 1–36 (2015)
13. Sehrawat, P., Rohil, H.: Taxonomy of swarm optimization. Int. J. Adv. Res. Comput. Sci. Softw. Eng. **3**(8), 1400–1406 (2013)
14. Pathania, K., Rana, A.: Fruit fly optimization solution for missile optimization attack drown problem. J. Comput. Technol. **1**(2), 1–6 (2016)
15. Pan, W.: A new fruit fly optimization algorithm: taking the financial distress model as an example. Knowl.-Based Syst. **26**(2), 69–74 (2012)
16. Li, X., Yin, M.: Parameter estimation for chaotic systems by hybrid differential evolution algorithm and artificial bee colony algorithm. Nonlinear Dyn. **77**(1–2), 61–71 (2014)
17. Li, H., Bai, P., Xue, J., Zhu, J., Zhang, H.: Parameter estimation of chaotic systems using fireworks algorithm. Adv. Swarm Comput. Intell. **9141**, 457–467 (2015)
18. Gao, W., Zhang, Z., Chong, Y.: Chaotic system parameter identification based on firefly optimization. Appl. Mech. Mater. **347–350**, 3821–3826 (2013)
19. Xiang-Tao, L., Ming-Hao, Y.: Parameter estimation for chaotic systems using the cuckoo search algorithm with an orthogonal learning method. Chin. Phys. B **21**(5), 1–6 (2012)
20. He, Q., Wang, L., Liu, B.: Parameter estimation for chaotic systems by particle swarm optimization. Chaos Solitons Fractals **34**(2), 654–661 (2007)
21. Li, L., Yang, Y., Peng, H., Wang, X.: Parameters identification of chaotic systems via chaotic ant swarm. Chaos Solitons Fractals **28**(5), 1204–1211 (2006)
22. Hao, D., Rong, L., Ke, L.: Parameters identification of chaotic systems based on artificial bee colony algorithm combined with cuckoo search strategy. Sci. China Technol. Sci. **61**(3), 417–426 (2018)
23. Niu, J., Zhong, W., Liang, Y., Luo, N., Qian, F.: Fruit fly optimization algorithm based on differential evolution and its application on gasification process operation optimization. Knowl.-Based Syst. **88**, 253–263 (2015)
24. Iscan, H., Gunduz, M.: Parameter analysis on fruit fly optimization algorithm. J. Comput. Commun. **2**(4), 137–141 (2014)
25. Jiang, T., Wang, J.: Study on path planning method for mobile robot based on fruit fly optimization algorithm. Appl. Mech. Mater. **536–537**, 970–973 (2014)

Optimal Shortest Path in Mobile Ad-Hoc Network Based on Fruit Fly Optimization Algorithm

Saad M. Darwish[1], Amr Elmasry[2], and Shaymaa H. Ibrahim[1(✉)]

[1] Department of Information Technology, Institute of Graduate Studies
and Research, Alexandria University, Alexandria, Egypt
saad.darwish@alexu-igsr.edu.eg,
igsr.shaymaahassan@alexu.edu.eg
[2] Department of Computer Engineering and Systems, Faculty of Engineering,
Alexandria University, Alexandria, Egypt
elmasry@alexu.edu.eg

Abstract. Mobile ad hoc network (MANET) can be defined as a self-configuring group of mobile devices or nodes. These nodes able to change locations and configure themselves rapidly, freely and dynamically, thus routing is big challenging in mobile ad hoc network. The important role of routing is selecting the route for transferring the data from the source node to destination node in network efficiently based on the routing table, which contains routing metrics such as cost, load, distance, and delay. To solve this problem used many shortest path search techniques. Based on network performance, the routing algorithms find the best path to take. In this paper proposed the fruit fly optimization algorithm (FOA) to find the optimal shortest route in a mobile ad hoc network. The simulation results show that the performance of fruit fly optimization algorithm is better than the classical algorithm (Dijkstra Algorithm) in terms of scalability and complexity time.

Keywords: Mobile Ad-hoc networks (MANET) · Routing protocol ·
Shortest route · Fruit fly optimization algorithm (FOA)

1 Introduction

Mobile ad hoc network (MANET) can be defined as a wireless network without using a fixed and centralized infrastructure, consisting of mobile nodes [1]. In fact, ad hoc nodes should be able to detect the other nodes and sharing of information, data, and services with each other. Therefore, the ad-hoc networks can be used in areas with incomplete communication infrastructure or inconvenient [2]. Basically, the MANET can be implemented for communication at the time of disaster, fire rescue operations, or other communication that requires fast deployable communications because of the nodes which are able to work as a router or host. The mobile nodes move randomly and freely, thus the network topology in MANET is dynamic [3]. In MANET the nodes have low power batteries, which can be difficult to recharge and replace after the nodes

© Springer Nature Switzerland AG 2020
A. E. Hassanien et al. (Eds.): AMLTA 2019, AISC 921, pp. 91–101, 2020.
https://doi.org/10.1007/978-3-030-14118-9_10

have been deployed and limited bandwidth. Hence, it is necessary to consider the specific factors of the networks in the process of developing the protocols.

Furthermore, routing is the process of transferring packets of data from a source to a destination in the network. Actually, the concept of routing involves two activities, namely, the routing protocol and routing metrics [4]. Subsequently, the main routing protocols in MANET are proactive and reactive protocols. In proactive routing protocols, each node records the latest routing information in a routing table and each node can update periodically the routing table with all other nodes. On the other hand, the reactive routing protocols determine routes only if needed to forward packets of data. Therefore, there is no need for the route information to update periodically [5]. As for the routing problems in MANET, serious issues like transmission delay and high energy consumption would occur if the optimal path has not been defined for transmitting data from a source to a destination by these nodes. Thus, it is needful for a routing optimization algorithm to solve this problem, particularly in a large size network, because the computational effort grows exponentially with the number of nodes and links in the networks [6].

Routing algorithms in routers work to determine mathematically the optimal path for the packet to its destination based on routing tables [7]. As a matter of fact, routing tables contain the total route information for the packet, such as load, cost, delay, and distance, which has an impact on the performance of the network based on routing metrics [8]. In addition, a routing algorithm requires a time-constraint service to select a path from a source to a destination. However, in MANET the topologies are more frequently changed, and the nodes cannot properly transmit data to the destinations on time if the routing algorithm cannot provide a time-constraint service. Thus, the optimal solution should be found by the routing algorithm within a reasonable time. There are many routing algorithms such as classical algorithms and optimization algorithms to solve this problem.

The classical algorithm work with real-valued edge-lengths using only comparisons and additions shall like Dijkstra and Bellman-Ford [9]. The classical optimization can find only the local optimum. The drawbacks of Dijkstra's algorithm are that it consumes a lot of time in the searching process and cannot obtain the right shortest path. On the other hand, Bellman-Ford's algorithm disadvantages are that changes in network topology are reflected slowly since updates are spread node-by-node [10]. By definition, optimization algorithms are search algorithms based on mechanics of natural selection and natural genetics utilized for obtaining global the maximum or minimum of an optimization problem [11]. They are based on survival of fittest among string structures with a structured yet randomized information exchange to form search algorithms with some of the innovative flairs of human search [12]. There are many types of optimization algorithms such as Bee Colony Optimization (BCO), Ant Colony Optimization (ACO), Firefly Algorithm (FA), Cuckoo Search Algorithm (CSA) and Fruit Fly Optimization Algorithm (FOA).

This paper proposes a fruit fly optimization algorithm (FOA) as a routing algorithm to solve the routing problem in MANET by finding the optimal shortest path between the source and the destination node. In practice, FOA has some advantages such as it is more scalable, it has a minimum computational process, and it can quickly search the feature space for optimal or near-optimal feature subset minimizing a given objective function within a reasonable time. Subsequently, this improves the data delivery performance and speeds up the computational time in the routing system [13, 14].

The remainder of this paper is organized as follows. In Sect. 2 depicted the related works. The detailed description of the proposed system has been made in Sect. 3. Section 4 explained the simulation and results. Section 5 ended with conclusions.

2 Literature Survey

Recently, the routing algorithm in MANET has become an active area of research to solve the routing problems in network communications. In this section, some of these algorithms are briefly reviewed. In [15], the authors have used the Ant Colony Optimization (ACO) technique in conjunction with queuing network analysis to minimize the overall end-to-end delay of routing packets from the source to destination in MANET. Basically, the ACO algorithm is derived from the food searching behavior of ants to select the shortest path in networks. Therefore, the system can be described as: the ACO uses the routing table information of each node to make decisions around which a node must be taken at the next hop. Then the trip time to reach the desired neighbor is computed. Actually, the queue works to organize the data packets according to the arrival time. The main advantage of this system is its ability to minimize the end-to-end delay in the network and adapt with the current topology of the network effectively. However, the principal limitation facing this technique is that it has a high complexity time for a large number of nodes.

Kumari and Kannammal [16] adopted a modified artificial bee colony optimization algorithm (MABCO) to solve dynamic routing problem for MANET. The Modified Bee Colony Optimization algorithm is a hybrid routing algorithm which is based on the Artificial Bee colony algorithm and contains both reactive and proactive elements. This algorithm depends on Bee behavior to find the shortest path, a reactive element to keep the routing information of only these nodes involved in communication, and a proactive element to make the nodes exchange routing information while the communication is going on. The main advantage of this system is the ability of identifying multiple stable paths from the source to the destination node in a dynamic network. The drawback of this system is the complexity because it is based on a lot of techniques.

Jang [17] adopted a tabu search algorithm to determine an optimal path from a source to a destination in mobile ad-hoc networks. For finding an optimal solution, the tabu search generates an initial solution which is inserted into a tabu list as a better solution. Then, the newest element is added to the tabu list and the oldest one is removed when the tabu list is full during sequent iterations. In each iteration, the set of neighbors (nodes) of the current solution is established by the neighborhood generating operations. Finally, only the node with the highest value is chosen as the new best solution of the next iteration. As a matter of fact, the proposed algorithm determines an optimal path within a minimum execution time. However, it is affected by its neighborhood generation, move selection mechanisms, and its parameters, which make the routing system in large network slow.

Another interesting approach is to be found in [18], where the authors introduced a combination between firefly algorithm and queuing network analysis to select the shortest route for transferring the packets of data from a source node to the destination node. Firefly algorithm is based on the characteristics of the firefly behavior such as attractiveness and brightness to select the next node, which can be used to find the shortest path in MANETs with minimum response time. The response time is the total time that the transaction spends waiting to be serviced and while being serviced. The affecting factors of time response are the number of hops, arrival rate and service time. Queuing network analysis can be applied as the model to calculate the taken response times to send packets of data in MANET. This system able to handle the routing problem with minimizing the response time which considered as advantages. However, the more hops, the much longer time it takes to detect routes.

Moreover, Persis et al. [19] suggested a firefly algorithm (FA) for enhancing the reliability and scalability in the network. This algorithm is based on brightness and attractiveness there in firefly behavior to select the next node for each fly then choosing the different paths, which are collected from the flies. Actually, these paths are called a non-dominated (Pareto-optimal) set of paths. Indeed, these paths are able to convert data packet from source to destination when the communication fails in the shortest path. What's more, this algorithm is able to transfer data from source to destination within an acceptable response time in large network size; hence, this makes the system more scalability. Moreover, the algorithm has a good converge rate. Nevertheless, the drawback of this system is that it may not find the shortest path because it is based on the parameters of firefly to select the neighbor node.

Likewise, the authors in [20] presented the genetic algorithmic (GA) to solve the shortest path routing problem. The GA is the encoding the routing path with a series of positive integers that represent the IDs of nodes through which the path passes. Then, it determines the solutions according to the fitness function and accurately evaluates the quality of a solution to choose the optimal path which has the least cost. Addressing the captured results in this technique achieves a good efficiency to network in terms of throughput, delay, and packet loss. However, this approach makes the routing system consume time considerably.

Accordingly, this paper proposes using FOA to address the problem of routing in MANET from source node to destination node, affected by the random movement of the mobile nodes in the network. FOA is used to find the shortest path. Different from related routing algorithms, the proposed approach is considered fast and easy to implement. Unquestionably, this reveals that the routing system possesses a high scalability and ability to reduce complexity time.

3   Proposed Methodology

3.1   MANET Routing Problem Formulation

Finding the shortest path from source to the destination node in MANETS is one of the network optimization problems. A mobile network topology can be represented as a simple connected graph $G = (V, E)$, where V and E denote the set vertices containing N nodes and the set of edges (i, j) that collectively connects the source node to the destination node. To transmit data from a source to a destination, each node has to send data through some intermediate nodes, due to the transmission range of each node. The state of the links can vary according to the link metrics such as distance, load, cost, and delay between communication nodes. There are present some assumption to apply to the proposed algorithm like every node can bi-directionally communicate with neighboring nodes, and has the same data processing capabilities. The aim is solving the routing optimization problem by finding the optimal shortest path, which has a total minimum link metrics. To design the algorithm, we assume every node knows, a priori, the information of all the nodes and links of the networks by using another management protocol [5]. The metrics of each link is determined by the distance, load, cost, and delay between the source and destination nodes. Therefore, the routing optimization problem in this paper can be formulated as a combinatorial optimization problem minimizing the objective function as follows:

$$Objective\ Function\ (F) = De_T + D_T + L_T + C_T \tag{1}$$

$$Minimize\ Delay\ (De_T) = \sum\nolimits_{(i,j)\in E} de_{ij} \tag{2}$$

$$Minimize\ Distance\ (D_T) = \sum\nolimits_{(i,j)\in E} d_{ij} \tag{3}$$

$$Minimize\ Load\ (L_T) = \sum\nolimits_{(i,j)\in E} l_{ij} \tag{4}$$

$$Minimize\ Cost\ (C_T) = \sum\nolimits_{(i,j)\in E} c_{ij} \tag{5}$$

where,
de_{ij} is the delay of the link (i, j)
d_{ij} is the distance of the link (i, j)
l_{ij} is the load of the link (i, j)
c_{ij} is the cost of the link (i, j)

3.2 Basic Fruit Fly Optimization Algorithm

Fruit fly optimization algorithm is a recent nature meta-heuristic algorithm, developed by pan [13]. Fruit fly swarm in nature exhibits social behavior that uses collective intelligence to perform their essential activities (see Fig. 1). The fruit fly is superior to other species in osphresis and vision [21]. The food finding process of fruit fly has two phases: (i) the flies use osphresis organ to smell the food source and flies towards that direction; (ii) Upon nearing the food source, the flies use its sensitive vision to find food and fruit flies' flocking location and flies towards that direction. The main steps of FOA can be concluded as follows (see Fig. 2) [22, 23]: (i) a swarm of m fruit flies is initially placed in source node S; (ii) each fly moves from source node to next node (neighbor location) randomly; (iii) calculate the smell concentration values of fruit flies according to the objective function (1); (iv) find the path of fruit fly, which has a lowest objective value by using vision; (v) repeat iterations until the fruit fly reach to the destination node (food location).

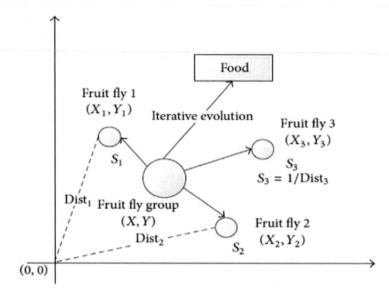

Fig. 1. Food searching for fruit fly.

4 Experimental Results

In this section, the fruit fly algorithm was applied to optimize the routing problems with different network topologies, by varying the number of nodes, edges and the link configurations randomly. The performance of the fruit fly algorithm is compared with the classical algorithm (Dijkstra's algorithm). Simulation experiments are carried out in MATLAB on a computer with Intel® Core Core™i5 – 2.40 GHz 2.40 GHz and 4 GB memory. The topologies are called problems A, B and C. Each problem contains some nodes and links, as is shown in Table 1. Moreover, the parameters of the algorithm that are used in the experiments are shown in Table 2. We measured the routing objective function (see Fig. 3) and complexity time (see Fig. 4) of the FOA with the number of iterations = $N-1$ and the number of fireflies: 5, 10 and 20.

Generally, the probability of finding the optimal solution increases if the swarm size increases in FOA algorithm. We observe that irrespective of the swarm size in the small size network the results of the minimum routing objective function are similar (see Fig. 3). This means that the fruit fly optimization algorithm can find an optimal solution in the small size network through a small swarm size. On the other hand, by increasing the number of nodes in the network, we note that the FOA with the larger swarm size finds an optimal solution with better performance. However, the minimum routing objective function of the FOA with swarm size = 10 is similar to that of the FOA with swarm size = 20. The average complexity time as a function of three problems of the proposed fruit fly optimization algorithm for 10 runs (see Fig. 4). The average complexity time of the fruit fly optimization algorithm increases by increasing the number of the swarm size for all problems. The average complexity time and the routing objective function of the nodes are trade-off factors, so these two factors must

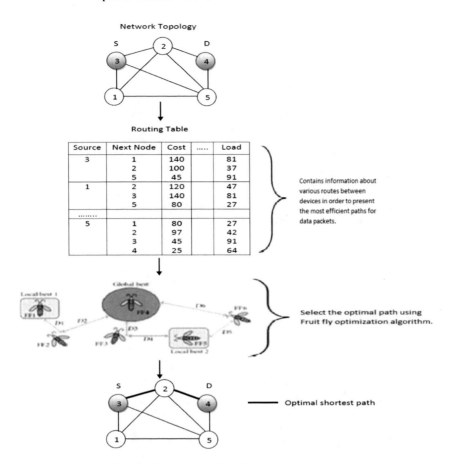

Fig. 2. The framework of the proposed system.

Table 1. Problems for the experiment

Problem	# of nodes	# of edges
A	5	8
B	10	22
C	15	28

Table 2. The parameters of the fruit fly optimization algorithm

Parameters	Values
Population size	5/10/20
Number of iterations	$N-1$

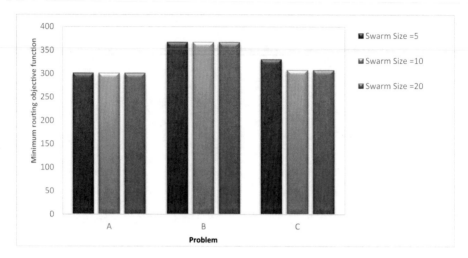

Fig. 3. Minimum routing objective function

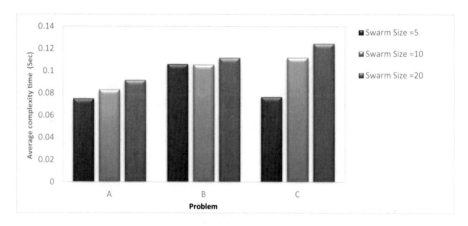

Fig. 4. Average routing complexity time

be simultaneously considered and balanced by the routing algorithm. Therefore, we will choose the swarm size = 10 because it has a lower average complexity time than swarm size = 20. Figure 5 shows comparisons of the average complexity time for 10 runs for the FOA and classical algorithms in the given problems. The results show that the FOA takes a shorter time. This is because the FOA reaches the optimal solution faster than another algorithm.

The scalability of the algorithms is tested with large-sized networks. The convergence to the local optima in a routing optimization problem is prevented in fruit fly optimization algorithm since the individual agent in the swarm searches path based on its own smell concentration and the smell concentration of other agents in the swarm that measured according to (1). Thereby fruit fly optimization algorithm is able to find the optimal path from the solution space. The complexity time taken by the classical and

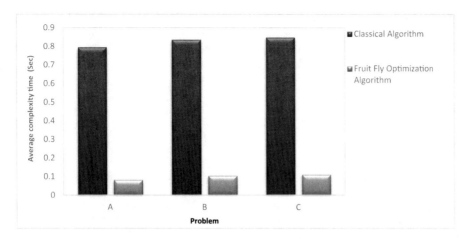

Fig. 5. Average routing complexity time

fruit fly optimization algorithms is presented in Table 3 to study the convergence time of these algorithms for different networks considered in the study. It is observed from the study that the classical algorithm of searching the shortest path in a network with more than 15 nodes encounter the maximum limit of the simulation period. However, it is found that the FOA algorithm is able to handle large network with less complexity time. Though the FOA algorithm does not guarantee to find the shortest path but will achieve an acceptable solution (shortest optimal path) in a reasonable time.

Table 3. The convergence time of the fruit fly optimization and classical algorithms

Topology configuration		Convergence time	
# of nodes	# of edges	Fruit fly optimization algorithm	Classical algorithm
5	8	0.07518	0.79554
10	21	0.10530	0.83643
15	28	0.10779	0.84837
20	99	0.13865	∞
40	158	0.22450	∞
60	302	0.22684	∞
80	454	0.23154	∞

5   Conclusion

In this paper, we proposed a routing optimization algorithm to efficiently determine an optimal path from a source to a destination in mobile ad-hoc networks. The proposed algorithm was designed by using the FOA algorithm, which is a typical meta-heuristic (optimization) algorithm. We evaluated the performance of the algorithms by carrying out the experiments by varying the number of nodes and links, and we compared the proposed algorithm with the classical algorithm (Dijkstra algorithm) in terms of the scalability and the average complexity time for the routing problem. The comparison results showed that the FOA outperforms classical algorithm in terms of the average complexity time and scalability and it is suitable for adapting the routing optimization problem.

References

1. Raza, N., Aftab, M., Akbar, M., Ashraf, O., Irfan, M.: Mobile ad-hoc networks applications and its challenges. Commun. Netw. **8**(3), 131–136 (2016)
2. Ghosekar, P., Katkar, G., Ghorpade, P.: Mobile ad-hoc networking: imperatives and challenges. Int. J. Comput. Appl. Spec. Issue MANETs **3**, 153–158 (2010)
3. Jayakumar, G., Gopinath, G.: Ad-hoc mobile wireless networks routing protocols – a review. J. Comput. Sci. **3**(8), 574–582 (2007)
4. Sharma, A., Gupta, A.: A survey on path weight-based routing over wireless mesh networks. Int. J. Innovations Eng. Technol. **3**(4), 15–20 (2014)
5. Devarajan, K., Padmathilagam, V.: Diversified optimization techniques for routing protocols in mobile ad-hoc wireless networks. J. Eng. Appl. Sci. **10**(12), 5229–5239 (2015)
6. Kushwaha, S., Kumar, A., Kumar, N.: Routing protocols and challenges faced in ad-hoc wireless networks. Adv. Electron. Electric Eng. **4**(2), 207–212 (2014)
7. Manjunath, M., Manjaiah, D.: PAR: petal ant routing algorithm for mobile ad hoc network. Int. J. Comput. Netw. Commun. **7**(2), 45–58 (2015)
8. Sumitha, J.: Routing algorithms in networks. Res. J. Recent Sci. **3**(ISC-2013), 1–3 (2014)
9. Pettie, S., Ramachandran, V., Sridhar, S.: Experimental evaluation of a new shortest path algorithm. In: LNCS, vol. 2409, pp. 126–142 (2002)
10. AbuRyash, H., Tamimi, A.: Comparison studies for different shortest path algorithms. Int. J. Comput. Appl. **14**(8), 5879–5986 (2015)
11. Roy, D., Das, S., Ghosh, S.: Comparative analysis of genetic algorithm and classical algorithms in fractional programming. In: Advanced Computing and Systems for Security, vol. 396, pp. 249–270 (2015)
12. Vasiljevic, D.: Classical and Evolutionary Algorithms in the Optimization of Optical Systems, 1st edn. Kluwer Academic Publishers, London (2002)
13. Pan, W.: A new fruit fly optimization algorithm: taking the financial distress model as an example. Knowl.-Based Syst. **26**(2), 69–74 (2012)
14. Allah, R.: Hybridization of fruit fly optimization algorithm and firefly algorithm for solving nonlinear programming problems. Int. J. Swarm Intell. Evol. Comput. **5**(2), 1–10 (2016)
15. Kaur, S., Sawhney, R., Vohra, R.: MANET link performance parameters using ant colony optimization approach. Int. J. Comput. Appl. **47**(8), 40–45 (2012)
16. Kumari, E., Kannammal, A.: Dynamic shortest path routing in mobile ad-hoc networks using modified artificial bee colony optimization algorithm. Int. J. Comput. Sci. Inf. Technol. **5**(6), 7423–7426 (2014)
17. Jang, K.: A tabu search algorithm for routing optimization in mobile ad-hoc networks. Telecommun. Syst. **51**(2–3), 177–191 (2012)
18. Zakaria, A., Saman, M., Nor, A., Hassan, H.: Finding shortest routing solution in mobile ad hoc networks using firefly algorithm and queuing network analysis. J. Technol. **77**(18), 17–22 (2015)

19. Persis, D., Robert, T.: Reliable mobile ad-hoc network routing using firefly algorithm. Intell. Syst. Appl. **8**(5), 10–18 (2016)
20. Biradar, A., Thool, R.: Effectiveness of genetic algorithm in reactive protocols for MANET. Int. J. Eng. Res. Technol. **2**(7), 1757–1761 (2013)
21. Shan, D., Cao, G., Dong, H.: LGMS-FOA: An improved fruit fly optimization algorithm for solving optimization problems. Hindawi Publishing Corporation Math. Probl. Eng. **2013**, 1–10 (2013)
22. Jiang, T., Wang, J.: Study on path planning method for mobile robot based on fruit fly optimization algorithm. Appl. Mech. Mater. **536–537**, 970–973 (2014)
23. Iscan, H., Gunduz, M.: Parameter analysis on fruit fly optimization algorithm. J. Comput. Commun. **2**(4), 137–141 (2014)

Swarm Optimization for Solving Load Balancing in Cloud Computing

Aya A. Salah Farrag$^{(\boxtimes)}$, Safia Abbas Mohamad,
and El Sayed M. El-Horbaty

Ain Shams University, Cairo, Egypt
aya.salah@cis.asu.edu.eg

Abstract. Cloud computing is the new paradigm of representing computing capabilities as a service. With its facility of resource sharing and being cost-effective, it exists in every domain of life, enhancing their functionality and adding new opportunities to it. Accordingly, the focus on solving its dilemmas like load balancing becomes more challenging and the research in swarm-based algorithms to find optimal results has been expanding. This paper discusses the use of two swarm algorithms including Ant-Lion optimizer (ALO) and Grey wolf optimizer (GWO) in task scheduling of the Cloud Computing environment. Additionally, compare the results with commonly known swarm algorithms: Particle Swarm Optimization (PSO) and Firefly Algorithm (FFA). The results show the ALO and GWO are a strong adversary to Particle Swarm Optimization (PSO), and better than Firefly (FFA) and they have potential in load balancing.

Keywords: Cloud computing · Load balancing · Task scheduling · Ant-Lion optimizer · Swarm intelligence

1 Introduction

Cloud computing is the process of leasing computing capacities to others without the need to have experience in the IT field. This process is done by cloud service providers (CSP) like Amazon, Google, Microsoft, and others. Additionally, it has been dominating and transforming the business environment in the last few years due to many factors: (1) reduced cost of storages, (2) assurance of data not lost, (3) cloud can handle buying, maintaining and upgrading hardware and high technical Support for it, (4) data availability any time, (5) additional resource can be added in near real-time and with ease [18]. In Retail business, cloud gives an equal opportunity to any retailer big or small to use services. These services explore big data analytics and social media analysis in order to connect with their customers and stay updated about their preferences example: Amazon, Walmart, and The Home Depot. Also, cloud reduces time to deploy ideas of new products in the market [16]. While in entertainment industry, cloud-powered entertainment enables small entertainment companies to reach global audiences and scale up their operations with minimal cost, for example: Netflix, which has experienced long-term growth as a result of its early adoption of cloud technology and become a major entertainment house [17].

© Springer Nature Switzerland AG 2020
A. E. Hassanien et al. (Eds.): AMLTA 2019, AISC 921, pp. 102–113, 2020.
https://doi.org/10.1007/978-3-030-14118-9_11

Cloud Computing confers services in 4 different forms: Infrastructure as a service (IaaS), Platform as a service (PaaS), Software as service (SaaS) and Mobile "backend" as a service (MBaaS). Quality of service is one of the main challenges of cloud computing which are: (1) Security and Privacy, (2) Portability, (3) Reliability and availability and (4) Quality of service (QoS). It is the process of maintaining proper management of resources in order to fulfill the Service Level Agreements (SLAs) between the cloud providers and the cloud users [7]. Considering the massive demand for handling Cloud Computing challenges, research has been done in this area especially in load balancing.

Load balancing distributes the load over servers to keep the system steady without over-loaded or under-loaded ones which maximizes resource utilization. The load can be network, memory and CPU loads. It is considered an NP-hard problem. In order to solve it, many researches have been done using heuristic and Metaheuristic Algorithms. Existing Cloud Systems EC2 Amazon and Google cloud App engine use Cron service as a scheduler a time-based job scheduler. Although it is fast and good still need improving in load balancing.

The rest of this paper is organized as follows: Sect. 2 presents the related work. The proposed algorithms ALO and GWO are explained in Sect. 3. Section 4 illustrates the implementation and experimental results. Finally, Sect. 6 contains the conclusion and future work.

2 Related Work

Load balancing of any cloud system is dependent on its scheduler either task scheduler or resource scheduler. Research on load balancing assist in improving one of these elements: (1) Makespan, (2) Response time, (3) Migration time, (4) Energy Consumption, (5) throughput and (6) Cost [7]. Load balancing as shown in Fig. 1 is branched to 2 types of work: Static Load Balancing (SLB) and Dynamic Load Balancing (DLB). Static Load Balancing runs from the start with prior knowledge of the system, while Dynamic Load Balancing (DLB) depends on in progress of the system as it runs when overload state or rebalance occurs [8]. Therefore, many researched scheduling algorithms either heuristics or meta-heuristics to improve load balancing.

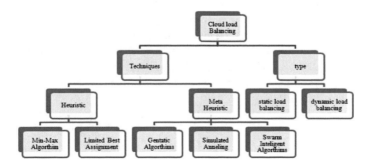

Fig. 1. Load balancing specifics

Many research done in heuristic Algorithms to find optimum scheduler as in [7, 9, 10]. In [7], it compared some of heuristic Algorithms: Minimum Completion Time (MCT), Minimum Execution Time (MET), Min-Min and Min-Max in by minimizing the makespan and energy Consumption. The results presented that MCT were exceeding others. While in [9], it modified Optimal Cost Scheduling algorithm as resource scheduling to maximize throughput and minimize response time, the results were better than the original algorithm and round robin. Furthermore, in [10] it combined the Min-Min and Min-Max in Enhanced load balancing Min-Min Algorithm to minimize the makespan which gave better results than Min-Min.

Other Researchers approached the problem by using Metaheuristic Algorithms specifically swarm intelligent Algorithms for their known high exploration of search space than traditional Algorithms Example: PSO in [13, 14] and FFA in [11, 12]. In [14], it used PSO as task scheduling to optimize both energy and processing time. It was tested on a large scale of 20 datacenters against Best Resource Selection (BRS) and Random Scheduling Algorithm (RSA) providing better results. Also in [13], it joined PSO with Cuckoo search (CS) to reduce the makespan, by comparison to PSO and Random Allocation (RA). Moreover, in [11], it proposed a combination of Firefly algorithm along with the fuzzy logic as resource scheduling to address both certainty and uncertainty time cases during load assignment across the virtual machines. On a small scale of 3 clients with each running a maximum of 2 VMs, it gave better results than a Genetic Algorithm to reduce completion time and number of migrated tasks. In the meantime [12] proposed FFA as job scheduler to minimize completion time. It gave better results than Ant Colony Optimizer (ACO) even when an increasing number of jobs from 100 to 400. With new metaheuristic algorithms discovered every day, researchers still explore them to find a better algorithm for load balancing. Example to that, the use of Endocrine algorithm in [1] with PSO to transfer the extra tasks form overloaded VMs to under-loaded ones using the enhanced feedbacking. This algorithm decreases the time span during the VM migration compared to traditional load balancing techniques which increase the Quality Of Service. However, it is needed to be checked on a large scale and with the algorithm in the paper [2]. Another example is the Lion Optimization Algorithm in [3] that is used a task scheduler to minimize the makespan. It provides a very high utilization of resources when compared with PSO and GA but results show the cost of LOA is between PSO and GA although the difference is not great. In [4] Orthogonal Taguchi Based-Cat Swarm Optimization is examined as task scheduler to minimize total the makespan. It outperformed Min-Max, PSO with Linear Descending Inertia Weight (PSO-LDIW) and Hybrid PSO with Simulated Annealing (HPSO-SA) algorithms in comparison. Although, it may also be compared with PSO itself. Moreover, Pollination Based Optimization (PBO) in [5] is used to minimize the total job completion time and the makespan. In comparison with GA using different numbers of VMS, it gives better results but needs to be compared to swarm algorithmic like PSO.

3 Swarm Optimization: Short Reviews

a. Ant Lion Optimizer Algorithm

Ant-Lion Optimizer (ALO) [15] is a new proposed swarm intelligent Algorithm for solving optimization problems. In this paper we intend to use ALO in task scheduling with its 2 variations to reduce the makespan. Mainly the ALO work as shown in the following Pseudo code (Fig. 2).

Initialize the position of n ants and n antlions randomly, where n is number of search agents
Calculate the fitness of antlions
Find the best fit antlion and assume it as the elite
While (the end criterion is not satisfied)
    Update search boundaries LB, UB
    For every ant
        Select an antlion using Roulette wheel on antlion fitness
        Create normalized random walk towards selected antlion and elite using Eq. (1) and Eq. (2)
        Update the position of ant
        Calculate ant fitness
    End for
    Replace an antlion position with its corresponding ant position it if becomes fitter
    Update elite if an antlion becomes fitter than the elite
End While
Return elite

Fig. 2. The pseudo code of ALO

ALO depends on creating a random walk of an ant_i and scales it within the boundaries of an ant $lion_j$. Any random walk on n number of iterations is calculated by adding either a step forward or backward randomly for each iteration, using the following equations.

$$RW_i = \left[RW_i^1, \ldots RW_i^k, \ldots RW_i^n \right] \tag{1}$$

such that $RW_i^k = RW_i^{k-1} \pm 1$; $RW_i^1 = 0$; n is number of iterations; $RW_i r$ and o m walk of ant_i. To use the walk in k^{th} iteration, it should firstly be normalized by equation:

$$NRW_i^k = \frac{\left(RW_i^k - a \right)}{(b - a)} \tag{2}$$

Where a is the min value in RW_i; b is the max value in RW_i.

In this paper, we propose 3 versions for calculating random walk as it consumes the most time:

- ALO code: it creates Random walk for each ant around each Ant-Lion which is the original algorithm.
- ALO2: creates one random walk for each ant around all ant lions.
- ALO RW: creates one random walk for each ant around all ant lions where the normalized random walk takes min and max of the iteration of all walks.

b. Grey Wolf Optimizer Algorithm

The other used algorithm is Grey wolf optimizer (GWO) [6]. It simulates the behavior of a wolf pack when chasing the pray, where there is hierarchy of the best in alpha (α), beta (β), delta (δ) and omega. The main phases of grey wolf hunting are as follows:

- Tracking, chasing, and approaching the prey.
- Pursuing, encircling, and harassing the prey until it stops moving.
- Attack towards the prey.

The algorithm simulates the part of encircling the prey in which wolfs position themselves according to the prey using the following equations:

$$D = \left| C * X(t)_p - X(t) \right| \tag{3}$$

$$X(t+1) = X(t)_p - A * D \tag{4}$$

Where D is the distance between the prey and wolf, t is the current iteration, $X(t)_p$ is the current position of the prey and $X(t)$ is the position of the wolf. A and C are coefficients and they are calculated as follows:

$$A = 2 * a * r1 - a$$
$$C = 2 * r2$$

Where r1 and r2 are random numbers in range [0 1] and a is linearly decreased from 2 to 0 over the course of iterations.

The other wolfs who are omega (the last rank in wolf hierarchy) position themselves around the prey according to their relative distance to alpha, beta and delta. Using the Eqs. 3 and 4 to calculate positions between the superior wolfs and omega, getting the omega position from the average of the 3 positions (Fig. 3).

$$D_\alpha = \left| C1 * X(t)_\alpha - X(t) \right|, D_\beta = \left| C2 * X(t)_\beta - X(t) \right|, D_\delta = \left| C3 * X(t)_\delta - X(t) \right|$$

$$X1 = X(t)_\alpha - A1 * D_\alpha, X2 = X(t)_\beta - A2 * D_\beta, X3 = X(t)_\delta - A3 * D_\delta$$

$$X(t+1) = \frac{X1 + X2 + X3}{3} \tag{5}$$

```
Initialize the position of n grey wolfs, where n is number of search agents.
Calculate coefficients
Calculate the fitness of grey wolfs.
Find the Alpha = best fit grey wolfs , Beta= second best and delta= third best.
While (the end criterion is not satisfied)
        For every wolf
                Update the position of wolf using Eq.(3) ,Eq.(4) and Eq.(5)
        End For
        Calculate All grey wolfs fitness.
        Update coefficients.
        Update Alpha, Beta and delta.
End While
Return Alpha
```

Fig. 3. The pseudo code of GWO

4   Experimental Results and Discussion

In this paper, we will compare the proposed algorithms with commonly known particle swarm optimizer and firefly algorithms. We will perform task scheduling by minimizing the makespan of a task which is calculated by:

$$TT = CT + WT \tag{6}$$

where TT is task Total time, CT is computation time and WT is waiting time.

The computation time and waiting time are calculated by:

$$CT = \frac{Tlength}{pecount * mip} \tag{7}$$

Where $Tlength$ is the task length, $pecount$ is number of processors in the VM and mip processors' computation power

$$WT_i^j = \sum_{1}^{i-1} CT_x^j \tag{8}$$

Where WT_i^j is the waiting time for task i in VM j and the summation of computation time of all tasks before task i in the same VM.

This work is done using a CloudSim extension called cloud reports which is a graphic tool that provides an easy-to-use user interface. Cloudsim is a simulation that simulates distributed computing environments based on the Cloud Computing paradigm. The implementation parameters can be shown in the following Table 1).

Table 1. Cloudsim parameters

Entity	Parameter	Values
Cloudlet	No of cloudlets	50
	length	50000
Virtual machine	No of VMs	20–30
	RAM	512 MB
	MIPS	1000
	Size	1000
	bandwidth	100000
	Policy type	Dynamic workload
	VMM	Xen
	Operating system	Linux
	No of CPUs	1
Host	No of hosts	6–7
	RAM	40000 MB
	Storage	1000000
	No of CPUs	4
	Bandwidth	10000000
	Policy type	Time shared
Data center	No of data center	1–2

5 Experimental Results

The simulation runs in 4 phases in each phase either increase the Datacenter or the number of VMs. The output of each phase is gathered in 3 tables and their respective figures. Two of these tables represent the standard deviation of usage on each run for each algorithm. The first table represents the standard deviation of usage of computation power in VM while the second shows the standard deviation of usage of computation power in DC. The third table demonstrates the total number of migration.

When executing the simulation by 20 VMs and 1 DC: ALO2 then PSO then GWO show better results on VMs and Datacenter computation usage while ALO_RW was the worst. All nearly have same results in VM migrations but there are slight differences that showed ALO_RW was the best one (Figs. 4, 5 and 6) and (Tables 2, 3 and 4).

Table 2. DC1 VM20 SD of CPU usage in VM

SD of cpu_VM	Run1	Run2	Run3	Run4	Run5
GWO	31.76478	33.9595	29.87974	28.56407	36.83937
FFA	48.12437	61.50484	33.84811	38.85762	35.48642
ALO_RW	44.75811	29.42873	47.08909	61.71134	40.32328
Alo_Code	37.48412	33.29953	69.10363	38.47589	38.20173
ALO2	27.24153	21.94493	24.36388	27.08552	26.72349
PSO	27.25623	24.78393	32.48366	31.77519	39.12427

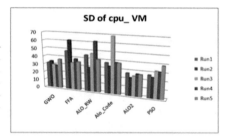

Fig. 4. DC1 VM20 SD of CPU usage in VM

Table 3. DC1 VM20 SD of CPU usage in DC

SD of Data Center_cpu	Run1	Run2	Run3	Run4	Run5
GWO	20.27925	21.05865	21.01944	18.88357	22.7781S
FFA	30.48978	25.17829	16.50816	22.71673	23.14078
ALORW	30.97252	21.15109	29.27881	25.3573	27.12319
Alo_Code	19.05455	21.28423	24.48056	22.87071	19.15162
ALO2	19.95746	11.60298	22.34225	21.47835	20.52657
PSO	11.26937	17.24932	22.23323	20.07091	18.13639

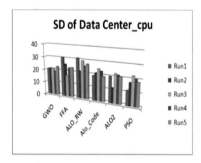

Fig. 5. DC1 VM20 SD of CPU usage in DC

Table 4. DC1 VM20 no. of VM Migration

No_VM_migrations	Run1	Run2	Run3	Run4	Run5
GWO	9	9	9	9	9
FFA	0	0	13	7	9
ALO_RW	0	9	9	0	9
Alo_code	15	9	0	9	5
ALO2	9	18	9	9	6
PSO	16	9	9	9	9

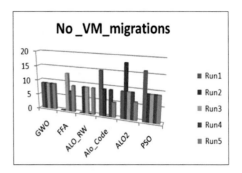

Fig. 6. DC1 VM20 no. of VM migration

When executing the simulation by 30 VMs and 1 DC (heaver loads): PSO then GWO then ALO2 are better than other (by slight difference between GWO and ALO2). On the other hand in migration, ALO_code presents better results (Figs. 7, 8 and 9) and (Tables 5, 6 and 7).

Table 5. DC1 VM30 SD of CPU usage in VM

SD of cpu_WM	Run1	Run2	Run3	Run4	Run5
GWO	32.30573	29.13538	31.83463	1S.09538	18.96382
FFA	60.9622	75.89679	24.97902	35.4948	32.83849
ALOSW	30.38869	18.68975	46.34659	59.23425	33.49068
Alo_Code	33.40604	33.4675	33.87311	61.69866	22.53788
A102	20.92942	30.33282	16.78599	27.4278	32.82376
PSO	20.63722	21.96876	18.17439	36.75326	16.42242

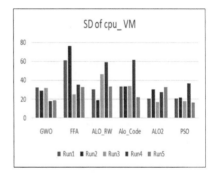

Fig. 7. DC1 VM30 SD of CPU usage in VM

Table 6. DC1 VM30 SD of CPU usage in DC

SD of Data Center_cpu	Run1	Run2	Run3	Run4	Run5
GWO	18.29314	14.71278	20.85847	7.040134	10.66249
FFA	24.46083	18.33993	12.36582	15.89244	17.63956
ALO_RW	16.78698	8.668745	25.76343	16.67581	17.04012
Alo_Code	22.66732	13.2516	15.30663	24.33215	8.270616
ALO2	8.655109	16.66486	15.30547	17.04671	19.77596
PSO	9.763257	9.688258	9.268105	20.31082	6.327539

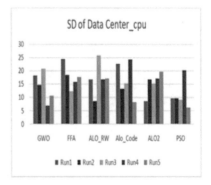

Fig. 8. DC1 VM30 SD of CPU usage in DC

After increasing the Data Centers and executing by 20VMs again: same best results as 1 Datacenter the results in computation usage was ALO2 then PSO then GWO. However in migration PSO was exceeding other GWO (Figs. 10, 11 and 12) and (Tables 8, 9 and 10).

Table 7. DC1 VM30 no. of migration of VM

No_ VM_migrations	Run1	Run2	Run3	Run4	Run 5
GWO	19	20	18	6	17
EFA	14	20	13	20	11
ALO_RW	11	20	14	19	24
AID Code	10	20	15	18	12
AL02	18	17	23	26	14
PSO	21	21	23	17	9

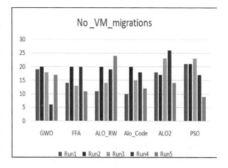

Fig. 9. DC1 VM30 no. of VM migration

Table 8. DC2 VM20 SD of CPU usage in VM

SD of cpu_VM	Run1	Run2	Run3	Run4	Run5
GWO	26.29297	18.09513	31.25691	21.52236	61.17653
FFA	41.78447	43.4154	40.89945	43.54235	39.12448
ALO RW	35.183	35.34739	30.2832	35.86092	34.83202
Alo Code	37.59097	38.26039	35.84564	36.85643	40.79685
AL02	17.56814	24.76555	19.02509	25.14942	22.15782
PSO	18.87558	17.82042	25.10197	25.60706	24.25797

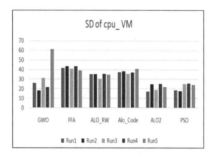

Fig. 10. DC2 VM20 SD of CPU usage in VM

Table 9. DC2 VM20 SD of CPU usage in DC

SD of Data Center_cpu	Run1	Run2	Run3	Run4	Run5
GWO	24.79656	26.50617	24.26489	22.21517	23.055
FFA	23.25695	27.12224	20.28459	25.26751	26.94233
ALO_RW	25.71981	24.29372	24.27936	23.0798	23.06714
Alo_Code	27.73939	18.03406	25.68396	23.81939	27.27785
AL02	23.30203	18.1769	23.06629	24.74092	19.78312
PSO	26.64778	23.52516	24.77898	24.75648	23.96061

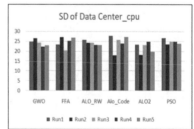

Fig. 11. DC2 VM20 SD of CPU usage in DC

Table 10. DC2 VM20 no of VM migration

No VM_ migrations	Runl	Run2	Run3	Run4	Run5
GWO	0	0	0	1	0
FFA	7	0	7	0	0
ALO_RW	0	0	0	0	5
Alo_Code	0	7	0	0	0
AL02	0	8	1	0	7
PSO	0	0	0	0	0

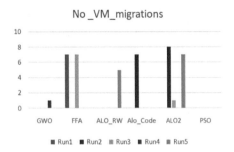

Fig. 12. DC2 VM20 no. of VM migration

At Last execution of simulation by 30VM and 2 DC: Although PSO is better in VM computation usage all nearly same in data center usage. In migration, ALO_RW then PSO have shown best results (Figs. 13, 14 and 15) and (Tables 11, 12 and 13).

Table 11. DC2 VM30 SD of CPU usage in VM

SD of cpu_VM	Runl	Run2	Run3	Run4	Run5
GWO	28.16634	36.00718	31.95295	33.2945	35.27935
FFA	60.74031	62.52765	40.43097	59.96365	42.41207
ALO_ RW	40.3276	41.62672	40.69102	35.88247	41.12091
Alo_ Code	38.75683	32.3281	41.80591	36.41324	41.18081
ALO2	32.86981	27.19764	55.53004	38.46025	37.29164
PSO	25.8028	29.04608	24.74904	31.18723	36.16269

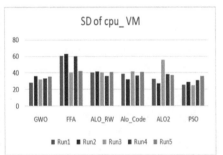

Fig. 13. DC2 VM30 SD of CPU usage in VM

Table 12. DC2 VM30 SD of CPU usage in DC

SD of Data Center_cpu	Runl	Run2	Run3	Run4	Run5
GWO	24.62795	23.10596	24.05538	25.97371	24.17005
FFA	21.68535	21.87335	28.69571	21.03657	18.63528
ALO_RW	24.57089	24.91588	25.34471	21.93462	26.85649
Alo_code	25.69677	25.51584	23.10923	22.24396	26.60429
ALO2	24.58847	26.69446	22.87614	25.57418	25.12961
PSO	17.66904	27.8129	22.10652	24.76325	25.44173

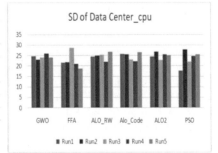

Fig. 14. DC2 VM30 SD of CPU usage in DC

Table 13. DC2 VM30 No of VM Migration

No VM_migrations	Run1	Run2	Run 3	Run 4	Run5
GWO	9	20	9	9	14
FFA	8	10	9	5	16
ALO_RW	16	9	8	5	9
Alo_Code	9	9	5	18	9
AL02	11	21	9	9	15
PSO	14	9	9	9	7

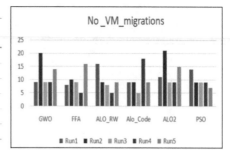

Fig. 15. DC2 VM30 no. of VM migration

6 Conclusion and Future Work

Load balancing is an important issue to stable cloud computing. Any load balancing needs a satisfactory scheduler to work efficiently. Thus many algorithms were researched including newly introduced intelligent algorithms. This Paper presented 3 versions of ALO and GWO to be used as Task scheduler to reduce the makespan. It compares the results with PSO and FFA. Although ALO2 and GWO were much better than FFA in reducing the makespan, they were as good as PSO in results and some-times ALO2 exceeds them. In VM migration, the results were different with every phase of simulation and sometimes close concluding that they all nearly same.

For future work the proposed algorithms are needed to be experimented by different factors of load balancing like reducing energy Consumption and compare its results to surveyed intelligent algorithms on different scales of cloud computing environment.

References

1. Aslanzadeh, S., Chaczko, Z.: Load balancing optimization in cloud computing: applying endocrine-particale swarm optimization. In: IEEE International Conference 2015 Electro/Information Technology (EIT), pp. 165–169. IEEE (2015)
2. Ramezani, F., Lu, J., Hussain, F.K.: Task-based system load balancing in cloud computing using particle swarm optimization. Int. J. Parallel Prog. **42**(5), 739–754 (2014)
3. Almezeini, N., Hafez, A.: Task Scheduling in Cloud Computing using Lion Optimization Algorithm. Algorithms **5**, 7 (2017)
4. Gabi, D., Ismail, A.S., Zainal, A., Zakaria, Z.: Solving task scheduling problem in cloud computing environment using orthogonal taguchi-cat algorithm. Int. J. Electr. Comput. Eng. (IJECE) **7**(3), 1489–1497 (2017)
5. Pathak, P., Mahajan, K.: A pollination based optimization for load balancing task scheduling in cloud computing. Int. J. Adv. Res. Comput. Sci. **25**(10) (2017)
6. Mirjalili, S., Mirjalili, S.M., Lewis, A.: Grey wolf optimizer. Adv. Eng. Softw. **69**, 46–61 (2014)
7. Mishra, S.K., Sahoo, B., Parida, P.P.: Load balancing in cloud computing: a big picture. J. King Saud Univ.-Comput. Inf. Sci. (2018)

8. Alam, M., Khan, Z.A.: Issues and challenges of load balancing algorithm in cloud computing environment. Indian J. Sci. Technol. **10**(25), 1–12 (2017)
9. Kaur, S., Sharma, S.: load balancing in cloud computing with enhanced optimal cost scheduling algorithm. Imp. J. Interdisc. Res. **2**(9), 1460–1466 (2016)
10. Patel, G., Mehta, R., Bhoi, U.: Enhanced load balanced min-min algorithm for static meta task scheduling in cloud computing. Procedia Comput. Sci. **57**, 545–553 (2015)
11. Susila, N., Chandramathi, S., Kishore, R.: A fuzzy-based firefly algorithm for dynamic load balancing in cloud computing environment. J. Emerg. Technol. Web Intell. **6**(4), 35–40 (2014)
12. Kaur, J., Bhardwaj, V.: A novel approach of task scheduling for cloud computing using adaptive firefly. Int. J. Comput. Appl. **147**(12), 9–13 (2016)
13. Al-maamari, A., Omara, F.A.: Task scheduling using hybrid algorithm in cloud computing environments. J. Comput. Eng. (IOSR-JCE) **17**(3), 96–106 (2015)
14. Jena, R.K.: Multi objective task scheduling in cloud environment using nested PSO framework. Procedia Comput. Sci. **57**, 1219–1227 (2015)
15. Mirjalili, S.: The ant lion optimizer. Adv. Eng. Softw. **83**, 80–98 (2015)
16. Mishra, S.K.: How has cloud computing affected the retail business. PCQuest, 5 October 2018. www.pcquest.com/cloud-computing-affected-retail-business/. Accessed 5 Oct 2018
17. Ryan: The Industries Most Affected by the Evolution of Cloud Computing. UTG, www.utgsolutions.com/the-industries-most-affected-by-the-evolution-of-cloud-computing. Accessed 5 Oct 2018
18. Cloud Computing – Allcenta Inc. http://allcenta.com/cloud-computing/. Accessed 5 Oct 2018

The Influence of New Energy Access on Load Peaks and Troughs Based on Optimization Techniques

Weibao Zhang$^{(\boxtimes)}$, Hong Gang, Baozhong Gan, and Qianhui Gang

Jinzhou Power Supply Branch, State Grid Liaoning Electric Power Supply Co.
Ltd., Jinzhou, China
zhangweibao@163.com

Abstract. Power is the basis of the continuous development of modern science and technology, the important guarantee of improving human life quality and the lifeblood of the national economy and security. Traditional power system generates electricity with the method of centralized power generation and then transmits electricity to the users with the network of electricity transmission. Developed countries and developing countries have the increased demand on energy. Therefore, how to solve the contradiction between increased demands for the energy and the traditional energy depleted by people increasingly will be the competitive field between countries. Meanwhile, it will be an important guarantee of the governments to maintain or improve the world influence and one of the keys to the continuous development of human civilization. In order to solve the contradiction between the distributed new energy power generation and the security of power system, we should maximize the utility of distributed new energy power generation to bring the technical, economic and social benefits, such as we can improve the flexibility and reliability of power system operation to meet people about the growing demand for electricity consumption and sustainable development in the future.

Keywords: Power system · Distributed new energy · Sustainable development

1 Introduction

New energy power generation and the difference between the traditional power generations: first, the power of new energy power station is an artificial force change and scheduling, the hair power natural environment has strong links with the outside world; the external performance is strong degeneration. Its power factors such as temperature, humidity, day and night, season are changing [1]. Second, the new energy power generation compared with the traditional power generation, its cost is lower, and emissions of greenhouse gases in the process of generating and harmful substances are very few, even not. Third, compared with traditional energy sources, new energy generally has the characteristics of less pollution and large reserves, which is of great significance to solve the serious environmental pollution problems and resource exhaustion problems in the world today. This chapter will focus on the above characteristics of photovoltaic power and wind power characteristic analysis research, based

© Springer Nature Switzerland AG 2020
A. E. Hassanien et al. (Eds.): AMLTA 2019, AISC 921, pp. 114–120, 2020.
https://doi.org/10.1007/978-3-030-14118-9_12

on the analysis of new energy power generation peak in cutting, voltage stability, reactive power support and the influence of the net energy loss, etc. [2].

2 Photovoltaic Power Generation

Photovoltaic power generation systems are designed around the Pv cells. A typical piece of photovoltaic battery output power is less than 5 W, is about 0.5 V, dc voltage so photovoltaic cells must be in series, to produce enough electricity for high power applications. Pv cells can be regarded as a special p-n junction, its work on the project mainly use diode equation described, its typical equivalent circuit diagram in Fig. 1.

$$I_L = I_{ph} - I_D - I_{sh} \tag{1}$$

$$I_D = I_o\left[\exp\left(\frac{qU_D}{AkT}\right) - 1\right] \tag{2}$$

$$I_{sh} = \frac{U_D}{R_{sh}} \tag{3}$$

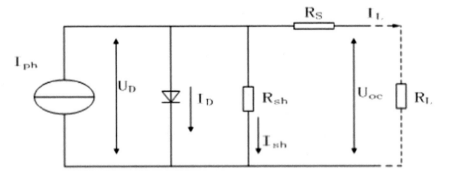

Fig. 1. The equivalent circuit of the photovoltaic battery

I_o: photovoltaic cells within the PN section of equivalent diode reverse saturation current

q: electron charge, 1.6×10^{-19} C

k: the Boltzmann constant, 1.38×10 to 18 erg/K

T: photovoltaic cells absolute temperature of the surroundings

A: The curve of constant: photovoltaic battery internal PN junction.

Power output characteristics of photovoltaic cells will be affected by the irradiance and temperature. Photovoltaic cells in the I-v curve of the fixed temperature 25 °C and PV curve. Can be seen that: fixed temperature, i-v curve can move up as light intensity increases with the whole, shows that short circuit current ISC rise with the increase of light intensity; At the same time, the open circuit voltage of battery increase slightly with the increase of light intensity. Can be seen from the photovoltaic cells have the output voltage and current limit, namely photovoltaic cells run even in the case of short circuit and short circuit phenomenon will not occur [3].

3 The Selection of Optimization Algorithm

New energy grid runtime through some optimization algorithm controls the size of the capacity of the new energy power generation access networks, so as to make the new energy power generation in power supply system to achieve the best balance between the stability of the economy and safety [4]. Looking for system optimal equilibrium is the essential connotation of the system of optimal power flow. The concept of optimal power flow first put forward by John Casteen pettier French engineer in the early 1960s, its core content is in the guarantee system of various constraint conditions under the condition of established, many of the variables in the control system by some means value target of the system to achieve the desired minimum or maximum. As a result of the optimal power flow including content is numerous, widely involved economic optimal control and scheduling of the system operation safety, power grid planning, etc.; optimal power flow, therefore, have been proposed since got rapid development, many excellent algorithms have been proposed [5]. Several kinds of the common optimization algorithm are summarized as follows:

3.1 Linear Programming Algorithm

Linear programming algorithm for the history of the development cycle is very long, is generally believed that Dantzig [11], he is the founder of the linear programming. Simplex algorithm is a classical algorithm for linear programming because its rules are simple and it can solve the linear programming problem well. Since then, the algorithm of linear programming problem can be divided into two categories: one is the traditional way along the feasible edge of the feasible domain, from one pole to the other pole, and finally the optimal solution of the programming problem is obtained. The other kind is not according to the traditional iterative method, that is, not strictly according to the direction of feasible falling (rising) edge, but from the internal or external of the feasible region, while most algorithms choose to travel from the feasible region, so as to finally find the optimal solution [6]. Linear programming algorithm is the main content has a standard linear programming, simplex algorithm, duality theory, and polynomial algorithm. Linear programming algorithm in the field of the simple convex set problem has good convergence, but when faced with the problem of a convex set, the algorithm of the solution set is often not the optimal solution. But the mathematical model of the modern power system is a typical nonlinear problem, so the adaptive linear programming algorithm is very poor, this kind of method of more used in decoupling system. Its outstanding characteristic is fast convergence speed.

3.2 Intelligent Algorithm

Optimization is in the field of intelligent algorithm mainly including genetic algorithm, ant colony algorithm, simulated annealing algorithm, and artificial neural network algorithm, etc. The essence of these algorithms are must establish the object of study the research of the feasible region of the objective function, and then randomly selected a feasible solution in the feasible region and its objective function value calculated, after the calculation and then randomly selected a feasible solution and calculate the function value, so constantly within the feasible solution domain search possible extremism. For simple function optimization problems, the classical algorithm is more effective and can obtain the exact optimal solution of the function. However, for complex functions and combinatorial optimization problems with non-linear and multi-extreme characteristics, the classical algorithm is often powerless [7]. Algorithm thought the main characteristic of this algorithm is simple and easy to operate, but because of this is fatal disadvantage of this kind of algorithm, the feasible solution of usually in order to guarantee the feasible region is the real extremism, such algorithms tend to divide the feasible region of the search space is large, cannot consider the existing computer calculation speed and memory space, its search effort is staggering. In order to improve the efficiency of the search, the scholar has offered a lot about a new method of heuristic search, but its performance and application scope is very limited. Because of this reason, the use of such methods is in the field and a long time [8].

3.3 Newton's Method

This algorithm is Newton method for unconstrained extreme value problem, the improved methods of traditional extremism problems need every step Newton's method for calculating haze matrix and its inverse matrix, the computing speed and memory size is a tall order. But by some scholars study of China's power system matrix is sparse, through certain way makes haze matrix of a highly sparse matrix, and then on the basis of the inequality constraints into equality constraints and the introduction of penalty function factor, through traditional Newton method for optimum each iteration variable amount of correction [9]. Newton algorithm is the main advantage of with quadratic convergence speed, can be found through a small amount of iterative optimal solution. The algorithm is the weakness of inequality constraints and the decisive set is often a complex polyhedron, in the process of calculation need to adopt the method of step by step to test to determine the calculation way, thereby increasing the amount of calculation. Newton method is an iterative algorithm, each step needs to solve the inverse matrix of the Hessian matrix of the objective function, the calculation is more complex.

3.4 The Original - Dual Method

The original dual method belongs to an important branch of the interior-point method; it has strong practicability and appealing. The basis of the algorithm framework is established by a Megiddo, Monteria, and others, then to the complementary problem by Kojima. The basic idea of the primal-dual method is to get the basic admissible solution of the original problem. The common method is to introduce artificial variables into the original problem and change the objective function into the negative value of the sum of the artificial variables. Then the objective function is maximized and the optimal basic admissible solution is removed from the artificial variable. This solution is the original problem. If the dual problem has an admissible solution and the basic admissible solution of the original problem satisfies the complementary relaxation condition, then the basic admissible solution of the original problem becomes the optimal basic admissible solution [10]. The original - dual algorithm can easily be applied to nonlinear convex programming, nonlinear complementarily, variation inequality, semidefinite optimization and second-order cone optimization problems such as field, has a similar approach in the best polynomial complex boundaries. This article will adopt this kind of method, which is the core of the system optimization research method, algorithm detailed steps are as follows.

4 The Influence of New Energy Power Generation on the System

A new energy access system, the aim is to ensure that the user power supply under the premise of safe and reliable as far as possible to reduce fossil fuel power generation plants. But the new energy power generation is influenced by natural factors on the timeline, part scheduling belongs to. Ideally, the network load are in a peak period of time, the new energy power just in peak power generation period; Network load is in a valley value time, generating new energy power also come in a valley. Reality is not the case, however, found in large amounts of data test research: new energy access system's influence on the system load "is the peak effect", "negative peak effect" and "peak effect". Simulation analysis, the graph "is the peak effect" wind power is set to 0.5 times the initial curve, photovoltaic power generation is set to 0.15 times the initial curve, set up the wind power in 1–15 and 46–60 time no electricity. To simplify the problem, it is assumed that the fossil fuel power plants generating a lower limit for the minimum load. The Fig. 2 shows that the fossil fuel power generation capacity in the trough time doesn't change, but in the peak period of time due to the participation of new energy power generation capacity to reduce. The end result is a peak value and the trough of fossil fuel power plants in the value of the difference is reduced, therefore the spinning reserve capacity can reach the purpose of saving fossil fuel with the decrease of the appropriate, the new energy access is a good way of new energy access.

Figure 2 as a "peak effect", the wind power is set to 0.8 times the initial curve, solar power is set to 0.5 times the initial curve, wind power and solar energy power set in the 1–15 and 46–60 h don't generate electricity. The figure shows that fossil fuels in the trough time don't change, in the peak period of time due to the participation of new energy power generation to reduce, but as a result of new energy power generation access networks too ambassador to fossil fuel power generation capacity in part-time limit through the phenomenon happens, the safe running of the system is under threat.

A single type of new energy power generation has the distinctive features of greater power fluctuation. Photovoltaic power generation in a day, for example, the rule is the lowest altitude of the sun, photovoltaic maximum power generation, and photovoltaic

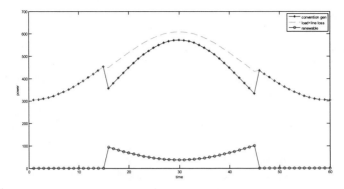

Fig. 2. Positive peak effect of new energy power generation access

panels at night without electricity. If the system of photovoltaic power generation capacity is large, then systematically power output will have occurred the phenomenon of volatility, real-time power control system will become very difficult. In order to overcome the single type of new energy power fluctuation big shortcoming, many scholars put forward many new energy hybrid power concepts. These theories with wind-light complementary research most, but is important to emphasize that most research papers are hypothetical wind power at night full power, reducing power generation in the daytime or not electricity, wind power can compensate for photovoltaic energy gaps in the night, photovoltaic power generation can be made of the energy gaps of wind power in the daytime. This scenario has certain rationality, in most of the wind field is suitable; But the field data tests show that the part of the wind field day wind resources may be the equilibrium constant, basic remain in a level power; There is also a part of the wind field wind resources may be the biggest in the daytime and night minimum, these obviously cannot adopt the assumptions mentioned above. Figure 3 is a photovoltaic power generation curve and composite picture of wind power curve under different conditions. Time delay axis curve of wind power and photovoltaic power generation curve geometry of phase center axis, the vertical axis represents active power. By changing the center axis of the phase difference is to change the new energy maximum power generation time position on the timeline.

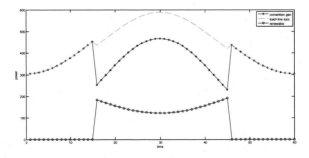

Fig. 3. Overload peak effect of new energy power generation access

5 Conclusion

From the above analysis, the new energy access to the system capacity will be affected by the natural environment and valley load of fossil fuel power plants, reserves and a variety of new energy mix various factors, the combination of them form diversified characteristics. After the new energy is connected to the system, the distribution system will change from the original single-source radiant network to the user-connected and multi-electric weak-loop network, and the size and direction of the load will be further affected. Soon the analysis of new energy power generation's influence on the actual system, can't take all kinds of the new energy power generation situation analysis of the combination of simple, should analyze their concrete composite structure, understand their portfolio is a good or bad combination.

References

1. Anindita, R., Kedare, B., Santanu, B.: Optimum sizing of wind-battery systems incorporating resource uncertainty. Appl. Energy **87**(8), 2712–2727 (2010)
2. Yuan, Y., Li, Q., Wang, W.: Optimal operation strategy of energy storage unit in wind power integration based on stochastic programming. IET Renew. Power Gener. **5**(2), 194–201 (2011)
3. Chen, Z., Ding, M., Su, J.: Modeling and control for large capacity battery energy storage system. In: 2011 4th International Conference on Electric Utility Deregulation and Restructuring and Power Technologies (DRPT), vol. **62**(6), pp. 1429–1436. IEEE (2011)
4. Hu, X., Tseng, K., Srinivasan, M.: Optimization of battery energy storage system with supercapacitor for renewable energy (2008)
5. Li, J.: Lanar cardinal spline curves with minimum strain energy, vol. **4**(3), pp. 2–4 (2018)
6. Tomura, Y., Nakagawa, T.: Adjust the energy supply and demand system for using renewable energy effectively. In: Proceedings of the Annual Conference of the Japan Institute of Energy, vol. **26**(10), pp. 859–862 (2003)
7. Mandal, J., Sinha, A.: Artificial neural network based hourly load forecasting for decentralized load management. In: Proceedings of the 1995 International Conference on Energy Management and Power Delivery (EMPD 1995), vol. **21**(1), pp. 61–66. IEEE (1995)
8. Junmeng, C., Ronghou, L.: Biomass Energy Engineering Research Center, School of Agriculture and Biology, Shanghai Jiao Tong University, Shanghai 201101, People's Republic of China. New distributed activation energy model and its application to pyrolysis kinetics of some types of biomass (2010)
9. Hu, Y.: Distributed energy transaction pattern and block chain based architecture design. In: Proceedings of the 2017 2nd International Conference on Energy, Power and Electrical Engineering (EPEE 2017), vol. **5**(25), pp. 31–40. Science and Engineering Research Center (2017)
10. Li, M.-J., Zhao, W., Chen, X., Tao, W.-Q.: Economic analysis of a new class of vanadium redox-flow battery for medium and large-scale energy storage in commercial applications with renewable energy. Appl. Therm. Eng. **9**(22), 114–128 (2017)
11. Dantzig, G.B.: General convex objective forms. In: Arrow, K.J., Karlin, S., Suppes, P. (eds.) Mathematical Models in the Social Sciences, 1959. Proceedings of the First Stanford Symposium. Stanford Mathematical Studies in the Social Sciences, vol. IV, pp. 151–158. Stanford University Press, Stanford (1960)

Multi-objective Solution of Traveling Salesman Problem with Time

Ibrahim A. Hameed[(✉)]

Department of ICT and Natural Sciences,
Faculty of Information Technology and Electrical Engineering,
Norwegian University of Science and Technology, Trondheim, Norway
ibib@ntnu.no
https://www.ntnu.edu/employees/ibib

Abstract. The traveling salesman problem (TSP) is a challenging problem in combinatorial optimization. No general method of solution is known, and the problem is NP-hard. In this paper, we consider the multi-objective TSP which encompasses the optimization of two conflicting and competing objectives: here the dual minimization of the total travel distance and total travel time at various traffic flow conditions. It is well known that travellers can experience extra travel time during peak hours (i.e., congestion conditions) compared to free flow conditions (i.e., uncongested conditions), therefore and under some conditions, minimizing traveled time could conflict and compete with travel distance and vice versa. This problem has been studied in the form of a single objective problem, where either the two objectives have been combined in a single objective function or one of the objectives has been treated as a constraint. The purpose of this paper is to find a set of non-dominated solutions (i.e., the sequence of cities) using the notion of Pareto optimality where none of the objective functions can be improved in value without degrading one or more of the other objective values. The traveller then has the chance to choose a solution that fits his/her needs at each congestion level. In this paper, a multi-objective genetic algorithm (MOGA) for searching for efficient solutions is investigated. Here, an initial population composed of an approximation to the extreme supported efficient solutions is generated. A Pareto local search is then applied to all solutions of the initial population. The method is applied to a simulated problem and to a real-world problem where distances and real estimates of the travel duration for multiple origins and destinations for specific transport modes are obtained from Google Maps Platform using a Google Distance Matrix API. Results show that solving a TSP as a multi-objective optimization problem can provide more realistic solutions. The proposed approach can be used for recommending routes based on variable duration matrix and cost.

Keywords: TSP · Optimization · Genetic algorithms

© Springer Nature Switzerland AG 2020
A. E. Hassanien et al. (Eds.): AMLTA 2019, AISC 921, pp. 121–132, 2020.
https://doi.org/10.1007/978-3-030-14118-9_13

1   Introduction

A travelling salesman problem (TSP) may be stated as follows: A salesman is required to visit each of the n given cities once and only once, starting from any city of origin and return to the original place of departure. Given, the distances between each pair of cities, What route, or tour, should he/she choose in order to minimize the total distance traveled? Instead of distance, other notions such as time, cost, etc., can be considered as well [1]. For n cities, there is a finite number of possible tours to consider, namely $\frac{1}{2}(n-1)!$, the problem is then all about how to obtain a reasonably efficient algorithm for finding an optimal solution in a reasonable time frame that is good enough for solving the problem at hand.

Here, the solution of travelling salesman problem as a multi-objective optimization problem with variable travel time (TSPT) is investigated. The TSPT is a generalization of the single-objective optimization problem (SOP) where only the total travel distance is minimized. The problem can be described as graph:

$$G = (V, E)$$

where $V = \{v_1, ..., v_n\}$ is a set of n nodes and E is a set of edges. Here, we associate a distance d_{ij} and time t_{ij} with each edge $(v_i, v_j) \in E$, respectively [2]. An example with $n = 5$ is shown in Fig. 1. The TSPT is composed of determining a tour of V that minimizes the total cost. Cost, here, is measured both in terms of the tour length and the tour time. Resulting in, under some circumstances, two conflicting objectives: (1) to minimize the length of the tour; (2) to minimize the total travel time. Minimizing distance and time could be conflict under some traffic circumstances such as traffic congestion due to reconstruction or at peak time. From the perspective of single-objective optimization, both objectives are combined or aggregated in a single objective function and the goal is, then, to find a solution that minimizes both the tour's length and time [1].

A multi-objective optimization problem (MOP), on the other hand, is defined as follows [2]:

$$(MOP) = \begin{cases} min\ \boldsymbol{F}(\boldsymbol{x}) = (f_1(\boldsymbol{x}), f_2(\boldsymbol{x}), \dots, f_m(\boldsymbol{x})) \\ s.t.\ \ \boldsymbol{x} \in D, \end{cases} \tag{1}$$

where $\boldsymbol{F}(\boldsymbol{x})$ is the objective vector; $m \geq 2$ is the number of objective functions; $\boldsymbol{x} = (x_1, x_2, \dots, x_n)$ is the decision variable vector where n is the number of cities; $\boldsymbol{x}$ is a permutation of integers from 1 through n that minimizes $\boldsymbol{F}(\boldsymbol{x})$. D, is the feasible solution space. The set $O = \boldsymbol{F}(D)$ corresponds to the feasible solutions in the objective space, and $\boldsymbol{y} = (y_1, y_2, \dots, y_m)$, where $y_i = f_i(\boldsymbol{x})$, is a solution. A MOP solution is the set of the non-dominated solutions called the Pareto-set (PS). For the TSPT problem under consideration, $m = 2$ where $f_1(\boldsymbol{x}) =$ travel distance and $f_2(\boldsymbol{x}) =$ travel time.

2 Problem Definition

The TSP is a problem defined by a set of cities and the distances between each city pair. The problem is to find a circuit that goes through each city once and that ends where it starts. Consider the following set of cities shown in Fig. 1:

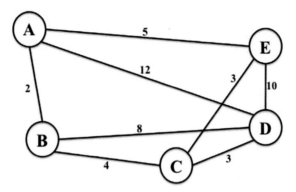

Fig. 1. A graph with weights (i.e., travel distance) on its edges as a single objective optimization problem.

The problem lies in finding a minimal path passing through all vertices once. For example, the path $P_1 = \{A, B, C, D, E, A\}$ and the path $P_2 = \{A, B, C, E, D, A\}$ pass all the vertices but P_1 has a total length of 24 and P_2 has a total length of 31. A Hamiltonian cycle is a cycle in a graph passing through all the vertices once [3]. The problem of finding a Hamiltonian cycle in a graph is NP-complete. As a multi-objective optimization problem, other notions such as time, cost, etc., can be considered as well as it is shown in Fig. 2 where travel time between cities is considered. Now, but P_1 has a total cost of (24, 23) and P_2 has a total cost of (31, 29). The mathematical model of the TSPT can be expressed as:

$$min \begin{cases} f_1(\boldsymbol{x}) = \sum_{i,j \in x} d_{ij} \\ f_2(\boldsymbol{x}) = \sum_{i,j \in x} t_{ij} \end{cases} \tag{2}$$

where d_{ij} and t_{ij} are the travel distance and time between city pair i and j, respectively. $\boldsymbol{x}$ is a route permutation of n. $f_1(\boldsymbol{x})$ and $f_2(\boldsymbol{x})$ are the total traveled distance and time for route $f_1(\boldsymbol{x})$.

3 Genetic Algorithms for Multi-objective TSP

The concept of genetic algorithms (GA) was developed by Holland and his colleagues in the 1960s and 1970s [4]. In GA terminology, a solution vector $x \in \boldsymbol{X}$ is called an individual or a chromosome. Chromosomes consist of discrete units

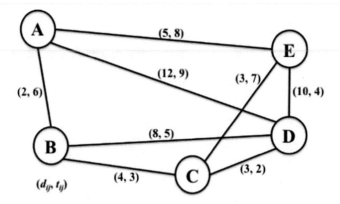

Fig. 2. A graph with weights (i.e., travel distance and time) on its edges as a multi-objective optimization problem.

called genes. Each gene represents one or more features of the chromosome. In the original implementation of GA by Holland, genes are assumed to be binary numbers. In later implementations, more gene types have been introduced. A chromosome corresponds to a unique solution x in the solution space. This requires a mapping mechanism between the solution space and the chromosomes. This mapping is called an encoding. In fact, GA works on the encoding of a problem, not on the problem itself. GA operates with a collection of chromosomes, called a population. The population is normally randomly initialized. In case of TSP, it is normally initialized with a random permutation. As the search evolves, the population converges, meaning that it is dominated by a single solution.

GA use two operators to generate new solutions from existing ones: *crossover* and *mutation*. The crossover operator is the most important operator of GA. In crossover, generally two chromosomes, called parents, are combined together to form new chromosomes, called offspring. The parents are selected among existing chromosomes in the population with preference towards fitness so that offspring is expected to inherit good genes which make the parents fitter. By iteratively applying the crossover operator, genes of good chromosomes are expected to appear more frequently in the population, eventually leading to convergence to an overall good solution. Crossover exploits the search space. The mutation operator introduces random changes into characteristics of chromosomes. Mutation is generally applied at the gene level. In typical GA implementations, the mutation rate (probability of changing the properties of a gene) is very small, typically less than 1%. Therefore, the new chromosome produced by mutation will not be very different from the original one. Mutation plays a critical role in GA. As discussed earlier, crossover leads the population to converge by making the chromosomes in the population alike. Mutation reintroduces genetic diversity back into the population and assists the search escape from local optima by exploring the search space.

Reproduction involves selection of chromosomes for the next generation. In the most general case, the fitness of an individual determines the probability of its survival for the next generation. There are different selection procedures in GA depending on how the fitness values are used. Proportional selection, ranking, and tournament selection are the most popular selection procedures. In many real-life problems, objectives under consideration conflict with each other. Hence, optimizing x with respect to a single objective often results in unacceptable results with respect to the other objectives. Therefore, a perfect multi-objective solution that simultaneously optimizes each objective function is almost impossible. A reasonable solution to a multi-objective problem is to investigate a set of solutions, each of which satisfies the objectives at an acceptable level without being dominated by any other solution.

Being a population-based approach, GA are well suited to solve multi-objective optimization problems. A generic single-objective GA can be modified to find a set of multiple non-dominated solutions in a single run. The ability of GA to simultaneously search different regions of a solution space makes it possible to find a diverse set of solutions for difficult problems with non-convex, discontinuous, and multi-modal solutions spaces. The crossover operator of GA may exploit structures of good solutions with respect to different objectives to create new non-dominated solutions in unexplored parts of the Pareto front. In addition, most multi-objective GA does not require the user to prioritize, scale, or weigh objectives. Therefore, GA has been the most popular heuristic approach to multi-objective design and optimization problems. Multi-object traveling salesman problem with time (MOT-SPT) requires the selection of a route that makes a balance between distance assignment and time assignment of the route. In GAMOP, the chromosome is composed of a random permutation of n. A route is considered valid if it satisfies the condition that each city is visited only once and that no city is missing [5]. Crossover and mutation for permutation operators are used [6, 7].

4   Results

4.1   Simulated Problem

In this section a TSPT with $n = 10$ cities is simulated. The travel distance and time between each city pair are generated in random. The problem run for 100 iterations, the mean running time is 10.95 s with a standard deviation of 0.66 s. The running time and the mean running time for 100 iterations is shown in Fig. 3. Minimum, mean, and maximum travel distance for 100 iterations is shown in Fig. 4. Minimum, mean, and maximum travel time for 100 iterations is shown in Fig. 5. The Pareto front of the two objective functions and the Pareto-set of the non-dominated solutions is shown in Fig. 6. The route that best minimizes the travel distance regardless of the time is shown in Fig. 7 (left), for this solution, the value of the travel distance is 2.56 while the value of the travel time is 4.88. The route that best minimizes the travel time regardless of the distance is shown in Fig. 7 (right), for this solution, the value of the travel time is 1.70 while the value of the travel distance is 4.58. For the other non-dominated solutions that are located on the Pareto front, none of the objective functions can be improved in value without degrading some of the other objective values.

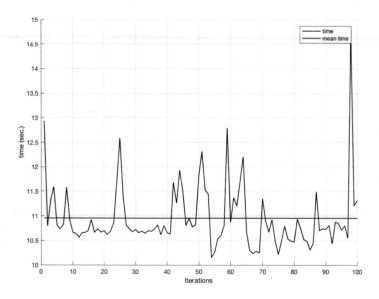

Fig. 3. Running time and mean running time for 100 iterations of the simulated problem.

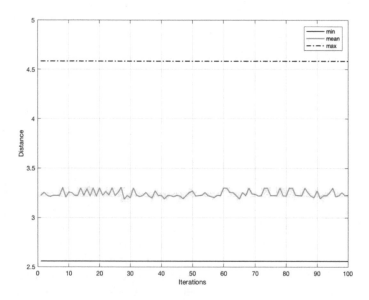

Fig. 4. Minimum, mean, and maximum travelled distance for 100 iterations of the simulated problem.

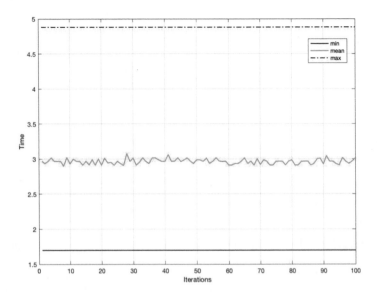

Fig. 5. Minimum, mean, and maximum travelled time for 100 iterations of the simulated problem.

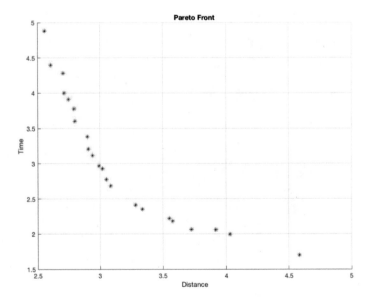

Fig. 6. Pareto front and Pareto-set of solutions of the simulated problem.

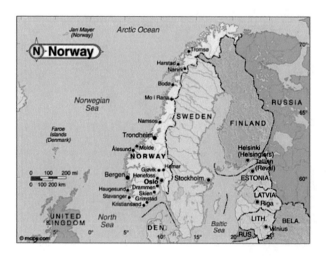

Fig. 7. Optimal route for distance (left) and time (right) of the simulated problem.

4.2 Real-World Problem

In this section, travel distance and travel time between each pair of cities of the following set of 17 cities located in Norway: *Nordkapp, Hammerfest, Tromso, Harstad, Narvik, Bodo, Mo i Rana, Trondhiem, Alesund, Floro, Bergen, Hamar, Oslo, Drammen, Fridrikstad, Stavanger,* and *Kristiansand,* are obtained using Google Maps API.[1] A map of Norway is shown in Fig. 8. Nordkapp is the depot city. The total travel distance of the given city tour is 6625.6 Km and total travel time for a car driving is 97.16 h.

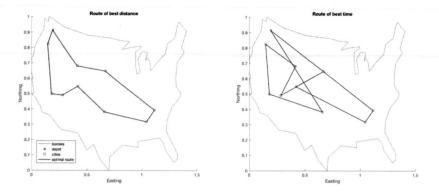

Fig. 8. Map of Norway.

The problem run for 100 iterations, the mean running time is 5.36 s with a standard deviation of 1.61 s. The running time and the mean running time for 100

---

[1] https://developers.google.com/maps/.

iterations is shown in Fig. 9. Minimum, mean, and maximum travel distance for 100 iterations is shown in Fig. 10. Minimum, mean, and maximum travel time for 100 iterations is shown in Fig. 11. The Pareto front of the two objective functions and the Pareto-set of the non-dominated solutions is shown in Fig. 12. The route that best minimizes the travel distance regardless of the time is shown in Fig. 13 (left), for this solution, the value of the travel distance is 5651.50 Km while the value of the travel time is 86.60 h. The route that best minimizes the travel time regardless of the distance is shown in Fig. 13 (right), for this solution, the value of the travel time is 86.18 h while the value of the travel distance is 5692.87. For the other non-dominated solutions that are located on the Pareto front, none of the objective functions can be improved in value without degrading some of the other objective values. For this problem, all initial solutions converged into two boundary solutions, as it is shown in Fig. 12. Both optimized solutions are much better than the default route in terms of travel distance and time.

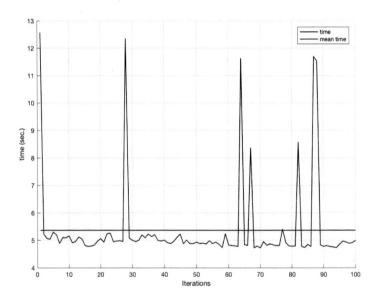

Fig. 9. Running time and mean running time for 100 iterations of the real-world problem.

Best Route for Travel Distance Minimization: Alesund, Trondhiem, Mo i Rana, Bodo, Harstad, Narvik, Tromso, Nordkapp, Hammerfest, Hamar, Oslo, Drammen, Fridrikstad, Kristiansand, Stavanger, Bergen, Floro. As it is shown in Fig. 13 (left).

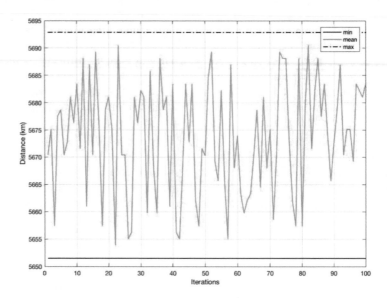

Fig. 10. Minimum, mean, and maximum travelled distance for 100 iterations of the real-world problem.

Best Route for Travel Time Minimization: Alesund, Trondhiem, Mo i Rana, Bodo, Harstad, Narvik, Tromso, Nordkapp, Hammerfest, Hamar, Oslo, Fridrikstad, Drammen, Kristiansand, Stavanger, Bergen, Floro. As it is shown in Fig. 13 (right).

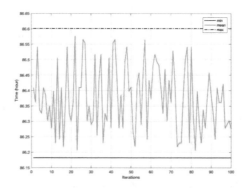

Fig. 11. Minimum, mean, and maximum travelled time for 100 iterations of the real-world problem.

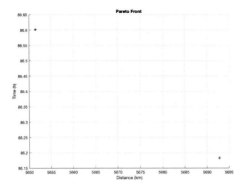

Fig. 12. Pareto front and Pareto-set of solutions for the simulated problem of the real-world problem.

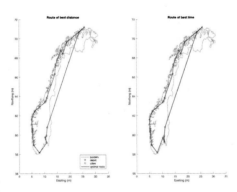

Fig. 13. Optimal route for distance (left) and time (right) of the real-world problem.

5   Conclusions and Future Work

The MOTSP presented in this paper is a problem defined by a set of cities, the travel distances and the travel times between each city pair. The problem is to find a cycle that goes through each city once and that ends where it starts. The problem is solved using multi-objective genetic algorithm (MOGA). In this paper, real-time estimates of travel duration between cities for specific transport modes and at certain time of the day are imported from Goggle Maps platform using specific application programming interfaces (APIs). The approach presented in this paper can be used for providing accurate estimates and recommendations under various traffic conditions in real-time. As a future work, real-time remaining useful life (RUL) estimates of roads will be included.

References

1. Lin, S.: Computer solutions of the traveling salesman problem. Bell Labs Tech. J. **44**(10), 2245–2269 (1965)
2. Jozefowiez, N., Glover, F., Laguna, M.: Multi-objective meta-heuristics for the traveling salesman problem with profits. J. Math. Model. Algorithms **7**(2), 177–195 (2008)
3. Weisstein, E.W.: Traveling salesman problem. Mathworld-a wolfram web resources (2010). http://mathworld.wolfram.com/TravelingSalemanProblem.html
4. Holland, J.H.: Adaptation in Natural and Artificial Systems. University of Michigan Press, Ann Arbor (1975)
5. Shi, L., Li, Z.: An improved pareto genetic algorithm for multi-objective TSP. In: The Proceeding of International Conference Fifth International Conference on Natural Computation, Tianjin, pp. 585–588 (2009)
6. Whitley, D., Yoo, N.W.: Modeling simple genetic algorithms for permutation problems. In: FOGA, vol. 1994, pp. 163–184 (1994)
7. Abdullah, K., David, W.C., Alice, E.S.: Multi-objective optimization using genetic algorithms: a tutorial. Reliab. Eng. Syst. Safety Spec. Issue - Genet. Algorithms Reliab. **91**(9), 992–1007 (2006)

Game Theory Based Solver for Dynamic Vehicle Routing Problem

Saad M. Darwish and Bassem E. Abdel-Samee[✉]

Department of Information Technology,
Institute of Graduate Studies and Research, University of Alexandria,
163 Horyya Avenue, Elshatby, P.O. box 832, Alexandria 21526, Egypt
saad.darwish@alex-igsr.edu.eg,
igsr.bassem.ezzat@alexu.edu.eg

Abstract. In recent decades, research concern in vehicle routing has progressively more concentrated on dynamic and stochastic approaches in solving sophisticated dynamic vehicle routing problems (DVRP) due to its importance in helping potential customers to better manage their applications and to provide e-freight and e-commerce systems similarly. Many techniques were introduced in the DVRP field to solve part of its challenges since it's hard to achieve equilibrium between finding feasible set of tours, minimum total travel time, shortest routing path and the capability of redirect a stirring vehicle to a new demand for additional savings; due to the fact that each one can be achieved at the expense of the other and combining them does not give an ideal result which requires a solution to track the optimum route during changes. In this paper, the game theory (GT) is integrated with Ant Colony Optimization algorithms (ACO) to adjust the attractiveness of arc and the pheromone level as it considers them competing players that prefer one according to the expected payoff. In addition, GT can be adapted inside the algorithm to facilitate the optimization, as it is considered a powerful technique in decision making and to find optimal solutions to conditions of conflict and cooperation. Experimental results show that the integration of GT with ACO algorithm improves the system performance in tackling DVRPs.

Keywords: Dynamic vehicle routing problem · Ant colony system ·
Game theory

1 Introduction

As a result of the remarkable advance in communication and information technologies that allows information to be obtained and processed in real time, dynamic vehicle routing problem (DVRP) which occasionally mentioned as online vehicle routing problems have recently arisen [1]. The DVRP can easily be modeled as a sequence of static VRP-like instances [2–5]. Most of early methods faced problems of how to deal with other dynamic constraint, the reality in real-time applications due to convergence time is unknown or to limit the problem only with time windows [6–8]. Moreover, existing systems neglect the time factor which is a significant aspect in solving the

© Springer Nature Switzerland AG 2020
A. E. Hassanien et al. (Eds.): AMLTA 2019, AISC 921, pp. 133–142, 2020.
https://doi.org/10.1007/978-3-030-14118-9_14

problem, uses stochastic methods which provide the only guarantee of global optimal solution which decreases once the problem size rises [9].

Also, these systems have several constraints to solve a problem characterized by multiple solutions and a great amount of variables. In addition to the mentioned faced problems above, there are still some problems that face the DVRP which can depend on several factors such as (1) time invested searching for better decisions. (2) the optimal routing paths. (3) cost reduction of materials. (4) rejection of new online customers. Therefore, a strong relationship can be observed between DVRP and static VRP through consequently modelling the DVRP as a sequence of static VRP- like stances. In particular, each static VRP will encompass all customers known at this time but requisite to be served [10].

In general, VRP can be classified to static or dynamic, and it can be deterministic or stochastic. All four combinations exist, and by type of problem it can be classified into: (a) Static and deterministic: if all inputs are known with certainty and there are no stochastic inputs. (b) Static and stochastic: if the priori route is determined before it is known which customers will be there or not. (c) Dynamic and deterministic: This is the type of problem we have in this research. If the evolving inputs are dynamically known but not stochastic information. (d) Dynamic and stochastic: if some probabilistic information is known about the inputs which dynamically evolve, and routes are updated whenever these inputs change in time [9].

To improve the routing process for vehicles, in this work, Game Theory (GT) is integrated with Ant Colony System (ACS). It adjusts the parameters of the main values of ACS which are the attractiveness of arc and the pheromone level and consider them competing players who are preferred according to the anticipated payoff of each. The suggested system will utilize the GT to facilitate the optimization-based in decision making and to determine the values of variables for activating the path with the best features. Moreover, GT focuses more on the dynamics of the strategy changes besides learning over time. A player might try something out. If it works well, they'll keep doing it. If it doesn't they might try out something else. They settle on a strategy that seems to do well in most circumstances.

The rest of this paper is organized as follows: Sect. 2 defines some of the recent related works. The detailed description of the proposed system has been made in Sect. 3. In Sect. 4, the results and discussions on the dataset are given. Finally, the conclusion is annotated in Sect. 5.

2   Literature Review

Research on DVRPs has grown considerably over the last three decades which aimed to solve it. But, it varies from one study to another depending on the constraints of the problem and the type of DVRP. The fundamental obstacle of these techniques is their processing time which increases with the problem size. Forcing them to use their techniques in small-scale problems only [4]. In recent years, genetic algorithms have received much attention from researchers. For instance, in 2018, Xu et al. presented an improved genetic algorithm with penalty aspect [12]. This approach possessed diversity, global searching ability, improvement of path selection and a statistical analysis to demonstrate the efficiency of the algorithm. Yet, it focuses only on solving DVRP with pickup and delivery issues without addressing other problem variables. Moreover, all algorithm parameters need fine-tuning which is often trial and error. Besides, there is no guarantee for the strong performance of the algorithm in large problems instances. Concurrently, in 2013, Ghannadpour et al. [13] integrated genetic algorithm with fuzzy time windows. That was aimed at finding a feasible spot to insert new customers without re-scheduling the customers already in the solution. Yet, the processing time and search space increased.

In a recent study [15], the work explores using particle swarm optimization algorithm for solving DVRP. For this purpose, the optimization is performed by a set of particles communication with each other with retention of location and velocity for each. This scheme costs less in computational time and less memory is needed. Also features by very efficient global search and simplicity in implementation. Besides, this technique initially has a population of random solutions. This algorithm lacks the local search ability and slow convergence in refined search stage. Moreover, the information about new requests can't be restructured during the algorithm's run process.

Montemanni et al. developed [4], an approach grounded on the idea of allocating the working day into equal time slices and to postpone newly received orders during a time slice to the end of it. This system can handle the dynamic travel time updates based on real-time traffic information. Also, it provides a solution to cover all known orders. Yet, it can only provide a probabilistic guarantee of global optimal solution which will reduce as the problem size rises. Besides, it is not reliable in real-time applications as a result of the convergence time is unknown. In addition, this technique has a problem to explore long edges nodes found in route network due to the evaporation of pheromone.

Although DVRPs have been studied for a few decades, there is still room to make it more efficient and practical in real applications. According to the aforementioned review, it can be found that past studies were primarily devoted to (1) time invested searching for better decisions, (2) the optimal routing paths, and (3) cost reduction of materials. However, to the best of our knowledge, slight consideration has been paid to advising new algorithm to facilitate the optimization-based in decision making and to determine the values of variables for activating the path with the best features. More information regarding the current DVRP approaches can be found in [5, 11–20].

3   Proposed Methodology

This paper proposes a new approach integrates GT with ACS grounded on the notion of dividing the working day into equal time slices and postpone each new order arrived at the end of it. The proposed system aimed to lessen the total travel time, provides a solution covering all known orders and facilitating the optimization-based in decision making for activating the path with the best features. Moreover, this system focuses more on the dynamics of the problem by changing the strategy of players. As mentioned before about dynamic problems types this approach is located under the dynamic and deterministic type which characterized by that all inputs of problem dynamically known and there is no stochastic information. The main diagram of the

suggested system is illustrated in Fig. 1. The following subsections describe in detail
the main elements and steps of the system.

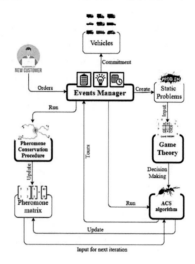

Fig. 1. The proposed game theory-based routing system

3.1   Events Manager

Events manager acts as an interface between the architecture and the external world.
So, it collects the new orders, holds over already served orders, determines the position
of the vehicles and finally commits orders to the drivers. The creation of the static
problem took place during the first time slice which is the beginning of the working day
to deal with orders received from the last working day. The next problems will consider
all orders received at the beginning of the time slice, and which haven't been committed yet [1]. In such problems, every vehicle starts from the location of the last
customer committed to it, with a starting time congruous to the end of the servicing
time for this customer. In addition, with a capacity congruous to the residual capacity of
the vehicle, after it has served all customers previously committed; see [4] for more
details. According to the end of each time slice, the best solution found for the congruous static problem is tested and orders with a computational time starting within the
next time slice and commitment time added to it which are committed to the respective
drivers [10, 11]. An exception to the previous commitment strategy is appeared by the
return journeys to the depot which can be represented only in two circumstances: if the
current time is larger than or equal the cutoff time and all clients have been served, or
the vehicle has used all its capacity [6].

3.2   Game Theory Integration

The role of game theory takes place after the creation of static problems instances by
events manager. After the creation of the static problems and the identification of all

data related to the customer from where it is, the time is taken to reach it, the time it takes to service it and the quantity of capacity required. So, the system has now a database for each customer in the network service and only requires the decision to serve it in the shortest way saves time and costs [4]. That's why the game theory was used as a formal study of decision-making where two players must make choices that potentially affect the interests of the other [23–25]. Therefore, decision-making has been splitted into two phases as follows:

Players Strategy Formation: Before proceeding in the players' strategy formation, the first is to identify the players which are the main components of the ACS algorithm inputs in the next stage to be as follows: Attractiveness of arc acts as player 1 and pheromone level of arc acts as player 2. Each player in this game has two points of interest or profit, which will be earned when the game ends, this is called the reward or benefit, which in turn measures the degree of satisfaction that the individual player can get from the conflicting position [21, 22]. For each player of the game, the choices obtainable to them are called strategies. The game represents an explanation of strategic interactions that include the constraints on the action that a player can take and the player's interests but does not specify the actions that the players do take. Formal definition of game theory can be found in [23–25].

Decision Making Procedure: Although all the components of the game mentioned in the previous step rests the most significant step is to make the decision and that by imposing six features for each arc connects the customer need to be served with the current location of the vehicle. These features are the number of clients to be served, the number of vehicles available, vehicle capacity, travel distance, travel time and the traffic capacity [19]. Each of the two players will then ennoble three of these features, which will calculate the different β values as input in ant colony system, which is a parameter regulating the relative significance of the arc versus the actual attractiveness of the same arc. Thus, β values become within each player's strategy on which the expected payoff is based before selecting each arc [25]. Also, it should consider that the type of game theory strategy proposed in this research is classified as competitive games which are known as zero-sum games of strategy. This kind of games is characterized by the conflict of the players' interests and because of the pay-off add up to zero for each outcome of a fair game or to another constant if the game is prejudiced [24].

3.3 ACS Algorithm

In this proposed system the ACS employs ants as simple computational agents that individually and iteratively make solutions for the problem. This approach has been applied to similar VRP problems and provided the best-known solutions for several benchmarks [4, 18]. As shown below the ACS algorithm entails three steps starting with the probability of ant moves through the update of the pheromone matrix as input for the next iteration, ending with finding the best-known solutions and submit them for the events manager to commit to the vehicles.

Probability of Movement: Every ant k calculates certain set of practical expansions to its current partial solution then selects one of these probabilities due to a probability distribution clarified as follows: For ant k the probability P_{ij}^k for visiting customer j after customer i which the last visited customer relies on the combination of values: (1) attractiveness η_{ij} of arc (i, j) which depends on the travel time between client i & customer j. (2) Pheromone level τ_{ij} of arc (i, j) which indicates how it was a good solution to visit customer j after i. In addition, all pheromone trails updated at each iteration while the good solutions arcs linked with high pheromone levels [4, 17]. The following formula represents the probability distribution using a set $\mathcal{F}_i^k$ which includes feasible customers the ant can extend to them by its partial solution.

$$p_{ij}^k = \begin{cases} \frac{\tau_{ij}(\eta_{ij})^\beta}{\sum_{l \in \mathcal{F}_i^k}(\tau_{il}(\eta_{il})^\beta)} & \text{if } j \in \mathcal{F}_i^k \\ 0 & \text{otherwise} \end{cases} \tag{1}$$

when the sum exceeds all possible moves, β parameter controls the proportional importance of trail τ_{ij} of arc (i, j) against the actual attractiveness η_{ij} of same arc. For this purpose, P_{ij}^k is a trade-off between the obvious desirability and information from the past [4].

Updating Pheromone Matrix: After ant k moves from customer i to j, there is a local updating on the pheromone matrix according to below rule:

$$\tau_{ij} = (1 - \rho)\tau_{ij} + \rho\tau_0 \tag{2}$$

while τ_0 is the starting value of trails which is defined previously by the user which used for the first static VRP and for the entries of pheromone matrix including new customers in the subsequent problems. Also, ρ $(0 \leq \rho \leq 1)$ is a parameter used to adjust the pheromone evaporation which occurred in real ants' case [4]. Moreover, an interesting side of the local updating is that while edge nodes are visited by ants, the previous formula makes the trail concentration decrease, making them less and less attractive and preferring to explore the not yet visited edges and diversity of solution generation.

Finding Best Known Solution: A parameter used to adjust the maximum computation time for this local search, t_{ls}, has to be defined by the user. Once the ants of the colony have been completed their computation, the best-known solution is used to globally adjust the pheromone trail. In this way, this preferred route is saved in the pheromone trail matrix and next iteration of ants will use this information to generate new solutions in an approximate of this preferred route [4, 10]. Where *CostBest* is the total travel time of solution *BestSol*, the best tour is produced by the algorithm since the beginning of the computation. This process is repeated until a termination case occurred. The pheromone matrix will be updated as:

$$\tau_{ij} = (1 - \rho)\tau_{ij} + \frac{\rho}{costBest}\forall(i,j) \in BestSol \tag{3}$$

Pheromone Conservation Procedure: After a time slice is finished and the respective static problem has been solved by the ACS, the pheromone matrix still contains encrypted information about features of good solutions for these problems. Precisely, pairs of customers which are visited in sequence in good solutions will have high values in the congruous entries of the pheromone matrix [4, 6]. This process prevents optimization to restart each time from scratch and heavily participate in the good performance aimed by this system. For this purpose, a parameter γ_r used to adjust pheromone conservation process in case each pair of customers which appear both in the old and in the new static problem. The pheromone matrix entry is started to the following value:

$$\tau_{ij} = (1 - \gamma_r)\tau_{ij}^{old} + \gamma_r\tau_0 \tag{4}$$

since τ_{ij}^{old} is the value of τ_{ij} appeared in the old static problem. Actually, pheromone values are not completely configured, but an effect of old values remains [4]. The entries of the modern pheromone matrix congruous to pairs of customers including new customers are initialized to τ_0; see [4] for more details.

4 Experimental Results

This section is dedicated to the experimental evaluation of the proposed GT based routing system. This software is written in MATLAB 2017b, and it was run on Hp ProBook core i5/2.3 GHZ- Windows 10 professional. The first group of experiments is conducted to investigate the comparison of the solution quality between the proposed system and other works. As we see, the proposed algorithm is competitive. It is better than both ACS [4] and adaptive ant colony system [18] on 10 benchmark instances, and it provides the shortest average for the traveled distance as shown in Table 1. The algorithm in [18] is built on the concept that it minimizes the total distance. These results allow us to say that the proposed algorithm is effective and shows viability to generate high-quality solutions. One explanation of this results is that the suggested system utilizes game theory approach as a decision making to weighing between the main ant colony parameters: the attractiveness of the arc and the level of the pheromone in the same arc and consider them as competitive players who prefer one according to the expected payoff.

The second set of experiments was to analyze the impact of a dynamic strategy when new events/demands are added into the schedule during operations and to extend the criterion of distance in the literature. For these instances used in Table 2. There are no strong time windows. It is therefore not penalizing, for the distance criterion, to delay an order to a driver. Moreover, delaying the orders, as in the time slice strategy, allows decision-making to occur when more information is available and therefore allows for superior decisions. The time slices methodology seems to be a better strategy when the final goal is to reduce distance especially when it adjusted to 25 time slices [2, 4]. So, it can be seen that the system could absorb an unexpected event or as an opportunity to reduce the time on the ground, therefore saving costs. Also, it is always possible to organize the trips so that each customer is served on time.

Table 1. Numerical results for the proposed system compared to other approaches

Problem	ACS		Adaptive ACS		The proposed approach	
	Best	Avg.	Best	Avg.	Best	Avg.
c50	631.30	681.86	601.78	623.09	571.691	591.93
c120	1416.45	1525.15	1272.65	1451.60	1215.38	1386.27
c199	1771.04	1844.82	1717.31	1757.02	1665.79	1407.30
tai75a	1843.08	1945.20	1832.84	1880.87	1796.18	1843.25
tai75b	1535.43	1704.06	1456.97	1477.15	1369.55	1388.52
tai 100a	2375.92	2428.38	2257.05	2331.28	2099.05	2168.09
tai 100b	2283.97	2347.90	2203.63	2317.30	2093.44	2201.43
tai 150a	3644.78	3840.18	3436.40	3595.40	3298.94	3451.58
tai 150b	3166.88	3327.47	3060.02	3095.61	2937.62	2971.78
f71	311.18	348.69	311.33	320.00	300.43	308.8
f134	15135.51	16083.56	15557.82	16030.53	16335.711	16832.05

Table 2. Results of new events added dynamically

Problem	Criterion: pheromone of arc (distance)	Criterion: attractiveness of arc
c50	665.57	55.19
c120	1288.58	78.88
c199	1774.71	35.79
f71	300.7	27.35
f134	15717.14	428.73
tai 75a	1833.03	106.03
tai 75b	1503.74	70.3
tai 100a	2510.26	105.69
tai 100b	2256.282	90.24
tai 150a	38764.19	124.52
tai 150b	3571.42	111.77

5   Conclusion

A dynamic vehicle routing algorithm based on game theory integrated with ant colony system has been introduced in this paper. Such an algorithm adopted consideration as a robust technique in decision making to find good optimal solutions for DVRPs. In addition, GT adjusts the β value according to the used instances in which it is a parameter regulating the weighing between ant colony main two elements the attractiveness of the arc and the level of the pheromone in the same arc and consider them competitive players who prefer one according to the expected payoff. Moreover, GT can be adapted within the algorithm to facilitate optimization and to find optimal solutions to conflict situations and collaboration. Furthermore, experimental results display that the proposed algorithm can fulfill good results in real problems. To set a plan for future works, the algorithm can be upgraded to use multi-objective optimization method contrary to what was used as a single-objective fitness function. Also, more enhancements can take place in game theory players strategies to improve the payoff expected.

References

1. Pillac, V., Gendreau, M., Guéret, C., Medaglia, A.: A review of dynamic vehicle routing problems. Eur. J. Oper. Res. **225**(1), 1–11 (2013)
2. Schyns, M.: An ant colony system for responsive dynamic vehicle routing. Eur. J. Oper. Res. **245**(3), 704–718 (2015)
3. Créput, J.C., Hajjam, A., Koukam, A., Kuhn, O.: Self-organizing maps in population based metaheuristic to the dynamic vehicle routing problem. J. Comb. Optim. **24**(4), 437–458 (2012)
4. Montemanni, R., Gambardella, L., Rizzoli, A., Donati, A.: Ant colony system for a dynamic vehicle routing problem. J. Comb. Optim. **10**(4), 327–343 (2005)
5. Ferrucci, F., Bock, S., Gendreau, M.: A pro-active real-time control approach for dynamic vehicle routing problems dealing with the delivery of urgent goods. Eur. J. Oper. Res. **225**(1), 130–141 (2013)
6. Yu, B., Ma, N., Cai, W., Li, T., Yuan, X., Yao, B.: Improved ant colony optimisation for the dynamic multi-depot vehicle routing problem. Int. J. Logist. Res. Appl. **16**(2), 144–157 (2013)
7. Mendoza, J., Castanier, B., Guéret, C., Medaglia, A., Velasco, N.: A memetic algorithm for the multi-compartment vehicle routing problem with stochastic demands. Comput. Oper. Res. **37**(11), 1886–1898 (2010)
8. Gromicho, J., van Hoorn, J., Kok, A., Schutten, J.: Restricted dynamic programming: a flexible framework for solving realistic VRPs. Comput. Oper. Res. **39**(5), 902–909 (2012)
9. Psaraftis, H., Wen, M., Kontovas, C.: Dynamic vehicle routing problems: three decades and counting. Netw. Int. J. **67**(1), 3–31 (2016)
10. Razavi, M., Eshlaghy, A.: Using an ant colony approach for solving capacitated vehicle routing problem with time windows. Res. J. Recent Sci. **4**(2), 2277–2502 (2015)
11. Pamucar, D., Ćirović, G.: Vehicle route selection with an adaptive neuro fuzzy inference system in uncertainty conditions. Decision Mak.: Appl. Manag. Eng. **1**(1), 13–37 (2018)
12. Xu, H., Duan, F., Pu, P.: Solving dynamic vehicle routing problem using enhanced genetic algorithm with penalty factors. Int. J. Perform. Eng. **14**(4), 611–620 (2018)
13. Ghannadpour, S., Noori, S., Tavakkoli-Moghaddam, R., Ghoseiri, K.: A multi-objective dynamic vehicle routing problem with fuzzy time windows: model, solution and application. Appl. Soft Comput. **14**(18), 504–527 (2014)
14. Ng, K., Lee, C., Zhang, S.Z., Wu, K., Ho, W.: Multiple colonies artificial bee colony algorithm for a capacitated vehicle routing problem and re-routing strategies under time-dependent traffic congestion. Comput. Ind. Eng. **109**(13), 151–168 (2017)
15. Okulewicz, M., Mańdziuk, J.: Application of particle swarm optimization algorithm to dynamic vehicle routing problem. In: International Proceedings on Proceedings of Artificial Intelligence and Soft Computing, pp. 547–558. Springer, Berlin (2013)
16. Kuo, R., Wibowo, B., Zulvia, F.: Application of a fuzzy ant colony system to solve the dynamic vehicle routing problem with uncertain service time. Appl. Math. Model. **40**(23–24), 9990–10001 (2016)

17. Yang, Z., Emmerich, M., Bäck, T.: Ant based solver for dynamic vehicle routing problem with time windows and multiple priorities. In: Proceedings of Conference on Evolutionary Computation, pp. 2813–2819. IEEE, Japan (2015)
18. Ouaddi, K., Benadada, Y., Mhada, F.: Multi period dynamic vehicles routing problem: literature review, modelization and resolution. In: 3rd International Proceedings on Proceedings of Logistics Operations Management, pp. 1–8. IEEE, Morocco (2016)
19. Mavrovouniotis, M., Yang, S.: Ant algorithms with immigrants schemes for the dynamic vehicle routing problem. Inf. Sci. **294**(32), 456–477 (2015)
20. Elhassania, M., Ahmed, E., Jaouad, B.: Application of an ant colony system to optimize the total distance and the customers response time for the real time vehicle routing problem. In: 3rd International Proceedings on Proceedings of Logistics Operations Management, pp. 1–6. IEEE, Morocco (2016)
21. Shah, I., Jan, S., Khan, I., Qamar, S.: An overview of game theory and its applications in communication networks. Int. J. Multi. Sci. Eng. **3**(4), 5–11 (2012)
22. Leyton-Brown, K., Shoham, Y.: Essentials of game theory: a concise multidisciplinary introduction. Synth. Lect. Artif. Intell. Mach. Learn. **2**(1), 1–88 (2008)
23. Banjanovic-Mehmedovic, L., Halilovic, E., Bosankic, I., Kantardzic, M., Kasapovic, S.: Autonomous vehicle-to-vehicle decision making in roundabout using game theory. Int. J. Comput. Sci. Appl. **7**(8), 292–298 (2016)
24. Gattami, A., Al Alam, A., Johansson, K., Tomlin, C.: Establishing safety for heavy duty vehicle platooning: a game theoretical approach. In: 18th International Proceedings on Proceedings of Automatic Control, vol. 44, no. 1, pp. 3818–3823. Elsevier, Italy (2011)
25. Inujima, W., Nakano, K., Hosokawa, S.: Multi-robot coordination using switching of methods for deriving equilibrium in game theory. Trans. Comput. Inf. Technol. **8**(2), 174–181 (2013)

A Hybridization of Sine Cosine Algorithm with Steady State Genetic Algorithm for Engineering Design Problems

M. A. El-Shorbagy[1,2], M. A. Farag[2(✉)], A. A. Mousa[2,3], and I. M. El-Desoky[2]

[1] Department of Mathematics, College of Science and Humanities Studies, Prince Sattam Bin Abdulaziz University, Al-Kharj, Kingdom of Saudi Arabia
[2] Department of Basic Engineering Science, Faculty of Engineering, Menoufia University, Shebin El-Kom, Egypt
mai.farag2015@gmail.com
[3] Mathematics and Statistics Department, Faculty of Science, Taif University, Taif, Kingdom of Saudi Arabia

Abstract. Sine Cosine Algorithm (SCA), a newly proposed optimization approach, has gained the interest of researchers to solve the optimization problems in different fields due to its efficiency and simplicity. As well as a genetic algorithm (GA) has proved its robustness in solving a large variety of complex optimization problems. In this paper, a hybridization of SCA with steady state genetic algorithm (SSGA) is proposed to solve engineering design problems. This approach integrates the merits of exploration capability of SCA and exploitation capability of SSGA to avoid exposure to early convergence, speed up the search process and quick the convergence to best results in a reasonable time. The proposed approach incorporates concepts from SSGA and SCA and generates individuals in a new generation by crossover and mutation operations of SSGA and also by mechanisms of SCA. Efficiency of the proposed algorithm is evaluated using two complex engineering design problems to verify its validity and reliability. Results show that the proposed approach has superior performance compared to other optimizations techniques.

Keywords: Sine Cosine Algorithm · Steady state genetic algorithm · Engineering design problems · Optimization algorithms

1 Introduction

Engineering design problem is considered very significant in the engineering field to obtain more efficient and accurate designs shapes. These problems are formulated generally as nonlinearly constrained optimization problems; where it can be defined as [1]:

© Springer Nature Switzerland AG 2020
A. E. Hassanien et al. (Eds.): AMLTA 2019, AISC 921, pp. 143–155, 2020.
https://doi.org/10.1007/978-3-030-14118-9_15

Min: $F(y)$

Subject to: $y \in S$

$$S = \left\{ \begin{array}{ll} g_k(y) \leq 0, & k = 1, 2, \ldots, p \\ h_j(y) = 0, & j = 1, 2, \ldots, m \end{array} \right\}$$

$$y_n^l \leq y_n \leq y_n^u \quad n = 1, 2, \ldots, N$$

(1)

where $F(y)$ is the objective function, $y = [y_1, y_2, \ldots, Y_n]^T$ are denotes solution and y_n are the decision variables. Each decision variable y_n is bounded by its upper and lower limits $[y_n^l, y_n^u]$. Where $h_j(y)$ and $g_k(y)$ are equality and inequality constraint functions, and $S \subset R^n$ is the feasible space which contains all the possible y that can be used to satisfy all the constraints.

Considerably, the engineering design problems are usually difficult to solve because of its complexity nature in the constraints; which leading to a feasible region (non-convex, discontinuity, and mixing continuity-discontinuity, etc.) that may be limited to a thin subset of the search area. In realistic, choosing the solving approach for any design problem requires knowledge and extensive experience about this problem. Nevertheless, there is no guarantee for obtaining an optimal or even sub-optimal solution [2].

There are different techniques in literature, which address these optimization problems. Generally, these techniques can be classified mainly into: deterministic and meta-heuristic. Over the last two decades, a growing interest has been observed in meta-heuristic techniques due to the drawbacks of the deterministic techniques such as: (1) requiring a good starting point, (2) the search spaces is continuous and (3) use gradient information [3].

Meta-heuristic techniques deal with optimization problems as black boxes. This means that the mathematical models' derivation is not required because these algorithms change only the inputs and adjust the system model for minimizing or maximizing it's outputs [4]. Meta-heuristic techniques like genetic algorithms (GA) [5], shuffled frog leaping algorithms (SFLA) [6], particle swarm optimization (PSO) [7], monkey algorithms [8], artificial bee colony (ABC) [9], gray wolf optimization (GWO) [10], sine cosine algorithm (SCA) [11], algorithms inspired by fish schools [12] and Bats [13, 51], cuckoo birds [14], and moth-flam [50] are some of the recently popular meta-heuristic techniques to solve global optimization problems. These techniques are characterized by the high flexibility, which is mean that meta-heuristic techniques are readily applicable to complex test as well as real life problems in different fields [15].

GA belongs to the meta-heuristic techniques; where it simulates the mechanics of natural selection, evolution and genetics and is introduced as an effective global algorithm to solve the complex nonlinear problems [5]. GA evaluates simultaneously many solutions in the search area, which give it more chance to obtain the global solution [16–18]. But, for large and complex systems, GA still has many defects such as being difficult to ensure the converge of the most global optimal solution, the loss of the best chromosome in group which leads to degradation the process of evolution [19] and permuting convergence which is one of the most important defects in GA is known to increase the iterations number to reach to the global optimum. To avoid these defects a lot of opinions were introduced such as: adding new genetic operator, adjusting the control parameters, improving the structure of the algorithm [20, 21] or introducing hybridization of it with other techniques of meta-heuristics [22, 23]. GA has several different models such as generational genetic algorithm (GGA) and steady state Genetic algorithm (SSGA) [24]. Both of GGA and SSGA follow the general model of GAs, but they differ by that SSGA generates small number of chromosomes only in each iteration, while in GGA a large portion of its population is replaced.

More recently, Mirjalili [11] proposed SCA, which is one of the meta-heuristic algorithms that used in the literature to solve optimization problems. The solutions SCA is updated according to the trigonometric sine and cosine functions; where it oscillate around the best solution. SCA is proved that it has an efficient performance in comparison with other meta-heuristic techniques. But in SCA, an acceptable solution cannot always be found every time due to it hasn't an internal memory [25]. During its procedure, SCA never maintains the set of solutions that may have a chance to converge at global optima.

This paper presents the performance of a very recently proposed a hybridization of SCA with SSGA (Hybrid SCA-SSGA) on a class of constrained design optimization problems. The distinct feature of this algorithm is that it integrates the merits of exploration capability of SCA and exploitation capability of SSGA to avoid exposure to early convergence, speed up the search process and quick the convergence to best results in a reasonable time. The hybridization of SCA with SSGA aims to improve their performance where SSGA operators makes the population more diverse and SSGA memory helps SCA to keep track of possible solutions and then converge to global solution. The performance of the algorithm is investigated on two engineering constrained design problems are solved to test its performance. The experimental results show that the hybrid SCA-SSGA has high performance compared with the other optimizations techniques.

The rest of this paper is presented as follows. Section 2 briefly introduces SSGA. The SCA is described in Sect. 3. While, Sect. 4 gives the details of the proposed hybrid SCA-SSGA algorithm. Experimental results are reported and analyzed in Sect. 5. Finally, Sect. 6 includes the conclusion.

2 Steady State Genetic Algorithm (SSGA)

Holland in 1970s introduced the GA as an optimization algorithm to obtain a global or near-global optimal solution [5]. The process of GA mimics natural evolutionary phenomena: Darwinian theory of biological evolution and genetic inheritance [26]. GA can be applied in several different models to solve any problem such as SSGA and GGA [24].

SSGA initiates with an initial random solutions (population) of a given size denoted by N_{pop}. SSGA generates an additional population, of size pop_{new} which generated by crossover and mutation of a specific individuals chosen from the base population. The offspring which created newly is evaluated and then are combined with the base population. The combined populations, of size $(pop_{new} + N_{pop})$ chromosomes are ranked according to their fitness value. After ranking process, N_{pop} worst chromosomes in the ranking set are replaced so as to restore the population to its base size. The chromosomes number (N_{pop}) which to be removed, is specified by the user and defines the amount of overlap between two consecutive generations. Once the new population has been produced, the termination condition is checked, if it satisfied, the algorithm stop, otherwise the evolution continues to complete the new population [27].

3   Sine Cosine Algorithm (SCA)

SCA is a new meta-heuristic approach based on the forms of sine and cosine functions, was proposed in 2016 [11]. SCA is a population based technique, begins by generating a set of random solution. These random solutions are repeatedly evolved toward better solutions over the course of the optimization process. Through this process, the search space is maintained by modifying the range of the sine and cosine based on their mathematical forms. These loops are repeated until the termination criterion is met. The mathematical formula of SCA can be described as follow:

$$y_i^{k+1} = \begin{cases} y_i^k + r_1 * \sin(r_2) * \left| r_3 D_i^k - y_i^k \right| r_4 < 0.5 \\ y_i^k + r_1 * \cos(r_2) * \left| r_3 D_i^k - y_i^k \right| r_4 \geq 0.5 \end{cases} ; \qquad (2)$$

where y_i^k is the current solution in i^{th} dimension at the kth iteration, r_1, r_2, r_3, r_4 are the random numbers and are updated in each iteration, D_i is the destination point in the ith dimension [28].

As shown in the above equation (Eq. (2)), SCA uses 4 variables r_1, r_2, r_3, r_4 to tune. r_1 decided where the movement direction of the search agent which could be either in the area between the destination and solution or outside it by this equation:

$$r_1 = c\left(1 - \frac{iter}{ITER_{max}}\right) ; \qquad (3)$$

where c is constant, $ITER_{max}$ is the maximum number of iterations and $iter$ is current iteration. The parameter r_2 is tell how far the movement might be outwards or towards the destination. r_3 assigns a random weight for destination and $r_4 \in [0, 1]$ decided whether sine or cosine formula is going to be used.

4   The Hybridization of SCA with SSGA

In this paper, a hybridization of SCA with SSGA (Hybrid SCA-SSCA) is proposed to solve engineering design problems. This approach integrates the merits of the exploration capability of SCA and exploitation capability of SSGA to avoid exposure to early convergence, accelerate the seeking operation and quick the convergence to the best results in a reasonable time. Hybrid SCA-SSCA combines concepts of SSGA and SCA and generates solutions in the new generation by SSGA operators and mechanisms of SCA. The features of hybridization of SCA with SSGA are:

- The internal memory of GA keeps track of possible solutions which have a potential to converge to global optima.
- The SSGA operators make the population more diverse and thus more immune to be stuck in local optima.
- The exploration capability for SCA works to accelerate access to optimal solutions.

The basic steps of Hybrid SCA-SSCA can be described as follows:

Step 1: Generate randomly the initial candidate solutions of SSGA to create an initial population.

Step 2: Define the initial reference point, as the proposed approach requires at least one feasible reference point to complete the algorithm procedure [29].

Step 3: The infeasible solutions are repaired; where the repair technique which used in the proposed algorithm is taken from [30].

Step 4: Evaluate the objective functions for all individual.

Step 5: Generating a new generation from the current population by using SSGA operators and SCA algorithms as:

　Selection: Four chromosomes are selected randomly.

　Ranking: Rank the four chromosomes according to their objective function value.

　Classification in order to applying SSGA and SCA: The SSGA operators are applied to the first two individuals which makes the population more diverse and thus more immune to be stuck in local optima where SCA is applied to the other two individuals.

Step 6: Repeat the above two steps (selection, classification) until the certain number of individuals (pop_{new}) taken from the base population are evolved by SSGA and SCA and generate additional population of size pop_{new}.

Step 7: Combine the new off-springs pop_{new} and the base population N_{pop} to generate the new generation have the same size of base population by choosing the best solutions of offspring and parents population.

Step 8: The approach is stopped, when the termination condition (maximum number of iterations) has been reached.

5   Numerical Results

To examine the performance of the proposed approach, the hybrid SCA-SSGA was applied to solve two engineering optimization design problems: design of a speed reduce and design of tension/compression spring [31, 37] to verify the validity and reliability of proposed approach in solving engineering applications. The proposed algorithm was programmed in Matlab (2016b). The proposed algorithm, as any meta-heuristics approach, includes several parameters that effect on its performance. The algorithm parameters are the size of population = 50, Generation gap = 0.9, Mutation rate = 0.1,

Crossover rate = 0.95 and the termination criteria is 100 iterations. In addition, the types of crossover and mutation operators are single point crossover and real-value mutation respectively.

Due to the stochastic characteristics of hybrid SCA-SSGA, the algorithm solves all the problems 30 times. The best values, worst values, the mean values of best values and the standard deviation (SD) are saved after every run.

5.1 Engineering Design Problem 1: Design of a Speed Reducer

In this design problem, the weight of speed reducer in small propeller-type aircraft is minimized. this problem subject to some constraints such as: stresses in the shafts, transverse deflections of the shafts, bending stress of the gear teeth and surface stress. This problem is very important optimization problem; where it aims to lower operating cost and material savings [31]. The diagrammatic representation of the problem and its design variables are given in Fig. 1. As shown in the figure, this problem has 7 design parameters, which are:

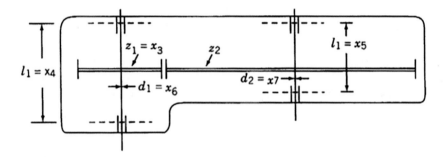

Fig. 1. Speed reducer [31]

- b: The face width,
- z: Number of teeth in the pinion,
- m: Module of teeth,
- l_1: Length of the first shaft between bearings,
- d_1: The diameter of the first shaft,
- l_2: Length of the second shaft between bearings,
- d_2: The diameter of second shaft.

These design variables can be formulated as: $(b, m, z, l_1, l_2, d_1, d_2)^T = (x_1, x_2, x_3, x_4, x_5, x_6, x_7)^T$. All of the mentioned design variables are continuous except the number of teeth in the pinion, which is an integer variable. The mathematical model of this problem is formulated as [31–36]:

$$\text{Min } [0.7854x_1x_2^2(-43.0934 + 14.933x_3 + 3.3333x_3^2)$$
$$+ 7.477(x_6^3 + x_7^3) - 1.508(x_6^2 + x_7^2) + 0.7854(x_4x_6^2 + x_5x_7^2)]$$

Subject to:
$$\frac{27}{x_1x_2^2x_3} - 1 \leq 0, \quad \frac{3947.5}{x_1x_2^2x_3^2} - 1 \leq 0, \quad \frac{1.93x_4^3}{x_2x_6^4x_3} - 1 \leq 0,$$

$$\frac{1.93x_5^3}{x_2x_3x_7^4} - 1 \leq 0, \quad \frac{x_1}{12x_2} - 1 \leq 0, \quad \frac{x_2x_3}{40} - 1 \leq 0,$$

$$\frac{5x_2}{x_1} - 1 \leq 0, \quad \frac{1.9 + 1.5x_6}{x_4} - 1 \leq 0, \quad \frac{1.9 + 1.1x_7}{x_5} - 1 \leq 0,$$:

$$\frac{\sqrt{\left(\frac{745x_4}{x_2x_3}\right)^2 + 16900000}}{110x_6^3} - 1 \leq 0, \quad \frac{\sqrt{\left(\frac{745x_5}{x_2x_3}\right)^2 + 15750000}}{85x_7^3} - 1 \leq 0,$$

$$2.6 \leq x_1 \leq 3.6, \qquad 0.7 \leq x_2 \leq 0.8, \qquad 17 \leq x_2 \leq 28,$$
$$7.3 \leq x_4 \leq 8.3, \qquad 7.8 \leq x_5 \leq 8.3, \qquad 2.9 \leq x_6 \leq 3.9,$$
$$5 \leq x_7 \leq 5.5.$$

This engineering problem is optimized by previously different algorithms such as: cuckoo search (CS) [31], upgraded firefly algorithm (UFA) [32], constrained-iterative topographical global optimization (C-ITGO) [33], Artificial bee colony algorithm (ABCA) [34]. Accelerated PSO (APSO) [35] and improved accelerated PSO (IAPSO) [36]. Table 1 shows a comparison between the best solutions for speed reducer problem that obtained from some of different approaches. While Table 2 presents the comparison of statistical results of the methods mentioned above and the proposed hybrid SCA-SSGA.

As shown in Tables 1 and 2, the proposed algorithm obtained the best solution between all comparison approaches, where it obtained the minimum objective value for the speed reducer problem. While in terms of the number of function evaluations (NFEs), the proposed algorithm found the best solution faster than other reported techniques by using 1450 NEFs, except CS algorithm and C-ITGO algorithm which uses 500 and 856 function evaluations to find its best solution respectively. The algorithms: UFA, IAPSO and C-ITGO have been come in the second rank in terms of the best solution among the seven approaches. But C-ITGO outperforms UFA, IAFSO in terms of NFEs to get the best solution, while ABCA is the most powerful approach to this design problem because its SD is zero. The results have shown that SCA-SSGA hybrids has greater competitiveness to other optimization methods that recently used in the literature to solve this design problem.

5.2 Engineering Design Problem 2: Design of Tension/Compression Spring Design Problem

The objective of this engineering problem is to design a spring with minimum-mass to can carry a given axial load without causing failure in material while satisfying two performance requirements [37]. Mathematically the objective function of this problem

Table 1. The best solutions obtained from different approaches for speed reducer problem

Methods	x_1	x_2	x_3	x_4	x_5	x_6	x_7	$F(x)$
Proposed algorithm	3.5000	0.70000	17.0000	7.300010634	7.8000699551	3.350214689	5.286683	2996.340656
UFA	3.50000	0.7000	17.000	7.300000	7.80000	3.350214666	5.286683	2996.348164
CS	3.5015	0.7	17.0000	7.6050	7.8181	3.352	5.2875	3000.981
IAPSO	3.49999	0.7	17.000	7.3	7.8	120	5.286683	2996.348

is the minimization of the weight of a tension/compression spring. In addition, there are four constraints on the surge frequency, the minimum deflection, the limits on outside diameter on design variables and shear stress. Among of these constraints, there one constraint is linear and three nonlinear.

The diagrammatic representation of the problem and its design variables are shown in Fig. 2. As shown in the figure, this problem has 3 design parameters, which are:

- The number of active coils (P)
- The mean diameter of the coil (D)
- Wire diameter (d)

The mathematical formulation of this problem:

$$\text{Min } [x_1^2 x_2 (x_3 + 2)]$$

$$\text{Subject to: } 1 - \frac{x_2^3 x_3}{71785 x_1^4} \le 0, 1 - \frac{140.45}{x_3 x_2^2} \le 0,$$

$$\frac{1}{5108 x_1^2} + \frac{4 x_2^2 - x_1 x_2}{12566 (x_1^3 x_2 - x_1^4)} - 1 \le 0,$$

$$\frac{x_1 + x_2}{1.5} - 1 \le 0,$$

$$0.05 \le x_1 \le 2, \, 0.25 \le x_2 \le 1.3, \, 2 \le x_3 \le 15.$$

This problem has been solved previously by different optimization methods such as: improved fruit fly optimization algorithm (IFFOA) [38], improved harmony search algorithm (IHS) [39], iterative dynamic diversity evolutionary algorithm (IDDEA) [40], salp swarm algorithm (SSA) [41], effective co-evolutionary differential evolution algorithm (CDE) [42], accelerating adaptive trade-off model using shrinking space technique (AATM) [43], inspired grey wolf optimizer (IGWO) [44], differential evolution with level comparison (DELC) [45], co-evolutionary particle swarm optimization (CPSO) [46], $(\mu + \lambda)$-evolutionary strategy [47], Mine blast algorithm (MBA) [48] and the grey wolf optimizer algorithm (GWO) [49]. The results of these algorithms, as well as the best result obtained by our algorithm are presented in Table 3. In addition, Table 4 gives the statistical results (best values, worst values, mean values and SD) that obtained by the proposed approach over 30 independent runs to evaluate the convergence speed, as well as the stability and solution accuracy of the algorithm.

Table 2. Statistical results obtained from the different methods for speed reducer problem

Methods	Best values	Mean values	Worst values	SD	NFEs
Proposed algorithm	2996.3406561098	2996.34128380294	2996.3431567554	9.7254e−04	1486
CS (2013)	3000.9810	3007.1997	3009	4.9634	500
IPSO (2016)	2996.34816497	2996.34816497	2996.34816497	6.88E−13	6000
UFA (2018)	2996.3481649760	2996.34816498	2996.3481649986	4.51E−09	3000
C-ITGO (2018)	2996.34816497	2996.34816497	2996.34816497	7.5E−13	856
ABCA (2012)	2997.058	2997.058	N/A	0.0	30000
APSO (2010)	3177.530771	3855.581557	4677.005187	473.767	30,000

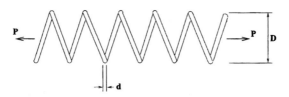

Fig. 2. Tension/compression spring design problem

Table 3. The best solutions obtained from different approaches for tension/compression spring design problem

Methods	x_1	x_2	x_3	Best $F(x)$
Proposed algorithm	0.051685	0.356629	11.294175	0.012665
IHS (2007)	0.051154	0.349871	.076432	0.0126706
AATM (2009)	0.051813	0.3569041	11.119252	0.0126682
IDDEA (2014)	0.051600	0.354581	11.415327	0.0126654
GWO (2014)	0.051690	0.356737	11.288850	0.0126660
CDE(2007)	0.051609	0.354714	11.410831	0.0126702
SSA (2017)	0.051207	0.345215	12.004032	0.0126763
IAFOA (2018)	0.051657	0.355939	11.334999	0.0126655
IGWO (2018)	0.05170	0.356983	11.275600	0.0126670
DELC (2010)	0.051689	0.356717	11.288965	0.012665
CPSO (2007)	0.051728	0.357644	11.244543	0.0126747
MBA (2013)	0.051656	0.355940	11.344665	0.012665
$(\mu + \lambda)$-ES (2005)	N/A	N/A	N/A	0.012689

As shown in the Tables 3 and 4, The proposed algorithm, IAFOA, DELC and MBA are the only algorithms that obtained the best solution. All other algorithms give poor performance relatively with high values of NFEs. The NFEs for Hybrid SCA-SSGA is less than ten times the NFEs for the competing algorithm, MBA, which has 7650 NFEs compared with 736 NFEs in Hybrid SCA-SSGA. In addition, the proposed algorithm obtained small SD (2.3579e−05) among most of the approaches. All runs of the hybrid SCA-SSGA converged to the best solution quickly, which show that our approach has better results than any other algorithm that solved this problem.

Table 4. Statistical results obtained from the different methods for tension/compression spring design problem

Methods	Best value	Mean value	Worst value	SD	NFEs
Proposed algorithm	0.012665	0.01268011	0.012716860	2.3579e−05	736
MBA	0.012665	0.012713	0.012900	6.30E−05	7650
DELC	0.012665233	0.012665267	0.0126655750	1.3E−07	20,000
CPSO	0.0129240	0.0127300	0.0126747	5.20E−04	240,000
CDE	0.0126702	0.012703	0.012790	2.7e−5	240,000
IAFOA	0.012665468	0.012688	0.012673	3.015E−06	N/A
GWO	0.0126660	0.012721	0.012691	2.314E−05	N/A
(μ + λ)-ES	0.012689	0.013165	0.014078	3.9E−4	30,000

6 Conclusions

A hybrid SCA-SSGA algorithm has been presented in this paper, as a new technique to solve engineering design optimization problems. This approach integrate the merits exploitation capability in SSGA with the exploration capability in SCA to avoid exposure to early convergence, speed up the search process and quick the convergence to best results in a reasonable time. In addition, there is a diversity in the population due to the SSGA operators. This algorithm has been tested by using two well-studied engineering design problems. The simulation results proved that the hybrid SCA-SSGA outperforms other methods, which verify the reliability and validity of our approach in solving real life engineering design problems. In summary, the proposed hybrid algorithm: SCA-SSGA is successful, and suitable for solving the engineering design problems. Although the analytical results demonstrate the superiority of the proposed algorithm in solving engineering design optimization problems, further complex experiments with high dimensions should be conducted in the future. In addition, we are planning to design the multi-objective of the Hybrid SCA-SSGA.

References

1. Rao, S.S.: Engineering Optimization: Theory and Practice. Wiley, Hoboken (2009)
2. Yang, X.S., Hossein Gandomi, A.: Bat algorithm: a novel approach for global engineering optimization. Eng. Comput. **29**(5), 464–483 (2012)
3. Rizk-Allah, R.M.: Hybridizing sine cosine algorithm with multi-orthogonal search strategy for engineering design problems. J. Comput. Des. Eng. **5**(2), 249–273 (2018)
4. Droste, S., Jansen, T., Wegener, I.: Upper and lower bounds for randomized search heuristics in black-box optimization. Theor. Comput. Syst. **39**(4), 525–544 (2006)
5. Holland, J.: Adaptation in Natural and Artificial Systems: An Introductory Analysis with Application to Biology, Control and Artificial Intelligence. MIT Press, Cambridge (1975)
6. He, B., Che, L., Liu, C.: Novel hybrid shuffled frog leaping and differential evolution algorithm. Jisuanji Gongcheng yu Yingyong (Comput. Eng. Appl.) **47**(18), 4–8 (2011)

7. Eberhart, R., Kennedy, J.: A new optimizer using particle swarm theory. In: 1995 Proceedings of the Sixth International Symposium on Micro Machine and Human Science, MHS 1995, pp. 39–43. IEEE, October 1995
8. Lenin, K., Reddy, B.R., Kalavathi, M.S.: Modified monkey optimization algorithm for solving optimal reactive power dispatch problem. Indones. J. Electr. Eng. Inform. (IJEEI) 3 (2), 55–62 (2015)
9. Zhou, Y., Wang, J., Gao, S., Yang, X., Yin, J.: Improving artificial bee colony algorithm with historical archive. In: Gong, M., Pan, L., Song, T., Zhang, G. (eds.) Bio-Inspired Computing-Theories and Applications, pp. 185–190. Springer, Singapore (2016)
10. Xu, H., Liu, X., Su, J.: An improved grey wolf optimizer algorithm integrated with cuckoo Search. In: 2017 9th IEEE International Conference on Intelligent Data Acquisition and Advanced Computing Systems: Technology and Applications (IDAACS), vol. 1, pp. 490–493. IEEE, September 2017
11. Mirjalili, S.: SCA: a sine cosine algorithm for solving optimization problems. Knowl.-Based Syst. 96, 120–133 (2016)
12. Monteiro-Filho, J.B., Albuquerque, I.M.C., Neto, F.L.: Fish School Search Algorithm for Constrained Optimization. arXiv preprint arXiv:1707.06169 (2017)
13. Mirjalili, S., Mirjalili, S.M., Yang, X.S.: Binary bat algorithm. Neural Comput. Appl. 25(3–4), 663–681 (2014)
14. Wang, H., Wang, W., Sun, H., Cui, Z., Rahnamayan, S., Zeng, S.: A new cuckoo search algorithm with hybrid strategies for flow shop scheduling problems. Soft. Comput. 21(15), 4297–4307 (2017)
15. Hoos, H.H., Stützle, T.: Stochastic local search: Foundations and applications. Elsevier, Amsterdam (2004)
16. Mousa, A.A., El-Shorbagy, M.A., Farag, M.A.: K-means-clustering based evolutionary algorithm for multi-objective resource allocation problems. Appl. Math 11(6), 1681–1692 (2017)
17. Farag, M.A., El-Shorbagy, M.A., El-Desoky, I.M., El-Sawy, A.A., Mousa, A.A.: Binary-real coded genetic algorithm based k-Means clustering for unit commitment problem. Appl. Math. 6(11), 1873 (2015)
18. Hussein, M.A., EL-Sawy, A.A., Zaki, E.S.M., Mousa, A.A.: Genetic algorithm and rough sets based hybrid approach for economic environmental dispatch of power systems. Br. J. Math. Comput. Sci. 4(20), 2978 (2014)
19. Renner, G., Ekárt, A.: Genetic algorithms in computer aided design. Comput. Aided Des. 35 (8), 709–726 (2003)
20. Iqbal, S., Hoque, M.T.: hGRGA: a scalable genetic algorithm using homologous gene schema replacement. Swarm Evol. Comput. 34, 33–49 (2017)
21. Lozano, M., Herrera, F., Cano, J.R.: Replacement strategies to preserve useful diversity in steady-state genetic algorithms. Inf. Sci. 178(23), 4421–4433 (2008)
22. Farag, M.A., El-Shorbagy, M.A., El-Desoky, I.M., El-Sawy, A.A., Mousa, A.A.: Genetic algorithm based on k-means-clustering technique for multi-objective resource allocation problems. Br. J. Math. Comput. Sci. 8(1), 80–96 (2015)
23. Elattar, E.E.: A hybrid genetic algorithm and bacterial foraging approach for dynamic economic dispatch problem. Int. J. Electr. Power Energy Syst. 69, 18–26 (2015)
24. Altiparmak, F., Gen, M., Lin, L., Karaoglan, I.: A steady-state genetic algorithm for multi-product supply chain network design. Comput. Ind. Eng. 56(2), 521–537 (2009)
25. Nenavath, H., Jatoth, R. K., Das, S.: A synergy of the sine-cosine algorithm and particle swarm optimizer for improved global optimization and object tracking. Swarm and Evolutionary Computation (2018)

26. Gen, M., Cheng, R.: Genetic Algorithms and Engineering Optimization, vol. 7. Wiley, Hoboken (2000)
27. Martorell, S., Carlos, S., Sanchez, A., Serradell, V.: Constrained optimization of test intervals using a steady-state genetic algorithm. Reliab. Eng. Syst. Saf. **67**(3), 215–232 (2000)
28. Ekiz, S.: Solving constrained optimization problems with sine-cosine algorithm. Period. Eng. Nat. Sci. (PEN) **5**(3), 378–386 (2017)
29. Osman, M.S., Abo-Sinna, M.A., Mousa, A.A.: A solution to the optimal power flow using genetic algorithm. Appl. Math. Comput. **155**(2), 391–405 (2004)
30. Mousa, A.A., Kotb, K.A.: A hybrid optimization technique coupling an evolutionary and a local search algorithm for economic emission load dispatch problem. Appl. Math. **2**(07), 890 (2011)
31. Gandomi, A.H., Yang, X.S., Alavi, A.H.: Cuckoo search algorithm: a metaheuristic approach to solve structural optimization problems. Engineering with Computers **29**(1), 17–35 (2013)
32. Brajević, I., Ignjatović, J.: An upgraded firefly algorithm with feasibility-based rules for constrained engineering optimization problems. J. Intell. Manuf. 1–30 (2018)
33. Ferreira, M.P., Rocha, M.L., Neto, A.J.S., Sacco, W.F.: A constrained ITGO heuristic applied to engineering optimization. Expert Syst. Appl. **110**, 106–124 (2018)
34. Akay, B., Karaboga, D.: Artificial bee colony algorithm for large-scale problems and engineering design optimization. J. Intell. Manuf. **23**(4), 1001–1014 (2012)
35. Yang, X.S.: Engineering Optimization: An Introduction with Metaheuristic Applications. Wiley, Hoboken (2010)
36. Guedria, N.B.: Improved accelerated PSO algorithm for mechanical engineering optimization problems. Appl. Soft Comput. **40**, 455–467 (2016)
37. Arora, J.S.: Introduction to Optimum Design. McGraw-Hill Book Company, New York (1989)
38. Wu, L., Liu, Q., Tian, X., Zhang, J., Xiao, W.: A new improved fruit fly optimization algorithm IAFOA and its application to solve engineering optimization problems. Knowl.-Based Syst. **144**, 153–173 (2018)
39. Mahdavi, M., Fesanghary, M., Damangir, E.: An improved harmony search algorithm for solving optimization problems. Appl. Math. Comput. **188**(2), 1567–1579 (2007)
40. Wei-Shang, G.A.O., Cheng, S.H.A.O.: Iterative dynamic diversity evolutionary algorithm for constrained optimization. Acta Automatica Sin. **40**(11), 2469–2479 (2014)
41. Mirjalili, S., Gandomi, A.H., Mirjalili, S.Z., Saremi, S., Faris, H., Mirjalili, S.M.: Salp swarm algorithm: a bio-inspired optimizer for engineering design problems. Adv. Eng. Softw. **114**, 163–191 (2017)
42. Huang, F.Z., Wang, L., He, Q.: An effective co-evolutionary differential evolution for constrained optimization. Appl. Math. Comput. **186**(1), 340–356 (2007)
43. Wang, Y., Cai, Z., Zhou, Y.: Accelerating adaptive trade-off model using shrinking space technique for constrained evolutionary optimization. Int. J. Numer. Meth. Eng. **77**(11), 1501–1534 (2009)
44. Long, W., Jiao, J., Liang, X., Tang, M.: Inspired grey wolf optimizer for solving large-scale function optimization problems. Appl. Math. Model. **60**, 112–126 (2018)
45. Wang, L., Li, L.P.: An effective differential evolution with level comparison for constrained engineering design. Struct. Multidiscip. Optim. **41**(6), 947–963 (2010)
46. He, Q., Wang, L.: An effective co-evolutionary particle swarm optimization for constrained engineering design problems. Eng. Appl. Artif. Intell. **20**(1), 89–99 (2007)

47. Mezura-Montes, E., Coello, C.A.C.: Useful infeasible solutions in engineering optimization with evolutionary algorithms. In: Mexican International Conference on Artificial Intelligence, pp. 652–662. Springer, Berlin, Heidelberg, November 2005
48. Sadollah, A., Bahreininejad, A., Eskandar, H., Hamdi, M.: Mine blast algorithm: a new population based algorithm for solving constrained engineering optimization problems. Appl. Soft Comput. **13**(5), 2592–2612 (2013)
49. Mirjalili, S., Mirjalili, S.M., Lewis, A.: Grey wolf optimizer. Adv. Eng. Softw. **69**, 46–61 (2014)
50. El Aziz, M.A., Ewees, A.A., Hassanien, A.E.: Whale optimization algorithm and moth-flame optimization for multilevel thresholding image segmentation. Expert Syst. Appl. **83**, 242–256 (2017)
51. Tharwat, A., Hassanien, A.E., Elnaghi, B.E.: A ba-based algorithm for parameter optimization of support vector machine. Pattern Recogn. Lett. **93**, 13–22 (2017)

Machine Learning in Biomedical

MolecRank: A Specificity-Based Network Analysis Algorithm
Ranking Therapeutic Molecules in the Bibliome

Ahmed Abdeen Hamed[1]([⊠]), Agata Leszczynska[2], and Mark Schreiber[1,3]

[1] Merck & Co., Inc., Boston, MA, USA
ahmed.hamed1@merck.com
[2] MSD, Prague, Czech Republic
[3] Kaleido Biosciences, Bedford, MA, USA

Abstract. Biomedical scientists often search databases of therapeutic molecules to answer a set of drug-related queries. In this paper, we present a novel network algorithm called MolecRank that is specialized in searching and ranking molecules using a biomedical literature. Starting with a disease-related set of publications (e.g., depression), a feature extraction step is performed to identify the biological features associated with the drugs of study. The MolecRank is a network centrality algorithm that is specialized in deriving a rank when *specificity* is in question. The algorithm's promise is two folds (a) an interesting search-and-rank tool that demonstrated its importance in the drug discovery research, (b) a theoretical network centrality measure that is based on the notion of specificity. We performed our experiments against a depression-related literature dataset. The results shows an interesting order that is significantly different from well-advertised drugs (e.g., Cymbalta#10 though well-advertised). We conclude that not all well-advertised drugs are most specific. This striking evidence highlights the significance of specificity as an important measure in discovering new drugs.

Keywords: Specificity centrality · Ranking algorithms ·
Therapeutic molecules · Literature mining

1 Introduction

The problem of ranking certain biological terms according to their association with other entities is well-established. Such a problem is manifested in various studies. The literature is rich of various algorithmic approaches taken in the biological ranking space [5]. For instance, gene-ranking a crucial task in bioinformatics has been previously investigated. (1) Bragazzi et al., presented an algorithmic approach for gene-ranking that leads to a list of few strong candidate genes [1], and (2) Winter et al., conducted a study that ranks gene features using a modified version of Google's PageRank, which they called NetRank [2]. Protein ranking is another problem that has also been researched; Weston et. al,

© Springer Nature Switzerland AG 2020
A. E. Hassanien et al. (Eds.): AMLTA 2019, AISC 921, pp. 159–168, 2020.
https://doi.org/10.1007/978-3-030-14118-9_16

demonstrated the importance of protein ranking in a network constructed from a sequence database [3]. Other scientists ranked biological features as sets to provide answers to various scientific questions. Wren et al., mined PubMed for biological features (genes, proteins, phenotypes, chemicals, and diseases), grouped them as sets of cohesions, and ranked them to satisfy scientific needs [4]. Studying biological ranking is an inherently network problem. Therefore, various researchers have framed their problems as biological networks. This representation offers tools such centrality measures for the purpose of identifying the most influential elements in the network [6,7].

From a drug discovery point of view, ranking molecules in a biological network holds great promise, especially one that integrates network biology and polypharmacology [8]. Bodnarchuk et al., presented their computational simulation technique for ranking chemical molecules [9]. In this paper, we address the molecule ranking problem based on two foundations: (1) a large network of biological features which we extracted from the biomedical literature, and (2) querying the network as a knowledge graph to answer a given question using SPARQL queries. The ranking algorithm we present here is a post-processing step that takes place over the result of the question. To date, graph databases do not provide rankings based on the specificity notion that is required for identifying significant molecules in this manner. Such a complex approach: (mining → transforming → persisting → querying → ranking) is worthwhile because it is flexible. It can answer an unlimited number of questions and for various purposes (drug discovery, drug design, decision support, hypothesis testing, etc).

This brief introduction constitutes the heuristic and the computational framework for measuring specificity algorithmically as a network analysis measure. To motivate the work, we present the following simple mockup scenario. For a given disease, suppose we found that four molecules were mentioned in publications in conjunction with the four biological feature types (gene, RNA, protein and cell type). Typically, some molecules exhibit stronger binding with a given gene, RNA, or a protein instance than others. As mentioned above, the stronger the binding, the stronger the specificity. Here we presented a weighted network that models the binding specificity aspects and captures its essence. Figure 1 shows how the four molecules are linked to specific instances of genes, proteins, RNA, and cell types. In a real example, a molecule may bind with an RNA instance, while another may not. Considering the figure, one might observe that molec3 has an one exact entity in each feature type (i.e, breast cancer, breast, brca1, eml4). Others have one exact instance in some of the features but missing information in other features: (e.g., molec1 is missing information in the gene feature, and molec4 is missing information in the cell-type and the gene features). Another very important observation: some molecules have more than a single instance in the same feature type (e.g., molec2 has two instances in the protein feature while molec5 has four instances in the protein feature). By the notion of specificity, molec3 is the most specific (because of the connection to one single instance), and the most informative (because there is information in each feature type). This grants molec3 the first rank based on these two heuristic factors. The second in the rank is molec1, since what we know is specific (single instances in each existing feature). Molec1 less informative than molec3, given that we are missing gene information. Both molec2 and molec5 are the least informative but also the least specific. However, molec2 (i.e., connected with three instances) is more specific than molec5 (connected with three instances). These observations drive the ranking to the following order (molec3, molce1, molec4, molec2, molec5). The full description of the example is also summarized in Table 1.

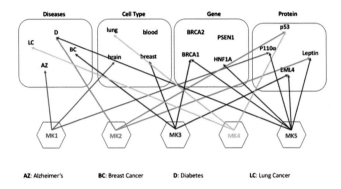

Fig. 1. Shows a motivating example for explaining the notion of specificity using a limited number of biological features (e.g. gene, protein, cell type and disease), and their association with therapeutic molecules mentions in a given set of publications. The hexagon shapes at the bottom represents five molecules each one is encoded with a color to better visualize the outgoing connections to the corresponding biological features.

Table 1. Shows a summary of the preliminary example. There are four column features in for each molecule. The last column shows the order of each molecule according to the notion of specificity (to be fully described onward). According to this notion we have the following order: molec3 is ranked#1, molec1, molec4, molec2, and then molec5.

MoleculeID	Disease	Cell type	Gene	Protein	MolecRank
Molec1	Alzheimer's	Brain	–	p53	2
Molec2	Diabetes	–	–	{p53, leptin}	4
Molec3	Breast cancer	Breast	brca1	eml4	1
Molec4	Lung cancer	Lung	–	p53	3
Molec5	Diabetes	–	–	{p53, p11$alpha$, eml4, leptin}	5

2   Methods

Using means of machine learning and Natural Language Processing (NLP) tools (e.g., ABNER [21]), we extracted the mentions of biological entities of each of the abstracts. This produced the features required. Representing the entire set of features as a network offers an attractive theoretical model that can be computationally explored. Representing the resulting features in the form of Resource Descriptor Framework (RDF) [23] triples offers a mechanism necessary to construct the network as visualized in Fig. 2. Each triple is uniquely described in the information that it communicates (subject→predicate→object). For instance, to express some knowledge about breast cancer it can be done as follows: ("breast cancer", is a, disease type), ("breast cancer", is mentioned in, PMID:28042876), and ("breast cancer", co-occurredwith, "BT-549").

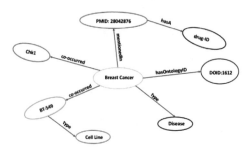

Fig. 2. Shows a graph resulting from the biological features extracted from PMID:28042876 and expressed as RDF triples. The following triples can be easily traced in the diagram: ("breast cancer", is a, disease type), ("breast cancer", is mentioned in, PMID:28042876), and ("breast cancer", co-occurredwith, "BT-549").

2.1 MolecRank Algorithm and Heuristics

Here we present the heuristic that empowers our network analysis algorithm to generate molecule ranks of a given search query. This algorithm is designed based on the notion of molecular specificity. The notion starts with a molecule and how it is connected to the various feature types. A molecule is considered specific if it is attached to one instance in each feature type. Contrarily, the more feature instances of the same type a molecule is linked to, the more general and less significant. This specificity notion is not be to confused with the knowledge-gain which entails: the more we know about the molecule (even if it is not specific enough), the more it contributes to the rank as shown in Fig. 1 and Table 1. Following are the steps that describe MolecRank, and the corresponding pseudocode described in Algorithm 1.

2.2 Experiments

The example shown in Fig. 1 above was a demonstration of how a set of biological features have co-occurred in a set of publications. Now, we need to show a query result for the algorithm to work on. As discussed earlier, the ranking is

---

Algorithm 1: SpecRank Pseudocode: SpecRank: A Molecule Ranking Algorithm for Biomedical Knowledge Networks

---

Data: A SPARQL query result in CSV format
Input: A bipartite network representation of the SPARQL result
Output: Sorted list of drugs in a descending order
Result: Ranked D's based on their computed scores

1 **for** $D_i \in G$ **do**
2 | Generate neighbors' adjacency matrices S and C ;
3 **end**
4 **for** $e \in S$ **do**
5 | **if** $e > 1$ **then**
6 | $e \leftarrow 0$, otherwise $e \leftarrow 1$;
7 | **end**
8 **end**
9 **for** $e \in C$ **do**
10 | **if** $e > 0$ **then**
11 | $e \leftarrow 1$, otherwise $e \leftarrow 0$;
12 | **end**
13 **end**
14 Calculate the product of S and the weights of each feature ;
15 Calculate the row sum of each matrix S and C ;
16 Calculate specificity Ψ and knowledge-gain Ω ;
17 Combine Ψ and Ω as in equation: 3 ;

---

a post-processing step after the query is issued. Here, we show how the algorithm operates on a weighted network that is hypothetically generated from a query result. Figure 3 the final network as a weighted bipartite graph. The figure shows the drugs as blue bigger nodes). All nodes are labeled (instance and type concatenated) to make it easier for the reader to understand and trace the algorithm steps. The edges are also labeled using co-occurrence frequencies, which is visualized with different thicknesses accordingly.

Producing the adjacency matrices (specificity and knowledge-gain) is a straightforward process: For each drug source, all neighbors are accessed. The source is prohibited to connect to a drug since the graph is bipartite. Therefore, we are guaranteed to have neighbors of all types except for drug nodes. Starting with the *Specificity Matrix*: For each drug, calculate the count of how many links to the neighbors. For each type, if the count is $= 1$, a value of "1" is placed in the type field, otherwise, a "0" value is placed. For instance D1's adjacency record would be (D1$\rightarrow$ DNA:0, RNA:1, PROT:0, CTYPE:0, CLINE:0). When applying the feature frequency values, we get the following considering the feature frequency weights, which captures the specificity strength, it produces (D1$\rightarrow$ DNA:0, RNA:0.05, PROT:0, CTYPE:0, CLINE:0) as in Eq. 1. Clearly this is because the RNA instance "rna1" is specific which has a value of "1" in the specificity matrix. When multiplying by the strength of such specificity, presented as a weight, it produces a value of "0.05". The RNA type instance is

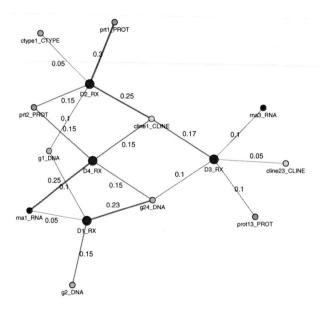

Fig. 3. Shows a simulated network of molecules and their corresponding biological features. Each feature type is colored differently. Molecules are colored in blue and displayed in a fairly larger size to highlight the importance.

the only specific feature according to our heuristic. The remaining features are not specific, hence their corresponding elements are values of "0".

Calculating the *Knowledge-gain Matrix* is treated differently. The data demonstrates that some drug have association with more than one instance of the same type. An example of that is D1 where it links to three different DNA instances (g1, g2, g24). This makes D1 not specific. However, this knowledge we gain must be factored in. The algorithm treats such instances probabilistically as in Eq. 2. This may result in marginal contributions but it does address the issue of how much information we gained, which eventually contribute to the final rank of the molecule significantly. The two measures (specificity and knowledge-gain) are represented as row vectors respectively. This step produces a single score for each feature type in each vector. Calculating the final ranks using the two vectors is a mechanical step. The final ranks for each drug are: [('D4': 1.4), ('D2', 0.845), ('D3', 0.609), ('D1', 0.103)].

$$specificity \quad \rightarrow \Psi_i = [\forall_{i=1} \mid (\psi_i + \epsilon)] \tag{1}$$

$$knowledge\_gain \quad \rightarrow \Omega(d_i) = \Pr_{(w_{i1})} \times \Pr_{(w_{i2})} \times \Pr_{(w_{i3})} \tag{2}$$

$$MolecRank \quad \rightarrow \sum_{i=1}^{n} \Theta_i = [\Psi_i + \Omega_i] \tag{3}$$

3   Discussion

Earlier, we mentioned the analysis of a depression-related PubMed dataset of 400,000+ abstracts. In this section we discuss the centrality benchmarking we have done against MolecRank. Table 2 shows the correlation analysis of MolecRank against the various centrality measures. The analysis shows that interesting patterns, which are consistent with the centrality measures. While degree, closeness, and eccentricity (the lowest "negative" correlation of −0.60), and eccentricity have a positive correlation (0.55). Most of the co-relation values, negative or positive are above (0.50) and they range from (0.55–0.60) regardless of the sign. Centrality measures are designed to explore network structures in various ways (edge connectivity and weights). For example, the degree centrality counts the neighbors of the each drug in a brute force manner. However, the specificity measure selectively counts the edges of the neighbors according to the singularity and the strength of the bond. This means, the more neighbors connected with edges to each given drug, the less the specificity. However, the information of the neighbors are not entirely lost since it was captured by the notion of knowledge-gain (Eq. 2 shows that it was computed by the multiplications of the probabilities of how the neighbors of connected). Since the algorithm favors the specificity over the knowledge-gain, the specificity always triumph. This explains why there is (−0.60) of negative correlation(the compliment of 0.40 is lost due accounting for the knowledge-gain probabilities). The same notion is also applied for closeness, betweenness, and eccentricity since the two measures consider the neighbors of the vertices in question (i.e., the drug). It is interesting to see that the absolute values all ranged around (0.55–60) which indeed demonstrates the spirit of the algorithm in capturing the quality of the connection as opposed to the bare frequencies without losing the connectivity information embedded in the network. MolecRank could potential be a favorite measure that summarizes all other measures. Clearly, the algorithm can be used not only for ranking molecules but can also be used in any theoretical context where specificity is in question. As for the actual ranks of the drug mentions in the dataset, here discuss the top-10 listed in Table 3. Indeed, there are some interesting surprises. Though Cymbalta is ranked number 1 in popularity on the drugs.com database, it claims the 10th position on the MolecRank list. On the other hand, a drug such as Protriptyline is ranked as #1. There are a few factors to consider in this analysis, which are yet to be incorporated in the next future phase: (1) The interaction with alcohol, (2) the impact on pregnancy, (3) Controlled Substance Act. Without keeping these factors in mind, it is not directly possible to link the popularity of a drug to its specificity. However, learning that Protriptyline is manufactured by companies such as Pfizer and Sanofi among a few other pharmaceutical companies, one may conclude a higher confidence in the drug's ability to treat depression. As for not being as popular as Cymbalta, that could be to various factors such as marketing and media advertisements. Table 3 also shows the order of each of the top-10 drugs according to MolecRank against the various centrality measures above.

Table 2. Shows the correlations among MolecRank scores for each drug and how much correlation it has with each centrality measure.

Centrality	MolecRank-Deg	MolecRank-Clos	MolecRank-Bet	MolecRank-ECC
Correlation	−0.58	−0.62	−0.59	0.55

Table 3. Shows the top-10 drugs ranked using the specificity-centrality to rank the drugs and the corresponding order using the various centrality measures

RX_DRUG	SPEC	DEG	CLOS	BET	ECC
protriptyline	1	40	41	36	26
Ludiomil	2	49	55	51	28
Symbyax	3	56	60	61	20
amoxapine	4	33	36	34	48
Sinequan	5	63	68	65	8
Parnate	6	44	61	48	21
methylphenidate	7	19	19	18	53
Trintellix	8	53	56	52	10
Surmontil	9	70	72	70	1
Cymbalta	10	42	48	41	3

4   Future Directions

The authors wish to extend the scope of the study in various directions: (1) studying other types of disease (e.g., Alzheimer's, cancer, and diabetes) in addition to depression, (2) incorporating real-world data from various sources to further enrich the networks (clinical trials, news, social media, etc), and (3) taking into consideration the adverse drug reactions and updating the ranking accordingly. We hope by addressing such directions, we offer a more significance steps towards discovering and designing new drugs for generations to come.

Acknowledgement. The authors would like thank Greg Temsi, Ramiro Barrantes for their valuable discussions. The authors also greatly appreciate the tremendous feedback on this work giving by Dr. Barabasi and his lab members. We also thank Dr. Karin Verspoor of University of Melbourne for the valuable discussions.

References

1. Bragazzi, N.L., Nicolini, C.: A leader genes approach-based tool for molecular genomics: from gene-ranking to gene-network systems biology and biotargets predictions. J. Comput. Sci. Syst. Biol. **6**, 165–176 (2013)
2. Winter, C., Kristiansen, G., Kersting, S., Roy, J., Aust, D., Knösel, T., Rümmele, P., Jahnke, B., Hentrich, V., Rückert, F., Niedergethmann, M., Weichert, W., Bahra, M., Schlitt, H.J., Settmacher, U., Friess, H., Büchler, M., Saeger, H.-D., Schroeder, M., Pilarsky, C., Grützmann, R.: Google goes cancer: improving outcome prediction for cancer patients by network-based ranking of marker genes. PLOS Comput. Biol. **8**(5), 1–16 (2012)
3. Weston, J., Elisseeff, A., Zhou, D., Leslie, C.S., Noble, W.S.: Protein ranking: from local to global structure in the protein similarity network. Proc. Nat. Acad. Sci. U. S. A. **101**(17), 6559–6563 (2004)
4. Wren, J.D., Garner, H.R.: Shared relationship analysis: ranking set cohesion and commonalities within a literature-derived relationship network. Bioinformatics **20**(2), 191–198 (2004)
5. Chen, J., Jagannatha, A.N., Fodeh, S.J., Yu, H.: Ranking medical terms to support expansion of lay language resources for patient comprehension of electronic health record notes: adapted distant supervision approach. JMIR Med. Inform. **5**(4), e42 (2017)
6. Koschützki, D., Schwöbbermeyer, H., Schreiber, F.: Ranking of network elements based on functional substructures. J. Theoret. Biol. **248**(3), 471–479 (2007)
7. Junker, B.H., Koschützki, D., Schreiber, F.: Exploration of biological network centralities with CentiBIN. BMC Bioinform. **7**(1), 219 (2006)
8. Hopkins, A.L.: Network pharmacology: the next paradigm in drug discovery. Nat. Chem. Biol. **4**(11), 682–690 (2008)
9. Bodnarchuk, M.S., Heyes, D.M., Dini, D., Chahine, S., Edwards, S.: Role of deprotonation free energies in pKa prediction and molecule ranking. J. Chem. Theory Comput. **10**(6), 2537–2545 (2014)
10. Koshland, D.E.: Application of a theory of enzyme specificity to protein synthesis. Proc. Nat. Acad. Sci. **44**(2), 98–104 (1958)
11. Lehninger, A., Nelson, D.L., Cox, M.M.: Lehninger Principles of Biochemistry, 5th edn. W. H. Freeman, San Francisco (2008)
12. Wood, E.J.: Harper's Biochemistry 24th edition by R.K. Murray, D.K. Granner, P.A. Mayes and V.W Rodwell. pp 868. Appleton & Lange, Stamford, CT (1996). £ 28.95 isbn 0-8385-3612-3. Biochem. Educ. **24**(4), 237–237 (1996)
13. Hu, L., Fawcett, J.P., Gu, J.: Protein target discovery of drug and its reactive intermediate metabolite by using proteomic strategy. Acta Pharm. Sin. B **2**(2), 126–136 (2012)
14. Hefti, F.F.: Requirements for a lead compound to become a clinical candidate. BMC Neurosci. **9**(3), S7 (2008)
15. Degterev, A., Maki, J.L., Yuan, J.: Activity and specificity of necrostatin-1, small-molecule inhibitor of rip1 kinase. Cell Death Differ. **20**(2), 366 (2013)
16. Eaton, B.E., Gold, L., Zichi, D.A.: Let's get specific: the relationship between specificity and affinity. Chem. Biol. **2**(10), 633–638 (1995)
17. Radhakrishnan, M.L., Tidor, B.: Specificity in molecular design: a physical framework for probing the determinants of binding specificity and promiscuity in a biological environment. J. Phys. Chem. B **111**(47), 13419–13435 (2007)

18. Strovel, J., Sittampalam, S., Coussens, N.P., Hughes, M., Inglese, J., Kurtz, A., Andalibi, A., Patton, L., Austin, C., Baltezor, M., et al.: Early drug discovery and development guidelines: for academic researchers, collaborators, and start-up companies (2016)

19. Hartley, J.A., Lown, J.W., Mattes, W.B., Kohn, K.W.: Dna sequence specificity of antitumor agents: oncogenes as possible targets for cancer therapy. Acta Oncol. **27**(5), 503–510 (1988)

20. Timchenko, L.T., Timchenko, N.A., Caskey, C.T., Roberts, R.: Novel proteins with binding specificity for DNA CTG repeats and RNA CUG repeats: implications for myotonic dystrophy. Hum. Mol. Genet. **5**(1), 115–121 (1996)

21. Settles, B.: ABNER: an open source tool for automatically tagging genes, proteins, and other entity names in text. Bioinformatics **21**(14), 3191–3192 (2005)

22. Carpenter, B.: Lingpipe for 99.99% recall of gene mentions. In: Proceedings of the Second BioCreative Challenge Evaluation Workshop, vol. 23, pp. 307–309 (2007)

23. Candan, K.S., Liu, H., Suvarna, R.: Resource description framework: metadata and its applications. SIGKDD Explor. Newsl. **3**(1), 6–19 (2001)

24. Shannon, C.E.: Prediction and entropy of printed English. Bell Labs Tech. J. **30**(1), 50–64 (1951)

25. Koschützki, D., Schreiber, F.: Centrality analysis methods for biological networks and their application to gene regulatory networks. Gene Regul. Syst. Biol. **2**, 193 (2008)

26. Jeong, H., Mason, S.P., Barabási, A.-L., Oltvai, Z.N.: Lethality and centrality in protein networks. Nature **411**(6833), 41–42 (2001)

27. Koschützki, D., Lehmann, K.A., Peeters, L., Richter, S., Tenfelde-Podehl, D., Zlotowski, O.: Centrality Indices, pp. 16–61. Springer, Berlin (2005)

28. Freeman, L.C.: Centrality in social networks conceptual clarification. Soc. Netw. **1**(3), 215–239 (1978)

29. Opsahl, T., Agneessens, F., Skvoretz, J.: Node centrality in weighted networks: generalizing degree and shortest paths. Soc. Netw. **32**(3), 245–251 (2010)

30. Zhou, Q., Womer, F.Y., Kong, L., Wu, F., Jiang, X., Zhou, Y., Wang, D., Bai, C., Chang, M., Fan, G., et al.: Trait-related cortical-subcortical dissociation in bipolar disorder: analysis of network degree centrality. J. Clin. Psychiatry **78**(5), 584–591 (2017)

31. Costenbader, E., Valente, T.W.: The stability of centrality measures when networks are sampled. Soc. Netw. **25**(4), 283–307 (2003)

32. Page, L., Brin, S., Motwani, R., Winograd, T.: The pagerank citation ranking: bringing order to the web. Technical report, Stanford InfoLab (1999)

33. Pretto, L.: A theoretical analysis of google's pagerank. In: String Processing and Information Retrieval, pp. 125–136. Springer (2002)

Detecting Epileptic Seizures Using Abe Entropy, Line Length and SVM Classifier

Aya Naser[✉], Manal Tantawi, Howida Shedeed,
and Mohamed F. Tolba

Scientific Computing Department, FCIS-Ain Shames University, Cairo, Egypt
aya.naser@cis.asu.edu.eg

Abstract. Epilepsy is a 4th prevalent neurological disorder which affects the individuals in all ages around the world. Epilepsy disorder is characterized by the abnormal movements of human muscles, called seizure, as a result of the abnormality in the brain electrical activity. The electroencephalogram (EEG) can serve as a powerful tool for detecting Epilepsy. In this paper, the most commonly used Andrzejak database is utilized for building an automated system for epilepsy detection. Digital Wavelet Transform (DWT) is applied on the segmented EEG signals to extract the five EEG sub-bands (delta, theta, alpha, beta, and gamma). Approximation and Abe entropies along with line length are calculated for the extracted sub-bands. Support Vector Machine (SVM) classifier with Radial Basis Function (RBF) kernel function is used to distinguish between three classes: (1) normal, (2) interictal (seizure free interval), and (3) ictal (during seizure). The best accuracies achieved are 93.75%, 98.75% and 98.125% for normal, interictal and ictal classes respectively. These accuracies are achieved using the combination of both Abe entropy and line length features together.

Keywords: Electroencephalogram (EEG) · Epilepsy · Seizure · Entropies ·
Line length · Digital Wavelet Transform (DWT) ·
Support Vector Machine (SVM)

1 Introduction

Epilepsy is a common disorder that affects the most important vital organ of humans (the brain) which controls the regularity of human muscles and nerves. Approximately 50 millions of all ages around the world suffer from epilepsy and roughly 70% of the epileptic patients respond to the treatment [1]. Epilepsy disorder is characterized by recurrent and unpredictable interruptions of human functions as a result of the abnormality of the brain electrical activity called epileptic seizure [2]. There are no fixed symptoms for the epileptic seizure, some patients lose their consciousness while others suffer from body shaking [2]. These attacks may threaten the patient's life [1]. The position of the abnormal brain activities which may be in specific area or include the whole brain is related to the seizure symptoms [2]. These seizures attack the patient suddenly and the reason for epilepsy is still ambiguous [2]. Thus, any forward step towards diagnosing and treating this disease is highly demanded.

© Springer Nature Switzerland AG 2020
A. E. Hassanien et al. (Eds.): AMLTA 2019, AISC 921, pp. 169–178, 2020.
https://doi.org/10.1007/978-3-030-14118-9_17

EEG signals demonstrate the spontaneous electrical activity of the brain. It is an indispensable diagnostic tool in neurological clinics. It is mostly used for diagnosing diseases such as epilepsy, sleep disorders, coma and brain death [3]. It is measured by placing a set of electrodes on the scalp [3]. There are no specific criteria to analyze EEG signals and the visual scanning is inaccurate, insufficient, and time consuming [4]. Therefore, a robust automated system to diagnose epilepsy is required. The beginning of the automated systems was in the second half of the twentieth century. Their promising results encouraged more and more studies to be emerged [2–16]. However, due to the difficulty of discriminating interictal class (out of seizure time), most of the existing studies consider only two classes: normal and ictal (during seizure time).

In this paper, the three classes: normal, interictal and ictal are investigated. Abe entropy is suggested as feature for discriminating between the three classes. The efficiency of Abe entropy has been compared to approximation entropy (APEN) and line length features which have been successfully used for discriminating between normal and ictal classes in the existing studies. Moreover, combinations of line length & APEN and line length & Abe were examined for improving results of the three classes. Support Vector Machines (SVM) classifier has been utilized for recognizing the three classes. Andrzejak EEG database [16] was considered for training and testing purposes.

The remaining of this paper is organized as follows: Sect. 2 discusses the existing related studies, the details of the proposed methodology are presented in Sect. 3, Sect. 4 provides the results and finally, the conclusion is provided in Sect. 5.

2 Related Work

Recently, many studies [3, 5–9] were developed for epilepsy detection and they provide very promising results. In this section, a summary for the key studies is given as follows:

Srinivasan et al. [5] divided each EEG signal of two sets from Dr. R. Andrzejak database [16] (one for normal and the other for ictal) into N segments. For each segment, approximation entropy (APEN) was calculated to measure the regularity and the predictability of the signal. Elman Network (EN) classifier (a special type of recurrent neural networks) and the Probabilistic Neural Network classifier (PNN) with two middle layers called radial basis and competitive layers were used to recognize normal and ictal classes. The achieved accuracy was 99.3% to 100% using EN, and 98% to 100% accuracy using PNN.

In order to extract the EEG signal sub-bands, Guo et al. [6] applied a four-level Digital Wavelet Transform (DWT) to EEG signals from Dr. R. Andrzejak database [16]. Line length feature was calculated for the extracted five sub-bands (delta, theta, alpha, beta, and gamma) to measure their irregularity. Multi-Layer Perceptron (MLP) artificial neural network was utilized for classification process. Three classification problems were examined: (1) normal versus ictal (Z-S), (2) normal and interictal versus ictal (ZNF-S), and (3) normal and interictal versus ictal (ZONF-S). An accuracy of 99.6%, 97.75% and 97.77% were achieved for the three problems respectively.

Later on, Guo et al. [7] examined three multi-wavelets to extract the five sub-bands (delta, theta, alpha, beta, and gamma): (1) Gernoimo–Hardin–Massopust (GHM), (2) Chui–Lian (CL), and (3) SA4. Approximation entropy was calculated for each sub-band. The calculated features were fed to three-layer MLP neural network with Bayesian regularization back propagation training algorithm. Two classification problems were examined: (1) normal versus ictal (Z-S), and (2) normal and interictal versus ictal (ZNF-S). Experiments were carried out using on Dr. R. Andrzejak database [16]. An accuracy of 99.85% was achieved using GHM and 98.27% was achieved using SA4 for the two classification problems respectively.

Long-duration EEG data were collected by Shen et al. [3]. Digital band pass filter (0.1–70 Hz), a notch filter (60 Hz), and a Daubechies 4 wavelet filter were applied to both unipolar and bipolar collected data. Approximation entropy and other statistical features were computed for each channel. SVM classifier with different kernel function was utilized for classification process. They achieved 98.91% accuracy with grid SVM which distinguishes between normal, interictal, and ictal classes. However, the testing data was biased toward the normal class (normal test samples are greater than other two classes).

The continuous 23-channel EEG signals from CHB-BIT database [17] were segmented by Ibrahim et al. [8] to get two-second EEG segments. Digital Wavelet Transform (DWT) was applied to extract four sub-bands from each segment. Shannon entropy was calculated for each sub-band and for the original EEG segment. Moreover, the standard deviation is calculated for the original segment. The K Nearest Neighbor (KNN) classifier was utilized for classification between normal and ictal classes. The best result achieved was 94.5% accuracy.

Gajic et al. [9] applied the Digital Wavelet Transform (DWT) to extract the five sub-bands from the EEG signals. Energy, entropy, mean and standard deviation were computed for each sub-band and also the standard deviation of each sub-band after it was reconstructed again resulting in a 25-feature vector. The feature vector was reduced by Karhunen-Loeve Expansion method. Divide and conquer strategy was utilized for classification between three sets of Dr. R. Andrzejak database [16]. This strategy was applied by using two quadratic classifiers: the first one distinguishes between normal class and the others, while the other one distinguishes between interictal and ictal classes. The best accuracy achieved was 99%.

From the mentioned studies we can observe that: (1) most of the exiting studies ignore the interictal class and discriminate only between normal and ictal classes, on the other hand, some studies consider it as normal or ictal [6, 7], (2) Approximation entropy (APEN) has been successfully used for classifying normal and ictal classes, (3) R. Andrzejak database [16] is the most commonly used database for the task in hand, and (4) most of the existing studies considered only some of the datasets of Andrzejak database, not all of them. In this study, the whole database is used for distinguishing between the three classes: normal, interictal and ictal.

3   Methodology

The proposed method in this paper consists of four main steps: (1) segmentation, (2) EEG sub-bands extraction, (3) features extraction, and (4) classification (see Fig. 1). These four steps are discussed in the following subsections.

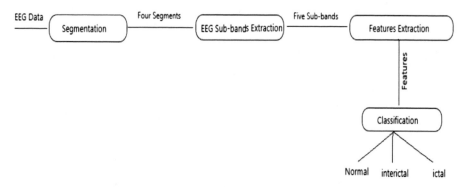

Fig. 1. The proposed methodology EEG based epilepsy detection

3.1   Segmentation

The EEG signal is segmented to N segments. Following Srinivasan et al. [5], the best value of N is four. Figure 2 shows the segmentation of an EEG signal of 4097 samples to four segments, each one of them is 1024 samples.

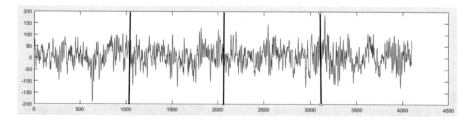

Fig. 2. Segmentation of the EEG signal to four segments

3.2   EEG Sub-bands Extraction

In this work, the wavelet analysis is used to extract the five primary EEG sub-bands: Delta (0–4 Hz), Theta (4–8 Hz), Alpha (8–16 Hz), Beta (16–32 Hz) and Gamma (32–64 Hz). The wavelet transform is an efficient analysis tool for dealing with the non-stationary signals such as EEG signals. It performs a multi-resolution analysis. It provides high time (low frequency) resolution at high frequencies. On the other hand, it provides high frequency (low time) resolution at low frequencies [6].

Digital Wavelet Transform (DWT) is performed by passing the input signal by series of low and high pass filters. DWT decomposes the original EEG signal into approximation (A) and details (D). The decomposition is repeated through levels on the previous approximation parts (see Fig. 3). In this work, seven-level decomposition is needed using db4 mother wavelet (fourth wavelet in Daubechies family) Srinivasan et al. [5]. Consequently, the approximation part (A7) and details part (D4–D7) are used to represent the five sub-bands: Delta (A7), Theta (D7), Alpha (D6), Beta (D5) and Gamma (D4).

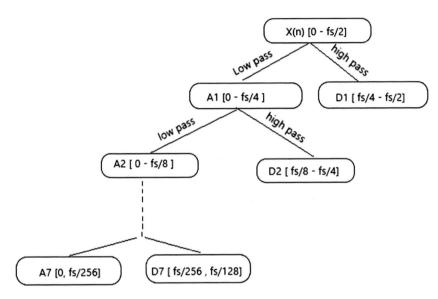

Fig. 3. Seven-level digital wavelet decomposition (A: Approximation Part, D: Details part, and Fs sampling frequency).

3.3   Feature Extraction

Approximation entropy, Abe entropy and line length are computed for each of the five sub-bands extracted from an EEG segment as follows:

Approximation Entropy (APEN)
APEN is a common feature which was developed to analyze medical data [19]. It can provide an indicator to the regularity and the predictability of time series data. Unlike other features, it works efficiently with small data and is uninfluenced to noise [5]. Hence, small values of APEN represent the regularity in the data, but the large values represent the irregularity. The following steps show how to calculate it:

1. Let the input vector X has N coefficients: [X(1), X(2), X(3), X(4),.......... X(n)]
2. Generate the embedded dimensional vector: [X(i), X(i + 1), ..., X(N−M + 1)]
3. Calculate the distance: $max_{k=1,2,....m}|X(i+k-1) - X(j+k-1)|$

4. Calculate the stimulatory calculation: $\theta^m(r) = \frac{1}{N-M+1}\sum_{i=1}^{N-M+1}\ln\left(\frac{N^m(i)}{N-M+1}\right)$ where m is the window length and r is the tolerance
5. Increase M to M + 1 and repeat steps from 2 to 4
6. $APEN(M, r, N) = \theta^m(r) - \theta^{m+1}(r)$.

Abe Entropy
Abe entropy [18] is one of the measurements that generalize the well-known Boltzmann–Gibbs entropy. In addition, it is asymmetric modification of the Tsallis entropy. It is calculated as follows:

$$S_q^{abe} = -\sum_i \frac{p_i^q - p_i^{q^{-1}}}{q - q^{-1}} \tag{1}$$

where p is wavelet coefficient and q is the order.

Line Length
Line Length [19] feature is derived from the fractal dimension. However, it is computationally more efficient and it was successfully used in epileptic seizure detection [19]. Its value increases as the variations of the signal increases and vice versa. Its value is calculated by adding the amplitude difference between every consecutive samples of the EEG signal.

$$L = \frac{1}{N-1}\sum_{i=1}^{N-1} abs(X_{i+1} - X_i) \tag{2}$$

where X_{i+1} and X_i are two consecutive wavelet coefficients and N is the total number of coefficients.

4 Classification

Support Vector Machines (SVM) classifier with RBF kernel function is used to discriminate between three classes: normal, interictal, and ictal. SVM is the most commonly used classifier in biomedical classification. It tends to maximize the gap between different classes using set of vectors defined from the training data called support vectors. For Linear classification problems, it uses the set of support vectors for building hyperplanes (Eqs. 3 and 4) to separate between two different classes.
    The hyperplane equations are:

$$W^T X_i + b = +1 \, for \, W^T X_i + b > +1 \tag{3}$$

$$W^T X_i + b = -1 \, for \, W^T X_i + b < -1 \tag{4}$$

where

  W^T is the transpose of the weight vector
  W, X_i is the input data, and b is bais.

However, for non-linear classification problem, SVM uses kernel function (Eq. 5) such as radial basis function (Eq. 6) first to map the nonlinear data to linear space.

The kernel function equation:

$$K(X_i, X_j) = (X_i, X_j) \tag{5}$$

where the X_i is the feature vector and X_j is the kernel center.

The radial bases function:

$$K(X_i, X_j) = \exp(\frac{-\|X_i - X_j\|^2}{2\sigma^2})) \tag{6}$$

where $\|X_i - X_j\|^2$ is the square euclidean distance between X_i and X_j and σ is the width of kernel function.

Regarding the task in hand, one versus one classification strategy has been utilized for solving our three-class classification problem. Hence, three binary SVM classifiers are used; the first one discriminates between the normal and the interictal classes, the second one discriminates between the normal and ictal classes and finally, the third one classifies between the inter ictal and ictal classes (see Fig. 4). During testing, the final decision is drawn by majority voting. However, if majority voting fails, the winner will be the classifier which provides the maximum confidence score.

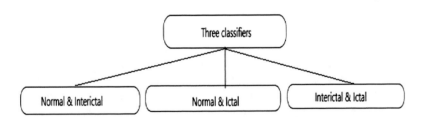

Fig. 4. One versus one classification strategy

5   Results

The performance of the proposed method is evaluated using Dr. R. Andrzejak database [10]. It consists of five datasets. Sets Z and O were collected form healthy volunteers in relaxed state with eye open (Set Z) and eye close (set O). Sets F and N were collected from interictal volunteers (out of seizure time). Finally, set S which was collected form ictal (during seizure) patients. Each set includes 100 EEG signal channels of 23.6 s duration. The length of each signal is 4097 samples [5]. In this paper, four classification problems are considered: Z-F-S, Z-N-S, O-F-S, and O-N-S. Each dataset is divided randomly into 60% training and 40% test.

After segmenting the EEG signals, the five EEG sub-bands are extracted from each segment as mentioned in Sect. 3.2. Features are computed for each sub-band. Features of the five sub-bands are concatenated together to form the final feature vector. Three binary SVM classifiers are trained in the way discussed in Sect. 4 to form a three-class SVM classifier. First, the considered features are evaluated separately. Table 1 shows the accuracy achieved by each feature for the four considered cases: (1) Z-F-S; (2) O-F-S; (3) Z-N-S; (4) O-N-S.

Table 1. Results achieved by the SVM classifier with APEN, Abe, and line length separately

Feature name	Case	Normal accuracy	Interictal accuracy	Ictal accuracy	Average accuracy
APEN	Z-F-S	62.5	49.375	65	58.9583
	O-F-S	28.75	65	63.75	52.5
	Z-N-S	53.75	50	66.875	56.875
	O-N-S	28.75	61.25	70.625	53.5417
Abe	Z-F-S	83.125	92.5	63.75	79.7917
	O-F-S	76.875	92.5	75.625	81.667
	Z-N-S	79.375	95.625	78.125	84.375
	O-N-S	76.25	95.625	86.875	86.25
Line length	Z-F-S	73.75	90	74.375	79.375
	O-F-S	75	86.875	61.25	74.375
	Z-N-S	70	66.25	73.75	70
	O-N-S	65	73.125	60	66.0417

As shown in Table 1, the best results are achieved by Abe entropy for all cases followed by line length. Although the APEN provides above 95% for normal and ictal classes in existing studies, it provides the worst results when interictal class is considered. The best values for RBF (kernel function) sigma are found empirically to be 1.5, 0.8, and 1 for APEN entropy, Abe entropy, and Line length, respectively.

Second, the combination of each type of entropies and line length features is examined with the SVM classifier. Table 2 shows the results achieved by the combination of APEN entropy & line length and the combination of the Abe entropy & line length features. By comparing Tables 1 and 2, it is clear that the considered combinations provide better results than each feature solely. Moreover, the results reveal the superiority of combining both Abe and line length features together. The best values for RBF (kernel function) sigma are found empirically to be 2.5 and 2 for the two combinations, respectively.

Table 2. The accuracy of combining each entropy and line length features using SVM classifier

Feature name	Case	Normal accuracy	Interictal accuracy	Ictal accuracy	Average accuracy
APEN and line length features	Z-F-S	80.625	88.125	75	81.25
	O-F-S	69.375	86.25	60.625	72.0833
	Z-N-S	76.25	76.25	81.875	78.125
	O-N-S	58.75	81.25	61.25	67.0833
Abe and line length features	Z-F-S	86.875	93.125	82.5	87.5
	O-F-S	93.75	93.125	87.5	91.4583
	Z-N-S	89.375	98.75	93.125	93.75
	O-N-S	93.75	98.75	98.125	96.875

6 Conclusion

To conclude, a method for epilepsy detection from EEG signals is proposed. The database presented by Dr. R. Andrzejak [10] is utilized for training and testing. The whole database with its five datasets (Z, O, F, N and S) is considered. The EEG signals are segmented in four segments with length of 1024 samples. The five EEG sub-bands (delta, theta, alpha, beta, and gamma) are extracted from each EEG segment using DWT. The proposed method suggests the use of Abe entropy along with line length as two features for discriminating the three classes: normal, interictal and ictal. The two features are fed into a three-class SVM classifier which is constructed by three binary SVM classifiers trained using one-versus-one strategy. The proposed method is evaluated by two experiments. First, the Abe entropy is evaluated against APEN entropy and line length separately. The results show the superiority of Abe entropy in providing the best results for the four cases (Z-F-S, O-F-S, Z-N-S and O-N-S). Second, two combinations are evaluated; the first one includes APEN entropy & line length and the other one includes Abe entropy & line length. The Abe entropy and line length combination provides the best results for all cases. In the future, we are looking forward to proving the efficacy of the proposed method on other databases.

References

1. WHO Homepage. http://www.who.int/news-room/fact-sheets/detail/epilepsy. Accessed Oct 2018
2. Naser, A., Tantawi, M., Shedeed, H.A., Tolba, M.F.: EEG based epilepsy detection using approximation entropy and different classification strategies. In: 8th International Conference on Intelligent Computing and Information Systems (ICICIS), pp. 92–97. IEEE, Egypt (2017)
3. Shen, C.P., Chan, C.M., Lin, F.S., Chiu, M.J., Lin, J.W., Kao, J.H., Chen, C.P., Lai, F.: Epileptic seizure detection for multichannel EEG signals with support vector machines. In: 11th International Conference on Bioinformatics and Bioengineering (BIBE), pp. 39–43. IEEE (2011)

4. Kiymik, M.K., Subasi, A., Ozcalık, H.R.: Neural networks with periodogram and autoregressive spectral analysis methods in detection of epileptic seizure. J. Med. Syst. **28**(6), 511–522 (2004)
5. Srinivasan, V., Eswaran, C., Sriraam, N.: Approximate entropy-based epileptic EEG detection using artificial neural networks. IEEE Trans. Inf. Technol. Biomed. **11**(3), 288–295 (2007)
6. Guo, L., Rivero, D., Dorado, J., Rabunal, J.R., Pazos, A.: Automatic epileptic seizure detection in EEGs based on line length feature and artificial neural networks. J. Neurosci. Methods **191**(1), 101–109 (2010)
7. Guo, L., Rivero, D., Pazos, A.: Epileptic seizure detection using multiwavelet transform based approximate entropy and artificial neural networks. J. Neurosci. Methods **193**(1), 156–163 (2010)
8. Ibrahim, S.W., Majzoub, S.: EEG-based epileptic seizures detection with adaptive learning capability. Int. J. Electr. Eng. Inf. **9**(4), 813–824 (2017)
9. Gajic, D., Djurovic, Z., Di Gennaro, S., Gustafsson, F.: Classification of EEG signals for detection of epileptic seizures based on wavelets and statistical pattern recognition. Biomed. Eng.: Appl. Basis Commun. **26**(02), 1450021 (2014)
10. Husain, S.J., Rao, K.S.: Epileptic seizures classification from EEG signals using neural networks. In: 2012 International Conference on Information and Network Technology (ICINT 2012), vol. 37, pp. 269–273, April 2012
11. Juarez-Guerra, E., Alarcon-Aquino, V., Gomez-Gil, P.: Epilepsy seizure detection in EEG signals using wavelet transforms and neural networks. In: New Trends in Networking, Computing, E-learning, Systems Sciences, and Engineering, pp. 261–269. Springer, Cham (2015)
12. Kannathal, N., Choo, M.L., Acharya, U.R., Sadasivan, P.K.: Entropies for detection of epilepsy in EEG. Comput. Methods Programs Biomed. **2**(81), 193 (2006). Comput. Methods Programs Biomed. **80**, 187–194 (2005)
13. Subasi, A.: Automatic detection of epileptic seizure using dynamic fuzzy neural networks. Expert Syst. Appl. **31**(2), 320–328 (2006)
14. Wang, D., Miao, D., Xie, C.: Best basis-based wavelet packet entropy feature extraction and hierarchical EEG classification for epileptic detection. Expert Syst. Appl. **38**(11), 14314–14320 (2011)
15. Nigam, V.P., Graupe, D.: A neural-network-based detection of epilepsy. Neurol. Res. **26**(1), 55–60 (2004)
16. Andrzejak, R.G., Lehnertz, K., Mormann, F., Rieke, C., David, P., Elger, C.E.: Indications of nonlinear deterministic and finite-dimensional structures in time series of brain electrical activity: dependence on recording region and brain state. Phys. Rev. E **64**(6), 061907 (2001)
17. CHB-BIT database. https://physionet.org/pn6/chbmit/
18. Beck, C.: Generalised information and entropy measures in physics. Contemp. Phys. **50**(4), 495–510 (2009)
19. Koolen, N., Jansen, K., Vervisch, J., Matic, V., De Vos, M., Naulaers, G., Van Huffel, S.: Line length as a robust method to detect high-activity events: automated burst detection in premature EEG recordings. Clin. Neurophysiol. **125**(10), 1985–1994 (2014)

Analyzing Electrooculography (EOG) for Eye Movement Detection

Radwa Reda[✉], Manal Tantawi[✉], Howida shedeed[✉],
and Mohamed F. Tolba[✉]

Scientific Computing Department, Faculty of Computer and Information Science,
FCIS-Ain Shams University, Cairo, Egypt
radwareda2012@gmail, manalmt2012@hotmail.com,
dr_howia@yahoo.com, Fahmytolba@gmail.com

Abstract. Although the cognitive parts of their brains are intact, some individual scan only interact with the outside environment through eye movements. Those people suffer from severe motor disabilities preventing them from moving all their limbs. Recently, Human Computer Interfaces (HCI) has emerged to help these people by providing them a new way for communication. These interfaces are based on detecting eye movements. Electro-oculogram (EOG) records eye movements through few electrodes placed around the eyes vertically and horizontally. In this paper, EOG vertical and horizontal signals are analyzed to detect four eye movements (left, right, up and down) along with blinking. Three statistical features are extracted from filtered EOG signals. Extracted features from horizontal and vertical EOG signals are concatenated to form final feature vector. K Nearest Neighbor (KNN), Linear Discriminant Analysis (LDA), Multinomial Logistic Regression (MLR), Naïve Bayes (NB), Decision Trees and Support Vector Machines (SVM) are six classifiers that are evaluated in this study. The results reveal the superiority of SVM Classifier in providing the best average accuracy.

Keywords: Electro-oculogram (EOG) · Human computer interface ·
Statistical features · Support Vector Machine

1 Introduction

There are many patients all over the world who suffer from motor neuromuscular dysfunction that leads to weakness of the voluntary muscles and thus impede their limbs from performing their vital functions. Unfortunately, the numbers of these patients are increasing as a result of the existence of many diseases that cause this severe motor (temporary or permanent) disability. One of the most common diseases is a Amyotrophic LateralSclerosis (ALS) that causes loss of the nerve cells that control movement [1, 2]. In early ALS, the muscles of the hands, fingers, and legs become weak and thin. As time passes, the symptoms get worse until voluntary breathing becomes difficult in the end stage, this progressive weakness is leading to death due to respiratory failure.

© Springer Nature Switzerland AG 2020
A. E. Hassanien et al. (Eds.): AMLTA 2019, AISC 921, pp. 179–189, 2020.
https://doi.org/10.1007/978-3-030-14118-9_18

According to the Amyotrophic Lateral Sclerosis Society (LALS), between 5,660 and 6,400 new cases are diagnosed per year and the motor paralysis in most cases (90% of cases) occur without any genetic history or any injury in the family [3–5].

As Also, many other diseases that cause motor paralysis of limbs because of several environmental factors were associated with injury such as exposure to poisons found in some food, smoking [6, 7], exposure to heavy metals(arsenic, mercury and lead), electromagnetic waves, insecticides and viruses such as Guillain–Barré Syndrome (GBS) [8], and Multiple Sclerosis [9]. The result of all these diseases, known or unknown is the inability of patients to communicate with their external environment. Hence, they are unable to perform daily life functions or express themselves.

However, although the limbs of those patients are paralyzed completely, eye movements are usually spared until the final stage. Thus, many patients with these diseases use communication supporting tools based on eye movement, which helps them through interfaces to take action [10]. Human Computer/Machine Interfaces (HCI/HMI) have emerged in the last few decades to act as an assistant for those patients. These interfaces help patients to overcome their disabilities and improve their quality of life by providing them a facility to communicate with their surrounding through eye movements. Most of the interfaces consider four Eye movements: left, right, up and down movements along with blinking. On the other hand, few interfaces consider also the four corners: up left, up right, down left and down right. Detection of eye movements can allow user for example to select an action displayed on a screen, write text using virtual keyboard or move a wheel chair, etc. [10].

An Electro-oculogram (EOG) is a bio-potentialsignal that measures the potential difference between the retina and the cornea [11]. The eyeball can be modeled as a dipole with the positive cornea in the front and the negative retinain the back.

As shown in Fig. 1, positive or negative pulses will be generated when the eyes roll upward or downward. The amplitude of pulse will be increased with the increment of rolling angle, and the width of the positive (negative) pulse is proportional to the duration of the eyeball rolling process [12, 13].Simultaneously, the EOG is recorded in the two directions horizontal and vertical using five electrodes (two vertical, two horizontal and one for ground) placed around the eyes.

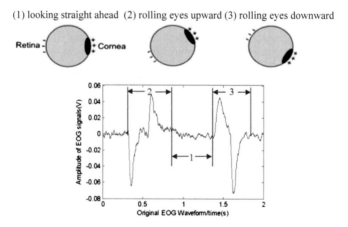

Fig. 1. Eye upward and downward movements and the corresponding waveform

In this paper, EOG signals are considered for detecting the four main eye movements (left, right, up and down) along with blinking. Both horizontal and vertical signals are filtered to maintain only the EOG band. From each signal (horizontal or vertical), three statistical features (mean, variance and energy) are extracted. Horizontal and vertical features are concatenated to form a final feature vector of six features. Six classifiers are investigated in this study: K-Nearest Neighbor (KNN), Linear Discriminate Analysis (LDA), Multinomial Logistic Regression (MLR), Naïve Bayes (NB), Decision Trees and Support Vector Machines (SVM). All experiments were carried out using the dataset provided by Usakli and Gurkan [14].

The remaining of this paper is organized as follows. Section 2 discusses the existing studies, the proposed methodology is provided in Sects. 3 and 4 presents the achieved results, and finally the conclusion is provided in Sect. 5.

2 Related Work

Recently, HCI becomes the hotspot of bio-based HCI research. Many studies have emerged in the last two decades for EOG based HCI that depends on classifying the eye blink along with the four directional eye movements or the eight directional eye movements. Some of these key studies can be summarized as follows. Deng et al. [15] developed an eye-tracking system that classifies four different directions of eye movement. Analog signals are converted to digital signals to be used as control signals for HCI. Fuzzy distinction rules were considered for classification. System performance was tested by measuring accuracy of eye direction detection. Eye direction was used to control a virtual device on computer screen such as TV system. The system has achieved 90% accuracy.

Merino et al. [16] developed a system to detect eye movement based on the EOG signal. They used Ag/AgCl sensors for EOG acquisition. Since EOG signal information is mainly contained in low frequencies, band pass filter with a range between 0.1 and 30 Hz was used to filter the signals. Noise was further removed by an averaging filter. They used an algorithm that depends on a threshold value of the amplitude to distinguish between a deliberate movement and a non-deliberate one. Moreover, the derivative of EOG signal was utilized to detect eye movements that can have three different values: a value greater than 0, a value smaller than 0 and a value equal to 0. The values that are not equal to 0 are related to initial and final edges of a pulse. The value equal to 0 is associated with the area between edges. A pulse was classified as a blink if the width of this area was smaller than 250 ms. For the vertical signal, up movement was detected when initial edge is positive and final edge is negative and the opposite for down movement. In the same way, the right and left movements can be detected from the horizontal signal. Average hit rate achieved was 94.11%.

Nathan et al. [17] designed an EOG based typing system. It used a virtual keyboard for typing letters on a monitor using EOG patterns. EOG signals were filtered using a 2–30 Hz band pass filter. If the amplitude is greater than a threshold, then the amplitude of the signal is checked within the next 0.078 s. If it is less (greater) than $-50 \mu V$ then it is identified as up (down) movement. The left and the right movements were classified in the same way but from the horizontal signal. The system achieved typing speed of

15 letters/min and an accuracy of 95.2%. Usakli and Gurkan in [14] developed and realized a virtual keyboard that allowed the user to write messages based on EOG signals. Five Ag/AgCl electrodes were used for EOG acquisition. Band pass filter with a range between 0.1 and 30 Hz was used to filter the signals. After filtering and amplification, EOG signals are digitized and transferred to PC. Nearest Neighborhood algorithm was employed to distinguish eye movements in four directions as well as blinking for typing in a virtual keyboard. By using this virtual keyboard, the user could type with a speed of 5 letters/25 s and an accuracy of 95% [18].

Aungsakul et al. [19] proposed 14 features extracted from specific characteristics of two directional EOG signals(vertical and horizontal) such as the maximum peak and valley amplitude values (PAV and VAV), and the maximum peak and valley position values (PAP and VAP), etc. the testing results for the eye movement detection system in eight directions revealed that the differences in mean features between the movements were statistically significant for ten features ($p < 0.0001$) including PAV, VAV, PAP, and VAP. Moreover, the classification accuracies based on a distinction feature rule from the proposed features approached 100% for some subject testing [20]. However, the number of features used in the algorithm was quite high and the distinction feature rule was quite complicated.

It can be observed that most of the existing studies considered only one classifier in their evaluation. In this study, for more robustness and reliability, statistical features are considered with different classifiers to classify the four eye movement directions along with blinking from both vertical and horizontal EOG signals.

3   The Proposed Method

The proposed approach includes three main stages namely preprocessing, feature extraction and classification as shown in Fig. 2. The following subsections provide a detailed discussion for each stage.

Fig. 2. Proposed approach for eye movement detection

3.1   Preprocessing

In general, the acquired eye signals are contaminated by noise and artifacts. Hence, preprocessing is needed to avoid the eye blinks, small involuntary EOG movements, background noise and power interference. First, eye signals are filtered using Butterworth band pass filter to keep only the EOG band [0.5–20]. Figure 3 shows an eye signal before and after filtering. Second, in order to reduce computation, signals are down sampled such that each signal becomes only 50 samples.

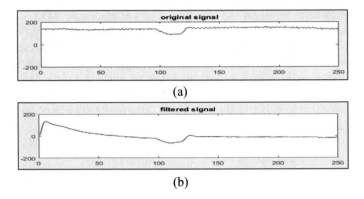

Fig. 3. EOG signal (a) before filtering and (b) after filtering

3.2 Feature Extraction

Statistical features have been successfully used in literature for representing biological signals in time domain. In this paper, three features are computed from each eye signal (horizontal or vertical), which are mean, standard deviation, energy. The features are calculated as follows:

$$\mu_i = \sum_{i=1}^{N} X_i \tag{1}$$

$$\sigma_i = \sum_{i=1}^{N} (X_i - \mu_i)^2 \tag{2}$$

$$E = \sum_{i=1}^{N} X_i^2 \tag{3}$$

Where (X_i) is eye signal, (μ_i) is the mean of signals, (σ_i) standerd deviationof signals and (E) energy of signals. Subsequently, the extracted features from both horizontal and vertical eye signals are concatenated to form final feature vector which consists of six features (three horizontal and three vertical).

3.3 Classification

Regarding the classification phase, the extracted feature vectors from the previous stage are fed into a classifier for training and testing purposes. Six classifiers are examined in this study to discriminate between five eye movements: left, right, up, down and blinking. The six classifiers are support K-Nearest Neighbors (KNN), Linear Discriminate Analysis (LDA), Multinomial Logistic Regression (MLR), Naive Bayes (NB), Decision Trees and Support Vector Machines (SVM). Each classifier can be discussed briefly as follows:

K-Nearest Neighbor (KNN) [21]: is considered among the oldest non-parametric classification algorithms. To classify an unknown example, the distance from that example to every other training example is measured. The k smallest distances are identified, the most represented class by these k nearest neighbours is considered the output class label.

Linear Discriminate Analysis (LDA) [22]: The basic idea of LDA is to find a transformation that best distinguishes between classes and the classification is then done in the transformed space based on some metric such as Euclidean distance. Mathematically, the implementation of a typical LDA is performed via scatter matrix analysis. In multi-classes case, a natural extension of linear discriminate exists using multiple discriminate analyses. As in two-class case, transformation from high dimensional space to low dimensional space is done under condition that the transformation maximizes the ratio of intra-class scatter to the inter-class scatter. The maximization should be performed among several competing classes. For n classes: intra-class matrix (within-class scatter)

$$S_w = \sum\nolimits_{i=1}^{n} S_i$$
$$\text{Where } S_i = \sum\nolimits_{X \in w_i} \left(x - x_i^-\right)\left(x - x_i^-\right)^T \tag{4}$$

Inter-class matrix (between-class scatter)

$$S_b - \sum\nolimits_{i=1}^{n} m_i\left(x_i^- - x^-\right)\left(x_i^- - x^-\right)^T \tag{5}$$

Where $\left(x_i^-\right)$ is the mean for each class and (x^-) is total mean vector given by $x^- = \frac{1}{m}\sum_{i=1}^{n} m_i x_i^-$, (w_i). The class separation direction and $\left(\vec{w}m_i\right)$ is the number of training samples for each class.

Multinomial Logistic Regression (MLR) [23]: generalizes logistic regression to multiclass problems. Given a set of independent variables, the probabilities of the various possible outcomes of a categorically distributed dependent variable are predicted. MLR can be defined by a one layer neural network [24]. The target class labels are encoded via one-hot encoding. Then, for the prediction step after learning the model, the "argmax," the index in the output vector with the highest value is returned as the class label. The main benefit of MLRis that it provides true probability where the sum of all network outputs is 1.

Naive Bays (NB) [25]: is one of simple "probabilistic classifiers" family based on applying Bayes' theorem with strong (naive) independence assumptions between the features. It is based upon the principle of maximum a posteriori (MAP). This approach can be extended to the case of more than two classes represented as vectors of feature values and the class labels are drawn from some finite set. For some types of probability models, Naive Bays classifiers can be trained using supervised learning, given a problem instance to be classified, represented by a vector $x = (x1,....xn)$ representing some n features (independent variables), then it assigns to this probabilities for each class (C_k):

$$p(C_k|x1,\ldots,xn) \tag{6}$$

Where (K) is number class. The problem with the above formulation is that if the number of features n is large, the model is reformulated to make it more tractable. The conditional probability can be decomposed as:

$$(\text{Posteriori})p(C_k|x) = \frac{p(C_k)p(x|C_k)}{p(x)} \tag{7}$$

Decision Trees [26]: the training data are split by the tree according to the values of the considered features with the aim to give the best possible generalization. The algorithm can work on binary or multiclass classification problems. The tree is a flowchart-like structure where each internal node refers to a "test" on feature, each branch refers to the result of the test, and finally each leaf node refers to a class label. Classification rules are referred by the paths from root to leaf. Hence, the decision tree can berepresented as decision rules where the result is the content of the leaf node and the conditions along the path form a conjunction in the if clause.

Support Vector Machine (SVM): SVMs are mainly two-class classifiers. SVM can efficiently generalize, since it reduces the complexity of the learning model. SVM constructs a hyperplane that maximizes the margin between various classes. The adjacent points of various classes are called support vectors [27]. The decision function is built to be used in testing as follows:

$$f(x) = \sum_{i=1}^{n} \alpha_i - (1/2) \sum_{i=1 \, j=1}^{n} \alpha_i \alpha_j y_i y_j K(x_i, x_j) \tag{8}$$

where $\left\{ (x_i, y_j), i = 1 \ldots N \right\}$ are the features vector, $a_i \alpha_i$ is the Lagrange multiple, $K(x, y) = (x^T y + 1)^d$ is the kernel function and (d) degree of polynomial kernel. Multi-class SVM can be constructed from binary SVM classifiers using one-versus-all classification strategy [27].

4   Experimental Results

Experiments are conducted using the dataset provided by Usakli and Gurkan [14]. The dataset includes data for the five considered classes: left, right, up, down and blinking. For each class, 20 pair of signals (horizontal & vertical) is provided. EOG signals were sampled at 176 Hz. In addition, the length of each signal is 251 samples (approximately 1.5 s). The data for each class is divided into four groups. Three groups for training and the fourth one for testing resulting in four trails (each group of each class appears once as a testing group). Tables 1, 2, 3, 4, 5 and 6 show the results achieved by each considered classifier. In addition, Fig. 4 provides a summary for the average achieved results using the six classifiers. Regarding SVM, MLR, KNN and LDA classifiers, the best results are achieved by using 2nd order band pass filter. On the other hand, the best results are achieved by using 5th order band pass filter using NB and Decision Trees classifiers. The accuracy of each class is calculated by the following equation:

$$\text{Accuracy} = \frac{\text{The number of samples correctly classified}}{\text{Total number of testing cases for specific class}} * 100 \qquad (9)$$

Table 1. The achieved accuracies using KNN classifier

Trials	Down	Up	Left	Right	Blink	Average Accuracy
Trial 1	100%	60%	100%	100%	100%	**92%**
Trial 2	80%	60%	80%	60%	100%	**76%**
Trial 3	80%	60%	100%	60%	100%	**80%**
Trial 4	60%	50%	100%	80%	100%	**78%**
Total The average accuracy: 81.5%						

Table 2. The achieved accuracies using LDA classifier

Trials	Down	Up	Left	Right	Blink	Average Accuracy
Trial 1	100%	80%	60%	80%	80%	**80%**
Trial 2	100%	60%	100%	80%	50%	**78%**
Trial 3	80%	65%	80%	80%	65%	**74%**
Trial 4	80%	60%	100%	100%	100%	**88%**
Total The average accuracy: 80%						

Table 3. The achieved accuracies using MLR classifier

Trials	Down	Up	Left	Right	Blink	Average Accuracy
Trial 1	100%	50%	60%	40%	50%	**60%**
Trial 2	100%	60%	100%	80%	80%	**84%**
Trial 3	60%	100%	80%	80%	100%	**84%**
Trial 4	80%	60	100%	100%	100%	**88%**
Total The average accuracy: 79%						

Table 4. The achieved accuracies using Naive Bayes classifier

Trials	Down	Up	Left	Right	Blink	**Average Accuracy**
Trial 1	100%	80%	80%	60%	50%	**74%**
Trial 2	80%	100%	50%	80%	100%	**82%**
Trial 3	80%	100%	80%	100%	60%	**84%**
Trial 4	80%	60%	100%	100%	100%	**88%**
Total The average accuracy: 82%						

Table 5. The achieved accuracies using Decision Trees classifier

Trials	Down	Up	Left	Right	Blink	Average Accuracy
Trial 1	100%	50%	100%	80%	60%	**78%**
Trial 2	100%	50%	50%	40%	60%	**60%**
Trial 3	100%	80%	80%	80%	60%	**80%**
Trial 4	80%	80%	100%	80%	100%	**88%**
Total The average accuracy: 76.5%						

Table 6. The achieved accuracies using SVM classifier

Trials	Down	Up	Left	Right	Blink	Average Accuracy
Trial 1	100%	100%	90%	85%	100%	**95%**
Trial 2	85%	90%	100%	100%	85%	**92%**
Trial 3	85%	100%	100%	85%	100%	**94%**
Trial 4	100%	100%	100%	80%	100%	**96%**
Total The average accuracy: 94.25%						

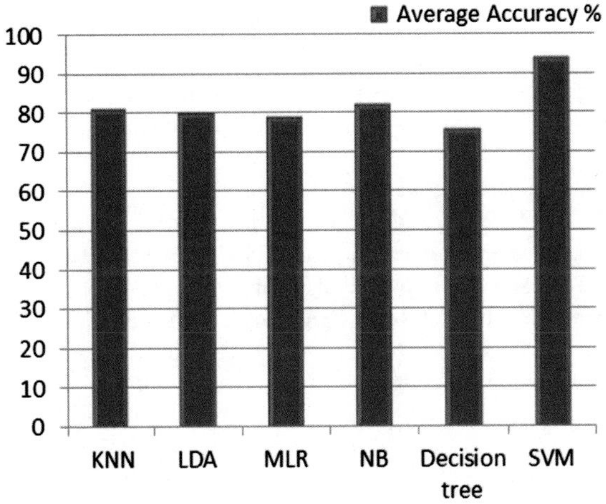

Fig. 4. Summarizes the results achieved by the six classifiers.

5   Conclusion

To sum up, this paper presents a method for detecting eye movements based on EOG signals. Five classes are considered: left, right, up, down and blinking. The EOG signals are filtered and then three statistical features: mean, standard deviation and energy are extracted for each EOG signal. Features of both vertical and horizontal EOG signals are concatenated to form a final feature vector of six features. Six classifiers have been examined. The results have revealed the superiority of the multi-class SVM classifier. The best results achieved for the four trails are 95%, 92%, 94% and 96%. In future, we are looking forward to testing the proposed approach with other bigger datasets.

References

1. Mitchell, J.D., Borasio, G.D.: Amyotrophic lateral sclerosis. Lancet **369**, 2031–2041 (2007)
2. Shaw, P.J.: Molecular and cellular pathways of neurodegeneration in motor neurone disease. J. Neurol. Neurosurg. Psychiatry **76**, 1046–1057 (2005)
3. Beleza-Meireles, A., Al-Chalabi, A.: Genetic studies of amyotrophic lateral sclerosis: controversies and perspectives. Amyotroph Lateral Scler. **10**(1–14), 27 (2009)
4. Dion, P.A., Daoud, H., Rouleau, G.A.: Genetics of motor neuron disorders: new insights into pathogenic mechanisms. Nat. Rev. Genet. **10**, 769–782 (2009)
5. Alonso, A., Logroscino, G., Hernan, M.A.: Smoking and the risk of amyotrophic lateral sclerosis: a systematic review and meta-analysis. J. Neurol. Neurosurg. Psychiatry **81**(11), 1249–1252 (2010)
6. Wang, H., O'reilly, E.J., Weisskopf, M.G., et al.: Smoking and risk of amyotrophic lateral sclerosis: a pooled analysis of 5 prospective cohorts. Arch. Neurol. **68**, 207–213 (2011)
7. Miller, R.G., Mitchell, J.D., Moore, D.H.: Riluzole for amyotrophic lateral sclerosis (ALS)/motor neuron disease (MND). Cochrane Database Syst. Rev. **3**, CD001447 (2012)
8. Sejvar, J.J., Baughman, A.L., Wise, M., Morgan, O.W.: Population incidence of Guillain-Barré syndrome: a systematic review and meta-analysis. Neuroepidemiology. **36**(2), 123–133 (2011). Accessed 04 Dec 2014
9. Milo, R., Kahana, E.: Multiple sclerosis: geoepidemiology, genetics and the environment. Autoimmun. Rev. **9**(5), A387–A394 (2010)
10. Majaranta, P.; Räihä, K.-J.: Twenty years of eye typing: Systems and design issues. In: Proceedings of the 2002 symposium on Eye tracking research & applications, pp. 15–22. ACM (2002)
11. Webster, J.: Medical Instrumentation: Application and Design. Wiley, Hoboken (2009)
12. Lv, Z., Wu, X.P, Li, M., Zhang, D.X.: Development of a human computer Interface system using EOG. Ministry of Education China, of Anhui University, Hefei, Anhui (2009)
13. Ya, T.O., Asumi, M.K.: Development of an input operation for the amyotrophic lateral sclerosis communication tool utilizing EOG. Med. Biol. Eng. **43**(1), 172–178 (2005). (in Japanese)
14. Usakli, A.B., Gurkan, S.: Design of a novel efficient human–computer interfacean electrooculagram based virtual keyboard. IEEE Trans. Instrum. Meas. **59**, 2099–2108 (2010)
15. Deng, L.Y., Hsu, C.L., Lin, TCh., Tuan, J.S., Chang, S.M.: EOG based human-computer interface system development. Int. J. Expert Syst. Appl. **37**(4), 3337–3343 (2010)
16. Merino, M., Rivera, O., Gomez, I., Molina, A., Dorrenzoro, E.: A method of EOG signal processing to detect the direction of eye movements. In: First International Conference of Sensor Device Technologies and Applications, Italy. pp. 100–105, July 18–25 (2010)
17. Nathan, D.S., Vinod. A.P., Thomas, P.K.: An electrooculogram based assistive communication system with improved speed and accuracy using multi-directional eye movements. In: 35th International Conference on Telecommunications and Signal Processing (TSP), pp. 554–558, Prague, Czech Republic (2012)
18. Usakli, A.B., Gurkan, S., Aloise, F., Vecchiato, G., Babiloni, F.: On the use of electrooculogram for efficient human computer interfaces. Comput. Intell. Neurosci. (2010)
19. Aungsakul, S., et al.: Evaluating feature extraction methods of electroocculography (EOG) signal for human-computer interface. In: Proceedings, 3rd International Science, Social Science, Engineering and Energy Conference, Nakhon Pathom, vol. 32, pp. 246–252, Thailand (2012)

20. Aungsakul, S., et al.: Robust eye movement recognition using EOG signal for human-computer interface. In: Proceedings, 2nd International Conference on Software Engineering and Computer Systems, Kuantan, vol. 180, pp. 714–723, Malaysia (2011)
21. Altman, N.S.: An introduction to kernel and nearest-neighbor nonparametric regression. Am. Stat. **46**(3), 175–185 (1992)
22. Johnson, A., Wichern, D.W.: Applied Multivariate Statistical Analysis. Prentice Hall, Upper Saddle River (1988)
23. Greene, W.H.: Econometric Analysis, 7th (edn.), pp. 803–806. Pearson Education, Boston (2012). ISBN 978-0-273-75356-8
24. Bishop, C.M.: Pattern Recognition and Machine Learning, pp. 206–209. Springer, New York (2006)
25. Mozina, M., Demsar, J., Kattan, M., Zupan, B.: Nomograms for visualization of naive bayesian classifier (PDF). In: Proceedings of PKDD 2004, pp. 337–348 (2004)
26. Quinlan, J.R.: Simplifying decision trees. Int. J. Man-Mach. Stud. **27**(3), 221 (1987)
27. Byun, H., Lee, S.-W.: Applications of support vector machines for pattern recognition: a survey. In: Pattern recognition with support vector machines, pp. 571–591(2002)

Analysis of Classification Methods for Gene Expression Data

Lamiaa Zakaria, Hala M. Ebeid$^{(\boxtimes)}$ (iD), Sayed Dahshan,
and Mohamed F. Tolba (iD)

Scientific Computing, Faculty of Computer and Information Sciences,
Ain Shams University, Cairo, Egypt
zakarialamiaa@gmail.com,
{halam,fahmytolba}@cis.asu.edu.eg

Abstract. The discovery of diseases at a molecular level is a great challenge for researchers in the field of bioinformatics and cancer classification. Understanding the genes that contribute to the cancer malady is a great challenge to many researchers. Cancer classification based on the molecular level investigation has gained the interest of researches as it provides a systematic, accurate and objective diagnosis for different cancer types. This Paper aims to present some classification methods for gene expression data. We compared the efficiency of three different classification methods; support vector machines, k-nearest neighbor and random forest. Two publicly available gene expression data sets were used in the classifications; Freije and Philips dataset. By performing the classification methods, results revealed that the best performance was achieved by using support vector machine classifier for both datasets comparing with other used classifiers.

Keywords: Gene expression · Classification

1 Introduction

Cancer is considered one of the deadliest genetic maladies of the human genome and has been the research interest in the field of medicine for the past decades. The World Cancer Report described cancer as a global problem because it affects the whole greater population. It was projected by the same report that cancer incidence will increase to 20 million new cases by 2025 [1]. There are several published researches for cancer classification that used new approaches for the molecular level. The important goal in cancer research is to identify the specific genes that contribute to cancer disease. Microarray has been extensively adapted to profile gene expression data of tumors and applied to cancer classification, but its success largely depends on the tools of data mining. Because, among too many gene expression data, only a part gives distinct expression levels for different disease types. Our study is about proposing a classification method for the gene expression data to achieve the best performance and accurate result.

© Springer Nature Switzerland AG 2020
A. E. Hassanien et al. (Eds.): AMLTA 2019, AISC 921, pp. 190–199, 2020.
https://doi.org/10.1007/978-3-030-14118-9_19

Several researchers have been studying the problem of cancer classification using data mining methods, machine learning algorithms and statistical methods to reach an efficient analysis for gene expression profiles [2, 3]. Studying the characteristics of thousands of genes simultaneously offered deep insight into cancer classification problem. Classifying gene expression data remains a difficult problem and an active research area due to their native nature of the data size "high dimensional low sample size" thereby resulting in great challenges to existing classification methods. It introduced an abundant amount of data ready to be explored. It has also been applied in a wide range of applications such as drug discovery, cancer prediction and diagnosis which is a very important issue for cancer treatment. DNA microarray is one of the fastest-growing new technologies that has empowered the study of gene expression in such a way that scientists can now measure the expression levels of large numbers of genes in a single experiment rather than performing many experiments and gathering data [4].

In this paper, we present a comparative study of classification methods on gene expression data using two different datasets. We find out two gene expression data sets that shared exactly the same genes for the same disease. Thus, we can study the stability of different methods. We study the impact of changing the number of used features in the classification methods, and how the change in the numbers of features effect in the classification accuracy.

This paper is organized as follows. In Sect. 2, we present the related work for classification methods using gene expression data. Section 3 describes the classification methods; KNN, SVM, and random forest. The description of datasets is given in Sect. 4.1 and Sect. 4.2 presents the results for our experiment. The conclusion is presented in Sect. 5.

2 Related Work

Microarray technologies enable the measurement of the molecular signatures of cancer cells. These data allow different types of analyses, such as (1) the identification of differentially expressed genes [5], which could indicate possible gene targets for more detailed molecular studies or drug treatments and (2) the building of classifiers, with machine learning techniques, which could be used to improve the diagnosis of patients with cancer [6]. Bioinformatics has been proposing novel clustering methods that take intrinsic characteristics of gene expression data into accounts, such as noise and high-dimensionality, to improve the classification [7, 8].

Golub et al. [9] presented an effective method to identify a predictive gene subset for cancer classification. They used a neighborhood analysis in selecting a gene subset that can distinguish between the two cancer types: AML and ALL, based on a separation measure similar to t-statistic. In their experiments, the leukemia dataset contained 6,817 gene expression levels in 72 samples. The training samples were used to select a gene subset that can distinguish between the two cancer types: AML and ALL. The fifty most predictive genes identified by the training samples were validated, to classify the test samples.

Yuanyuan et al. [10] presented the Genetic algorithm/k-nearest neighbors (GA/KNN) method [11–13] for pan-cancer classification. GA/KNN employs a genetic algorithm (GA) as the gene/feature selection engine and the K-Nearest Neighbors (KNN) algorithm as the classification tool. They used UNC RNASeqV2 level 3 expression data sets from the TCGA data for 9096 patients representing 31 tumor types. In conclusion, they were able to identify many sets of 20 genes that could correctly classify more than 90% of the samples from 31 different tumor types in a validation set.

Zhong et al. [14] compared between their proposed method Indexes B/SVM and other classification methods like SWKC/SVM and SVM-RFE. They proposed a new gene selection method for classification using SVMs. They used two dataset Colon cancer dataset and Leukemia dataset. Finally, the proposed B/SVM method outperforms the other two in terms of small average misclassification rate (1.1%) and high average recovery rate (95.7%). This means that B/SVM appears to be more effective and has more power in finding marker genes than SVM-RFE and SWKC/SVM.

3   Classification Methods

There are different classification algorithms such as SVM, KNN, and RF that have been applied to predict and classify different tumor types using gene expression data in recent researches.

3.1   Support Vector Machines (SVMs)

SVM is a supervised machine learning algorithm that can be employed for both classification and regression purposes. SVM is a powerful method for building a classifier. It aims to create a decision boundary between two classes that enable the prediction of labels from one or more feature vectors. Using non-linear kernel function (such as a radial basis function or Polynomial kernel) enables the SVM to deal with higher dimensional data and non-linear models.

In a non-linear problem, a kernel function used to add additional dimensions to the raw data and thus make it a linear problem in the resulting higher dimensional space. The Kernel defines as:

$$K\ (x,\ y) = \langle f(x),\ f(y) \rangle \tag{1}$$

where K is the kernel function while x and y are n-dimensional inputs. f is used to map the input from n-dimensional to m-dimensional space. $< x,\ y >$ denotes the dot product. With kernel functions, we could calculate the scalar product between two data points in a higher dimensional space without explicitly calculating the mapping from the input space to the higher dimensional space. The Radial Basis Function (RBF) kernel defines as:

$$KRBF(x,\ y) = \exp\left(-\gamma ||x\ -\ y||^2\right) \tag{2}$$

Where $\gamma > 0$ is a free parameter. The choice of kernel function among other factors could greatly affect the performance of an SVM model. However, there is no way to figure out which kernel would do the best for a specific pattern recognition problem. The only way to choose the best kernel is through trials.

3.2 K-Nearest Neighbor (KNN)

KNN is one of the simplest classification algorithms but widely used machine learning algorithms. In this algorithm, "distance" is used to classify a new sample from its neighbors. KNN is considered a non-parametric lazy learning algorithm. Non-parametric technique means that it does not make any assumptions on the underlying data distribution. Lazy algorithm means that there is no explicit training phase. For high-dimensional datasets, dimension reduction is usually performed prior to applying the KNN algorithm in order to avoid the effects of the curse of dimensionality [17, 20]. In this case, we are given some data points for training and also a new unlabeled data for testing. We aim to find the class label for the new point. The algorithm has different behavior based on k. The process of the KNN algorithm to classify sample X is:

- Suppose there are j training categories $C_1, C_2, ..., C_j$ and the sum of the training samples is N after feature reduction, they become m-dimension feature vector.
- Make sample X to be the same feature vector of the form $(X_1, X_2, ..., X_m)$, as all training samples.
- Calculate the similarities between all training samples and X. Taking the ith sample di $(d_{i1}, d_{i2}, ..., d_{im})$ as an example, the similarity SIM(X, di) is as following:

$$SIM(X, d_i) = \frac{\sum_{j=1}^{n} X_j . d_{ij}}{\sqrt{\left(\sum_{j=1}^{m} X_j\right)^2} \sqrt{\left(\sum_{j=1}^{m} d_{ij}\right)^2}} \tag{3}$$

Choose k samples which are larger from N similarities of SIM(X, di), (i = 1, 2... N), and treat them as a KNN collection of X. Then, calculate the probability of X that belongs to each category respectively with the following formula.

$$P(X, C_i) = \sum_{d} SIM(X, d_i) . y(d_i, C_i) \tag{4}$$

Where $y(d_i, C_i)$ is a category attribute function:

$$y(d_i, C_i) = \begin{cases} 1, d_i \in C_i \\ 0, d_i \notin C_i \end{cases} \tag{5}$$

Judge sample X to be the category which has the largest $P(X, C_j)$.

3.3   Random Forest (RF)

RF is a classifier that uses a collection of decision trees, each of which is a simple classifier learn a random sample from the data. The classification of a new sample is done by majority voting of the decision trees. Given N training samples, each having M features, each tree is grown as follows: First, n instances are sampled at random from the training set. This sample is the training set of the tree. Then, at each node, m $\ll$ M of the features are selected at random. The best split on these m features is used to branch at the node. Although each individual tree growing by the random forest algorithm can severely over-fit the data, the whole random forest is very resistant to over-fitting, owing to the effect of averaging over many different trees. From this respect, the larger the number of trees, the better. Random forest gains many advantages of decision trees as well as achieves better results through the usage of bagging on samples, random subsets of variables, and a majority voting scheme. Random forest is a classification algorithm well suited for microarray data, it reports superior performance even when most predictive variables are noisy.

4   Experimental Results

This section presents experimental results and is divided into three subsections. The first subsection summarizes the two gene expression datasets. In the second subsection, we present the performance criteria that were used with the classification methods. In the third subsection, we have the result of the three classification methods which are SVM, KNN, and RF to the gene expression data.

4.1   Data Sets

In Freije dataset, all patients for primary brain cancers between 1996 and 2003 undergoing surgical treatment at the University of California. Fresh-frozen samples were obtained at the time of initial surgical treatment. Only grade III and IV gliomas were included in this study. Patient ages at diagnosis varied from 18 to 82 years. There were 46 females and 28 males [18].

In Phillips dataset, fresh-frozen samples were obtained at the time of initial surgical resection from patients with ages greater than twenty without prior therapy. Institutional Review Board/Human Subjects approval was obtained for these retrospective laboratory studies at the University of California, San Francisco and M.D. Anderson Cancer Center [19]. The two data sets Freije and Philips are summarized in Table 1.

After selecting the top 20 genes, we applied three different classification methods SVM, KNN, and random forest, and obtained the accuracy on the test set.

Table 1. Summary of the information of the two datasets.

Dataset	Samples	Tumor Type	Samples in each type	Features
Freije [13]	74	Grade IV	50	12287
		Grade III	24	
Philips [14]	98	Grade IV	75	12287
		Grade III	23	

4.2   Performance Criteria

There are several ways to measure the performance of classification methods. We will use the accuracy percentage to measure the classification methods performance. Accuracy refers to the percentage of correct predictions made by the model when compared with the actual class labels in the test data.

$$\text{Accuracy} = \frac{TP + TN}{TP + TN + FP + FN} \tag{6}$$

Where TP is the number of positive samples that are classified as positive in the test set; FP is the number of negative samples that are classified as positive. FN is the number of positive samples that are classified as negative, and TN is the number of negative samples that are classified as negative. Also, we plot the receiver operating characteristic (ROC) curve to show the performance of the classification model. The ROC plots true positive rate (TP/(TP + FN)) vs. false positive rate (FPL(FP + TN)). After that, the area under the ROC curve (AUC) measures from (0,0) to (1,1).

4.3   Results

The methodology is simply divided into two phases: (i) the feature (genes) selection that will be used for training and testing and (ii) evaluation of the effectiveness of feature selection methods using different classifiers. Feature selection was performed on the training set only. Several numbers of genes were tested like (20, 50, 100, 150, 200, and 250) and top 20 genes that have the highest score were selected because it performed well, consumed less time, and required less memory compared to others. After selecting the top 20 genes, we applied three different classification methods SVM, KNN, and random forest.

SVM Classifier: The ability of SVMs to deal with high dimensional data such as gene expression made this method the first choice for classification. The linear kernel is tested. Figures 1 and 2 show the classification accuracy of SVM using Freije dataset and Philips dataset. Figure 1 draws the number of genes against the accuracy using the SVM classifier for Freije dataset. There is no change in classification accuracy after the number of genes equals 24. Figure 2 draws the number of genes against the accuracy using the SVM classifier for Philips dataset. We notice that the accuracy of classification increases as the number of genes increase.

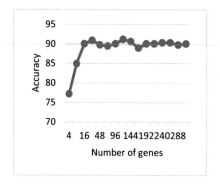

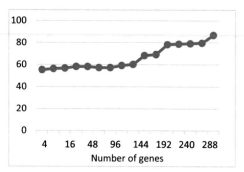

Fig. 1. Performance of SVM using Freije dataset

Fig. 2. Performance of SVM using Philips dataset

KNN Classifier: The simplicity is its main advantage. Euclidian distance is used as the distance metric, and to break the tie. K is set to a value from 1 to 5. The K reports the best cross-validation accuracy is selected. Figure 3 shows the accuracy of classification for different K-values. In Fig. 3, we noticed that the best accuracy was achieved when we use the value of K equal to 3. Table 2 gives the accuracy of KNN classification method using a different number of features.

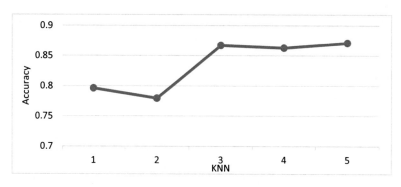

Fig. 3. Performance of KNN for a different K-values using Freije dataset.

Table 2. Performance of KNN using a different number of features.

Accuracy	Freije Dataset			Phillips Dataset		
KNN Feature (%)	1	3	5	1	3	5
60%	90.54	91.59	91.24	76.53	79.59	82.65
40%	90.54	90.54	90.54	82.65	84. 65	83.71
20%	77.02	83.78	82.43	84.69	84.69	83.73

Random Forest (RF): The random forest is used in the experiments with all genes and without replacement. Table 3 shows the impact in classification accuracy depending on using different numbers of tree size. Figure 4 shows the ROC curve for SVM classifier when applied over Freije and Phillips dataset where the area under the ROC curve (AUC) = 0.9334 and 0.8519 respectively. Figure 5 shows ROC curves for KNN

Table 3. Performance of random forest for different size of trees using Freije dataset

Tree size	Accuracy (%)
20	87.96
50	87.79

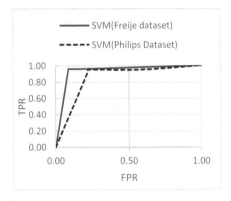

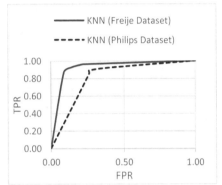

Fig. 4. ROC curve for SVM using Freije and Phillips dataset.

Fig. 5. ROC curve for KNN using Freije and Phillips dataset.

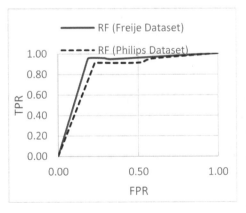

Fig. 6. ROC curve for RF using Freije and Phillips dataset.

Table 4. The best performance with SVM, RF, and KNN

	Freije	Phillips
SVM	91.8%	86.7%.
KNN (K = 3)	91.5%	84.6%
Random forest (T = 20)	87.9%	49.1%

classifier where AUC = 0.9228 using Freije dataset and AUC = 0.8137 using Phillips dataset. Figure 6 shows the ROC curves for RF classifier where AUC = 0. 0.8784 using Freije dataset and AUC = 0. 0.8336 using Phillips dataset. Table 4 shows the best accuracy for each classification methods, namely SVM, KNN, and RF using two gene expression datasets; Freije and Philips for all genes.

5   Conclusions and Future Work

This paper conducted an empirical study on various classification methods in order to observe their performance on gene expression data in terms of using two different datasets. Thus, the classification methods are potentially useful to discover the unknown disease classes in more detail with the results of classification in the field of bioinformatics and cancer classification. We applied these methods to two available datasets and compared how these methods performed in test datasets. The used classification methods applied to two datasets without any feature selections. In both datasets, SVM performed better than other classification methods. SVM classification had the best accuracy in the Freije data set with accuracy better than Philips data set. KNN achieved the best performance when K equals 3. Random forest reports the minimum accuracy in the Freije data. In the future, we plan to study the effect of doing feature selection and classification altogether. Another possible way is to combine the feature selection in the classification learning process.

References

1. Stewart, B.W., Wild, C.P.: World Cancer report 2014. In: International Agency for Research on Cancer (IARC), World Health Organization (WHO). WHO Press, Switzerland (2014)
2. Wang, J.J.-Y., Bensmail, H., Gao, X.: Multiple graph regularized nonnegative matrix factorization. Pattern Recogn. 46(10), 2840–2847 (2013)
3. Wang, J.J.-Y., Wang, X., Gao, X.: Non-negative matrix factorization by maximizing correntropy for cancer clustering. BMC Bioinform. 14, 107–118 (2013)
4. Wang, J.-Y., Almasri, I., Gao, X.: Adaptive graph regularized nonnegative matrix factorization via feature selection. In: 21st International Conference on Pattern Recognition (ICPR), pp. 963–966 (2012)
5. Tusher, V.G., Tibshirani, R., Chu, G.: Significance analysis of microarrays applied to the ionizing radiation response. Proc. Natl. Acad. Sci. U.S.A. 98(9), 5116–5121 (2001)
6. Spang, R.: Diagnostic signatures from microarrays: a bioinformatics concept for personalized medicine. BIOSILICO 1, 64–68 (2003)
7. Brunet, J.P., Tamayo, P., Golub, T.R., Mesirov, J.P.: Metagenes and molecular pattern discovery using matrix factorization. Proc. Natl. Acad. Sci. U.S.A. 101(12), 4164–4169 (2004)
8. McLachlan, G.J., Bean, R., Peel, D.: A mixture model-based approach to the clustering of microarray expression data. Bioinformatics 18(3), 413–422 (2002)
9. Golub, T.R., Slonim, D.K., Tamayo, P., Huard, C., Gaasenbeek, M., Mesirov, J.P., Coller, H., Loh, M.L., Downing, J.R., Caligiuri, M.A., Bloomfield, C.D., Lander, E.S.: Molecular classification of cancer: class discovery and class prediction by gene expression monitoring. Science 286(5439), 531–537 (1999)

10. Li, Y., Kang, K., Krahn, J.M., Croutwater, N., Lee, K., Umbach, D.M., Li, L.: A comprehensive genomic pan-cancer classification using the Cancer Genome Atlas gene expression data. BMC Genom. 18(1), 508 (2017)
11. Li, L., Weinberg, C.R., Darden, T., Pedersen, L.G.: Gene selection for sample classification based on gene expression data: study of sensitivity to choice of parameters of the GA/KNN method. Bioinformatics 17(12), 1131–1142 (2001)
12. Li, L., Darden, T.A., Weinberg, C.R., Levine, A.J., Pedersen, L.G.: Gene assessment and sample classification for gene expression data using a genetic algorithm/k-nearest neighbor method. Comb. Chem. High Throughput Screen. 4(8), 727–739 (2001)
13. Singha, R.K., Sivabalakrishnan, M.: Feature selection of gene expression data for cancer classification: a review. Procedia Comput. Sci. 50, 52–57 (2015)
14. Zhong, W., Lu, X., Wu, J.: Feature selection for cancer classification using microarray gene expression data. Biostat. Biometr. 1(2), 1–7 (2017)
15. Li, T., Zhang, C., Ogihara, M.A.: comparative study of feature selection and multiclass classification methods for tissue classification based on gene expression. Bioinformatics 20(15), 2429–2437 (2004)
16. Wu, X., Kumar, V., Ross Quinlan, J., Ghosh, J., Yang, Q., Motoda, H., McLachlan, G.J., Ng, A., Liu, B., Yu, P.S., Zhou, Z.H., Steinbach, M., Hand, D.J., Steinberg, D.: Top 10 algorithms in data mining. Knowl. Inf. Syst. 14(1), 1–37 (2008)
17. Nello, C., Taylor, J.S.: An Introduction to support vector machines and other kernel-based learning methods. Cambridge Univ. Press 22(2), 204–210 (2001)
18. The Freije dataset. http://www.ncbi.nlm.nih.gov/geo/query/acc.cgi?acc=GSE4271. last accessed 10 Aug 2018
19. The Phillips dataset. http://www.ncbi.nlm.nih.gov/geo/query/acc.cgi?acc=GSE4412. last accessed 10 Aug 2018
20. Schlkopf, B., Tsuda, K., Vert, J.P.: Kernel methods in computational biology. MIT Press series on Computational Molecular Biology, Berlin (2003)

Using Eye Movement to Assess Auditory Attention

Alaa Bakry[1], Radwa Al-khatib[1], Randa Negm[1], Eslam Sabra[1],
Mohamed Maher[1], Zainab Mohamed[1], Doaa Shawky[2(✉)],
and Ashraf Badawi[1]

[1] Center for Learning Technologies, University of Science and Technology,
Zewail City, Giza, Egypt
{s-alaa.mobarak, s-radwa.alkhatib,
s-randa.negm, s-eslamalaa, s-moa.maher,
zrajab, abadawi}@zewailcity.edu.eg
[2] Engineering Mathematics Department, Faculty of Engineering,
Cairo University, Giza, Egypt
doaashawky@staff.cu.edu.eg

Abstract. Eye movement has been found to be one of the factors that highly affect attention. In this paper, a study for detecting the influence of eye movement on attention is presented. Forty-three participants attended four sessions introducing different auditory stimuli while wearing a 14-channel wireless headset that collects their EEG signals. The participants were asked to fix their eyes in two of the sessions and they were allowed to move them freely in the other two. Their attention during the sessions was estimated using questionnaires that assess the information they were able to gain after the sessions. Different classifiers were trained to predict the attention scores when subjects were freely moving their eyes or fixing them. Among the trained classifiers, K nearest-neighbors classifiers yielded the best classification accuracy, which varied with the addition of the eye-movement features from about 72% to 87%. Thus, the obtained results show that there is an effect of eye movement on the gained attention. Hence, it is possible to detect attention of subjects using their eye movement patterns.

Keywords: Attention measurement · EEG · Eye movement · Eye gaze

1 Introduction

Utilization of several biosensors to detect attention and various cognitive skills is not a novel approach [1]. In many situations such as a delivered lecture, a conducted workshop, or even a meeting, knowing the level of attention and engagement of the participants is of utmost importance as it can enable the instructor/coordinator to modify the presented content to cater for the attendees and ensure maximum benefit and delivery of the presented information [2, 3]. Thus, having a real time and automated processing application that can give insights about attendees' attention through merely data collected from a camera is the aim of this study.

© Springer Nature Switzerland AG 2020
A. E. Hassanien et al. (Eds.): AMLTA 2019, AISC 921, pp. 200–208, 2020.
https://doi.org/10.1007/978-3-030-14118-9_20

Many research groups have measured attention with different tools and analysis techniques including EEG to know the attention state of learners and team members. Electroencephalography (EEG) is the electrophysiological method used in order to record electrical activity of certain brain areas. To measure various attention states, the authors in [3] have built a classification model that is able to characterize EEG signals of 24 participants to attentive and inattentive during instruction. In another relevant study, the relation between changes in spectral analysis of EEG signals for 18 subjects and sustained visual attention is analyzed in a real classroom setting. Changes in delta, theta and beta power values are found to be features that can characterize sustained visual attention [1]. Also, in [4], four attention states have been classified using artificial neural networks with classification accuracies range from 56.5–79.75%. To monitor task engagement in previous studies, researchers used EEG signals and data from the camera to predict user's emotional state and attentiveness. When EEG biofeedback is used to detect low attention level, it has significant effects on overall attention [5]. Eye tracking is a process of measuring a person's point of gaze or the motion of their eyes relative to the position of their head. Previous studies tried to extract features of eye gaze and facial expressions to estimate attentional states in various conditions [6, 7]. Eye gaze, head pose and facial action information are the main cues for engagement prediction [8, 9]. Kawahara et al. [10] analyzed the conversation in two different situations using a human robot in the first experiment and a person in the second one. Eye-gaze is found to be the most critical behavior in engagement detection and in expecting the mental state during conversations compared to nodding, laughing and backchannels especially in noisy environment [10]. Other studies investigated whether task-evoked eye movements during an auditory task can predict the magnitude of the activity. They used different looped melodies and recorded vertical and horizontal gaze displacements to find that attentive listening is associated with changes in eye movements that are independent of visual stimuli or visual demands [8]. In addition, a group of researchers investigated the attention of 29 students in a psychology lecture through features of head pose, eye gaze and facial expressions, extracted from a YI 4k action camera [11]. To monitor task engagement, in previous studies, researchers used EEG signals and data from the camera to predict user's emotional state and attentiveness [12, 13]. EEG was used to obtain an objective index of neural face discrimination during different facial expression to be able to correlate with eye movements [14]. When EEG biofeedback is used to detect low attention level, it has significant effects on overall attention [15].

Thus, a lot of research is going on to come up with a viable model of attentional states based on EEG waves and facial expressions. To achieve this objective, in this paper, an approach for detecting the effect of eye movement on auditory attention is presented. In the conducted experiments, participants are subjected to an auditory stimulus, then they fill a questionnaire that aims to test their attention through direct questions related to the presented content. The collected data is then analyzed and a classifier that is trained to detect attention scores is built.

The rest of the paper is organized as follows. In Sect. 2, the experimental setup and feature extraction are presented. In Sect. 3, the results of the proposed approach are discussed. Finally, in Sect. 4, conclusions and directions for the future work are outlined.

2   Materials and Methods

To test the relation between eye movement and auditory attention, and achieve the aspired result, we used the following materials: (1) EPOC + EEG model 1.1, EMOTIV scientific contextual EEG headset (2) EMOTIVPRO software (3) laptop webcam with video quality of 720p, ratio of 16:9, and 30 frames per second (4) Free access Google Forms for recording participants responses (5) Bluetooth headphones (6) Matlab version 2016a (7) Bio-true multi-purpose solution for hydrating the electrodes.

2.1   Experimental Setup

For the experiments to be done, a post was made on the group of Zewail City students informing them about the experiment purposes. Those who wished to participate filled in a form where information about their mental health, long term medication, age, field of study, presence of skin allergies, and means of communication, were collected. From those responses 43 participants were chosen according to the inclusion criteria, and each was assigned a randomly generated code of two numbers to be used for all labeling in the experiment to ensure privacy. A consent form and time log form were sent to the prospective participants to know the procedure and choose time to do the experiment. The participant would then show up on the agreed time, sign the consent form, do the experiment, and receive an incentive. The participants were of age 18 years old to 22 years; 17 of them were females and the rest of the 43 were males from different educational backgrounds encompassing the 7 majors of the University of Science and Technology at Zewail City. The experimental analysis steps are shown in Fig. 1.

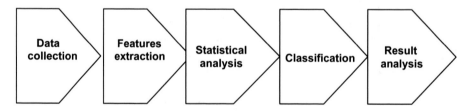

Fig. 1. The experimental analysis

2.1.1   Task Description
In the following subsections, the details of each of the four experiments will be presented.

2.1.1.1 Experiment 1 Description
The EPOC headset was placed to each participant after being well hydrated and connected to the EmotivPRO software to record the EEG signals. The connectivity was checked and the baseline recorded. The participant would sit in front of the laptop having the webcam to record the video. Then, he/she would use the headphone to hear

a 6.37 min recording from an audiobook about atmosphere. The participant was asked to move his/her eyes freely while hearing the record. After the record was done, the video was stopped and saved as well as the EEG data for that section. Next, the participant was asked to answer 6 true/false questions based on the information heard in the audio; for instance: "The four layers of the atmosphere include the: troposphere, biosphere, stratosphere and hydrosphere." True or False? Experiment 2 then starts.

2.1.1.2 Experiment 2 Description
Another EEG and video records were formed for the participant, while he/she was hearing a 6.21 min audio record for Quran. Participants were asked again to move their eyes freely while hearing the record. Finally, the participant solved another 6 true/false questions about the Ayat mentioned in the audio; for instance: "was this the first Ayah the reader started with in the audio *Ayah was placed in Arabic*" True or False?

2.1.1.3 Experiment 3 Description
The third EEG and video records were formed for the participant, while he/she was hearing a 6.22 min recording from an audiobook about geology and fossil formation. This time participants were asked to choose a point anywhere in the room (taking in consideration the visibility of their eyes to the camera) and fix their eyes to it while hearing the record. Again, the participant solved another 6 true/false questions about the information mentioned in the audio; for instance: "Estimates say about 80 species in ten thousand have made it into fossil record." True or False?

2.1.1.4 Experiment 4 Description
The final EEG and video records were formed for the participant, while he/she was hearing a 6.28 min recording from the same surah of the Quran. Again, participants were asked to choose a point anywhere in the room (taking in consideration the visibility of their eyes to the camera) and fix their eyes to it while hearing the record. Finally, the participant solved another 6 true/false questions about the ayat heard in the audio. Then, to ensure considering bias, each participant has to answer another questionnaire at the end of the full experiment to address his knowledge and interest in the heard topics (atmosphere, and geology) on a scale of 5, having 0 as no previous knowledge/totally not interested, and 5 as have an advanced level of knowledge about the topic/extremely interested in the topic. Their memorization of the surah, from which the Quran audio was obtained, was assessed as well through a yes/no question. Each participant ended up having 4 videos and 4 CSV files containing the recorded data of each experiment.

2.1.2   EEG Acquisition
EEG signals were recorded through the 14 channels of the EPOC+ headset connected to the Emotiv software, and data was downloaded as CSV files (4 for each participant, 1 for each experiment) and EDF files.

2.1.3   Feature Extraction

Since human thoughts and activities come from neurons' electrical behaviors and to measure these signals and hence their reflection on the psychological state, an EEG measuring tool was used, which is a headset with 16 sensors- 14 channel (AF3, F7, F3, FC5, T7, P7, O1, O2, P8, T8, FC6, F4, F8, AF4) with two reference channels, where these sensors have 128 sample/s. The waves that represent human state from being very active and alert (Gamma waves (30–60 Hz) & Beta waves (12–30 Hz)) going to relaxation and meditation (Alpha waves (8–12 Hz)) to deep meditation and being asleep (Theta waves (4–8 Hz)) up to being in a deep sleep (Delta and Infra- low waves) [16, 17]. Alpha is considered as the transition state between being awake and falling asleep which may represent in our experiment the losing attention stage. Since our study focuses on brain activity during lectures, we excluded the Delta and Infra- low measurements expecting it will have a very low power and thus its effect will be negligible. Each of the six brain waves has its own characteristics and significations representing a specific consciousness level [17]. First, Filtration is done before the feature extraction; the DC offset is removed to be able to analyze the AC signal (oscillations). Second, the outliers were removed which makes the signal more representative as outliers change the statistical data misleadingly. Then feature extraction is divided into two main categories.

Features that were extracted in time domain which include the mean, maximum, minimum, and standard deviation. In addition to features that were extracted in the frequency domain including normalized power for the average wave, Gamma, Beta, Alpha and Theta. This step was done to know the power of each mind state compared to the normal power of the overall combined signal and compared to each one's baseline to know the probability of the state and also to solve the problem of individual differences. All features were collected for each participant, which resulted in 14 channel × 9 features making overall 126 feature.

3   Classifications Results and Discussion

In the conducted analysis, the attention scores were mapped to attentive (1) and inattentive (0) in three different ways. First, the scores that are greater than 3 points (out of 6) are considered as attentive (Atten1). Second, only the scores that are greater than 4 are considered as attentive (Atten2). Finally, only those that are greater than 5 are considered as attentive (Atten3). It is worth-mentioning that the analysis was conducted using the classification Apps in Matlab, with the default parameters settings. In addition, for each experiment, the analysis was conducted on all EEG features, when adding the eye-movement information as a predictor to the model, and when applying principal component analysis (PCA) as a feature selection. Only the results for the best performing classifiers were reported in each case.

3.1   Effect of Eye Movement on Attention in Experiment 1 and Experiment 3

To assess the ability of the used EEG features on predicting attention when subjected to auditory scientific information, different classifiers were trained and the classification accuracy and confusion matrix are presented. The confusion matrix is composed of two rows and two columns with cells a_{ij}, where i and j denotes the row and column numbers, respectively. Moreover, a_{11} indicates the number of correctly-identified inattentive participants, a_{12} indicates the mistakenly-identified inattentive participants, a_{21} indicates the mistakenly-identified attentive participants, and finally, a_{22} indicates the correctly-identified attentive participants. Table 1 shows the best classifiers and their performance measures for each attention scores. As shown in the table, K-nearest neighbor (KNN) classifier yielded the best results. Moreover, although the classification accuracies enhance when increasing the number of data points with attentive scores, the confusion matrices show that the results are biased and that the predicting ability of inattentive class is very low.

Table 1. Best classifiers for attention with all EEG features

Target	Classifier	Accuracy	Confusion Matrix
Atten1	Coarse KNN and other classifiers	71.1%	0 | 22
			0 | 54
Atten2	Cubic KNN, PCA applied	72.4%	54 | 0
			21 | 1
Atten3	Cubic KNN	89.5%	0 | 8
			0 | 68

To enhance the results, the two-sample t-test was applied to choose only the set of EEG features that are relevant to eye-movement. Thus, the set of EEG features with eye movement is tested against the same set with eye fixation. Only the features with p-value smaller than 0.05 were selected and used in building the classification models. Table 2 shows the set of EEG features that were highly affected by eye-movement together with their p-values. As shown in the table, most features are related to channel 9, which is located near the cerebellum. Cerebellum is responsible for movement coordination [19]. Moreover, Table 3 presents the best classifiers together with their performance measures when only the set of EEG features relevant to eye-movement were selected. As shown in the table, the obtained results have been improved.

Table 2. EEG features relevant to eye-movement

EEG Feature	p-value
Average Power for channel 9	0.006
Power in Theta range for channel 9	0.017
Power in Alpha range for channel 5	0.048
Power in Alpha range for channel 9	0.014
Power in Beta range for channel 9	0.017
Power in Gamma range for channel 9	0.015
Mean value of channel 9	0.045
Standard deviation of values of channel 14	0.007

Table 3. Best classifiers for attention using only EEG features affected by eye movement

Target	Classifier	Accuracy	Confusion Matrix
Atten1	Weighted KNN, with eye movement information added to the predictors	73.7%	7 | 14 2 | 49
Atten2	Weighted KNN, with PCA	78.7%	45 | 5 8 | 14
Atten3	Medium KNN and several other classifiers	85.3%	0 | 8 0 | 64

In addition, the best obtained results are those when weighted KNN classifier is used with PCA. It should be noted that Atten2 usually yields the best confusion matrix, since it provides an approximate balance between the two classes to be detected.

3.2 Effect of Eye Movement on Attention in Experiment 2 and Experiment 4

Similar analyses were conducted when different auditory stimuli (Quran) were presented to the participants. Table 4 presents the best classifiers and their performance measures for each attention score. As shown in the table, KNN yielded the best accuracy, however, as shown in the confusion matrix, the true positive and false negative rates are low. In addition, it is worth-mentioning that when the t-test was applied to choose only the set of relevant features, no features with p-value smaller than 0.05 were found.

Table 4. Best classifiers for attention with all EEG features

Target	Classifier	Accuracy	Confusion Matrix
Atten1	Weighted KNN, without eye movement information	57.5%	10 | 23 8 | 32
Atten2	Coarse KNN	74%	54 | 0 19 | 0
Atten3	Coarse KNN	78.1%	0 | 16 0 | 57

4 Conclusions and Future Work

In this paper, an approach for predicting the effect of eye movement on auditory attention is presented. The EEG signals of the participants are collected while experiencing auditory information sessions. Their attention is then calculated based on their responses to a questionnaire that measures the information content they were able to gain during the sessions. Different classification models were trained on the collected data with the EEG features as the predictors and the attention scores as the target.

The obtained results show that it is possible to detect the auditory attention using eye-movement. However, a more reliable method for detecting whether the eyes of participants are fixed or not is needed. As a future work, we intend to add this information to further enhance the obtained results. In addition, more participants will be included to be able to obtain a more robust model for detecting attention based on subjects' eye-movement patterns.

References

1. Liu, N., Chiang, C., Chu, H.: Recognizing the degree of human attention using EEG signals from mobile sensors. Sensors **13**(8), 10273–10286 (2013)
2. Shawky, D., Badawi, A.: A reinforcement learning-based adaptive learning system. In: International Conference on Advanced Machine Learning Technologies and Applications, pp. 221–231 (2018)
3. Shawky, D., Badawi, A.: Towards a personalized learning experience using reinforcement learning. In: Machine Learning Paradigms: Theory and Application, pp. 169–187 (2019)
4. Mohammadpour, M., Mozaffari, S.: Classification of EEG-based attention for brain computer interface. In: 3rd Iranian Conference on Intelligent Systems and Signal Processing (ICSPIS) (2017)
5. Ko, L., Komarov, O., Hairston, W., Jung, T., Lin, C.: Sustained attention in real classroom settings: an EEG study. Front. Hum. Neurosci. **11**, 388 (2017)
6. Ghosh, P., Mazumder, A., Bhattacharyya, S., Tibarewala, D.: An EEG study on working memory and cognition. In: Proceedings of the 2nd International Conference on Perception and Machine Intelligence - PerMIn 2015 (2015)
7. Pavlov, Y., Kotchoubey, B.: EEG correlates of working memory performance in females. BMC Neurosci. **18**(1), 26 (2017)
8. Braga, R., Fu, R., Seemungal, B., Wise, R., Leech, R.: Eye movements during auditory attention predict individual differences in dorsal attention network activity. Front. Hum. Neurosci. **10**, 2016 (2016)
9. Kwok, C.K.: Understanding user engagement level during tasks via facial responses, eye gaze and mouse movements. PhD Thesis, Hong Kong polytechnic University (2017)
10. Putze, F., Küster, D., Annerer-Walcher, S., Benedek, M.: Dozing off or thinking hard? In: Proceedings of the 2018 on International Conference on Multimodal Interaction (2018)
11. Kawahara, T., Inoue, K., Lala, D., Takanashi, K.: Audio-visual conversation analysis by smart posterboard and humanoid robot. In: IEEE International Conference on Acoustics, Speech and Signal Processing (ICASSP) (2018). https://doi.org/10.1109/icassp.2018.8461470
12. Rajavenkatanarayanan, A., Babu, A., Tsiakas, K., Makedon, F.: Monitoring task engagement using facial expressions and body postures. In: Proceedings of The 3rd International Workshop on Interactive and Spatial Computing - IWISC 2018 (2018). https://doi.org/10.1145/3191801.3191816
13. Li, Y., Li, X., Ratcliffe, M., Liu, L., Qi, Y., Liu, Q.: A real-time EEG-based BCI system for attention recognition in ubiquitous environment. In: Proceedings of 2011 International Workshop on Ubiquitous Affective Awareness and Intelligent Interaction - UAAII 2011 (2011). https://doi.org/10.1145/2030092.2030099
14. Niu, X., Han, H., Zeng, J., Sun, X., Shan, S., Huang, Y., et al.: Automatic engagement prediction with GAP feature. In: Proceedings of the 2018 on International Conference on Multimodal Interaction - ICMI 2018 (2018). https://doi.org/10.1145/3242969.3264982

15. Stacchi, L., Ramon, M., Lao, J., Caldara, R.: Neural representations of faces are tuned to eye movements. J. Neurophysiol. **101**(5), 2581–2600 (2018). https://doi.org/10.1101/402263
16. Sun, J., Yeh, K.: The effects of attention monitoring with EEG biofeedback on university students' attention and self-efficacy: The case of anti-phishing instructional materials. Comput. Educ. **106**, 73–82 (2017). https://doi.org/10.1016/j.compedu.2016.12.003
17. Mohamed, Z., El Halaby, M., Said, T., Shawky, D., Badawi, A.: Characterizing focused attention and working memory using EEG. Sensors **18**(11), 3743 (2018)
18. Mindvalley Blog: This is How Brain Waves Contribute to the State of Mind (2018). https://blog.mindvalley.com/brain-waves/?utm_source=google_blog. Accessed 29 Nov 2018
19. Thach, W.T., Goodkin, H.P., Keating, J.G.: The cerebellum and the adaptive coordination of movement. Annu. Rev. Neurosci. **15**(1), 403–442 (1992)

Facilitating Classroom Orchestration Using EEG to Detect the Cognitive States of Learners

Zainab Mohamed[1], Mohamed El Halaby[2], Tamer Said[1],
Doaa Shawky[3(✉)], and Ashraf Badawi[1]

[1] Center for Learning Technologies, University of Science and Technology,
Zewail City, Giza, Egypt
{zrajab,tsaid,abadawi}@zewailcity.edu.eg
[2] Mathematics Department, Faculty of Science, Cairo University, Giza, Egypt
halaby@sci.cu.edu.eg
[3] Engineering Mathematics Department, Faculty of Engineering,
Cairo University, Giza, Egypt
doaashawky@staff.cu.edu.eg

Abstract. Technology, when successfully integrated in a classroom environment can help redefine and facilitate the role of the teacher. Classroom orchestration is an approach to Technology Enhanced Learning that emphasizes attention to the challenges of classroom use of technology, with a particular focus on supporting teachers' roles. The automatic detection of learners' cognitive profiles is an important step towards adaptive learning, where the learning material are adapted to match that of the learners in order to enhance the learning outcome. Electroencephalogram (EEG) is a methodology that monitors the electric activity in the brain. It has been utilized in several applications including, for example, detecting the subject's emotional and cognitive states. In this paper, an approach for detecting two basic cognitive skills that affect learning using EEG signals is proposed. These skills include focused attention and working memory. The proposed approach consists of the following main steps. First, subjects undergo a cognitive assessment test that stimulates and measures their full cognitive profiles while putting on a 14-channel wearable EEG headset. Second, only the scores of the two cognitive skills aforementioned above are extracted and used to encode the two targets for a classification problem. Third, the collected EEG data are analyzed and a number of time and frequency-domain features are extracted. Fourth, several classifiers were trained to be able to correctly classify and predict three levels (low, average, and high) of the measured cognitive skills. The classification accuracies that were obtained for the focused attention and working memory were 90% and 87%, respectively, which indicates the suitability of the proposed approach for the detection of these two skills. This could be used as a first step towards adaptive learning where adaptation is to be done according to the predicted levels of focused attention and working memory.

Keywords: Cognitive skills measurement · Electroencephalography ·
Short-time fourier transform · Classification

© Springer Nature Switzerland AG 2020
A. E. Hassanien et al. (Eds.): AMLTA 2019, AISC 921, pp. 209–217, 2020.
https://doi.org/10.1007/978-3-030-14118-9_21

1 Introduction

Detecting cognitive states and skills is an important step towards adaptive learning, in which the learning material and pace are adjusted to match some collected data about learners during a learning task [1, 2]. In addition to the invasive approach in which learners are asked to provide some information about their learning sessions, an automated non-invasive one is also possible. A promising methodology for the automated collection of data during a mental task includes the use of bio-sensors that could measure subjects' emotions, attention, and engagement in a non-intrusive way in order to not interfere with her/his learning process.

In this paper, an approach for measuring the basic cognitive skills that affect the outcomes of the learning process using bio-signals is presented. The ultimate goal is to provide the learning management systems (or instructors) with information about how these skills vary from one learner to another to help personalize the learning process. The measured cognitive skills include focused attention (FA) and working memory (WM). FA means that all learners' activities involve active cognitive processes such as problem-solving and critical thinking [3]. On the other hand, WM is a type of short-term memory that allows learners to store and manipulate temporary information [4, 5].

As presented in [5], the cognitive skills of learners play a major role in determining how well they perceive the knowledge being presented in a learning task, although the learning outcome depends on many other personal and affective factors of learners such as their motivation and interpersonal skills [6–8].

After the advancements in bio-sensors, the brain activity can be considered as one of the bio-signals that could be measured in a convenient, non-intrusive way. The electroencephalogram (EEG) is the brain electrical activity measured by mounting electrodes on the scalp, where the activity of millions of cortical neurons, produces an electrical field which can be measured from the human scalp [9]. Analyzing EEG can be effectively used in measuring emotions and some other psychological states of subjects, especially after the advancements in the technology and the availability of convenient wireless devices.

Utilization of several biosensors to detect various cognitive skills and the application in education is not a novel approach. Many studies have tackled the same problem with different tools and analysis techniques. In regard to attention, the authors in [10] have built a classification model that is able to classify EEG signals of 24 participants to attentive and inattentive while listening to instructions. Support vector machines are used, and the best achieved accuracy is 76.8%. In another relevant study [11], the relation between changes in spectral analysis of EEG signals for 18 subjects and sustained visual attention is analyzed in a real classroom setting. Changes in power values of some frequency bands (alpha, theta, beta, and gamma [12]) are found to be features that can characterize sustained visual attention. Also in [13], four attention states have been classified using artificial neural networks with classification accuracies that range from 56.5–79.75%. Moving to working memory, the authors in [14] have built a support vector machines classifier that is able to discriminate between working memory and recognition cognitive states from EEG data with a classification accuracy of 79%. In addition, the work in [15] utilizes the characteristics of power values in different frequency bands for EEG data during working memory tasks that vary in type and complexity. Changes in theta to beta EEG frequency bands are found to indicate the performance of WM.

In this paper, the presented approach in [16] is further improved. The approach includes the following main steps. First, the subjects put on a wireless headset that measures EEG signals from 14 different electrodes distributed across the scalp. Second, they undergo a scientifically-validated cognitive assessment test. This step helps elicit and stimulate the signals related to the cognitive skills that need to be measured. Third, the collected EEG signals are analyzed to find out whether there are relationships between the extracted EEG features and the measured cognitive skills.

The rest of the paper is organized as follows. In Sect. 2, the experimental study is presented. In Sect. 3, the results of the proposed approach are discussed. Finally, in Sect. 4, conclusions and directions for the future work are outlined.

2 Experimental Study

2.1 Data Used

The data set that was collected in [16] is used. This includes the EEG signals for 86 undergraduate students (72 males, 14 females) with ages ranging from 18 to 23 years were recorded using 14-channel wireless Emotiv EPOC EEG headset while the participants underwent a scientifically-validated cognitive assessment battery for around 40 min. The test is available at https://www.cognifit.com/. It consists of a series of fixed activities that stimulate perception, memory, attention, and other cognitive states. The outcome of interest represents the given test scores or measures of the cognitive skills for each subject. The cognitive skills measured include 23 basic skills and 5 compound skills. The five compound skills include memory, attention, coordination, perception, and reasoning. Each of these 5 compound skills are composed of a set of basic skills. For instance, the score of perception is the average of the scores of six basic skills; namely recognition, auditory perception, spatial perception, visual scanning, visual perception, and estimation. On the other hand, focused attention (FA) and working memory (WM) belong to the basic skills that are directly measured and whose scores are directly reflected in the test. The numeric scores given to cognitive skills vary from 0 to 800 with a categorization of the profile to low (score < 200), moderate $(200 \leq \text{score} < 400)$ and high (score ≥ 400). Thus, in the conducted experiments, FA and WM, scores were encoded to these three levels. Further details about the experimental setup can be found in [16].

2.2 Feature Extraction and Analysis

As a preprocessing step, first, the DC offset was removed from the EEG signals, then the statistical outliers were removed. To be able to get information about the dynamics of the collected EEG data, features were extracted from time as well as frequency domains. Figure 1 shows the first 1000 samples of the raw time series EEG signal for one subject. In addition, Fig. 2 shows the event-related potentials (ERP) of the 14 channels for one subject.

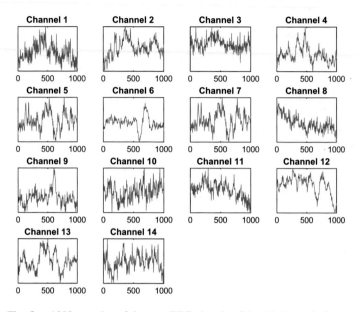

Fig. 1. The first 1000 samples of the raw EEG signals of the 14 channels for one subject

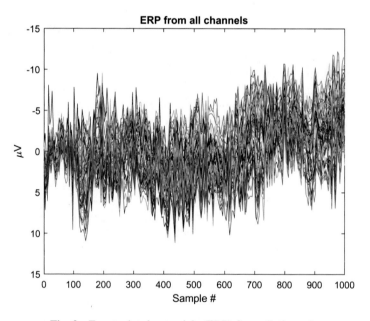

Fig. 2. Event-related potentials (ERP) from all channels

After the preprocessing stages, in total, 15 time domain features were calculated for each channel. The extracted features are mainly statistical ones (e.g., mean, standard deviation, skewness, kurtosis,.., etc.). In addition, the Hjorth mobility and complexity parameters are calculated as given by (1) and (2) [17], respectively:

$$\text{Mobility} = \sqrt{\frac{\text{var}\left(\frac{dx(t)}{dt}\right)}{\text{var}(x(t))}} \tag{1}$$

$$\text{Complexity} = \frac{\text{Mobility}\left(\frac{dx(t)}{dt}\right)}{\text{Mobility}(x(t))} \tag{2}$$

Moreover, frequency-domain features were extracted. Those include five power values of four frequency bands for each of the 14-channel EEG signals. These frequency bands include theta (4–8 Hz), alpha (8–12 Hz), beta (12–30 Hz), and gamma (30–60 Hz) bands, in addition to the total average power in each channel. Thus, in total, 280 features were calculated for each subject. Figures 3, 4 and 5 show the average persistent spectrum [18] for the subjects with low, moderate, and high scores for FA, respectively. As shown in the figures, there are clear variations between the power values for each class of the FA scores.

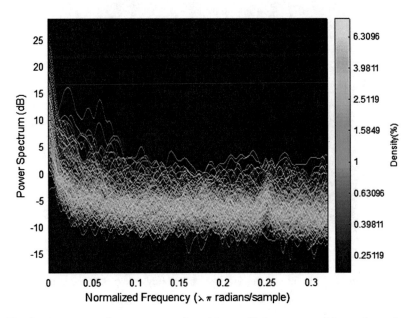

Fig. 3. Average persistent spectrum for subjects with low scores of focused attention

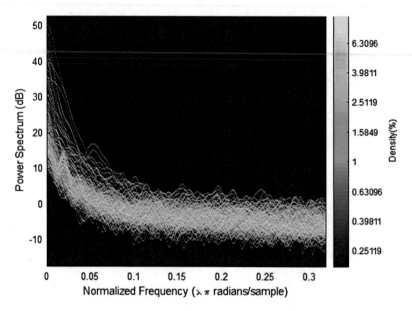

Fig. 4. Average persistent spectrum for subjects with moderate scores of focused attention

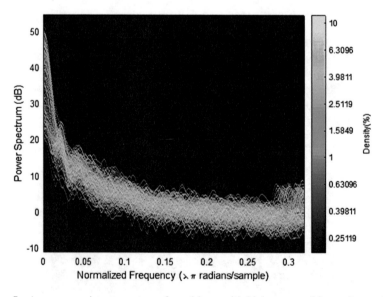

Fig. 5. Average persistent spectrum for subjects with high scores of focused attention

3   Classification Results and Discussion

To be able to correctly classify the collected EEG signals into one of the 3 classes (i.e., low, average, and high) of each target, many classifiers were implemented. However, the classifiers which were able to give the best results include neural network (NN) and linear support vector classifier (SVC). The models were implemented using scikit-learn in Python. The NN consists of two layers, each of size 40 with logistic activation function. To avoid over-fitting, L2 penalty regularization with a parameter of 0.001 was used. As for the linear SVC, the penalty used is L2 (parameter = 0.1) with the squared-hinge-function loss. In addition, the maximum number of iterations was set to 1000. It is worth mentioning that the hyper-parameters' values were chosen as a result of Grid search optimization in the python library sklearn.

Furthermore, to reduce the features' dimensions, meta-transformation feature selection was applied using a logistic regression classifier with k-fold cross validation ($k = 10$) and the mean score is obtained. Best obtained results are summarized in Table 1, together with the confusion matrices for the used classifiers. As can be shown in the table, the best obtained accuracy for WM was 87% using NN. Meanwhile, linear SVC yielded the best accuracy (90%) for the detection of FA. Other classifiers, however, were not able to provide good performance on the used data set. It is worth mentioning that the selected features in the best models include mostly the spectral domain features, in addition to Hjorth mobility and complexity values from the time-domain.

Table 1.   Obtained results for best performing classifiers.

	K-fold cross-validation mean		Confusion-matrix	
	FA	WM	FA	WM
NN	0.86	0.87	[[10 0 6] [0 7 3] [2 0 58]]	[[37 0 2] [3 8 3] [2 0 31]]
Linear SVC	0.90	0.85	[[13 1 2] [0 9 1] [4 0 56]]	[[35 2 2] [4 6 4] [3 0 30]]

4   Conclusions and Future Work

In this paper, an approach for the prediction of focused attention and working-memory using EEG is proposed. EEG signals were recorded while the subjects undertook a cognitive test that stimulated these cognitive skills. The collected EEG signals were analyzed in the time and frequency domains to extract a set of 280 features, which were then used to train different classifiers. The built models were able to detect the three levels of FA and WM with good accuracy. The obtained good results show the suitability of the proposed approach to predict the three levels (i.e., low, average, and high) of focused attention and working memory of learners using a wireless 14-channel EEG headset.

As a future work, to avoid feature extraction and selection, using the raw EEG for the detection of the levels of FA and WM will be examined. For this purpose, deep learning approaches such as convolutional and recurrent neural networks can be used.

Acknowledgments. This research was supported by Information Technology Industry Development Agency (ITIDA) and Smart & Creative Solutions (SMACRS).

References

1. Badawi, A., Shawky, D.: The need for a paradigm shift in CSCL tools. In: Computing Conference, pp. 1131–1135. IEEE (2017)
2. Shawky, D., Badawi, A.: Reinforcement learning-based adaptive learning system. In: 3rd International Conference on Advanced Machine Learning Technologies and Applications (2018)
3. Kearsley, G., Shneiderman, B.: Engagement theory: a framework for technology-based teaching and learning. Educ. Technol. **38**(5), 20–23 (1998)
4. Baddeley, A.: Working memory: theories, models, and controversies. Annu. Rev. Psychol. **63**, 1–29 (2012)
5. Shuell, T.J.: Cognitive conceptions of learning. Rev. Educ. Res. **56**(4), 411–436 (1986)
6. Shawky, D., Badawi, A., Said, T., Hozayin, R.: Affordances of computer-supported collaborative learning platforms: a systematic review. In: International Conference on Interactive Collaborative Learning (ICL), pp. 633–651. IEEE (2014)
7. Said, T., Shawky, D., Badawi, A.: Identifying knowledge-building phases in computer supported collaborative learning: a review. In: International Conference on Interactive Collaborative Learning (ICL), pp. 608–614. IEEE (2015)
8. Fahmy, A., Said, Y., Shawky, D., Badawi, A.: Collaborate-it: a tool for promoting knowledge building in face-to-face collaborative learning. In: 15th International Conference on Information Technology Based Higher Education and Training (ITHET), pp. 1–6. IEEE (2016)
9. Savelainen, A.: An introduction to EEG artifacts. In: Independent Research Projects in Applied Mathematics (2010)
10. Liu, N.-H., Chiang, C.-Y., Chu, H.-C.: Recognizing the degree of human attention using EEG signals from mobile sensors. Sensors **13**(8), 10273–10286 (2013)
11. Ko, L.-W., Komarov, O., Hairston, W.D., Jung, T.-P., Lin, C.-T.: Sustained attention in real classroom settings: An eeg study. Front. Hum. Neurosci. **11**, 388 (2017)
12. Klimesch, W.: EEG alpha and theta oscillations reflect cognitive and memory performance: a review and analysis. Brain Res. Rev. **29**(2–3), 169–195 (1999)
13. Mohammadpour, M., Mozaffari, S.: Classification of EEG-based attention for brain computer interface. In: 3rd Iranian Conference on Intelligent Systems and Signal Processing (ICSPIS). IEEE (2017)
14. Ghosh, P., Mazumder, A., Bhattacharyya, S., Tibarewala, D.N.: An EEG study on working memory and cognition. In: Proceedings of the 2nd International Conference on Perception and Machine Intelligence, pp. 21–26. ACM (2015)
15. Pavlov, Y.G., Kotchoubey, B.: EEG correlates of working memory performance in females. BMC Neurosci. **18**(1), 26 (2017)

16. Mohamed, Z., El Halaby, M., Said, T., Shawky, D., Badawi, A.: Characterizing focused attention and working memory using EEG. Sensors **18**(11), 3743 (2018)
17. Hjorth, B.: EEG analysis based on time domain properties. Electroencephalogr. Clin. Neurophysiol. **29**(3), 306–310 (1970)
18. Lascari, L.D., Pera, R.J.: Wireless radio system optimization by persistent spectrum analysis. Google Patents (2015)
19. Djamal, E.C., Pangestu, D.P., Dewi, D.A.: EEG-based recognition of attention state using wavelet and support vector machine. In: 2016 International Seminar on Intelligent Technology and Its Applications (ISITIA), pp. 139–144. IEEE (2016)

A New Nano-robots Control Strategy for Killing Cancer Cells Using Quorum Sensing Technique and Directed Particle Swarm Optimization Algorithm

Doaa Ezzat$^{(\boxtimes)}$, Safaa Amin$^{(\boxtimes)}$, Howida A. Shedeed$^{(\boxtimes)}$, and Mohamed F. Tolba$^{(\boxtimes)}$

Scientific Computing Department, Faculty of Computer and Information Sciences, Ain Shams University, Cairo, Egypt
doaa.ezzat@gmail.com, ahmed_andeel76@hotmail.com, dr_howida@cis.asu.edu.eg, fahmytolba@gmail.com

Abstract. Nowadays cancer is considered one of the most killing diseases. Traditional cancer therapy leads to dangerous side effects on healthy tissues. A recent direction has been proposed to overcome these side effects. This direction is to use Nano-robots to deliver drugs directly to tumor cells without harming healthy ones. In this paper, we propose a new Nano-robots control strategy that uses Directed Particle Swarm Optimization (DPSO) algorithm for delivering Nano-robots to the cancer area. A Quorum Sensing (QS) algorithm is also used in this strategy to control drug concentration in the cancer area. The results show that using the proposed control strategy increases the rate of killing cancer cells efficiently. This study also proposes to use a certain number of Nano-robots for destroying 100 cancer cells.

Keywords: Nano-robots · Cancer treatment · Quorum Sensing · Directed PSO

1 Introduction

It was noticed that the number of deaths caused by cancer increased considerably in the last decades [1]. Cancer treatment is done traditionally by two ways: chemo-therapy and radio-therapy. Both of these two ways have harmful side effects on healthy tissues. To decrease these side effects, scientists proposed to use Nano-robots to deliver drugs directly to tumor area without affecting healthy tissues.

A Nano-robot is a Nano-scale robot programmed to perform a specific task. The size of this Nano-robot begins from 0.1 to 100 nm (a nanometer = 10^{-9} m). A Nano-robot generally has sensors, actuators, battery, data transmission module and on-board computer. The Nano-robots are injected in the blood vessel and try to find the tumor area. This area can be recognized by its pH value. During their journey to the target area, the Nano-robots need to control their movement using appropriate method like a swarm intelligence algorithm [22]. After reaching the target, the Nano-robots release the drug and kill only the cancer cells. Inside the tumor area, they also need an appropriate algorithm to control the release rate of the drug. The drug release must be

© Springer Nature Switzerland AG 2020
A. E. Hassanien et al. (Eds.): AMLTA 2019, AISC 921, pp. 218–226, 2020.
https://doi.org/10.1007/978-3-030-14118-9_22

controlled properly because if the drug dose is too small, it will not be enough to destroy the cancer cells. And if the drug dose is too large, it will harm the healthy cells. And in both cases, the expensive drug is wasted [2].

Particle Swarm Optimization (PSO) algorithm is one of the most famous swarm intelligence algorithms used in delivering Nano-robots to the target area. This technique was inspired from birds and fish. It was proposed by Kennedy and Eberhart [3]. It uses a swarm of particles flying in the solution space to get the optimal solution. Each particle represents one solution in the search space and has a position and a velocity. Each particle updates its position using the best position found by this particle and the best position found by the whole swarm.

Due to its simplicity and efficiency, PSO became a suitable solution for the problem of delivering a group of Nano-robots to the tumor area [4–6]. Many modifications have been made to PSO to enhance its ability to solve this problem. One of these modifications is Directed PSO (DPSO) algorithm [7]. This modification makes PSO more efficient in delivering a swarm of Nano-robots to the tumor area. It was proved experimentally that when using DPSO, all Nano-robots can reach the target area after a very few iterations [7].

This paper introduces a new control strategy that uses DPSO in delivering Nano-robots to cancer area. After reaching the target area, the proposed control strategy uses a Quorum Sensing algorithm to control drug concentration in this area. This algorithm simulates the quorum sensing behavior of bacteria. The Quorum Sensing algorithm can estimate the number of Nano-robots in the target area and then determine when to release the drug and how much dose has to be released to efficiently kill the cancer cells [8].

Experiments have shown that the proposed control strategy can speed up the process of eradicating cancer cells. This paper also provides an optimal number of Nano-robots that can destroy 100 tumor cells.

2 Related Work

A lot of work is made in the area of using Nano-robots for detecting and treating diseases [4–15]. Cavalcanti et al. introduced many Nano-robot architectures for medical applications [10–15]. They proposed new hardware architecture to diagnose brain aneurysm [10] and diabetes control [11]. In [12], the authors proposed the architecture for Nano-robots with sensing capability to identify medical target. These Nano-robots has the ability to move with six-degree-of-freedom, and each Nano-robot has a sensor for measuring chemical concentration without any communications between Nano-robots.

Lewis and Bekey [16] used swarm intelligence algorithms and chemical signaling techniques to remove tumors using Nano-robots. Chandrasekaran and Hougen [17] presented control strategies to apply swarm intelligence in medical Nano-robots; where Nano-robots can sense the chemical signals, make decisions for their next moves, and then reach the target region, which has the highest or the lowest chemical concentration. Also, there was no communication among Nano-robots.

In 2013, Zhang, Li and Guo [18] introduced a cooperative control technique for Nano-robots that can sense the pH value for delivering drugs to the cancer cells. They proposed a distributed control method that achieves cooperation by robot-to-robot communications. In their control method, the Nano-robots cooperate with each other to trace the gradient of pH values, and then reach the center of the tumor area which has the lowest pH value.

In 2014, Ahmed et al. [4, 19–21] proposed a new control strategy for Nano-robots to deliver drugs to the tumor area. They introduced a modification in the PSO algorithm (MPSO) to deliver a swarm of Nano-robots to the tumor region. They also made a modification in the obstacle avoidance algorithm (MOA) for avoiding collision with blood cells. Finally, they combined the MPSO with the MOA algorithms to make a novel control strategy that controls the movement of the Nano-robots in the blood vessel.

In 2015, a Nano-robot control strategy based on Quorum Sensing was presented [8] to control the number of Nano-robots in the target area and the drug release rate needed to kill the tumor cells. This strategy used chemical signals released by Nano-robots to estimate the number of Nano-robots in a certain position. This control strategy used Bacteria Foraging Optimization Algorithm (BFOA) for delivering Nano-robots to cancer area.

In 2016, the researchers presented an Improved Bacterial Foraging Optimization Algorithm (IBFOA) to reach and kill tumor cells using Nano-robots cooperation [5]. In 2017, an innovative computing-inspired bio-detection framework called touchable computing was proposed for cancer detection [6]. In the same year, a Q-learning algorithm was used to optimize the path travelled by Nano-robots [23]. This algorithm decreases the length of the path used by Nano-robots to reach their target.

3   The Proposed Control Strategy

3.1   Directed PSO Algorithm

Directed PSO (DPSO) algorithm represents a modification in PSO algorithm [7]. The main purpose of this algorithm is to efficiently deliver a swarm of Nano-robots to the cancer area. In this algorithm, Nano-robots use the traditional PSO to update their positions until one of them reaches the target. Then the other Nano-robots in the swarm direct themselves toward the position of the one that reached this target. Then they start to move in this direction with constant steps until the whole swarm becomes inside the target region.

Each Nano-robot detects if it reaches the tumor area or not by sensing the pH value at its current position. The pH value is 7.4 at normal cells. This value decreases continuously until the Nano-robot reaches the tumor area. In this area, the pH value is less than or equals to 5.7 [18].

DPSO can considerably reduce the time needed to deliver all Nano-robots to the cancer area, because it needs only one or two PSO iterations for delivering at least one Nano-robot to the target area. Then it needs a small number of direct steps to deliver all other Nano-robots to this area. Also, using DPSO guarantees the reachability of the target by all Nano-robots [7].

3.2   Quorum Sensing Algorithm

Quorum Sensing is a method that enables bacteria to communicate and coordinate using chemical signals [8]. Each bacterium is a single-celled microscopic organism and can produce light. When a bacterium is alone, it doesn't produce light. When some bacteria are aggregated together and reach a certain cell number, all the bacteria generate light simultaneously.

The bacteria can achieve this feature using chemical language. Each bacterium releases a chemical substance, called auto-inducer, to inform others about its position. If the bacterium is alone, the concentration of this substance is low and the chemical molecules just float away. On the other hand, when the bacteria are aggregated to reach a certain number, they are all participating in producing these chemical molecules. So, the concentration of these molecules proportionally increases with the number of bacteria. When the concentration reaches a certain amount, it tells the surrounding bacteria how many neighbors there are, and then all the bacteria generate light [8].

We can apply the Quorum Sensing method when using Nano-robots for killing cancer cells [8]. The main purpose of this method is to control the number of Nano-robots and drug release rate in the cancer area. Each Nano-robot represents a bacterium and releases a kind of chemical called Ai (Like auto-inducer) when it reaches the tumor region. Like bacteria, the concentration of Ai indicates the Nano-robots number inside this region. When a Nano-robot reaches the target region, and finds that the concentration of Ai exceeds an appropriate value, then it moves away to maintain the number of Nano-robots in this region at a certain threshold, and the other Nano-robots, that already exist in this region, decide when to release the drug and how much drug to be released to kill the cancer cells.

The concentration of Ai is affected by some factors, including diffusion and evaporation [8]. At some iteration j, when a Nano-robot i detects that it is inside the cancer area, it measures the Ai concentration. If the concentration of Ai is greater than the supper-threshold value h, it indicates that the cancer area is too crowded, and this Nano-robot will move away from this area. Otherwise, it will release $\Delta\tau(i,j)$ doses of Ai. $\Delta\tau(i,j)$ is an increment of Ai, which can be expressed as follows:

$$\Delta\tau(i,j) = \begin{cases} Q, & \Delta\tau(i,j) \leq h \\ 0, & otherwise \end{cases} \tag{1}$$

The Ai concentration is updated by the following formula:

$$\tau(i,j) = (1-\sigma)(1-\rho)\tau(i,j-1) + \Delta\tau(i,j) + m \tag{2}$$

Where σ is the Ai diffusion rate, ρ is the Ai evaporation rate, and m is the Ai diffusion constant that comes from other locations [8]. If the Ai concentration value $\tau(i,j)$ is between the closed interval $[l, h]$, where l is the lower threshold value for Ai concentration, it indicates that the number of Nano-robots in cancer area is sufficient, and the Nano-robot will stay at the target and release drug to kill cancer cells. This behavior is inspired from the quorum sensing of bacteria.

To control the release rate of drug, we used a method based on the concentration of Ai [8]. At iteration j, a Nano-robot i detects the concentration of drug $d(i,j)$ which is expressed as follows:

$$d(i,j) = (1 - \omega)d(i,j-1) + \Delta d(i,j) + n \tag{3}$$

Where $\Delta d(i,j)$ is the drug dose released by a Nano-robot i at step j. And n is the drug diffusion constant that comes from other locations [8]. When the concentration of Ai is l, the drug dose released by the Nano-robot is d_l. And when the concentration of Ai is h, the drug dose released by the Nano-robot is d_h. The drug dose released by an individual Nano-robot is inversely proportional to the concentration of Ai [8]. If the concentration of Ai t is in the open interval (l, h), the drug dose released by the Nano-robot is Δd, and can be expressed as follows:

$$\Delta d = \frac{(t - l)(d_h - d_l)}{(h - l)} + d_l \tag{4}$$

3.3 The Proposed Control Strategy

The proposed control strategy combines the DPSO algorithm with the Quorum Sensing algorithm (discussed in the previous two subsections) to form a new control strategy

Table 1. The proposed control strategy.

Begin
1.   Initialize Nano-robots with random positions
2.   **For** each Nano-robot i **do**
3.   **If** the Nano-robot i didn't reach the target area
4.   **Then If** there is no Nano-robots in the target area
5.   **Then** update the position of the ith Nano-robot with PSO
6.   **Else**
7.   Perform a direct step toward the global best position
8.   **End If**
9.   **Else If** $\tau(i,j) > h$
10.   **Then** the ith Nano-robot moves away from the target area
11.   **Else**
12.   **Then** update the Ai concentration by equations (1) and (2)
13.   **If** $l \leq \tau(i,j) \leq h$
14.   **Then** release drug dose Δd and update the drug concentration using equations (3) and (4)
15.   **End If**
16.   **End If**
17.   **End For**
End

that can efficiently deliver Nano-robots to the cancer area and enable them to efficiently destroy cancer cells. The proposed control strategy is shown in Table 1.

Experiments have proved that this control strategy reduces the number of iterations needed to kill cancer cells in the target region. The next section shows the results of these experiments in details.

4 Simulation and Results

We implemented a program using visual C++ to simulate the movements of the Nano-robots in the blood vessel. Initially Nano-robots are located randomly in the deployment area nearby the tumor region. The initial state of the Nano-robots is shown in Fig. 1. Simulations are performed using 10, 20 and 30 Nano-robots and repeated experiments 10 times each. Two different control strategies are applied in these simulations. The first control strategy is Strategy A. This strategy is the proposed control strategy. It uses DPSO for delivering Nano-robots to cancer area. The second one is Strategy B. This strategy is the control strategy proposed in [8]. It uses Bacterial Foraging Optimization Algorithm (BFOA) for delivering Nano-robots to cancer area. Both strategies use the same Quorum Sensing algorithm to control the drug concentration in the target area.

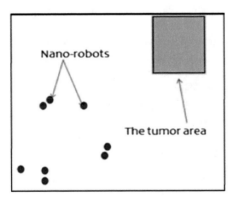

Fig. 1. The initial state of the Nano-robots

The BFOA, used in the Strategy B, is inspired from the behavior of bacteria. In this algorithm, each Nano-robot takes a unit step toward a random direction, then it will take another unit step toward the same direction if the fitness value is improved; otherwise it will make another random step [5].

We put $h = 0.8$, $l = 0.5$, $\sigma = 0.6$, $\rho = 0.5$, $d_l = 0.8$, $d_h = 0.5$, $\omega = 0.5$. Using Strategy A, we conducted 10 experiments with 10 Nano-robots, 10 experiments with 20 Nano-robots and 10 experiments with 30 Nano-robots. Each time we took the average number of iterations needed to kill 100 cancer cells. The results are shown in Table 2.

For Strategy B, we conducted 10 experiments with 10 Nano-robots, 10 experiments with 20 Nano-robots and 10 experiments with 30 Nano-robots. The average number of iterations needed to kill 100 cancer cells is calculated each time. Table 3 shows the averaged results.

Table 2. The results of Strategy A.

The number of Nano-robots	The average number of iterations
10	35
20	23
30	24

Table 3. The results of Strategy B.

The number of Nano-robots	The average number of iterations
10	46
20	45
30	36

From these results we can conclude that using the proposed strategy (Strategy A) significantly decreases the time needed to kill cancer cells. For instance, with 20 Nano-robots, the average number of iterations is reduced by about a half when using Strategy A instead of Strategy B. This improvement results from two reasons: the first is that Strategy A uses DPSO for delivering Nano-robots to the target area. And it was proved that DPSO delivers Nano-robots to the target in a very short time [7]. So, the total number of iterations needed to complete the process of destroying cancer cells is also reduced. The second reason is that the Quorum Sensing algorithm used in Strategy A needs the aggregation of Nano-robots to initiate the drug release. This aggregation is easily done by DPSO, used in Strategy A, because once a Nano-robot reaches

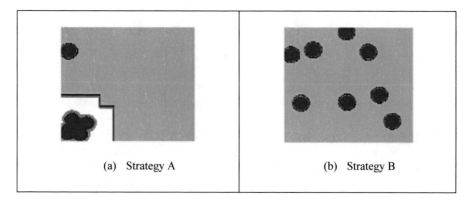

(a)   Strategy A   (b)   Strategy B

Fig. 2. The simulator state when using Strategy A and Strategy B.

the target, it attracts the others to its position. So, the drug release starts quickly and this reduces the time needed to kill the cancer cells. Figure 2 shows that using DPSO in Strategy A aggregates the Nano-robots but using BFOA in Strategy B scatters the Nano-robots. The cancer cells are colored with pink and when a cancer cell is killed, it is colored with white.

Using a relatively small number of Nano-robots, 10 for example, with Strategy B, extremely slows down the process of killing cancer cells. We found that in 60% of the experiments, that uses Strategy B with 10 Nano-robots, no cancer cells were killed even after expending 100 iterations. On the other hand, this situation occurred in only 10% of the experiments that uses Strategy A with 10 Nano-robots. In other words, Strategy A always outperforms Strategy B even if a small number of Nano-robots are used.

Also, we observed that the optimal number of Nano-robots needed to kill 100 cancer cells using Strategy A is 20. That is because before 20 the number of aggregated Nano-robots is not sufficient to start releasing the drug. And after 20 the target area becomes too crowded and the extra Nano-robots leave this area without any contribution in the drug release.

5   Conclusions and Future Work

It was proposed recently to use Nano-robots for delivering drugs to cancer cells without harming the healthy ones. Many algorithms were introduced in this field. Some of these algorithms focused on delivering Nano-robots to the target area, and some of them focused on controlling the drug release in this area. In this paper, we introduce a new control strategy that uses the Directed PSO algorithm for delivering Nano-robots to the cancer area and the Quorum Sensing algorithm for controlling the drug release after reaching the cancer area. Experiments proved that the proposed strategy efficiently kills the cancer cells. This paper also suggests an optimal number of Nano-robots to kill certain number of cancer cells.

In the future, we need to extend this work to treat more than one cancer area. We also need to modify the proposed strategy to deal with obstacles that may be encountered by Nano-robots on their way to tumor environment.

References

1. Devasena Umai, R., Brindha Devi, P., Thiruchelvi, R.: A review on DNA nanobots - a new technique for cancer treatment. Asian J. Pharm. Clin. Res. **11**(6), 61–64 (2018)
2. Zhao, Q., Li, M., Luo, J.: Relationship among reaction rate, release rate and efficiency of nanomachine-based targeted drug delivery. Technol. Health Care **25**, 1119–1130 (2017)
3. Eberhart, R., Kennedy, J.: A new optimizer using particle swarm theory. In: Proceedings of the Sixth International Symposium on Micro Machine and Human Science, pp. 39–43 (1995)
4. Ahmed, S.: Nano-Robotics Control for Biomedical Applications (Unpublished Doctoral Dissertation). Ain Shams University, Cairo, Egypt (2014)

5. Cao, J., Li, M., Wang, H., Huang, L., Zhao, Y.: An improved bacterial foraging algorithm with cooperative learning for eradicating cancer cells using nanorobots. In: Proceedings of the IEEE International Conference on Robotics and Biomimetics, Qingdao, China, pp. 1141–1146, 3–7 December 2016

6. Chen, Y., Shi, S., Yao, X., Nakano, T.: Touchable computing: computing-inspired bio-detection. IEEE Trans. Nanobiosci. **16**(8), 810–821 (2017)

7. Ezzat, D., Amin, S., Shedeed, H.A., Tolba, M.F.: Directed particle swarm optimization technique for delivering nano-robots to cancer cells. Accepted in the 13th IEEE International Conference on Computer Engineering and Systems (ICCES 2018) (2018)

8. Zhao, Q.Y., Li, M., Luo, J., Li, Y., Dou, L.H.: A quorum sensing algorithm to control nanorobot population and drug concentration in cancer area. In: IEEE International Conference on Robotics and Biomimetics, ZhuHai, China, pp. 42–47 (2015)

9. Ummat, A., Sharma, G., Mavroidis, C., Dubey, A.: Bio-nanorobotics: state of the art and future challenges. In: Biomedical Engineering Handbook. CRC Press, London (2005)

10. Cavalcanti, A., Shirinzadeh, B., Fukuda, T., Ikeda, S.: Nanorobot for brain aneurysm. Int. J. Robot. Res. **28**(4), 558–570 (2009)

11. Cavalcanti, A., Shirinzadeh, B., Kretly, L.: Medical nanorobotics for diabetes control. Nanomedicine **4**, 127–138 (2008)

12. Cavalcanti, A., Shirinzadeh, B., Freitas, R., Hogg, T.: Nanorobot architecture for medical target identification. Nanotechnology **19**, 1–15 (2008)

13. Cavalcanti, A., Freitas, R.: Nanorobotics control design: a collective behavior approach for medicine. IEEE Trans. Nanobiosci. **4**, 133–140 (2005)

14. Cavalcanti, A.: Assembly automation with evolutionary nanorobots and sensor-based control applied to nanomedicine. IEEE Trans. Nanotechnol. **2**, 82–87 (2003)

15. Cavalcanti, A., Freitas, R.: Autonomous multi-robot sensor-based cooperation for nanomedicine. Int. J. Nonlinear. Sci. Numer. Simul. **3**, 743–746 (2002)

16. Lewis, M., Bekey, G.: The behavioral self-organization of nanorobots using local rules. In: Proceedings of IEEE International Conference on Intelligent Robots and Systems, pp. 1333–1338 (1992)

17. Chandrasekaran, S., Hougen, D.: Swarm intelligence for cooperation of bio-nano robots using quorum sensing. In: Bio Micro and Nanosystems Conference, San Francisco, p. 104 (2006)

18. Zhang, S., Li, S., Guo, Y.: Cooperative control design for nano-robots in drug delivery. In: Guo, Y. (ed.) Selected Topics in Micro/Nano-Robotics for Biomedical Applications, pp. 101–123. Springer, New York (2013)

19. Ahmed, S. Amin, S.E., Alarif, T.: Simulation for the motion of nanorobots in human blood stream environment. In: Proceedings of Scientific Cooperation International Workshops on Electrical and Computer Engineering Subfields, pp. 70–75. Koc University, Istanbul/Turkey (2014)

20. Ahmed, S., Amin, S.E., Alarif, T.: Efficient cooperative control system for pH sensitive nanorobots in drug delivery. Int. J. Comput. Appl. **103**(1), 39–43 (2014)

21. Ahmed, S., Amin, S.E., Alarif, T.: Assessment of applying path planning technique to nanorobots in a human blood environment. In: Proceedings of the UKSim-AMSS 8th European Modeling Symposium on Mathematical Modeling and Computer Simulation, Pisa, Italy, pp. 45–51, 21–23 October 2014

22. Hassanien, A.E., Alamry, E.: Swarm Intelligence: Principles, Advances, and Applications. CRC – Taylor & Francis Group (2015). ISBN 9781498741064 - CAT# K26721

23. Lambe, A.: Reinforcement learning for optimal path length of nanobots using dynamic programming. In: Proceedings of IEEE International Conference on Industrial and Information Systems, Peradeniya, Sri Lanka (2017)

Data Mining: Technology and Applications

Mining Student Information System Records to Predict Students' Academic Performance

Amjad Abu Saa[1], Mostafa Al-Emran[2]([✉]) [ID], and Khaled Shaalan[1] [ID]

[1] Faculty of Engineering and IT, The British University in Dubai, Dubai, UAE
a.abusaa@ajman.ac.ae, khaled.shaalan@buid.ac.ae
[2] Faculty of Computer Systems and Software Engineering,
Universiti Malaysia Pahang, Gambang, Malaysia
malemran@buc.edu.om

Abstract. Educational Data Mining (EDM) is an emerging field that is concerned with mining and exploring the useful patterns in educational data. The main objective of this study is to predict the students' academic performance based on a new dataset extracted from a student information system. The dataset was extracted from a private university in the United Arab of Emirates (UAE). The dataset includes 34 attributes and 56,000 records related to students' information. The empirical results indicated that the Random Forest (RF) algorithm was the most appropriate data mining technique used to predict the students' academic performance. It is also revealed that the most important attributes that have a direct effect on the students' academic performance are belonged to four main categories, namely students' demographics, student previous performance information, course and instructor information, and student general information. The evidence from this study would assist the higher educational institutions by allowing the instructors and students to identify the weaknesses and factors affecting the students' performance, and act as an early warning system for predicting the students' failures and low academic performance.

Keywords: Educational data mining · Students' performance ·
Performance prediction

1 Introduction

In the past two decades, we have witnessed a vast technological advancement in the area of computers and information systems [1, 2]. Educational institutions took the advantage of this advancement and employed it to digitalize most of its educational information, and transactions [3, 4]. Currently, it is very hard to find an educational institution without a student information system, where most of the students' information resides [5]. Student information systems store a huge amount of data about students, such as student's demographics, courses' information, instructors' information, students' class attendance, students' grades, among many others. There is a high potential in those systems that could enable them to be used in educational data mining (EDM) to predict students' performance based on their data in the student information

© Springer Nature Switzerland AG 2020
A. E. Hassanien et al. (Eds.): AMLTA 2019, AISC 921, pp. 229–239, 2020.
https://doi.org/10.1007/978-3-030-14118-9_23

system [6]. This study addresses this matter, where we try to collect as many students' information as we can from an information system of a private university in the United Arab of Emirates (UAE), and we try to analyze and predict the student performance based on the retrieved data. In addition, we try to identify the most appropriate data mining technique for use in prediction. Analyzing and predicting the students' academic performance has been an emerging topic in the past two decades, and has been an interest for many researchers and scholars. With no exception, this study focuses on this subject. The most important aspect of this study is to learn from the past data and to help instructors and university management to understand the factors affecting their students' performance. This would assist them to focus on low performing students and help them to overcome their weaknesses and improve their academic performance [7].

The main goal of the present study is to build a predictive model based on student information system's data and select the most appropriate data mining techniques and algorithms to predict the students' academic performance. This research uses a list of 34 attributes of data extracted from a student information system of a private university in the UAE. It is worth mentioning that the selected university has a challenging environment for research. Also, it has a variety of students coming from more than 100 nations. Therefore, our research falls in such a challenging environment with various students' backgrounds and diversities. The present study addresses the following two research questions:

RQ 1: What is the most appropriate data mining technique/algorithm that could achieve the best results for predicting the students' academic performance?
RQ 2: What are the main predictors of students' academic performance among the attributes selected from the student information system?

2   Related Work

A study conducted by [8] studied 17 attributes of students, in which, nine of them were extracted from a student information system of a high school. The studied attributes are related to the students' scores in specific subjects, students' attendance, among the other attributes. It was found that the GPA in secondary school and the number of students enrolled in the group/class were the most significant attributes and had a large impact on the students' performance in a high school. Another study carried out by [9] conducted an EDM research on admission data trying to build an early warning system by predicting low performing students as early as possible and provide them with the potential opportunities to improve their performance. The study had a simple dataset in which it consists of eight attributes only. All the attributes were related to the students' high school marks as well as the first and second year of university. All data were collected from the university's student information system where the study took place. The study used multiple classification techniques to model their system, including Decision Trees, Rule Induction, K-NN, Naïve Bayes, etc. The results showed that all the attributes used in data mining had a significant impact on the students' academic performance; suggesting that all previous grades and scores achieved by the student are considered as successful predictors of a future student grade.

Fernandes and his colleagues [10] conducted a predictive analysis of the academic performance of public school students. The authors studied the effect of multiple students' attributes collected from two different datasets acquired at the beginning of the semester. The second dataset included the same attributes as the first dataset in addition to other attributes that became available after two months of the study, such as grades, absences, and subjects' information. The results showed that after adding these attributes, the importance of attributes from the first dataset became less significant. However, not all attributes of the second dataset bypassed the importance of the first dataset's attributes. Finally, the significant attributes according to their scale of importance include grades, student neighborhood, school, school subjects, absences, student city, and age. A study conducted by [11] employed a comparative research of predictive models of students' academic performance. The study was very specific in which it attempted to predict the students' performance on a specific course (i.e., Engineering Dynamics). The study focused on a limited number of attributes extracted from the student information system. Specifically, the attributes included students' last GPA, grades on related subjects, and scores of mid-term exams of the engineering dynamics course. The results showed that there were insignificant differences between the four DM models applied to the data when applied to all students. However, when applied to one student only, SVM was the most appropriate predictive model to be used. Additionally, it was found that the GPA is the only effective predictor variable among the other attributes used in the modeling when it is applied to all students. However, when predicting the individual student's performance, all attributes make a difference.

3 Dataset

3.1 Data Collection

A new dataset has been collected from a private university in the UAE. The university has more than 7000 active students and more than 35,000 graduates over the past 30 years. The data was extracted directly from the university database; a Microsoft SQL Server relational database. The dataset was limited to the past five years, from 2013 to 2018 including summer semesters. Besides, it was also limited to only regular academic courses (i.e., special courses like training, orientation, and studios, were not considered). The overall dataset size extracted from the database includes 231,782 records related to students' academic history. The dataset of this study has 34 attributes in total within four types (i.e., Demographics, Course and Instructor Information, Student General Information, and Student Previous Performance Information).

3.2 Data Preprocessing

Data preprocessing is a critical procedure in data mining research [12–14]. In this study, some data preprocessing steps were performed on the data, which were divided into two phases. The following subsections describe these two phases and the related processes in detail.

3.2.1   Phase 1

The first phase of data preprocessing was performed during the data extraction from the database by the SQL query. Some attributes were generated based on other fields as follows:

- Age: it was generated based on the difference between the student date of birth and the start date of the semester of the record.
- High School Merit: it was generated based on the student high school marks.
- Low Performer: it was labeled for students with high school mark less than 75%.
- High Performer: it was labeled for students with high school mark between 75% and 85%.
- Excellent: it was labeled for students with high school mark equal to or greater than 85%.
- Student and Instructor Genders: it was generated by comparing the student and instructor gender. If they were the same, it would be "True", otherwise, "False".
- Student and Instructor Nationalities: it was generated by comparing the student and instructor nationalities. If they were the same, it would be "True", otherwise, "False".
- Section Size Nominal: this attribute is based on the "Section Size" attribute, where:
  - Small: was given to classes of size less than 20 students.
  - Average: was given to classes of size 20 to 39 students.
  - Large: was given to classes of size 40 to 64 students.
  - Huge: was given to classes of size equal to or greater than 65 students.
- Class Timing: was calculated based on the class start and end times, where:
  - Morning: was given to classes starting from the early morning and ending no later than 1:30 PM.
  - Evening: was given to classes starting from 1:30 PM and ending after any time after that.
  - Mixed: was given to classes starting in the morning and ending after 1:30 PM.
- Math Average: is the average of the accumulated marks of a student in all mathematics subjects in which s/he studied in the university, where:
  - Failure: is for averages less than 60 marks.
  - Low Performer: is for averages equal to or greater than 60 and less than 75 marks.
  - High Performer: is for averages equal to or greater than 75 marks.
- Physics Average: is the average of the accumulated marks of a student in all physics subjects in which s/he studied in the university, where:
  - Failure: is for averages less than 60 marks.
  - Low Performer: is for averages equal to or greater than 60 and less than 75 marks.
  - High Performer: is for averages equal to or greater than 75 marks.
- Prerequisite Average: is the average of the accumulated marks of a student in all the prerequisites of the subject, in which, s/he is studying, where:
  - Failure: is for averages less than 60 marks.
  - Low Performer: is for averages equal to or greater than 60 and less than 75 marks.
  - High Performer: is for averages equal to or greater than 75 marks.

- University Requirements Average: is the average of the accumulated marks of a student in all university requirements subjects in which s/he studied in the university, where:
  - Failure: is for averages less than 60 marks.
  - Low Performer: is for averages equal to or greater than 60 and less than 75 marks.
  - High Performer: is for averages equal to or greater than 75 marks.
- Class (label): is also calculated based on the student final marks for the course in which s/he is studying, where:
  - Failure: is for marks less than 60.
  - Low Performer: is for marks equal to or greater than 60 and less than 75.
  - High Performer: is for marks equal to or greater than 75.

3.2.2 Phase 2

The second phase was undertaken using the RapidMiner software with an educational license. In that, all the data mining and preprocessing tasks were performed. It is worth mentioning that the dataset used in this study is imbalanced, in which most of the examples are of class "High Performer". This would mislead the performance measures in general and the accuracy measure in particular. Consequently, it was decided to use a balanced sample of the data where all the class examples are of equal size. Therefore, we have used the under-sampling technique to achieve this goal, in which the sample size of each class was downsized to match the size of the least class dataset. As a result, the total sample size became 56,544 (18,848 × 3), which is still good enough for a classification data mining study.

4 Data Mining Implementation and Results

4.1 Data Mining Approaches and Algorithms

According to the findings of the systematic literature review conducted by [15], it is pointed out that the most frequently used data mining techniques by the EDM community are Decision Trees (DT), Naive Bayes (NB), Artificial Neural Networks (ANN), Support Vector Machine (SVM), and Regression. Therefore, this study concentrates on these techniques in terms of selecting the DM algorithms. Table 1 represents the DM algorithms selected and applied in this study. Unfortunately, we could not use the SVM algorithms as it requires only numerical attributes, whereas our data has a mixture of nominal and numerical attributes.

Table 1. Selected data mining algorithms

Algorithm category	Algorithm name	Abbreviation
Decision Tree	Decision Tree (RapidMiner's default)	DT
Decision Tree	Random Forest	RF
Decision Tree	Gradient Boosted Trees	GBT
Artificial Neural Networks	Deep Learning	DL
Naïve Bayes	Naïve Bayes	NB
Regression	Logistic Regression	LR
Regression	Generalized Linear Model	GLM

4.2   Data Mining Implementation

Before finalizing the data mining models, it is crucial to test multiple settings and parameters for each predictive algorithm. Similarly, we have tested our data mining models with many combinations of settings and parameters before finalizing them into the list provided in Table 2; keeping in mind that the list mentions only the changes to default values.

Table 2. Algorithms parameters

Algorithm	Settings/Parameters				
DT	Criterion: Information Gain	Confidence: 0.05	Minimal gain: 0.001	Minimal leaf size: 1	
RF	Criterion: Information Gain	Confidence: 0.05	Minimal gain: 0.001	Minimal leaf size: 1	Number of Trees: 50
GBT	Learning Rate: 1.0				
DL	No change				
NB					
LR					
GLM					

Furthermore, the performance validation was generated using 10-fold cross-validation. The data were split into ten folds, in which nine-folds were reserved for the training and one-fold for the testing. The resulted accuracy is the average of the accuracies of the ten runs. In addition, we have used the Precision and Recall as performance measures in order to get a broader analysis of the algorithms performance.

4.3   Results

Table 3 summarizes the performance of the data mining algorithms executed on our selected dataset. As it can be seen from Table 3, the RF algorithm had the highest Accuracy score over the other algorithms; however, it does not have the best scores in terms of Recalls and Precisions classes. The RF is shown to have the best Recall of predicting "Failure" class, and the best Precision for "Low Performer" class. This indicates that RF is the best at predicting "Failure" class than other algorithms, and it was the second significant one in predicting the "High Performer" and "Low Performer" classes as well. On the other hand, a surprising result is noticed in the DL results. Although the data were consistently balanced across the three classes, the DL algorithm was powerful enough to predict 98% of "Low Performer" class records, but very weak in predicting the other two classes. Nevertheless, the DL is still a very powerful data mining algorithm and has a high potential to perform better in other data mining studies.

Table 3. The performance of seven data mining algorithms

Algorithm	Accuracy %	Recall %			Precision %		
		High performer	Low performer	Failure	High performer	Low performer	Failure
DT	68.49	60.43	62.16	82.89	63.74	56.82	86.52
RF	**75.52**	71.41	66.87	**88.28**	71.53	**66.04**	89.26
GBT	72.45	69.23	64.44	83.67	69.19	62.35	86.62
DL	57.33	10.55	**98.05**	63.39	**88.44**	43.86	**98.28**
NB	66.56	67.57	56.08	76.02	64.08	56.15	80.30
LR	72.88	70.26	64.12	84.26	70.18	63.02	85.85
GLM	58.70	**76.10**	32.86	67.13	52.30	47.83	78.25

Although "Failure" was the least class in the dataset before under-sampling the data, another interesting observation is that the majority of algorithms had a perfect Recall for "Failure" class over the other two classes. This is due to the successful dataset balancing procedure. However, we tried to increase the sample size of the other two classes slightly, in the hope that their Recall might increase, but we did not notice any significant improvements in the Recalls nor the overall Accuracies. Furthermore, it was noticed that the least predicted class among the three classes was "Low Performer" where most of the algorithms faced difficulties in predicting it. This might be reasoned to the fact that "Low Performer" is a middle class between "Failure" and "High Performer", in which most of the noise usually exists.

Additionally, the procedure of comparing the generated results with a baseline is usually practiced by researchers to emphasize their results. Since this study is based on a new dataset, we have selected the default model as our baseline which is described as a model that predicts the class of all examples in a dataset as the class of its mode. This creates a baseline measure that we can compare our results with. In our case, since the dataset is balanced, and we have three classes, the baseline Accuracy of the default model is easily calculated as 33.33%. Therefore, it is confirmed that the Accuracies of all the models generated in this study outperform the baseline model significantly. Overall, we can answer our first research question by stating that the most appropriate data mining algorithm for predicting the students' academic performance is the Random Forest (RF).

In terms of the second research question, Table 4 shows the top 15 most important attributes according to the information gain. As it can be noticed from Table 4, the most repeated attribute type is the "Student Previous Performance Information", in which it has eight records out of 15. This result indicates that the most important attribute type that mainly predicts a student's academic performance is the student's previous grades and general performance. This finding comes in line with the findings of other prior studies carried out by [8–10, 15–19].

Table 4. The most important attributes according to the Information Gain

Attribute type	Attribute	Importance
Demographics	High School Name	0.3842
Student Previous Performance Information	University Requirements Average	0.3016
Course and Instructor Information	Course Name	0.1939
Student Previous Performance Information	Attendance Warning	0.1164
Student Previous Performance Information	Number of Absences	0.0959
Student Previous Performance Information	Prerequisite Average	0.0742
Demographics	Student Program	0.0567
Student Previous Performance Information	High School Merit	0.0510
Student Previous Performance Information	High School Percentage	0.0503
Student Previous Performance Information	Math Average	0.0488
Student General Information	Has Discount	0.0382
Demographics	Student Nationality	0.0332
Course and Instructor Information	Offering College	0.0332
Demographics	Gender	0.0325
Student Previous Performance Information	Physics Average	0.0244

Surprisingly, the first and most important attribute was "High School Name" which indicates that a student is going to perform very well or very bad in college based on the high school that s/he attended. However, this attribute had a huge number of distinct values in the dataset, in which it includes more than 3200 high schools. Additionally, it was noted that there are other three demographical information that were identified as more important than the "High School Name", namely "Student Program", "Student Nationality", and "Gender". This indicates that the demographical information holds an important value for predicting the students' academic performance. This finding is consistent with the findings of existing studies conducted by [8, 10, 17–24]. It is also evident that the difficulty of a program affects the students' academic performance, either negatively or positively. On the other hand, the student nationality was also flagged as relevant to the prediction of students' academic performance in which students coming from certain countries performed better than others. Moreover, it was evident from the prediction models that males and females have significant differences in their performance.

Comparatively, including the course names in the dataset seemed a bad idea at first, but after reaching out to the results, it was concluded that some courses in the university are harder than others, and some of them are easier. Hence, the prediction model confirms this issue and uses this attribute effectively to predict the students' performance accordingly. This also seems very important in order to warn the students studying these courses in the future, in which they might face difficulties in getting higher grades and provide them with the potential opportunities to increase their efforts accordingly. It was also learned that the student who has a discount from the university would perform better in his/her study. This finding is in line with the findings concluded by [21].

Given the above analysis and results, we can now answer our second research question. The most important attributes and the main predictors of students' academic performance are categorized under four types, namely Students' Demographics [8, 10, 17, 24], Student Previous Performance Information [8–10, 15–19], Course and Instructor Information [16, 25–27], and Student General Information [21].

5   Conclusion and Future Works

The main objective of this study is to predict the students' academic performance based on a new dataset extracted from a student information system. Different data mining algorithms were applied on the extracted dataset. The results indicated that the Random Forest (RF) algorithm was the most appropriate data mining technique used to predict the students' academic performance. The results also pointed out that the most important attributes that have a direct effect on the students' academic performance are the student's high school, university requirements grades, some specific courses, attendance warnings, number of absences, the course's prerequisites average, student program, high school grades and percentage, grades in math, having a discount, nationality, gender, and physics grades. These attributes were belonged to four main types, namely students' demographics, student previous performance information, course and instructor information, and student general information. Future studies can rely on this set of attributes instead of wasting time extracting unnecessary information about the students. The evidence from this study would assist the higher educational institutions by allowing the instructors and students to identify the weaknesses and factors affecting the students' performance, and act as an early warning system for predicting the students' failures and low academic performance.

This study is subject to two limitations. First, the study was limited to the data extracted from a student information system only. Further research could replicate this study by extracting data from other relevant systems such as eLearning systems in order to improve the prediction levels of students' performance. Second, the data collection in this study was limited to the past five years only while extracting the dataset. Future studies are encouraged to include more historical records in order to leverage the prediction levels of students' performance.

References

1. Mhamdi, C., Al-Emran, M., Salloum, S.A.: Text mining and analytics: a case study from news channels posts on Facebook, vol. 740 (2018)
2. Al-Emran, M.: Hierarchical reinforcement learning: a survey. Int. J. Comput. Digit. Syst. **4** (2), 137–143 (2015)
3. Salloum, S.A., Al-Emran, M., Abdallah, S., Shaalan, K.: Analyzing the Arab gulf newspapers using text mining techniques. In: International Conference on Advanced Intelligent Systems and Informatics, pp. 396–405 (2017)
4. Salloum, S.A., Al-Emran, M., Monem, A.A., Shaalan, K.: Using text mining techniques for extracting information from research articles. In: Studies in Computational Intelligence, vol. 740. Springer (2018)
5. Romero, C., Ventura, S.: Educational data mining: a review of the state of the art. IEEE Trans. Syst. Man Cybern. Part C Appl. Rev. **40**, 601–618 (2010)
6. Kiron, D., et al.: Analytics: the widening divide. MIT Sloan Manag. Rev. **53**, 1 (2012)
7. Bienkowski, M., Feng, M., Means, B.: Enhancing teaching and learning through educational data mining and learning analytics: an issue brief. DC SRI Int., Washington (2012)
8. Márquez-Vera, C., Cano, A., Romero, C., Noaman, A.Y.M., Mousa Fardoun, H., Ventura, S.: Early dropout prediction using data mining: a case study with high school students. Expert Syst. **33**, 107–124 (2016)
9. Asif, R., Merceron, A., Ali, S.A., Haider, N.G.: Analyzing undergraduate students' performance using educational data mining. Comput. Educ. **113**, 177–194 (2017)
10. Fernandes, E., Holanda, M., Victorino, M., Borges, V., Carvalho, R., Van Erven, G.: Educational data mining: predictive analysis of academic performance of public school students in the capital of Brazil. J. Bus. Res. **94**, 335–343 (2018)
11. Huang, S., Fang, N.: Predicting student academic performance in an engineering dynamics course: a comparison of four types of predictive mathematical models. Comput. Educ. **61**, 133–145 (2013)
12. Salloum, S.A., Al-Emran, M., Shaalan, K.: Mining social media text: extracting knowledge from Facebook. Int. J. Comput. Digit. Syst. **6**(2), 73–81 (2017)
13. Salloum, S.A., Al-Emran, M., Shaalan, K.: Mining text in news channels: a case study from Facebook. Int. J. Inf. Technol. Lang. Stud. **1**(1), 1–9 (2017)
14. Salloum, S.A., Mhamdi, C., Al-Emran, M., Shaalan, K.: Analysis and classification of Arabic newspapers' Facebook pages using text mining techniques. Int. J. Inf. Technol. Lang. Stud. **1**(2), 8–17 (2017)
15. Shahiri, A.M., Husain, W., Rashid, N.A.: A review on predicting student's performance using data mining techniques. Procedia Comput. Sci. **72**, 414–422 (2015)
16. Gómez-Rey, P., Fernández-Navarro, F., Barberà, E.: Ordinal regression by a gravitational model in the field of educational data mining. Expert Syst. **33**, 161–175 (2016)
17. Márquez-Vera, C., Cano, A., Romero, C., Ventura, S.: Predicting student failure at school using genetic programming and different data mining approaches with high dimensional and imbalanced data. Appl. Intell **38**, 315–330 (2013)
18. Kotsiantis, S.B.: Use of machine learning techniques for educational proposes: a decision support system for forecasting students' grades. Artif. Intell. Rev. **37**, 331–344 (2012)
19. Anuradha, T., Velmurugan, T.: A comparative analysis on the evaluation of classification algorithms in the prediction of students performance. Indian J. Sci. Technol. (2015)

20. Abazeed, A., Khder, M.: A classification and prediction model for student's performance in University level. J. Comput. Sci. **13**, 228–233 (2017)
21. Abu Saa, A.: Educational data mining & students' performance prediction. Int. J. Adv. Comput. Sci. Appl. **7**, 212–220 (2016)
22. Badr El Din Ahmed, A., Sayed Elaraby, I.: Data mining: a prediction for student's performance using classification method. World J. Comput. Appl. Technol. **2**, 43–47 (2014)
23. Araque, F., Roldán, C., Salguero, A.: Factors influencing University drop out rates. Comput. Educ. **53**, 563–574 (2009)
24. Inan, F.A., Yukselturk, E., Grant, M.M.: Profiling potential dropout students by individual characteristics in an online certificate program. Int. J. Instr. Media **36**, 163–177 (2009)
25. Jiang, Y.H., Javaad, S.S., Golab, L.: Data mining of undergraduate course evaluations. Inform. Educ. **15**, 85 (2016)
26. Hung, J., Hsu, Y.-C., Rice, K.: Integrating data mining in program evaluation of K-12 online education. Educ. Technol. Soc. **15**, 27–41 (2012)
27. Costantini, P., Linting, M., Porzio, G.C.: Mining performance data through nonlinear PCA with optimal scaling. Appl. Stoch. Model. Bus. Ind. **26**, 85–101 (2010)

Bayesian Classification of Personal Histories - An application to the Obesity Epidemic

Christopher R. Stephens[1]([✉]), José Antonio Borras Gutiérrez[2], and Hugo Flores[3,4]

[1] C3 - Centro de Ciencias de la Complejidad, Instituto de Ciencias Nucleares, Universidad Nacional Autónoma de México, 04510 Mexico City, Mexico
stephens@nucleares.unam.mx

[2] Facultad de Ciencias, Universidad Nacional Autónoma de México, 04510 Mexico City, Mexico

[3] TechMileage, 3295 N Drinkwater Blvd, Suite 13, Scottsdale, AZ 85251, USA

[4] C3 - Centro de Ciencias de la Complejidad, Universidad Nacional Autonoma de Mexico, 04510 Mexico City, Mexico

Abstract. Bayesian classifiers are an important tool in machine learning. The Naive Bayes classifier in particular has shown itself to be a robust, computationally efficient and transparent approximation. However, due to its strong assumption of feature independence it is often too easily discarded. Understanding under what circumstances feature correlations are likely to occur and how to diagnose them is an important first step in improving the Naive Bayes approximation. Furthermore, combining features can lead to enhanced approximations. In this paper we show the benefits of a Generalised Bayes Approximation that accounts for feature correlations while, at the same time, showing that an important area where feature correlations are ubiquitous is in time series that represent aspects of human behaviour. Using data that represent historical patterns of exercise we show how more predictive and more insightful models for obesity can be constructed using a Generalised Bayes approximation that combines historical features, thereby capturing the idea of human habits. In particular, by analysing feature combinations we show that abandoning good past exercise habits is more correlated with obesity than never having had them in the first place.

Keywords: Naive Bayes approximation · Generalised Bayes · Time series · History · Obesity · Exercise

1 Introduction

The Naive Bayes approximation (NBA) and associated classifier (NBC) are widely used and offer robust performance across a large spectrum of problem domains (Domingos and Pazzani 1996). It depends, however, on a very strong

© Springer Nature Switzerland AG 2020
A. E. Hassanien et al. (Eds.): AMLTA 2019, AISC 921, pp. 240–249, 2020.
https://doi.org/10.1007/978-3-030-14118-9_24

assumption - independence among features. This has led many to criticise its use, claiming that it has been superseded by much more sophisticated classifiers. Often this criticism is superficial, based only on the fact that it is based on a strong assumption. It remains, however, a robust, computationally efficient and transparent algorithm that leads to surprisingly good results. The reasons why it performs so well, in spite of the strong independence assumption, are now much better understood. Importantly, however, one can use diagnostics (Stephens et al. 2017) to determine when the correlation between features is sufficiently strong so as to indicate that the distinct features would be better considered as one joint feature.

Having a mathematical machinery for generalising the NBA in hand, it is important to determine which problem domains and/or variable types typically contain correlated features. In this paper, we will show how going beyond the NBA naturally occurs in situations where features are to be extracted out of time series where correlations are a common occurrence. Furthermore, time series - "histories" - that represent human behaviour, where the notions of homeostasis, habit and instinct play important roles, would be expected to show a high degree of correlation, as is manifest in the difficulties in fomenting positive behaviour change in the context of lifestyle diseases such as obesity and type-2 diabetes (Teixeira et al. 2015). In particular, we will present a computationally simple, illustrative and highly relevant example, where the relationship between obesity and historical patterns of exercise is considered, showing that the degree of correlation is also a function of educational level. Obesity is currently one of the world's major health crises (WHO 2000) associated with a host of chronic-degenerative diseases. Two major risk factors are malnutrition and sedentarism (Church et al. 2011).

2   Naive and Generalised Bayesian Classifiers

In this section we will consider the NBA and a simple class of generalisations - the Generalised Bayes Approximation (GBA) - wherein the complete factorisation associated with the NBA is not a prerequisite. The advantage is that we may look for approximations that better respect correlations that exist between features in the feature set, while the disadvantage is that there are very many potential factorisations to check. In both cases it is important to be able to have a base evaluation of the potential predictive value of features or feature combinations.

As a potential feature selection method and as a means to gauge the degree of predictability inherent in a feature X, we use a binomial test to determine the degree of statistical significance of the relation between the class of interest C and X relative to a null hypothesis. The corresponding statistical diagnostic ε is given by

$$\varepsilon = \frac{N_X(P(C|X) - P(C))}{\sqrt{N_X P(C)(1 - P(C))}} \tag{1}$$

where the null hypothesis $P(C) = N_C/N$ is associated with the class probability independent of the conditioning variable X. Here, N_C is the number of elements in the class and N the overall sample size. $P(C|X) = N_{CX}/N_X$, where N_{CX} is the number of sample elements associated with the feature X that are also in the class C, while N_X is the number of sample elements associated with the feature X. The numerator of Eq. (1) is therefore the difference between the actual number of elements in the class C and with feature X relative to the expected number if the null hypothesis were valid. The denominator is just the standard deviation of the binomial distribution. Values of ε allow hypothesis testing in the standard way. When the binomial distribution may be approximated by a normal distribution, $|\varepsilon| > 1.96$ corresponds to the 95% confidence interval that the observed data is not consistent with the null hypothesis.

The NBA is set up in the standard fashion: We wish to determine $P(C|\mathbf{X})$, where $\mathbf{X} = X_1 X_2, \ldots X_M$ represents a vector of features. Bayes theorem relates the posterior probability $P(C|\mathbf{X})$, associated with the information in the features $\mathbf{X}$, to a prior probability $P(C)$ in the absence of the information

$$P(C|\mathbf{X}) = \frac{P(\mathbf{X}|C)P(C)}{P(\mathbf{X})} \qquad (2)$$

where $P(\mathbf{X}|C)$ is the likelihood of observing the evidence $\mathbf{X}$ given the class C and $P(\mathbf{X})$ is the probability of the evidence $\mathbf{X}$. We now assume independence of the features X_i, which yields $P(\mathbf{X}|C) = \prod_{i=1}^{M} P(X_i|C)$. In order to eliminate $P(\mathbf{X})$ it is convenient to consider the ratio $P(C|\mathbf{X})/P(\overline{C}|\mathbf{X})$, where $\overline{C}$ is the set complement of C, i.e., those not in the class C. The likelihood $P(\mathbf{X}|\overline{C})$ is also factorised to yield a score function

$$S(C|\mathbf{X}) = \ln\left(\frac{P(C|\mathbf{X})}{P(\overline{C}|\mathbf{X})}\right) = \sum_{i=1}^{M} S(X_i) + \ln(P(C)/P(\overline{C})) \qquad (3)$$

where

$$S(X_i) = \ln\left[\frac{P(X_i|C)}{P(X_i|\overline{C})}\right] = \ln\left[\frac{\frac{N_{CX_i}}{N_c}}{\frac{N_{X_i} - N_{CX_i}}{N - N_C}}\right] \qquad (4)$$

Viewing $S(C|\mathbf{X})$ as a classifier, if $S(C|\mathbf{X}) > 0$, an element corresponding to the feature set $\mathbf{X}$ is placed in the class C, and, on the contrary, if $S(C|\mathbf{X}) < 0$. A class of generalisations of the NBA can be posited from a generalisation of the factorisation of the likelihood $P(\mathbf{X}|C) = \prod_{i=1}^{M} P(X_i|C)$ to $P(\mathbf{X}|C) = \prod_{i=1}^{M_\xi} P(X_\xi|C)$ (Stephens et al. 2017; Kononenko 1991), where, now, instead of X_ξ representing a single feature value, it can represent a combination of feature values. M_ξ is the number of feature combinations in a given factorisation. As a simple example, consider the case of three binary features - X_1, X_2 and X_3 - and a corresponding likelihood $P(X_1 X_2 X_3|C)$. In the NBA, $P(X_1 X_2 X_3|C) = P(X_1|C)P(X_2|C)P(X_3|C)$. However, this is only one potential factorisation of the likelihood. Another is $P(X_1 X_2 X_3|C) = P(X_1 X_2|C)P(X_3|C)$, or $P(X_1 X_2 X_3|C) = P(X_1|C)P(X_2 X_3|C)$ etc. Of course, we may view these latter factorisations as another version of the NBA, where the composite variable is treated as a single variable but now with a different cardinality. For instance, for two binary variables X_1 and X_2 we may consider the single composite variable $\xi_1 = X_1 X_2$ with now $2 \times 2 = 4$ possible values - 00, 01, 10 and 11.

The question then is: Which feature combinations should be put together? Viewing feature combinations as single features with a higher cardinality, our diagnostic ε, or simple generalisations thereof, can be utilised for this task (Stephens et al. 2017). We may then ask which factorisation leads to better performance and/or greater insight.

3 Data and the Prediction Problem

The data is associated with 1,076 academics, non-academics and students of the Universidad Nacional Autónoma de México (UNAM) using protocols approved by the Ethics Committee of the Facultad de Medicina of the UNAM, approved under the project FM/DI/023/2014, and with the aim of developing a deeper understanding of the time evolution of risk factors for obesity and metabolic disease. The age range of the population was 23–85 years, with an average age of $\mu = 47$ with standard deviation $\sigma = 12.6$. 21% were obese, corresponding to a Body Mass Index (BMI) > 30, where $BMI = weight/height^2$.

The problem is associated with two prediction classes: (i) current obesity status for a participant, as measured using BMI; and (ii) academic/non-academic, a binary variable denoting if a participant's job description was associated with the academic versus non-academic sector. The class variable $C = 1$ for obese or academic in the two distinct models.

Although an overall goal of the study is to predict risk factors for obesity, previous analysis had shown a very important difference - greater than 150% - in obesity rates between academics and non-academics and therefore an important subsidiary objective was to understand what differentiated academics from non-academics.

Although over 2000 distinct data fields were collected on this population, here we will focus, for the purpose of illustration, on 5 data variables that correspond to a participant's self-reported history of participation in exercise. The explicit variables are: *ejer30, ejer20, ejer10, ejer5, ejer1, ejer0*, where the value of each variable is the number of hours of exercise reported by the participant as being typical of their exercise pattern currently (*ejer0*), 1 year ago (*ejer1*), 5 years ago (*ejer5*), 10 years ago (*ejer10*), 20 years ago (*ejer20*) and 30 years ago (*ejer30*). Participants could also report a "not applicable" status to account for the fact, for example, that they were not of sufficient age to self-report an exercise status. Thus, the overall goal is to predict obesity or academic status according to participants' self-reported exercise history. In the first case, we wish to understand whether certain exercise histories are more predictive of actual obesity and, given that academics display a much lower overall obesity rate, determine if they exhibited significantly different exercise histories relative to non-academics.

As exercise was reported in hours/week, few participants had exactly the same history. As we wished to use Bayesian classifiers we therefore coarse grained the values considering a binary classification for each $ejer_i = 0,\ 1$, where 0 signifies less than the minimum recommended weekly number of exercise hours and 1 signifies greater than that number. We used 2.5 h (WHO 2018) as the recommended minimum amount of aerobic activity of moderate intensity.

With this coarse graining, a participant j could be represented by a sequence $\mathbf{X}^j = a_0^j a_1^j a_2^j a_3^j a_4^j a_5^j$ that represented their physical activity pattern history. Attribute a_0 corresponds to exercise 30 years ago, while a_5 refers to their current exercise pattern. $a_i^j = A,\ B$ represents a person who at the time period i exercised more (A) or less (B) than the recommended amount. For example, the sequence $\mathbf{X}^j = BBBABB$ corresponds to a person who was exercising less than the recommended amount 30, 20 and 10 years ago; exercising more than the recommended amount 5 years ago and less than that amount 1 year ago and currently. As mentioned, some participants may not be old enough to report exercise for the older time periods. Their histories can be represented either by shorter sequences or using another symbol N to represent the non-applicability of that time period. Of the 1076 exercise histories: 658 contain 6 applicable characters $(a_0 \to a_5)$, 281 contain 5 $(a_1 \to a_5)$, 128 contain 4 $(a_2 \to a_5)$ and 9 correspond to other cases. Note that there are $2^6 = 64$ distinct exercise histories, each one representing a specific feature. The advantage of considering full histories is that it allows us to explore the particularities of a given history. However, it potentially obfuscates the search for interesting patterns in different histories. For instance, an exercise history $ABABAB$ is unlikely to occur. However, if we are interested in the relation between obesity now and current exercise pattern then this is a sequence that would be examined relative to others, such as $AAAAAB$ or $BBBBB$. All these histories have in common that currently the corresponding persons are doing less exercise than recommended.

From the point of view of a classifier, there are two natural interpretations of a history $\mathbf{X}$. One is to consider the entire history as a single composite variable with a cardinality of 64, while another is to consider it as 6 single variables with cardinality 2. Thus, we may imagine a likelihood $P(\mathbf{X}|C)$ associated with a history as not being factorised, or also consider it as $P(\mathbf{X}|C) = P(a_0|C)P(a_1|C)P(a_2|C)P(a_3|C)P(a_4|C)P(a_5|C)$ - the NBA to this history. Of course, we may also consider any feature combination of up to 6 variables. Thus, for instance, we may consider $P(\xi|C) = P(a_i a_j|C)$. This would be the probability that given the class attribute the observed exercise pattern is described by $a_i a_j$. For example, for $a_i = a_2 = A$ and $a_j = a_5 = B$ we have that $P(a_i a_j|C)$ represents the probability that an obese person would be associated with the exercise pattern where currently they exercise less than the minimum recommended amount whereas 10 years ago they were exercising more than the minimum recommended amount.

4   Profiling Histories

We restrict attention to those 658 histories that contain 6 applicable characters A/B. Of these, 33% were academics and 25% were obese. For this group we calculate the values of ε, using Eq. (1), for each exercise history pattern, and also the

Table 1. This table shows the 5 history schemata of highest ε values, and also the 5 lowest, for the obesity class. We also show the score contributions for each history.

History	ϵ	N_x	N_{cx}	%	Score
A*A*BB	3.56	94	38	40.43	0.73
AAA*B	3.55	91	37	40.66	0.74
AA**BB	3.53	113	44	38.94	0.67
AA**B*	3.40	131	49	37.40	0.60
A***BB	3.23	137	50	36.50	0.57
*A***A	−3.27	157	21	13.38	−0.75
***AAA	−3.27	157	21	13.38	−0.75
AA**AA	−3.51	103	10	9.71	−1.11
*A**AA	−3.61	134	15	11.19	−0.95
****AA	−3.76	193	25	12.95	−0.79

corresponding values of the score function $S(C|\mathbf{X})$, using Eq. (4). This is done for the two classes - obesity and academic. $N_{C\mathbf{X}}$ represents the number of persons in the class C and with the exercise history $\mathbf{X}$, while $N_{\mathbf{X}}$ represents the number of persons with that exercise history.

As many exercise patterns are associated with a small number of participants, thereby reducing the potential statistical significance of any relation between the class of interest and that exercise pattern, we consider a further coarse graining, introducing a notation familiar in Evolutionary Computation (Holland 1992), that of a schema associated with a wildcard symbol *. Thus, * at any point in the sequence means we "don't care" about its value. For example, the sequence $* * * * *B$ represents those participants who are doing less than the recommended amount of exercise currently, independent of their previous history. In this case In terms of probabilities

$$P(* * * * *B) = \sum_{a_0=A,\,B} \sum_{a_1=A,\,B} \sum_{a_2=A,\,B} \sum_{a_3=A,\,B} \sum_{a_4=A,\,B} P(a_0 a_1 a_2 a_3 a_4 B) \quad (5)$$

represents a coarse-grained marginal probability relative to the original fine-grained set of probabilities on specific histories.

In Table 1 we see the top and bottom 5 ranked schemata in terms of ε. Note that all show a statistically significant relation beyond the 95% confidence interval ($p < 0.05\,|\varepsilon > 1.96|$). The most striking conclusion to draw from these results is that obesity is much more linked to patterns of exercise such that the corresponding persons exercised more than the minimum recommended other than in the recent past but currently and in the recent past - one year - they exercised less than the recommended amount. On the contrary, the highest probabilities to not be obese are associated with patterns of exercise that show no significant change through time. Simply put, in terms of habits, obesity is much linked to the loss of a good habit while not being obese is associated with maintaining a good habit. Besides

Table 2. This table shows the 5 history schemata of highest ε values, and also the 5 lowest, for the academic class. We also show the score contributions for each history.

History	ϵ	N_x	N_{cx}	%	Score
*A***A	5.55	157	85	54.14	0.86
*A**AA	5.21	134	73	54.48	0.88
*AA**A	5.13	135	73	54.07	0.86
*A*A*A	5.06	129	70	54.26	0.87
AA	4.97	165	85	51.52	0.76
*BBB**	−4.32	197	37	18.78	−0.77
***BB*	−4.40	267	55	20.60	−0.65
**BBB*	−4.41	207	39	18.84	−0.76
***BBB	−4.41	245	49	20.00	−0.69
***B*B	−4.55	260	52	20.00	−0.69

the fact that we can quantify this fact the notable conclusion is that changing from good to bad habits is seen to be much more linked to obesity than maintaining a bad habit. The latter would manifest itself through the appearance of schemata such as $*BB * *B$, $BB * *BB$ etc.

Turning to the *academic* class, we can perform a similar analysis to determine which patterns of exercise most differentiate between academics and non-academics. In Table 2 we see again the list of the 5 highest and 5 lowest values of ε. As with obesity there is a very sharp differentiation between the class and its complement. Academics are clearly distinguished by a higher incidence of good exercise habits while non-academics are distinguished by a higher incidence of bad exercise habits.

5   Model Performance Using Naive Bayes Vs. Generalised Naive Bayes

We will now show how different models using different feature combinations lead to different classifier performance. In this analysis we used histories with both 6 and 5 features. We can combine together the schemata from these two different history types into a hyperschemata using the symbol # to denote that we don't care if the corresponding end symbol came from a history with 5 or 6 defining features. In other words, # can represent A, B or N, where N always occurs to the left, i.e., earlier, of the A and B symbols. Joining those persons with either 5 or 6 symbols we have 939 individuals. We divide this population into a training set (626 individuals) and a test set (313 individuals). In Table 3 we see the most important feature combinations, which is similar to that of Table 1. We now consider the predictive value and performance of models based on different factorisations of the histories. We will consider histories of the form

Table 3. The top 5 and bottom 5 ranked histories with highest and lowest values for ε as determined from the training set

History	ϵ	N_x	N_{cx}	%	Score
#A***B	2.98	184	59	32.07	0.47
#A*B*B	2.81	84	30	35.71	0.63
#*A**B	2.79	177	56	31.64	0.45
#*A*BB	2.77	123	41	33.33	0.52
#****B	2.70	392	112	28.57	0.30
#*A**A	−2.93	160	21	13.13	−0.67
#A**AA	−2.95	121	14	11.57	−0.82
#**A*A	−2.98	167	22	13.17	−0.67
#****A	−3.50	234	31	13.25	−0.66
#***AA	−3.55	185	22	11.89	−0.79

$\#a_1 a_2 a_3 a_4 a_5$. In other words we will consider 32 models associated with different exercise histories up to 20 years in the past. A model in this setting is just a score function associated with a given factorisation or coarse graining of the history. We can then compare the relative performance of each model. In particular,

1. *Naive Bayes*: Here, the score is the sum of scores for each time period. Thus, for example, $S_{NB}(\#AABBB) = S(a_0 = \#) + S(a_1 = A) + S(a_2 = A) + S(a_3 = B) + S(a_4 = B) + S(a_5 = B)$
2. *Generalised Bayes*: Here, the score is taken to just come from the considered feature combination. For example: $S(a_0 = \#, a_1 = A, a_2 = A, a_3 = B, a_4 = B, a_5 = B)$

Model performance may be measured via multiple metrics. For instance, one may use the AUC associated with the ROC. The ROC curves for the models with $a_0 = \#$ y $a_n = \{A, B\}$, with $n > 0$ for both the NBA and the generalised approximation can be seen in Figs. 1a and b, respectively.

In Table 4 we see a comparison of the overall model performance. We can clearly see that models that account for the time correlations within the histories lead to better predictive performance. This is a mathematical manifestation of the fact that people are creatures of habit. It is easier to maintain a habit than to change it and it is significantly easier to change a good habit into a bad one rather than vice versa. Finally, in Table 5 we note how more detailed historical information leads to more accurate Generalized Bayes models with a 15% improvement in AUC for a single variable NB model ($\# * * * *L$) corresponding to only the most recent data, versus a model that contains information that contains the maximal amount of historical information ($\#LLLLL$).

Table 4. Model performance comparison for three models: the NB model, a *Generalised Bayes* model with histories #*LLLLL* and a *Generalised Bayes* model with coarse grained histories #*L ∗ ∗LL*. The column *threshold* refers to the score threshold used for the model to classify predictions. N and P are the number in the no class and class respectively ($N = 246$, $P = 67$), while TN and TP are the number of true negatives and true positives associated with each model. PPV is the *positive predictive value*, defined as $PPV = TP/(TP + FP)$ where FP means the number of false positives. $x(1 - TN/N)$ and $y(TP/P)$ are the specificity and sensitivity respectively. Dist is the distance of the point on the ROC curve farthest from the diagonal line corresponding to a random model and Area is the AUC.

ScoreType	Best	Model	Threshold	TN	TP	PPV	x(1-TN/N)	y (TP/P)	dist	Area
Generalised	PPV	#L**LL	0.27	186	32	0.36	0.24	0.48	0.17	0.66
Generalised	dist	#LLLLL	−0.48	106	59	0.30	0.57	0.88	0.22	0.70
Generalised	Area	#LLLLL	−0.48	106	59	0.30	0.57	0.88	0.22	0.70
NB	-	#LLLLL	−0.78	82	60	0.27	0.67	0.90	0.16	0.62

Table 5. Model performance comparison for an obesity classifier using different degrees of historical information included as Generalized Bayes feature combinations.

Model	Threshold	PPV	x(1-TN/N)	y (TP/P)	dist	Area
#****L	−0.17	0.26	0.60	0.79	0.13	0.59
#***LL	−0.39	0.27	0.67	0.90	0.16	0.60
#**LLL	−0.52	0.27	0.67	0.90	0.16	0.61
#*LLLL	−0.49	0.28	0.60	0.85	0.18	0.67
#LLLLL	−0.48	0.30	0.5	0.88	0.22	0.70

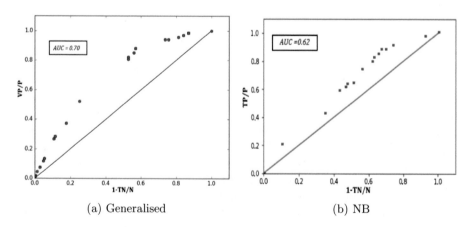

(a) Generalised   (b) NB

Fig. 1. ROC curve with AUC value for (a) the generalised model with $a_0 = \#$, $a_n = A, B$ with $n > 0$ and (b) Naive Bayes.

6   Conclusion

There are multiple generalisations of the NBA that lead to enhanced performance. In all such generalisations feature combinations are considered so as to go beyond the full factorisation of the NBA. Here we have argued that a classifier based analysis of systems with time dependence is an area where correlations in time occur frequently and naturally. In such circumstances the NBA can, and normally will, be suboptimal. We have shown that a generalisation of the NBA that accounts for historical time correlations leads to enhanced classifier performance. We have also shown that it leads to greater insight.

We believe that the formalism shown here, in the context of a simple but highly relevant example - obesity and historical exercise patterns - can be extremely useful in classification problems where features are represented by time series and particularly when there are correlations within the time series. This is clearly the case when human behaviour is concerned. We have also shown that an analysis considering histories as composite features also leads to greater insight. In particular, we showed that obesity/non-obesity was associated with very particular and distinct feature sets; obesity being more associated with having good exercise habits then losing them than with always having had bad exercise habits. We also showed that academics exhibited healthier exercise patterns than non-academics, a conclusion that is consistent with the observation that academics exhibit a much lower incidence of obesity.

Acknowledgements. This work was supported by CONACyT Fronteras grant 1093.

References

Domingos, P., Pazzani, M.: Beyond independence: conditions for the optimality of the simple Bayesian Classifier. In: Proceedings of the Thirteenth International Conference on Machine Learning, pp. 105–112. Morgan Kaufmann (1996)

Stephens, C.R., Huerta, H.F., Linares, A.R.: Machine Learning, vol. 107, Issue 2, pp. 397–441 (2017)

Teixeira, P.J., Carraca, E.V., Marques, M.M., et al.: Successful behavior change in obesity interventions in adults: a systematic review of self-regulation mediators. BMC Med. **13**, 84 (2015). https://doi.org/10.1186/s12916-015-0323-6

World Health Organization: Obesity: preventing and managing the global epidemic (No. 894). World Health Organization (2000)

Church, T.S., Thomas, D.M., et al.: Trends over 5 decades in US occupation-related physical activity and their associations with obesity. PloS one **6**(5), e19657 (2011)

Kononenko, I.: Semi-naive Bayesian classifier. In: Proceedings of the Sixth European Working Session on Learning, pp. 206–219. Springer, Berlin (1991)

World Health Organisation: Global recommendations on physical activity for health (2018). http://www.who.int/dietphysicalactivity/factsheet_recommendations/en/

Holland, J.: Adaptation in Natural and Artificial Systems, 2nd edn. MIT Press, Cambridge (1992)

Poverty and Its Relation to Crime and the Environment: Applying Spatial Data Mining to Enhance Evidence-Based Policy

Christopher R. Stephens[1,2(✉)], Oliver López-Corona[1,3,4],
Ricardo David Ruíz[1,6], and Walter Martínez Santana[1,5]

[1] C3 - Centro de Ciencias de la Complejidad, y Universidad Nacional Autónoma de México, 04510 Mexico City, México
[2] Instituto de Ciencias Nucleares, Universidad Nacional Autónoma de México, 04510 Mexico City, México
stephens@nucleares.unam.mx
[3] Cátedras CONACyT, Comisionado en Comisión Nacional para el Conocimiento y Uso de la Biodiversidad (CONABIO), Mexico City, México
[4] Red ambiente y Sostenibilidad, Instituto de Ecología A.C., Xalapa, México
[5] Instituto Tecnológico Autónomo de México (ITAM), Mexico City, Mexico
[6] TechMileage, 3295 N Drinkwater Blvd, Suite 13, Scottsdale 85251, USA

Abstract. The Data Revolution provides an unprecedented opportunity for enhancing evidence-based decision making in the area of public policy. Machine learning techniques will play an increasingly important role in knowledge extraction in data bases associated with important social phenomena such as poverty, crime and environmental degradation. As much of the corresponding data is spatio-temporal it is important to develop spatial data mining methodologies to attack these problems. In this paper, we will use spatial data mining techniques to analyze the relation between poverty and crime and poverty and environmental integrity in two bespoke data sets. We will show that the role and relation of poverty is measurable but is highly complex and heterogeneous.

Keywords: Spatial data mining · Naive Bayes · Poverty · Complexity · Crime · Ecosystem integrity · Niche

1 Introduction

The Data Revolution of the last few decades is leading to unprecedented opportunities for advances in Evidence Based Policy making (EBP). Huge data sets, representing data in multiple formats - structured, unstructured, spatio-temporal etc. - are becoming more widely available. Indeed, our ability to generate data is far outstripping our ability to analyze and extract the knowledge from it that may aid in decision making processes. Two important, related areas of EBP are

© Springer Nature Switzerland AG 2020
A. E. Hassanien et al. (Eds.): AMLTA 2019, AISC 921, pp. 250–260, 2020.
https://doi.org/10.1007/978-3-030-14118-9_25

that of poverty and sustainability. In this context, the Sustainable Development Goals (SDGs) (Cf 2015) agreed to by the United Nations aim to: "Eradicate extreme poverty, combat inequality and injustice and solve climate change".

These SDGs form part of a wider group that divide up into 3 categories: **Poverty (multidimensional), Peace**, and **Environment**. Poverty is a highly multi-dimensional state that both impacts on, and is related to, Peace and Environment. It is to data that we must look to help us understand these relations, and to use that knowledge to improve public policy and make progress towards the above-stated SDGs.

However, although data is a necessary condition for this to occur, it is not sufficient. Rather, this data must first be turned into knowledge. It is there that machine learning plays an important role. In this paper, using machine learning techniques, we will analyze three data sets in order to better understand the complex, multi-factorial relation between poverty, peace and environment. For the former, we will use an official, multi-dimensional poverty measurement used in Mexico by the National Council for the Evaluation of Social Development Policy (CONEVAL). For peace, we will use a set of crime data from the municipality of General de Escobedo, Nuevo Leon, Mexico, while for environment we will consider an Ecosystem Integrity index developed by the Consejo Nacional de Biodiversidad (CONABIO). An important goal of this paper is to show how variables linked to poverty relate to crime and the environment and, particularly, to show how these complex relations may be analyzed and quantified and used as a motor for EBP.

For example, many questions that can be posed are relevant to understanding both crime patterns and also the risk factors associated with different crimes. Are all populations equally at risk for a given crime? What are the underlying socio-demographic and socio-economic factors, such as poverty dimensions, associated with a given crime? Recently, data mining techniques have been brought to bear on this type of problem (Chen et al. 2004; Nath 2006; McCue 2014; Estivill-Castro and Lee 2001; Keyvanpour et al. 2011).

In the case of Ecosystem Integrity one may ask whether there is a coupling between poverty and adverse environmental impacts due to the use of economically high impact technologies; or if regions with high indigenous population percentages are in higher levels of Ecosystem Integrity. It is also important to better understand the impact of road construction on poverty, while at the same time gauging any adverse environmental impacts.

2 Data and Methods

2.1 Poverty in Mexico: A Multidimensional Perspective

Poverty may be measured by governments as a means to provide a continuous assessment of how its policies affect the conditions of the poor. This is part and parcel of a government's quest to be accountable for its actions and to provide accurate information on a central socio-economic problem (Foster 2007).

In general, one may think of poverty as a state in which a person does not have access to sufficient resources to achieve a minimum level of living. The general framework for measuring poverty is based on the seminal work of (Sen 1976) in which he proposed three steps: (1) select the space or spaces in which poverty is to be evaluated; (2) Identify the poor with the help of a cutoff for each space, as well as a rule for deciding when a person is poor; and (3) aggregate the resulting data using some form of overall poverty index. Of course, this general framework for measuring poverty allows a great deal of leeway in selecting a specific measurement procedure (Foster 2007).

In Mexico, poverty measurement is carried out by the CONEVAL, which considers two main approaches for analyzing the multidimensional character of poverty: welfare and rights. The first relates to the basic unsatisfied needs of the poor; while the second is associated with the existence of fundamental, inalienable rights of the poor. From this perspective, poverty constitutes, in itself, a denial of human rights.

In this framework the CONEVAL establishes that "a person is in a situation of multidimensional poverty if you have not guaranteed the exercise of at least one of your rights for social development, and if your income is insufficient to acquire the goods and services you need to meet your needs". The way the distinct dimensions of poverty are measured can be found at official site. Some of the most relevant variables are: (i) per-capita household income, or the total income of a household divided by the number of persons in the household; (ii) the population with a lack of adequate access to education; (iii) the population with a lack of access to health services; (iv) the population with a lack of access to social security; (v) the population with a lack of access to adequate living conditions.

In order to evaluate these poverty metrics, CONEVAL uses official data from the INEGI, which uses a biennial polling exercise with spatial resolution at the state scale; and from a 10-year based census with spatial resolution at the municipality level. In this work we used CONEVAL reports for 2010 (municipality scale) and several others from INEGI. All this data was previously collected and curated by the CONABIO in its Geoportal.

2.2   Crime

Crime and security are issues that are almost always at the forefront of the public's mind. There are a wide variety of crimes that can afflict a society or a certain population: from business robbery, home robbery and domestic violence to homicides and kidnappings. In this paper, we consider three crime types with the greatest incidence in the municipality of General de Escobedo in Nuevo Leon, México. The three types are: domestic violence (DV), business robbery (BR), and burglary (HR). Criminal data were obtained from police officers who attended the crimes.

The considered crime data is at a more fine grained scale than that of the CONEVAL poverty data, associated with AGEBs (Basic Geo-statistical Areas) and provided by INEGI (INEGI 2015). The data consists of 188 socio-economic and socio-demographic variables within 121 AGEBs associated with the General Escobedo municipality. The variables are grouped into population, migration, indigenous population, disability, educational characteristics, economic characteristics, health, marital status, religion and housing.

We performed several different types of analysis, ranging from basic exploratory analysis, using simple statistics, to a more sophisticated classification model using a Naive Bayesian classifier. The classifier was used to determine risk profiles for different crime types and perform a spatial risk analysis at the level of AGEBs using, for example, heat maps as a visualization tool.

2.3 Ecosystem Integrity

The ability to estimate ecosystem integrity is key for the design, implementation and evaluation of practically any action toward the achievement of the SDGs. For this reason, there is an urgent need to develop new analytical approaches that help balance multiple, and often contradictory, public development goals. Given the increasing pressure of human activities on ecosystems, it is necessary to evaluate their health or the condition in which they are found.

To address this problem, a Sistema Nacional de Monitoreo de la Biodiversidad (SNMB) strategy was developed in Mexico based on the conceptual framework of ecosystem integrity. This approach makes it possible to estimate, in a practical way, the correspondence between attributes of biodiversity and their organization in functional ecosystems, which in turn allows us to design a systematic and rigorous, although also complex, strategy to acquire the data and perform the required computations.

The approach followed in the SNMB recognizes that ecosystem integrity is not easily measurable directly. An analogy would be that of doctors inferring the health status of patients from a series of symptoms and clinical data. In this sense, ecosystem integrity is equivalent to a type of "ecological health". The current Ecosystem Integrity Index takes into account field data, collected by national agencies of Forestry and National Parks, as well as several remote sensing variables from LANDSAT series. All this data are used by a Bayesian Neural Network to assess the Index, which at this moment is available at a scale of $1\,\mathrm{km}^2$.

2.4 Statistical Analysis

For any of our data sets, from a classification point of view, we are interested in one or more classes, C, and the influence of predictors on class membership. We will use a binomial test for the identification of features, X_i, that correlate in a significant way with class membership. We will then use the Naive Bayes approximation to construct predictive models. Explicitly we use the statistical diagnostic in Eq. (1).

$$\varepsilon\left(C|X_i\right) = \frac{N_{X_i}\left(P\left(C|X_i\right) - P\left(C\right)\right)}{\left(N_{X_i}P\left(C\right)\left(1 - P\left(C\right)\right)\right)^{1/2}} \tag{1}$$

where $P(C|X_i) = \frac{N_{CX_i}}{N_{X_i}}$, $P(C) = \frac{N_C}{N_t}$, N_{X_i} = number of times that the feature value X_i appears, N_C = number events of class C, N_t = the total number of records and N_{CX_i} = is the number of co-occurrences of C and X_i. ε is a binomial test that determines the statistical significance of the observed distribution of co-occurrences of the class C and feature X_i relative to the null hypothesis that they are uncorrelated. The associated distribution is binomial but may be approximated in most circumstances by a normal distribution, in which case, $|\varepsilon(C|X_i)| > 1.96$ corresponds to the 95% confidence interval that the observed co-occurrence of C and X_i would not have occurred by chance.

Having determined those features which are correlated with the class C we may construct a classifier (score function)

$$S(\mathbf{X}) = \ln \frac{P(C|\mathbf{X})}{P(\bar{C}|\mathbf{X})} \tag{2}$$

where $\bar{C}$ is the complement of the class C and $\mathbf{X}$ is the vector of relevant features. As $P(C|\mathbf{X})$ is potentially a high-dimensional conditional probability it cannot be estimated from data directly. We can proceed however using Bayes rule and adopting the Naive Bayes approximation. Using Bayes rule

$$P(C|\mathbf{X}) = \frac{P(\mathbf{X}|C)P(C)}{P(\mathbf{X})} \tag{3}$$

we wish to determine the likelihood function $P(\mathbf{X}|C)$. As due to the high-dimensional nature of the feature space, these likelihoods cannot be determined directly we assume the Naive Bayes approximation, wherein the features X_i are considered uncorrelated, therefore

$$P(\mathbf{X}|C) = \prod_i P(X_i|C) \qquad P(\mathbf{X}|\bar{C}) = \prod_i P(X_i|\bar{C}) \tag{4}$$

We thus obtain for our classifier

$$S(X) = \sum_i \ln \frac{P(X_i|C)}{P(X_i|\bar{C})} + \ln \frac{P(C)}{P(\bar{C})} \tag{5}$$

where $\ln \frac{P(C)}{P(\bar{C})}$ is a constant independent of the features X_i. A high positive score value indicates a higher probability of belonging to the class C, while a high negative score indicates a low probability of belonging to the class.

In the case of crime, the class variables are the incidences of the three crimes - DV, BR and HR. However, as the provided data base consisted only of crime events there is no natural "non-crime" class complement. There are several ways to overcome this. In the case of BR for instance, if one had access to a data base with all businesses in the municipality one could determine which businesses had not been robbed and these businesses would form $\bar{C}$. In the present case however we will consider C to be a subset of crimes within a wider set. For instance, BR could be the class C and then $\bar{C}$ is the class of all robberies that were not BR.

In this way we will characterize and profile crimes relative to one another, determining, for instance, what are the particular predictive drivers of BR relative to other types of robbery. The features X_i are census-type variables by AGEB. In this case, for any variable, such as "Male population 0–2 years", there exists a value for each AGEB. The AGEBs are ranked with respect to this variable and the ranked list divided into deciles. Thus, decile 10 corresponds to that 10% of AGEBs in the data with the highest value of the corresponding variable, while decile 1 corresponds to the 10% of AGEBs with the lowest values. In the case of Ecosystem Integrity, we will take the class variable to be membership in the 10% of municipalities that exhibit the worst ecosystemic integrity. For the features a similar coarse-graining as to that for crime is carried out except that in this case the coarse graining is at the level of municipalities not AGEBs.

3 Results

3.1 Poverty and Crime

We will first show results for those predictive drivers most associated with each crime using the ε diagnostic (1). For DV, Table 1 shows the 10 variables with the highest ε values. As $\varepsilon = 1.96$ corresponds to the 95% confidence interval we can see that these factors are all statistically significant. The variable "male population 15 years and over with primary school as last level of studies completed" with the coarse grained value 8 has the highest ε. Value 8 indicates the AGEBs with a high proportion of males 15 years and over who have a maximum education level corresponding to primary education. These tendencies in

Table 1. Most significant risk factors for domestic violence.

Variable	Value	Epsilon
Population 15 years old and over with unfinished secondary school	6	9.0692
Private inhabited house with car property	1	9.0940
Private inhabited house with fixed telephone line	1	9.1247
Male population from 0 to 2 years	6	9.2146
Population from 12 years old and over	3	9.3088
Male population from 15 years old and over with primary incomplete	7	9.3385
Female population aged 15 years old and over with primary school finished	7	9.3845
Female illiterate population from 15 years old or more	7	9.6971
Male illiterate population from 15 years old or more	7	9.9840
Male population from 15 years old and over with primary school completed	8	10.0506

Table 2. Confirmatory risk factors for domestic violence.

Variable	Value	Epsilon
Private inhabited houses with one bedroom	6	7.7072
Population with no health service	6	7.8016
Population from 12 to 14 years old who do not attend school	6	7.8171
Male population born in another entity	8	7.9968
Private inhabited houses with two bedrooms or more	2	8.0916
Private inhabited houses with a computer	1	8.4723
Private inhabited houses with piped water	2	8.4739
Occupants per room average in private inhabited houses	6	8.6375
Illiterate population from 15 years old or more	7	8.8522
Population 0–2 years old	7	9.0056

the risk factors can also be confirmed by considering the next ten most relevant risk factors in terms of their ε values, as seen in Table 2.

Fig. 1. Heat map regarding the incidence of DV in each AGEB (red color indicates a higher incidence, white color means little or no presence).

In Fig. 1 we show a heat map based on the Naive Bayes score function regarding the incidence of DV in each AGEB (red color indicates a higher incidence, white color means little or no presence). Thus, the methodology not only predicts overall risk but also identifies in a transparent fashion key drivers. In table 3 we see the 10 variables associated with the most negative values of ε. This can be compared and contrasted with Table 1.

Table 3. Variables with most negative correlation with domestic violence.

Variable	Value	Epsilon
Male population from 18 years old and over with post-basic education	8	−9.3353
Private inhabited houses with fixed telephone line	8	−8.9527
People from 18 to 24 years old attending school	7	−8.8008
Population from 18 years old and over with post-basic education	8	−8.7268
Female population from 18 years old and over with post-basic education	8	−8.7268
Female population from 15 years old and over with secondary school incomplete	1	−8.6027
Population without religion	1	−8.4974
Private inhabited houses with radio	8	−8.4072
Private inhabited houses with a washing machine	8	−8.4072
Female population from15 years old and over with secondary school complete	1	−7.9293

3.2 Poverty and Ecosystem Integrity

Using the same analysis but using the environmental data set in relation to Ecosystem Integrity (in place of Domestic Violence) we found (See Table 4) that higher values ε correspond to state number, investment in health services, altitude, and highway access. In Fig. 2 we see the performance of a predictive model for Ecosystem Integrity based on the Naive Bayes approximation Table 4.

4 Discussion and Conclusions

We can readily infer from Table 1 that low educational attainment is an important factor when considering DV, as does the related category of illiteracy. The fact that the lowest deciles (1) of "Private inhabited house with car or telephone line" appear also strongly indicates the preponderance of DV crimes in areas of poor economic development. The presence of infants also seems to be a risk factor. The overall profile is that of economically stressed households - insufficient education to obtain decent employment and infants to be cared for. All these factors may be considered as related to the overall concept of poverty.

Table 2 confirms this, where the importance of economic level is notable. The appearance of the variable "male population born in another entity" with a value of 8, indicates that a substantial fraction of the male population has emigrated to the municipality. Average number of occupants per room in private occupied houses and private inhabited houses with a bedroom both have value 6, which shows a tendency to have many people living in a small space, while the variable private inhabited houses with two bedrooms and more, with value 2, shows

Table 4. Variables with the most significant ε for Ecosystem Integrity.

Variable	Value	Epsilon
Mins access to highway 20 thousand population	1	13.437
Mins access to highway 250 thousand population	1	12.032
State key	29	8.115
Number of medical units per municipality	1	6.436
Mean altitude	10	6.294
Municipality's jungle percentage	0	5.548
Mins access to highway 250 thousand population	2	5.136
Health service transfer	8	4.973
Environmental service transfer	4	4.890
Health spending per capita	5	4.842
Environmental service transfer	3	4.707
Municipality's forest percentage	0	3.987
Number of schools	10	3.858
Mean altitude	9	3.686

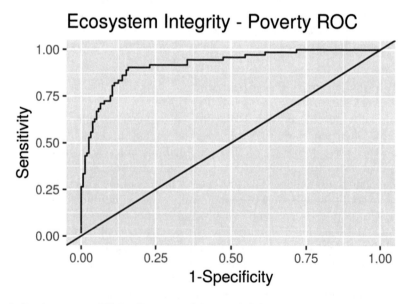

Fig. 2. Performance of Naive Bayes model via a ROC curve for predicting Ecosystem Integrity.

that the majority of dwellings where there is a high incidence of DV are houses with no more than two bedrooms. Thus, we conclude that having more people living in reduced inhabitable space, in a precarious economic situation and of low education level leads to high rates of DV.

Table 3 shows the variables least associated with DV. For example, the variable "private inhabited houses with fixed telephone line" with a value of 8 and $\varepsilon = -8.95$ is the opposite of the variable with the same name in table 1 but with value 1 and $\varepsilon = 9.124$. In Table 3 we see that variables which involve education are those that indicate a higher education level than those with only basic education. All these variables have a value of 8 or 7 and belong to the coarse-graining that brings a greater number of people who have post-basic education. On the other hand, other education variables - incomplete or complete basic education with a value of 1 - indicate a coarse-grained value with the least number of people of this type. The variable "private houses inhabited with a washing machine" with a value of 8 indicates a socio-economic level better than those described in the variables of Table 1. This indicates that the incidence of DV is less than expected as the level of education rises, the same applies to economic level.

A further analysis for HR and BR have also been carried out. In this case the poverty related variables that play such a significant role in the case of DV do not play such a role in HR and BR. Rather, the socio-economic profile of these crimes is that they occur in relatively better off neighbourhoods. This however is a function of where the crime takes place not a profile of the perpetrator.

Turning now to the relation between poverty and Ecosystem Integrity, we see that an important indicator of a degraded ecosystem is how many minutes on average any given center of population is from the nearest highway. This is an indication that transport infrastructure has a very negative impact on ecosystem integrity. We also see that a large number of schools is highly correlated with a high degree of ecosystem degradation. In general, we may observe that proxies for poverty are not very predictive of ecosystem degradation.

Overall, we see that the role of poverty in crime and Ecosystem Integrity is both complex and heterogeneous, wherein its impact and role is quite different from one area to another. For instance, each considered crime has its own unique "poverty profile", where in DV social marginalization and poverty seem to be strongly causative elements, whereas this is not the case for Ecosystem Integrity. Clearly there is enormous potential for data mining and machine learning in the application of EBP.

References

Cf, Objetivos de Desarrollo Sostenible. Transforming our world: the 2030 agenda for sustainable development (2015)

Chen, H., et al.: Crime data mining: a general framework and some examples. Computer **37**(4), 50–56 (2004). https://doi.org/10.1109/MC.2004.1297301. ISSN: 0018-9162

Estivill-Castro, V., Lee, I.: Data mining techniques for autonomous exploration of large volumes of geo-referenced crime data. In: Proceedings of the 6th International Conference on Geocomputation (2001)

Foster, J.E.: A report on Mexican multidimensional poverty measurement (2007)

INEGI. Instituto Nacional de Estadística y Geografía (2015). http://www.inegi. org.mx/sistemas/consulta%5C_resultados/ageb%5C_urb2010.%5C%5C%20aspx? c=28111

Keyvanpour, M.R., Javideh, M., Ebrahimi, M.R.: Detecting and investigating crime by means of data mining: a general crime matching framework. In: Procedia Computer Science 3.0. World Conference on Information Technology, pp. 872– 880 (2011). https://doi.org/10.1016/j.procs.2010.12.143. http://www.sciencedirect. com/science/article/pii/S1877050910005181. ISSN: 1877-0509

McCue, C.: Data Mining and Predictive Analysis: Intelligence Gathering and Crime Analysis. Butterworth-Heinemann, Oxford (2014)

Nath, S.V.: Crime pattern detection using data mining. In: 2006 IEEE/WIC/ACM International Conference on Web Intelligence and Intelligent Agent Technology Workshops, 2006. WI- Poverty, Crime and the Environment 11 IAT 2006 Workshops, pp. 41–44, December 2006. https://doi.org/10.1109/WI-IATW.2006.55

Sen, A.: Poverty: an ordinal approach to measurement. Econometrica **44**(2), 219–231 (1976)

Identifying Different Types of Biclustering Patterns Using a Correlation-Based Dilated Biclusters Algorithm

Mahmoud Mounir[1]($\boxtimes$), Mohamed Hamdy[1],
and Mohamed Essam Khalifa[2]

[1] Information Systems Department,
Faculty of Computer and Information Sciences,
Ain Shams University, Cairo, Egypt
{mahmoud.mounir,m.hamdy}@cis.asu.edu.eg
[2] Basic Sciences Department, Faculty of Computer and Information Sciences,
Ain Shams University, Cairo, Egypt
esskhalifa@cis.asu.edu.eg

Abstract. An essential step in the analysis of gene expression profiles is the identification of sets of co-regulated genes or genes tend to be active under only subsets of experimental conditions or participate in multiple cellular processes or functions. Biclustering is a non-supervised technique exceeds the traditional clustering techniques because it can find groups of both genes and conditions simultaneously. In this paper, we proposed a biclustering algorithm called Correlation-Based Dilated Biclusters CBDB to find sets of biclusters with correlated gene expression patterns. This algorithm has many phases starting with the preprocessing phase, determination of elementary biclusters, then the dilation phase depending on a heuristic searching approach with Pearson correlation coefficient as a measure of coherency, after that, the removal phase to exclude sets of genes and conditions that show low level of coherency, finally, the elimination of duplicated and overlapped biclusters phase. This approach showed reasonable results on both synthetic and real datasets compared with other correlation-based biclustering techniques.

Keywords: Clustering · Biclustering · Microarrays ·
Gene expression profiles · Correlated patterns

1 Introduction

Gene expression represents the amount of mRNA obtained during the transcription process inside the cell, which will be translated to proteins. These proteins have the main responsibility in the function of any cell [1].

DNA microarrays are tools used to observe the gene expression levels of a large number of genes in different cells under different experimental conditions. Analyzing the gene expression can allow the extraction of huge number of information, one of the major tasks in this analysis is finding groups of genes that tend to have similar expression levels under some conditions; this task is known as biclustering of gene

A. E. Hassanien et al. (Eds.): AMLTA 2019, AISC 921, pp. 261–271, 2020.
https://doi.org/10.1007/978-3-030-14118-9_26

expression data. Discovering the relations between genes according to their expression levels gives an indication that these genes may participate in the same cellular processes or the share same cellular functions [2, 3]. A gene or condition may belong to more one group or none, which is supported by biological believes that every gene may participate in many biological processes and each experimental condition may be active in many genes [4]. Biclustering [5, 6] identifies groups of genes that show similar activity patterns under a specific subset of the experimental conditions. Starting with an *(n X m)* data matrix *A* where n represents number of genes and m is the number of experimental conditions. Each value in this matrix (a_{ij}) represents the amount of gene expression level for specific gene *(i)* under condition *(j)*. The goal of the biclustering process is to partition the data into biclusters, each of which is a sub-matrix *(X, Y)*, with *I* rows *(X ε N)*, *Y* columns *(Y ε M)* that satisfies some homogeneity or coherency characteristics [7, 8].

The rest of this paper is organized as follows: the related work is summarized in Sect. 2. The proposed correlation based dilated biclustering algorithm (CBDB) is presented in Sect. 3. Results obtained after applying the proposed algorithm on both synthetic and real datasets are discussed in Sect. 4. Finally, conclusions from this work are presented in Sect. 5.

2 Related Work

Most of the existing biclustering algorithms start with searching for initial random sets of biclusters, which varies each time when applying the same algorithm to the same microarray dataset. However, the initial biclusters identification is a necessary step to overcome the computational complexity of biclustering. A few studies were proposed to deal with this issue. Erten and Sozdinler [9] proposed a localization method depending on reordering of rows and columns in the original data matrix to have patterns close enough to each other. Another crucial issue to improve the performance of biclustering algorithms is finding the best way to score the coherency of biclusters. Most of biclustering methods use Mean-Square Residual (MSR) distance as a measure of coherency [10]. However, it was proved by Aguilar-Ruiz [11] that MSR has limitations in finding scaling patterns of biclusters. To cope with these issues the correlation coefficient can be used as a measure of similarity instead of MSR. Alloco et al. [12] proved that if the correlation coefficient between pairs of genes is larger than 0.84, the probability that such genes may share the same function or regulated by the same transcription factor exceeds 50%. Bahattacharya and De [13] proposed the Bi-Correlation Clustering Algorithm (BCCA), they proved that BCCA can find number of significantly enriched biological terms compared to non-correlated biclustering algorithms. Yun et al. [14] presented the BIclustering by Correlated and Large number of Individual Clustered seeds (BICLIC), it starts by finding set of biclusters as seeds, expanding these seeds and keep adding genes with high correlation and similar conditions. On the other hand, Qualitative Biclustering Algorithm (QUBIC) [15] converts the original microarray data matrix to a discretized matrix called the representing matrix. Bentham et al. [16] proposed the Maximal Correlated Biclustering (MCbiclust) to select specific sets of genes and experimental conditions. However, MCbiclust was able to identify large biclusters and complex networks, it cannot handle weak signaled biclusters. Henriques et al. [17] proposed another biclustering technique based on pattern mining called Biclustering based on Pattern Mining Software (BicPAMS) that includes parameters related to bicluster's coherency, structure, quality, and efficiency.

3   The Proposed Algorithm

Our proposed algorithm Correlated-Based Dilated Biclusters (CBDB) has six main phases, starting by a preprocessing phase, followed by a single-dimension clustering phase, then finding elementary biclusters which will dilate later to larger biclusters, the dilation of these elementary biclusters to a set probable biclusters, the removal of less correlated genes and experimental conditions, and finally, the checking and removal of duplicated biclusters to find the final desired biclusters. Figure 1 shows a complete workflow of the proposed CBDB technique.

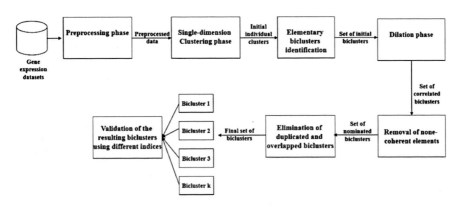

Fig. 1. A complete workflow of the proposed CBDB algorithm

3.1   Preprocessing Phase

In the preprocessing phase, gene expression dataset is examined to search for missing values of gene expression levels, if found, it is swapped with the average over all rows value for all genes. Another important processing step is the standardization of data if needed. The platform of gene expression microarray used to measure the expression values of genes can affect these values and can vary significantly. Standardization process of the gene expression values in each biclusters is needed to scale these values to a mutual range.

3.2   Single-Dimension Clustering Phase

In single-dimension clustering phase, an ($n \times m$) microarray data matrix is converted to a number (m) of one-dimension ($n \times 1$) gene matrices. A single-dimension clustering is applied to find genes with similar expression levels in each condition. In order to achieve this process, an adapted version of DBSCAN algorithm is used to find similar genes in each condition. For each gene vector, starting with an arbitrary unvisited gene, retrieving all neighborhood within the value of (eps) for this gene. If this gene has number of neighbors larger than the predefined threshold value (MinPts), this gene is defined as a cluster. Otherwise, this gene is considered as a noise. In an upcoming phase, this noise gene may be a neighbor of another gene and hence a part of another cluster.

Later, the clustered genes in each condition are labeled with a number as a cluster label. Genes with similar experimental condition are labeled with the same cluster labels. Now, we have a set of (m) cluster label vectors, which will be converted to its original matrix form again of ($n \times m$) dimension.

3.3   Elementary Biclusters Identification Phase

Now, we have a set of grouped genes within each condition, genes with the same cluster labels in each of (m) conditions are grouped together to form a set of elementary or initial biclusters. Genes in the elementary biclusters in each condition are labeled with the same cluster label, in another condition these genes may be labeled with the same or another cluster label. Phases 2 and 3 are illustrated in Algorithm 1.

3.4   Dilation of Elementary Biclusters Phase

The dilation process proceeds in two ways, by gene and by condition. In gene dilation, Pearson correlation coefficient is calculated between gene vector and the elementary bicluster to find set of genes to be added to the elementary bicluster. These genes are added in a decreasing order based on value of Pearson correlation coefficient, this process stops when the Pearson correlation coefficient within elements of the same bicluster is less than the predefined threshold value (δ). The same process is done in condition dilation, but condition by condition rather than genes. Set of resulting biclusters are called dilated biclusters DB.

3.5   Removal of None-Coherent Elements Phase

In the dilated biclusters DB, not all genes show correlated patterns over all experimental conditions in, but at least it is guaranteed that all genes and conditions are correlated with the initial elementary bicluster.

Algorithm 1. Elementary biclusters dentification

Input   ***Gene Expression Data Matrix***
Output   ***Set of elementary biclusters***
Begin
    Convert the original gene expression data matrix
    (n X m) to a number of (m) one-dimension *(n X 1)* matrices.
    While GI != n
    Set CL = 0, L = 0 , CGS = Null
    for each unvisited gene g in set of genes G = {g_1, g_2, ..., g_n}
        mark g as visited
        N = getNeighbors (g, eps)
        if sizeof (N) < MinPts
          mark P as NOISE
        else
          Set CL = CL + 1
          expandCluster(g, N, CL, eps, MinPts)
          Set L = L + 1 and   CGS = CGS U g
          Label (g) = CL
        end if
      nd for
      end while
      for each (m) condition
        Combine conditions which are labeled with the same cluster index
    End for
end
expandCluster(g, N, CL, eps, MinPts)
Begin
    add g to cluster CL
    for each gene g' in N
      if g' is not visited
        mark g' as visited
        N' = getNeighbors(g', eps)
        if sizeof(N') >= MinPts
            N = N joined with N'
        end if
      end if
    end for
end

While maintaining an average Pearson correlation coefficient (PCC) that are greater than or equal to the predefined threshold value (δ), less coherent genes and conditions are identified in the dilated biclusters. The condition with the least correlation value is eliminated from the dilated bicluster and the increase in correlation of the bicluster is computed and its value is scored. Similarly, the gene with the least coherency value is eliminated from the dilated bicluster and the increase in the correlation is computed and scored. Comparing values of increase in both cases, the one with larger value will be eliminated from the dilated bicluster. Keep repeating this process while the value of correlation within the elements of dilated bicluster is less than or equal to the (δ), the result is a correlated bicluster matrix. A complete Pseudo code for this phase is illustrated in Algorithm 2.

3.6   Elimination of Duplicated and Overlapped Biclusters Phase

Different elementary biclusters may dilate to the same biclusters, in addition to that biclusters may include other biclusters of smaller size. Therefore, this phase of checking and removal of duplicated genes is essential. An overlapping control strategy is applied to prevent overlapping between biclusters. After this process, the remaining biclusters are defined to be the desired set of unique-correlated non-overlapped biclusters.

Algorithm 2.   Removal of Less Correlated Genes or Conditions

---

Input   *Set of Elementary Biclusters*
Output   *Set of Candidate Biclusters*
Begin
    for each dilated bicluster DB
        Calculate PCC between DB and each gene vector as GenePCC
        if GenePCC $>= \delta$
            Append gene vector to the DB in a descending order
        end if
        do the same for condition
    end for
    for each dilated bicluster DB
        Calculate the average Pearson correlation coefficient AvgPCC for DB
        While AvgPCC $< \delta$
            Remove the appended gene vector with the least GenePCC
            Calculate the overall Pearson correlation coefficient as AllGenePCC
            do the same for the appended condition vector and calculate AllCondPCC
            if AllGenePCC > AllCondPCC
                Remove from the dilated bicluster the gene vector
            else
                Remove from the dilated bicluster the condition vector
            end if
        end while
    end for
        Return the resulting bicluster as candidate bicluster CB
end

4   Results and Discussion

In this section, the performance of the proposed (CBDB) algorithm was compared with the performance of three other know correlation based biclustering algorithms BCCA, CPB, and QUBIC. This comparison has two phases. In the first phase, CBDB were tested on synthetic datasets to test and prove the ability of the algorithm to cover all correlated biclusters with different patterns. Secondly, real microarray datasets were used to show that CBDB can find large number of different correlated biclusters.

4.1   Synthetic Datasets

A number of 10 (100 × 10) data matrices are generated for each type of correlated patterns. Values in this matrix are random values in normal distribution with mean. For each type of patterns, there are ten data matrices with a total number of 30 for all types of patterns shifting, scaling, and shifting-scaling patterns.

4.2   Performance Indices

Different performance indices were proposed in [18], the accuracy of the biclustering algorithms compared by using the average match score proposed in [19]. The match score (MS) is defined in Eq. 1.

$$MS_g(G_1, G_2) = \frac{|G1 \cap G2|}{|G1 \cup G2|} \tag{1}$$

where $|G1 \cap G2|$ is the number of genes in the intersection of both sets of genes and $|G1 \cup G2|$ represents the total number of genes in the two sets of genes.

Let BC_1 and BC_2 be two sets of biclusters, the Sensitivity of BC_1 with respect to BC_2 is given by Eq. 2.

$$Sensitivity(BC_1, BC_2) = \frac{\sum_{(G1,C1) \in BC1} Max(G2, C2 \in BC2) MS_g(G1, G2)}{|BC1|} \tag{2}$$

Therefore, if BC_1 and BC_2 represent true bicluster and identified bicluster respectively, the *Sensitivity* is a measure of how the biclustering algorithm can recover all biclusters in the dataset. On the other hand, if we swap BC_1 and BC_2 in Eq. 2, we will get the *specificity*, which is a measure of similarity level for all generated biclusters compared to the true biclusters. The correlation threshold δ was set to 0.9 and minimum number of rows and columns in each bicluster were set to 5 (Table 1).

Table 1. (a) The average *Sensitivity* and (b) The average *Specificity* in each correlated pattern for each biclustering algorithm using synthetic datasets

(a)				(b)			
Algorithm	Shifting	Scaling	Shifting-Scaling	Algorithm	Shifting	Scaling	Shifting-Scaling
QUBIC	0.527	0.243	0.513	QUBIC	0.053	0.052	0.139
BCCA	0.174	0.213	0.182	BCCA	0.083	0.137	0.113
BICLIC	1	1	0.921	BICLIC	0.732	0.874	0.761
CBDB	1	1	1	CBDB	0.995	0.973	0.929

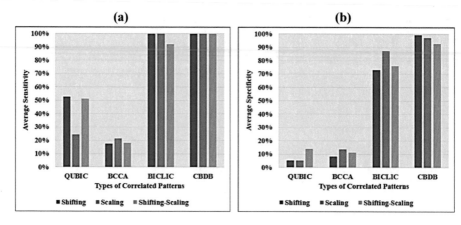

Fig. 2. A comparison of (a) the average Sensitivity (b) and the average Specificity in each correlated pattern for each biclustering algorithm using synthetic datasets

CBDB and BICLIC can recover most of the biclusters in all types of correlated patterns, BCAA performed very poor here, it may be due to its inability to detect small size biclusters, while QUBIC performance was poor particularly in scaling patterns as QUBIC works on a discretized version of the dataset (Fig. 2).

Moreover, the value of *Specificity* for CBDB was very high, which indicates that biclusters obtained by CBDB was too much similar to the those true biclusters. On the other hand, nearly 90% of the biclusters extracted by BCAA and QUBIC are not related to true biclusters. However, BICLIC performs well in *Sensitivity*, about 25% of the biclusters obtained when applying this algorithm is not relevant to the true biclusters in the original datasets, this may be as a result of the absence of the overlapping control strategy in BICLIC algorithm.

4.3   Real Experimental Datasets

The yeast saccharomyces cerevisiae dataset [8] represents the expression level of yeast under different stress conditions. Total number of 2654 genes and 48 induced oxidative stress conditions are represented in this yeast dataset.

In Table 2(a), Although the average size of biclusters obtained from BCAA and BICLIC biclustering algorithms are larger than that's of CBDB, most of these biclusters were overlapped.

In Table 2(b), the final set of biclusters by CBDB covered nearly all sets of genes, conditions, and 98% of cells. Almost all genes and conditions were included in at least a bicluster. BCCA can extract biclusters that covered most of genes and 70% of conditions, however, it can only cover a small part of cells in yeast dataset, because most of biclusters found were overlapped.

Table 2. (a) Number and average size of biclusters (b) gene, condition, and cell coverages obtained by biclustering algorithms using real yeast stress dataset

(a)   **(b)**

						Coverage	
Algorithm	No. of Biclusters	Average Size		Algorithm	Gene	Condition	Cell
QUBIC	511	204.7		QUBIC	0.748	0.687	0.182
BCCA	1967	701.3		BCCA	0.658	1	0.273
BICLIC	2853	1023.8		BICLIC	0.823	1	0.418
CBDB	3506	620.1		CBDB	1	1	0.986

Although average size of biclusters found by BICLIC was 1023.8, it can only cover about 80% of genes and 40% of cells in the original data matrix. QUBIC scored the minimum value of cell coverage as it can only find sets of up-regulated or down-regulated of genes or conditions which represented only 18% of all cells in the data matrix (Fig. 3).

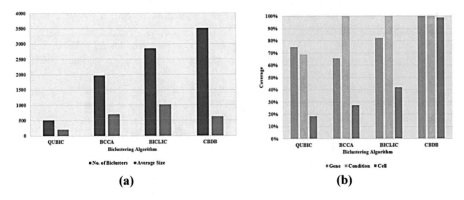

Fig. 3. A comparison between (a) number of and size of biclusters (b) different coverage types of biclusters obtained by each biclustering algorithm using yeast stress dataset

5   Conclusion

In this paper, a biclustering algorithm CBDB was proposed, CBDB used initial single-dimension clustering to find sets of elementary biclusters, which will dilate later to larger sets of true biclusters. This algorithm uses Pearson correlation coefficient as a scoring function for coherency between elements of the same bicluster. CBDB outperforms different correlation-based biclustering techniques when applying to both synthetic and real datasets by using different indices. CBDB has a limitation in identifying overlapping biclusters as it removes all of them in the last phase, which can be handled later by limiting the elimination process. In the future, we may use another clustering technique in the second phase such as bipartite graph-based techniques to combine the second and third phases and get the set of elementary biclusters in one phase. Another enhancement may be to apply CBDB to real cancer datasets with clinical information to emphasize the ability of the algorithm to identify meaningful sets of genes that may affect the diagnosis and prognosis of the disease.

References

1. Dziuda, D.: Data Mining for Genomics and Proteomics: Analysis of Gene and Protein Expression Data, 1st edn. Wiley, New York (2010)
2. Dumancas, G., Adrianto, I., Bello, G., Dozmorov, M.: Current developments in machine learning techniques in biological data mining. Bioinform. Biol. Insights. **11** (2017)
3. Chen, J., Lonardi, S.: Biological Data Mining. In: Chapman and Hall/CRC Data Mining and Knowledge Discovery Series, 1st edn. CRC Press (2017)
4. Iswarya Lakshmi, K., Chandran, C.: Biclustering approaches for prediction of class discovery from gene expression data. In: Proceeding of International Seminar on Emerging Trends and Innovative Technologies in Biological Sciences (2011)
5. Beatriz, P., Raúl, G., Aguilar-Ruiz, J.: Biclustering on expression data: a review. J. Biomed. Inform. **57**, 163–180 (2015)
6. Mounir, M., Hamdy, M.: On biclustering of gene expression data. In: IEEE Seventh International Conference on Intelligent Computing and Information Systems (ICICIS), Cairo, Egypt, pp. 641–648 (2015)
7. Madeira, S., Oliveira, A.: Biclustering algorithms for biological data analysis: a survey. IEEE/ACM Comput. Biol. Bioinf. **1**, 24–45 (2004)
8. Ben Saber, H., Elloumi, M.: A new study on biclustering tools, biclusters validation and evaluation functions. Int. J. Comput. Sci. Eng. Surv. (IJCSES) **6**(1), 1–13 (2015)
9. Erten, C., Sözdinler, M.: Improving performances of suboptimal greedy iterative biclustering heuristics via localization. Bioinformatics **26**, 2594–2600 (2010)
10. Denittoa, M., Farinellia, A., Figueiredob, M., Bicego, M.: A biclustering approach based on factor graphs and the max-sum algorithm. Pattern Recogn. **62**, 114–124 (2017)
11. Aguilar-Ruiz, J.: Shifting and scaling patterns from gene expression data. Bioinformatics **21**, 3840–3845 (2005)
12. Allocco, D., Kohane, I.S., Butte, A.J.: Quantifying the relationship between co-expression, co-regulation and gene function. BMC Bioinformatics **5**, 18 (2004)
13. Bhattacharya, A., De, K.: Bi-correlation clustering algorithm for determining a set of co-regulated genes. Bioinformatics **25**, 2795–2801 (2009)
14. Yun, T., Yi, G.S.: Biclustering for the comprehensive search of correlated gene expression patterns using clustered seed expansion. BMC Genomics **14**(1), 144 (2013)
15. Zhang, Y., Xie, J., Yang, J., Fennell, A., Zhang, C., Ma, Q.: QUBIC: a bioconductor package for qualitative biclustering analysis of gene co-expression data. Bioinformatics **33** (3), 450–452 (2017)

16. Bentham, R., Bryson, K., Szabadkai, G.: MCbiclust: a novel algorithm to discover large-scale functionally related gene sets from massive transcriptomics data collections. Nucleic Acids Res. **45**(15), 8712–8730 (2017)
17. Henriques, R., Ferreira, F., Madeira, S.: BicPAMS: software for biological data analysis with pattern-based biclustering. BMC Bioinf. **18**(1), 82 (2017)
18. Eren, K., Deveci, M., Küçüktunç, O., Çatalyürek, Ü.: A comparative analysis of biclustering algorithms for gene expression data. Brief. Bioinform. **14**, 279–292 (2012)
19. Rodrigo, S., Luis, Q., Roberto, T.: Methods to bicluster validation and comparison in microarray data. In: The Proceeding of 8th International Conference in Intelligent Data Engineering and Automated Learning - IDEAL 2007, 16–19 December, Birmingham, UK, pp. 780–789 (2007)

Reduction of Variations Using Chemometric Model Transfer: A Case Study Using FT-NIR Miniaturized Sensors

Mohamed Hossam[1](✉), Amr Wassal[2](✉),
and M. Watheq El-Kharashi[1](✉)

[1] Computer and Systems Engineering Department, Ain Shams University,
Cairo 11517, Egypt
{mohamed.hossam,watheq.elkharashi}@eng.asu.edu.eg
[2] Computer Engineering Department, Cairo University, Cairo 12613, Egypt
wassal@eng.cu.edu.eg

Abstract. The aim of this paper is to study the unit-to-unit variations in miniaturized Fourier Transform Near-InfraRed (FT-NIR) spectral sensors and the effects of these variations on a classification model developed on a single reference calibration sensor. The paper introduces a simple technique to transfer a classification model from the reference calibration sensor to any other target sensor taking into account variations that might occur. The unit-to-unit variations of the sensors usually result from changes in the signal to noise ratio (SNR) of the sensor due to changes in the mode of operation, variations due to aging, variations due to production tolerances, or changes that occur due to the setup and usage scenario such as scanning through a different medium. To prove the effectiveness of the model transfer technique, we use a Gaussian process classification (GPC) model developed using spectral data from ultra-high temperature (UHT) pasteurized milk with different levels of fat content. The model aims to classify the milk samples based on their fat content. After the model is developed, three experiments are held to mimic each type of the variations and to test how far this will influence the GPC model accuracy after applying the transfer technique.

Keywords: FT-NIR · Gaussian process classification · Milk · Model transfer · Unit-to-unit variation

1 Introduction

Spectroscopy is the discipline of studying material composition using different types of electromagnetic radiation. The instruments used for measuring materials' radiations to study them are called spectrometers or spectrophotometers, in case of visible light radiation. Spectrometers usually measures a certain range of the electromagnetic spectrum. In this paper, spectrometers working in the near infrared (NIR) range are being used. One of the most important advantages in using NIR spectrometers for materials inspection is that they are very fast and non-destructive compared to many other methods especially the traditional chemical methods which are much slower and require sample preparation destroying it in the process. In NIR spectroscopy, the

© Springer Nature Switzerland AG 2020
A. E. Hassanien et al. (Eds.): AMLTA 2019, AISC 921, pp. 272–280, 2020.
https://doi.org/10.1007/978-3-030-14118-9_27

sample is shined with light in the NIR range, the sample molecules vibrate as a result of absorbing some of the wavelengths with different rates while transmitting or reflecting the rest of the light. The received light in the NIR range has overtones of the wavelengths that identify major chemical functional groups. However, these overtones are crowded and overlap making them difficult to discriminate. Thus, it becomes challenging to identify or to quantify materials from their NIR spectrum only. This is where the importance of using machine learning techniques to resolve the ambiguous nature of the NIR spectra based on a prior calibration emerges.

With the advent of miniaturized spectrometers, spectroscopy is liberated from the confinements of the laboratories and is now available to be used directly in the field. However, these mass-produced spectrometers also suffer from unit-to-unit variations that need to be accounted for in order for the chemometrics calibration models to behave in the same way on all spectrometer units despite of variations between them. In practice, there are more types of variations that the model needs to accommodate, namely; variations in the same spectrometer due to heating, aging or other factors, variations due to changes in the measuring conditions, and the unit-to-unit variations among different spectrometers. Any of these three types of variations can be very intense requiring a new and costly model calibration for this specific unit or it can be minimal requiring only some tuning of the existing model using a process called Calibration Transfer.

2 Calibration Transfer

The main purpose of this step in developing spectroscopy applications is to offer a standardization method between the sensor unit or instrument used in the model calibration process and any other units that will use the same model. Originally, calibration transfer used to transfer the models among large benchtop spectrometers with large differences in their specifications. In our case, the differences between spectrometers are minimal since they are all of the same model and brand, having the same design and coming out of the same production line, however, suffering from component and manufacturing variations and tolerances. More importantly, the transfer technique needs to be simple and fast to cope with the large number of the miniaturized sensors coming out of the production line in our case.

To transfer a model from sensor A to sensor B, some spectra readings are taken using both sensors from samples having the same nature as the calibration samples and cover most of the samples variations. Next, dimensional reduction is applied to the data in the same way applied in the calibration model development process. Finally, the data points are transferred from sensor A to sensor B using the linear transformation given in Eq. (1)

$$X_b' = AX_b + B \tag{1}$$

where X_b is the original data points of sensor B and X_b' is the transferred points, while A and B are the coefficients of the linear equation. The coefficients A and B are calculated using the least squares method as per Eq. (2)

$$e = \sum \left[X_a - X_b' \right]^2$$

$$e = \sum [X_a - (AX_b + B)]^2 \qquad (2)$$

where X_a is the data points of sensor A and e is the error between X_a and X_b' which should be minimized with respect to A and B. The coefficients A and B are the output of this step and for any new sensor to be transferred to sensor A these steps should be repeated to get new coefficients associated with the new senor. Afterwards, for every new reading taken with the new sensor B, Eq. (1) is applied before using the model calibrated for sensor A.

3 Experiments and Results

3.1 Samples

The samples used to perform the experiments and to build the model were bought from local grocery store and consisted of 11 different UHT milk packs. The 11 packs were distributed as follows:

- 5 samples of full-fat milk
- 4 samples of skimmed milk
- 2 samples of half-and-half milk.

3.2 Instrumentations

The NeoSpectra micro DVK [1, 2] miniaturized FT-NIR spectrometers from Si-ware Systems based in Cairo, Egypt, were used. These spectrometers are based on Michelson interferometers [3] implemented using micro-electro-mechanical-systems (MEMS) technology. NeoSpectra micro works in the range from 1350 nm to 2550 nm with a resolution of 16 nm at wavelength 1550 nm. All experiments were held using at least one of the three available spectrometer sensors or all three together. The three sensors were labeled D20, D39 and D48. In all experiments, samples are measured in reflection mode and each sample is measured 5 times with different exposure spots on the sample.

3.3 Classification Model

The Classification model is developed with the spectral data measured using sensor D48. That data is plotted in Fig. 1a. The measurement scan time was set to 10 s and taken through a quartz cuvette of a 10 mm path length. Spectra are treated using a 2nd order detrending function, which is a well-known preprocessing technique [4], given in Eq. (3) to remove any trends and scattering effects in the data.

$$F(\lambda) = a\lambda^2 + b\lambda + c \qquad (3)$$

$F(\lambda)$ is assumed to be the trend in the spectra which is a function of the wavelength (λ). The quadratic coefficients of the function, a, b and c, are calculated separately for each spectrum. Afterwards, partial least squares discriminant analysis (PLS-DA) [5, 6] is applied to reduce the data dimensionality to two latent variables. PLS-DA is similar to the principle component analysis (PCA) [7, 8] but unlike the PCA, it aims to maximize the correlation between latent variables and response variables while PCA tries to maximize the variance in latent variables. We develop two Gaussian Process Classification (GPC) [9] models using the latent variables resulting from the PLS-DA run. The first model is a one-dimensional GPC which uses the first latent variable only as shown in Fig. 1b. The second model uses the two latent variables together in a two-dimensional GPC model as shown in Fig. 1c. The models are validated with a leave-one-sample-out cross validation and both of them show 100% classification accuracy. These two models will be used in the following experiments to test their performance under different variations.

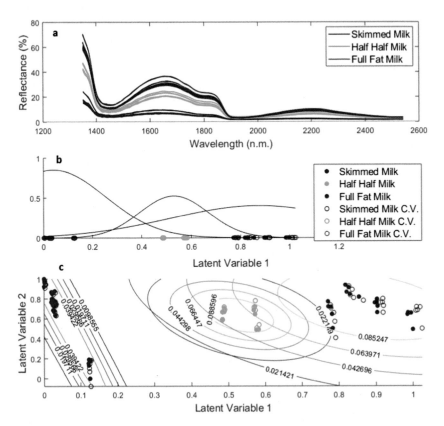

Fig. 1. (a) Raw spectra of scan time 10 s which are measured using the D48 sensor through quartz cuvette of a 10 mm path length. (b) The scores of the 1st latent variable where solid dots represent calibration data, hollow dots represent cross validation data and black lines represent the probability of each class across the space. (c) The scores of both 1st and 2nd latent variables.

3.4   Experiment 1

The purpose of this experiment is to study the effect of changing the signal-to-noise ratio (SNR) within a single spectrometer sensor. The effect of changing SNR can be simulated by changing the scan time duration for the sample measurement. As the time increases, the SNR increases. All samples are measured using sensor D48 with a scan time of 2 s and a scan time of 5 s through a quartz cuvette of a 10 mm path length. These two test sets are presented to the model developed using spectra collected with 10 s scan time on the same sensor D48. From Fig. 2a and b, it is clear that changing the SNR has minimal effect on the classification accuracy of the model, however, it may still be beneficial to test the calibration transfer procedure on this data as shown in Fig. 2c and d.

3.5   Experiment 2

In this experiment, we were keen on testing the effect of the variations between different sensors. We used the three sensors D20, D39 and D48 to measure all the samples using a scan time duration of 2 s and through the regular cuvette we used in the prior experiment. Sensors D20 and D48 should be the closest to each other as both of them passed production testing with similar margins while sensor D39 has passed with a smaller margin from some production tests. From Fig. 3a, it is clear that testing with the 1D GPC model D39 achieves a performance poorer than that of D20 while D48 shows the best performance since it is the sensor unit for which the model was developed. On the other hand, testing with the 2D GPC model shows a performance degradation for both D20 and D39, as shown in Fig. 3b. It can be seen that there is a larger spread in the 2nd latent variable in these cases. Based on these results, calibration transfer is a must and it shows considerable improvements as shown in Fig. 3c and d and Table 1.

Table 1. The classification accuracy for the test data measured using the three sensors

Sensor	1D GPC model		2D GPC model	
	Accuracy before standardization	Accuracy after standardization	Accuracy before standardization	Accuracy after standardization
D48	100%	100%	100%	100%
D20	100%	100%	87.27%	98.18%
D39	90.9%	100%	83.64%	96.36%

3.6   Experiment 3

This experiment focuses mainly on studying the variations that may occur because of changing the container's material. Since the model is developed using milk spectra measured through a quartz cuvette, the test data is measured through a different container which is a glass beaker in this experiment. Similar to prior experiments, the same three sensors are used and their scan time is set to 2 s. From Fig. 4a and b, it is clear that changing the measuring medium or container has a great impact on the model classification accuracy even for sensor D48. After applying the calibration transfer procedure, the spread of the data decreases noticeably and the three classes become more confined as shown in Fig. 4c and d. Both models benefit from standardization but the 1D GPC model achieves a better classification accuracy as shown in Table 2.

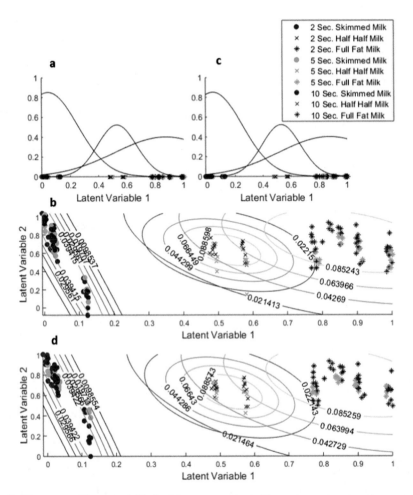

Fig. 2 The scores of latent variable 1 of the test set using milk spectra across changing scan time are plotted (a) before and (c) after linear standardization where the black lines represent the 1D GPC model. The scores for 1st and 2nd latent variables for the 2D GPC model (b) before and (d) after linear standardization respectively.

Table 2. The classification accuracy for the test data measured through a glass beaker

Sensor	1D GPC model		2D GPC model	
	Accuracy before standardization	Accuracy after standardization	Accuracy before standardization	Accuracy after standardization
D48	45.45%	100%	50.9%	89.1%
D20	90.9%	92.27%	96.36%	90.9%
D39	47.27%	96.36%	81.81%	90.9%

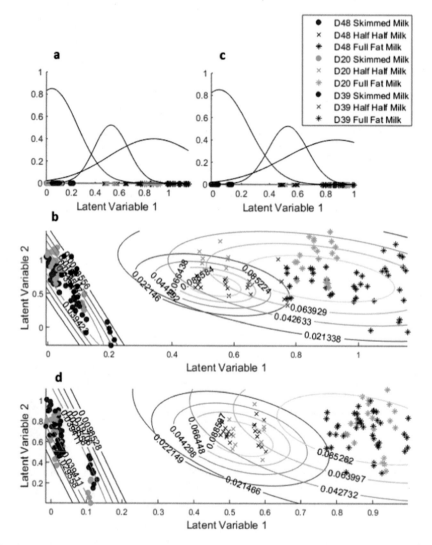

Fig. 3. The scores of latent variable 1 of the test set using milk spectra measured with 3 different sensors, (a) before and (c) after linear standardization where the black lines represent the 1D GPC model. The scores for 1st and 2nd latent variables for the 2D GPC model (b) before and (d) after linear standardization respectively.

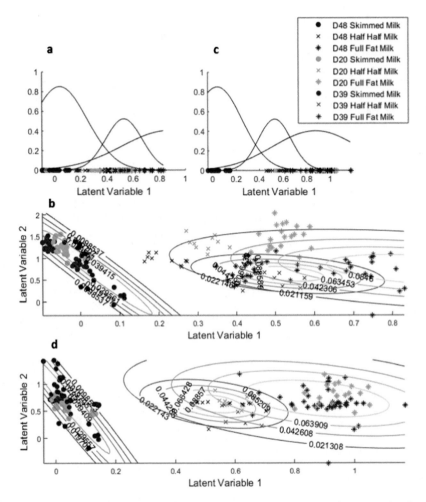

Fig. 4. The scores of latent variable 1 of the test set using milk spectra through a glass beaker instead of a quartz cuvette (a) before and (c) after linear standardization where the black lines represent the 1D GPC model. The scores for 1st and 2nd latent variables for the 2D GPC model (b) before and (d) after linear standardization respectively.

4   Conclusion

In this paper, a calibration transfer process has been presented to overcome the variations that appear more often in miniaturized spectral sensors. The proposed technique does not help with variations due to degradation in SNR since the model does not suffer much from this variation as was demonstrated in experiment 1. On the other hand, unit-to-unit variations and variations due to changing the measuring medium are the greatest beneficiaries from calibration transfer. This was demonstrated in experiments 2 and 3 where the model accuracy jumped from poor classification to 100% accuracy for some cases and over 90% for the rest of the cases. Finally across all the three experiments, the one-dimensional GPC model always shows better or similar results than the two-dimensional GPC model.

Acknowledgment. We would like to express our appreciation to Si-Ware Systems for supporting this research with their state-of-the-art spectrometers and allowing us to use their facilities and laboratories.

References

1. Sabry, Y.M., Khalil, D.A.M., Medhat, M., Haddara, H., Saadany, B., Hassan, K.: Si Ware Systems. Integrated spectral unit. U.S. Patent Application 15/623,961 (2017)
2. Anwar, M., Medhat, M., Mortada, B., El Shater, A.O., Seif, M.G., Nagy, M., Saadany, B.A., Hafez, A.N.: Si Ware Systems. Self calibration for mirror positioning in optical MEMS interferometers. U.S. Patent 9,658,107 (2017)
3. Griffiths, P.R., De Haseth, J.A.: Fourier Transform Infrared Spectrometry, vol. 171. Wiley, Hoboken (2007)
4. Rinnan, Å., van den Berg, F., Engelsen, S.B.: Review of the most common pre-processing techniques for near-infrared spectra. TrAC Trends Anal. Chem. **28**(10), 1201–1222 (2009)
5. Pérez-Enciso, M., Tenenhaus, M.: Prediction of clinical outcome with microarray data: a partial least squares discriminant analysis (PLS-DA) approach. Hum. Genet. **112**(5–6), 581–592 (2003)
6. Ma, J., Pu, H., Sun, D.W., Gao, W., Qu, J.H., Ma, K.Y.: Application of Vis–NIR hyperspectral imaging in classification between fresh and frozen-thawed pork Longissimus Dorsi muscles. Int. J. Refrig. **50**, 10–18 (2015)
7. Eriksson, L.: Introduction to Multi- and Megavariate Data Analysis Using Projection Methods (PCA & PLS). Umetrics AB, Umeå (1999)
8. Næs, T., Isaksson, T., Fearn, T., Davies, T.: A User Friendly Guide to Multivariate Calibration and Classification. NIR publications, Chichester (2002)
9. Bernardo, J., Berger, J., Dawid, A., Smith, A.: Regression and classification using Gaussian process priors. Bayesian Stat. **6**, 475 (1998)

Predicting Drug Target Interaction by Integrating Drug Fingerprint and Drug Side Effect Using Machine Learning

Abdelrahman Saad⊙, Fahima A. Maghraby$^{(\boxtimes)}$,
and Yasser M. Omar$^{(\boxtimes)}$

Arab Academy for Science, Technology and Maritime Transport (AASTMT),
Cairo, Egypt
{abdelrahman.saad,fahima,yasser.omar}@aast.edu

Abstract. Drug discovery is an important step before drug development. Drug discovery is the process of identifying, testing a drug before medical use. Drugs are used to cure diseases by interacting with the target, which is the protein in the human cells. Many resources are wasted (cost and time) on lab experiments to discover drugs and its application. Yet machine learning enhanced the process of drug discovery and the prediction of drug-target interaction, which helped in predicting new drugs and finding more applications for old drugs. Predicting drug-target interaction starting by studying the nature of drugs and its properties. Most of the datasets existing are drugs, targets and their interactions datasets. We compiled our dataset to include side effect as drug feature. The dataset contains 400 drugs, 794 targets and 3990 side effects. In this study, a machine-learning model is implemented using three different classifiers: Decision Tree, Random Forest (RF) and K-Nearest Neighbors (K-NN) for classification. Drug fingerprint and side effect were used as input features to train our model. Three different experiments were conducted using fingerprint, side effect and both fingerprint and side effect. Results showed improvement in prediction when integrating both drug fingerprint and side effect. K-NN scored best results in the three experiment with an average accuracy of 94.69%.

Keywords: Drug · DTI · Fingerprint · Side effect · Target

1 Introduction

Drug-Target Interaction (DTI) prediction is important in medical field especially in drug discovery. Simply drug represents compounds and target represents proteins, which interact with drugs to cure a certain disease [1, 2]. According to DrugBank database which is a unique bioinformatics and cheminformatics resource that contains both drug and target valuable information, the total number of discovered drugs are 10562 while the only approved are 3254 having known interactions with targets [3]. Nowadays the number of chemical compounds (drugs formed of) reached approximately 35 million [4] where they could be nominated to be drug candidates. The combination of each drug and target to determine interaction will require a long time and expensive experiments [5].

© Springer Nature Switzerland AG 2020
A. E. Hassanien et al. (Eds.): AMLTA 2019, AISC 921, pp. 281–290, 2020.
https://doi.org/10.1007/978-3-030-14118-9_28

Computational modeling and algorithms are used to predict DTI to save time and cost like machine learning which is better than the traditional docking simulation that requires three-dimensional structures of targets which is not available for all existing targets [6].

The rest of the paper is organized as follows: Sect. 2 introduces the related work of drug-target interaction. Section 3 discusses constructing dataset, input features, used machine learning techniques and performance measures. Section 4 presents proposed framework. Section 5 discusses the experimental results. Finally, Sect. 6 presents the conclusion and future work.

2 Literature Review

Identifying drug and target interaction is not only important to discover new drugs, but also discovering new targets for the existing old drugs [7].

In 2015 Shi et al. [1], used the enhanced similarity measures and super target clustering to predict the interaction between drug and target. They focused on handling the missing interactions; the dataset was collected from KEGG BRITE [8], BRENDA [9], SuperTarget & Matador [10] and DrugBank [11] databases, which contained 932 drugs, 989 targets and 5127 drug interactions. First, they calculated the similarity between new drugs as well the similarity between known targets, they clustered the targets in-group called super target group and then they built their model by matching drug and target using confidence score. Results from their proposed algorithm were high in terms of AUC (Area Under Curve of Receiver Operating Characteristic) compared to Kernelized Bayesian Matrix Factorization (KBMF2K) and Weighted Nearest Neighbor (WNN) algorithm.

In 2016 Coelho et al. [12], predicted DTI by combining Support Vector Machine (SVM) and RF as a two-classification model, one was to predict drug taking into consideration the type of target and the other without considering the type of target. The dataset was collected from DrugBank [13] and study by Yamanishi et al. [14] which contained 927 drugs, 1370 targets and 5127 drug interactions. Results were better in SVM model considering the type of target (protein) in terms of AUC. They also proposed that integrating of network centrality metrics, in parallel to the expansion of the chemical and proteomic spaces would enhance drug-target interaction.

In 2016 Galeano et al. [15], introduced chemical similarity prediction, their idea was based on the assumption of similar chemical structures of drugs, can predict the closeness of their target. The dataset was collected from Biogrid [16] and DrugBank [13] databases, which contained 9336 drugs, 4612 targets. They started by building chemical similarity network where each node represents a drug and the weight of the edges is the Tanimoto chemical similarity to calculate the similarity between drugs, then building an interactome which represents protein-protein interaction to identify the relationship between targets and finally identify similarities between the two networks to predict similar targets. Similarity reached 85% in terms of AUC they suggested that integrating side-effect similarity would enhance the similarity percentage.

In 2016 Hao et al. [17], predicted drug-target interactions by dual-network integrated logistic matrix factorization (DNILMF) based on similar drugs or target may help in the accuracy of the predictions for their neighbors. They constructed a new DTI dataset with PubChem CID, which contained 829 drugs, 733 targets and 3688 interactions. First, they built drug and target profiles (kernel construction), followed by kernel calculation, then diffused two classes of similarity and finally proposed a DNILMF algorithm to perform predictions. DNILMF achieved better results than a Neighborhood Regularized Logistic Matrix Factorization (NRLMF); they said that their model could be improved using genetic algorithm.

In 2017, Sinha et al. [18] adopted machine learning techniques (Naïve Bayes, Decision Tree, Random Forest and Support Vector Machine) to predict a certain kind of protein called Leishmania donovani membrane to predict whether this protein will be a drug target candidate or vaccine candidate. They prepared 28 known proteins [19] and used the four classifiers then evaluated the classifiers and used the best classifier. Finally, they trained the best classifier with 37 unknown proteins and identified drug targets, vaccine candidates among unknown proteins. Naive Bayes achieved the best result for this kind of problem with accuracy 76.17%.

In 2017, Li et al. [20] predicted DTI networks based on drug chemical structure and protein sequences. Their objective was to enhance the accuracy of SVM by introducing Discriminative Vector Machine (DVM), which focused on enhancing the accuracy of feature extraction and classification. The dataset was collected from KEGG BRITE [8], BRENDA [9], SuperTarget & Matador [21] and DrugBank [11] databases, which contained 932 drugs, 989 targets and 5127 drug interactions (Gold standard dataset). First, they started by feature extraction, fingerprint as a feature vector for drug and Local Binary Pattern (LBP) operator as a feature vector for the target. Second, they applied the Principle Component Analysis (PCA) to extract the discriminatory feature information. Third, they trained the two classifiers. Finally, they tested the two classifiers and obtained the results. Results were significantly enhanced using DVM in terms of average accuracy, precision and sensitivity. Actually, there was a problem that DVM would spend a lot of time finding the representative vector; they proposed that using multi-dimensional indexing techniques would speed up the search process.

We can conclude from the previous researches that there are two approaches, the first approach-building network (graphs) where drug molecules and target proteins are nodes and the known drug–target interactions are edges between these nodes. Therefore, predicting new drug–target interaction is equivalent to predicting new edges [22].

The second approach is a machine learning approach for predicting drug target interaction (DTI) by building a classifier based on known interactions. Each drug–target pair is marked positive if they interact and marked negative if there is no interaction and can be represented by a feature vector. All recent papers represented drug features as descriptors generated from its chemical structure but did not considered drug side effect as a feature, while targets' features are derived from its protein sequence [23].

In this study, drug fingerprint and drug side effect, which is our contribution in this research, will be used as features along with the drug and its known interaction with the targets. Model is implemented using machine learning algorithms.

3   Materials and Methods

This section introduces the materials and methods used in this paper.

3.1   Dataset Construction

The first dataset was downloaded from drug central which is online drug compendium created and maintained by Division of translational informatics at the University of New Mexico in collaboration with the IDG the dataset contains 2376 drugs, 1938 targets and 14521 known interactions between drug and target.

The second dataset was downloaded from spider 4.1, which is a drug-side effect resource, which contains information about drugs, its side effects, side effect frequency and side effect classifications. The dataset contains 1430 drugs, 5868 side effect and 139756 drug side effect pairs.

We joined two datasets-using MYSQL database using PubChem CID, which is a unique identifier for each drug (compound) as a primary key. The output was two tables. The first table contains drugs with their corresponding targets and the second table contains drugs with their corresponding side effect as shown in Table 1.

Table 1. Summarizes features in datasets

No. of features	Drugs	Targets	Side-effect
First dataset	2376	1938	–
Second dataset	1430	–	5868
Compiled dataset	400	794	3990

3.2   Fingerprint and Side Effect as Features

Every drug (compound) has a fingerprint. Every molecule contains information representing it like number of atoms and bonds this information encoded as a fingerprint of same length and type in binary bits. Not necessarily that each drug has a unique fingerprint, maybe two drugs have the same finger-print or near to it which can be used as a powerful and effective feature in classification and structural similarity clustering [24].

Every drug has side effects, which can affect patients using these drugs causing unpleasant symptoms e.g. skin irritation and nausea. Many drugs share many side effects, which can help in predicting targets [25]. Drug side effect similarity is a measurement framework to see how two or more drugs similar in their side effects. There is a strong relationship between drug side effect similarity and their indications (diseases treated by drug) where a group of some drugs treating certain diseases tend to have similar side effects.

3.3   Random Forest (RF)

RF is a supervised classification technique used in different classification purposes [26]. As proposed by Breman, a collection of tree-structured classifiers (bagged decision trees) introduces ensemble learning using multiple machine learning algorithms and put them together to create a bigger machine algorithm. In RF, we build multiple trees. Randomly subset of data is built by each tree and based on bootstrap aggregating algorithm [27]. After that, each tree is established based on splitting criteria e.g. Gini. Features are classified by trees and vote for the tree class. The forest chooses the classification having the most votes of all other trees.

Decision trees is built in a tree-structured form to classify instances by sorting them down from the root node to the leaf node. A root node is defined to have no-incoming nodes but maybe zero or several outgoing nodes. While the internal node has one incoming node and two or more outgoing nodes and the node, which has one incoming node and no outgoing nodes is called leaf or terminal node [28]. This technique is very useful to our problem as each drug interacts with one target or more and that drug has multiple side effects. Each leaf node is a class label while root node and internal node are rules to test the value of some features. Entropy is one of the criteria used to build a decision tree, which is a common measure, used in information theory measures the randomness in data using Eq. (1)

$$Entropy(S) = -\sum_{c \in C} p_c Log_2 p_c \tag{1}$$

Consider a set S of examples with C many classes where p_c is the probability that an element of S belongs to class c. Entropy measures degree of uncertainty where dominant classes have less entropy (low certainty) while classes that have the same or equal probabilities have high entropy (high uncertainty).

The other criteria is the information gain, which measures the effectiveness of an attribute to classify data simply by reducing entropy as in Eq. (2)

$$Information\ Gain(S, F) = Entropy(S) - \sum_{f \in F} \frac{|S_f|}{|S|} Entropy(S_f) \tag{2}$$

Consider element of S consists of a set of features number of elements of S with feature F having value f. therefore Information Gain(S, f) is the expected reduction in entropy by knowing the value of F.

3.4   K-Nearest Neighbors (K-NN)

K-NN is called a non-parametric method. Unlike other supervised learning algorithms, K-NN does not learn from training data, it simply uses the training data at the test time to make predictions. K-NN algorithm computes distances of the test instance from each of the training instances based on the type of features. For Real-valued features were

$(x_i \in R^d)$ Euclidean distance is commonly used by calculating the square root of the sum of the squared differences between a new point (x_i) and an existing point (x_j) across all input attributes j using Eq. (3)

$$Euclidean\ Distance\left(X_i, X_j\right) = \sqrt{\sum_{m=1}^{D}(X_{im} - X_{jm})^2} \tag{3}$$

For Binary-valued features, hamming distance is applied using Eq. (4)

$$Hamming\ Distance\left(X_i, X_j\right) = \sum_{m=1}^{D}(X_{im} \neq X_{jm}) \tag{4}$$

3.5   Performance Measures

The performance of the three models was evaluated using Accuracy, Sensitivity, Specificity and Precision as shown below:

$$Accuracy = \frac{TP + TN}{TP + TN + FP + FN} \tag{5}$$

$$Sensitivity = \frac{TP}{TP + FN} \tag{6}$$

$$Specificity = \frac{TN}{TN + FP} \tag{7}$$

$$Precision = \frac{TP}{TP + FP} \tag{8}$$

In our classification problem, TP means true positive (drug-target pairs are interacting), TN means true negative (drug-target pairs are not interacting), FP (predicted negative drug target pairs to be interacting) and FN (predicting positive drug target pairs not to be interacting) where positive means interacting while negative is not interacting.

4   Proposed Model

The aim of the proposed model is to predict DTI using three different machine-learning techniques as shown in Fig. 1.

The first step is the preprocessing phase where the two datasets were joined containing 400 drugs in common as mentioned before in Table 1. Features in two output tables such as drug name, drug side effect and target name are only just names so these features were labeled (given random numbers) then were hot encoded (0s and 1s) to replace text with numbers so that machine can understand features for training.

To generate drug fingerprint, we first downloaded Structure Data File (SDF) for 400 drugs from National Center for Biotechnology Information (NCBI) that contains features about each drug such as molecular weight, number of bonds and number of atoms. Using R language, special library called ChemmineR we created a database contain SDF file for each drug then generated drug fingerprint a fixed binary length of 20 fixed characters enclosing drug features.

Finally, constructing the drug-side effect matrix where each drug (D1) has multiple side effects (Sn) as shown in Fig. 2, adding to it the fingerprint generated for each drug and the other matrix was drug-target matrix where it is the matrix of interaction in which a drug (D1) interacts with multiple targets (Tn) as shown in Fig. 3.

The second step is the classification phase where we trained the data using decision tree, RF and K-NN. We trained the three classifiers three different times. First, using drug's fingerprint only, second using drug's side effect and third using both against the drug target matrix. In case of using drug fingerprint, a new drug can be classified to be interacting or not interacting after training data where new drug is considered interacting when it is similar to the trained drug fingerprint or near to it. Also considered when integrating drug fingerprint and drug side effect a new drug can be classified to be interacting or not interacting after training data where new drug is considered interacting when this drug has a certain fingerprint and side effect.

The third step we tested our model by splitting dataset in 70% training and 30% testing.

The Final step, is to compare results using fingerprint only, side effect only and using both side effect and fingerprint as features.

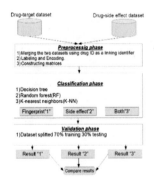

Fig. 1. Drug-target interaction model

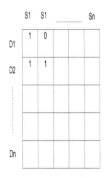

Fig. 2 Drug side effect matrix

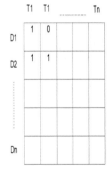

Fig. 3 Drug-target matrix

5   Experiment Results and Discussion

In this section, three different experiments were conducted using drug fingerprint only, using drug side effect only and finally using both.

The first experiment was training our classifiers on drug's fingerprint. The purpose of this experiment is to classify drugs have similar or near fingerprints and determine their interaction. The carried experiments showed that K-NN achieved the higher accuracy with 95.6% while the higher sensitivity (rate of predicting interacting drugs) was by the decision tree with 37.38%. K-NN has lowest precision ratio with 28.24% due to the number of predicted interacting drugs is less than the non-interacting one.

The second experiment goal is to classify drugs based on their side effects and identify their interaction with the targets. K-NN achieved the higher accuracy and sensitivity with 91.28% and 35.27% respectively. Decision tree has the lowest accuracy with 89.97%.

The third experiment purpose was classifying drugs based on their fingerprints and side effects. Integrating the two features enhanced the accuracy. K-NN achieved the higher accuracy with 97.63% and higher sensitivity of 90%. RF recorded the lowest precision ratio with 64.1%.

To sum up, integrating both drug fingerprint and side effect enhanced the accuracy, sensitivity and specificity of the three classifiers. As it helped the classifiers to increase the true positive cases (predict the positive drug target pairs) and lowered both false positive (predicted negative drug target pairs to be interacting) and false negative (predicting positive drug target pairs not to be interacting). Accuracy increased by an average 5.12% in case of decision tree, 4.41% in case of K-NN and 4.93% in case of random forest as shown in Figs. 4, 5 and 6.

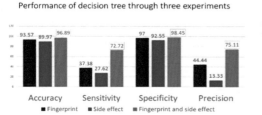

Fig. 4. Decision tree performance

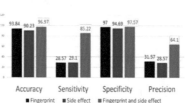

Fig. 5. Random forest performance

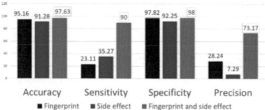

Fig. 6. K-NN performance

6   Conclusion

Drugs (compounds) play a vital role in curing diseases through interacting with targets (proteins). Therefore, studying drug feature will help in developing drug discovery, which will help to find application for many drugs in turn. Many papers used features by generating drug descriptors and fingerprint but none of them used drug side effect as a feature. In this study, we integrated both drug fingerprint and side effect by compiling a new dataset, which enhanced the result of the prediction. The results showed that using both drug fingerprint and side effect together as features enhanced the performance of prediction. In the future, we can try other different classification techniques to enhance the prediction performance in addition to increasing number of drugs and targets families.

References

1. Shi, J.-Y., Yiu, S.-M., Li, Y., Leung, H.C.M., Chin, F.Y.L.: Predicting drug–target interaction for new drugs using enhanced similarity measures and super-target clustering. Methods **83**, 98–104 (2015)
2. Lan, C., Chandrasekarany, S., Huan, J.: A distributed and privatized framework for drug-target interaction prediction. In: International Conference on Bioinformatics and Biomedicine (BIBM), pp. 731–734. IEEE (2016)
3. Statistics: DrugBank. https://www.drugbank.ca/stats. Accessed Nov 2018
4. Bolton, E., Wang, Y., Thiessen, P., Bryant, S.: PubChem: integrated platform of small molecules and biological activities. Ann. Rep. Comput. Chem. **4**, 217–241 (2008)
5. Hurle, M., Yang, L., Xie, Q., Rajpal, D., Sanseau, P., Agarwal, P.: Computational drug repositioning: from data to therapeutics. Clin. Pharmacol. Ther. **93**(4), 335–341 (2013)
6. Chen, X., Yan, C., Zhang, X., Zhang, X., Dai, F., Yin, J., Zhang, Y.: Drug–target interaction prediction: databases, web servers and computational models. Briefings Bioinf. **17**(4), 696–712 (2015)
7. Li, H., Gao, Z., Kang, L., Zhang, H., Yang, K., Yu, K., Luo, X., Zhu, W., Chen, K., Shen, J., Wang, X., Jiang, H.: TarFisDock: a web server for identifying drug targets with docking approach. Nucleic Acids Res. **34**(Web Server), W219–W224 (2006)
8. Kanehisa, M.: From genomics to chemical genomics: new developments in KEGG. Nucleic Acids Res. **34**(90001), D354–D357 (2006)
9. Schomburg, I.: BRENDA, the enzyme database: updates and major new developments. Nucleic Acids Res. **32**(90001), 431D–433D (2004)
10. Kuhn, M., Szklarczyk, D., Franceschini, A., Mering, C., Jensen, L., Bork, P.: STITCH 3: zooming in on protein-chemical interactions. Nucleic Acids Res. **40**(D1), D876–D880 (2011)
11. Wishart, D.S., Knox, C., Guo, A.C., Cheng, D., Shrivastava, S., Tzur, D., Gautam, B., Hassanali, M.: DrugBank: a knowledgebase for drugs, drug actions and drug targets. Nucleic Acids Res. **36**(suppl_1), D901–D906 (2007)
12. Coelho, E., Oliveira, J., Arrais, J.: Ensemble-based methodology for the prediction of drug-target interactions. In: 29th International Symposium on Computer-Based Medical Systems (CBMS), pp. 36–41. IEEE (2016)
13. Wishart, D.S.: DrugBank: a comprehensive resource for in silico drug discovery and exploration. Nucleic Acids Res. **34**(90001), D668–D672 (2006)

14. Yamanishi, Y., Kotera, M., Kanehisa, M., Goto, S.: Drug-target interaction prediction from chemical, genomic and pharmacological data in an integrated framework. Bioinformatics **26** (12), i246–i254 (2010)
15. Galeano, D., Paccanaro, A.: Drug targets prediction using chemical similarity. In: XLII Latin American Computing Conference (CLEI), pp. 1–7. IEEE (2016)
16. Stark, C.: BioGRID: a general repository for interaction datasets. Nucleic acids Res. **34** (suppl 1), D535–D539 (2006)
17. Hao, M., Bryant, S., Wang, Y.: Predicting drug-target interactions by dual-network integrated logistic matrix factorization. Sci. Rep. **7**(1), 40376 (2017)
18. Sinha, A., Singh, P., Prakash, A., Pal, D., Dube, A., Kumar, A.: Putative drug and vaccine target identification in leishmania donovani membrane proteins using naïve bayes probabilistic classifier. IEEE/ACM Trans. Comput. Biol. Bioinform. **14**, 204–211 (2017)
19. Kumar, A., Misra, P., Sisodia, B., Shasany, A., Sundar, S., Dube, A.: Proteomic analyses of membrane enriched proteins of Leishmania donovani Indian clinical isolate by mass spectrometry. Parasitol. Int. **64**(4), 36–42 (2015)
20. Li, Z., Han, P., You, Z.-H., Li, X., Zhang, Y., Yu, H., Nie, R., Chen, X.: In silico prediction of drug-target interaction networks based on drug chemical structure and protein sequences. Sci. Rep. **7**(1) (2017)
21. Gunther, S., Kuhn, M., Dunkel, M., Campillos, M., Senger, C., Petsalaki, E., Ahmed, J., Urdiales, E.G., Gewiess, A., Jensen, L.J., Schneider, R., Skoblo, R., Russell, R.B., Bourne, P.E., Bork, P., Preissner, R.: SuperTarget and matador: resources for exploring drug-target relationships. Nucleic Acids Res. **36**(Database), D919–D922 (2007)
22. Azuaje, F., Zhang, L., Devaux, Y., Wagner, D.: Drug-target network in myocardial infarction reveals multiple side effects of unrelated drugs. Sci. Rep. **1**(1), 52 (2011)
23. Cao, D.-S., Liu, S., Xu, Q.-S., Lu, H.-M., Huang, J.-H., Hu, Q.-N., Liang, Y.-Z.: Large-scale prediction of drug–target interactions using protein sequences and drug topological structures. Anal. Chim. Acta **752**, 1–10 (2012)
24. Cao, D.-S., Hu, Q.-N., Xu, Q.-S., Yang, Y.-N., Zhao, J.-C., Lu, H.-M., Zhang, L.-X., Liang, Y.-Z.: In silico classification of human maximum recommended daily dose based on modified random forest and substructure fingerprint. Anal. Chim. Acta **692**(1–2), 50–56 (2011)
25. Campillos, M., Kuhn, M., Gavin, A., Jensen, L., Bork, P.: Drug target identification using side-effect similarity. Science **321**(5886), 263–266 (2008)
26. Fayz, S., Rizka, M., Maghraby, F.: Cervical cancer diagnosis using random forest classifier with SMOTE and feature reduction techniques. IEEE Access 1 (2018)
27. Wu, Y., Wang, H., Wu, F.: Automatic classification of pulmonary tuberculosis and sarcoidosis based on random forest. In: 10th International Congress on Image and Signal Processing, BioMedical Engineering and Informatics (CISP-BMEI). IEEE (2017)
28. Bombara, G., Vasile, C.-I., Penedo, F., Yasuoka, H., Beltaz, C.: A decision tree approach to data classification using signal temporal logic. In: Proceedings of the 19th International Conference on Hybrid Systems: Computation and Control - HSCC 2016 (2016)

Statistical Insights and Association Mining for Terrorist Attacks in Egypt

Nour Eldeen M. Khalifa[1,3(✉)] ⓘ, Mohamed Hamed N. Taha[1,3] ⓘ,
Sarah Hamed N. Taha[2] ⓘ, and Aboul Ella Hassanien[1,3] ⓘ

[1] Information Technology Department, Faculty of Computers and Information,
Cairo University, Giza, Egypt
{nourmahmoud, mnasrtaha, aboitcairo}@cu.edu.eg
[2] Forensic Medicine and Clinical Toxicology Department, Faculty of Medicine,
Cairo University, Giza, Egypt
snasrtaha@cu.edu.eg
[3] Scientific Research Group in Egypt (SRGE), Giza, Egypt
http://www.egyptscience.net

Abstract. Terrorist attacks are the most significant challenging issue for the humankind across the world, which need the whole attention of all nations. No doubt, that the terrorism attacks in Egypt negatively affect its economy and decreases its opportunities for foreign investments in the last decade. In this paper, statistical techniques will be applied to terrorism attacks occurred in Egypt in the last four decades. The used database in this study is the Global Terrorism Database (GTD), which is published by the University of Maryland. Statistical insights graphs will be presented in the paper. In addition, association-mining algorithms will be used to find the frequent hidden patterns in the database. Those algorithms help in generating composite rules by mining the database, which generates deeper meaningful insights. The statistical insights and association mining will help to understand the nature of terrorist attacks in Egypt. Moreover, it will give the decision-makers the knowledge to overcome and step ahead the terrorist attacks before occurring. This will reflect on Egypt economy positively and increases the opportunities to have better investments and encourage the return of tourism.

Keywords: Terrorism attacks · Egypt · Association mining ·
Statistical insights

1 Introduction

Terrorism is a serious threat to international peace and security. No nation can consider itself immune from the risks of terrorism. No nation can disengage itself from the efforts to combat terrorism. Researchers want to solve the problems effectively and efficiently [1, 2]. Terrorism is classified as terror all activities which are committed to the main goal of killing people, destroying the infrastructure, the economy, etc. [3]. Some activities target specific persons and others are committed without any specific target and hence are more dangerous leading to more casualties and destruction. Persons who commit such activities

© Springer Nature Switzerland AG 2020
A. E. Hassanien et al. (Eds.): AMLTA 2019, AISC 921, pp. 291–300, 2020.
https://doi.org/10.1007/978-3-030-14118-9_29

are known as terrorists. They are dangerous people who are the main source of threat to the whole society locally and to the humanity at large [4].

In the Global Terrorism Index annual report in January 2017 [5], the report named Egypt as one of the ten most terrorism-struck countries in the world. Terrorism in Egypt has targeted government police, officials, tourists and the normal citizens. Most of these attacks have been linked to Islamic extremism. Terrorist attacks increased in the 1990s when the Islamist movement al-Gama'a al-Islamiyya targeted high-level with the politics leaders and killed hundreds [1]. The attacks also increased in 2010s after revolutions in 2011 and 2013. The Egyptian government's counter-terrorism campaign has resulted in a reduction in the number of terrorist attacks on the Egyptian mainland.

The rest of the paper is organized as follows. Section 2 is a quick review of the used methods and related works. Section 3 illustrates the Global Terrorism Database (GTD). Statistical insights will be presented in Sect. 4, while Sect. 5 introduces the association mining results on Egypt GTD extracted data. In Sect. 6, a discussion of research findings will be outlined. Finally, Sect. 7 concludes the paper and list the possible future work.

2 Literature Reviews

Understanding the nature of data and retrieving a useful information from it will help directly to take corrective actions and decisions by government decision makers to overcome the terrorist attacks. Computers algorithms and techniques with its protentional speed in calculations and accuracy contribute to understand the data and thus generate useful information about it.

2.1 Statistical Analysis

One of the popular computers techniques is statistical analysis. Statistical analysis is a component of data analytics. It includes inspecting and collecting all data sample in a group of items from which samples can be pulled. In statistics, a sample is defined as a representative selection pulled from a total population [6].

Most of the time, the statistical analysis is not enough to draw a conclusion about the collected data. More deep insights are indeed needed to find relationships between the collected data items. The term "mining" is commonly used nowadays in researches that add an extra layer of analysis over the data to deeply understand it and thus having solid useful information.

2.2 Association Mining

Association mining or "Association rule mining" is a procedure, which is meant to find frequent patterns, correlations, associations, or causal structures from database. Association rule mining can be described as a data mining process. It involves finding the rules and relations between set of items. The techniques that are used for discovering association rules from the data have focused on finding relationships between items. Therefore, the association rules of one type describe a local pattern which can be easily interpreted and communicated [7].

The two important basic measures of association rules are support and confidence as explained in detail in [7]. If there are two items, then support is defined as the ratio of occurrence of that two items and a total number of transactions. Minimum support is that if rules which have supported greater than a user-defined support.

Confidence is defined as the possibility of seeing the rule's consequent under the condition that the transactions also contain the antecedent. In addition, minimum confidence is one in which rules have a confidence greater than a user-defined confidence.

Let $I = \{i_1, i_2, i_3, \ldots i_n\}$ be a set of n attributes called items and $D = \{t_1, t_2, \ldots t_n\}$ be the set of transactions. It is called a dataset. Every transaction, t_i in D has a unique transaction ID, and it consists of a subset of item sets in I. A rule can be defined as an implication, $X \rightarrow Y$ where X and Y are subsets of $I(X, Y \subseteq I)$, and they have no element in common, i.e., $X \cap Y$. X and Y are the antecedent and the consequent of the rule, respectively.

The support of an itemset X, supp(X) is the proportion of transaction in the database in which the item X appears. It signifies the popularity of an itemset. The support and the confidence values are calculated according to Eqs. (1) and (2).

$$supp(X) = \frac{Number\ of\ transaction\ in\ which\ X\ appears}{Total\ number\ of\ transactions} \tag{1}$$

$$conf(X \rightarrow Y) = \frac{supp(X \cup Y)}{supp(X)} \tag{2}$$

2.3 Related Works

The report in [8] presents the global terrorism index which depends on Global Terrorism Database. The newest annual report is released in 2017, this report provides a comprehensive summary of the key global trends and patterns in terrorism over the last 17 years in covering the period from the beginning of 2000 to the end of 2016.

In [9], the authors introduce visual analytics of terrorism in India based on GTD. They extracted data was from online news articles and tweets and integrated with GTD to produce useful analytics.

The work presented in [10] concerning about terrorist attacks in the Middle East. The attacks were analyzed using methods of network science, statistical methods, geographic information science, and artificial neural networks. The authors also, used the GTD.

The previous mentioned works used the GTD to generate statistical technical reports or present visual analytics globally or for a certain country or region. According to literature reviews, this study is considered one of the first trials to address Egypt as a country for terrorist attacks and uses association-mining algorithms to generate deeper meaningful insights.

3 Global Terrorism Database (GTD)

The Global Terrorism Database (GTD) [8] is an open-source database including information on terrorist events around the world from 1970 through 2016 (with annual updates planned for future releases). GTD is collected and collated by the National Consortium for the Study of Terrorism and Responses to Terrorism (START); a Department of Homeland Security Centre of Excellence led by the University of Maryland. The GTD is the most comprehensive global dataset on terrorist activity and has now codified over 170,000 terrorist incidents.

The database contains (1) basic information on the type of terrorist attack, including location, name of group taking responsibility, and number of deaths and injuries; (2) detailed information on the fate of the terrorists or terrorist group claiming responsibility; (3) detailed information on terrorist events involving hostages; and (4) detailed information on terrorist events involving skyjackings.

4 Statistical Insights for EGYPT Terrorist Attacks

The extracted data from GTD for EGYPT was reviewed and numerically coded using programming language C++. The coding phase is a necessary step in order to generate the required statistics and generating rule in the association mining phase.

The first statistical insight graph is the distribution of the terrorist attacks over the years between 1970 to 2016. Figure 1 illustrates the distribution of the terrorist attacks over the years. The figure also demonstrates that there are two peaks or periods of terrorism attacks that EGYPT suffered from it mostly.

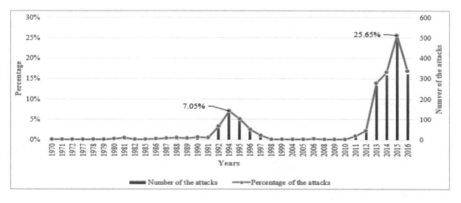

Fig. 1. The distribution of the terrorist attacks over the years

The first period is the nineties from 1990 to 1998 with the total percentage of 20.22% and the highest number of attacks in 1994. This period the rise of the Islamist movement al-Gama'a al-Islamiyya when the Islamist movement al-Gama'a al-Islamiyya targeted politics leaders, police and tourists, killing hundreds. The second period is the revolution period from 2011 to 2016 with 76.18% of the total attacks on Egypt and the highest number of attacks in 2015. Egypt had two revolutions in 2011 and 2013. After both revolutions, there was instability period due to the absence of police and the absence of legal authority on Egypt including the presidency chair and parliament, which reflected on the number of terrorist attacks on this period. The two peaks possessed around 96.39% of the total percentage of terrorist attacks on Egypt.

Figure 2 shows the distribution of terrorist attacks on targets. It shows that the police and military possessed 55% of total terrorist attacks. It is an expected percentage as the military and police main objective is to stand against them and defend the national security of their country. If the terrorists manage to attack police and military successfully, their terror messages to citizens will be delivered. The citizens will feel unsecure and thus begin to make plans to have a secure life through immigration or any other way.

Figure 3 presents the distribution of successful/not successful attacks and the distribution of suicidal attacks/not suicidal attacks.

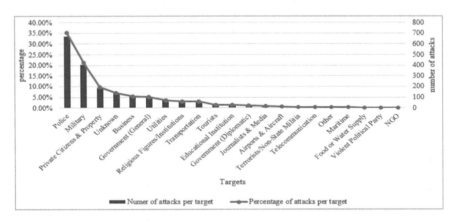

Fig. 2. The distribution of the terrorist attacks on targets

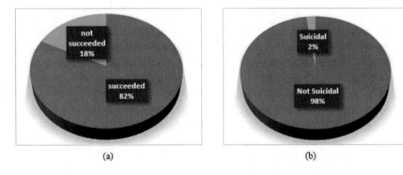

(a)   (b)

Fig. 3. (a) The distribution of successful/not successful attacks – (b) The distribution of suicidal attacks/not suicidal attacks

The subfigure (a) in Fig. 3 shows that 82% of terrorist attacks in Egypt was successful, it is a sad fact, but it reflects that the terrorists are well organized. It also reflects that the security procedures and countermeasures failed to predict terrorist attacks and then manage to stop it before happening. The sub-figure (b) illustrates that only 2% of terrorist attacks were suicidal while 98% is not.

Figure 4 demonstrates a complex insight, which presents the distribution of the target of the attack with attack type versus the number of attacks. The figure shows that the police category was targeted using armed assault attack with 16.46% of the total percentage of the attacks. Meanwhile, the Military category was targeted using bombing/explosion with 11.29%. The using of explosion with the military because of the nature of military regulations to protect important sensitive buildings.

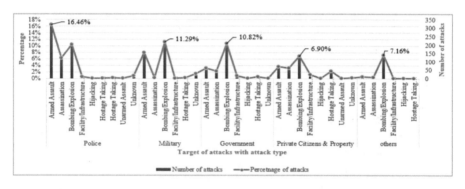

Fig. 4. The distribution of the target of the attack with attack type versus the number of attacks

5   Association Mining on Egypt Extracted Data from GTD

Association rule is a rule-based machine learning method for discovering interesting relations between variables in datasets. It is intended to identify strong rules discovered in the dataset using some measures of interestingness.

In this study, two well-known algorithms have been used on the extracted data for Egypt. The two association mining algorithms are apriori and fp-growth. Both algorithms were implemented using C++. They produced the same results. The only difference was the running time. The apriori and fp-growth are meant to be run on a large dataset contains thousands of records. However, the extracted data for Egypt from GTD has only contained 1914 records, so the running time can be ignored.

Both algorithms generate 12324 rules with different confidence and support values. Figure 5 illustrates the visualization of the 12324 rules generated by apriori and fp- growth on the collected data. Figure 5 also demonstrates that most of the rules concentrated on low support and confidence values in the lower left corner of the graph thus, it can be ignored for its low values.

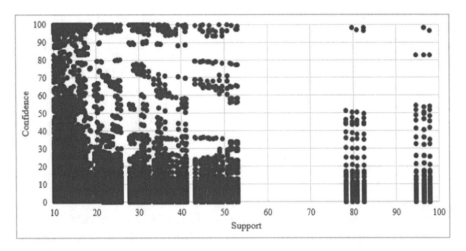

Fig. 5. Visualization of the generated rules generated by apriori and fp-growth algorithms on the extracted data of Egypt from GTD.

The top five rules generated by apriori and fp-growth algorithms based on confidence values are presented in Table 1. The table illustrates interesting deep insights such as rule number 1 {attack_type: Bombing/Explosion, gname: Unknown} → {nationality: Egypt} which indicates that if the terrorist attack occurred using Bombing/Explosion and the group terrorism name is Unknown then the nationality of the terrorist is Egyptian with confidence 99.8208%. The rule number 5 {nationality: Egypt, year: 2013} → {success: yes} shows that 99.5327% success rate of the terrorist attack if the nationality of the terrorist is Egyptian and the year of the attack is 2013.

Table 1. Top five generated rules based on confidence values

No.	Rule	Confidence
1	{attack_type: Bombing/Explosion, gname: Unknown} → {nationality: Egypt}	99.8208%
2	{year: 2013} → {nationality: Egypt}	99.6109%
3	{nationality: Egypt, target_type: Military} → {gname: Unknown}	99.5763%
4	{nationality: Egypt} → {target_type: Police}	99.5536%
5	{nationality: Egypt, year: 2013} → {success: yes}	99.5327%

Table 2 presents the top five rules based on both confidence and support values ordered by support values with a minimum threshold of 50% in both confidence and support values. In the following table, the support value of 100% was ignored, as it did not present a useful rule.

Table 2. Top five rules based on both confidence and support values ordered by support values.

No.	Rule	Support	Confidence
1	{nationality: Egypt} → {suicide: No}	97.9101%	96.6916%
2	{Number_of_kills: 1} → {suicide: No}	97.9101%	51.8143%
3	{nationality: Egypt} → {success: yes}	96.5517%	82.5758%
4	{Number_of_kills: 0} → {nationality: Egypt}	96.5517%	50.7035%
5	{gname: Unknown, nationality: Egypt} →{suicide: No}	94.6708%	54.3046%

Therefore, Table 2 showed that many frequent items on the extracted data dominate the support values. The dominant frequent item will be the most the attacks are occurred by Egyptian with no suicidal act, kills none or one person, they did not belong to any terrorist group and managed to do their terrorist attack successfully.

Table 3 presents the top five rules based on the failure of terrorist attacks. The table shows that with support values 17.5549% of the total records. If a number of the kills was zero or the group name is unknown, or the used weapon is a firearm or the target is unknown or the attack occurred in Cairo; then the terrorist attack is mostly to end with failure.

Table 3. Top five rules based on the failure of the terrorist attack.

No.	Rule	Support	Confidence
1	{number_of_kills: 0} → {success: No}	17.5549%	83.3333%
2	{gname: Unknown} → {success: No}	17.5549%	68.75%
3	{weapon_type: Firearms} → {success: No}	17.5549%	26.7857%
4	{target_type: Unknown} → {success: No}	17.5549%	24.1071%
5	{city: Cairo} → {success: No}	17.5549%	12.5%

6   Discussion

This study may help decision makers who oversee national security in Egypt in the elimination process of terrorism in Egypt. The study generated useful insights, which accurately describes the terrorist attack, occurred in Egypt in the last four decades. In this section, a summary of highlighted insight will be outlined as follow:

1. Egypt suffered from 2 periods of terrorism; the first period was between 1990 to 1998 and the second one was between 2011 to 2016.
2. The highest number of attacks occurred in 2015 and 1994.
3. Most of terrorist attacks occurred in summer within Sinai and Cairo.
4. The successful rate of terrorist attacks was 82%.
5. The bombing/explosion was the most common way used by terrorists.
6. Police and military personnel were the most valuable target by terrorists.
7. 1848 of total 1914 terrorist attacks on Egypt was operated by Egyptians.
8. A round 50% of the terrorist attacks ended with no killing.

9. The police personnel were always attacked by firearms while military personnel were attacked by bombing weapons.
10. The most frequents terrorist attack properties would be occurred by Egyptian with no suicidal act, kills none or one person, did not belong to any terrorist group and managed to do their terrorist attack successfully.

From the previously generated insights, an extra layer of explanation can be added to clarify them.

1. Egypt suffered from 2 period of terrorism due to the instability of the authorities with the new president for the country. Therefore, stability and peace are very important in Egypt.
2. Most terrorist attacks occurred in Sinai as most of it is desert and lies close to the border. Therefore, Sinai needs more investment to be inhabited by citizens and more security is indeed needed which is happing now by current leaders in Egypt.
3. There is a need to control the manufacturing of homemade bombs and explosives and they are the most commonly used weapon by terrorists
4. More Security Intelligence to unveil the names of terrorist group in Egypt as 53% of the terrorist attacks is unknown.
5. More financial resources are needed for police and military administration in Egypt to protect their personnel from terrorists, as they are the highest target for them.
6. Poor education and economic pressure are one of the reasons why youth have been manipulated by terrorist groups to join them. Egypt is on it is way to improve education and economy.

7 Conclusion and Future Works

The issue of terrorism is one of the main concerns in the recent world; it threats the peace all over the world. In Egypt, terrorism hold down the development and prosperity progress. It consumes lives, assets and financial resources. This study highlights and analyses terrorist attacks in Egypt that occurred in last the four decades. The source of data used in the study is the Global Terrorism Database (GTD). GTD is an online open source database published by the University of Maryland in the USA. According to the literature review, this study is considered on first trials to address Egypt for terrorism attacks with different computer techniques such as statistical analysis and association mining algorithms. The study generated statistical insight graphs and more than 12000 association rules. This study may help decision makers who oversee national security in Egypt to eliminate terrorism in Egypt. Poor education and economic pressure are one of the reasons why youth have been manipulated by terrorist groups to join them. Egypt on it is way to improve education and economy.

One of the future works to extend the current study is to include more databases of terrorism such as the American Terrorism Study (ATS), Extremist Crime Database (ECDB) in one database and extract the records of Egypt. The unified terrorism database will generate more solid insights concerning Egypt terrorist attacks.

References

1. Murphy, C.: Passion for Islam: Shaping the Modern Middle East: the Egyptian Experience. Simon and Schuster, New York City (2002)
2. Ganor, B.: Terrorism in the twenty-first century. In: Shapira, S.C., Hammond, J.S., Cole, L.A. (eds.) Essentials of Terror Medicine, pp. 13–26. Springer, New York (2009)
3. Elhajj, A., et al.: Estimating the importance of terrorists in a terror network. In: Özyer, T., Erdem, Z., Rokne, J., Khoury, S. (eds.) Mining Social Networks and Security Informatics, pp. 267–283. Springer, Dordrecht (2013)
4. Qin, J., Xu, J.J., Hu, D., Sageman, M., Chen, H.: Analyzing terrorist networks: a case study of the global salafi jihad network. In: Intelligence and Security Informatics, pp. 287–304 (2005)
5. Institute for Economics and Peace: Global Terrorism Index 2016: Measuring and Understanding the Impact of Terrorism. The Institute for Economics and Peace (IEP) (2016)
6. Kotb, M.A., Elmahdy, H.N., Khalifa, N.E.D.M., El-Deen, M.H.N., Lotfi, M.A.N.: Pediatric online evidence-based medicine assignment is a novel effective enjoyable undergraduate medical teaching tool: a SQUIRE compliant study. Medicine 94(29), e1178 (2015)
7. Solanki, S.K., Patel, J.T.: A survey on association rule mining. In: 2015 Fifth International Conference on Advanced Computing and Communication Technologies, pp. 212–216 (2015)
8. Institute for Economics and Peace: Global Terrorism Index 2017: Measuring and Understanding the Impact of Terrorism. The Institute for Economics and Peace (IEP) (2017)
9. Hegde, L.V., Sreelakshmi, N., Mahesh, K.: Visual analytics of terrorism data. In: 2016 IEEE International Conference on Cloud Computing in Emerging Markets (CCEM), pp. 90–94 (2016)
10. Li, Z., Sun, D., Chen, H., Huang, S.-Y.: Identifying the socio-spatial dynamics of terrorist attacks in the middle east. In: 2016 IEEE Conference on Intelligence and Security Informatics (ISI), pp. 175–180 (2016)

Regression with Support Vector Machines and VGG Neural Networks

Su Wu[1], Chang Liu[1], Ziheng Wang[2], Shaozhi Wu[3(✉)], and Kai Xiao[4]

[1] School of Information and Software Engineering,
University of Electronic Science and Technology of China, No. 4, Section 2,
North Jianshe Road, Chengdu, China
ml8781078969@163.com, luucas799@gmail.com
[2] School of Aerospace Engineering and Applied Mechanics, Tongji University,
Shanghai, China
wangzhl364@hotmail.com
[3] School of Computer Science and Engineering,
University of Electronic Science and Technology of China, No. 2006,
Xiyuan Road, Chengdu High Tech Zone (West District), Chengdu, China
wszfrank@uestc.edu.cn
[4] School of Electronic Information and Electrical Engineering,
Shanghai Jiao Tong University, 800 Dongchuan Road, Shanghai, China
showkey@sjtu.edu.cn

Abstract. In the area of machine learning, classification tasks have been well studied, while another important application of regression is not of the same level. In this paper, we propose parameterization and settings obtained from multiple experiments for traditional supervised machine learning of Support Vector Machine (SVM) and recently widely used deep unsupervised learning technology of Convolutional Neural Networks (CNN) based on Visual Geometry Group (VGGNet). In this study, different dataset used for regression task have been adopted. We have experimented on six data types obtained from the UCI Machine Learning Repository, and one converted handwritten image dataset from the MNIST. Accuracy of the regression results generated by the proposed models are validated with statistical methods of Mean Absolute Error (MAE) and R_square, i.e. coefficient of determination. Experimental results demonstrate that VGG has clear advantages over SVM in the cases of image recognition and attributes with strong correlation, and SVM performs better in the cases of discrete, irregular and weak correlation data than. By comparing the three kernel functions of SVM, it is found that in most cases, Rbt kernel function performs more effectively than Linear and Poly ones.

Keywords: Regression · VGG · SVM · R_square · Kernel function · Supervised and unsupervised learning

© Springer Nature Switzerland AG 2020
A. E. Hassanien et al. (Eds.): AMLTA 2019, AISC 921, pp. 301–311, 2020.
https://doi.org/10.1007/978-3-030-14118-9_30

1 Introduction

With the gradual rise of artificial intelligence, more and more machine learning algorithms and networks have emerged. Faced with so many different algorithms and networks, how to choose the appropriate learning network or algorithm has become a very difficult problem. In terms of classification, people have obtained good research results. In 1984, Clancey provided a useful framework to solve the classification problem [1]. In 1995, Meehl, use answered a classification problem in psychopathology [5]. Joachims give us a method to learn text classifiers in 1998 [3]. And in 2012, Krizhevsky, Sutskever and Hinton give a method for the classification of image [4]. But in another task area of machine learning regression, people have not obtained research results at the same level as classification. In this paper, we consulted some information of two major classifications of machine learning: supervised learning and unsupervised learning. In 2017, Kotsiantis, Zaharakis and Pintelas sourted out most supervised learning algorithms and their application examples [2] for people. Gentleman and Carey explored the use of unsupervised machine learning in 2008 [21]. We select two algorithms that are representative in the field of supervised learning and unsupervised learning, and use them to test different data sets to provide people with a solution or suggestion to solve the problem.

The UCI Machine Learning Repository [8] is a database for machine learning proposed by the University of California Irvine. There are currently 335 data sets in the database. The UCI dataset is a commonly used standard test data set. The MNIST handwritten image dataset is from the National Institute of Standards and Technology (NIST) [9]. It consists of numbers written by 250 different people, there are 60,000 pictures in the train set and 10,000 pictures in the test set. This handwritten dataset is widely used for machine learning of pictures. We selected 6 different test data from the UCI dataset and MNIST handwritten dataset as testing data. And in order to get the most realistic results, we did not pre-process any of these data to improve the correlation between their properties. In terms of testing the network, we chose SVM and VGGNet networks [7] as the presented test network. The VGG model is the second in the 2014 ILSVRC competition, the first place is googLeNet. But the VGG model performs better than googLeNet in multiple migration learning tasks. We combine the VGGNet networks structure with the traditional convolutional neural network (CNN) [6] to adjust the ConvNets architecture of the network. SVM is an optimized linear classifier proposed by Corinna and Vapnik in 1995 [10]. The principle is based on linear separability. It can be used for linear classification and used so-called kernel techniques, whose inputs are implicitly mapped into high-dimensional feature spaces for efficient nonlinear classification. So in our work, we have tested the three commonly used SVM kernel functions. In order to evaluate results, we choose MAE and R_Square to evaluate the accuracy of the prediction data. In statistics, MAE and R_Square are usually used as indicators for determining the regression of the model. We use the Pearson correlation coefficient in verifying the correlation between data attributes. It is a linear correlation coefficient that is a statistic that reflects the degree of linear correlation between two variables.

The rest of the paper is structured as follows: Sect. 2 detailed explanation test experiments and results. Section 3 is to discuss the analysis of the data set and the results of the experiment. Section 4 provides conclusion through the experimental results.

2 Test Experiments and Results

2.1 Experimental Data

In this paper, we used six types of data from UCI Machine Learning Repository, including Beijing PM2.5, Concrete Compressive Strength, Energy Efficiency, Parkinsons Telemonitoring, Appliances Energy Prediction, Geographical Original Of Music. And one from the MNSIT dataset. The number of attributes of these data and the relationship between them are different. Below we will give a brief introduction to these data [8]:

(1) In the BeijingPM2.5 data set, there are 13 attributes. This hourly data set contains the PM2.5 data of US Embassy in Beijing. Meanwhile, meteorological data from Beijing Capital International Airport are also included.

(2) In the Appliances Energy Prediction data set, it includes 29 attributes. The house temperature and humidity conditions were monitored with a ZigBee wireless sensor network. The energy data was logged every 10 min with m-bus energy meters. Two random variables have been included in the data set for testing the regression models and to filter out non predictive attributes.

(3) In the Concrete Compressive Prediction data set, there are a total of 9 attributes, and the attribute breakdown 8 quantitative input variables, and 1 quantitative output variable.

(4) For Energy Efficiency data set, The dataset comprises 768 samples. The dataset contains eight attributes and two responses. The aim is to use the eight features to predict each of the two responses.

(5) In the Geographical Original of Music data set, there are 68 attributes for training and two attributes (longitude and latitude) for prediction. These 68 features are audio features extracted from the waveform file and normalized to these features.

(6) Parkinsons Telemonitoring data set includes 26 attributes. This dataset is composed of a range of biomedical voice measurements from 42 people with early-stage Parkinson's disease recruited to a six-month trial of a telemonitoring device for remote symptom progression monitoring. Columns in the table contain subject number, subject age, subject gender, time interval from baseline recruitment date, motor UPDRS, total UPDRS, and 16 biomedical voice measures. And the main aim of the data is to predict the motor and total UPDRS scores ('motor_UPDRS' and 'total_UPDRS') from the 16 voice measures.

(7) For the MNIST handwritten image dataset, we preprocessed the image to reduce the length and width of all images by half. Then, 7 pictures with the same content, such as 7 pictures with the content "1", and two blank pictures of the same size are stitched into a 3 * 3 picture. (two blank pictures are respectively at the beginning and the end of the figure), As shown in Fig. 1. We treat the lable of the image as a normal distribution, turning lable into a continuous value rather than discrete (Table 1).

Fig. 1. Some examples of MNIST handwritten image create

Table 1. Experimental data

BeijingPM2.5	13 attributes (include 1 predicted attribute)
Appliances Energy Prediction	29 attributes (include 2 predicted attributes)
Concrete Compressive Prediction	9 attributes (include 1 predicted attribute)
Energy Efficiency	10 attributes (include 2 predicted attributes)
Geographical Original of Music	70 attributes (include 2 predicted attributes)
Parkinsons Telemonitoring	26 attributes (include 16 biomedical voice measures and 2 predicted attributes)
MNIST	1000 pictures

2.2   Experimental Network Structure and Result

Network Structure and Method. Since different test data have different number of features, this has a certain impact on convolution and pooling operations, especially on the number of ConvNet network layers. So the ConvNet architecture used by each data is different. In order to reduce the gap in the evaluation of the final prediction results due to differences in the ConvNet architecture, we have tried several times on each data by adjusting the number of convolutional and pooling layers, expecting to get a model with high computational efficiency and better results. We use MAE (Mean Absolute Error) and R_square as test indicators.

$$\text{MAE} = \frac{1}{m} \sum_{i=1}^{m} |y_i - \hat{y}_i| \tag{1}$$

$$\text{R\_square} = 1 - \frac{\sum_i (y_i - \hat{y}_i)^2}{\sum_i (y_i - \bar{y}_i)^2} \tag{2}$$

Equation (1). m is the total number, i is a subscript, indicating the number, y_i is the independent variable, $\hat{y}_i$ is the predicted value, and $\bar{y}_i$ is the average value of y_i. Because we calculate each of the predicted data separately, that is, each calculation has only one independent variable, the effect of using the coefficient of determination here is exactly the same as the effect of using the correction coefficient. It can be seen from the formula that R_square can better reflect the fit of a model [20]. Therefore, we use R_square as the main test indicator for good and bad, and MAE as an auxiliary test indicator. On the full connection layer, we have a total of 5 layers, namely FC 2048, FC 500, FC 100, FC 20, FC 1. Through many experiments, the computational efficiency and accuracy of the model are considered comprehensively. For Energy Efficiency, we used 6 convolution layers and 1 pooling layer. In Geographical Original of Music, we used 10 convolutional layers and 3 pooling layers and stripped out four unrelated variables. In the predictions for Concrete Compressive Prediction and Beijing PM2.5 and Appliance Energy Prediction and Parkinsons Telemonitoring, we used 8 convolutional layers and 1 pooling layer. For the MNIST data set, we use 8 convolutional layers and 4 pooling layers (Tables 2 and 3).

Table 2. ConvNet layers

	BJPM2.5	EE	CCP	GOM	AEP	PT	MNIST
Conv	8	6	8	10	8	8	8
Pooling	1	1	1	3	1	1	4

Table 3. Experiment results

Name	VGG_MAE	VGG_R2	LN_MAE	LN_R2	PL_MAE	PL_R2	Rbf_MAE	Rbf_R2
1	21.9	−1.875	0.558	0.224	0.556	0.194	0.501	0.346
2_Y1	1.182	0.907	2.343	0.878	0.175	0.926	0.248	0.793
2_Y2	1.311	0.862	2.110	0.889	0.212	0.877	0.190	0.916
3	3.132	0.838	0.450	0.656	0.387	0.728	0.281	0.843
4	10.037	0.149	0.623	0.129	0.654	−0.031	0.570	0.228
5	15.553	−0.091	0.700	0.260	0.500	0.437	0.378	0.745
6	13.001	−0.179	0.755	0.173	0.551	0.479	0.378	0.745
7	0.235	0.972	0.866	−3.830	2.067	−93.065	0.407	0.684

1: BeijingPM2.5, 2: Energy Efficiency, 3: Concrete Compressive Prediction, 4: Geographical Original of Music, 5: Appliance Energy Prediction, 6: Parkinsons Telemonitoring, 7: MNIST

The Result of VGG vs SVM. As can be seen from Fig. 2, the effect of SVM in 1 is obviously better than VGG, but the overall prediction effect of both is poor. The results of SVM and VGG on the prediction of 2 are similar and the accuracy is high. For the prediction of 3, both of the results of SVM and VGG are good, but VGG is better than SVM. In the prediction of 4, both VGG and SVM are not effective, but the overall results of SVM are better than VGG. For the prediction of data 5 and data 6, the results are almost the same: SVM performance is much better than VGG. However, in terms of the performance of the prediction of data 7, VGG has obvious advantages over SVM.

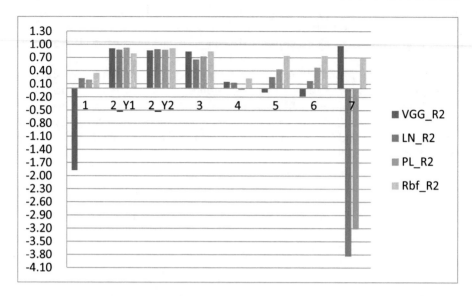

Fig. 2. The result of R_Square. We changed the poor result data of −93.065 to −4 to make the whole chart clearer. In fact, the −4 represents −93.065

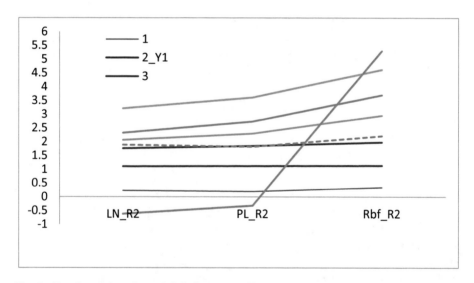

Fig. 3. Results of three kernel functions. 1: BeijingPM2.5, 2: Energy Efficiency, 3: Concrete Compressive Prediction, 4: Geographical Original of Music, 5: Appliance Energy Prediction, 6: Parkinsons Telemonitoring, 7: MNIST

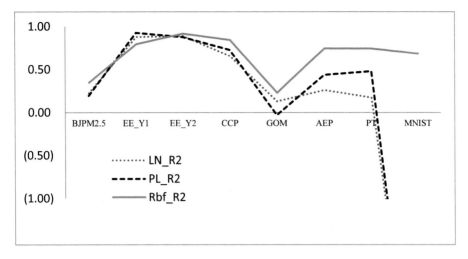

Fig. 4. Linear VS Poly VS Rbf.

We can see from Fig. 3 that for most data predictions, Rbf performs better than Linear and Poly. It can also be seen from Fig. 4 that Rbf outperforms Linear and Poly on most datasets, especially in the prediction of MNIST datasets. Rbf is several times more accurate than the other two (Table 5).

In order to measure the correlation between each feature and the results that need to be predicted, we use Pearson correlation coefficient method as the measurement index.

$$R_{x,y} = \frac{n \sum x_i y_i - \sum x_i \sum y_i}{\sqrt{n \sum x_i^2 - (\sum [x_i])^2} \sqrt{n \sum y_i^2 - (\sum [y_i])^2}} \tag{3}$$

Formula 2. x, y are two independent variables is the total, i is a subscript, indicating the number. As can be seen from Fig. 5 and Table 4, in Appliance Energy Prediction, Geographical Original of Music and Parkinsons Telemonitoring datasets, there is no moderate or higher intensity relationship between the feature and the value to be predicted, which is basically weak or negative weak correlation. However, in Energy Efficiency and Concrete Compr. In the essive Prediction dataset, the proportion of features with negative correlation and weak correlation is greatly reduced, and the proportion of features with medium and strong correlation is increased. In BeijingPM2.5 data set, 71.43% of the features and the predicted features show weak correlation or no correlation or even negative correlation.

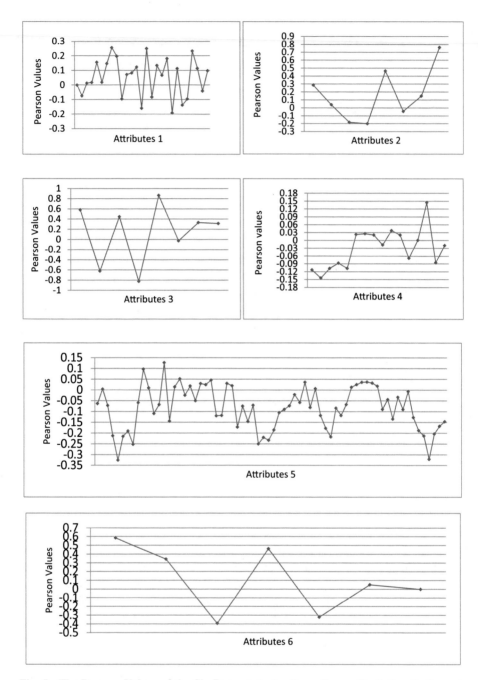

Fig. 5. The Pearson Values of the Six Dataset. 1: Appliance Energy Prediction, 2: Concrete Compressive Prediction, 3: Energy Efficiency, 4: Geographical Original of Music, 5: Parkinsons Telemonitoring, 6: BeijingPM2.5

Table 4. The correlation ratio of each data.

Correlation grade	Extremely weak or irrelevant	Weak	Moderate	Strong	Extremely strong
1	0.8846	0.0385	0	0	0
2	0.50	0.250	0.1250	0.1250	0
3	0.1250	0.250	0.250	0.1250	0.250
4	1.0	0	0	0	0
5	0.8382	0.1618	0	0	0
6	0.2857	0.4286	0.2857	0	0

Comment: The ratio = Characteristic numbers of the same grade/Total number of characteristics. (1: Appliance Energy Prediction, 2: Concrete Compressive Prediction, 3: Energy Efficiency, 4: Geographical Original of Music, 5: Parkinsons Telemonitoring, 6: BeijingPM2.5)

Table 5. Pearson correlation coefficient grade.

Grade	Value
Extremely weak or irrelevant correlation	$0 \leq |x| < 0.2$
Weak correlation	$0.2 \leq |x| < 0.4$
Moderate correlation	$0.4 \leq |x| < 0.6$
Strong correlation	$0.6 \leq |x| < 0.8$
Extremely correlation	$0.8 \leq |x| \leq 1$

3   Discussion

In the paper [11], the author uses various classical and non-parametric statistical analysis tools to systematically study the correlation strength between each input variable and each output variable in order to identify the strongest correlation input variable. The statistical tools used here indicate that RC, wall area and roof area appear mostly associated with HL and CL. This can also be seen from chart 4 and chart 5. We can also find similar correlations for features in the Concrete Compressive Prediction data set [12]. In the Appliance Energy Prediction data set, it contains the temperature, humidity, energy consumption of electric lights and atmospheric data from local weather stations (including wind speed, humidity, visibility, dew point) [13]. From the above experimental results, the correlation between each feature and the predicted data is very weak, so we use VGG network to train the data is not good, on the contrary, we use the RBF kernel function of SVM to get a better result. The same situation also appears in the Geographical Original of Music and Parkinsons Telemonitoring datasets. In the Beijing PM2.5 data set, we found that all features are weather characteristics. Moreover, the mutual influence of these features is not immediately significant, a feature change will not cause the rest of the features to change immediately, So you will find that there are many data in it, although one data has changed, the rest of the data has not changed in a few hours. And the concentration of PM2.5 has changed a lot during these few hours [14]. And we know that the concentration of PM2.5 is not only related to the weather, but also closely related to the number of cars driving at that time, the amount of exhaust gas emitted, and the amount of exhaust gas emitted by local factories. So these will result in a small amount of features and one-sidedness of the training model. Although dropout is used, there is still an over-fitting phenomenon that causes the prediction results to be highly biased [15], neither VGG nor SVM can get a good result. For Geographical Original Of Music and Parkinsons Telemonitoring data sets, audio features are used. In GOM, the author associates audio with geographical location. Music in different places has local characteristics. Therefore, the track analysis of music in different areas can be obtained these music originated from roughly that place. But the author only uses the central towns of that area as a symbol of regional division; this will cause two more similar but different music to be divided into a central town [16]. In the Parkinsons Telemonitoring data set, because Parkinson's patients have certain dysphonia. The voices of Parkinson's patients have certain audio characteristics [17]. All of them are early Parkinson's patients in this dataset, but some early Parkinson's patients' voice characteristics are not very different from those of ordinary people in terms of sound tracks. The results of regression prediction using VGG are not good, but the effect of SVM's Rbf kernel function has changed significantly [18, 19]. In the MNIST handwritten image dataset, convolution has an inherent advantage in the processing of images, and it is easier to extract the features of the image, which makes VGGNet's prediction results much better than SVM [9].

4   Conclusions

From the above experiments, we can see that in Appliance Energy Prediction dataset, nearly 88.46% of the features show extremely weak or no correlation with the final predicted Eigen values. In Geographical Original of Music dataset, nearly 100% of the features show extremely weak or irrelevant correlation. In the Parkinsons Telemonitoring dataset, 83.82% of the features are extremely weakly correlated or uncorrelated with the final predicted Eigen values, and nearly 16.18% of the features show weak correlation. In Concrete Compressive Prediction, 50% of the features are weakly correlated or uncorrelated with the final predicted feature, nearly 25% of the features show weak correlation, 12.5% of the features show a moderated correlation and 12.5% of the features show strong correlation. And in Energy Efficiency datasets, 25% of the features show extremely strong correlation, 12.5% of the features show a strong correlation and 25% of the features show moderated correlation. 71.43% of the features are weakly correlated or unrelated to the predicted features in the BeijingPM2.5 dataset. We can see that VGGNet outperforms SVM in image recognition and continuous, strongly correlated data sets, but performs worse than SVM on discrete, single, weakly correlated data sets. Under most data sets, Rbf performs better than Linear and Poly, especially in image processing.

References

1. Clancey, W.J.: Classification Problem Solving. Stanford University, Stanford (1984)
2. Kotsiantis, S.B., Zaharakis, I., Pintelas, P.: Supervised machine learning: a review of classification techniques. Emerg. Artif. Intell. Appl. Comput. Eng. **160**, 3–24 (2017)
3. Joachims, T.: Text categorization with support vector machines: learning with many relevant features. In: European Conference on Machine Learning, pp. 137–142. Springer, Heidelberg (1998)
4. Krizhevsky, A., Sutskever, I., Hinton, G.E.: ImageNet classification with deep convolutional neural networks. In: Advances in Neural Information Processing Systems, pp. 1097–1105 (2012)
5. Meehl, P.E.: Bootstraps taxometrics: solving the classification problem in psychopathology. Am. Psychol. **50**(4), 266–275 (1995)
6. LeCun, Y., Boser, B., Denker, J.S., Henderson, D., Howard, R.E., Hubbard, W., Jackel, L.D.: Backpropagation applied to handwritten zip code. Neural Comput. **1**(4), 541–551 (1998)
7. Simonyan, K., Zisserman, A.: Very Deep Convolutional Networks for Large-Scale Image Recognition (2015). arXiv preprint: arXiv:1409.1556v6
8. UCI machine learning repository. http://www.ics.uci.edu/~mlearn/MLRepository.html
9. The MNIST Database of Handwritten Digits. http://yann.lecun.com/exdb/mnist/
10. Cortes, C., Vapnik, V.: Support-vector networks. Mach. Learn. **20**(3), 273–297 (1995)
11. Tsanas, A., Xifara, A.: Accurate quantitative estimation of energy performance of residential buildings using statistical machine learning tools. Energy Build. **49**, 560–567 (2012)
12. Yeh, I.-C.: Modeling of strength of high performance concrete using artificial neural networks. Cem. Concr. Res. **28**(12), 1797–1808 (1998)
13. Candanedo, L.M., Feldheim, V., Deramaix, D.: Data driven prediction models of energy use of appliances in a low-energy house. Energy Build. **140**, 81–97 (2017). ISSN 0378-7788
14. Liang, X., Zou, T., Guo, B., Li, S., Zhang, H., Zhang, S., Huang, H., Chen, S.X.: Assessing Beijing's PM2.5 pollution: severity, weather impact, APEC and winter heating. Proc. R. Soc. A **471**, 20150257 (2015)
15. Srivastava, N., Hinton, G., Krizhevsky, A., Sutskever, I., Salakhutdinov, R.: Dropout: a simple way to prevent neural networks from overfitting. J. Mach. Learn. Res. **15**(1), 1929–1958 (2014)
16. Zhou, F., Claire, Q., King, R.D.: Predicting the geographical origin of music. In: IEEE ICDM, pp. 1115–1120 (2014)
17. Tsanas, A., Little, M.A., McSharry, P.E., Ramig, L.O.: Accurate telemonitoring of Parkinson's disease progression by non-invasive speech tests. IEEE Trans. Biomed. Eng. **57**(4), 884–893 (2010)
18. Little, M.A., McSharry, P.E., Hunter, E.J., Spielman, J., Ramig, L.O.: Suitability of dysphonia measurements for telemonitoring of Parkinson's disease. IEEE Trans. Biomed. Eng. **56**(4), 1015–1022 (2009)
19. Little, M.A., McSharry, P.E., Roberts, S.J., Costello, D.A.E., Moroz, I.M.: Exploiting nonlinear recurrence and fractal scaling properties for voice disorder detection. BioMed. Eng. OnLine. **6**, 23 (2007)
20. Frost, J.: Regression analysis: how do I interpret R-squared and assess the goodness-of-fit. The Minitab Blog (2013)
21. Gentleman, R., Carey, V.J.: Unsupervised machine learning. In: Bioconductor Case Studies, pp. 137–157. Springer, New York (2008)

A Distributed Multi-source Feature Selection Using Spark

Bochra Zaghdoudi[1(✉)], Waad Bouaguel[1,2(✉)],
and Nadia Essoussi[1(✉)]

[1] LARODEC, ISG, University of Tunis, Tunis, Tunisia
bochrazaghdoudi@gmail.com, bouaguelwaad@gmail.com,
nadia.essoussi@isg.rnu.tn
[2] College of Business, University of Jeddah, Jeddah, Saudi Arabia

Abstract. Feature selection is one of the key problems in data pre-processing because it brings the immediate effects on the data mining algorithm. Using high-dimensional data sets, we can describe the data based on multiple sources, which corresponding to different knowledge sources. Multi-source feature selection is another topic relevant with large-scale data. Learning and selecting features from multiple data sources is becoming more common and much needed in many real-world applications. In this work, we propose a new multisource feature selection method based on traditional filters where data sources contain the same set of instances but different sets of features. This method is implemented using Spark as a powerful parallel framework for large-scale data processing. Conducted experiments approve the effectiveness of our approach in terms of execution time and where the classification accuracy is maintained.

Keywords: Feature selection · Multi-source features · Distributed · Large data · Spark

1 Introduction

In the last few years, data has become more and more larger in both number of instances and number of features in a variety of domains such as text categorization, healthcare, image retrieval, and social media. Large scale data present some challenges in data analysis and knowledge discovery process. Such data sets with a large number of features can contain many noisy, redundant and irrelevant features that will significantly demand more on the computational and memory storage requirements to provide a reliable analysis [1]. Thus, when performing machine learning or data mining techniques on high dimensional data, data reduction is a suitable solution in order to avoid the curse of dimensionality [2].

Dimensionality reduction techniques are divided into two main approaches: feature selection that select relevant features and remove the redundant ones and feature extraction that consists on creating a new set of features. The technique that we focus on in this work is the feature selection since it retains the original features and as it is useful to keep them for the interpretation of the learning model and for the knowledge extraction [3].

© Springer Nature Switzerland AG 2020
A. E. Hassanien et al. (Eds.): AMLTA 2019, AISC 921, pp. 312–320, 2020.
https://doi.org/10.1007/978-3-030-14118-9_31

The increasing availability of data provides motivations to analyze and extract useful knowledge coming from different sources to obtain a consistent result is one of the major problems in selecting features from multiple data sources. Analyzing and modeling multi-source data refers to the exploration of global models by aggregating different data-source results into a unified perspective to obtain a global source that could be very useful for improving the learning performance.

Multi-source data can be defined as a set of different data sources where each source contains a different type of feature sets and the same set of instances. Dealing with the heterogeneity of data from various perspectives can pose several issues. The most common problem is how to exploit this data and integrate it to have a global pattern while dealing with the redundancy of features to enhance the learning accuracy. Multi-source data can suffer from numerous types of redundant data. This redundancy can be located at the features level or even at the data sources level. The multi-source feature selection problem can be formulated as follows:

Given a set of H sources of data $(D_1, ..., D_H)$ containing the same set of instances $X = \{ x_1, ..., x_N \}$ where $x_i \in R$.

Learning and selecting features from multi-source data is becoming more common and much needed in many real-world applications. The need to analyze those data vary considerably from one application domain to another.

In the literature, various works for selecting features from a single data source have been well studied: CFS [7], χ^2 [4], $mRMR$ [5], $ReliefF$ [10], etc. In another hand, only few works have addressed the problem of multi-source feature selection with different definitions such as $GDCOV$ [6], $MSTDA$ [9], $MSFS$ and $KOFS$ proposed by [8]. However, most of the existing multi-source feature selection methods are not adapted to scale well when dealing with large data because they use the whole data set at once. Thus, they cannot handle the large amounts of data and may even become inapplicable. As a solution, the selection process may be parallelized by distributing the data subsets and assigning them to multiple processors in parallel and then combining them [11].

To deal with the multi-source feature selection problem using high-dimensional data sets, we propose a new distributed design for multi-source feature selection using Spark. For that, we present a new definition of the multi-source feature selection problem and the approach in which two steps of feature selection are processed in a distributed environment. Firstly, a feature selection technique is performed on each data source independently, then the outputs are combined into a single feature subset. Finally, another step of feature selection is processed on the output of the previous step to remove redundancies coming from the different sources to generate a new subset of relevant features. The experimental results on different data sets demonstrate important savings in selection time with acceptable performance.

The rest of the paper is organized as follows: we present in the Sect. 2 the background about the multi-source feature selection. Section 3 describes our proposed approach for the multi-source feature selection using Spark. In Sect. 4, we detail and discuss the experimental results provided by this study. We conclude finally with a conclusion that resumes our work and introduces the future works.

2   Related Work

Several researches were developed in the field of feature selection, but few of them involved data sets with multiple data sources.

In [6], Zhao et al. proposed an unsupervised feature selection framework for multi-source data using geometry-dependent covariance. In this study, the multisource feature selection task is defined as identifying relevant features from a target source basing on the information obtained from the other sources which jointly describing a set of N instances and each source with a different set of features. The approach consists of introducing a global geometric pattern that identify relevant data from a target source depending on local geometric patterns extracted from all the other sources. Features are selected through the use of the global geometric pattern in the covariance analysis that allows the creation of a geometrically dependent sample covariance matrix. Thus, the approach allows the selection of features with less redundancy and considers the interactions between features.

Guo et al. in [9] suggested a multi-source temporal data aggregation (*MSTDA*) model that includes feature selection step and data prediction step. The technique involves combining data from multiple sensors by eliminating redundancy, and reducing the number of transmissions that will transfer data to a base station to effectively increase network lifetime. The first step for aggregation is feature selection from multi-source data based on the *Particle Swarm Optimization (PSO)* method. Then, only extracted subsets are used as examples of data for the prediction step.

In [8] the multi-source feature selection task is introduced as the use of additional sources of information to enhance the size of samples considering the interactions between features of different sources. The authors describe two spectral-based feature selection approaches to select features from different data sources, *Multi-source feature selection (MSFS)* and *Knowledge-oriented multi-source feature selection (KOFS)*. MSFS method consists of extracting, from each knowledge source, a local sample similarity. Then selecting features is performed basing on a global similarity matrix constructed by combining local similarity matrices. The second framework, KOFS, is a rank aggregation-based method. This technique receives external knowledge from data sources and converts them into internal knowledge as first step. These internal knowledges are then ranked and aggregated into a final list.

The proposed studies in the field of multi-source feature selection have shown good results but they are not adapted to scale with high-dimensional data as they are designed to be applied in a sequential manner. Then it is interesting to parallelize feature selection methods to deal with large-scale data and adapting them to select features from multiple data sources.

3 Proposed Approach: Distributed Multi-source Feature Selection Using Spark (D-MSFSS)

In this study, we present a new distributed and scalable multi-source feature selection framework using the Spark computing model to enable the selection of relevant features from multiple large data sources while using standard feature selection methods.

3.1 Distributed Programming Using Spark

The significant increase in the dimensions of available data, presents several challenges especially when dealing with extremely large data bases. Therefore, a distributed programming is required. A variety of parallel techniques and frameworks have been proposed in the few last decades to cope with these problems such as *Hadoop' MapReduce* and *Apache Spark*.

Apache Hadoop is an open source, scalable and fault tolerant framework that is based on the MapReduce model and provides a distributed processing of big data sets across clusters of commodity hardware. However, Hadoop MapReduce use the disk to read and write every job which slows down the processing speed.

Recently, the Apache Spark framework has been emerged and designed to handle the Hadoop limitations: it is 100 times faster in memory and 10 times faster on disk than Hadoop thanks to its in-memory operations. It allows more capacities for treating huge data and presents a good alternative especially for machine learning problems. Its in-memory processing enhance the computation time. It is adapted to be used for complex computing operations, machine learning tasks and real-time applications [3].

To deal with large-scale data, Spark provides *Resilient Distributed Dataset (RDDs)* that are distributed between different nodes. RDDs are a data abstraction that represent distributed data sets or collections. They may be produced by HDFS, stored data or even by other processed RDDs. The two major types of operations used on RDDs are transformations and actions. Transformations consist of generating a new RDD from another processed one (e.g., *map, filter,...*), while actions aims to return the results obtained after transformations from a Spark cluster (e.g., *reduce, collect,...*).

As Spark is better suited for machine learning tasks [11], it is chosen in this work.

3.2 D-MSFSS

Given H distinct data sources $\{S_1, ..., S_H\}$, where the different sources contain the same set of n instances $X = \{x_1, ..., x_n\}$ but with different sets of features $F_i = \{f_{i1}, ..., f_{id}\}$. The feature selection is performed on each data source independently and all interactions between the sources will be ignored.

The flowchart in Fig. 1 details the different steps of our proposed approach.

Our approach consists of applying two steps of feature selection that operates in three main phases.

Firstly, the local feature selection phase, in which we perform a feature selection algorithm on each data source separately to select the most relevant features from each source.

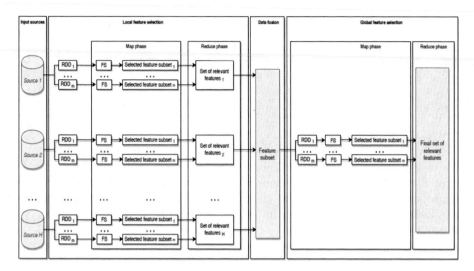

Fig. 1. The flowchart of the proposed approach

Then, in the fusion phase, the resulted subsets are combined to form a unified set of features obtained from the different data source.

However, one or more identical features may exist and selected from the different sources, which presents redundancy in the resulting feature set, for this reason, the global feature selection step is achieved to remove the redundancies that may coming from the different data sources by applying a feature selection algorithm on the new set of features.

4   Result and Discussion

Our approach is implemented using Apache Spark 2.3 with Python 3.5 language on a cluster of one master (with 4 GB of RAM and 1 CPU) and two workers (with 8 GB of RAM and 1 CPU for each worker).

To evaluate the performance of our distributed approach, we used two real data sets from the UCI machine learning repository and one simulated data set. The first one is the mfeat data set which represents a set of handwritten digit elements (from 0 to 9) and contains 2000 instances described by 649 features and six different sources. We also used the Gisette data set which is a two-class classification problem that contains a normalized representation of the four and nine patterns and adapted to be used in feature selection problems. It consists of 7000 instances and 5000 features divided into 8 sources. Finally, the ArtDB which is an artificial data set with 10 sources containing 100000 instances and 100000 features.

For the feature selection process, we will opt for filter methods due to their statistical scalability as they operate independently of the learning algorithm and have the advantage of being fast and less expensive in calculation. Three well known filters and supervised classifiers are chosen to test the efficiency of the proposal.

- **ReliefF:** the ReliefF algorithm [13] is a multivariate method that presents a more robust multi-class extension of the original Relief algorithm proposed by [12]. It consists of adding the possibility of handling incomplete and noisy data and capable of proceeding multi-class problems. ReliefF estimates features according to the way in which their values distinguish from one class to another in a local neighborhood.
- **Chi-square test:** is a statistical test proposed by [4] that uses the test of independence to test how the feature is independent of the class label. The algorithm selects a proper χ^2 value, search for the intervals of a numeric attribute and select features according to the characteristics of the data.
- **mRMR:** introduced in [5], is a method that guarantees maximum relevance of features with the class label and a minimum redundancy in each class. For that, it uses several statistical measures.

The evaluation is based on two criteria: (a) the processing time and (b) the learning accuracy. Then, we compare the parallel results with the sequential ones.

We evaluate our proposed approach based on two criteria: (a) the processing time and (b) the learning accuracy by taking advantage of the distributed processing. Then, we compare the parallel results with the sequential ones. We present and discuss the experimental results based on two strategies in order to study the impact of the method changes on the final results. The first strategy is to apply the same feature selection algorithm to the entire process (local and global feature selection steps). The second strategy is to apply a different feature selection algorithm for each data source.

All results are compared basing on two approaches: the sequential approach and the parallel approach. Table 1 reports the runtime consumed in seconds of the three feature selection algorithms performed on three multi-source data sets.

Table 1. Sequential vs. parallel selection time for each strategy

	χ^2		ReliefF		mRMR		Mixed FS	
	Sequential	Parallel	Sequential	Parallel	Sequential	Parallel	Sequential	Parallel
mfeat	0.097	0.0925	0.154	0.133	54.502	20.193	0.298	0.292
Gisette	1.379	1.21	1.37	1.113	76.009	27.14	3.796	3.682
ArtDB	27.198	12.936	32.537	9.158	282.899	32.11	48.749	23.234

We compare the selection time when processing data in a sequential way versus the parallel paradigm. For these experimental tests, the studied feature selection algorithms are: χ^2, relieff, mRMR and mixed methods for the same data set.

As expected, the time spent by the parallel version is radically reduced for all data sets. We can note that using the third data set (which contains 100000 features and 100000 instances) the reduction in execution time is much more observable. This fact proves the capacity of the distributed approach when dealing with large data sets. Note that the mRMR algorithm is noticeably higher than the other approaches and ReliefF have the lowest runtime. Also, we can notice that the use of mixed methods for the same data set provides an acceptable time comparing to the time required by mRMR algorithm in both of sequential and parallel approaches.

For evaluating the classification accuracy, we test the KNN, LR and NB classifiers with the sequential and parallel approaches. Figures 2, 3, 4 and 5 show the classification accuracy results obtained by the three classifiers for all data sets and feature selection methods. As can be seen, in general there were no differences in terms of accuracy, but we succeeded to reduce the runtime with maintaining the same accuracy as the sequential approach. The best results of the classification performance are achieved by the NB classifier for mRMR algorithm in Gisette and mfeat cases.

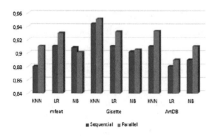

Fig. 2. Sequential vs. parallel accuracy using χ^2 algorithm

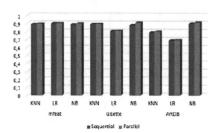

Fig. 3. Sequential vs. parallel accuracy using mRMR algorithm

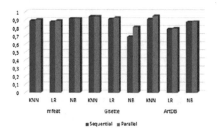

Fig. 4. Sequential vs. parallel accuracy using ReliefF algorithm

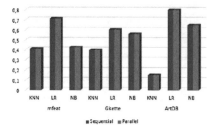

Fig. 5. Sequential vs. parallel accuracy using mixed methods

For the comparative study, we compared our proposed method with the existing multi-source feature selection works to distinguish our particularity in Table 2. The different studies are compared in terms of the used feature set, the instances set, if a target source is considered, the elimination of redundancy level, the supervision and the used aggregation method.

Table 2. Comparison of the existing multi-source feature selection works with our proposed method

Existing works	Same features set	Same instances set	Target source	Redundancy elimination from all data sources	Supervision	Aggregation
GDCOV [6]	No	Yes	X	No	Unsupervised	–
PSO [9]	No	No	–	Yes	Supervised	X
MSFS [8]	No	No	X	No	Supervised	–
KOFS [8]	No	No	X	Yes	Supervised/Unsupervised	X
D-MSFS	No	Yes	–	Yes	Supervised	X

5   Conclusion and Future Work

In this work, we presented a new distributed method for multi-source feature selection task using high-dimensional data sets. Experimental results approve that our proposed method can provides good and shorter runtime results when dealing with large data sets without modifying the classification accuracy. For future work, we will use for our proposed approach to deal with different data types. Also, we will adapt it to perform using Big data sets and evaluate it in real case studies.

References

1. Sutha, K.: A review of feature selection algorithms for data mining techniques. Int. J. Comput. Sci. Eng. **7**(6), 63 (2015)
2. Li, J., Cheng, K., Wang, S., Morstatter, F., Trevino, R.P., Tang, J., Liu, H.: Feature selection: a data perspective. ACM Comput. Surv. **50**(6), 94:1–94:45 (2017)
3. Ramirez-Gallego, S., Mouriño-Talín, H., Martinez-Rego, D., Bolón-Canedo, V., Manuel Benitez, J. M., Alonso-Betanzos, A., Herrera, F.: An information theoretic feature selection framework for big data under apache spark. CoRR, abs/1610.04154 (2016)
4. Liu, H., Setiono, R.: Chi2: Feature selection and discretization of numeric attributes, pp. 388–391 (1995)
5. Peng, H., Long, F., Ding, C.: Feature selection based on mutual information criteria of max-dependency, max-relevance, and min-redundancy. IEEE Trans. Pattern Anal. Mach. Intell. **27**(8), 1226–1238 (2005)
6. Zhao, Z., Liu, H.: Multi-source feature selection via geometry-dependent covariance analysis. In: Proceedings of the 2008 international conference on new challenges for feature selection in data mining and knowledge discovery, vol, 4, pp. 36–47 (2008)
7. Hall, M.A.: Correlation-based feature selection for machine learning, Technical Report (1999)

8. Zhao, Z.A., Liu, H.: Spectral feature selection for data mining. Chapman, Hall/CRC (2011)
9. Guo, W., Xiong, N., Vasilakos, A.V., Chen, G., Cheng, H.: Multi-source temporal data aggregation in wireless sensor networks. Wirel. Pers. Commun. 56(3), 359–370 (2011)
10. Yu, L., Liu, H.: Efficient feature selection via analysis of relevance and redundancy. J. Mach. Learn. Res. 5, 1205–1224 (2004)
11. Palma-Mendoza, R.-J., Rodriguez, D., de Marcos, L.: Distributed ReliefF-based feature selection in spark (2018)
12. Kira, K., Rendell, L.A.: The feature selection problem: traditional methods and a new algorithm, 2, 129–134 (1992)
13. Kononenko, I.: Estimating attributes: analysis and extensions of RELIEF, pp. 171–182 (1994)

The Classification of Multiple Power Quality Disturbances Based on Dynamic Event Tree and Support Vector Machine

Qiang Gao[1,2], Fenghou Pan[1(✉)], Feng Yuan[1], Jiayu Pan[1],
Jiannan Zhang[1], and Yunhua Zhang[1]

[1] State Grid Liaoning Electric Power Research Institute,
Shenyang 110003, China
panfenghou@dbdky.com
[2] Shanghai Jiao Tong University, Shanghai 200240, China

Abstract. It's difficult to classify multiple disturbances with the single disturbance classification method. This paper describes multiple disturbances and designs the classifier of multiple power quality (PQ) disturbances; the process of power quality disturbance classification can be divided into two-stage, feature extraction, and classification. This paper extracts features of disturbances with dq Transform, Wavelet Packet Transform (WPT) and S-Transform (ST), and combines them to reflect the characteristics of disturbances better. The design of binary tree Support Vector Machine (BT-SVM) with the concept of the class distance of the clustering analysis makes classifications intelligently. And dynamic event tree is proposed to make classifications of multiple disturbances. By these methods, disturbances can be classified fast and accurately. The results of simulation show that the classification method in this paper is able to classify multiple disturbances effectively.

Keywords: Multiple power quality · Disturbances dynamic event tree · Support Vector Machine

1 Introduction

The use of electronic equipment, high-voltage electronic devices, large capacitance, and nonlinear load has brought serious pollution to the power quality [1]. It causes losses to the user. Therefore, it's necessary to analyze power quality disturbances and make effective detection and classification to solve the power quality problems. A large number of researches have been conducted for single power quality disturbance [1–5], and great results have been achieved. However, power quality disturbances are often multiple, which means that there may be many kinds of disturbances at the same time. The single disturbance classification method is difficult to solve the problem of multiple disturbances. Therefore, the classification method of multiple disturbances based on dynamic event tree and Support Vector Machine (SVM) is designed in this paper, according to the combination of some features that are gotten through Wavelet Packet Transform (WPT), S-Transform (ST) and dq Transform.

© Springer Nature Switzerland AG 2020
A. E. Hassanien et al. (Eds.): AMLTA 2019, AISC 921, pp. 321–329, 2020.
https://doi.org/10.1007/978-3-030-14118-9_32

2 Multiple Power Quality Disturbances

Power quality disturbances can be divided into steady disturbances and transient. Steady disturbances include unbalanced three-phase voltage deviation, frequency deviation, voltage deviation, harmonic, voltage fluctuation and flicker [2]. Transient disturbances include voltage rising, voltage dropping, voltage interrupting, transient pulse and oscillation. However, in practice disturbance is not single; it may consist of more than two kinds of single disturbance.

Considering the randomness of disturbance and the influence of noise, several multiple disturbance models are established by using a mathematical model of computer and field data, which are shown in Fig. 1. To solve the above problems in the paper, a new feature extraction method is designed by using different transforms.

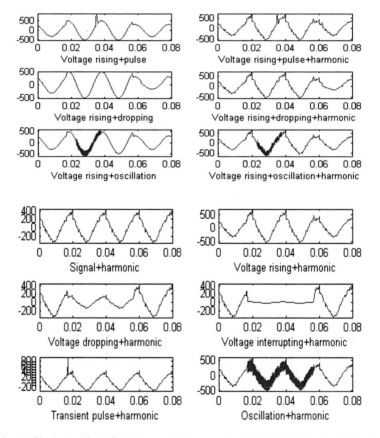

Fig. 1. (a) Single transient disturbance with harmonic; (b) Multiple transient disturbances

3   The Feature Extraction Method: Harmonic Detection

The first step of dq Transform is changing variables from three-phase voltage to two-phase orthogonal $\alpha - \beta$ system, and then transform to dq coordinate system [3], the formula is as follows:

$$\begin{bmatrix} v_d \\ v_q \end{bmatrix} = C_2 \begin{bmatrix} v_\alpha \\ v_\beta \end{bmatrix} = C_2 C_1 \begin{bmatrix} v_a \\ v_b \\ v_c \end{bmatrix}, C_1 = \sqrt{2/3} \begin{bmatrix} 1 & -1/2 & -1/2 \\ 0 & \sqrt{3}/2 & -\sqrt{3}/2 \end{bmatrix},$$

$$C_2 = \begin{bmatrix} \sin \omega t - \cos \omega t \\ -\cos \omega t; -\sin \omega t \end{bmatrix}$$

(1)

The sinωt and cosωt is sine and cosine signals and has the same phase with A phase voltage.

If power quality disturbance is symmetrical in a three-phase system, then the dq Transform becomes available. According to the features of the symmetrical three-phase power system that the phase difference of each phase is 120°, a virtual three-phase power system can be constructed. Set A phase voltage as va = $\sqrt{2}$Vsinωt, then the B and C phase voltage is got: vb = $\sqrt{2}$Vsin(ωt – π/3)–sinωt and vc = – vb– va, and V is the ideal phase voltage, then plug va, vb, vc into formula (1), then the disturbance features of dq Transform can be extracted which are shown in Fig. 2.

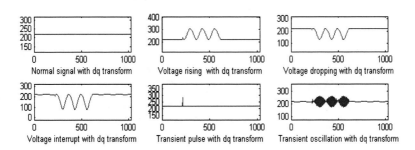

Fig. 2. Features of disturbances by dq transform

3.1   The Extraction Method Based on WPT

Wavelet Packet Transform is the development of the Wavelet Transform [3], it can decompose the high-frequency part of the signal, which can provide more disturbance features.

Power signal u(t) can be decomposed according to formula (2), where H1(k) indicates a high-pass filter coefficient, H0 (k) indicates a low-pass filter coefficient.

$$u_{2n}(t) = \sqrt{2} \sum_k H_1(k) u_n(2t - k), \ u_{2n-1}(t) = \sqrt{2} \sum_k H_0(k) u_n(2t - k) \qquad (2)$$

Wavelet packet decomposition process can be expressed as a binary tree structure, and the node in the binary tree is (j, p). The coefficients of the node (j + 1, p) in the wavelet packet decomposition can be described in the formula (3) and (4):

$$d_{j+1}^{2p}(k) = d_j^p(k) * \bar{h}_0(2k) = \sum_{m=-\infty}^{m=\infty} d_j^p(m) * h_0(m - 2k) \qquad (3)$$

$$d_{j+1}^{2p+1}(k) = d_j^p(k) * \bar{h}_1(2k) = \sum_{m=-\infty}^{m=\infty} d_j^p(m) * h_1(m - 2k) \qquad (4)$$

Wavelet packet can deal with low frequency and high frequency part, but the calculation may be complex, if all features of decomposition layers are selected, so the energy of node [1, 1], [2, 1], [3, 1], [3, 2], [3, 3], [3, 4], [3, 5], [3, 6], [3, 7] are used as features. Energy e(i) and relative energy $\bar{E}(i)$ of each node can be got by formula (5), di (k) represents the k wavelet coefficient of i node.

$$e(i) = \sum_{k=1}^{N} d_i^2(k), \ E_{\max} = \sum_{i=1}^{N} e(i), \ \bar{E}(i) = e(i)/E_{\max} \qquad (5)$$

The features $\bar{E}$ of disturbances are gotten based on the formula (6).

$$\bar{E} = [\bar{E}(1), \bar{E}(2), \bar{E}(3), \bar{E}(4), \bar{E}(5), \bar{E}(6), \bar{E}(7), \bar{E}(8), \bar{E}(9)] \qquad (6)$$

The features of disturbances gotten through WPT are shown in Fig. 3.

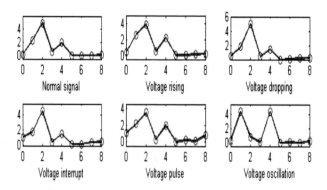

Fig. 3. Relative energy of power quality disturbances

3.2 The Extraction Method Based on S-Transform

S-Transform is used as disturbances detection algorithm. It's a reversible time-frequency analysis algorithm [4]. The one-dimensional continuous S-transform of signal x (t) is as followed:

$$S(\tau,f) = \int_{-\infty}^{\infty} x(t)w(\tau - t,f) \exp(-i2\pi ft)dt,$$

$$w(\tau - t,f) = |f|\Big/\sqrt{2\pi}^{*} \exp(-(t - \tau)^2 f^2\Big/2) \tag{7}$$

Where w $(\tau - t, f)$ is Gaussian window function, the τ controls the timeline position of Gaussian window function, f means frequency, and i is an imaginary number. The x [k · T] is the discrete time series of signal x (t). T is the sampling interval. N is the total number of sampling points. The discrete form of the formula is shown as followed:

$$S[jT, n/NT] = \sum_{m=0}^{N-1} X[(m+n)/NT] \exp(-2\pi^2 m^2/n^2) \exp(i2\pi mj/N) \tag{8}$$

The j is time, and n is frequency. After the S transform of x[k·T], get the two-dimensional complex time-frequency matrix [5]. The matrix columns correspond to the discrete frequency, and its row corresponds to the sampling time point and local spectrum.

This paper adopts the amplitude's sum of each line of the S matrix (ST1) and fundamental frequency amplitude curve of the S matrix (ST2) to extract features. ST1 and ST2 can be represented by formula (9), and N is the line's total number of the matrix, A is the S matrix. The f0 corresponds to the fundamental frequency 50 Hz.

$$ST_1(f) = 1/N \sum_{t=0}^{N} A(t,f),\ ST_2(t) = A(t,f_0) \tag{9}$$

The maximum value (EA1) of ST1 and the maximum value (EA2) of ST2 are used as features of disturbances, which are shown in Fig. 4.

The EA1 of the normal signal is defined as the standard value, upper and lower thresholds are defined respectively. If the EA1 of some disturbance is greater or less than the threshold, the value of EA1 is different. Finally, features of all kinds of disturbances can be obtained, which are shown in Table 1.

This paper adopts the value in the highest frequency line of S matrix as a sign that whether multiple disturbance happens. If multiple disturbances happened, more than two mutations appear, which are shown in Fig. 5.

4 The Design of Classifier Single Disturbance Classification

A classification method for power quality multiple disturbances should also be suitable for the single power quality disturbance. And SVM has the advantages of avoiding local minima and dealing with the curse of dimensionality. It has good scalability and classification accuracy and has the advantage to solve the small sample and nonlinear problems [6, 7].

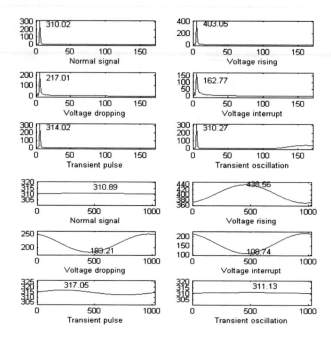

Fig. 4. (a) The EA1 of various disturbances; (b) The EA2 of various disturbances

Table 1. Features of power quality disturbance

Signal	EA₁	EA₂
Voltage dropping	−1	−1
Voltage rising	2	2
Voltage interrupt	−2	−2
Transient pulse	0	1
Transient oscillation	0	0

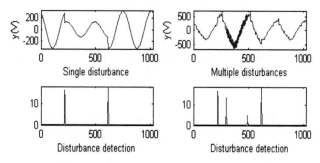

Fig. 5. Features of the disturbance signal

WPT and ST are used to extract features. The features from wavelet packet decomposition and S matrix are combined to obtain the combination feature $\bar{E}$, and $\bar{E} = [\bar{E}(i), EA1, EA2]$, $\bar{E}(i)$ is relative energy value obtained by WPT, $i = 1, 2, \ldots, 9$, EA1, EA2 are obtained by S-Transform.

This design adopts binary tree Support Vector Machine (BT-SVM). K classifiers just need to construct K-1 classifiers with BT-SVM. Classifiers of this method can be less than others. A BT-SVM needs to select an appropriate tree structure. The basic idea of classification is the concept of the class distance of the clustering analysis. Calculate the shortest distance between different classes in the eigenvectors space. After getting the distance, the structure can be designed. There are 5 kinds of disturbances. They are voltage pulse, oscillation, dropping, rising and interrupt. The structure of BT-SVM (SVM classifier 1) is shown in Fig. 6.

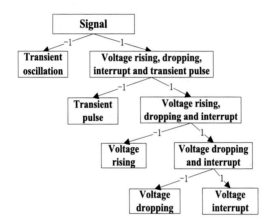

Fig. 6. Structure of SVM classifier 1

4.1 Multiple Disturbances Classification

The features of single disturbance with harmonic obtained by WPT are different, so a new binary tree classifier is needed to improve the identification accuracy [8]. Similarly, features are obtained by WPT and ST, the design of classifier (SVM classifier 2) also uses clustering analysis, and the difference from SVM classifier 1 is that SVM classifier 2 calculates the distance between classes firstly, and then gets classes together which have the nearest distance and separates them which have the far distance. SVM classifier 2 is shown in Fig. 7.

The traverse binary tree structure for classifying disturbances combined by two or more transient disturbances is shown in Fig. 8.

Dynamic event tree classifier of multiple disturbances is shown in Fig. 9. It consists of a trunk, fixed branch, and dynamic branch. The disturbance detection is the trunk, and multiple disturbance detections, harmonic detection and exit are fixed branches, SVM classifier 1, SVM classifier 2 and multiple disturbances traverse binary tree are dynamic branches.

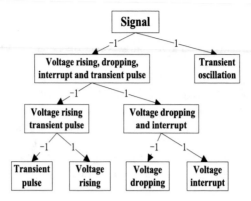

Fig. 7. Structure of SVM sub-classifier 2

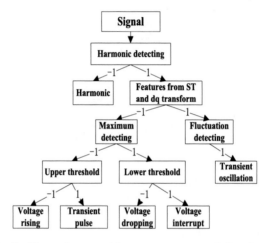

Fig. 8. Chart of traverse binary tree structure of disturbances

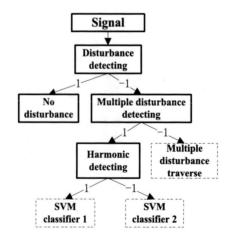

Fig. 9. Chart of the dynamic event tree structure

Table 2. The results of multiple disturbances classifier

Type of power quality disturbance	Accuracy/%
Voltage dropping	93%
Voltage rising	93.5%
Voltage interrupt	96.5%
Transient pulse	96%
Transient oscillation	94%
Average precision	94.5%

Multiple disturbances are detected by ST corresponding to the number of mutations when the number of mutations is greater than 2, dynamic event tree produces multiple disturbances traversal branches to classify disturbances.

The experimental result for multiple disturbances classifier designed by this paper is showed in Table 2.

5 Conclusion

Multiple disturbances are established and several typical multiple disturbances are enumerated. The method which combines WPT and ST is adapted to extract features and changes the monotonicity of extracting features. Finally, dynamic event tree classifier is established to classify all kinds of disturbances intelligently. Experiments show that disturbances can be classified in multiple disturbances in different cases.

References

1. Xiao, X.: Power quality analysis and control. China Electric Power Press Beijing **44**(2), 45–47 (2004). (in Chinese)
2. Zhang, X.: The research on the method of power quality disturbances detection and classification. North China Electric Power Univ. Beijing **1**(1), 2–4 (2004). (in Chinese)
3. Zhou, L., Guan, C., Lu, W.: Application of multi-label classification method to the categorization of multiple power quality disturbances. Proc. CSEE **31**(4), 45–47 (2011)
4. Li, Y.: Detection and analysis of transient power quality based on the S transform. Northeastern University, Shenyang, **1**(1), 1–2 (2009). (in Chinese)
5. Ma, Z., Li, P., Yang, Y., et al.: Power quality detecting based on wavelet multi-resolution method. J. North China Electric Power Univ. **30**(13), 6–9(2003). (in Chinese)
6. Qiao, Z., Sun, W.: A multi-class classifier based on the SVM decision tree. Comput. Appl. Soft. **11**(227), 27–30 (2009). (in Chinese)
7. Katsaprakakis, D.A.J., Christakis, D.G., Zervos, A.: A power-quality measure. IEEE Trans. Power Deliv. **23**(2), 553–561 (2008)
8. Tong, W., Song, X., Song, J.: Detection and classification of power quality disturbances based on wavelet packet decomposition and support vector machines. In: Proceedings of the 8th International Conference on Signal Processing, vol. 4, pp. 3015–3018 (2006). (in Chinese)

Power Cable Fault Diagnosis Based on Wavelet Analysis and Neural Network

Minghang Jiao[1(✉)], Yang Gao[2], Xuemin Leng[2], Yangqun Ou[1],
and Lin Zhang[1]

[1] State Grid Liaoyang Electric Power Supply Co., Ltd., Shenyang, China
1013511462@qq.com, 245954159@qq.com, 945859784@qq.com
[2] Shenyang Institute of Engineering, Shenyang, China
gaoyangsie@163.com, 522450260@qq.com

Abstract. With the widespread use of cables, the problem of power system fault diagnosis is becoming more and more serious. As we all know, the sudden blackout caused by cable fault will bring serious threat to the life and property safety of users, and even cause adverse social impact. Avoiding losses caused by cable failures is popular. To diagnose the cable faults is vital to guarantee the safe and steady operation of power transmission line. The combination of wavelet analysis and neural network is adopted as the fault diagnosis method to realize the accurate identification. Wavelet packet decomposition is used for feature extraction of cable fault signals which are input vectors after normalization processing. Radial basis function (RBF) network structure is built and relevant practice and test of cable fault diagnosis are conducted select 8 sets of samples for testing. By selecting different failover resistance values, the target output of the first four groups is 0.9, and the actual output is also 0.9; the target output of the last four groups is 0.1, the actual output is also 0.1, and the target output and actual output are The error between the two is basically zero, which also indicates that the RBF network has good fault discrimination. According to the test result, it shows that this method can be effectively achieved in cable faults diagnosis.

Keywords: Cable fault diagnosis · Wavelet analysis · Neural network

1 Introduction

Due to cable buried in the ground, when a failure occurs, if the fault is not identified and located accurately in time, it will not be able to guarantee safety and reliability of the power supply. Through the analysis of power cable grounding fault features, there is a certain mapping relation between the fault characteristic signal and different transition resistance in the neutral point via small resistance grounding system [1]. Wavelet analysis not only has the adaptive ability for the signal, and can well solve the problem between time and frequency, so it is the necessary tools for signal analysis and fault detection [2, 3, 8–12]. This paper uses a method combining the wavelet analysis and the neural network to realize the cable grounding fault diagnosis. Through data processing and simulation, it verifies this method can satisfy the requirement of power

© Springer Nature Switzerland AG 2020
A. E. Hassanien et al. (Eds.): AMLTA 2019, AISC 921, pp. 330–338, 2020.
https://doi.org/10.1007/978-3-030-14118-9_33

supply system and can effectively diagnose the fault. This method not only improves the security and reliability of power supply system, and well restrains the influence of interim resistance to cable fault location.

First of all, this article introduced Mechanism Analysis of Water Tree Branch and Electric Tree. At the same time, this paper also introduces the theoretical part of neural network, and also extends the application of neural network in cable faults, it also shows that RBF network have good fault discrimination.

2 Mechanism Analysis of Water Tree Branch and Electric Tree

Studies have shown that water tree is a further factor in triggering electrical tree branches, once the formation of electric branches, cable insulation can be caused in the short term breakdown, lead to insulation incidents.

The water-bearing tree cable has: rectification effect, low frequency or high-frequency harmonic effect, polarity effect and nonlinear low-frequency loss characteristic. In the cable, the area where the water tree appears is mainly the protrusion of the interface between the semiconducting layer and the insulating layer and the internal defects (porosity, impurities, etc.) of the insulating layer, where the electric field intensity is high, point.

There are many factors affecting water tree growth, such as field strength, frequency, temperature, and material structure and the type and concentration of salt solution and so on. It is found that the frequency of the applied electric field is an important factor to accelerate the aging of the insulation material, which has great influence on the growth rate of water tree. For example, water trees are seldom found in insulating materials at DC voltages. The effect of the electric field frequency on the water tree provides us with an important message that the fatigue of the insulating material is a crucial factor in triggering the water tree.

Electrical tree branches are an electrical-induced cracking phenomenon that occurs in XLPE cables, which occur in the local regions of the polymer domain, due to impurities, bubbles and other defects caused by local electric field concentration caused by the local breakdown, and the formation of branch-like discharge destruction of the channel, its shape and branches similar to the name. For XLPE cable insulation. Electrical tree branches appear to the insulation thickness of all the breakdown time of the saddle is short, which is the electrical tree branches and water tree distinction important feature. It is generally believed that the process of electric tree branches from initiation to growth is an extremely complex phenomenon of electrical corrosion, including charge injection, partial discharge, partial pressure, local high temperature, electro-mechanical force, physical deformation, chemical decomposition, within the complex process. Different types of media, different states, microstructure differences have increased the initiation and development of electrical tree randomness.

According to the conduction characteristics of the electrical tree channel, the electrical branches in the XLPE cable can be divided into conductive type, three types of electric and hybrid, respectively, corresponding to different growth mechanism, mainly depends on the media aggregation and pressure frequency and so on.

The electrical branches in the uniform crystalline state are conductive, and the non-conductive electrical tree branches are non-conductive. The frequency will promote the development of the non-conductive electrical tree branches, but not the initiation and growth of the conductive branches. Significantly affected. Non-conductive type of electric tree branch initiation process is weak; it will slowly develop into a mixed type. Previously, researchers in the laboratory using the needle plate electrode accelerated aging of the XLPE cable found that the initiation and growth of electrical tree branches and many factors, including voltage and frequency, temperature, defects, mechanical stress.

3   Theory of Wavelet Packet Analysis

Wavelet analysis can devise the frequency bands into multi-levels, put the high-frequency signal into more detailed wavelet decomposition, matches the corresponding frequency bands based on the characteristic signals analyzed, thus enhances the signal analysis ability, which has good time-frequency features [4, 15].

Taking a three-layer wavelet packet decomposition tree as an example to understand what is the wavelet packet analysis as is shown in Fig. 1.

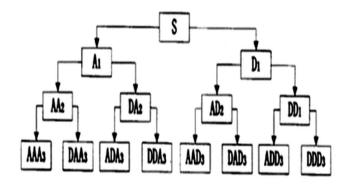

Fig. 1. The structure of three layers wavelet packet decomposition tree

It can be found from the Fig. 1, the wavelet packet is an extension of orthogonal decomposition step by step, and the relationship between its decomposition for three layers is shown as:

$$S = AAA_3 + DAA_3 + ADA_3 + DDA_3 + AAD_3 + DAD_3 + ADD_3 + DDD_3 \tag{1}$$

In the multi-resolution analysis, define the scale function and wavelet basis function dual scaling equation is:

$$\begin{cases} \Phi(t) = \sum_{k \in Z} h_k \Phi(2t - k) \\ \Psi(t) = \sum_{k \in Z} g_k \Phi(2t - k) \end{cases} \tag{2}$$

Among them, the sequence $\{h_k\}$ and $\{g_k\}$ are conjugate filters of discrimination analysis. In order to define the wavelet packet, define the $u_0(t)$ and $u_1(t)$ express respectively $\Phi(t)$ and $\Psi(t)$. Through further derivation, the double scale equation of the function $u_n(t)$ [5]:

$$\begin{cases} u_{2n}(t) = \sqrt{2} \sum_{k \in Z} h(k) u_n(2t - k) \\ u_{2n+1}(t) = \sqrt{2} \sum_{k \in Z} g(k) u_n(2t - k) \end{cases} \tag{3}$$

Among them, the coefficient of $h(k)$ and $g(k)$ have orthogonal relation, define the closure spaces of function $u_n(t)$ and $u_{2n}(t)$. The equivalent expression of formula 2 is as follows:

$$U_{j+1}^n = U_j^n \oplus U_j^{2n+1} j \in Z, n \in Z_+ \tag{4}$$

The sequence $\{u_n(t)\}$ is called orthogonal wavelet packet build by the basis function $\Phi(t)$.

4 The Extraction of Characteristic Signals for Cable Fault

In order to match the characteristics signals, the selection of wavelet basis function is very necessary, the most widely used in engineering and is relatively mature wave system is db N. Through theoretical derivation, different serial number of the small wave have different energy concentration, for the convenience of signal analysis, it usually select high energy concentration wavelet as basis function, so we choose db5 wavelet [6].

The layer number of wavelet packet decomposition is directly related to the width of the band, if the layer is overmuch, the bandwidth will be too narrow, and the corresponding sampling data will be too little, which does not favor the fault point positioning; On the contrary, if the layers are too little, it will cause the bandwidth too big, and cause more interfering signals, which will reduce the reliability of the diagnosis of fault point. Through comprehensive analysis and comparison, we set the layer number of wavelet packet decomposition 5 layers, so it is convenient for the division of frequency band and the extraction of feature signal [7, 8].

In order to realize the fault accurate positioning, we set a 5 layer wavelet packet decomposition aiming at the zero-sequence current of the faulted cable according to different cable fault location, the energy band diagram of which is shown in Fig. 2. It can be obviously found from the Fig. 2, the energy of 17th frequency band is biggest, the fault information of which is also the most abundant, so choose the 17th band as the featured band, and extract the frequency band of wavelet packet decomposition coefficient as the characteristic signal.

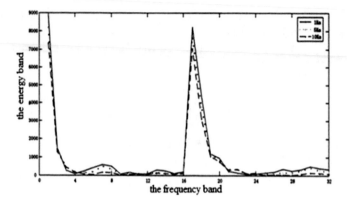

Fig. 2. Energy diagram of zero-sequence current

5 The Theory of Neural Network

Neural network [13, 14] mostly uses the layered structure, according to the presence of feedback between layer and layer it can be divided into feedforward and feedback network. Among them, the typical feedforward networks are BP network and RBF network, and the main forms of feedback network are the Hopfield net and recursive neural network. This paper chooses RBF network because it is superior to the BP network in convergence speed, the ability to recognize and forecast precision. Take a radial basis function (RBF) neurons as the column, the model of which is shown in Fig. 3.

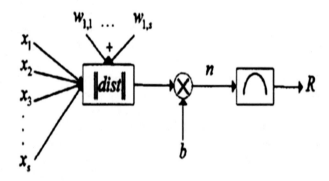

Fig. 3. Model of radial basis function neuron

Figure 3 express the distance between the input vector x and the weight vector w, b express the threshold, the role of which is to adjust the sensitivity of neurons. The expression of the hidden layer output is:

$$R = \left(\frac{-\left(\| x_i - w_{1,i} \| \times b - c \right)^2}{2\sigma^2} \right) (i = 1, \ldots, s) \tag{5}$$

It can be seen from the radial basis function curve and the model that when the distance between the input vector and weight vector is reduced, the output of the radial basis function (RBF) increases, on the other hand, the output of hidden layer nodes generated will be reduced. It reflects the function has a strong capacity of local approximation.

In the RBF network, it realizes the nonlinear mapping relationship between the input signal and the hidden layer, and it realizes the linear mapping from the hidden layer to the output layer, the expression of which is:

$$y_j = \sum_{i=1}^{P} w_{ij} R_i \, (j = 1, \ldots, m) \tag{6}$$

Accounting for subsequent analysis, here we cite a constant spread to replace the threshold, which is called the distribution density of the radial basis function. Distribution density decides the size of the neuron response area, when the distribution density is big enough, the response range will be wide, so the smoothness of each neuron is better, but if it beyond a certain limit, it will cause the cross leads of neuron area, which cannot meet the precision demand.

6 Neural Network in Cable Fault Diagnosis

The basic principle of cable fault diagnosis is shown in Fig. 4. Among them, the 1first two steps are mainly getting through the ATP - EMTP simulation, using the conversion tool to convert the format from p14 to mat, so it is convenient for data preprocessing. Then use wavelet packet analysis to get the extraction of feature signal, the role of which directly affects the fault diagnosis. Finally is the construction of the neural network and through the training sample and test, sample pretreatment to realize the cable fault diagnosis effectively.

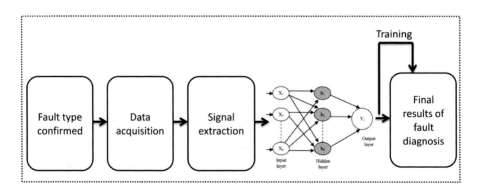

Fig. 4. The fundamental cable fault diagnosis

In this paper, the cable fault identification is based on a zero-sequence current of the circuit fault feature. As is shown in Fig. 5, the zero-sequence current waveform between the normal line and single-phase earth fault line is quite different. The zero-sequence current of the normal line is almost zero, but the zero-sequence current signal of the fault line changes due to impedance characteristics, time-varying characteristics and many other factors, and through this feature, it will be very convenient for cable fault recognition.

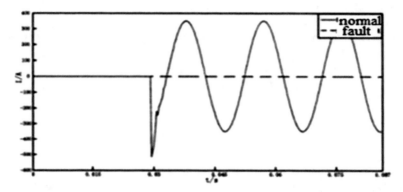

Fig. 5. The zero-sequence current waveform of normal line and single-phase grounding fault line

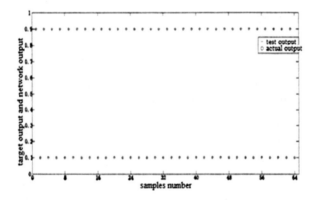

Fig. 6. The training results of cable fault recognition subnet

Select zero-sequence current of normal line and single-phase grounding fault line as the signal. Get the energy spectrums of the 4th layer sample data through wavelet packet decomposition of zero sequence current using db5. Take the energy spectrum obtained by wavelet packet decomposition as the characteristics vector of the neural network input after normalization. Here select 64 sets of data as training samples, including 32 group for fault samples and 32 groups for the normal samples, with a " + " mark sample target output, with a "0" indicates the actual output, and the output is shown in Fig. 6.

Table 1. Test results

Samples	Test number	Transition resistance	Test output	Actual output
Fault test samples	1	1 Ω	0.90	0.9000
	2	10 Ω	0.90	0.9000
	3	100 Ω	0.90	0.9000
	4	1000 Ω	0.90	0.9000
Normal test samples	5	–	0.10	0.1000
	6	–	0.10	0.1000
	7	–	0.10	0.1000
	8	–	0.10	0.1000

In order to train the effect of the network fault recognition, select 8 set of samples for testing, as is shown in Table 1. By choosing the different value of fault transition resistance, it can be found that the influence of the transition resistance on cable fault recognition is not big, the error between the target output and the actual output is basically zero, and it also shows that RBF network has good fault discrimination.

7 Conclusion

This paper first introduces the theory of wavelet packet and the basic knowledge of neural network theory, and analyses the structure of RBF network, based on which adopted a method that combined wavelet analysis and neural network for cable fault identification. Using wavelet packet analysis to extract the cable fault characteristic signal as the input feature vectors of neural network, training samples and testing, according to the training error curve and the test result, it shows that this method that based on wavelet analysis and RBF neural network can play an important role in cable fault detection and diagnosis.

References

1. Bao, Y.Sh.: Partial discharge of power cable on-line monitoring and fault diagnosis. Beijing Jiaotong University (2012). (in Chinese)
2. Yuan, Y.L., Zhou, H., Dong, J.: On-line monitoring and fault diagnosis technology of high voltage power cable sheath current (in Chinese)
3. Zhai, L., Wen, Y.K.: Power cable fault diagnosis and analysis. In: Technology and Innovation, vol. 22, pp. 114–116 (2015). (in Chinese)
4. Liu, H.: Cable fault diagnosis theory and key technology research. Huazhong University of Science and Technology (2012). (in Chinese)
5. Zhang, G., Ye, J.H.: Intelligent fault diagnosis and management system for power cable. Power Syst. Telecommun. **02**, 56–60 (2011). (in Chinese)
6. Gao, Q.S., Yang, J.: Review of fault diagnosis of power cables. Guizhou Electric Power Technol. **05**, 54–58 (2016). (in Chinese)

7. Zhang, Z.Ch.: Partial discharge monitoring of power cables and insulation fault diagnosis. The Hubei University of Technology (2013). (in Chinese)
8. Fu, J.P.: Underground pipe network on-line monitoring and fault diagnosis system. University of Electronic Science and Technology (2013). (in Chinese)
9. Zhu, G.M.: Research on fault location method of power cable-based on wavelet neural network. Guangdong Electric Power (2013). (in Chinese)
10. Zhan, L.H.: Research on calculation of power cable core temperature based on BP neural network. Technology Information (2016). (in Chinese)
11. Yang, X.H.: Pattern recognition of partial discharge of XLPE power cable-based on BP artificial neural network. High Voltage Electrical Appliance (2013). (in Chinese)
12. Wang, Y.X.: Simulation analysis of partial discharge circuit model of power cable-based on capacitance sensor technology. Wire and Cable (2017). (in Chinese)
13. Press, W.H., Teukolsky, S.A., Vetterling, W.T., Flannery, B.P.: Section 3.7.1. radial basis function interpolation. Numerical Recipes: The Art of Scientific Computing, 3rd edn. Cambridge University Press, New York (2007)
14. Bishop, C.M.: Neural networks for pattern recognition. Clarendon Press, Oxford (1995)
15. Graps, A.: An introduction to wavelets. J. IEEE Comput. Sci. Eng. 2(2), 50–61 (1995)

Predictive Control of Superheated Steam Temperature of Molten Salt Heat Storage System

Zhi Wang[(✉)]

Technical Services Department, Liaoning CPI Power Station Combustion
Engineering Research Center Co., Ltd., Shenyang, China
33732880@qq.com

Abstract. With the development of the economy, the demand for energy in
factories is increasing, and the demand for the development of electric power
industry is also increasing. Higher requirements for various capabilities and
targets in the process of generating electricity are put forward. As a result, the
capacity of the unit becomes larger and larger, the controlled object becomes
more and more complex, and the control of the superheated steam temperature
in the power plant is more and more difficult to meet. In order to overcome the
control problems such as non-linearity, large inertia, large parameter change,
and large operating condition change, the process principle of heat exchanger
system in the energy storage system is studied, and a multi-model predictive
control algorithm is proposed. The control problem of the nonlinear industrial
system is solved. The simulation results show that the method can suppress the
superheated steam temperature fluctuation and reduce the dynamic error and
steady-state error of the superheated steam temperature.

Keywords: Molten salt thermal system · Super-heated steam ·
Multiple model control · Automatic control system

1 Introduction

The molten salt thermal storage power generation system stores heat by using molten
salt as a medium and generates power by releasing it when necessary. It consists of a
number of heliostat arrays that reflect solar radiation to the top of the heat sink, which
absorbs heat to produce steam directly or generates heat through heat, and the alternating
current drives the turbine generator to convert solar energy into electrical energy.

Since the solar radiation itself varies with the seasons, the characteristics of different time and discontinuous changes during the day are greatly affected by the
weather, so the quality control of the thermal power generation system is the key to
thermal power generation. However, the tower solar thermal power generation system
is a multivariate, large variable, and there are obvious nonlinear and lag complex
industrial processes. The system is mainly controlled objects (such as heating steam
drums, energy storage systems) difficult to obtain accurate models through traditional
control methods, the traditional control method can better control the thermal system,
and greatly improve the various quality parameters of solar energy. The control of the

© Springer Nature Switzerland AG 2020
A. E. Hassanien et al. (Eds.): AMLTA 2019, AISC 921, pp. 339–345, 2020.
https://doi.org/10.1007/978-3-030-14118-9_34

thermal power tower system is very difficult, which makes people have to introduce the process principle and control method in detail. For example, the literature [1] adopts the parameter self-tuning control method to achieve constant control of the output oil temperature and proposes a general prediction based on the nonlinear model. The control algorithm is applied to the solar power system [2].

At present, China's research on tower solar thermal power generation system is still in its infancy, and a very mature control scheme has not yet been formed. The polymer tower solar thermal power generation system has a very large working range and has nonlinearity, large inertia, large hysteresis, and time-varying characteristics. Therefore, the adaptive control of a single model is not ideal, which can lead to large transient errors. The multi-model control method based on comprehensive decomposition can effectively overcome the above problems [3].

Predictive control is widely used in the field of process control. Multi-model predictive control is combined with many scholars to solve the nonlinear chemical reactor model, waste treatment and superheated steam temperature control and other nonlinear process control problems [4]. Multi-model predictive control (MMPC) is the basic form of nonlinear predictive control, and multi-linear local models are used to describe the same nonlinear object. The control design is a local model controller. The global output is switched or weighted. The advantage is that most of the control parameters are calculated offline. It is difficult to ensure the model switching time and maintain the stability of the model switching [5].

In this paper, the principle of heat transfer in molten salt storage is studied, and a thermal power generation system based on multi-model predictive control is proposed.

2   Model Predictive Control Method and Implementation

2.1   Multi-model Predictive Control Algorithm Structure

The multi-model predictive control design principle is to use the multi-model decomposition synthesis strategy to divide the nonlinear space into multiple sub-spaces, and design each sub-space and its corresponding single-controller mathematical model to design an effective controller scheduling scheme [6].

A multi-model predictive control algorithm is used in the offline design method. According to different synthesis methods, it is divided into weight control and handover control. This paper uses multi-model weighted control. The controller uses a predictive controller and the multi-mode switching algorithm uses a weighting algorithm. The characteristics of the algorithm are divided into n fixed sub-models, and the offline controller is designed according to the sub-model respectively. Then, in the control process, the output is controlled according to a certain weighting synthesis rule. The linear weighted switching method can be used for smooth and continuous model switching and can ensure that the model is stable and computationally intensive.

2.2   Predictive Control

The successful application of predictive control in industrial control fields such as the chemical industry and the aerospace industry has caused widespread concern in the field of industrial control. The predictive control algorithm is a new computer-optimized control algorithm based on the model, rolling implementation and feedback correction. Due to the excellent performance of predictive control, many scholars at home and abroad have carried out a series of researches on the application of predictive control technology in thermoelectric power generation process control. For example, the GPC algorithm based on the parametric model makes the system design more flexible by introducing the level of not equal to the prediction and control, with model prediction, rolling optimization and feedback correction online. Because it adopts the minimized parameter model, it can reduce the calculation of the predictive control algorithm by using less parameter than the non-parametric model. And because of the multi-step prediction based on object output, the pure lag of object is not sensitive, which is very applicable to industrial process control [7–9].

The generalized pre-test control system is used for the control of the CARIMA model.

$$A(q^{-1})y(t) = B(q^{-1})u(t-1) + C(q^{-1})\xi(t)/\Delta \tag{1}$$

In order to ensure the system output prediction to track the reference trajectory in the best way and limit the changing amplitude of the input signal, the performance index of GPC is defined as:

$$\min J(t) = \left\{ E\sum pj = [y(t+j) - w(t+j)]^2 + \sum Nj = \lambda(j)[\Delta u(t+j-1)] \right\}^2 \tag{2}$$

The optimal control law is obtained

$$\Delta U = (G^T G + lI)^{-1} G^T (W - f) \tag{3}$$

Clearly

$$\Delta u = [1, 0, .., 0]_{1\times p} \Delta U \tag{4}$$

2.3   Multi-model Switching Strategy

This article is based on multi-step prediction error calculating performance indicators, performance indexes of each model of the current time and the ratio of performance index summation as the weighting coefficient; calculate overall control of the current moment.

At the kth moment, ei(k) = y(k) − ymi (k) represents the error between the actual output and the output of the "I" sub model.

2.4   Multi-model Predictive Control Algorithm Steps

The specific steps of the predictive control algorithm based on multi-model weighting are shown below.

- Step 1 collect data to establish local sub-models of nonlinear objects.
- Step 2 designs the corresponding GPC controller for local models.
- Step 3 to k times the input data and historical input and output data, using type (4) and (5) the local control of the current time increment Δ u.
- Step 4: using formula (6) and (7), the weighted value of the local control quantity is calculated by the weighted criterion.
- Step 5 use Eq. (8) to calculate the control amount and make use of the actual object.
- Step 6: k = k + 1, return step 3.

3   Simulation and Analysis

3.1   General Characteristics of Superheated Steam

The main steam temperature of the main steam is mainly composed of the main steam flow, the pressure of the main steam and the temperature of the main steam. The main steam flow has the most significant influence on the characteristics. In load rises, superheated steam flow rate and flow velocity increases, and the system of the static gain, delay time and time constant is less, will greatly change the object properties, and the main steam temperature system is typical multiple model [7]. The change of load is the most important factor influencing the model, and the parameter size of the steam temperature process model is closely related to the change of unit load. Load rises, superheated steam flow rate and flow rate will increase accordingly, with the same change of warm water flow static gain, the time constant and model reduction will subsequently change monotonously, and vice versa [10]. Therefore, the working interval can be divided according to the change of load.

Compared with the traditional conventional coal-fired boiler, polymer solar thermal power generation system in water/steam heat exchanger with conventional boiler thermal system strongly nonlinear, inertia and difficult to control similar properties, such as delayed in superheated steam temperature control has a lot of cans draw lessons from. Therefore, this article will design the control strategy is applied to boiler superheated steam temperature system in simulation experiment, for after tower solar thermal system heat exchanger of superheated steam temperature control to provide a technical reference for export [11].

The parameters of the superheated steam temperature object model change with the change of load, although the change is not strictly linear change, but can be approximated as a linear variable. Assuming that the model parameter can be approximated as a linear function of load, the error caused by this hypothesis is small, which is acceptable in the simulation research and practical operation [12]. Based on this assumption, the transfer function of an operating point can be obtained through the linear interpolation of the local model.

3.2   Control Results Under Fixed Load

Based on the linear interpolation of the superheated steam temperature object model with a load of 100%, the model set of the sub-model under this condition is obtained. The simulation results of system control are shown in Fig. 1 respectively.

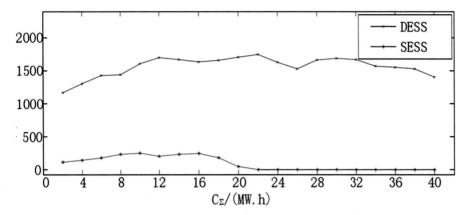

Fig. 1. System output at 100% load.

3.3   Control Results Under Variable Load

If the simulation system has an initial load of 37% and the simulation time is greater than 500s, the system load becomes 100%. The increment curve of the system output and control of variable load is shown in Figs. 2 and 3 respectively. The simulation results show that the method can meet the control requirements of the variable load and prevent the excessive fluctuation of the overheating steam temperature and reduce the dynamic and steady-state deviation of the superheated steam temperature.

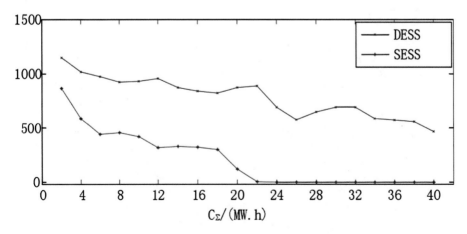

Fig. 2. Variable load system output curve.

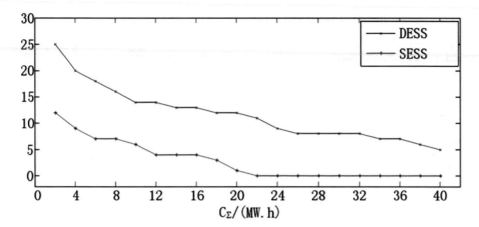

Fig. 3. Variable load controller incremental curve.

4   Conclusion

Based on the multi-model control strategy of the decomposed synthesis strategy, the generalized predictive control method is adopted to study the multi-model predictive control method for the common class of highly nonlinear time-delay systems. A local model family is established within the hot steam temperature condition, and the corresponding sub-gpc controller is designed. The control increment is calculated by tracking the variable weight controller of the working condition. Simulation results show that the method has good control effect, and can satisfy the control requirement of the constant load and varying load, to prevent overheating steam temperature fluctuations, significantly reduce the dynamic and steady state of superheated steam temperature deviation.

References

1. Camacho, E.F., Rubio, F.R., Hughes, F.M.: Control of a solar power plant with for field. J. Control Systems (1992) Self tuning distributed collector field. In: Maxwell, J.C.: A Treatise on Electricity and Magnetism, 3rd ed., vol. 2, pp. 68–73. Clarendon, Oxford (1892)
2. Camacho, E.F., Berenguer M.: Application of generalized predictive control to a solar power plant. In: Proeedings of the Third IEEE Conference on Control Applica-dons. Ulasgow, UK, pp. 1657–1662 (1994)
3. Yunguang, Z., Tong, S., Yiguo, L., et al.: Multi-model switching based UPC and its application to super-heated steam temperature systems. East China Electric Power, **37**(1), 164–168 (2009) (in press). (in Chinese)
4. Yorozu, Y., Hirano, M., Oka, K., Tagawa, Y.: Electron spectroscopy studies on magneto-optical media and plastic substrate interface. IEEE Transl. J. Magn. Japan, 2, 740–741 (1987) (Digests 9th Annual Conf. Magnetics Japan, p. 301, 1982)
5. Richard, T.: The Global CSP Market-its Industry Structure and Decision Mechanisms. Hanburg, pp. 45–46 (2003)

6. Herrmann, U., Kelly, B., Price, H.: Two-tank molten salt storage for parabolic 'dough solar power plants. Energy **29**, 883–893 (2004)
7. Tiu, W., Bavidson, J.H., Mantell, S.C., et al.: Natural convection from a horizontal tube heat exchanger immersed in a tilted enclosure. ASME J. Solar Energy **125**(1), 67–75 (2003)
8. Liu, W., Davidson, J.H., Kulacki, F.A.: Natural convection tube a tube bundle in a thin inclined enclosure. ASME J. Solar Energy **126**(2), 702–709 (2004)
9. Haltiwangez, J., Dauidson, J.H.: Discharge of a thermal storage tank using an immersed heat exchanger with an annular baffle. Solar Energy **83**, 193–201 (2009)
10. Luo, W.G.: J. Syst. Simulation. **19**(24), 5750–5754 (2007)
11. Qiu, X.Z.: International Conference on Advanced Computer Control, pp. 742–746 (2009)
12. Yao, J.F.: Adv. Mater. Res. **516–517**(1), 232–238 (2012)
13. Shang, M.: Optimal algorithm for scheduling large divisible workload on heterogeneous system. Appl. Math. Model. **32**(9), 1682–1695 (2007)
14. Lipton, Z.: Learning to diagnose with LSTM recurrent neural networks. Comput. Sci. **28**(10), 135–149 (2015)
15. Hua, J., Yi, H., Zhang, Q.: Minimizing accumulative memory load cost on multi-core DSPs with multi-level memory. J. Syst. Architect. **59**(7), 389–399 (2013). (in Chinese)
16. Jingnan, Y., Jiani, G., Laxmi, B.: Fair link striping with FIFO delivery on heterogeneous channels. Comput. Commun. **31**(14), 3427–3437 (2008)
17. Richard, S., Mikael, A., Lennart, S.: Approaching wind power forecast deviations with internal ex-ante self-balancing. Energy **57**, 06–115 (2013)
18. Pramod, K., Abhishek, M., Kamal, S.: Benchmarking the clustering algorithms for multiprocessor environments using dynamic priority of modules. Appl. Math. Model. **36**(12), 6243–6263 (2012)
19. Chen, S., Wen, J., Zhang, R.: GRU-RNN based question answering over knowledge base. In: Proceedings of International Conference on Systems Knowledge Graph and Semantic Computing. Springer Singapore, pp. 80–91 (2016)

Control and Chaotic Systems

Optimal Proportional Integral Derivative (PID) Controller Design for Smart Irrigation Mobile Robot with Soil Moisture Sensor

Ahmad Taher Azar[1,2(✉)], Hossam Hassan Ammar[1],
Gabriel de Brito Silva[1], and Mohd Saiful Akmal Bin Razali[1]

[1] School of Engineering and Applied Sciences, Nile University,
6th of October City, Egypt
ahmad_t_azar@ieee.org, hhassan@nu.edu.eg,
{gabriel.silva, saifulakmal.razali}@eu4m.eu
[2] Faculty of Computers and Information, Benha University, Banha, Egypt

Abstract. Uncertainty on the condition of the weather always give a major headache to the agricultural industry as the cultivated plant that is grown on a large scale commercially rely on the condition of the weather. Therefore, to reduce the interdependency on the weather itself, a recommendation to develop a prototypic mobile robot for smart irrigation is submitted. Smart irrigation system is an essential tool from yield point of view and scarcity of the water. This smart irrigation system adopts a soil moisture sensor to measure the moisture content of the soil and automatically provide a signal to switches the water pump when the power is on. This mobile robot uses a tracked vehicle as the chassis of the robot whereby two parallel forces of motion are controlled to create a linear and rotation motion of the wheel. Besides, ultrasonic sensor is also assembled on this prototypic design to avoid any object or obstacles during the movement of the mobile robot. The novelty of this study is to demonstrate that the smart irrigation mobile robot can be used, not only to provide water to the crops but also, to reduce water usage. In this research, 2-Degree of Freedom Proportional Integral Derivative (2-DOF PID) controller is proposed for smart irrigation mobile robot after comparison between optimal 1-Degree of Freedom Proportional Integral Derivative (1-DOF PID) to acquire the best controller. Further investigation of controller effort, disturbance rejection and reference tracking are implemented to validate the capability of the proposal within the parameter of the system itself.

Keywords: Optimal PID · 1-DOF PID · 2-DOF PID ·
Smart irrigation mobile robot · Soil moisture sensor

1 Introduction

Manual irrigation system is easy to do and inexpensive. However, this kind of irrigation system is using more water which ends up provide more waste to the agricultural industry [1]. Mass irrigation is one of the methods that also commonly used to water the plant [2]. This method however represents massive losses since the amount of water

© Springer Nature Switzerland AG 2020
A. E. Hassanien et al. (Eds.): AMLTA 2019, AISC 921, pp. 349–359, 2020.
https://doi.org/10.1007/978-3-030-14118-9_35

given to the crops always more than it should. Due to increasing water cost, some countries such as United States is taxing the water consumption. Agricultural industry players really need to start concerning more on the conversation of the water as the demand of the water is forecasted to be increase in the future. Besides, as water is brought into the system manually, this thereof requires high labor input [3]. Moreover, it is important to check the systems regularly to improve the production and avoid water loss on the crops. Moreover, human labor is becoming more and more expensive [4]. Due to this reason, if there is no effort being invested in optimization of these resources, agricultural industry players really need to spend more money in this process. It is without a doubt, that technology is probably the only solution to reduce cost and prevent loss of more resources. With the advanced development of technology especially in automation industry and smart irrigation system, automatic irrigation or smart irrigation system is becoming a trend lately [5]. Contrary to a traditional irrigation system, the smart irrigation one regulates the water which is provided according to the needs of the soil and crops [6]. The feedback mechanism of a smart irrigation system is a soil moisture sensor. Soil moisture sensor is an essential parameter in monitoring growth of the plant. Since soil moisture sensor determines the amount of liquid content (commonly water) of the soil, most of the researchers start to design a device that will determine the moisture level of the soil that will trigger a water irrigation system to release and gives enough water for the plants to reach their full growth [7].

Normally, automatic irrigation has typical features that should be take note. The range of area that required to irrigate probably includes several acres of land which need a lot of efforts to cover all of it [8, 9]. Besides, the points to irrigate and sample by moisture sensor spread around which another difficulty that need to be handle in a timely manner [10]. The parameters to sample include temperature, soil moisture and air humidity. Moreover, farming activity that carry the wire and soil near to each other is very difficult to design [11]. The poor circumstance including high moisture due to air humidity and immensely high temperatures due to the sunlight provides a dilemma and obstacle for this kind of smart devices to sustain for a quite longer time [12]. Another factor that need to be considered also, is that most of the workers in the agricultural industry are uneducated and only know a basic knowledge to operate and maintain the device [13]. Therefore, this paper is presented to design and manufacture a smart irrigation mobile robot with soil moisture sensor. This paper introduces a simulation analysis of our designed smart irrigation mobile robot by using 1-DOF. After that, the results of 1-DOF will be compared with 2-DOF. The proposed method in this paper attempt to produce the best output performances. From the proposed controller, the results that give a high stability and a fast time response is captured. Besides, this paper will discuss a development of a program that will process the data from the soil moisture sensor and control the whole irrigation system by implementing several optimal PID controller design such as the Integral of Squared Error (ISE), Integral of Absolute Error (IAE), Integral of Time Multiply Squared Error (ITSE), Integral of Time multiply Absolute Error (ITAE) and Ziegler-Nichols stability margin tuning (ZN). The obtained results are very promising.

The rest of the paper is organized as follows: While in Sect. 2, the mathematical model is introduced and explained. In Sect. 3, the schematic and hardware design are well explained and in Sect. 4, the controller design is discussed. The results and discussion are shown in Sect. 5. In addition, the conclusion is written to conclude this article in Sect. 6.

2 Mathematical Modeling

The mathematical modeling including the kinematic model, dynamic modeling, mathematical modelling of a DC motor and Two-Degree-of-Freedom Control has been presented in our previous study under Chapter 2: Mathematical Model from Equation 1 until Equation 35 [14].

3 Schematic Design

Figure 1 shows a sub-system diagram of smart irrigation mobile robot. The system is sub-divided to three sub-systems which are water irrigation system, mobile robot movement system and object avoidance system.

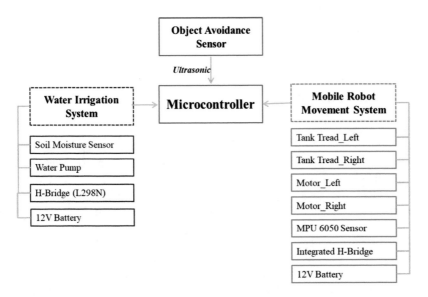

Fig. 1. Sub-system diagram of robotic infusion stand.

For this study, microcontroller which is NexusDuino 60003 will reads angular velocity data from the MPU 6050. Then, PWM signals will be sends to the integrated H-Bridge to change that particular signals to the form of mechanical motion [15]. Meanwhile, NexusDuino 60003 will receive the voltage data from the soil moisture sensor and transmits the PWM signals to the H-Bridge L298N to start supplying the water via water pump. This H-Bridge L298N is powered by 3 V Nickel-Metal-Hydride (Ni-MH) batteries that are 12 V in series connection. Besides, ultrasonic distance sensor is installed on this overall system for safety purposes (Fig. 2).

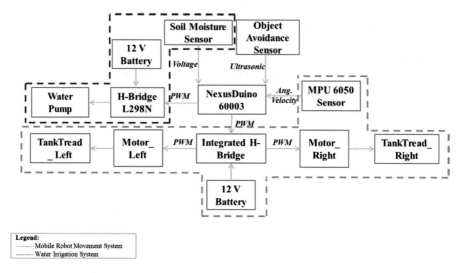

Fig. 2. Schematic diagram of smart irrigation mobile robot

4   Controller Design

The design target of the system is to control the smart irrigation mobile robot with a tuned PID controller. Figure 3 describes a full block diagram of this smart irrigation mobile robot while Figs. 4 and 5 shows a diagram for motor subsystem left and right. The controllers are separated into Controller 1 and also Controller 2. It should be noted that, Controller 1 is a rotation movement in y-axis whereas Controller 2 is translational movement in x-axis. Besides, 2 set point speed for each motor (motor left and motor right) are take into consideration. Then, this data is used in the subsystem of the motor right speed control. By applying a PID Controller, an alteration method to gain analog results with digital means can be created. This step is commonly defined as PWM in order to have charge of the PWM output and gain the transfer function of the motor [16]. Smart irrigation mobile robot can be said at its physical state, meaning it will move to the position that is being designated.

5   Results and Discussion

By implementing the system identification, the transfer function of each DC motors is noted, and this transfer function will be applied to integrate the actual transfer function with a kinematic model of the entire system. The main aims are the complete PID with

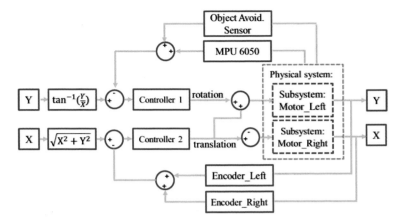

Fig. 3. Block diagram for smart irrigation mobile robot

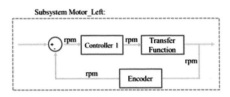

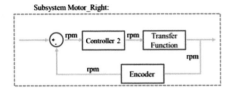

Fig. 4. Block diagram for subsystem motor (Left)

Fig. 5. Block diagram for subsystem motor (Right)

One Degree of Freedom (1-DOF) and the PID with Two Degrees of Freedom (2-DOF). By evaluation each of the transfer function, the correlation can be made for both, in order to comprehend the behavior and advantage of each one of its.

Figure 6 shows the input and output signal for Left_Side of the robot while Fig. 7 shows the best fit simulated output for Left_Side which is 81.46% that is higher compared to other trial fits. the same way was applied on the right side and the best fit simulated output for Right_Side was is 74.6%.

With the estimation of each one separately, the transfer function of the Left_Side (Eq. 1) and The Right_Side (Eq. 2) are as follows:

$$\frac{0.06174s + 0.06785}{s^2 + 0.08144s + 0.1011} \tag{1}$$

$$\frac{0.9154s^2 + 0.008741s + 0.004563}{s^3 + 4.939s^2 + 0.03201s + 0.00603} \tag{2}$$

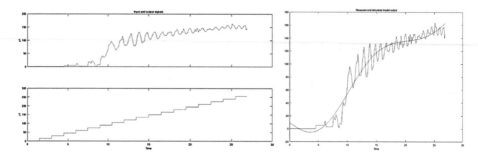

Fig. 6. Input and output signals for Left_Side   **Fig. 7.** Best fits simulated Left_Side

The first trial was optimal 1-DOF PID controller. Upon receiving acceptable values in the determinants to a good compensator, the values of K_P, K_i, K_d and the transfer function of the compensator itself are exported to the MATLAB workspace. Figure 8 shows the performance criterion based on the optimal 1-DOF PID controller design. From this figure, ISE and ITSE show a very promising value for this Left_Side while Fig. 9 shows the disturbance rejection of ISE and ITSE.

Table 1 below shows the transient response values for different optimal 1-DOF PID controller for Left_Side

Table 1. Transient response values for optimal 1-DOF PID controllers

Optimal 1-DOF PID	Steady state error, e_{ss}	Overshoot, M_P (%)	Rise time, $T_r(s)$	Settling time, $t_s(s)$
ISE	0.0351	34.6514	0.2481	2.0834
IAE	0.0671	9.3890	10.5170	82.8251
ITSE	0.1359	37.3618	0.7894	6.8788
ITAE	0.0960	7.4433	12.1070	47.5708
Z-N	0.1096	0.7514	40.1078	94.6071

For optimal 1-DOF PID controllers, ISE and ITSE are the only optimal 1-DOF PID controller method that will reach the step value in the good real time compared with IAE, ITAE and Z-N which have a very high-rise time and settling time. All possible 2-DOF controllers are simulated, and from the analysis, the behavior of the controllers is similar to each other whereby all are in the acceptable range to be implemented. Since, there is no rigorous requirement to follow for the chosen controller, the easiest controller that can be implemented is chosen (Figs. 10 and 11).

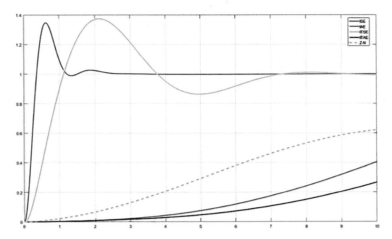

Fig. 8. Performance criterion for Left_Side

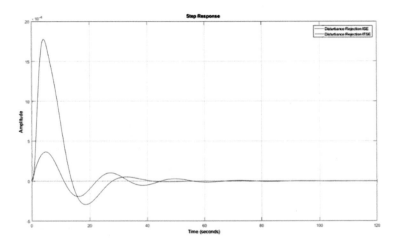

Fig. 9. Disturbance rejection for Motor_Left

The 2-DOF PID controller has lower overshoot which is 12% whereas the 1-DOF PID has 34.7% percentage of overshoot. The 2-DOF PID which is a kind of modified form of PID control is recommended to minimize the limitation of PID controllers. This 2-DOF PID has a capability to perform fast disturbance rejection by not having any meaningful increment on the factor of overshoot. In addition, this controller can be utilized to alleviate changed affected in the input signal.

The Same techniques are used to Control the Right_Side and Fig. 12 shows the performance criterion based on the optimal 1-DOF PID controller design. From this figure, ITAE and Z-N show a very promising value for this Right_Side while Fig. 13 shows the disturbance rejection of ITAE and Z-N (Table 2).

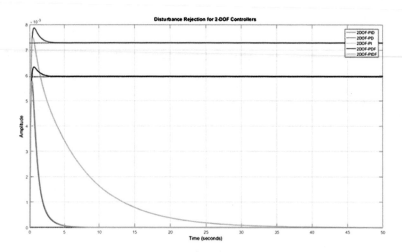

Fig. 10. Comparison of disturbance rejection of 2-DOF controllers for Left_Side

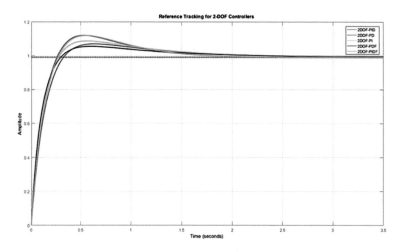

Fig. 11. Comparison of reference tracking of 2-DOF controllers for Left_Side

For 1-DOF PID controllers, ITAE and Z-N are the most suitable optimal 1-DOF PID controller method to be implemented on the hardware compared with ISE, IAE and ITSE which have a very high percentage of maximum overshoot and steady state error. The only acceptable 2-DOF controllers are PID and PIDF controllers. PIDF controller has no percentage of overshoot while the PID controller has 16% percentage of overshoot (Figs. 14 and 15).

The 2-DOF PIDF controller has no percentage of overshoot while the 1-DOF PID has 37.4% percentage of overshoot. Besides, the steady state error of 2-DOF PIDF controller is lower than the 2-DOF PID controller.

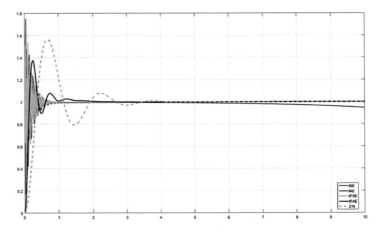

Fig. 12. Performance criterion for Right_Side

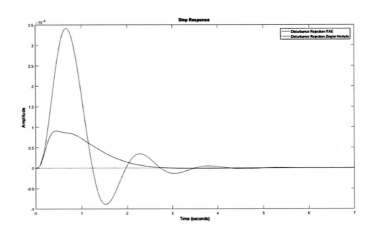

Fig. 13. Disturbance rejection for Right_Side

Table 2. Transient response values for Optimal 1-DOF PID controllers

Optimal 1-DOF PID	Steady state error, e_{ss}	Overshoot, M_P (%)	Rise time, $T_r(s)$	Settling time, $t_s(s)$
ISE	0.8371	91.3136	0.0052	0.6743
IAE	0.5862	75.1090	0.0174	0.7005
ITSE	0.8725	93.3089	0.0041	0.6837
ITAE	0.1016	37.4261	0.1025	1.3735
Z-N	0.2087	56.6868	0.2445	3.1867

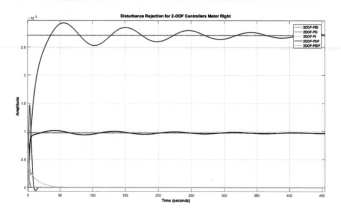

Fig. 14. Comparison of disturbance rejection of 2-DOF controllers for Right_Side

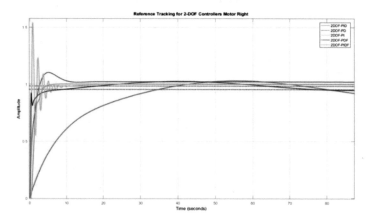

Fig. 15. Comparison of reference tracking between of 2-DOF controllers for Right_Side

6 Conclusion

In conclusion, our smart irrigation mobile robot is working as what we desired. Our mobile robot can move forward and in a rotation movement to the desired destination. Once the robot at the desired destination, the soil moisture sensor will measure the condition of the soil and the water pump will react accordingly to the condition of the soil. Besides, here are some recommendations for the future work: Automation of the soil moisture sensor mechanism by installing a motor with proper mechanism that capable to insert in and out the probes to the soil. Implement a machine vision knowledge by adding a camera that able to monitor daily watering of the crops. Adding Bluetooth mechanism to make wireless control of the mobile robot. Adding and mixing of fertilizer to water which may flow at the crops that can assist the crops to grow faster to expand the scope of the system in a large scale.

References

1. Maisiri, N., Senzanje, A., Rockstrom, J., Twomlow, S.J.: On farm evaluation of the effect of low cost drip irrigation on water and crop productivity compared to conventional surface irrigation system. Phys. Chem. Earth Parts A/B/C **30**(11–16), 783–791 (2005)
2. Wiesner, C.J.: Climate, irrigation and agriculture (1970)
3. Bigot, Y., Bigot, Y., Binswanger, H.P.: Agricultural Mechanization and the Evolution of Farming Systems in Sub-Saharan Africa. Johns Hopkins University Press, Baltimore (1987)
4. Li, H., Li, L., Wu, B., Xiong, Y.: The end of cheap Chinese labor. J. Econ. Perspect. **26**(4), 57–74 (2012)
5. Ćulibrk, D., Vukobratovic, D., Minic, V., Fernandez, M.A., Osuna, J.A., Crnojevic, V.: Sensing Technologies for Precision Irrigation. Springer, New York (2014)
6. Kumar, A., Kamal, K., Arshad, M.O., Mathavan, S., Vadamala, T.: Smart irrigation using low-cost moisture sensors and XBee-based communication. In: 2014 IEEE Global Humanitarian Technology Conference (GHTC), pp. 333–337. IEEE, October 2014
7. Aguilar, R.B., Ecija, E.B., Medalla, M.M., Morales, R.F.N., Platon, C.A.P., Rodrigo, J.A., Caldo, R.B.: Automatic soil moisture sensing water irrigation system with water level indicator. Adv. Sci. Lett. **23**(5), 4505–4508 (2017)
8. Ji, X., Tang, F.: The study and development of system for automatic irrigation. Irrig. Drain. **21**(4), 25–27 (2002)
9. Cui, Y.: Technology and Application of Water Saving Irrigation, pp. 345–349. Chemical Industry Press, Beijing (2005)
10. Liu, G., Sun, J.: The development and application of automatic system for irrigation management. Irrig. Drain. **20**(1), 65–68 (2001)
11. Li, K., Mao, H., Li, B.: The development of automatic system for irrigation and fertilization. J. Jiangsu Univ. Sci. Technol. (Nat. Sci.) **22**(1), 12–15 (2001)
12. Yeo, T.L., Sun, T., Grattan, K.T.V.: Fibre-optic sensor technologies for humidity and moisture measurement. Sens. Actuators A Phys. **144**(2), 280–295 (2008)
13. Dillon, J.L., Hardaker, J.B.: Farm Management Research for Small Farmer Development, vol. 41. Food & Agriculture Org (1980)
14. Azar, A.T., Hassan, H., Razali, M.S.A.B., de Brito Silva, G., Ali, H.R.: Two-degree of freedom proportional integral derivative (2-DOF PID) controller for robotic infusion stand. In: International Conference on Advanced Intelligent Systems and Informatics, pp. 13–25. Springer, Cham, September 2018
15. Immega, G., Antonelli, K.: The KSI tentacle manipulator. In: Proceedings of the 1995 IEEE International Conference on Robotics and Automation, vol. 3, pp. 3149–3154. IEEE, May 1995
16. Adar, N.G., Kozan, R.: Comparison between real time PID and 2-DOF PID controller for 6-DOF robot arm. Acta Phys. Polonica A **130**(1), 269–271 (2016)

Adaptive Higher Order Sliding Mode Control for Robotic Manipulators with Matched and Mismatched Uncertainties

Ahmad Taher Azar[1]([✉]), Fernando E. Serrano[2],
Sundarapandian Vaidyanathan[3], and Hani Albalawi[4]

[1] Faculty of Computers and Information, Benha University, Benha, Egypt
ahmad_t_azar@ieee.org
[2] Universidad Tecnologica Centroamericana (UNITEC), Zona Jacaleapa,
Tegucigalpa, Honduras
serranofer@eclipso.eu
[3] Research and Development Centre, Vel Tech University, Avadi,
Chennai 600 062, Tamil Nadu, India
sundarvtu@gmail.com
[4] Electrical Engineering Department, University of Tabuk, Tabuk, Saudi Arabia
halbala@ut.edu.sa

Abstract. Robotic manipulators have been extensively used in industrial and other kinds of applications. Thus, it is important to design effective control strategies for tracking precision for robotic manipulators. In this work, an adaptive higher order sliding mode control for a robotic manipulator with matched and mismatched uncertainties is proposed. Matched uncertainties occur when they are found in the input of the system and mismatched uncertainties are found in the system parameters. Hence, an adaptive higher order sliding mode controller is designed when both matched and mismatched uncertainties are found. Considering that uncertainties yield unwanted effects in the controller design, sliding mode control provides a suitable control strategy for robotic manipulators when extreme tracking precision of the end effector is needed especially in a reduced task space. The design procedure starts with the dynamic model represented in the Euler-Lagrange form considering the uncertainties of the system and then by implementing a Lyapunov stability method and selecting an appropriate sliding surface suitable control. Finally, adaptive laws are obtained taking into account the matched and mismatched uncertainties in the system model. As a numerical example, the proposed control strategy is validated for trajectory tracking purposes of a five bar linkage mechanism.

Keywords: Robotics · Sliding mode control · Uncertainties ·
Adaptive control · Mechatronics

© Springer Nature Switzerland AG 2020
A. E. Hassanien et al. (Eds.): AMLTA 2019, AISC 921, pp. 360–369, 2020.
https://doi.org/10.1007/978-3-030-14118-9_36

1   Introduction

Robotic manipulators have been extensively used in different kinds of applications such as industrial, aerospace and medical applications. Hence, the controls for trajectory tracking of these mechanisms have needed a lot of effort of robotic and automatic control specialists since several decades ago. The precision of an efficient automatic control system is necessary to drive the end effector of a robotic manipulator on the way of a pre-specified trajectory for different tasks such as welding, painting and screwing among other activities. In the last decade some approaches involving strategies such as passivity based backstepping and sliding mode control have been implemented to improve the accuracy of the trajectory tracking of different mechanisms (Mondal and Mahanta 2011; Mobayen and Tchier 2018; Azar and Zhu 2015; Azar and Vaidyanathan 2015, 2016, 2018; Azar et al. 2018a,b). In this work, an adaptive higher order sliding mode controller is proposed for robotic manipulators with matched and mismatched uncertainties.

Matched uncertainties are those kind of uncertainties found in the input of a system while mismatched uncertainties are those that are found in the system parameters. For this reason, we consider both matched and mismatched uncertainties in order to design the adaptive higher order sliding mode providing a novel contribution in which a robust approach (Azar and Serrano 2017) is needed to reduce to zero the tracking error of the system for different kinds of robotic manipulators and mechanisms (Gutierrez et al. 2018). We note that the higher order sliding mode have been implemented in different kinds of complex systems. The higher order sliding mode technique consists in designing a sliding surface and then finding a control law that makes the $r-1$ higher order derivatives to approach zero where r is the sliding mode order. For this purpose, it is sometimes necessary to implement a super-twisting algorithm in order that the sliding variable and its r derivatives approach zero. Some examples of adaptive higher order sliding mode control can be found in (Han and Liu 2016a; Barth et al. 2018; Thomas et al. 2018). Besides, some applications of higher order sliding mode control in fields different from mechatronics and robotics are found in (Han and Liu 2016b) where this control strategy is implemented for trajectory tracking of an air cushion vehicle. Another example can be found in (Tiwari et al. 2016) where this sliding mode approach is designed for the attitude control of a rigid aircraft. Finally in (Tannuri and Agostinho 2010), a higher order sliding mode controller is implemented in the dynamic position system of an offshore vessel. In (Mondal and Mahanta 2013; Khan et al. 2011; Mohamed et al. 2018), some chattering reduction sliding mode strategies for multi-input multi-output MIMO systems and single-input single-output SISO systems are shown considering the matched and mismatched uncertainties. One of the most important conditions that are considered in this study for both kinds of uncertainties are the norm bounded characteristics of them and so it is important to establish some conditions as they appear in some studies such as (Köroğlu and Scherer 2005; Battilotti 1996; Adrot et al. 2004).

In this work, an adaptive higher order sliding mode controller is designed for robotic manipulators with matched and mismatched uncertainties. The controller design is done by implementing a super-twisting algorithm in order to obtain the control law and then the adaptive laws are obtained by a Lyapunov stability approach. It is important to consider that some conditions are taken into account for the matched and mismatched uncertainties, conditions that in this case is by assuming decreasing uncertainties, so in this way the conditions that the norm of the matched and mismatched uncertainties are upper bounded is overcame. Finally, the theoretical results obtained in this study are validated by a numerical experiment in which a five-bar linkage mechanism is used as an example and then the results obtained in this experiment are compared with the results obtained by (Zhao et al. 2018; Mujumdar et al. 2014).

2   Problem Formulation

Consider the following dynamical model of a robotic manipulator in an Euler-Lagrange form:

$$D(q(t))\ddot{q}(t) + C(q(t), \dot{q}(t))\dot{q}(t) + g(q(t)) = \tau(t) \tag{1}$$

where $q(t) \in \mathbb{R}^n$ is the position vector, $\tau(t) \in \mathbb{R}^n$ is the torque vector, $D(q(t)) \in \mathbb{R}^{n \times n}$ is the inertia matrix, $C(q(t), \dot{q}(t)) \in \mathbb{R}^{n \times n}$ is the Coriolis matrix and $g(q(t)) \in \mathbb{R}^n$ is the gravity vector. By making the following change of variables $x_1(t) = q(t)$, $x_2(t) = \dot{q}(t)$ and $X(t) = [x_1^T(t), x_2^T(t)]^T$, the following uncertain system is obtained:

$$\dot{X}(t) = F(X(t)) + G(X(t))[U + \Xi(X, t)] + P(X, t) \tag{2}$$

where

$$F(X(t)) = \begin{bmatrix} x_2(t) \\ -D^{-1}(x_1(t))C(x_1(t), x_2(t))x_2(t) - D^{-1}(x_1(t))g(x_1(t)) \end{bmatrix}$$

$$G(X(t)) = \begin{bmatrix} 0_{n \times n} & 0_{n \times n} \\ 0_{n \times n} & D^{-1}(x_1(t)) \end{bmatrix}$$

$$U = \begin{bmatrix} 0_n \\ \tau \end{bmatrix} \tag{3}$$

and $\Xi(X, t)$, $P(X, t)$ are the matched and mismatched uncertainties that are defined in Sect. 3. The control and adaptive laws of the system are obtained in Sect. 3 by considering some suitable conditions to overcome these issues.

3   Adaptive Higher Order Controller Design

In this section, an adaptive higher order controller for robotic manipulators is obtained by considering the matched and mismatched uncertainties in the input and system parameters respectively. It is important to note that these kinds of disturbance have not been studied well in the past for the design of sliding mode controllers for robotic manipulators. As explained before, due to the necessity of accuracy in trajectory tracking is important to take into account some issues related to input uncertainties and unmodeled dynamics that are not considered in previous studies. In this section, the control law is obtained by a super-twisting algorithm selecting an appropriate control law and then the adaptive gains are obtained by a Lyapunov stability approach dealing with the matched and mismatched system uncertainties. Before deriving the control and adaptive laws, we consider the following definition related to the matched and mismatched uncertainties.

Definition 1. *The uncertainties $\Xi(X,t)$ and $P(X,t)$ are class $\mathcal{L}$ functions, i.e. they are decreasing functions for which their norms meet the following conditions $lim_{t\to\infty}\|\Xi(X,t)\| = 0$ and $lim_{t\to\infty}\|P(X,t)\| = 0$.* $\qquad\square$

Before deriving the higher order sliding mode control, the error dynamics of the system and its derivative are given by the following equations:

$$e(t) = X(t) - X_d(t)$$
$$\dot{e}(t) = \dot{X}(t) - \dot{X}_d(t) \tag{4}$$

where $X_d(t)$ is the desired trajectory vector of the robotic manipulator.

The higher order sliding mode controller is obtained by using the super-twisting algorithm (Zhao et al. 2018; Mujumdar et al. 2014) as follows:

$$\dot{S}(t) = -K_1\|S(t)\|^{1/2}\mathrm{sign}(S(t)) + Z(t)$$
$$\dot{Z}(t) = -K_2\mathrm{sign}(S(t)) \tag{5}$$

where $K_1, K_2 \in \mathbb{R}$ are the sliding mode gains that will be adjusted by the adaptive algorithm.

The sliding variable $S(t)$ and its derivative $\dot{S}(t)$ are given by the following equations:

$$S(t) = Ke(t)$$
$$\dot{S}(t) = K\dot{e}(t) \tag{6}$$

where $K \in \mathbb{R}^{2n \times 2n}$ is the sliding variable gain matrix.

The error dynamics shown in (4) is defined as:

$$\dot{e}(t) = F(X(t)) + G(X(t))[U + \Xi(X,t)] + P(X,t) - \dot{X}_d(t) \tag{7}$$

Substituting (7) in (6) and then in (5), the following sliding mode control law is obtained:

$$U = -G^{-1}(X(t))F(X(t)) + G^{-1}(X(t))\dot{X}_d(t)$$
$$-G^{-1}(X(t))K^{-1}K_1\|S\|^{1/2}\,\mathrm{sign}(S(t)) \tag{8}$$

Substituting (8) in (7), the following sliding mode dynamics is obtained:

$$\dot{e}(t) = -K^{-1}K_1\|S(t)\|^{1/2}\,\mathrm{sign}(S(t)) + G(X(t))\Xi(X,t) + P(X,t)$$
$$\dot{Z}(t) = -K_2\,\mathrm{sign}(S(t)) \tag{9}$$

The following theorem is needed to find the adaptive laws for the adjustable gains K_1 and K_2 by selecting an appropriate Lyapunov functional and implementing the control law shown in (8).

Theorem 1. *The uncertain system shown in (1) is stabilized by the super-twisting sliding mode control law (8) and the following adaptive laws:*

$$\dot{K}_1(t) = \frac{1}{\alpha} e^T(t) K^{-1} \|S(t)\|^{1/2} sign(S(t))$$
$$\dot{K}_2(t) = \frac{1}{\beta} Z^T(t) sign(S(t)) \tag{10}$$

Proof 1. Consider the following Lyapunov function:

$$V(e, Z, K_1, K_2) = \frac{1}{2} e^T(t) e(t) + \frac{1}{2} Z^T(t) Z(t) + \frac{\beta}{2} K_2^2(t) + \frac{\alpha}{2} K_1^2 \tag{11}$$

where $\alpha, \beta \in \mathbb{R}^+$ are constants selected by the designer.

By finding the derivative of the Lyapunov function (11), the following result is obtained:

$$\dot{V}(e, Z, K_1, K_2) = e^T(t) \dot{e}(t) + Z^T(t) \dot{Z}(t) + \beta K_2(t) \dot{K}_2(t) + \alpha K_1(t) \dot{K}_1(t) \tag{12}$$

Using (9), we obtain

$$\begin{aligned} \dot{V}(e, Z, K_1, K_2) = &-e^T(t) K^{-1} K_1(t) \|S(t)\|^{1/2} \, sign(S(t)) + e^T(t) G(X(t)) \Xi(X, t) \\ &+e^T(t) P(X, t) - Z^T(t) K_2(t) \, sign(S(t)) + \beta K_2(t) \dot{K}_2(t) \\ &+\alpha K_1(t) \dot{K}_1 \end{aligned} \tag{13}$$

Thus, we select the adaptive laws as follows:

$$\dot{K}_1(t) = \frac{1}{\alpha} e^T(t) K^{-1} \|S(t)\|^{1/2} \, sign(S(t))$$
$$\dot{K}_2(t) = \frac{1}{\beta} Z^T(t) \, sign(S(t)) \tag{14}$$

By substituting (14) in (13), we obtain

$$\dot{V}(e, Z, K_1, K_2) = e^T(t) G(X, t) \Xi(X, t) + e^T(t) P(X, t) \tag{15}$$

Now taking the norm of (15) and by implementing Definition 1, the following conclusion is obtained

$$\dot{V}(e, Z, K_1, K_2) \leq \|e^T(t)\| \|G(X(t))\| \|\Xi(X, t)\| + \|e^T(t)\| \|P(X, t)\| \leq 0 \tag{16}$$

This completes the proof. □

Theorem 1 ensures that the closed-loop system drives the error variables $e(t)$ to zero while adjusting the respective parameters K_1 and K_2 implementing the control law obtained by the super-twisting algorithm shown in (8).

4   Numerical Experiment and Discussion

In this numerical example, the control strategy described in Sect. 3 is validated for trajectory tracking purposes of a five bar linkage mechanism (Spong et al. 2006). The parameters of this mechanism are $m_1 = m_2 = m_3 = m_4 = 0.5\ Kgs$, $l_1 = l_3 = 0.4\ m$, $l_2 = l_4 = 0.8\ m$, $l_{c1} = l_{c3} = 0.2\ m$, $l_{c2} = l_{c4} = 0.4\ m$, $I_1 = I_3 = 0.3\ kg.m^2$, $I_2 = I_4 = 0.4\ kg.m^2$ and $g = 9.81\ m/s^2$ while the mechanism is depicted in Fig. 1. The inertia matrix and the gravity vector components are shown below (Spong et al. 2006):

$$d_{11}(q(t)) = m_1 l_{c1}^2 + m_3 l_{c3}^2 + m_4 l_1^2 + I_1 + I_3$$
$$d_{12}(q(t)) = d_{21}(q(t)) = (m_3 l_2 l_{c3} - m_4 l_1 l_{c4})cos(q_2(t) - q_1(t))$$
$$d_{22}(q(t)) = m_2 l_{c2}^2 + m_3 l_2^2 + m_4 l_{c4}^2 + I_2 + I_4 \tag{17}$$

$$g_1(q(t)) = gcos(q_1(t))(m_1 l_{c1} + m_3 l_{c3} + m_4 l_1)$$
$$g_2(q(t)) = gcos(q_2(t))(m_2 l_{c2} + m_3 l_2 - m_4 l_{c4}) \tag{18}$$

It is important to note that the Coriolis matrix is zero for this experiment. The matched and mismatched uncertainties used in this experiment are given as follows:

$$\Xi(X,t) = 0.1e^{-10t}X(t)$$
$$P(X,t) = 0.1e^{-5t}X(t) \tag{19}$$

The controller parameters are $\alpha = \beta = 0.5$ and the gain matrix K is given below:

$$K = \begin{bmatrix} 10000 & 0 & 0 & 0 \\ 0 & 10000 & 0 & 0 \\ 0 & 0 & 10000 & 0 \\ 0 & 0 & 0 & 10000 \end{bmatrix} \tag{20}$$

The system initial conditions are $X(0) = [0,0,0,0]^T$. The reference variables $X_d(t)$ are shown below.

$$x_{d1}(t) = \begin{bmatrix} 0.1sin(t) \\ 0.1sin(t) \end{bmatrix}$$
$$x_{d2}(t) = \dot{x}_{d1}(t) \tag{21}$$

In Fig. 2, the trajectory of the variable X_{11} is shown. It is noticed that the trajectory is tracked accurately by the action of the proposed control strategy in comparison with the strategies shown in (Mujumdar et al. 2014; Zhao et al. 2018).

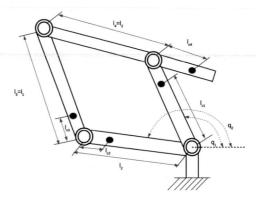

Fig. 1. Five bar linkage mechanism

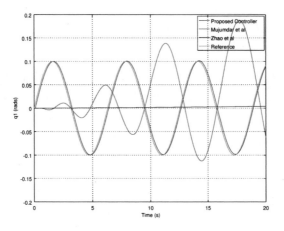

Fig. 2. Variable x_{11} trajectory

It can be also noticed in Fig. 3 that the error for the variable e_1 approaches zero. In Fig. 4, the input variable τ_1 is shown where it can be noticed that the control effort yielded by the action of the proposed controller is smaller than the control effort yielded by the approaches shown in (Mujumdar et al. 2014; Zhao et al. 2018).

The theoretical results obtained in this study prove that even when matched and mismatched uncertainties are found in robotic manipulators or another kind of mechanism, this issue can be overcome by an effective adaptive higher order sliding mode controller. It is important to notice that the chattering effect is reduced to zero by the proposed strategy and this is achieved by the adaptive gains that make the sliding variables to reach the desired value.

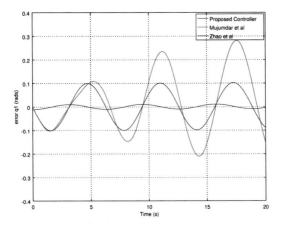

Fig. 3. Error variable e_1

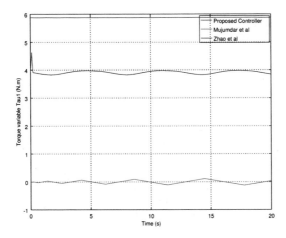

Fig. 4. Input variable τ_1

5 Conclusion and Future Work

In this paper, a higher order sliding mode controller for robotic manipulators and mechanisms is proposed. A novel feature of our work is that the matched and mismatched uncertainties are considered when they are found in the input and the system parameters respectively. To solve these problems, some conditions for these uncertainties were considered, and by using a Lyapunov stability approach and implementing a super-twisting algorithm. In the future, this research study will guide to design and analyze different kinds of sliding mode controllers that deal with other kinds of uncertainties.

References

Adrot, O., Shariari, K., Flaus, J.-M.: Estimation of bounded model uncertainties. IFAC Proc. **37**(15), 391–396 (2004)

Azar, A., Serrano, F.: Robust control for asynchronous switched nonlinear systems with time varying delays. Adv. Intell. Syst. Comput. **533**, 891–899 (2017)

Azar, A.T., Radwan, A.G., Vaidyanathan, S.: Fractional Order Systems. Academic Press, London (2018a)

Azar, A.T., Radwan, A.G., Vaidyanathan, S.: Mathematical Techniques of Fractional Order Systems. Academic Press, London (2018b)

Azar, A.T., Vaidyanathan, S.: Chaos Modeling and Control Systems Design. Springer, Heidelberg (2015)

Azar, A.T., Vaidyanathan, S.: Advances in Chaos Theory and Intelligent Control. Springer, Heidelberg (2016)

Azar, A.T., Vaidyanathan, S.: Advances in System Dynamics and Control. IGI Global, Hershey (2018)

Azar, A.T., Zhu, Q.: Advances and Applications in Sliding Mode Control. Springer, Heidelberg (2015)

Barth, A., Reger, J., Moreno, J.A.: Indirect adaptive control for higher order sliding mode. IFAC-PapersOnLine **51**(13), 591–596 (2018)

Battilotti, S.: Stabilization of nonlinear systems with norm bounded uncertainties. IFAC Proc. **29**(1), 2002–2007 (1996)

Gutierrez, I., Hernandez-Martinez, E., Oropeza, A., Keshtkar, S.: High-order sliding mode control for solar tracker manipulator. Mech. Mach. Sci. **54**, 235–243 (2018)

Han, Y., Liu, X.: Continuous higher-order sliding mode control with time-varying gain for a class of uncertain nonlinear systems. ISA Trans. **62**, 193–201 (2016a)

Han, Y., Liu, X.: Higher-order sliding mode control for trajectory tracking of air cushion vehicle. Optik - Int. J. Light Electron Opt. **127**(5), 2878–2886 (2016b)

Khan, Q., Bhatti, A.I., Ahmed, Q.: Dynamic integral sliding mode control of nonlinear siso systems with states dependent matched and mismatched uncertainties. IFAC Proc. **44**(1), 3932–3937 (2011)

Köroglu, H., Scherer, C.W.: Robust stability analysis for structured uncertainties with bounded variation rates. IFAC Proc. **38**(1), 179–184 (2005)

Mobayen, S., Tchier, F.: Robust global second-order sliding mode control with adaptive parameter-tuning law for perturbed dynamical systems. Trans. Inst. Meas. Contr. **40**(9), 2855–2867 (2018)

Mohamed, G., Sofiane, A., Nicolas, L.: Adaptive super twisting extended state observer based sliding mode control for diesel engine air path subject to matched and unmatched disturbance. Math. Comput. Simul. **151**, 111–130 (2018)

Mondal, S., Mahanta, C.: Nonlinear sliding surface based second order sliding mode controller for uncertain linear systems. Commun. Nonlinear Sci. Numer. Simul. **16**(9), 3760–3769 (2011)

Mondal, S., Mahanta, C.: Chattering free adaptive multivariable sliding mode controller for systems with matched and mismatched uncertainty. ISA Trans. **52**(3), 335–341 (2013)

Mujumdar, A., Kurode, S., Tamhane, B.: Control of two link flexible manipulator using higher order sliding modes and disturbance estimation. IFAC Proc. **47**(1), 95–102 (2014)

Spong, M., Hutchinson, S., Vidyasagar, M.: Robot Modeling and Control. Wiley, Hoboken (2006)

Tannuri, E.A., Agostinho, A.C.: Higher order sliding mode control applied to dynamic positioning systems. IFAC Proc. **43**(20), 132–137 (2010)

Thomas, M., Kamal, S., Bandyopadhyay, B., Vachhani, L.: Continuous higher order sliding mode control for a class of uncertain mimo nonlinear systems: an iss approach. Eur. J. Control **41**, 1–7 (2018)

Tiwari, P.M., Janardhanan, S., un Nabi, M.: Attitude control using higher order sliding mode. Aerosp. Sci. Technol. **54**, 108–113 (2016)

Zhao, Y., Huang, P., Zhang, F.: Dynamic modeling and super-twisting sliding mode control for tethered space robot. Acta Astronautica **143**, 310–321 (2018)

Robust Path Tracking of Mobile Robot Using Fractional Order PID Controller

Hossam Hassan Ammar[1] and Ahmad Taher Azar[1,2(✉)]

[1] School of Engineering and Applied Sciences, Nile University,
Sheikh Zayed District, Juhayna Square, 6th of October City, Giza 12588, Egypt
hhassan@nu.edu.eg, ahmad_t_azar@ieee.org
[2] Faculty of Computers and Information, Benha University, Benha, Egypt

Abstract. This paper represents the control of the Pioneer-3 Mobile Robot as a complex non-linear system which provides an object for research nonlinear system kinematics and dynamics analysis. In this paper, the system modeling and simulation is divided into two main parts. The first part is the modeling and simulation using MATLAB and the second part is the whole mechanical design and its characteristics as a function of the motor speed and the torque depending on the system using Virtual Robot Environment Program (V-REP). The study uses Proportional–Integral–Derivative (PID) and Fractional Order PID (FOPID) controllers to obtain a robust controller for the system. The linear velocity loop controls the robot wheels speeds using the motor speed feedback signal from the encoder. The angular velocity control loop keeps the robot always in the accepted angle boundary using a six-degree of freedom gyroscope and accelerometer as a feedback signal. A state space model is obtained considering some assumptions and simplifications. This paper also studies and compares the results of two controllers PID and FOPID controllers from analysis perspectives with different optimization methods. The results demonstrated that the FOPID controller is superior in performance to the traditional PID controller.

Keywords: Pioneer-3 Mobile Robot · PID · Fractional Order PID ·
Mobile robot modeling and simulation · Controller tuning and optimization ·
Path tracking · V-REP

1 Introduction and Problem Statement

Fractional calculus extends the ordinary calculus by extending the ordinary differential equations to fractional order differential equations, i.e. those having non-integer orders of derivatives and integrals (Podlubny 1999). Recently, there has been an explosion of interest in designing and using fractional-order controllers in different practical applications (Azar et al. 2018a, b; Li et al. 2017; Das et al. 2012). Podlubny (1994) introduced fractional order PID controller with integrator order λ and differentiator order μ which showed that the performance of Fractional dynamics performance becomes better with FOPID. The application fields of fractional calculus are increasingly widening, including areas such as electrical engineering, automation and control engineering, robotics, biomedical engineering and recently the renewable energy domain introduced (Azar et al. 2017; Azar et al. 2018c, d; Kumar et al. 2018).

© Springer Nature Switzerland AG 2020
A. E. Hassanien et al. (Eds.): AMLTA 2019, AISC 921, pp. 370–381, 2020.
https://doi.org/10.1007/978-3-030-14118-9_37

The main motivation for this research in fractional order operators and systems are their good performances, hereditary properties, and the recent advances in computer science and numerical tools. Because there are existing problems in the path-tracking and motion control; therefore, different trajectory tracking control approaches for (WMR) have been proposed so as to achieve the best performance for the wheeled mobile robot including high speed, high tracking accuracy (minimized tracking error), low energy consumption and smoothness of velocity control signal obtained (Kim and Oh 1999; Ammar et al. 2018; Azar and Serrano 2018). This research introduced the dynamic and kinematic model of Pioneer-3 Mobile Robot in an environment with path planning and path tracking methodology to calculate the distance and the angle. The mathematical model has to be covered in order to meet the destination. This paper presents an effective and more important controller based on Fractional Order PID (FOPID) controller which is used for controlling the wheeled mobile robot. Different tuning algorithms are used for numerical calculation of optimal FOPID controller gains which is used to adjust the linear velocity and the angular velocity for the wheeled mobile robot and give better performance. This manuscript is organized as follows: In Sect. 2 Modeling and Simulation of Pioneer-3 Mobile Robot is introduced. In Sect. 3, control of Pioneer-3 Mobile Robot is described. In Sect. 4, Simulation results and analysis are attained. Finally, in Sect. 5 concluding remarks with future directions are given.

2 Modeling and Simulation of Pioneer-3 Mobile Robot

Modeling and simulation of the mobile is very important section to study the behavior of the robot and compare the model of the robot with the real robot (Azar et al. 2018c, d). Figure 1 shows Pioneer-3 Mobile Robot model. In order to model the robot on MATLAB, the mathematical model that describe the robot kinematics and dynamics should be obtained as following: The kinematic model of the robot with differential chassis can be written in matrix form presented in Spyros (2013):

$$\begin{pmatrix} x^{\cdot} \\ y^{\cdot} \\ \phi^{\cdot} \end{pmatrix} = \begin{pmatrix} Cos(\phi) & 0 \\ Sin(\phi) & 0 \\ 0 & 1 \end{pmatrix} \begin{pmatrix} v \\ \omega \end{pmatrix} \tag{1}$$

where v is the overall linear velocity of robot, ω is the overall angular velocity of robot, x, y are position coordinates in global coordinate system and ϕ is orientation angle of the mobile robot.

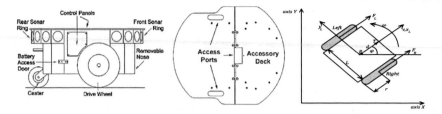

Fig. 1. Pioneer-3 Mobile Robot model

For implementation and control reasons, it is often required to control the robot by wheel's linear velocities v_R, v_L instead of the linear and angular velocity of the robot v, ω. The transformation between velocities

$$\begin{pmatrix} v_R \\ v_L \end{pmatrix} = \begin{pmatrix} 1 & \frac{L}{2} \\ 1 & \frac{-L}{2} \end{pmatrix} \begin{pmatrix} v \\ \omega \end{pmatrix} \quad \text{and} \quad \begin{pmatrix} v \\ \omega \end{pmatrix} = \begin{pmatrix} \frac{r}{2} & \frac{r}{2} \\ \frac{r}{L} & \frac{-r}{L} \end{pmatrix} \begin{pmatrix} \omega_R \\ \omega_L \end{pmatrix} \qquad (2)$$

The dynamic model which describes the mass m and moment of inertia J for each mathematical model for the mobile robot dynamic part can be described in state space. Where the state variables for the robot is defined as the manipulated inputs variable $U = (\omega_R \ \omega_L)$ and the output variable as $y = (v \ \phi)$. Then:

$$\dot{X} = AX + BU, \quad y = CX \qquad (3)$$

$$where \quad A = \begin{pmatrix} a_1 & 0 & 0 \\ 0 & 0 & 1 \\ 0 & 0 & a_2 \end{pmatrix} \quad B = \begin{pmatrix} b_1 & b_1 \\ 0 & 0 \\ b_2 & -b_2 \end{pmatrix} \quad C = \begin{pmatrix} 1 & 0 & 0 \\ 0 & 1 & 0 \end{pmatrix}$$

$$a_1 = \frac{-2C}{(Mr^2 + 2I_w)}, \quad a_2 = \frac{-2CI^2}{(I_v r^2 + 2I_w I^2)}, \quad b_1 = \frac{Kr}{(Mr^2 + 2I_w)}, \quad b_2 = \frac{-KrI}{(I_v r^2 + 2I_w I^2)}$$

The following Eq. (4) represents the relation between the input torques to the robot (ω_R and ω_L) and the output of the controller (μ_ω and μ_v):

$$\omega_R = \mu_v + \mu_\omega \quad \text{and} \quad \omega_L = \mu_v - \mu_\omega \qquad (4)$$

Where μ_v is the torque required for controlling the velocity of robot and μ_ω is the torque required for controlling the robot's azimuth.

3   Control of Pioneer-3 Mobile Robot

Applying the controller to the system ensures the best performance and increases the robustness of the system as shown in Fig. 2. Some different controllers have been designed and implemented starting from classical controllers like PI and PID developed by Azar et al. (2018d). The Fractional Order PID controller to visualize and evaluate the performance of the system is developed in this study.

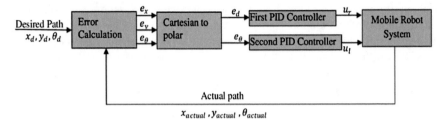

Fig. 2. Pioneer-3 Mobile Robot closed loop block diagram

3.1 Motion Planning and Path Tracking

Many methods can be implemented to navigate and control the motion planning of the robot form point to point where any path is a sequence of points (Spyros 2013). The configuration of the motion can be represented by $q = (x, y, \theta)$ where x, y is the coordinate of the robot in 2-D plane and θ is the direction angle measuring from the center of the robot to the destination point with respect to the x axis and target position $(X_t, Y_t,$ and $\beta)$ as shown in Fig. 3.

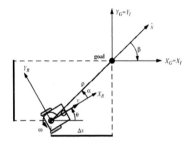

Fig. 3. Pioneer-3 Mobile Robot position control schematic

Point to Point motion planning method starts with maintain the distance error (Eq. 5) and angle error (Eq. 7).

$$D_{Error} = \sqrt{((X_2 - X_1)^2 + (Y_2 - Y_1)^2)}, \quad \beta = \tan^{-1}\frac{\Delta y}{\Delta x} \tag{5}$$

If $x_{ref} < x$ and $y_{ref} \geq 0$ then $\delta = \beta + \pi$
If $x_{ref} < x$ and $y_{ref} < 0$ then $\delta = \beta - \pi$
If $x_{ref} \geq x$ then $\delta = \beta$
and the orientation angle error α is calculated as:

$$\alpha_{error} = \delta - \theta \tag{6}$$

Figure 4 illustrate the closed loop control system of differential drive motion planning method of the mobile robot navigation which consists of two control loops. The inner control loop focuses on the speed of each wheel and the outer loop is responsible for the total motion of the robot and give the setpoints of wheels speed to the inner loop using MATLAB and V-REP (Ciszewski et al. 2017; Coppelia Robotics 2018) as shown in Fig. 5.

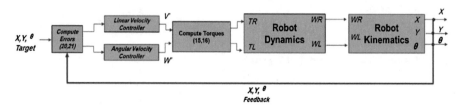

Fig. 4. Detailed path tracking closed loop control block diagram

The final transfer functions of the differential drive dynamical model between control signal (Input control signals $\mu_v(s)$ and $\mu_\omega(s)$) and the dynamical outputs (Linear Velocity V(s) and angular Velocity $\omega(s)$) are:

$$\frac{v(s)}{\mu_v(s)} = \frac{1.25}{s^2 + 1.48s + 3.86} e^{-0.15}, \quad \frac{\omega(s)}{\mu_\omega(s)} = \frac{1.28}{s^2 + 0.52s + 2.47} e^{-0.15} \quad (7)$$

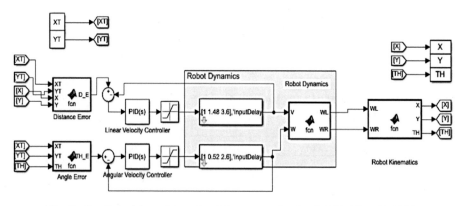

Fig. 5. Detailed path tracking closed loop control using MATLAB and V-REP

3.2 Classical PID Controller (One Degree and Two Degree of Freedom)

The controllers are configured to improve the guidance of the robot in case of angular velocity error between the direction of the robot and the goal direction while keeping the linear velocity error as constant and vice versa for the second controller. The mathematical model of one degree of freedom (1-DOF) PID controller design is given by:

$$c(t) = K_p e(t) + K_p K_i \int e(t)dt + K_p K_d \frac{de(t)}{dt} \tag{8}$$

The 2-DOF PID Feedforward type is used as shown in Fig. 6 where P(s) is the controlled process transfer function, $C_f(s)$ is the feedforward compensator transfer function, C(s) is the serial (or main) compensator transfer function, r is the set-point, d the load-disturbance, and y the controlled variable. In this case, C(s) and $C_f(s)$ are given by (Araki and Taguchi 2003):

$$C(s) = K_p \left\{ 1 + \frac{1}{T_i s} + T_D D(s) \right\}, \quad C_f(s) = -K_p \{ \alpha + \beta T_D D(s) \}, \quad D(s) = \frac{s}{1 + \tau s} \tag{9}$$

where α and β are controller weighting parameters ranging from 0 to 1, D(s) is the approximate derivative filter term and $\tau = T_d/K_d$

$$C(s) = K_p \{ 1 + \frac{1}{T_1 s} + T_D D(s) \}$$
$$C_f(s) = -K_P \{ \alpha + \beta T_D D(s) \} \tag{10}$$

Where the degree of freedom for any closed loop control system is the number of controller transfer functions adjusted independently.

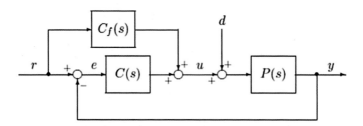

Fig. 6. Two-DOF PID Feedforward type (Araki and Taguchi 2003)

3.3 Fractional Order PID Controller (FOPID)

Podlubny (1994) introduced fractional order PID (FOPID) controller in 1994. Figure 7 illustrates the block diagram which represent the FOPID control structure.

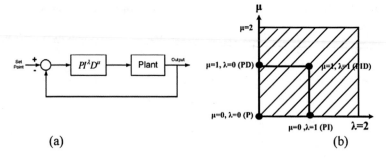

Fig. 7. (a) Fractional Order PID controller block diagram (b) Fractional PID controller converge

FOPID controller provides extra degree of freedom for not only the need of design controller gains (K_p, K_i, K_d) but also design orders of integral λ and derivative μ. The orders of integral and derivative are not necessarily integer but any real numbers. The transfer function of FOPID controller:

$$C_{FOPID}(s) = K_P + \frac{K_I}{s^{\lambda}} + K_D s^{\mu} \tag{11}$$

It is clear from Fig. 7(b) that by selecting $\lambda = 1$ and $\mu = 1$, a classical PID controller can be recovered. Using $\lambda = 1$, $\mu = 0$, and $\lambda = 0$, $\mu = 1$, corresponds to the conventional PI & PD controllers, respectively. All these classical types of PID controllers are special cases of the FOPID controller.

3.4   Controller Tuning Criteria

For designing controllers based on time domain, controllers aim to minimize different integral performance indices namely:

$$\text{Integral square error}\quad ISE = \int_0^{\infty} e^2 dt \tag{12}$$

$$\text{Integral absolute error}\quad IAE = \int_0^{\infty} |e| dt \tag{13}$$

$$\text{Integral time} - \text{square error}\quad ITSE = \int_0^{\infty} t^2 |e| dt \tag{14}$$

$$\text{Integral time} - \text{absolute error}\quad ITAE = \int_0^{\infty} t |e| dt \tag{15}$$

The tuning gains of PID and FOPID for linear and angular velocities using different tuning methods are summarized in Table 1. Where (Kp.1, Kd.1......μ.1) represent the linear velocity controller (Controller 1) tuning parameters and (Kp.2, Kd.2.....μ.2) represent the angular velocity controller (Controller 2) using different tuning methods.

4 Simulation Results and Analysis

Simulated results of the proposed strategy are presented on this section. A Classical PID and FOPID controller have been implemented as positioning controller (liner and angular velocities) with different tuning for the mobile robot into two stages:

Table 1. PID and FOPID controller gains using different tuning methods

Type of controller	Tuning criteria	Kp.1	Kp.2	Ki.1	Ki.2	Kd.1	Kd.2	λ.1	λ.2	μ.1	μ.2
PID	ISE	1.451	1.308	3.964	2.821	6.189	5.651	1	1	1	1
FOPID	ISE	2.325	2.105	8.621	7.361	2.961	1.860	0.801	0.903	0.897	0.754
PID	IAE	1.892	2.107	2.584	1.987	5.892	4.671	1	1	1	1
FOPID	IAE	3.325	3.605	9.621	6.361	3.961	2.860	0.781	0.852	0.922	0.678
PID	ITSE	1.652	1.498	4.014	3.621	5.729	5.781	1	1	1	1
FOPID	ITSE	4.325	2.901	8.534	7.578	3.022	2.587	0.907	0.789	0.789	0.891
PID	ITAE	2.583	1.889	4.164	3.021	6.029	5.754	1	1	1	1
FOPID	ITAE	4.325	2.105	8.621	7.361	2.961	1.860	0.801	0.903	0.897	0.754

Stage 1: Robot Dynamical Control and Simulation
The robot dynamic simulation results have been derived experimentally for a mobile robot using MATLAB based on the system transfer functions of linear and angular velocities introduced in Sect. 3.1 as shown in Figs. 8 and 9.

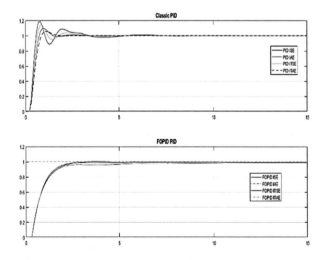

Fig. 8. Linear velocity control results using PID and FOPID

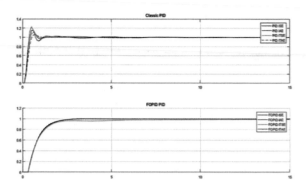

Fig. 9. Angular velocity control results using PID and FOPID

Figure 8 shows the closed loop step response of the robot linear velocity using both PID and FOPID with the different tuning methods illustrated in Table 1 with zero input angular velocity. Figure 9 also shows the closed loop step response of robot angular velocity using both PID and FOPID with the different tuning methods illustrated in Table 1 with zero input linear velocity. The results showed that in case of applying classical PID with linear and angular velocities, the overshot is larger than applying FOPID. Applying FOPID overcomes the overshot but gives slower response. Table 2 shows the performance results of the closed loop with input step using classical PID and FOPID with the different tuning criteria at the same conditions. For classical PID it's noted that tuning parameters using ITAE gave the best results of the perfomance. Tuning using ISE achieved the best results of the performance for FOPID tunning.

Table 2. Closed position control results using PID and FOPID

Type of controller	Tuning criteria	Rise time (sec)	Peak time (sec)	Overshoot (%)	Setting time (sec)
PID	ISE	0.9541	2.6863	24.4907	5.9068
FOPID	ISE	0.6512	0.6863	0	1.9068
PID	IAE	0.8546	2.6863	14.2905	4.8025
FOPID	IAE	0.6543	1.6863	0	2.0024
PID	ITSE	0.5548	2.6863	12.3302	3.9241
FOPID	ITSE	1.1522	0.9863	0	1.2056
PID	ITAE	0.6833	2.6863	9.5489	3.4512
FOPID	ITAE	0.7541	0.6863	0	2.8601

Stage 2: Path Planning Control and Simulation
Simulation and control results of the complete system introduced in Sect. 3.1 are divided into different tasks of the robot using PID-ITAE parameters and FOPID-ISE parameters as these controllers achieved the best performance results in stage 1.

Task 1: **Linear Trajectory**

This task is to test the mobile robot for tracking linear path with desired velocity = 0.5 [m/sec] and desired angle = 45°. The simulation result of the linear trajectory is shown in Fig. 10.

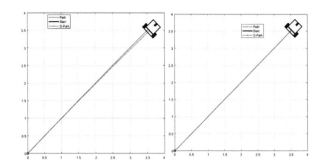

Fig. 10. Linear path tracking control results using PID and FOPID

Task 2: **Go-to-Goal**

Move from one point to point task has been simulated in order to test the performance of the controllers for the robot motion. PID-ITAE and FOPID-ISE have been implemented and simulated using MATLAB and V-REP to move from point [0, 0] to point [5, 5] as shown in Fig. 11.

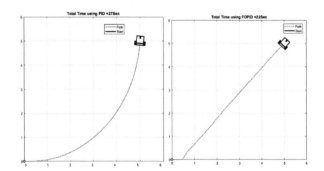

Fig. 11. Closed go-to-goal control results using MATLAB and V-REP for PID and FOPID

Task 3: **Infinity Path Tracking**

The infinity path has been used as the following:

$$Scale = \frac{20}{3 - Cos(2t)}, \quad X(t) = Scale \times Cos(t), \quad Y(t) = \frac{1}{2}Scale \times Sin(2t) \quad (16)$$

Where t = 0:0.1:7 s. PID-ITAE and FOPID-ISE which achieved the best performance, results have been implanted as shown in Fig. 12. The results demonstrated that FOPID is faster and more robust than PID.

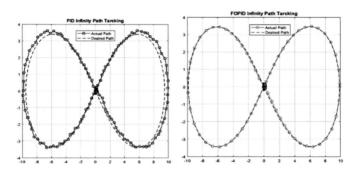

Fig. 12. Infinity path tracking results using MATLAB and V-REP for PID and FOPID

5   Conclusion and Future Work

In this paper, a model of the Pioneer-3 Mobile Robot is proposed as a complex nonlinear system to study the nonlinear system kinematics and dynamics analysis. This study compares the results of two controllers PID and FOPID controllers. The result analysis showed that FOPID controller works better as it has five different parameters to be tuned which is not possible in the case of classical PID controller. For higher order systems, the performance of classical PID controller deteriorates, whereas fractional PID controller provides better results. For a system with long time delay, a fractional PID controller provides better results than classical PID controller.

References

Ammar, H.H., Azar, A.T., Tembi, T.D., Tony, K., Sosa, A.: Design and implementation of Fuzzy PID controller into multi agent smart library system prototype. In: International Conference on Advanced Machine Learning Technologies and Applications, pp. 127–137. Springer, Cham (2018)

Araki, M., Taguchi, H.: Two-degree-of-freedom PID controllers. Int. J. Control Autom. Syst. **1** (4), 401–411 (2003)

Azar, A.T., Kumar, J., Kumar, V., Rana, K.P.S.: Control of a two link planar electrically-driven rigid robotic manipulator using fractional order SOFC. In: International Conference on Advanced Intelligent Systems and Informatics, pp. 57–68. Springer, Cham (2017)

Azar, A.T., Radwan, A.G., Vaidyanathan, S.: Mathematical Techniques of Fractional Order Systems. Elsevier, Amsterdam (2018a). ISBN 9780128135921

Azar, A.T., Radwan, A.G., Vaidyanathan, S.: Fractional Order Systems: Optimization, Control, Circuit Realizations and Applications. Elsevier, Amsterdam (2018b). ISBN 9780128161524

Azar, A.T., Ammar, H.H., Mliki, H.: Fuzzy logic controller with color vision system tracking for mobile manipulator robot. In: International Conference on Advanced Machine Learning Technologies and Applications, pp. 138–146. Springer, Cham (2018c)

Azar, A.T., Ammar, H.H., Barakat, M.H., Saleh, M.A., Abdelwahed, M.A.: Self-balancing robot modeling and control using two degree of freedom PID controller. In: International Conference on Advanced Intelligent Systems and Informatics, pp. 64–76. Springer, Cham (2018d)

Azar, A.T., Serrano, F.E.: Fractional order sliding mode PID controller/observer for continuous nonlinear switched systems with PSO parameter tuning. In: International Conference on Advanced Machine Learning Technologies and Applications, pp. 13–22. Springer, Cham (2018)

Ciszewski, M., Mitka, Ł., Buratowski, T., Giergiel, M.: Modeling and simulation of a tracked mobile inspection robot in Matlab and V-REP software. J. Autom. Mob. Robot. Intell. Syst. **11**(2), 5–11 (2017)

Coppelia Robotics (2018). http://www.coppeliarobotics.com/index.html. Accessed 04 Oct 2018

Das, S., Pan, I., Das, S.: Gupta A (2012) Improved model reduction and tuning of fractional-order $PI^\lambda D^\mu$ controllers for analytical rule extraction with genetic programming. ISA Trans. **51**(2), 237–261 (2012)

Kim, D.H., Oh, J.H.: Tracking control of a two-wheeled mobile robot using input-output linearization. Control Eng. Pract. **7**(3), 369–374 (1999)

Kumar, J., Azar, A.T., Kumar, V., Rana, K.P.S.: Design of fractional order fuzzy sliding mode controller for nonlinear complex systems. In: Mathematical Techniques of Fractional Order Systems, pp. 249–282 (2018)

Li, C., Zhang, N., Lai, X., Zhou, J., Xu, Y.: Design of a fractional-order PID controller for a pumped storage unit using a gravitational search algorithm based on the Cauchy and Gaussian mutation. Inf. Sci. **396**, 162–181 (2017)

Podlubny, I.: Fractional-order systems and fractional-order controllers. Institute of Experimental Physics, Slovak Academy of Sciences, Kosice (1994)

Podlubny, I.: Fractional Differential Equations. Academic Press, San Diego (1999)

Spyros, G.T.: Introduction to mobile robot control. Elsevier, Amsterdam (2013). ISBN 978-0124170490

Synchronization between a Novel Integer-Order Hyperchaotic System and a Fractional-Order Hyperchaotic System Using Tracking Control

Ayub Khan[1], Shikha Singh[2], and Ahmad Taher Azar[3,4(✉)]

[1] Department of Mathematics, Jamia Millia Islamia, New Delhi, India
akhan012@jmi.ac.in
[2] Department of Mathematics, Jesus and Mary College, University of Delhi,
New Delhi, India
sshikha7014@gmail.com
[3] Faculty of Computers and Information, Benha University, Banha, Egypt
ahmad_t_azar@ieee.org, ahmad.azar@fci.bu.edu.eg
[4] School of Engineering and Applied Sciences, Nile University Campus,
6th of October City, Giza, Egypt

Abstract. This manuscript investigates the synchronization between a novel integer order hyperchaotic system and a fractional order hyperchaotic system. The controllers are constructed using the technique of tracking controller and the stability theory of the linear fractional order system. Chaotic analysis of the introduced novel integer order hyperchaotic system is also investigated. The Lyapunov exponent, bifurcation diagram, Poincare section, Kaplan-Yorke dimension, equilibria and phase portraits are given to justify the chaotic nature of the system. Theoretical results are supported with the numerical simulations.

Keywords: Novel integer-order hyperchaotic system ·
Chaotic analysis · Fractional-order hyperchaotic system ·
Synchronization · Stability theory · Tracking control

1 Introduction

In the most recent decades researchers from all fields of natural sciences have studied phenomena that include nonlinear systems exhibiting chaotic behavior [2]. This is because of the fact that nonlinear systems demonstrate rich dynamics and have sensitive dependence on initial conditions. Chaotic systems are third order or higher order nonlinear differential equations with at least one positive lyapunov exponent. Hyperchaos is characterized as a chaotic system with more than one positive Lyapunov exponent [16], this infer that its dynamics are expanded in several different directions simultaneously. Thus, hyperchaotic systems have more complex dynamical behaviors than ordinary chaotic systems. Analysis of the dynamics of hyperchaotic systems received a great deal of

© Springer Nature Switzerland AG 2020
A. E. Hassanien et al. (Eds.): AMLTA 2019, AISC 921, pp. 382–391, 2020.
https://doi.org/10.1007/978-3-030-14118-9_38

interest in the recent past due to its applications in secure communication, information processing, biological systems, chemical reactions, neural network and other fields. So, more and more chaotic or hyperchaotic systems showing wide dynamical behavior were found [10,11]. In this manuscript, we constructed a novel 4D continuous autonomous hyperchaotic system having three equilibrium points by adding a feedback controller to the 3D chaotic system [13]. Further, some basic dynamical properties including Lyapunov exponent, bifurcations, Poincare section, Kaplan-Yorke dimension and phase portraits are illustrated by both theoretical analysis and computer simulations. The theory of fractional calculus is a 300-year-old topic which can trace back to Leibniz, Riemann, Liouville, Grünwald, and Letnikov [15]. However, the fractional calculus did not attract much attention for a long time. Nowadays, the past three decades have witnessed significant progress on fractional calculus, because the applications of fractional calculus were found in more and more scientific fields, covering mechanics, physics, engineering, informatics, and materials. Fractional differential equations have successfully modelled phenomena related to diffusion, visco-elastic systems, signal processing, dielectric polarization and so on [15]. Recently, the research on fractional-order chaotic systems has become an important issue [1]. However very few research work is done based on synchronization between the fractional-order chaotic system and the chaotic system of integer order. Looking for better strategies for chaos control and synchronization different types of methods have been developed for controlling chaos and synchronization of non identical and identical systems for instance adaptive control [5,6], active control [9], sliding mode control [7,8,12] etc. Recently, synchronization between integer-order and fractional-order chaotic systems began to attract much attention in recent years, where different methods and techniques have been proposed. For example, general control schemes have been described in [3].

This manuscript is organized as follows: In Sect. 1 introductory definitions of fractional derivatives are given. In Sect. 2 system description and chaotic analysis of the novel integer order hyperchaotic system is given. Also in this section system description of the fractional order hyperchaotic system is given. In Sect. 3 synchronization between a novel integer-order hyperchaotic system and a fractional-order hyperchaotic system using tracking control is performed. Finally in Sect. 4 concluding remarks are given.

2   System Description and Chaotic Analysis

2.1   Chaotic Analysis of a Novel Integer-Order Chaotic System

A novel continuous autonomous integer order hyperchaotic system formed by adding a linear feedback controller to the Lorenz chaotic system satisfying the two necessary condition [16] of hyperchaotic behavior is given as

$$\begin{cases} \dot{u}_1 = a(u_2 - u_1) \\ \dot{u}_2 = cu_1 - u_2 - u_1 u_3 - u_4 \\ \dot{u}_3 = -bu_3 + u_1 u_2 \\ \dot{u}_4 = ku_4 + u_2 \end{cases} \tag{1}$$

where $(u_1,\ u_2,\ u_3,\ u_4) \in \mathbb{R}^4$ are the state variables and $a,\ b,\ c,\ k \in \mathbb{R}$ are parameters.

2.1.1   Chaotic Attractor of the System

When the parameters $a = 10$, $b = 2$, $c = 28$, $k = 0.1$ and initial condition $(u_1,\ u_2,\ u_3,\ u_4) = (1,\ 2,\ 3,\ 4)$ are chosen then the system displays chaotic attractor as shown in Fig. 1. The corresponding Lyapunov exponents of the novel hyperchaotic attractor are $\gamma_1 = 0.59528$, $\gamma_2 = 0.17969$, $\gamma_3 = 0.018209$, $\gamma_4 = -13.6906$. The Kaplan-Yorke dimension is defined by

$$D = j + \sum_{i=1}^{j} \frac{\gamma_i}{|\gamma_{j+1}|}$$
$$= 3.0579$$

where j is the largest integer satisfying $\sum_{i=1}^{j} \gamma_j \geq 0$ and $\sum_{i=1}^{j+1} \gamma_j < 0$. Therefore Kaplan-Yorke dimension of the chaotic attractor is $D = 3.0579$ which means that the Lyapunov dimension of the chaotic attractor is fractional.

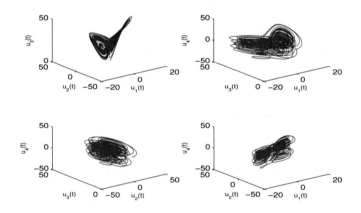

Fig. 1. 3D phase portrait of the 4D novel hyperchaotic system

2.1.2 Dissipation and Existence of Chaotic Attractor
The divergence of the novel integer order 4D hyperchaotic system (1) is

$$\nabla.V = \frac{\partial \dot{u}_1}{\partial u_1} + \frac{\partial \dot{u}_2}{\partial u_2} + \frac{\partial \dot{u}_3}{\partial u_3} + \frac{\partial \dot{u}_4}{\partial u_4}$$
$$= -a - 1 - b + k$$
$$= -12.9 < 0$$

So, the system (1) is dissipative system and the system converges by the index rate of $e^{-12.9}$. It means that a volume element V_0 is contracted into $V_0 e^{-12.9t}$ as time t. That is, each volume containing the system orbit shrinks to zero as $t \rightarrow 0$, at an exponential rate ∇V, which is independent of u_1, u_2, u_3, u_4. Consequently, all system orbits will ultimately be confined to a specific subset of zero volume and the asymptotic motion settles onto an attractor. Then the existence of attractor is proved.

2.1.3 Symmetry and Invariance
The novel hyperchaotic system (1) is invariant under the transformation $(u_1, u_2, u_3, u_4) \rightarrow (-u_1, -u_2, u_3, -u_4)$ this means the system (1) is symmetric about u_3-axis.

2.1.4 Poincare Map and Bifurcation Diagram
As an important analysis technique, the Poincare map reflects the periodic, quasi-periodic, chaotic and hyper-chaotic behavior of the system. When $a = 10$, $b = 2$, $c = 28$ and $k = 0.1$ one may take $u_1 = 0$ as the crossing section which shows the hyperchaotic behavior as shown in Fig. 2(a).

The bifurcation diagram of $|u_1|$ with respect to parameters c is shown in Fig. 2(b) which shows abundant and complex dynamical behaviors.

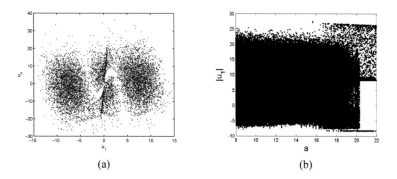

(a)   (b)

Fig. 2. (a) Poincare map on the crossing section $u_2 = 0$. (b) Bifurcation diagram of $|u_1|$ versus a

2.1.5   Equilibria and Stability

The equilibria of system (1) for the parameters $a = 10$, $b = 2$, $c = 28$, $k = 0.1$ is obtained by solving the following equations

$$
\begin{cases}
a(u_2 - u_1) = 0 \\
cu_1 - u_2 - u_1 u_3 - u_4 = 0 \\
-bu_3 + u_1 u_2 = 0 \\
ku_4 + u_2 = 0
\end{cases}
\tag{2}
$$

The system has three equilibrium points $E_0 = (0,0,0)$, $E_1 = (-86.0232, 8.6023, 8.6232, 37)$, $E_2 = (86.0232, -8.6023, -8.6232, 37)$.

Proposition 1. The equilibrium point E_0 is unstable.

Proof. The Jacobian matrix of system (1) at the equilibrium point E_0 is given by

$$
Jac = \begin{bmatrix}
-10 & 10 & 0 & 0 \\
28 & -1 & -0 & -1 \\
0 & 0 & -2 & 0 \\
0 & 1 & 0 & 0.1
\end{bmatrix}
\tag{3}
$$

The eigenvalues of the Jacobian Jac are $\lambda_1 = -22.8116$, $\lambda_2 = 11.7738$, $\lambda_3 = 0.1378$ and $\lambda_4 = -2$. This implies that the equilibrium point E_0 is unstable. This proof is completed.

Remark. In the same way, we can prove that E_1 and E_2 are also unstable equilibrium points.

2.2   Fractional-Order Hyperchaotic System

The fractional order hyperchaotic system [4] is given by

$$
\begin{cases}
\dfrac{d^q v_1}{dt^q} = -a_1 v_1 + a_2 v_2 \\
\dfrac{d^q v_2}{dt^q} = a_3 v_1 - v_1 v_3 - v_2 + v_4 \\
\dfrac{d^q v_3}{dt^q} = v_1^2 - a_4(v_1 + v_3) \\
\dfrac{d^q v_4}{dt^q} = -a_5 v_1
\end{cases}
\tag{4}
$$

where v_1, v_2, v_3, v_4 are the state variables, $q > 0$ is the fractional order and a_1, a_2, a_3, a_4, a_5 are the parameters.

In Fig. 3 3D phase portraits are displayed, it is clear that the system has a double-scroll hyperchaotic attractor.

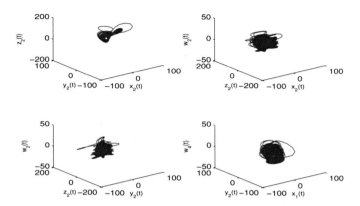

Fig. 3. 3D phase portrait of the fractional order hyperchaotic system

3 Synchronization between a Novel Integer-Order Hyperchaotic System and a Fractional-Order Hyperchaotic System Using Tracking Control

In this section, we perform the synchronization between the novel integer order hyperchaotic system and a fractional order hyperchaotic system using tracking control. We consider the novel integer order hyperchaotic system as the master system which is described in (1). Fractional order hyperchaotic system is considered as slave system which is described in (4). Now adding controller to the slave system we get

$$
\begin{bmatrix} \frac{d^q v_1}{dt^q} \\ \frac{d^q v_2}{dt^q} \\ \frac{d^q v_3}{dt^q} \\ \frac{d^q v_4}{dt^q} \end{bmatrix} = \begin{bmatrix} -a_1 v_1 + a_2 v_2 \\ a_3 v_1 - v_1 v_3 - v_2 + v_4 \\ v_1^2 - a_4(v_1 + v_3) \\ -a_5 v_1 \end{bmatrix} + \theta(u) + \Gamma(v, u) \tag{5}
$$

where $\theta(u(t))$ is the compensation controller and $\Gamma(v(t), u(t))$ is feedback controller. According to the methodology of tracking controller, the compensation controller is given by

$$
\theta(u(t)) = \frac{d^q u(t)}{dt^q} - f(u(t))
$$

where

$$
f(u(t)) = \begin{bmatrix} -a_1 u_1 + a_2 u_2 \\ a_3 u_1 - u_1 u_3 - u_2 + u_4 \\ u_1^2 - a_4(u_1 + u_3) \\ -a_5 u_1 \end{bmatrix}
$$

So, the controlled fractional order hyperchaotic system can be rewritten as

$$
\begin{bmatrix} \frac{d^q v_1}{dt^q} \\ \frac{d^q v_2}{dt^q} \\ \frac{d^q v_3}{dt^q} \\ \frac{d^q v_4}{dt^q} \end{bmatrix} = \begin{bmatrix} -a_1 v_1 + a_2 v_2 \\ a_3 v_1 - v_1 v_3 - v_2 + v_4 \\ v_1^2 - a_4(v_1 + v_3) \\ -a_5 v_1 \end{bmatrix} + \frac{d^q u(t)}{dt^q}
$$
$$
- \begin{bmatrix} -a_1 u_1 + a_2 u_2 \\ a_3 u_1 - u_1 u_3 - u_2 + u_4 \\ u_1^2 - a_4(u_1 + u_3) \\ -a_5 u_1 \end{bmatrix} + \Gamma(v(t), u(t)) \tag{6}
$$

The synchronization error $e \in \mathbb{R}^4$ for synchronization is defined as

$$
e_i = v_i - u_i, \quad where \quad i = 1, 2, 3, 4 \tag{7}
$$

The corresponding error dynamics is obtained as follows :

$$
\begin{bmatrix} \frac{d^q e_1}{dt^q} \\ \frac{d^q e_2}{dt^q} \\ \frac{d^q e_3}{dt^q} \\ \frac{d^q e_4}{dt^q} \end{bmatrix} = \begin{bmatrix} -a_1 e_1 + a_2 e_2 \\ a_3 e_1 - e_1 e_3 - e_1 u_3 - e_3 u_1 - e_2 + e_4 \\ e_1^2 - 2 e_1 u_1 - a_4(e_1 + e_3) \\ -a_5 e_1 \end{bmatrix} + \Gamma(v(t), u(t)) \tag{8}
$$

Now, the aim is find the suitable controller $\Gamma(v(t), u(t))$ in such a way that the error e tends to zero i.e. $\|e\| = \lim_{t\to\infty} \|v(t) - u(t)\| = 0$, so that the required synchronization takes place. Let $e(t) = \begin{bmatrix} e_1(t) \\ e_2(t) \end{bmatrix}$ where $e_1(t) = e_1$, $e_2(t) = (e_2, e_3, e_4)^T$. Then, the Eq. (8) can be rewritten as

$$
\begin{bmatrix} \frac{d^q e_1}{dt^q} \\ \frac{d^q e_2}{dt^q} \\ \frac{d^q e_3}{dt^q} \\ \frac{d^q e_4}{dt^q} \end{bmatrix} = \begin{bmatrix} B_1 e_1(t) + F_1(u(t), e_1(t), e_2(t)) \\ B_2 e_2(t) + F_{21}(u(t), e_1(t), e_2(t)) + F_{22}(u(t), e_2(t)) \end{bmatrix} \tag{9}
$$
$$
+ \Gamma(v(t), u(t))
$$

where $B_1 = -a_1$, $B_2 = \begin{bmatrix} -1 & 0 & 0 \\ 0 & -a_4 & 0 \\ 0 & 0 & 0 \end{bmatrix}$, $F_1(u(t), e_1(t), e_2(t)) = a_2 e_2$,

$$
F_{21}(u(t), e_1(t), e_2(t)) = \begin{bmatrix} a_3 e_1 - e_1 e_3 - e_1 u_3 \\ e_1^2 - 2 e_1 u_1 - a_4 e_1 \\ -a_5 e_1 \end{bmatrix}, F_{22}(u(t), e_2(t)) = \begin{bmatrix} -e_3 u_1 + e_4 \\ -a_4 e_3 \\ 0 \end{bmatrix}.
$$

It is easy to see that

$$
\lim_{e_1(t)\to 0} F_{21}(u(t), e_1(t), e_2(t)) = \lim_{e_1(t)\to 0} \begin{bmatrix} a_3 e_1 - e_1 e_3 - e_1 u_3 \\ e_1^2 - 2 e_1 u_1 - a_4 e_1 \\ -a_5 e_1 \end{bmatrix}
$$
$$
= 0
$$

For simplicity, we can rewrite (9) in the following form

$$\begin{bmatrix} \frac{d^q e_1}{dt^q} \\ \frac{d^q e_2}{dt^q} \end{bmatrix} = \begin{bmatrix} B_1 e_1(t) + F_1(u(t), e_1(t), e_2(t)) \\ B_2 e_2(t) + F_{22}(u(t), e_2(t)) \end{bmatrix} + \Gamma(v(t), u(t)) \qquad (10)$$

Designing the controller $\Gamma(v(t), u(t))$ as follows :

$$\Gamma(v(t), u(t)) = \begin{bmatrix} A_1 e_1(t) - F_1 \\ A_2 e_2(t) - F_{22} \end{bmatrix}$$

where A_1 and A_2 are matrices to be determined. Thus, we obtain

$$\begin{cases} \dfrac{d^q e_1}{dt^q} = (-a + A_1) e_1(t) \\[2mm] \dfrac{d^q e_2}{dt^q} = \left(\begin{bmatrix} -1 & 0 & 0 \\ 0 & -a_4 & 0 \\ 0 & 0 & 0 \end{bmatrix} + A_2 \right) e_2(t) + F_{21}(u(t), e_1(t), e_2(t)) \end{cases} \qquad (11)$$

The following Lemma and Theorem is used to carry out the numerical simulations.

Lemma [14]. The following autonomous system

$$D^q U = AU, U(0) = U_0$$

with $0 < q < 1$, $U \in \mathbb{R}^n, A \in \mathbb{R}^{n \times n}$, is asymptotically stable if and only if $|arg(\lambda)| > \frac{q\pi}{2}$ is satisfied for all eigenvalues (λ) of matrix A. Also, this system is stable if and only if $|arg(\lambda)| \geq \frac{q\pi}{2}$ is satisfied for all eigenvalues of matrix A and those critical eigenvalues which satisfy $|arg(\lambda)| > \frac{q\pi}{2}$ have geometric multiplicity one.

Theorem. If we choose appropriate matrices $A_1 \in \mathbb{R}$, $A_2 \in \mathbb{R}^{2 \times 2}$, so that $A_1 + a < 0$ and all eigenvalues of matrix

$$\left(\begin{bmatrix} -1 & 0 & 0 \\ 0 & -a_4 & 0 \\ 0 & 0 & 0 \end{bmatrix} + A_2 \right)$$ satisfies Lemma, then $\lim_{t \to \infty} e_i = \lim_{t \to \infty}(v_i(t) - u_i(t)) = 0$ $(i = 1, 2, 3, 4)$ is satisfied which means that master system (1) and slave system (4) achieve the required synchronization.

3.1   Numerical Simulations

Numerical simulation is performed to illustrate the validity and feasibility of the presented synchronization technique. The parameter are chosen so that the systems behave chaotically without the controller. The matrices A_1 and A_2 are chosen to be $A_1 = 9$, $A_2 = \begin{pmatrix} -1 & 0 & 0 \\ 0 & 3 & 0 \\ 0 & 0 & -3 \end{pmatrix}$. The convergence of error state variables in Fig. 4 shows that the synchronization between novel integer-order hyperchaotic system (1) and fractional-order hyperchaotic systems (4) is achieved when controllers are activated at $t > 0$.

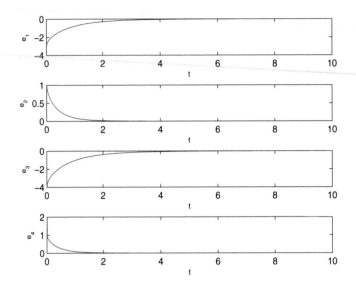

Fig. 4. Error dynamics among the drives and the response system with controllers activated for $t > 0$

4   Conclusion

In this manuscript, we designed a suitable controller by using stability criteria of the fractional-order system, based on the tracking control, so that the synchronization between novel integer-order hyperchaotic system and fractional-order hyperchaotic system is achieved. The analytical properties of novel systems are analyzed both theoretically and numerically. Finally, numerical simulations are displayed to show the viability of the methodology.

References

1. Azar, A.T., Ouannas, A., Singh, S.: Control of new type of fractional chaos synchronization. In: International Conference on Advanced Intelligent Systems and Informatics, pp. 47–56. Springer, Cham (2017)
2. Chen, G., Dong, X.: From chaos to order–perspectives and methodologies in controlling chaotic nonlinear dynamical systems. Int. J. Bifurcat. Chaos **3**(06), 1363–1409 (1993)
3. Gang-Quan, S., Zhi-Yong, S., Yan-Bin, Z.: A general method for synchronizing an integer-order chaotic system and a fractional-order chaotic system. Chin. Phys. B **20**(8), 080505 (2011)
4. Gao, Y., Liang, C., Wu, Q., Yuan, H.: A new fractional-order hyperchaotic system and its modified projective synchronization. Chaos Solitons Fractals **76**, 190–204 (2015)
5. Khan, A., Pal, R.: Adaptive hybrid function projective synchronization of chaotic space-tether system. Nonlinear Dyn. Syst. Theory **14**(1), 44–57 (2014)

6. Khan, A., Shikha, : Hybrid function projective synchronization of chaotic systems via adaptive control. Int. J. Dyn. Control **5**, 1–8 (2016)
7. Khan, A., Shikha, : Combination synchronization of Genesio time delay chaotic system via robust adaptive sliding mode control. Int. J. Dyn. Control **6**, 1–10 (2017a)
8. Khan, A., Shikha, : Combination synchronization of time-delay chaotic system via robust adaptive sliding mode control. Pramana **88**(6), 91 (2017b)
9. Khan, A., Shikha, : Increased and reduced order synchronisations between 5D and 6D hyperchaotic systems. Indian J. Ind. Appl. Math. **8**(1), 118–131 (2017c)
10. Khan, A., Shikha: Dynamical behavior and reduced-order combination synchronization of a novel chaotic system. Int. J. Dyn. Control **6**, 1–15 (n.d.)
11. Khan, A., Singh, S.: Chaotic analysis and combination-combination synchronization of a novel hyperchaotic system without any equilibria. Chin. J. Phys. **56**, 238–251 (2017)
12. Khan, A., Singh, S.: Generalization of combination-combination synchronization of n-dimensional time-delay chaotic system via robust adaptive sliding mode control. Math. Methods Appl. Sci. **41**(9), 3356–3369 (2018)
13. Lorenz, E.N.: Deterministic nonperiodic flow. J. Atmos. Sci. **20**(2), 130–141 (1963)
14. Matignon, D.: Stability results for fractional differential equations with applications to control processing. In: Computational Engineering in Systems Applications, vol. 2, pp. 963–968. IMACS, IEEE-SMC Lille, France (1996)
15. Podlubny, I.: Fractional Differential Equations: An Introduction to Fractional Derivatives, Fractional Differential Equations, to Methods of Their Solution and Some of Their Applications, vol. 198. Academic press, Cambridge (1998)
16. Rossler, O.: An equation for hyperchaos. Phys. Lett. A **71**(2–3), 155–157 (1979)

Design of Air-Cooled Control System
for Intelligent Transformer

Dantian Zhong[1]([⊠]), Qiang Gao[1], Jiayu Pan[2], Zhannan Guo[2],
and Maojun Wang[2]

[1] State Grid Chaoyang Electric Power Supply Company,
Chaoyang 122000, China
zhongdantian@dbdky.com
[2] State Grid Liaoning Electric Power Research Institute,
Shenyang 110006, China

Abstract. According to the functional requirements of smart substation, the problem of the traditional air-cooled control system using the electromagnets and programmable logic controller (PLC) is analyzed. The basic framework of intelligent transformer cooling control system is put forward, and based on advanced electronic technology the intelligent monitoring device is developed in this paper. The system could monitor the status of transformer cooling system, and upload the information to the integrated supervision and control system of smart substation. In addition, it can control the fan to make the transformer within a suitable temperature range. Generally the system could improve the efficiency of operation and maintenance, and ensure the safety and stable operation of the transformer.

Keywords: Air-cooled control system · Programmable logic controller (PLC) ·
Intelligent · Transformer

1 Introduction

At present, while the climatic conditions of the northeast region, load conditions of grid and the structure of the transformer are into account, in 220 kV substation the cooling method of 240/180 MVA transformers is almost using oil nature air forced (ONAF) mode. With the development of intelligent substation, intelligent transformer is emerged, but the air-cooled control system of intelligent transformer is still using electromagnetic or a programmable logic controller (PLC) control mode. These two control methods can only provide local control with low accuracy and remote control with private communications protocol, but there is not any monitoring function for the entire control system. The air-cooled control system like this does not meet the requirements of intelligent transformer, and it cannot be integrated into the integrated supervision and control system in the intelligent substation.

© Springer Nature Switzerland AG 2020
A. E. Hassanien et al. (Eds.): AMLTA 2019, AISC 921, pp. 392–399, 2020.
https://doi.org/10.1007/978-3-030-14118-9_39

2 Traditional Air-Cooled Control System

As shown in Fig. 1, the traditional air-cooled control system generally consists of the following components: oil temperature indicator, winding temperature indicator, current relay and electrical control of four parts. Current relay and electrical control are put in the air-cooled control cabinet which is near the transformer. Oil temperature indicator and winding temperature indicator are put on tank wall or placed in the terminal box. Manual and automatic control mode for air-cooled control system could be changed with switch in the cabinet. When the system is in manual mode, operating personnel can stop any group fan. When the system is in automatic mode, the control system will be running according to the oil temperature, winding temperature and load condition. When the temperature or load is too high up to the threshold, the fan will begin to run automatically. The basic data such as oil temperature, winding temperature and load is from the oil temperature indicator, winding temperature indicator and current relay, but the accuracy of the senses is too low, especially winding temperature. Winding temperature indicator consists of convert, heating element and elastic element. The convert is used to transfer the current signal which is proportional to load current to the heating element that is in the elastic element. When the temperature is changed, an additional displacement will be generated within the elastic element, thus produce an indicated value which is higher the oil temperature. Such indirect method is used in winding temperature indicator, so an average value of winding temperature could be taken. This control system mode is simple, run inflexible, and accuracy is poor [1, 2].

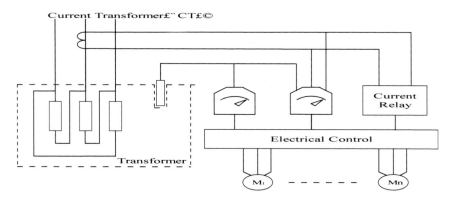

Fig. 1. Traditional air-cooled control system block diagram

In recent years, in order to adapt to the substation automation, many of the traditional air-cooled control system adds PLC module that provides remote control and alarm functions on the basis of electrical control part. When the air-cooled control system is running in automatic control mode, load current signal and contact signal of oil temperature and winding are collected by PLC module [3–5]. The PLC model would send start-stop control signal to fans according to the above signal. Remote alarm signal will be sent through contact when faults appear in the system. The contact also provides the control function. The biggest drawback of this control system is that IEC 61850 protocol is not supported, and cannot access to the integrated monitoring system of intelligent substation. In this situation when fans go wrong, fault details cannot be provided, it may delay troubleshooting and apparatus.

3   Design Scheme of Air-Cooled Control System in Intelligent Substation

With the construction of a new generation of intelligent substation, a higher requirement to air-cooled control system of intelligent transformer is put forward. The design scheme of air-cooled control system in intelligent substation is based on data sharing in the IEC 61850 standard. In the scheme the cooling control secondary circuit is optimized. The air-cooled control system of intelligent transformer is realized in the scheme, and the portion of the measurement and monitoring functions are taken into account. Further expansion the transformer monitoring system has been kept beforehand [6, 7].

The transformer status and control information transfer and sharing are fully taken into account in the design process of intelligent transformer air-cooled control system. While retaining all the functions of original PLC, the scheme of air-cooled control system adds the features of the new remote control and monitoring capabilities to adapt demand intelligent transformer. As shown in Fig. 2, in addition to oil temperature indicator, winding temperature indicator and current relay, the system optimized control secondary circuit, and added the transformer intelligent monitoring device. The more devices and unified electrical components are placed in intelligent component cabinet near the transformer.

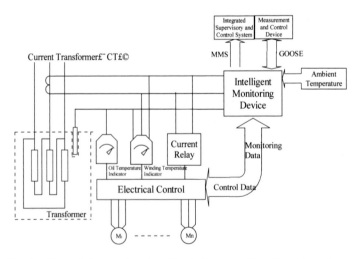

Fig. 2. Air-cooled control system block diagram of intelligent transformer

As shown in Fig. 3, intelligent component cabinet that is made of stainless steel uses before and after double-door form. On the left side of cabinet 19-in. rack structure is used. The intelligent monitoring device which is the core device in the air-cooled control system is put in the 19-in. rack. Furthermore other intelligent devices of transformer monitoring system could be put in the cabinet to extend functions such as neutral point current merging unit (MU), intelligent terminal and so on. Air-cooled electrical control part is on the right of cabinet. Some monitoring function including monitoring current of fans, status of power, cooling system operating status is added in the system. These monitoring values can still be displayed through the transparent window, which is in the front of cabinet to improve the control system operation and maintenance efficiency.

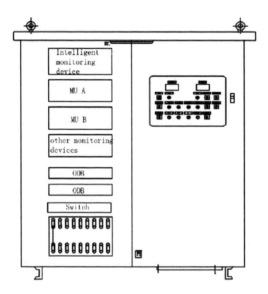

Fig. 3. Intelligent component cabinet schematic

Air-cooled control system of intelligent transformer is functionally divided into measurement, monitoring and control. Measurement and monitoring functions are mainly the fan motor current, the main power circuit voltage, the top oil temperature, load and other parameters monitoring, and can be integrated into monitoring platform of the integrated supervisory and control system; control functions can be converted by the two switches S1 and S2 selection, S1: local and remote (smart), S2: manual and automatic.

When the switch S1 is local place, the control mode is consistent with the traditional way. When the switch S1 is in remote (Smart) mode, the air-cooled control system will be controlled by the intelligent monitoring device, and can be also controlled by monitoring and control platform. The oil temperature, ambient temperature, load and other information will be collected by intelligent monitoring device. On the basis of these data the winding hot spot and cooling efficiency are estimated according to the transformer thermal load model (GB/T 1094.7-2008) [8, 9]. The cooler fan could be controlled through monitoring and evaluation results. The control logic is shown in Fig. 4. In addition, taking into account the operation and maintenance of the fan, the fan startup sequence is determined by the length time which is calculated from the transformer run in order to avoid frequent use of the fan, while another part is not used. In this way the life of fans could be extended.

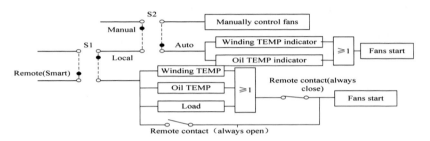

Fig. 4. Control logic diagram

When the system is in remote mode, the control information is sent to monitoring and control device through GOOSE service of IEC 61850. The integrated supervisory and control system in the intelligent substation also can receive the information, then the operating personnel could operate fans through the interface which is integrated into monitoring computer. But when the operating personnel want to stop fans, the monitoring and control device will subject to comprehensive judgments based on the oil temperature, winding temperature and load. If fans don't stop when operating personnel sent the command to stop fans, the interface will display the message that fans cannot be stopped because of present oil temperature, winding temperature and load. When the system is in remote mode, the monitoring information is sent to the integrated supervisory and control system through MMS service of IEC 61850. In the final all the information about air-cooled control system will be displayed in the integrated supervisory and control system, so remote alarm and fault diagnosis are realized. In addition, the air-cooled control system is also configured independently interface based on B/S structure to convenient operation and maintenance personnel to view the system's operation.

4   Design of Intelligent Monitoring Device

The intelligent monitoring device is the core device of air-cooled control system, the device uses a standard 19″ plug-in box construction, 4U high, with main control board, digital input board, digital output board, 4–20 mA signal acquisition board, temperature acquisition board, AC acquisition board, power board and the expansion board as shown in Fig. 5.

Digital input board, digital output board, 4–20 mA signal acquisition board, temperature acquisition board and AC acquisition board are all called as functional board.

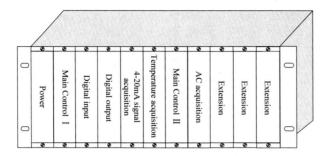

Fig. 5. Intelligent monitoring device

The hardware and software architecture which are based on ARM Cortex-M3 processor core and embedded real-time operating system μcos are all the same.

The hardware structure of function board can be divided into four parts: acquisition and processing module, signal analysis scheduling module, communication module and power module, as shown in Fig. 6. In next content the temperature acquisition board will be an example with detailed description.

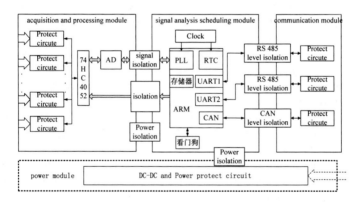

Fig. 6. Hardware block diagram of function board

The oil temperature is collected by PT100 (Resistance Temperature Detector, RTD) which is on the transformer. The temperature acquisition board configured with 6-channel PT100 acquisition interface can collect the amount of 6-channel temperature. This function is performed with four pieces of 74HC4052 (high-speed multi-channel selector). 74HC4052 enables digital control to select different analog channel switch, which has an enable pin. When the pin is on the enabled state, 74HC4052 may be turned in a two channels. Depending on channel selection capability of 74HC4052, analog signal input of AD (analog to digital) can be accessed by one channel at each time so as to prevent mutual interference between channels, but also to meet the multi-channel input function.

AD7793 which is made by ADI Corporation is selected as AD sampling chip. AD7793 includes a highly integrated, high functional diversity and anti-noise ability and other characteristics. It is very suitable for high precision and low power requirements of complex applications. PT100 generally uses three-wire connection. In order to compensate off the errors as a consequence of PT100 RTD lead wire resistance, two-way constant current source is needed to compensate circuit in the data collection process. The common-mode voltage is produced by the voltage between RL1 and RL2 and RL3 which is caused by two equal and constant current sources IOUT1 and IOUT2. This connection mode never affects the difference between the AD sampling, to ensure the accuracy of temperature acquisition. AD7793 exchanges data with ARM through the SPI bus. ARM will convert the voltage signal into temperature value according with PT100, which is needed to look up table to obtain temperature and send the result to main control board through the RS485 bus.

The main control board is the core of the intelligent monitoring device, which could handle large amounts of data. The board is responsible for communication with the integrated supervisory and control system, so ARM Cortex-A8 high speed processing chip and Linux system are used. The main clock of ARM Cortex-A8 is up to 720 MHz with low power consumption, low cost, high reliability and other characteristics. It is easy to expand by embedded industrial control bus interface, the rich resources of the on-chip bus interface (GPMC bus, MMC bus, SPI bus, I2C bus, UART) interfaces, and industrial peripheral resources. The board also has a large-capacity storage and dynamic non-volatile storage device to ensure efficient data processing and persistent. The main control board communicates with digital input board, digital output board, 4–20 mA signal acquisition board, temperature acquisition board and AC acquisition board through RS485 serial bus to form fast arithmetic processing and the overall deployment mechanism, which could meet the requirement of collection and control field signal command submitting. Further, the photoelectric conversion module is arranged to meet IEC 61850 Ethernet interface with the host station or other devices to communicate.

The intelligent monitoring device uses master-slave hot standby dual power design to ensure the stable operation of the internal power supply system. Each individual board is self-designed short-term recovery.

5 Conclusion

The existing form of transformer air-cooled control system cannot meet the need for rapid development of intelligent substation. In the situation, the design scheme of air-cooled control system for intelligent transformer is proposed. This solution retains the full functionality of the original control system. On this basis, the system with intelligent monitoring device as a core further improves the transformer monitoring, control and measurement functions, and integrates monitoring systems and intelligent substation seamless.

References

1. Kang, Z., Zhang, M.: A new design of transformer air-cooled control system. Power Syst. Clean Energy **26**(2), 23–25 (2010). (in Chinese)
2. Shi, Y., Zhai, Y., Yan, X.: Design of intelligent cooling control device for oil-immersed transformer. Transformer **44**(2), 71–73 (2007). (in Chinese)
3. Zhang, W., Gu, B., Gu, K.: Application of air-cooled control system based on PLC in large transformer. Zhejiang Electr. Power **1**(3), 19–21 (2010). (in Chinese)
4. Li, X., Zheng, X.: Application of PLC and frequency conversion technology to forced-air cooling system in large transformer. Transformer **44**(1), 61–64 (2007). (in Chinese)
5. Feng, X.: Design improvement for air cooler of secondary circuit of transformer. Transformer **44**(4), 73–75 (2007). (in Chinese)
6. Wu, Z., Li, Y., Hu, G.: Research on intelligent air-cooled control device used in the digital substation. Electr. Eng. **1**(6), 63–65 (2013)
7. Xian, R.: Strategy of temperature control on large-scale oil-dipped transformer. Distribution & Utilization **29**(2), 60–62 (2012)
8. Chen, Q., Li, H., Zhong, M.: Winding hot-point temperature calculation and analysis of control of transformer operation with winding. Transformer **47**(11), 15–18 (2010). (in Chinese)
9. Susa, D., Nordman, H.: A simple model for calculating transformer hot-spot temperature. IEEE Trans. Power Deliv. **24**(4), 1257–1265 (2009)

Quoting Model Strategy of Thermal Power Plant Considering Marginal Cost

Anlong Su[1(⊠)], Mingyang Zhu[2], Shunjiang Wang[1], Kai Gao[1], Jun Yuan[1], and Zhenjiang Lei[1]

[1] State Grid Liaoning Electric Power Co., Ltd., Shenyang 110000, China
418484168@qq.com
[2] Shenyang Institute of Engineering, Shenyang 110136, China

Abstract. In the bidding market of the electricity market, most thermal power enterprises compete on the basis of marginal cost. In order to meet the needs of reform and development, thermal power generation enterprises must set up a set of practical bidding strategies to maximize the interests of enterprises while tapping potential and increasing efficiency and reducing costs to provide the basis for their competition in the electricity market. This paper first explains the theoretical basis of marginal cost bidding function. Then the mathematical model of the quotation is established. Finally, a concrete algorithm based on the mixed strategy model is given. The optimal value of the expected return of the generator is obtained to form the basis of quotation for the generator.

Keywords: Marginal cost · Quotation function · Mixed strategy · Quotation basis

1 Introduction

Because electricity cannot be stored in large quantities and the particularity of the power industry determines that the electricity market is not a completely competitive market. Under the unified price mechanism, the power producers can obtain excess profits through strategic bidding [1]. Moreover, the strategic bidding of the power generation company always gains more profit than the marginal cost of the unit.

In the actual operation of the power market, the power producers can offer corresponding prices at different output levels when quoted [2]. The competition in the generation side electricity market is similar to supply function competition [3]. At present, China's power generation side electricity market has been established for a period of time, and the historical data are relatively sufficient, which has practical significance for analyzing the bidding strategy of power generation companies [4–6]. In this paper, a hybrid strategy model is proposed to represent the benefits of various possible combinations of power producers, thus providing optimal bidding strategies for power producers [7].

© Springer Nature Switzerland AG 2020
A. E. Hassanien et al. (Eds.): AMLTA 2019, AISC 921, pp. 400–405, 2020.
https://doi.org/10.1007/978-3-030-14118-9_40

2 The Theoretical Basis of the Quotation Function

Under the unified clearing price mechanism, for a power producer, according to its own bid price equal to the winning bid price, lower than the winning bid price or higher than the bid price, there are only three bidding results. However, below the clearance price, the winning bid is equal to the clearing price and the bidding failure. The probability of these three outcomes is very important for the launch of the quotation strategy [8].

2.1 Expected Return

Considering the i of any power plant C_i is used to represent the production cost of a power plant. Its bidding capacity is q_i. Its unit electricity is quoted at p_i (yuan/MWh). The probability that the quotation is successful and as a marginal unit is PR_1. The probability that the quotation is successful but the probability of loading position below the marginal unit is PR_2. At this time, the marginal price of the system is p. PR_1 and PR_2 can be expressed by the function of p_i. When the power supplier i is quoted as p_i the expected return is

$$\pi_i = [(p_i - C_i)PR_1 + (p - C_i)PR_2]q_i \tag{1}$$

2.2 Optimal Quotation

According to the optimal conditions, the optimal bid p_i^* with the largest expected revenue should satisfy the following conditions

$$\left.\frac{\partial \pi_i}{\partial p_i}\right|_{p_i'} = 0 \tag{2}$$

$$\left.\frac{\partial^2 \pi_i}{\partial p_i}\right|_{p_i'} < 0 \tag{3}$$

The optimal price of p_i is p_i^*

$$p_i^* = C_i + \frac{PR_1 + \frac{\partial PR_2}{\partial p_i}(p - C_i)}{-\frac{\partial PR_1}{\partial p_i}} \tag{4}$$

The bidding strategy described in (4) is based on the expected marginal price. A power plant will face great risks when the predicted bid price is very different from the actual bid price.

Obviously, p_i^* is a monotone subtraction function of p. Because p is the market marginal price of a power producer without marginal unit price, it is well known that $p_i < p$. The expected return of the power producer represented by the formula (1) can be written as follows

$$\pi_i = \left[(p_i - C_i)PR_1 + \int_{p_i}^{\infty} (p - C_i)f(p)dp \right] q_i \tag{5}$$

$f(p)$ is the probability density function of p.

Find the derivation of (5) and let $\frac{\partial \pi_i}{\partial p_i} = 0$.

p_i^* can be obtained as follows:

$$p_i^* = C_i + PR_1 / \left(f(p_i) - \frac{\partial PR_1}{\partial p_i} \right) \tag{6}$$

Which is available of

$$\frac{\partial PR_2}{\partial p_i} = -f(p_i) \tag{7}$$

If (7) is brought into (6), the probability of successful bidding is PR. It can be obtained that $PR = PR_1 + PR_2$.

p_i^* can be written as:

$$p_i^* = C_i + PR_1 / \left(-\frac{\partial PR}{\partial p_i} \right) \tag{8}$$

From $PR_1 > 0$ and $\frac{\partial PR}{\partial p_i} < 0$, we can see that under the uniform clearing price mechanism, the optimal quotation of a power plant is based on its cost plus a lifting amount.

3   Mathematical Quotation Model

3.1   Mathematical Model Establishment

In the electricity market where N power producers are involved, each generator has a quadratic cost function: $C_i(q_i) = a_i + b_i q_i + \frac{1}{2} c_i q_i^2$, $(i = 1, \ldots, N)$. And a_i, b_i, c_i are constant, and q_i is the power generation. Its marginal cost is $MC_i(q_i) = \frac{\partial C_i(q_i)}{\partial q_i} = b_i + c_i + q_i$, $(i = 1, \ldots, N)$. And the price function of the power producers can be set to

$$p_i = \alpha_i + \beta_i q_i, \ \alpha_i, \beta_i \geq 0 \tag{9}$$

The price curve of other power producers is p_k.

$$p_k = \alpha_k + \beta_k q_k, \ \alpha_k, \beta_k \geq 0, k \neq i, k = 1, \ldots, N \tag{10}$$

Assume

$$\beta_i = c_i \tag{11}$$

The power producers bid for the trading center. Under the unified clearing price mechanism, the trading center determines whether the generating unit can access the grid and the amount of electricity through the solution (12).

$$p_i = p_k = p \tag{12}$$

In order to solve the problem, this paper regards other power producers other than itself as a whole and records it as j.

$$\sum_{k=1, k \neq i}^{N} q_k = q_j \tag{13}$$

Its cumulative quotation function is

$$p_j = \alpha_j + c_j q_j, \; \alpha_j, c_j \geq 0 \tag{14}$$

3.2   Quotation Model

The market is assumed to be a rigid load market. All generators quoted together to become a cumulative supply function. The intersection point between the supply function and the demand function determines the market clearing price p.

$$Q = \frac{p_i - \alpha_i}{c_i} + \frac{p_j - \alpha_j}{c_j} \tag{15}$$

To bring (12) to (19), we can get the following equation:

$$p = \frac{Q + \frac{\alpha_i}{c_i} + \frac{\alpha_j}{c_j}}{\frac{1}{c_i} + \frac{1}{c_j}} = \frac{Q c_i c_j + \alpha_i c_j + \alpha_j c_i}{c_i + c_j} \tag{16}$$

It can be seen that the market clearing price can be expressed as a function of α_i. The bidding power of the generator is q_i.

$$q_i = \frac{p - \alpha_i}{c_i} = \frac{Q c_j - \alpha_i + \alpha_j}{c_i + c_j} \tag{17}$$

The profit of the generator can be expressed as π_i.

$$\pi_i(\alpha_i) = q_i p - C_i(q_i) = q_i p - \left(\alpha_i + b_i q_i + \frac{1}{2} c_i q_i^2 \right) \tag{18}$$

It can be seen that α_i is a function of the cost of the generator itself and the cost and cost of other generators.

4 Mixed Strategy Model

It is assumed that the power market only contains i and its competitor j, and two power producers have limited generating units. Power generation companies can obtain different cost coefficients b through internal unit commitment. Express these cost factors as the pure strategy of the power producers.

When the power plant i chooses a pure strategy x and power producer j chooses a pure strategy y, the profits they get are $\pi(i)_{xy}$ and $\pi(j)_{xy}$ respectively. Under the mixed strategy, the expected return E_i and E_j of the two generation companies can be expressed as:

$$E_i = \psi_1^T \Omega_i \psi_2 \tag{19}$$

$$E_i = \psi_1^T \Omega_j \psi_2 \tag{20}$$

Suppose the power producer i has two pure strategies, namely S_{a1} and S_{a2}. The probability of i using S_{a1} and S_{a2} is Pr_1 and $1 - \text{Pr}_1$ respectively. The power producer j has two pure strategies, namely S_{b1} and S_{b2}. The probability of j using S_{b1} and S_{b2} is Pr_2 and $1 - \text{Pr}_2$ respectively.

Therefore, the hybrid strategy of the generator i can be expressed as:

$$\psi_1 = \begin{bmatrix} \text{Pr}_1 \\ 1 - \text{Pr}_1 \end{bmatrix} \tag{21}$$

And the hybrid strategy of the generator j can be expressed as:

$$\psi_2 = \begin{bmatrix} \text{Pr}_2 \\ 1 - \text{Pr}_2 \end{bmatrix} \tag{22}$$

Through the comparative analysis of historical data, we can know the probability that the power producers j use various strategies to be relatively accurate.

5 Conclusion

Under the unified clearing price mechanism of the power market, power producers pursue the most profitable bidding strategy in the spot market. This paper derives the theoretical basis of the quotation of power producers. The results show that the optimal bidding strategy is a function of the cost of the power producer itself and the quotations and costs of other power producers. A hybrid strategy model is set up to show the benefits of various possible combinations of power producers in this paper. Finally, an example is given to illustrate the results.

References

1. Li, S.F.: Study on voltage and reactive power control strategy of substation based on nine-domain diagram. Qinghai Electr. Power **2**(5), 1–4+64 (2005)
2. Ren, X., Cheng, H., Liu, J.: Voltage and reactive power optimization control of substation based on taboo search algorithm. Relay **8**, 31–34+39 (2008)
3. Zhang, Y., Ren, Z., Liao, M., Li, F.: Optimal reactive compensation on tower of 10 kV long distribution feeder. Electr. Power **33**(9), 50–52 (2000)
4. Wang, L., Huang, C., Guo, S., Cao, G.: The design of the voltage reactive power comprehensive control device of substation based on fuzzy control theory. Relay **31**(8), 40–42 (2003)
5. Wu, Y., Yang, Q., Yu, Z.: Design for voltage and reactive power automatic control device in substation on DSP. Program. Controll. Fact. Autom. **9**(8), 101–103 (2008)
6. Mohsenian-Rad, A.H., Wong, V., Jatskevich, J., et al.: Optimal and autonomous incentive-based energy consumption scheduling algorithm for smart grid. In: Innovative Smart Grid Technologies (ISGT), pp. 1–6 (2010)
7. Lujano-Rojas, J.M., Monteiro, C., Dufo-Lopez, R.A.: Optimum residential load management strategy for real time pricing (RTP) demand response programs. Energ. Policy **45**(2), 671–679 (2012)
8. Gottwalt, S., Ketter, W., Block, C., et al.: Demand side management—a simulation of household behavior under variable prices. Energ. Policy **39**(12), 8163–8174 (2011)

Combination-Combination Anti-Synchronization of Four Fractional Order Identical Hyperchaotic Systems

Ayub Khan[1], Shikha Singh[2], and Ahmad Taher Azar[3,4($\boxtimes$)]

[1] Department of Mathematics, Jamia Millia Islamia, New Delhi, India
akhan012@jmi.ac.in
[2] Department of Mathematics, Jesus and Mary College,
University of Delhi, New Delhi, India
sshikha7014@gmail.com
[3] Faculty of Computers and Information, Benha University, Benha, Egypt
ahmad_t_azar@ieee.org, ahmad.azar@fci.bu.edu.eg
[4] School of Engineering and Applied Sciences, Nile University Campus,
6th of October City, Giza, Egypt

Abstract. In this manuscript, we investigate the methodology of combination-combination anti-synchronization of four identical fractional order hyperchaotic system. The methodology is implemented by considering a 4D fractional order hyperchaotic system. The controllers are constructed using adaptive control technique to ensure the combination-combination anti - synchronization. The synchronization schemes such as chaos control problem, projective anti-synchronization, combination anti-synchronization becomes the special cases of combination-combination anti-synchronization. The combination - combination scheme can additionally enhances the security of transmission of message signals. The theoretical results and numerical simulations are given to justify the validity and feasibility of the proposed control technique.

Keywords: Fractional-order hyperchaotic system ·
Combination-combination anti-synchronization · Stability theory ·
Active control

1 Introduction

The fractional dynamics is a established scientific idea. The study of fractional dynamics is a 300-year-old point which can follow back to Leibniz, Riemann, Liouville, Grünwald, and Letnikov [15]. However, the fractional calculus did not draw in much consideration for quite a while. It has attracted more researchers interest and has more broad application prospects due to its unique advantages. These days, the previous three decades have seen noteworthy advance on fractional calculus, in light of the fact that the utilizations of fractional calculus were found in an ever increasing number of scientific fields covering mechanics, material science, engineering, informatics and materials. Fractional differential equations have effectively displayed wonders identified with diffusion, visco-elastic

© Springer Nature Switzerland AG 2020
A. E. Hassanien et al. (Eds.): AMLTA 2019, AISC 921, pp. 406–414, 2020.
https://doi.org/10.1007/978-3-030-14118-9_41

systems, signal processing, dielectric polarization and so on [9]. More cases for fractional-order dynamical systems can be found in [1]. Besides, uses of fractional calculus have been accounted for in numerous territories, for example, image processing, signal processing and automatic control [16]. These and numerous other comparative illustrations splendidly elucidate the significance of thought and examination of fractional-order dynamical systems. Chaos synchronization is of awesome hypothetical noteworthiness and down to earth esteem and has made incredible commitments because of its potential application in numerous logical and designing fields amid the current years. In 1990, Pecora and Carroll [14] gave the synchronization of riotous frameworks utilizing the idea of drive and response system of integer-order chaotic systems. In 2003, Li et al. [10] demonstrated that fractional order chaotic systems can likewise be synchronized. In 2006, Lu et al. [11] demonstrated that the nonlinear control is utilized to synchronize two fractional-order chaotic systems. Searching for better methodologies for chaos control and synchronization distinctive sorts of techniques have been created for chaos and synchronization of non identical and identical systems for adaptive control [20], active control [4,5,18], sliding mode control [19] etc. Various synchronization techniques, such as complete synchronization (CS) [12], anti-synchronization generalized synchronization [22], lag synchronization [17], projective synchronization [13], hybrid synchronization [21], hybrid projective synchronization [8], hybrid function projective synchronization [3] and so forth, have been analysed for synchronizing the chaotic systems.

Combination-combination synchronization scheme [6,7] utilized as a part of this paper is summed up such that different types of synchronization scheme can be accomplished from it. Thus, combination-combination synchronization scheme is more adaptable and pertinent to this present reality frameworks. Motivated by above researches, in this manuscript, we investigates the combination-combination anti synchronization among identical fractional-order hyperchaotic systems.

This manuscript is organized as follows: In Sect. 2, methodology of combination-combination is described. In Sect. 3, system description of fractional order hyperchaotic systems is given. In Sect. 4, combination-combination anti-synchronization among four fractional order identical hyper-chaotic systems is performed. Finally in Sect. 5 concluding remarks are given.

2 Methodology

Methodology to perform combination-combination anti-synchronization for fractional order chaotic systems with two drive systems and two response systems is reported in this section.

The two fractional order drive systems are:

$$_cD^q u_1(t) = A_1 u_1 + f_1(u_1) \tag{1}$$

$$_cD^q u_2(t) = A_2 u_2 + f_2(u_2) \tag{2}$$

while the two fractional order response systems are:

$$_cD^q v_1(t) = B_1 v_1 + g_1(v_1) + \mu \tag{3}$$

$$_cD^q v_2(t) = B_2 v_2 + g_2(v_2) + \mu^* \tag{4}$$

where $u_1 = (u_{11}, u_{12}, ..., u_{1n})$, $u_2 = (u_{21}, u_{22}, ..., u_{2n})$, $v_1 = (v_{11}, v_{12}, ..., v_{1n})$ and $v_2 = (v_{21}, v_{22}, ..., v_{2n})$ are the state vectors of the drive and response systems respectively. $f_1, f_2, g_1, g_2 : \mathbb{R}^n \rightarrow \mathbb{R}^n$ are the continuous vector functions and $u = (u_1, u_2, ..., u_n)$, $u^* = (u_1^*, u_2^*, ..., u_n^*)$ are the controllers added to the response systems which are to be determined.

The error states to achieve combination-combination anti-synchronization is given as:

$$e = Cu_1 + Du_2 + Ev_1 + Fv_2 \tag{5}$$

where $e = (e_1, e_2, ..., e_3)^T$

Theorem 1. Anti-Synchronization in four identical fractional order hyperchaotic systems in the form (1), (2), (3), (4) and the error system in the form (5) by adaptive control technique will be attained, if we choose the control functions as:

$$U = E\mu + F\mu^* = -f_1 - f_2 - g_1 - g_2 - Ke \tag{6}$$

where $K = (k_1, k_2, k_3, k_4)$ is the control gain matrix, which satisfies $|arg(\lambda_i(A_1 - K))| > \frac{\alpha\pi}{2}$ for all eigenvalues of $(A_1 - K)$.
The following observation arises:

1. The constant matrices C, D, E, F are called the scaling matrices. In addition C, D, E, F can be extended to functional matrices of state variables u_1, u_2, v_1 and v_2.
2. If $E = 0$ or $F = 0$, then combination-combination synchronization problem becomes the combination synchronization problem.
3. If $C = 0, E = I, F = 0$ or $C = E = 0, F = I$ or $D = 0, E = I, F = 0$ or $D = E = 0, F = I$, then the combination synchronization becomes projective synchronization, where I is the $n \times n$ matrix.
4. If $C = D = E = 0$ or $C = D = F = 0$, then the combination synchronization becomes the chaos control problem.

3   Fractional-Order Hyperchaotic System

The fractional-order hyperchaotic system [2] is given by

$$\begin{cases} \dfrac{d^q u_1}{dt^q} = -a_1 u_1 + a_2 u_2 \\[2mm] \dfrac{d^q u_2}{dt^q} = a_3 u_1 - u_1 u_3 - u_2 + u_4 \\[2mm] \dfrac{d^q u_3}{dt^q} = u_1^2 - a_4(u_1 + u_3) \\[2mm] \dfrac{d^q u_4}{dt^q} = -a_5 u_1 \end{cases} \tag{7}$$

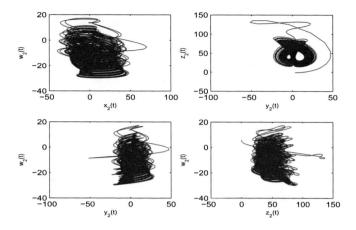

Fig. 1. 3D Phase Portrait of the fractional order hyperchaotic system

where u_1, u_2, u_3, u_4 are the state variables, $q > 0$ is the fractional order and a_1, a_2, a_3, a_4, a_5 are the parameters.

For the parameters values $a_1 = 25, a_2 = 60, a_3 = 40, a_4 = 4, a_5 = 5$ and $q = 0.95$ the four Lyapunov exponents of the system (3) are calculated as $\gamma_1 = 3.0057$, $\gamma_2 = 0.0304$, $\gamma_3 = -0.1631$ and $\gamma_4 = 46.1578$, respectively.

Therefore, the Kaplan-Yorke dimension of system is

$$D = j + \sum_{i=1}^{j} \frac{\gamma_i}{|\gamma_{j+1}|}$$
$$= 3.0622$$

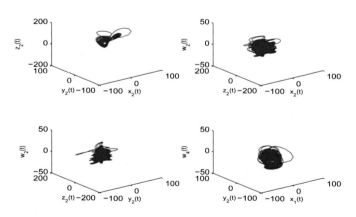

Fig. 2. 2D Phase Portrait of the fractional order hyperchaotic system

where j is the largest integer satisfying $\sum_{i=1}^{j} \gamma_j \geq 0$ and $\sum_{i=1}^{j+1} \gamma_j < 0$. The Lyapunov dimension is fractional, which implies system (3) is really a dissipative system. In Figs. 1 and 2, 3D phase portraits and 2D phase portraits are displayed respectively, it is clear that the system has a double-scroll hyperchaotic attractor.

4   Illustrative Example

To accomplish the combination-combination anti-synchronization, we consider the four identical fractional order hyperchaotic system described above (7). The two drive systems are:

$$
\begin{cases}
\dfrac{d^q u_{11}}{dt^q} = -a_1 u_{11} + a_2 u_{12} \\[2mm]
\dfrac{d^q u_{12}}{dt^q} = a_3 u_{11} - u_{11} u_{13} - u_{12} + u_{14} \\[2mm]
\dfrac{d^q u_{13}}{dt^q} = u_{11}^2 - a_4(u_{11} + u_{13}) \\[2mm]
\dfrac{d^q u_{14}}{dt^q} = -a_5 u_{11}
\end{cases}
\tag{8}
$$

$$
\begin{cases}
\dfrac{d^q u_{21}}{dt^q} = -b_1 u_{21} + b_2 u_{22} \\[2mm]
\dfrac{d^q u_{22}}{dt^q} = b_3 u_{21} - u_{21} u_{23} - u_{22} + u_{24} \\[2mm]
\dfrac{d^q u_{23}}{dt^q} = u_{21}^2 - b_4(u_{21} + u_{23}) \\[2mm]
\dfrac{d^q u_{24}}{dt^q} = -b_5 u_{21}
\end{cases}
\tag{9}
$$

while the two response systems are given as:

$$
\begin{cases}
\dfrac{d^q v_{11}}{dt^q} = -c_1 v_{11} + c_2 v_{12} + v_{11} + u_1 \\[2mm]
\dfrac{d^q v_{12}}{dt^q} = c_3 v_{11} - v_{11} v_{13} - v_{12} + v_{14} + u_2 \\[2mm]
\dfrac{d^q v_{13}}{dt^q} = v_{11}^2 - c_4(v_{11} + v_{13}) + u_3 \\[2mm]
\dfrac{d^q v_{14}}{dt^q} = -c_5 v_{11} + u_4
\end{cases}
\tag{10}
$$

$$
\begin{cases}
\dfrac{d^q v_{21}}{dt^q} = -d_1 v_{21} + d_2 v_{22} + u_1^* \\[2mm]
\dfrac{d^q v_{22}}{dt^q} = d_3 v_{21} - v_{21} v_{23} - v_{22} + v_{24} + u_2^* \\[2mm]
\dfrac{d^q v_{23}}{dt^q} = v_{21}^2 - d_4(v_{21} + v_{23}) + u_3^* \\[2mm]
\dfrac{d^q v_{24}}{dt^q} = -d_5 v_{21} + u_4^*
\end{cases}
\tag{11}
$$

where $A_1 = diag(-a_1, -1, -a_4, 0)$, $A_2 = diag(-b_1, -1, -b_4, 0)$, $B_1 = diag(-c_1, -1, -c_4, 0)$, $B_2 = diag(-d_1, -1, -d_4, 0)$

$$f_1 = \begin{bmatrix} a_2 u_{12} \\ a_3 u_{11} - u_{13} u_{11} + u_{14} \\ u_{11}^2 - a_4 u_{11} \\ -a_5 u_{11} \end{bmatrix}, f_2 = \begin{bmatrix} b_2 u_{22} \\ b_3 u_{21} - u_{23} u_{21} + u_{24} \\ u_{21}^2 - b_4 u_{21} \\ -b_5 u_{21} \end{bmatrix},$$

$$g_1 = \begin{bmatrix} c_2 v_{12} \\ c_3 v_{11} - v_{13} v_{11} + v_{14} \\ v_{11}^2 - c_4 v_{11} \\ -c_5 v_{11} \end{bmatrix}, g_2 = \begin{bmatrix} d_2 v_{22} \\ d_3 v_{21} - v_{23} v_{21} + v_{24} \\ v_{21}^2 - d_4 v_{21} \\ -d_5 v_{21} \end{bmatrix}.$$

For combination-combination anti-synchronization we choose $C = I$, $D = I$, $E = I$ and $F = I$ in this synchronization scheme.

The error system is obtained as follows:

$$\begin{cases} e_1 = u_{11} + u_{21} + v_{11} + v_{21} \\ e_2 = u_{12} + u_{22} + v_{12} + v_{22} \\ e_3 = u_{13} + u_{23} + v_{13} + v_{23} \\ e_4 = u_{14} + u_{24} + v_{14} + v_{24} \end{cases} \qquad (12)$$

Denote

$$\begin{cases} U_1 = u_1 + u_1^* \\ U_2 = u_2 + u_2^* \\ U_3 = u_3 + u_3^* \\ U_4 = u_4 + u_4^* \end{cases} \qquad (13)$$

By applying the Theorem 1, we get the control functions as follows:

$$\begin{cases} U_1 = -a_2 u_{12} - b_2 u_{22} - c_2 v_{12} - d_2 v_{22} - k_1 e_1 \\ U_2 = -(a_3 u_{11} - u_{13} u_{11} u_{14}) - (b_3 u_{21} - u_{21} u_{23} + u_{24}) \\ \quad - (c_3 v_{11} - v_{11} v_{13} + v_{14}) - (d_3 v_{21} - v_{21} v_{23} + v_{24}) - k_{12} e_2 \\ U_3 = -(u_{11}^2 - a_4 u_{11}) - (u_{21}^2 - b_4 u_{21}) - (v_{11}^2 - c_4 v_{11}) \\ \quad - (v_{21}^2 - d_4 v_{21}) - k_3 e_3 \\ U_4 = a_5 u_{11} + b_5 u_{21} + c_5 v_{11} + d_5 v_{21} - k_4 e_3 \end{cases} \qquad (14)$$

4.1 Numerical Simulations

Numerical simulations are performed to justify the efficiency of the above designed controllers. The parameters values $a_1 = b_1 = c_1 = d_1 = 25$, $a_2 = b_2 = c_2 = d_2 = 60$, $a_3 = b_3 = c_3 = d_3 = 40$, $a_4 = b_4 = c_4 = d_4 = 4$ and $a_5 = b_5 = c_5 = d_5 = 5$ are chosen so that system shows chaotic behavior in the absence of controllers as shown in Fig. 2.

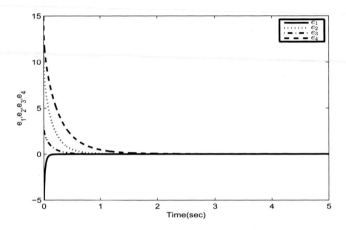

Fig. 3. Error dynamics among the drive and the response systems with controllers activated for $t > 0$

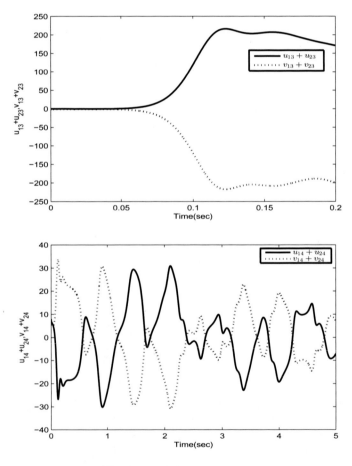

Fig. 4. Dynamics of the drive and the response state variables with controllers are activated for $t > 0$

The initial conditions of the drive systems and response systems are chosen as $(u_{11},\ u_{12},\ u_{13},\ u_{14}) = (1,\ 2,\ 3,\ 4)$, $(u_{21},\ u_{22},\ u_{23},\ u_{24}) = (-1,\ 2,\ -3,\ 4)$, $(v_{11},\ v_{12},\ v_{13},\ v_{14}) = (-3,\ 4,\ -1,\ 1)$, $(v_{21},\ v_{22},\ v_{23},\ v_{24}) = (-2,\ 3,\ -1,\ 6)$ for which the systems shows chaotic behaviour. The control gain matrix is chosen to be $K = diag(5, 4, 2, 3)$. The corresponding initial conditions for error system is obtained as $(e_1,\ e_2,\ e_3,\ e_4) = (-5,\ 11,\ 3,\ 15)$. The convergence of error state variables in Fig. 3 shows that the combination-combination synchronization among systems (8), (9), (10) and (11) is achieved when controllers are activated at $t > 0$. Figure 4 shows the trajectories of drive and response state variables when controller are activated at $t > 0$, this again confirms combination-combination anti-synchronization among systems (8), (9), (10) and (11).

5   Conclusion

Combination-combination anti-synchronization is accomplished by considering four identical 4D fractional-order hyperchaotic systems. Numerical simulations are performed to justify the legitimacy of the theoretical results discussed. Computational and scientific outcomes are in fantastic agreement. It is observed that synchronization scheme, for example, chaos control problem, projective anti-synchronization, combination anti-synchronization turns into the special cases of combination-combination anti-synchronization. The combination-combination anti-synchronization of a 4D hyperchaotic system presented in this manuscript may be used in information processing having many applications in physical, biological and social systems.

References

1. Azar, A.T., Serrano, F.E.: Fractional order sliding mode PID controller/observer for continuous nonlinear switched systems with PSO parameter tuning. In: International Conference on Advanced Machine Learning Technologies and Applications, pp. 13–22. Springer (2018)
2. Gao, Y., Liang, C., Wu, Q., Yuan, H.: A new fractional-order hyperchaotic system and its modified projective synchronization. Chaos Soliton. Fract. **76**, 190–204 (2015)
3. Khan, A., Pal, R.: Adaptive hybrid function projective synchronization of chaotic space-tether system. Nonlinear Dyn. Syst. Theor. **14**(1), 44–57 (2014)
4. Khan, A., Shikha, S.: Increased and reduced order synchronisations between 5D and 6D hyperchaotic systems. Indian J. Ind. Appl. Math. **8**(1), 118–131 (2017)
5. Khan, A., Shikha, S.: Mixed tracking and projective synchronization of 6D hyperchaotic system using active control. Int. J. Nonlinear Sci. **22**(1), 44–53 (2016)
6. Khan, A., Singh, S.: Chaotic analysis and combination-combination synchronization of a novel hyperchaotic system without any equilibria. Chinese J. Phys. (2017)
7. Khan, A., Singh, S.: Generalization of combination-combination synchronization of n-dimensional time-delay chaotic system via robust adaptive sliding mode control. Math. Method. Appl. Sci. (2018)
8. Khan, A., et al.: Hybrid function projective synchronization of chaotic systems via adaptive control. Int. J. Dyn. Control **5**(4), 1114–1121 (2017)

9. Koeller, R.: Applications of fractional calculus to the theory of viscoelasticity. J. Appl. Mech. **51**, 299–307 (1984). (ISSN 0021-8936)
10. Li, C., Liao, X., Yu, J.: Synchronization of fractional order chaotic systems. Phys. Rev. E **68**(6), 067203 (2003)
11. Lu, J.G.: Chaotic dynamics of the fractional-order lü system and its synchronization. Phys. Lett. A **354**(4), 305–311 (2006)
12. Mahmoud, G.M., Mahmoud, E.E.: Complete synchronization of chaotic complex nonlinear systems with uncertain parameters. Nonlinear Dyn. **62**(4), 875–882 (2010)
13. Mainieri, R., Rehacek, J.: Projective synchronization in three-dimensional chaotic systems. Phys. Rev. Lett. **82**(15), 3042 (1999)
14. Pecora, L.M., Carroll, T.L.: Synchronization in chaotic systems. Phys. Rev. Lett. **64**(8), 821 (1990)
15. Podlubny, I.: Fractional Differential Equations: An Introduction to Fractional Derivatives, Fractional Differential Equations, to Methods of their Solution and Some of their Applications, vol. 198. Academic press (1998)
16. Podlubny, I.: Fractional-order systems and PI/sup/spl lambda//D/sup/spl mu//-controllers. IEEE Trans. Autom. Control **44**(1), 208–214 (1999)
17. Rosenblum, M.G., Pikovsky, A.S., Kurths, J.: From phase to lag synchronization in coupled chaotic oscillators. Phys. Rev. Lett. **78**(22), 4193 (1997)
18. Singh, S., Azar, A.T., Zhu, Q.: Multi-switching master-slave synchronization of non-identical chaotic systems. In: Innovative Techniques and Applications of Modelling, Identification and Control, pp. 321–330. Springer (2018)
19. Singh, S., Azar, A.T., Ouannas, A., Zhu, Q., Zhang, W., Na, J.: Sliding mode control technique for multi-switching synchronization of chaotic systems. In: 9th International Conference on Modelling, Identification and Control (ICMIC) 2017, pp. 880–885. IEEE (2017)
20. Singh, S.V.S., Serrano, F.E., Sambas, A.: A novel hyperchaotic system with adaptive control, synchronization, and circuit simulation. In: Advances in System Dynamics and Control, p. 382 (2018)
21. Vaidyanathan, S., Azar, A.T.: Hybrid synchronization of identical chaotic systems using sliding mode control and an application to vaidyanathan chaotic systems. In: Advances and Applications in Sliding Mode Control Systems, pp. 549–569. Springer (2015)
22. Zheng, Z., Hu, G.: Generalized synchronization versus phase synchronization. Phys. Rev. E **62**(6), 7882 (2000)

A New Generalized Synchronization Scheme to Control Fractional Chaotic Systems with Non-identical Dimensions and Different Orders

Adel Ouannas[1], Giuseppe Grassi[2], and Ahmad Taher Azar[3,4(✉)]

[1] Department of Mathematics, University of Larbi Tebessi, 12002 Tebessa, Algeria
ouannas.a@yahoo.com
[2] Dipartimento Ingegneria Innovazione, Università del Salento, 73100 Lecce, Italy
giuseppe.grassi@unisalento.it
[3] Faculty of Computers and Information, Benha University, Benha, Egypt
ahmad_t_azar@ieee.org, ahmad.azar@fci.bu.edu.eg
[4] School of Engineering and Applied Sciences, Nile University,
Giza, Egypt

Abstract. This paper addresses the problem of generalized synchronization (GS) between fractional order chaotic systems. In this paper, we propose a new control strategy for a complex generalized synchronization (GS) scheme dedicated to non-identical fractional-order chaotic systems characterized by different dimensions. The proposed control parameters are nonlinear in nature. In order to ensure that the proposed scheme converge towards zero, we establish the asymptotic stability of the zero solution to the error system by means of the stability of linear fractional-order systems. In order to assess the validity of the findings, numerical results have been presented for a 3D master system and a 4D slave system. The fractional-order systems employed here are well known in the literature. Matlab simulation results have confirmed the convergence of the error in sufficient time.

Keywords: Chaos and hyperchaos · Generalized synchronization ·
Fractional order systems · Different dimensions ·
Fractional stability theory

1 Introduction

Recently, with the development of fractional-order algorithms, dynamics of fractional order systems have received much attention [1,2,4–6,8,10,11,14,18,31, 32]. Studying chaos in fractional-order dynamical systems is an interesting topic as well. It has been shown that many fractional-order dynamical systems behave chaotically with total order less then three [27,33]. Similar to nonlinear integer-order differential systems, nonlinear fractional-order dynamical systems can also be synchronized [3].

© Springer Nature Switzerland AG 2020
A. E. Hassanien et al. (Eds.): AMLTA 2019, AISC 921, pp. 415–424, 2020.
https://doi.org/10.1007/978-3-030-14118-9_42

Recently, study on synchronization of chaotic fractional order dynamical systems has starts to attract increasing attention of many researchers. Until now, a wide variety of fractional techniques have been used to design a synchronization control in fractional order chaotic systems [15–17, 19–26, 28, 30]. Among these, generalized synchronization (GS) is the most effective synchronization method that has been used widely to achieve the chaos synchronization. GS implies the establishment of functional relation between the master and the slave systems. It has received a great deal of attention for its universality in the recent years. Nowadays, numerous researches of GS in fractional-order chaotic systems have been done theoretically and experimentally [9, 12, 34].

Based on these considerations, the aim of this work is to present new control method to investigate a new complex generalized synchronization scheme between arbitrary fractional chaotic systems with different dimensions. Using nonlinear control law and stability theory of fractional-order systems, a theorem is proved, which enable synchronization between n-dimensional and m-dimensional fractional chaotic systems to be achieved. The complexity of the proposed scheme can be used to enhance security in communication and encryption. The outline of the rest of this paper is organized as follows. Section 2 provides a brief review of the fractional calculus and some results on the stability of fractional order systems are introduced. The main results are presented in Sect. 3. In Sect. 4, example of synchronization is given to validate the theoretical result derived in this paper. Section 5 is the brief conclusion.

2   Basic Concepts

Definition 1. *Caputo fractional derivative is defined as follows [7].*

$$D_t^p x\left(t\right) = J^{m-p} x^m\left(t\right), \tag{1}$$

where $p \in]0,1]$ and m is the first integer which is not less than p, x^m is the m-order derivative in the usual sense, and $J^q\left(q > 0\right)$ is the q-order Reimann-Liouville integral operator with expression:

$$J^q y\left(t\right) = \frac{1}{\Gamma\left(q\right)} \int\limits_0^t \left(t - \tau\right)^{q-1} y\left(\tau\right) d\tau, \tag{2}$$

where Γ denotes Gamma function.

Lemma 1 [29]. *Suppose $f(t)$ has a continuous kth derivative on $[0,t]$ ($k \in N$, $t > 0$), and let $p, q > 0$ be such that there exists some $\ell \in N$ with $\ell \le k$ and p, $p + q \in [\ell - 1, \ell]$. Then*

$$D_t^p D_t^q f\left(t\right) = D_t^{p+q} f\left(t\right) \tag{3}$$

Remark 1. *Note that the condition requiring the existence of the number ℓ with the above restrictions in the property is essential. In this work, we consider the case that $0 < p, q \le 1$, and $0 < p + q \le 1$. Apparently, under such conditions this property holds.*

Lemma 2. *The fractional-order linear system [13]*

$$D_t^p X(t) = AX(t), \tag{4}$$

where $D_t^p = [D_t^{p_1}, D_t^{p_2}, ..., D_t^{p_n}], 0 < p_i \leq 1, (i = 1, 2, ..., n), X(t) = (x_i(t))_{1 \leq i \leq n}$ *and* $A \in \mathbf{R}^{n \times n}$, *is asymptotically stable if all roots* λ *of the characteristic equation*

$$det\left(diag\left(\lambda^{Mp_1}, \lambda^{Mp_2}, ..., \lambda^{Mp_n}\right) - A\right) = 0, \tag{5}$$

satisfy $|\arg(\lambda)| > \frac{\pi}{2M}$, *where* M *is the least common multiple of the denominators of* p_i's.

3 Main Results

We consider the following fractional chaotic system as the master system

$$D_t^{p_i} x_i(t) = F_i(X(t)), \qquad i = 1, ..., n, \tag{6}$$

where $X(t) = (x_1(t), x_2(t), ..., x_n(t))^T$ is the state vector of the master system (6), $D_t^{p_i}$ is the Caputo fractional derivative of order $p_i, 0 < p_i \leq 1$ and $F_i : \mathbf{R}^n \to \mathbf{R}, (i = 1, ..., n)$, are nonlinear functions.

Also, consider the slave system as

$$D_t^{q_i} y_i(t) = \sum_{j=1}^m b_{ij} y_j(t) + G_i(Y(t)) + u_i, \qquad i = 1, ..., m, \tag{7}$$

where $Y(t) = (y_1(t), y_2(t), ..., y_m(t))^T$ is the state vector of the slave system (7), $B \in \mathbf{R}^{m \times m}, G_i : \mathbf{R}^m \to \mathbf{R}, (i = 1, ..., m)$, are nonlinear functions $0 < q_i \leq 1, D^{q_i}$ are the Caputo fractional derivatives of order q_i and $U = (u_i)_{1 \leq i \leq m}$ is a controllers to be determined.

Definition 2. *The master system (6) and the slave system (7) are said to be generalized synchronized, if there exists a controllers* $U = (u_i)_{1 \leq i \leq m}$ *and given differentiable functions* $\phi_i : \mathbf{R}^n \longrightarrow \mathbf{R}^m, 1 \leq i \leq m$, *such that the error*

$$e_i(t) = y_i(t) - \phi_i(X(t)), \quad i = 1, ..., m, \tag{8}$$

satisfies the $\lim_{t \longrightarrow +\infty} \|(e_1(t), e_2(t), ..., e_m(t))\| = 0$, *where* $\|.\|$ *is the Euclidian norm.*

We assume that $m \leq n$ and $q_i < p_i$ for $i = 1, ..., m$. Hence, we have the following result.

Theorem 1. *There exists a suitable feedback gain matrix* $L = (l_{ij})_{m \times m}$ *to realize GS between the master system (6) and the slave system (7) under the following controllers*

$$u_i = -\sum_{j=1}^m b_{ij} y_j(t) - G_i(Y(t)) + J^{p_i - q_i}\left(\sum_{j=1}^m (b_{ij} - l_{ij})(y_j(t) - \phi_j(X(t))) + D_t^{p_i} \phi_i(X(t))\right), \quad i = 1, ..., m. \tag{9}$$

Proof. By inserting the control law described by Eq. (9) into system (7), we can rewrite the slave system as follows

$$D_t^{q_i} y_i(t) = J^{p_i - q_i} \left(\sum_{j=1}^{m} (b_{ij} - l_{ij}) e_j(t) + D_t^{p_i} \phi_i(X(t)) \right), \quad 1 \le i \le m. \quad (10)$$

By applying the fractional derivative of order $p_i - q_i$ to both the left and right sides of Eq. (10), we obtain

$$D_t^{p_i - q_i} \left(D_t^{q_i} y_i(t) \right) = D_t^{p_i} y_i(t)$$

$$= D_t^{p_i - q_i} J^{p - q_i}(v_i) = \sum_{j=1}^{m} (b_{ij} - l_{ij}) e_j(t) + D_t^{p_i} \phi_i(X(t)), \quad 1 \le i \le m.$$

$$(11)$$

Note that $p_i - q_i$ satisfies $p_i - q_i \in [0, 1)$. The error system can be written as follows

$$D^{p_i} e_i(t) = D_t^{p_i} y_i(t) - D_t^{p_i} \phi_i(X(t)), \quad i = 1, ..., m. \quad (12)$$

From Eqs. (11)–(12), we get the error dynamical system as follows

$$D_t^{p_i} e_i(t) = \sum_{j=1}^{m} (b_{ij} - l_{ij}) e_j(t), \quad i = 1, ..., m. \quad (13)$$

Now, the error system (13) can be written in following compact form

$$D_t^{p} e(t) = (B - L) e(t), \quad (14)$$

where $B = (b_{ij})_{m \times m}$ and $L = (l_{ij})_{m \times m}$ is a control matrix. The control matrix K is chosen such that all roots λ of $\det \left(diag \left(\lambda^{Mp_1}, \lambda^{Mp_2}, ..., \lambda^{Mp_m} \right) + L - B \right) = 0$, satisfy $|\arg(\lambda)| > \frac{\pi}{2M}$, where M is the least common multiple of the denominators of p_i's. According to Lemma 2, we conclude that the zero solution of the error system (14) is globally asymptotically stable and therefore systems (6) and (7) are globally generalized synchronized.

4   Numerical Example

In this example, we assume that the fractional order permanent magnet synchronous motor model (PMSM) [36] is the master system and the controlled new fractional order hyperchaotic system, proposed by Wu et al. in [35], is the slave system. The master system is described by

$$D_t^{p_1} x_1 = -x_1 + x_2 x_3, \quad (15)$$
$$D_t^{p_2} x_2 = -x_2 - x_1 x_3 + a x_3,$$
$$D_t^{p_3} x_3 = b(x_2 - x_3),$$

where $(a, b) = (100, 10)$ and $p_1 = p_2 = p_3 = 0.95$. Fig. 1 show different chaotic attractors of the fractional order PMSM system (15).

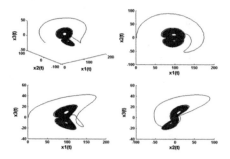

Fig. 1. Chaotic attractors of the master system (15).

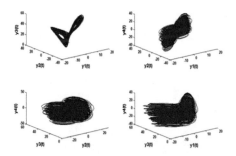

Fig. 2. Attractors of the slave system (16).

The slave system is given as follows

$$D_t^q y_1 = 10\,(y_2 - y_1) + u_1, \tag{16}$$
$$D_t^q y_2 = 28y_1 + y_2 - y_4 - y_1 y_3 + u_2,$$
$$D_t^q y_3 = y_1 y_2 - \frac{8}{3} y_3 + u_3,$$
$$D_t^q y_3 = 0.1 y_2 y_3 + u_4,$$

where u_i, $i = 1, 2, 3, 4$, are controllers. System (16), when $q = 0.94$, exhibits hyperchaotic behaviours without control as shown in Fig. 2.

In this case, the matrix $B = (b_{ij})_{4\times 4}$ and the nonlinear function $G = (G_i)_{1 \le i \le 4}$ of the slave system (16) are given by

$$B = \begin{pmatrix} -10 & 10 & 0 & 0 \\ 28 & 1 & 0 & -1 \\ 0 & 0 & -\frac{8}{3} & 0 \\ 0 & 0 & 0 & 0 \end{pmatrix}, \quad G = \begin{pmatrix} 0 \\ -y_1 y_3 \\ -y_1 y_2 \\ 0.1 y_2 y_3 \end{pmatrix}.$$

Define the errors of GS between the master system (15) and the slave system (16) by

$$(e_1, e_2, e_3, e_4)^T = (y_1, y_2, y_3, y_4)^T - \phi\,(x_1, x_2, x_3), \tag{17}$$

where

$$\phi\left(x_1, x_2, x_3\right) = \begin{bmatrix} \phi_1\left(x_1, x_2, x_3\right) \\ \phi_2\left(x_1, x_2, x_3\right) \\ \phi_3\left(x_1, x_2, x_3\right) \\ \phi_4\left(x_1, x_2, x_3\right) \end{bmatrix} = \begin{bmatrix} x_1 \\ x_2 \\ x_3 \\ x_1 x_2 x_3 \end{bmatrix}. \tag{18}$$

By applying our approach of GS, described in Sect. 3, the feedback gain matrix L is chosen as

$$L = \begin{pmatrix} 0 & 10 & 0 & 0 \\ 28 & 2 & 0 & -1 \\ 0 & 0 & 0 & 0 \\ 0 & 0 & 0 & 1 \end{pmatrix}, \tag{19}$$

and the controllers u_i, $i = 1, 2, 3, 4$, can be designed as

$$u_1 = -10\left(y_2 - y_1\right) + J^{0.01}\left[-10e_1 + D_t^{0.95} x_1\right], \tag{20}$$

$$u_2 = -28y_1 - y_2 + y_4 + y_1 y_3 + J^{0.01}\left[-e_2 + D_t^{0.95} x_2\right],$$

$$u_3 = \frac{8}{3}y_3 + y_1 y_2 + J^{0.01}\left[-\frac{8}{3}e_3 + D_t^{0.95} x_3\right],$$

$$u_4 = -0.1y_2 y_3 + J^{0.01}\left[-e_4 + D_t^{0.95} x_1 x_2 x_3\right].$$

It is easy to know that $B - L$ is a negative definite matrix. Then, according to Theorem 1, the master system (15) and the slave system (16) are globally generalized synchronized. In this case, the error system between system (15) and (16) can be written as follow:

$$D_t^{0.95} e_1 = -10e_1, \tag{21}$$

$$D_t^{0.95} e_2 = -e_2,$$

$$D_t^{0.95} e_3 = -\frac{8}{3}e_3$$

$$D_t^{0.95} e_4 = -e_4.$$

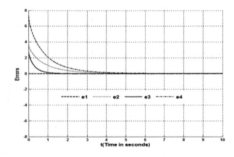

Fig. 3. Time evolution of GS errors of system (21).

For the purpose of numerical simulation, fractional Euler method has been used. The initial values of the master system and the slave systems are $[x_1(0), x_2(0), x_3(0)] = [2, -1, 1]$ and $[y_1(0), y_2(0), y_3(0), y_4(0)] = [2, 3, 4, 6]$, respectively, and the initial states of the error system are $[e_1(0), e_2(0), e_3(0), e_4(0)] = [0, 4, 3, 8]$. The error functions evolution, in this case, is shown in Fig. 3.

5   Discussion and Conclusion

In this work, we have shown that a new complex scheme of generalized synchronization for non-identical and different dimensions fractional-order chaotic systems can be controlled. We assumed a n-dimensional fractional-order master system and a m-dimensional fractional-order slave system with $m \leq n$. The main result of the study was obtained via controlling the linear part of the slave system. We presented a nonlinear control scheme whereby generalized synchronization is achieved for n-dimensional master system and a m-dimensional slave system in m-D. The stability of the zero solution, and consequently the convergence of the synchronization errors, was established by means of the stability of linear fractional-order systems. Numerical results have confirmed the findings of the study. Simulations were carried out on Matlab to ensure that the errors converge to zero subject to the proposed control laws. In addition, fractional chaotic systems have received considerable interest in the field of data, image, and video encryption. It is common to use a large set of random or pseudo-random keys in a secret or public key encryption scenario. The basic idea is that chaotic systems can be used instead of conventional algorithmic key generators. It has been claimed in the literature that the dependence of fractional chaotic systems on changes in the fractional order adds a new degree of freedom, which makes them more suitable for encryption purposes. It is the intention of the authors to put this claim to the test in a future study by comparing the synchronization scheme discussed herein with previous results in a data encryption scenario.

References

1. AbdelAty, A.M., Azar, A.T., Vaidyanathan, S., Ouannas, A., Radwan, A.G.: Chapter 14 - applications of continuous-time fractional order chaotic systems. In: Azar, A.T., Radwan, A.G., Vaidyanathan, S. (eds.) Mathematical Techniques of Fractional Order Systems, Advances in Nonlinear Dynamics and Chaos (ANDC), pp. 409–449. Elsevier (2018)
2. Azar, A.T., Serrano, F.E.: Fractional order sliding mode PID controller/observer for continuous nonlinear switched systems with PSO parameter tuning. In: Hassanien, A.E., Tolba, M.F., Elhoseny, M., Mostafa, M. (eds.) The International Conference on Advanced Machine Learning Technologies and Applications (AMLTA2018), pp. 13–22. Springer, Cham (2018)
3. Azar, A.T., Vaidyanathan, S., Ouannas, A.: Fractional Order Control and Synchronization of Chaotic Systems. Studies in Computational Intelligence, vol. 688. Springer, Berlin (2017)

4. Azar, A.T., Kumar, J., Kumar, V., Rana, K.P.S.: Control of a two link planar electrically-driven rigid robotic manipulator using fractional order SOFC. In: Hassanien, A.E., Shaalan, K., Gaber, T., Tolba, M.F. (eds.) Proceedings of the International Conference on Advanced Intelligent Systems and Informatics 2017, pp. 57–68. Springer, Cham (2018)

5. Azar, A.T., Radwan, A.G., Vaidyanathan, S. (eds.) Mathematical Techniques of Fractional Order Systems. Elsevier (2018)

6. Azar, A.T., Serranot, F.E., Vaidyanathan, S.: Chapter 10 - sliding mode stabilization and synchronization of fractional order complex chaotic and hyperchaotic systems. In: Azar, A.T., Radwan, A.G., Vaidyanathan, S. (eds.) Mathematical Techniques of Fractional Order Systems, Advances in Nonlinear Dynamics and Chaos (ANDC), pp. 283–317. Elsevier (2018)

7. Caputo, M.: Linear models of dissipation whose Q is almost frequency independent-II. Geophys. J. Roy. Astron. Soc. **13**(5), 529–539 (1967)

8. Ghoudelbourk, S., Dib, D., Omeiri, A., Azar, A.T.: MPPT control in wind energy conversion systems and the application of fractional control (piα) in pitch wind turbine. Int. J. Model. Ident. Control **26**(2), 140–151 (2016)

9. Hongtao, L., Zhen, W., Zongmin, Y., Ronghui, L.: Generalized synchronization and control for incommensurate fractional unified chaotic system and applications in secure communication. Kybernetika **48**(2), 190–205 (2012)

10. Khettab, K., Bensafia, Y., Bourouba, B., Azar, A.T.: Chapter 20 - enhanced fractional order indirect fuzzy adaptive synchronization of uncertain fractional chaotic systems based on the variable structure control: Robust h∞ design approach. In: Azar, A.T., Radwan, A.G., Vaidyanathan, S. (eds.) Mathematical Techniques of Fractional Order Systems, Advances in Nonlinear Dynamics and Chaos (ANDC), pp. 597–624. Elsevier (2018)

11. Kumar, J., Azar, A.T., Kumar, V., Rana, K.P.S.: Chapter 9 - design of fractional order fuzzy sliding mode controller for nonlinear complex systems. In: Azar, A.T., Radwan, A.G., Vaidyanathan, S. (eds.) Mathematical Techniques of Fractional Order Systems, Advances in Nonlinear Dynamics and Chaos (ANDC), pp. 249–282. Elsevier (2018)

12. Martinez-Guerra, R., Mata-Machuca, J.L.: Fractional generalized synchronization in a class of nonlinear fractional order systems. Nonlinear Dyn. **77**(4), 1237–1244 (2014)

13. Matignon, D.: Stability results for fractional differential equations with applications to control processing. In: In Computational Engineering in Systems Applications, pp. 963–968 (1996)

14. Meghni, B., Dib, D., Azar, A.T., Ghoudelbourk, S., Saadoun, A.: Robust Adaptive Supervisory Fractional Order Controller for Optimal Energy Management in Wind Turbine with Battery Storage, pp. 165–202. Springer, Cham (2017)

15. Ouannas, A., Al-sawalha, M.M., Ziar, T.: Fractional chaos synchronization schemes for different dimensional systems with non-identical fractional-orders via two scaling matrices. Optik **127**(20), 8410–8418 (2016a)

16. Ouannas, A., Azar, A.T., Vaidyanathan, S.: A robust method for new fractional hybrid chaos synchronization. Math. Method. Appl. Sci. **40**(5), 1804–1812 (2016b)

17. Ouannas, A., Abdelmalek, S., Bendoukha, S.: Coexistence of some chaos synchronization types in fractional-order differential equations. Electr. J. Differ. Equ. **128**, 1–15 (2017)

18. Ouannas, A., Azar, A.T., Vaidyanathan, S.: A new fractional hybrid chaos synchronisation. Int. J. Model. Ident. Control **27**(4), 314–322 (2017b)

19. Ouannas, A., Azar, A.T., Ziar, T., Radwan, A.G.: A Study on Coexistence of Different Types of Synchronization Between Different Dimensional Fractional Chaotic Systems, pp. 637–669. Springer, Cham (2017). https://doi.org/10.1007/978-3-319-50249-6_22
20. Ouannas, A., Azar, A.T., Ziar, T., Vaidyanathan, S.: Fractional Inverse Generalized Chaos Synchronization Between Different Dimensional Systems, pp. 525–551. Springer, Cham (2017). https://doi.org/10.1007/978-3-319-50249-6_18
21. Ouannas, A., Azar, A.T., Ziar, T., Vaidyanathan, S.: A New Method to Synchronize Fractional Chaotic Systems with Different Dimensions, pp. 581–611. Springer, Cham (2017). https://doi.org/10.1007/978-3-319-50249-6_20
22. Ouannas, A., Azar, A.T., Ziar, T., Vaidyanathan, S.: On New Fractional Inverse Matrix Projective Synchronization Schemes, pp. 497–524. Springer, Cham (2017f)
23. Ouannas, A., Odibat, Z., Alsaedi, A., Hobiny, A., Hayat, T.: Investigation of QS synchronization in coupled chaotic incommensurate fractional order systems. Chinese J. Phys. 56(5), 1940–1948 (2018a)
24. Ouannas, A., Wang, X., Pham, V.T., Grassi, G., Ziar, T.: Coexistence of identical synchronization, antiphase synchronization and inverse full state hybrid projective synchronization in different dimensional fractional-order chaotic systems. Adv. Differ. Equ. 2018(1), 35 (2018)
25. Pham, V., Gokul, P.M., Kapitaniak, T., Volos, C., Azar, A.T.: Chapter 16 - dynamics, synchronization and fractional order form of a chaotic system with infinite equilibria. In: Azar, A.T., Radwan, A.G., Vaidyanathan, S. (eds.) Mathematical Techniques of Fractional Order Systems, Advances in Nonlinear Dynamics and Chaos (ANDC), pp. 475–502. Elsevier (2018)
26. Pham, V.T., Vaidyanathan, S., Volos, C.K., Azar, A.T., Hoang, T.M., Van Yem, V.: A Three-Dimensional No-Equilibrium Chaotic System: Analysis, Synchronization and Its Fractional Order Form, pp. 449–470. Springer, Cham (2017)
27. Pham, V.T., Ouannas, A., Volos, C., Kapitaniak, T.: A simple fractional-order chaotic system without equilibrium and its synchronization. AEU Int. J. Electron. Commun. 86, 69–76 (2018b)
28. Shukla, M.K., Sharma, B.B., Azar, A.T.: Chapter 19 - control and synchronization of a fractional order hyperchaotic system via backstepping and active backstepping approach. In: Azar, A.T., Radwan, A.G., Vaidyanathan, S. (eds.) Mathematical Techniques of Fractional Order Systems, Advances in Nonlinear Dynamics and Chaos (ANDC), pp. 559–595. Elsevier (2018)
29. Si, G., Sun, Z., Zhang, Y., Chen, W.: Projective synchronization of different fractional-order chaotic systems with non-identical orders. Nonlinear Anal. Real World Appl. 13(4), 1761–1771 (2012)
30. Singh, S., Azar, A.T., Vaidyanathan, S., Ouannas, A., Bhat, M.A.: Chapter 11 - multiswitching synchronization of commensurate fractional order hyperchaotic systems via active control. In: Azar, A.T., Radwan, A.G., Vaidyanathan, S. (eds.) Mathematical Techniques of Fractional Order Systems, Advances in Nonlinear Dynamics and Chaos (ANDC), pp. 319–345. Elsevier (2018)
31. Soliman, N.S., Said, L.A., Azar, A.T., Madian, A.H., Radwan, A.G., Ounnas, A.: Fractional controllable multi-scroll v-shape attractor with parameters effect. In: 6th International Conference on Modern Circuits and Systems Technologies (MOCAST), pp. 1–4 (2017)
32. Tolba, M.F., AbdelAty, A.M., Saida, L.A., Elwakil, A.S., Azar, A.T., Madian, A.H., Radwan, A.G., Ounnas, A.: FPGA realization of caputo and grünwald-letnikov operators. In: 6th International Conference on Modern Circuits and Systems Technologies (MOCAST), pp. 1–4 (2017)

33. Wang, X., Ouannas, A., Pham, V.T., Abdolmohammadi, H.R.: A fractional-order form of a system with stable equilibria and its synchronization. Adv. Differ. Equ. **2018**(1), 20 (2018)
34. Wu, X., Lai, D., Lu, H.: Generalized synchronization of the fractional-order chaos in weighted complex dynamical networks with nonidentical nodes. Nonlinear Dyn. **69**(1), 667–683 (2012a)
35. Wu, X., Wang, H., Lu, H.: Modified generalized projective synchronization of a new fractional-order hyperchaotic system and its application to secure communication. Nonlinear Anal. Real World Appl. **13**(3), 1441–1450 (2012b)
36. Xue, W., Li, Y., Cang, S., Jia, H., Wang, Z.: Chaotic behavior and circuit implementation of a fractional-order permanent magnet synchronous motor model. J. Franklin Inst. **352**(7), 2887–2898 (2015)

Design of Reactive Voltage Automatic Control Device Based on Fuzzy Control

Qiang Zhang[1(✉)], Zhengdao Zhou[2], Yingjun Ju[3], Jianhan Jianhan[4], and Yong Liu[4]

[1] State Grid Liaoning Electric Power Research Institute,
Shenyang 110003, China
hexuanna505@qq.com
[2] Northeastern University, Shenyang 110819, China
[3] State Grid Liaoning Electric Power Supply Co., Ltd., Shenyang 110004, China
[4] Shanghai Proinvent Information Technology Co., Ltd.,
Shanghai 200240, China

Abstract. A method plan and the hardware realization of substation reactive power comprehensive control device are presented in this paper. The method of fuzzy control is used to control the voltage and reactive power of substation, and the structure of the compensator is designed as a whole. The function of the device is to improve power factor, reduce line loss, save electric energy and improve voltage quality. The controller mainly uses single chip microcomputer technology to collect and calculate the current, voltage, active power, reactive power and power factor of the circuit or distribution equipment. And device can effectively decrease day-adjusting times of the switching of on-load transformer and paralleled compensation capacitors in substation, and then improve the controlling ability of power system and its voltage quality remarkably. Based on the theory of area control, the switching power capacitor is controlled to realize the automatic compensation of reactive power, and the power factor is adjusted to the optimal state. Establishing reactive power automation network with AVC system.

Keywords: Fuzzy control · Reactive power compensation · Voltage regulation · Plant design

1 Introduction

In recent years, high economic growth has led to an increase in demand for electricity, structural power shortages in some parts of China, power shortages in some areas and the problem of switching and limiting power during peak hours [1]. With the increasing of high power nonlinear load, reactive power impact and harmonic pollution are on the rise [2]. The lack of effective reactive power regulation means makes the bus voltage change greatly, which leads to the increase of line loss of distribution system and the decrease of voltage qualification rate [3, 9]. For a long time in the construction and operation of power network in China, the reactive power compensation capacity is insufficient, the equipment is unreasonable, and the effect of reactive power compensation is not ideal [4, 10]. Especially in 10 kV distribution system, rapid, accurate, continuous compensation technology has not been applied.

© Springer Nature Switzerland AG 2020
A. E. Hassanien et al. (Eds.): AMLTA 2019, AISC 921, pp. 425–433, 2020.
https://doi.org/10.1007/978-3-030-14118-9_43

In the distribution network of 10 kV in our country, the reactive power compensation device is usually installed on the side of the special transformer, but the number of utility transformers is not enough, so that the compensation degree of the distribution network is not high, and the loss space of the network is large [5, 11]. Therefore, it is very important to install reactive power compensator on the distribution line. In the 10 kV distribution network, the shunt capacitor and other reactive power compensation devices are installed on the transmission line towers, which can further improve the power factor and reduce the network loss. Compared with the reactive power compensation scattered in the main transformer, the reactive power compensation on the installation and distribution lines has the characteristics of centralized compensation, high utilization of equipment and basic non-occupation of land, which makes up for the lack of reactive power on the low-voltage side of the transformer. Reduced reactive power on the line. Line compensation, especially for long lines and low power factor, can reduce line losses and improve power quality. Based on the fuzzy control strategy, the module of the automatic reactive power compensation device and the whole design of the device are designed this paper. The automatic control of reactive power and voltage for 10 kV distribution network is realized [6].

2 Traditional Reactive Power Compensation Control Strategy

At the end of the distribution line, the operating voltage is low, especially in heavy load, thin wire lines [7]. When the compensation capacitor is installed, the operating voltage can be increased, which leads to the question of how large the compensation capacitor is to be selected according to the requirement of increasing the voltage. In addition, in order to satisfy this constraint, the relationship between compensation capacity and network voltage increment must be obtained in order to meet the limit of the voltage rise of the network when the compensatory capacitance is installed in the normal network voltage line, and it is also necessary to find out the relationship between the compensation capacity and the voltage increment of the network [8].

The network voltage may be calculated by the following expression before installing compensating capacitors.

$$U_1 = U_2 + \frac{PR + QX}{U_2} \tag{1}$$

After the compensation capacitor is installed, the supply voltage U2 is not changed, the voltage U2 of the substation bus is raised to U2', and

$$U_1 = U_2' + \frac{PQ + (Q + Q_C)X}{U_2'} \tag{2}$$

$$\Delta U = U_2' - U_2 \approx \frac{Q_C X}{U_2'} \tag{3}$$

$$Q_C = \frac{U_2' \Delta U}{X} \tag{4}$$

U_2'–bus voltage value after capacitor input (kV). ΔU–Voltage Increment of bus Line with Capacitance input (kV). The total capacitance required for three phases is:

$$\sum Q_C = 3Q_C = 3\frac{U_{2L}'}{\sqrt{3}} \times \frac{\Delta U_L}{\sqrt{3}}$$
$$\times \frac{1}{X} = \frac{\Delta U_L \times U_{2L}'}{X} \tag{5}$$

At this point, the voltage and the increment of the voltage are based on the line voltage, while the expression of the single-phase compensation capacity is for the phase voltage.

Nine-domain diagram control strategy is a typical voltage and reactive power two parameter control strategy. It is based on the substation operation voltage and reactive power in three states: qualified, too high and too low. The traditional nine-domain diagram control strategy is to divide the voltage-reactive power two-dimensional coordinate plane into nine regions according to the upper and lower limits of the fixed voltage and reactive power (or power factor), as shown in the following Fig. 1.

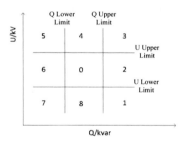

Fig. 1. The partition map of nine-domain diagram.

The traditional nine-domain diagram control has some defects of discriminating fuzzy regions, which coincides with the characteristics of self-adjusting, self-organizing and adaptive fuzzy control.

3 Fuzzy Control Strategy

Fuzzy control strategy is based on the traditional nine-domain diagram to introduce the influence of voltage state. By changing the fixed upper and lower limits of reactive power into fuzzy boundaries, a more widely used nine-domain diagram with fuzzy boundaries is formed. Compared with the traditional nine-domain diagram, the fuzzy reactive power boundary control strategy is more reasonable, and the unnecessary voltage fluctuation and gear adjustment caused by too many switching capacitors are avoided successfully. Figure (2) shows the improved partition map of nine-domain.

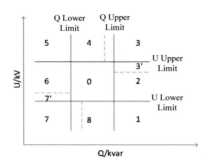

Fig. 2. The improved partition map of nine-domain diagram.

According to the requirements of the power system, it is possible to optimize the control by predicting whether the switching capacitor can achieve the optimal effect. If the switching capacitor can not only guarantee the qualified power factor and restrain harmonics, but also adjust the voltage, then the voltage can be adjusted by switching the capacitor, otherwise, the main variational connector can be adjusted. Based on the above analysis, an optimized improved control strategy can be obtained:

(1) If the capacitor does not cause the power factor to exceed the upper limit, priority is given to the capacitor. If the voltage after the capacitor is still not up to standard, then the main variation connector boosts.

(2) When the capacitor does not exceed the limit of voltage and power factor, only the capacitor is put on.

(3) When the voltage is high but the power factor is low, only the main varistor is adjusted to reduce the voltage, and the re-decision after operation depends on the operation.

(4) When the voltage is high and a set of capacitors cut will cause the power factor to exceed the lower limit, adjust the main variation connector.

(5) If the cut capacitor does not cause the power factor to exceed the lower limit, cut the capacitor first. If the voltage after capacitor cutting is not up to standard, adjust the main variation connector to lower voltage.

(6) The capacitor will not lower the voltage and power factor, only cut the capacitor.

(7) When the voltage is low but the power factor is too high, the voltage of the main varistor is increased, and the decision after operation depends on the operation condition.

(8) When the voltage is too low and the switching of a group of capacitors causes the power factor to exceed the upper limit, only the main transformer tap boost is adjusted.

(9) 3′ area: when it is detected that the voltage of the capacitor exceeds the upper limit, if there is a capacitor that can be put in, first adjust the main varistor to lower the voltage, and then check whether the capacitor needs to be put in. If other circumstances arise, re-decide. If there is no capacitor in the area, it can be maintained temporarily, without the operation of the main transformer joint.

(10) 3′ area: When it is detected that the cut capacitor causes the voltage to cross the lower limit, if there is a capacitor that can be removed, first adjust the main variational connector to raise the voltage, and then check whether the capacitor needs to be removed. If other circumstances arise, re-decide. If no capacitor is found to be resectable in the area, maintain the status quo for the time being without operating the main variational connector.

The control device should be judged synthetically according to voltage, reactive power, time and so on. Judging the current running area according to the real-time data, and then according to the predetermined control scheme. The switching of shunt compensation capacitors in closed loop control station and the adjustment of tap of on-load transformer should be based on the optimal control sequence and the minimum number of actions to make the operation point into the normal working area.

4 Design of Automatic Control Device for Reactive Power and Voltage

4.1 Design of Switch Control Program Module

Capacitor switching control module is the most important module in each module. In this paper, the capacitor switching control module is designed according to the voltage and reactive time compound control criterion, which includes two modules: control flow and capacitor switching control strategy.

(1) Capacitor switching control flow

In the control flow module, a two-stage voltage threshold is set, which is the upper and lower limits of the voltage respectively. The system can automatically judge and calculate the corresponding calibration value. If the voltage is over the upper limit, the lower limit is the power network fault, so the controller sends out tripping signal. If the voltage is between the upper limit and the upper limit, the capacitor is removed step by step to reduce the grid voltage, regardless of the calculated reactive power level and the voltage is detected once per excision stage. If the voltage is restored between the upper and lower limits, the capacitor will no longer be removed. If the voltage is still higher than the upper limit, then the removal continues step by step. If the capacitor is fully cut and the voltage is still above the upper limit, return to the main program and continue to detect the grid voltage, waiting for the grid voltage to return to its normal range. If the voltage is at the lower limit, the capacitor is put in step by step to increase the voltage, regardless of the size and nature of the reactive power, and the voltage is detected once per input stage.

In order to prevent instantaneous interference and switching oscillation, the switching and tripping signals of capacitors do not act immediately with one test result, but are detected again after a period of delay. If it is consistent with the last test result, the device will act. The flow chart of capacitor switching control is shown in Fig. 3.

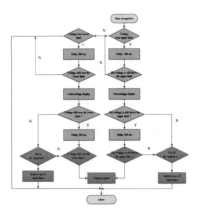

Fig. 3. The flowchart of Capacitor switching control

In the process of capacitor switching control, the protection functions of the device system and the switching control of the capacitor are mainly set up. The design of the program is mainly based on the principle of reactive power compensation. First, the reactive power deficit of the system is detected accurately, then the optimal switching of capacitors is carried out according to the size of the gap, so that the reactive power value can be reduced and the optimal configuration can be achieved. Therefore, after considering the factors such as reactive power and voltage of the system, the following switching control laws are obtained:

(1) Voltage range: The compensation device shall have an effective voltage range for normal operation. According to the actual demand, the relevant parameters are adjusted in advance, and the capacitor banks choose different switching combinations according to the switching values.
(2) Cyclic switching: When the compensation capacity is the same, the device follows the principle of cycle switching.
(3) Switching interval: Preset delay interval. In the time delay, only when the input or the cutting condition is satisfied, can the operation be performed, otherwise the detection can be restarted.

(2) Capacitor switching control strategy

Set Q as reactive power. If Q is perceptual reactive power, Q is positive sign. If Q is capacitive reactive power, then Q is negative sign. 'i' is a capacitor bank array combination number (i = 0, 1, 2, ···, 2n − 1). Qi is the capacitor bank capacity of the i combination. The detected reactive power is Qm. Set Qt is the capacity of a capacitor bank already in use. Under the assumption that a set of capacitors are not put into operation, the total reactive power missing in the system is calculated to be Qz (Qz = Qt + Qm). Qi, which satisfies the minimum condition (Qz − Qi) and (Qz − Qi) ≥ 0, is the capacity that should be put into. Accordingly, the optimal capacitance combination of input is determined. At the same time, according to the condition of the capacitor banks that have been put in, the best combination is selected for switching. The switching control flow chart for capacitor banks is shown in Fig. 4:

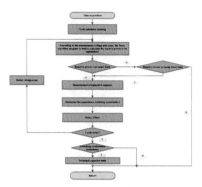

Fig. 4. Switching control strategy of capacitor

4.2 Integral Structure Design of Compensator

There is no uniform standard for line compensation devices. But the main parts are basically the same, including drop fuses, arresters, control parts, capacitors and some other switchgear. The device is shown in Fig. 5.

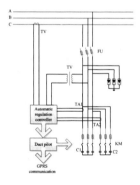

Fig. 5. Electrical schematic diagram of compensator

4.3 Program Module Design of Fuzzy Control Algorithm

Classical control theory and modern control theory have been widely used, but they all need to establish mathematical model of object. But in some complex control systems, it is very difficult to establish accurate mathematical models. The fuzzy control theory can solve this problem well, because they don't need a Pre-established mathematical model to control it. Due to the influence of various factors such as load change, line loss and power network fluctuation, the exact model of power system is difficult to establish, and it is suitable for fuzzy controller to control power system.

In order to improve the reaction speed of the controller and simplify the programming, the system classifies the fuzzy subsets of each input variable according to the control characteristics of the capacitive compensation control device. Then, according to the analysis of the law of capacitive compensation control, several single input systems are designed synthetically. The switching of capacitor banks is determined by the output of multiple independent fuzzy controllers, as shown in Fig. 6.

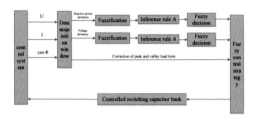

Fig. 6. Fuzzy control design diagram.

5   Conclusion

Reactive voltage intelligent controller is suitable for distribution network 10 kV feeder 10 thousand V and 400 V distribution. Combined with capacitor switching switch, reactive power dispersion compensator is formed. The function of the device is to improve power factor, reduce line loss, save electric energy and improve voltage quality. The controller mainly uses single chip microcomputer technology to collect and calculate the current, voltage, active power, reactive power and power factor of the circuit or distribution equipment. Based on the theory of area control, the switching power capacitor is controlled to realize the automatic compensation of reactive power, and the power factor is adjusted to the optimal state. Establishing reactive power automation network with AVC system.

References

1. Li, S.: Study on voltage and reactive power control strategy of substation based on nine-domain diagram. Qinghai Electr. Power **2**(3), 1–4 (2005). (in Chinese)
2. Ren, X., Cheng, H., Liu, J.: Voltage and reactive power optimization control of substation based on taboo search algorithm. Relay **8**(1), 31–34+39 (2008). (in Chinese)
3. Zhang, Y., Ren, Z., Liao, M., et al.: Optimal reactive compensation on tower of 10 kV long distribution feeder. Electr. Power **33**(9), 50–52 (2000). (in Chinese)
4. Wang, L., Huang, C., Guo, S., Cao, G.: The design of voltage & reactive power comprehensive control device of substation based on fuzzy control theory. Relay **31**(8), 40–42 (2003). (in Chinese)

5. Wu, Y., Yang, Q., Yu, Z.: Design for voltage and reactive power automatic control device in substation on DSP. Program. Control. Fact. Autom. **8**(5), 101–103 (2008). (in Chinese)
6. Yu, C.: Design of optimal reactive power compensation equipment based on genetic algorithm. Shandong Ind. Technol. **12**(6), 227–228 (2017). (in Chinese)
7. Shen, G., Ma, L.: A reactive power compensation method for distribution network. Electron. Test. **9**(8), 73–74 (2013). (in Chinese)
8. Zhang, X., Cao, W., Song, Y.: Comparative analysis of two optimal compensation methods for reactive power. Sci. Technol. Innov. Rev. **33**(5), 183–185 (2009). (in Chinese)
9. Gupta, T., Sambariya, D.K.: Optimal design of fuzzy logic controller for automatic voltage regulator. In: 2017 International Conference on Information, Communication, Instrumentation and Control (ICICIC), Indore, India, 17–19 August 2017
10. Wu, X., Wang, J., Yang, P., Piao, Z.: Fuzzy control on voltage/reactive power in electric power substation. In: Cao, B., Li, T.F., Zhang, C.Y. (eds.) Fuzzy Information and Engineering. Advances in Intelligent and Soft Computing, vol. 62. Springer, Heidelberg (2009)
11. Abusorrah, A.M.: Optimal power flow using adaptive fuzzy logic controllers. Math. Probl. Eng. **2013**, 7 (2013). https://doi.org/10.1155/2013/975170. Article ID 975170

Fractional-Order Control Scheme for Q-S Chaos Synchronization

Adel Ouannas[1], Giuseppe Grassi[2], and Ahmad Taher Azar[3,4(✉)]

[1] Department of Mathematics, University of Larbi Tebessi, 12002 Tebessa, Algeria
ouannas.a@yahoo.com
[2] Dipartimento Ingegneria Innovazione, Università del Salento, 73100 Lecce, Italy
giuseppe.grassi@unisalento.it
[3] Faculty of Computers and Information, Benha University, Banha, Egypt
ahmad_t_azar@ieee.org, ahmad.azar@fci.bu.edu.eg
[4] School of Engineering and Applied Sciences, Nile University, Giza, Egypt

Abstract. In this paper, a fast control scheme is presented for the problem of Q-S synchronization between fractional chaotic systems with different dimensions and orders. Using robust control law and Laplace transform, a synchronization approach is designed to achieve Q-S synchronization between n-D and m-D fractional-order chaotic systems in arbitrary dimension d. This paper provides further contribution to the topic of Q-S synchronization between fractional-order systems with different dimensions and introduces a general control scheme that can be applied to wide classes of fractional chaotic and hyperchaotic systems. Numerical example and simulations are used to show the effectiveness of the proposed approach.

Keywords: Fractional chaos · Q-S synchronization ·
Different dimensions · Laplace transform · Fast control

1   Introduction

The theory of fractional calculus is old as its inception can be attributed to two of the most prominent figures of modern calculus. Over the last decade, researchers have found interest in the theory as well as applications of fractional calculus [2–4]. Chaos refers to the high sensitivity of a dynamical system to small changes in the initial condition and chaotic systems have been around for a while [1]. One of the most studied aspects of chaos is the synchronization of chaotic systems, which aim to force a slave system to mimic the dynamics of a master one. Numerous synchronization types and control strategies have been proposed [1]. The theory related to this kind of systems makes them suitable for certain applications such as encryption [15,16]. In recent years, synchronization of fractional-order chaotic (hyperchaotic) systems became a subject of interest [8,9,11,13,14].

© Springer Nature Switzerland AG 2020
A. E. Hassanien et al. (Eds.): AMLTA 2019, AISC 921, pp. 434–441, 2020.
https://doi.org/10.1007/978-3-030-14118-9_44

Q-S synchronization is one of the most widely studied synchronization types. In this scheme, two functions Q and S are used to condition the slave and master states, respectively, such that different values for Q and S lead to different types of synchronization. The Q-S scheme was first proposed by Yan in [18] for continuous–time dynamical systems. The author, then, extended the Q-S scheme to discrete-time systems [17]. In the years that followed, several algorithms were proposed for the Q-S synchronization [10,12,17].

In this paper, we aim to propose control law for Q-S chaos synchronization relating to fractional systems with different dimensions and non–identical orders. The objective of this study is to establish the convergence of the proposed Q-S synchronization scheme for a pair of fractional systems by means of asymptotic stability results reported in the literature. The next section of this study, Sect. 2, will highlight some of the necessary concepts related to fractional calculus. Section 3 proposes the control law for the Q-S synchronization of two general fractional systems with non-identical dimensions and orders. Section 4 presents the numerical results. Finally, Sect. 5 provides a general summary of the main findings of this study.

2   Preliminaries

The Riemann-Liouville fractional integral operator of order $p \geq 0$ of function $f(t)$ is defined as,

$$J^p f(t) = \frac{1}{\Gamma(p)} \int_0^t (t - \tau)^{p-1} f(\tau) d\tau, \quad p > 0, \ t > 0. \quad (1)$$

where Γ denotes Gamma function. The Caputo derivative of $f(t)$ is defined as,

$$D_t^p f(t) = J^{m-p} \left(\frac{d^m}{dt^m} f(t) \right) = \frac{1}{\Gamma(m-p)} \int_0^t \frac{f^{(m)}(\tau)}{(t-\tau)^{\alpha-m+1}} d\tau, \quad (2)$$

for $m - 1 < \alpha \leq m$, $m \in \mathbb{N}$, $t > 0$. The fractional differential operator $D_t^\alpha f(t)$ is left-inverse (and not right-inverse) to the fractional integral operator J^α, i.e $D_t^\alpha J^p = I$ where I is the identity operator [5].

The Laplace transform of the Caputo fractional derivative rule reads [7]

$$L\left\{{}^C D_t^\alpha f(t)\right\} = s^\alpha F(s) - \sum_{k=0}^{n-1} s^{\alpha-k-1} f^{(k)}(0), \quad (\alpha > 0, \ n - 1 < \alpha \leq n). \quad (3)$$

Particularly, when $\alpha \in (0, 1]$, we have $L\left\{{}^C D_t^\alpha f(t)\right\} = s^\alpha F(s) - s^{\alpha-1} f(0)$. The Laplace transform of the Riemann-Liouville fractional integral rule satisfies

$$L\left\{J^\alpha f(t)\right\} = s^{-\alpha} F(s), \quad (\alpha > 0). \quad (4)$$

Caputo fractional derivative appears more suitable to be treated by the Laplace transform technique in that it requires the knowledge of the (bounded) initial values of the function and of its integer derivatives of order $k = 1, 2, \cdots, m-1$, in analogy with the case when $\alpha = n$.

3   System Description and Problem Formulation

Consider the following system

$$D_t^\alpha X(t) = F(X(t)), \tag{5}$$

where $X(t) = (x_1(t), x_2(t), ..., x_n(t))^T$ is the state vector of the master system (5), $F : \mathbf{R}^n \to \mathbf{R}^n$ is a nonlinear function, $D_t^\alpha = [D_t^{\alpha_1}, D_t^{\alpha_2}, ..., D_t^{\alpha_n}]$ is the Caputo fractional derivative and α_i, $i = 1, 2, ..., n$, are rational numbers between 0 and 1. Also, consider the slave system as

$$D_t^\beta Y(t) = G(Y(t)) + U, \tag{6}$$

where $Y(t) = (y_1(t), y_2(t), ..., y_m(t))^T$ is the state vector of the slave system (6), $G : \mathbf{R}^m \to \mathbf{R}^m$, $D_t^\beta = \left[D_t^{\beta_1}, D_t^{\beta_2}, ..., D_t^{\beta_m}\right]$ is the Caputo fractional derivative, β_i, $i = 1, 2, ..., m$, are rational numbers between 0 and 1 and $U = (u_i)_{1 \le i \le m}$ is a vector controller to be designed.

Definition 1. *The n-dimensional master system $X(t)$ and the m-dimensional slave system $Y(t)$ are said to be Q-S synchronization in dimension d, if there exists a controller $U = (u_i)_{1 \le i \le m}$, two differentiable functions $Q : \mathbf{R}^m \to \mathbf{R}^d$ and $S : \mathbf{R}^n \to \mathbf{R}^d$, such that the synchronization error*

$$e(t) = Q(Y(t)) - S(X(t)), \tag{7}$$

satisfies that $\lim_{t \to +\infty} \|e(t)\| = 0$.

We assume that $d \le m$.

4   Analytical Results

The error system (7), between the master system (5) and the slave system (6), can be derived as

$$\dot{e}(t) = \mathbf{D}Q(Y(t)) \times \dot{Y}(t) - \mathbf{D}S(X(t)) \times \dot{X}(t), \tag{8}$$

where $\mathbf{D}Q(Y(t)) = \left(\frac{\partial Q_i}{\partial y_j}\right)_{d \times m}$, $\mathbf{D}S(X(t)) = \left(\frac{\partial S_i}{\partial x_j}\right)_{d \times n}$ are the Jacobian matrices of the functions Q and S, respectively.

The error system (8) can be written as

$$\dot{e}(t) = M \times (\dot{y}_1, ..., \dot{y}_d)^T + N \times (\dot{y}_{d+1}, ..., \dot{y}_m)^T - \mathbf{D}S(X(t)) \times \dot{X}(t) \tag{9}$$

where $M = \left(\frac{\partial Q_i}{\partial y_j}\right)_{d \times d}$ and $N = \left(\frac{\partial Q_i}{\partial y_j}\right)_{(m-d) \times m}$.

To achieve the Q-S synchronization between systems (5)–(6), The control law $U = (u_i)_{1 \leq i \leq m}$ is selected as

$$(u_1, ..., u_d)^T = -\left(g_1(Y(t)), ..., g_d(Y(t))\right)^T + J^{1-\beta}\left[\mathbf{M}^{-1} \times R\right], \tag{10}$$

and

$$(u_{d+1}, ..., u_m)^T = -\left(g_{d+1}(Y(t)), ..., g_m(Y(t))\right)^T, \tag{11}$$

where $\mathbf{M}^{-1}$ is the inverse of $\left(\frac{\partial Q_i}{\partial y_j}\right)_{d \times d}$ and

$$R = -\mathbf{K}\left[Q(Y(t)) - S(X(t))\right] + \mathbf{D}S(X(t)) \times \dot{X}(t), \tag{12}$$

where $\mathbf{K} = \mathrm{diag}(k_1, k_2, ..., k_d)$ is a feedback gain matrix and k_i $(i = 1, ..., d)$ are positive numbers.

Now, by inserting Eqs. (10) and (11) into (9), we can rewrite the slave system as follows

$$\left(D_t^{\beta_1} y_1(t), ..., D_t^{\beta_d} y_d(t)\right)^T = J^{1-\beta}\left[\mathbf{M}^{-1} \times R\right], \tag{13}$$

and

$$D_t^{\beta_{d-1}} y_{d+1}(t) = ... = D_t^{\beta_m} y_m(t) = 0. \tag{14}$$

By Applying Laplace transform to (13) and letting $\mathbf{F}(s) = \mathbf{L}(y_1(t), ..., y_d(t))^T$, we obtain,

$$s^{\beta}\mathbf{F}(s) - s^{\beta-1}Y(0) = s^{\beta-1}\mathbf{L}\left(\mathbf{M}^{-1} \times R\right), \tag{15}$$

multiplying both the left-hand and right-hand sides of (15) by $s^{1-\beta}$ and applying the inverse Laplace transform to the result, we get the following equation

$$(\dot{y}_1(t), ..., \dot{y}_d(t))^T = \mathbf{M}^{-1} \times R. \tag{16}$$

By using Eqs. (14) and (16), the error system (8) can be written as follow

$$\dot{e}(t) = -Ke(t). \tag{17}$$

It is easy to schow that all solutions of error system (17) go to zero as $t \to \infty$. Now, we can state the following theorem.

Theorem 1. *The master system (5) and the slave system (6) are globally in dimension d subject to (10) and (11).*

5   Numerical Results

Consider the incommensurate fractional order financial system studied by Chen in [6].

$$D^{\alpha_1} x_1 = x_3 + (x_2 - a) x_1, \tag{18}$$
$$D^{\alpha_2} x_2 = 1 - b0.1x_2 - x_1^2,$$
$$D^{\alpha_3} x_3 = -x_1 - cx_3,$$

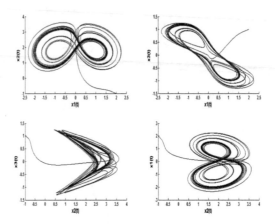

Fig. 1. Chaotic attractors in 2-D of the master system (18).

as a master system. This system exhibits chaotic behaviors when $(\alpha_1, \alpha_2, \alpha_3) = (0.97, 0.98, 0.99)$, $(a, b, c) = (1, 0.1, 1)$ and the initial point is $(2, -1, 1)$. Attractors of the master system (18) are shown in Fig. 1.

The slave system is

$$
\begin{aligned}
D^{\beta_1} y_1 &= 0.56y_1 - y_2 + u_1, & (19)\\
D^{\beta_2} y_2 &= y_1 - 0.1y_2 y_3^2 + u_2,\\
D^{\beta_3} y_3 &= 4y_2 - y_3 - 6y_4 + u_3,\\
D^{\beta_4} y_4 &= 0.5y_3 + 0.8y_4 + u_4,
\end{aligned}
$$

where y_1, y_2, y_3, y_4 are states and u_1, u_2, u_3, u_4 are synchronization controllers. This system, as shown in [19], exhibits hyperchaotic behavior when $(u_1, u_2, u_3, u_4) = (0, 0, 0, 0)$ and $(\beta_1, \beta_2, \beta_3, \beta_4) = (0.98, 0.98, 0.95, 0.95)$. Attractors of the uncontrolled system (19) are shown in Fig. 2.

Using the notations presented in Sect. 3, the errors between the master system (18) and the slave system (19) are defined as follows

$$
(e_1, e_2, e_3)^T = Q(y_1, y_2, y_3, y_4) - S(x_1, x_2, x_3), \tag{20}
$$

where

$$
Q(y_1, y_2, y_3, y_4) = (y_1, y_2 + y_4, 2y_3 + y_4^2), \tag{21}
$$

and

$$
S(x_1, x_2, x_3) = (x_1, x_3 + x_2, x_3 x_2). \tag{22}
$$

So,

$$
\mathbf{DQ} = \begin{pmatrix} 1 & 0 & 0 & 0 \\ 0 & 1 & 0 & 1 \\ 0 & 0 & 2 & 2y_4 \end{pmatrix}, \ \mathbf{DS} = \begin{pmatrix} 1 & 0 & 0 \\ 0 & 1 & 1 \\ 0 & x_3 & x_2 \end{pmatrix} \text{ and } \mathbf{M} = \begin{pmatrix} 1 & 0 & 0 \\ 0 & 1 & 0 \\ 0 & 0 & \frac{1}{2} \end{pmatrix}.
$$

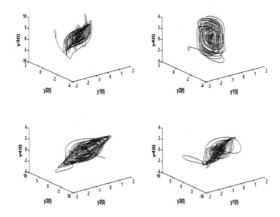

Fig. 2. Chaotic attractors in 3D of the uncontrolled slave system (19).

The matrix $\mathbf{K}$ can be chosen as

$$\mathbf{K} = \begin{pmatrix} 1 & 0 & 0 \\ 0 & 2 & 0 \\ 0 & 0 & 3 \end{pmatrix}, \tag{23}$$

and the vector controller $u = (u_1, u_2, u_3, u_4)^T$ can be obtained as

$$\begin{pmatrix} u_1 \\ u_2 \\ u_3 \\ u_4 \end{pmatrix} = \begin{pmatrix} -0.56y_1 + y_2 + J^{0.02}\left(-e_1 + \dot{x}_1\right) \\ -y_1 + 0.1y_2 y_3^2 + J^{0.02}\left(-2e_2 + \dot{x}_2 + \dot{x}_3\right) \\ -4y_2 + y_3 + 6y_4 + J^{0.05}\left(-\frac{3}{2}e_3 + \frac{1}{2}x_3\dot{x}_2 + \frac{1}{2}x_2\dot{x}_3\right) \\ -0.5y_3 - 0.8y_4 \end{pmatrix} \tag{24}$$

The numerical simulations of the error functions evolution are shown in Fig. 3.

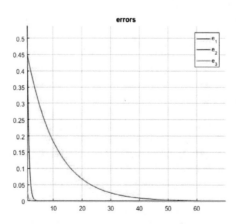

Fig. 3. Time evolution of synchronization errors. (19).

6   Conclusion

This paper has studied Q-S chaos synchronization between fractional-order systems with different dimensions and non-identical orders. The Q-S scheme aims for the general definition of the error towards zero in finite time, thereby covering a range of different synchronization types. We assume an n-dimensional master system and an m-dimensional slave system. Nonlinear control law has been derived and the asymptotic stability of their zero solution investigated through the stability theory of integer-order linear systems. The strategy correspond to the synchronization dimension d with $d \leq m$.

Computer simulation results have been presented, whereby a 4-dimensional fractional-order hyperchaotic slave system synchronized the 3-dimensional fractional-order chaotic master system. The proposed scheme was utilized to achieve 4-dimensional synchronization. The errors have been shown to converge towards zero in sufficient time. Our plan for future work includes an investigation of the applicability of the fractional Q-S synchronization proposed in this paper in encryption and secure communication.

References

1. Azar, A.T., Vaidyanathan, S.: Advances in Chaos Theory and Intelligent Control, vol. 337. Springer, Berlin (2016)
2. Azar, A.T., Vaidyanathan, S., Ouannas, A.: Fractional Order Control and Synchronization of Chaotic Systems, Studies in Computational Intelligence, vol. 688. Springer, Berlin (2017)
3. Azar, A.T., Radwan, A.G., Vaidyanathan, S. (eds.): Fractional Order Systems: Optimization, Control, Circuit Realizations and Applications. Elsevier, Amsterdam (2018)
4. Azar, A.T., Radwan, A.G., Vaidyanathan, S. (eds.): Mathematical Techniques of Fractional Order Systems. Elsevier, Amsterdam (2018)
5. Caputo, M.: Linear models of dissipation whose Q is almost frequency independent-II. Geophys. J. Roy. Astron. Soc. **13**(5), 529–539 (1967)
6. Chen, W.C.: Dynamics and control of a financial system with time-delayed feedbacks. Chaos Solitons Fractals **37**(4), 1198–1207 (2008)
7. Miller, K.S., Ross, B. (eds.): An Introduction to The Fractional Calculus and Fractional Differential Equations. Wiley, New York (1993)
8. Ouannas, A., Al-sawalha, M.M., Ziar, T.: Fractional chaos synchronization schemes for different dimensional systems with non-identical fractional-orders via two scaling matrices. Optik **127**(20), 8410–8418 (2016)
9. Ouannas, A., Azar, A.T., Vaidyanathan, S.: A robust method for new fractional hybrid chaos synchronization. Math. Methods Appl. Sci. **40**(5), 1804–1812 (2016)
10. Ouannas, A., Odibat, Z., Shawagfeh, N.: A new Q-S synchronization results for discrete chaotic systems. Differential Equations and Dynamical Systems (2016). https://doi.org/10.1007/s12591-016-0278-x
11. Ouannas, A., Abdelmalek, S., Bendoukha, S.: Coexistence of some chaos synchronization types in fractional-order differential equations. Electron J. Diff. Eqn. **128**, 1–15 (2017)

12. Ouannas, A., Azar, A.T., Vaidyanathan, S.: On a simple approach for Q-S synchronization of chaotic dynamical systems in continuous-time. Int. J. Comput. Sci. Math. **8**(1), 20–27 (2017)
13. Ouannas, A., Grassi, G., Ziar, T., Odibat, Z.: On a function projective synchronization scheme for non-identical fractional-order chaotic (hyperchaotic) systems with different dimensions and orders. Optik **136**, 513–523 (2017)
14. Ouannas, A., Odibat, Z., Hayat, T.: Fractional analysis of co-existence of some types of chaos synchronization. Chaos Solitons Fractals **105**, 215–223 (2017)
15. Pham, V.T., Ouannas, A., Volos, C., Kapitaniak, T.: A simple fractional-order chaotic system without equilibrium and its synchronization. AEU - Int. J. Electron. Commun. **86**, 69–76 (2018)
16. Wang, X., Ouannas, A., Pham, V.T., Abdolmohammadi, H.R.: A fractional-order form of a system with stable equilibria and its synchronization. Adv. Differ. Eqn. **2018**(1), 20 (2018)
17. Yan, Z.: Q-S synchronization in 3D Hénon-like map and generalized Hénon map via a scalar controller. Phys. Lett. A **342**(4), 309–317 (2005). https://doi.org/10.1016/j.physleta.2005.04.049. http://www.sciencedirect.com/science/article/pii/S0375960105006080
18. Yan, Z.: Q-S (lag or anticipated) synchronization backstepping scheme in a class of continuous-time hyperchaotic systems-a symbolic-numeric computation approach. Chaos **15**(2), 023902 (2005)
19. Zhou, P., Wei, L.J., Cheng, X.F.: A novel fractional-order hyperchaotic system and its synchronization. Chin. Phys. B **18**(7), 2674 (2009)

Path Planning Control for 3-Omni Fighting Robot Using PID and Fuzzy Logic Controller

MennaAllah Soliman[1], Ahmad Taher Azar[1,2(✉)],
Mahmood Abdallah Saleh[1], and Hossam Hassan Ammar[1]

[1] School of Engineering and Applied Sciences, Nile University Campus,
Sheikh Zayed District, Juhayna Square, 6th of October City, Giza 12588, Egypt
ahmad_t_azar@ieee.org
[2] Faculty of Computers and Information, Benha University, Benha, Egypt
ahmad.azar@fci.bu.edu.eg

Abstract. This paper addresses a comparison between some control methods of three Omni wheels firefighting robot due to the variety of maneuverability. To achieve path planning for firefighting robot to reach a specific point with the shortest path, a kinematics model of omni wheel robot is applied with some control algorithms based on PID controller, Fuzzy logic controller and self-tuned PID using fuzzy logic techniques. Hardware prototype has been tested to validate the simulation results.

Keywords: Fuzzy logic controller · Fuzzy-PID controller · 3-omni wheel · Kinematic model · Path planning · Mobile robot · Motion control

1 Introduction

Mobil robots have been recently become a core component of automation system as it involved in many aspects however in industrial application or everyday life. Motion control of mobile robots as path planning is an issue for academic researcher to reach optimization path for doing specific task. Omni wheel robot ease the control of robot path more than differential robot as omni wheels give the robot the flexibility and efficiency [1]. Studying the kinematics of omni wheel robot (3 or 4 wheel), analyzing motion and control have drawn much attention from researchers as it provides an alternative to differential one as it able to synchronize steering and linear motion in any direction [2]. Controlling the motion of mobile robot as well as omni wheel depend mainly on its kinematics by getting velocities needed for certain path in determined area and let the robot to follow these velocities (linear and angular) and transfer them to wheel motor pulse width. This method is an open loop control method as if there is any change in robot working area, the robot should be re-programed to update its path. Conventional controllers like PID are commonly used for controlling wheel motors of the robot but rarely on the whole robot motion to create a closed loop system able to be adapted to new working areas. Also, controlling of nonlinear system is difficult using conventional methods due to uncertainties in the system. Fuzzy set theory [3] which led to a new control method called Fuzzy Control which is able to cope with system uncertainties [4, 5]. The objective of this paper is to design a self-tuned firefighting robot to move from its original location to fire location in a closed loop path. The

© Springer Nature Switzerland AG 2020
A. E. Hassanien et al. (Eds.): AMLTA 2019, AISC 921, pp. 442–452, 2020.
https://doi.org/10.1007/978-3-030-14118-9_45

specific location of fire area should be defined and the robot will navigate to it automatically to achieve this task. The robot is designed to go through some possible paths like linear, circular and curved paths. Because good manipulation is one of the main challenges of the robot, so that, 3-Omni wheels have been chosen as for their variety and flexibility of possible movements. This paper is mainly focused on controlling the robot and mechanical mathematical model based on kinematics for motion control. The methodology of this solution is based on the PID controller, fuzzy controller and self-tuned PID controller. The model is implemented on MATLAB and the simulation result are verified using v-rep software. Hardware prototype has been tested also to validate the simulation results.

2   Mechanical Model: - Omni-Directional Wheeled Mobile Robot (WMR) Kinematic Modeling

Mathematical model of 3 omni wheel drive gets from kinematics model of robot for validating simulation results. First, the forward kinematics are derived where velocities of motors are the inputs, and the outputs are linear velocity (V_X, V_y) and angular velocity ($V\phi$) during motion on the path [6, 7]. As shown in Fig. 1, there is two frames to estimate kinematics of robot; world frame (x, y) and robot frame (x_r, y_r).

 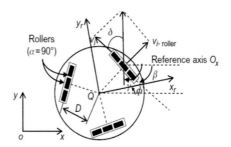

Fig. 1. Nexis 3 Omni wheel kit and robot frames

2.1   Forward Kinematics

Forward kinematics refers to the use of the kinematic equations of a robot to compute the position of the end-effector from specified values for the wheel velocities [8]. As in [8], the forward kinematics equations are:

$$Vx = V3 - V1\cos(\delta) - V2\cos(\delta) \tag{1}$$

$$Vy = V1\sin(\delta) - V2\sin(\delta) \tag{2}$$

$$V\varphi = \frac{V1}{L} + \frac{V2}{L} + \frac{V3}{L} \tag{3}$$

$$Vi(1, 2, 3) = w.r \tag{4}$$

Where V_x is the linear velocity in X-axis; V_y is the linear velocity in Y axis, $V\varphi$ is the angular velocity, $V_{1,2,3}$ are velocities of 3 wheels, δ is the angle of robot, L is the length from wheel to center of mass of Robot and r is the radius of the wheel.

2.2   Inverse Kinematics

Inverse kinematics makes use of the kinematics equations to determine the wheel velocities that provide a desired position for of the robot [9]. The inverse equations are [8]:

$$V1 = (-\sin(\delta)\cos(\delta)Vx + cos2(\delta)Vy + rV\varphi)/L \tag{5}$$

$$V2 = (-\sin(\delta + 120°)\cos(\delta)Vx + \cos(\delta + 120°)\cos(\delta)Vy + rV\varphi)/L \tag{6}$$

$$V2 = (-\sin(\delta + 240°)\cos(\delta)Vx + \cos(\delta + 240°)\cos(\delta)Vy + rV\varphi)/L \tag{7}$$

2.3   DC Motor Model and Parameters Estimation

The equivalent system for DC motor which consists of equivalent electric circuit of the armature and the free-body diagram or the rotor as shown in Fig. 2.

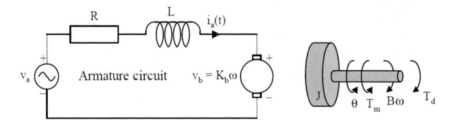

Fig. 2. DC motor free body diagram

The following equations are two transfer functions for Speed with input voltage and torque with input voltage.

$$G(s) = \frac{\dot{\theta}(s)}{V(s)} = \frac{0.2}{0.0422s + 0.2511} \tag{8}$$

$$P(s) = \frac{T(s)}{V(s)} = \frac{0.2s + 0.0004}{0.04222s + 0.2511} \tag{9}$$

Where V(s) is the speed, T(s) is the torque, and $\dot{\theta}$ is the position.

2.4   Linear and Angular Velocity Transfer Function

Using system identification techniques in MATLAB, the transfer function of the linear velocity system (output) and motor speed (input) is extracted as shown in Fig. 3 and Eq. (10). The transfer function of the robot angular velocity (output) and motor speed as (input) is also shown in Fig. 4 and Eq. (11). The transfer function of the estimated system has two poles and one zero with sampling time = 0.5 s.

$$G(s) = \frac{0.07996s + 5.313}{s^2 + 28.63s + 17.65} \tag{10}$$

$$G(s) = \frac{-0.1479s - 0.000969}{s^2 + 8.916s + 0.08298} \tag{11}$$

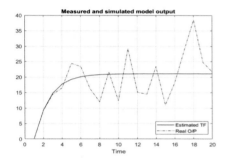

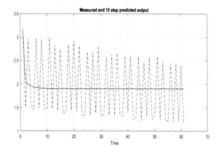

Fig. 3. Linear velocity step response   **Fig. 4.** Angular velocity step response

2.5   Fuzzy Model

The fuzzy model input is the pulse width modulation (PWM) or the speed of the three motors and the robot linear and angular speed as output. Each PWM is represented by 5 triangular memberships ranges from 0 to 255 based on Mamdani rule [3]. The PWM input voltage range is [−100, 0, 100] considering dead zone operation mode for DC motor as shown in Fig. 5. From experience of operation modes of 3 omnidirectional robot motion control, building rule base between 3 PWM as input and the output is linear and angular velocities to get robot model using fuzzy logic as shown in Fig. 6.

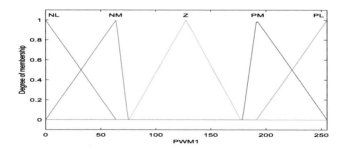

Fig. 5. Input PWM1 membership

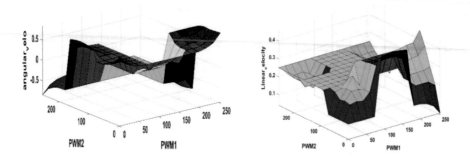

Fig. 6. PWM versus linear and angular velocities surface using fuzzy logic

3   Motion Control Methods

A cascaded model is designed to explain the inner loop and outer loop of control system of the robot as shown in Fig. 7.

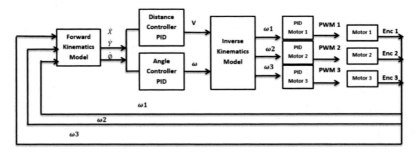

Fig. 7. Cascaded model for control modes

3.1   One Degree of Freedom PID Controller

The design target of the system is to control the Self-Balancing robot with a tuned Proportional-Integral-Derivative (PID) controller [10]. The overall control function can be expressed mathematically as:

$$u(t) = K_p e(t) + K_i \int_0^t e(t')dt' + K_d \frac{de(t)}{dt} \tag{12}$$

Where K_p, K_i and K_d all non-negative, denote the coefficients for the proportional, integral, and derivative terms respectively.

3.2 PID Tuning Used Symmetric Optimum Method (Loop Shaping)

A loop-shaping approach for tuning Proportional-Integral-Derivative (PID) controllers is used in this study. One of loop shaping techniques, known as symmetric optimum method [11], is used to verify the robot velocities response with a reduced overshoot and with a high stiffness degree. Also, this method is used to reach the desired value with zero error and to reject the constant load disturbances. This method, proposed by Kessler [11] considers the following generic process transfer function:

$$G(s) = \frac{K}{(\tau_f s + 1) \prod_{k=1}^{m} (\tau_a s + 1)} \tag{13}$$

$$Kp = \frac{1}{2} \left(\frac{\tau a}{K}\right)\left(\frac{1}{\tau f}\right) \tag{14}$$

$$Ti = 4\tau f \tag{15}$$

where τ_a is the system dominant poles and τ_f represents the process parasitic elements (including fast time constants and time delays).

3.3 Linear Velocity Controller Design

From Eqs. (13), (14) and (15) calculating Linear velocity PID parameters, yields $K_p = 25.9$, $K_i = 0.1185$, $K_d = 0$.

3.4 Angular Velocity Controller Design

From Eqs. (13), (14) and (15) calculating Angular velocity PID parameters, yields $K_p = 32.5$, $K_i = 0.0087$, $K_d = 0$.

4 Results and Discussion

4.1 Path of PID Controller

PID controller is applied for robot path tracking control to reach point (1,1) as target point. Figure 8 shows the result where robot approximately reach target point but with some distortion due to the system nonlinearity (slipping force of the ground – Input delay from 2 sensors encoder and IMU – Motors as nonlinear system).

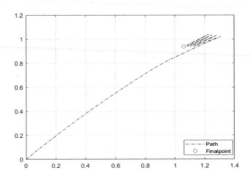

Fig. 8. PID controller for robot path tracking

4.2   FUZZY Logic Controller

After implementing the PID controller, it's noted that conventional control is sufficient for the robot system by considering the input delay, disturbance rejection and working at linear part of DC motor for moving robot in linear paths. For certain paths and better result, fuzzy logic controller (FLC) is used to adapt the robot system in any environment to self-tune and reject any kind of disturbance (slipping and torque) [12, 13]. As 3 Omni wheel robot is a nonlinear system, so that using FLC to control this system is better for more accurate paths using motion control on both linear and angular velocities. Mamdani rules for fuzzy logic technique is used to build the rule-based system between robot location and 3 PWM of wheel motor which effects the linear and angular velocities of the robot as shown in Fig. 9. Table 1 summarizes the rule based system where PL is Positive Large, PM is Positive Medium, Z is zero, NM is Negative Medium, and NL is Negative Large.

Table 1. Rule base of fuzzy logic controller

PWM1 - PWM2 - PWM3		Displacement				
		NL	NM	Z	PM	PL
Angle	NL	NL – NL - NL	PM – NM - Z	PL – PL - PL	NM – PM - Z	NL – PL - Z
	NM	PM – NM - Z	PM – NM - Z	PM – PM - PM	NM – PM - Z	NM – PM- Z
	Z	Z- NL- PL	Z – NM - PM	Z – Z - Z	Z – PM - NM	Z – PL - NL
	PM	NL – Z - PL	NM – Z - PM	NM – NM - NM	PM – Z - NM	PL – Z - NL
	PL	NM – Z - PM	NM – Z - PM	NL – NL - NL	PM – Z - NM	PM – Z - NM

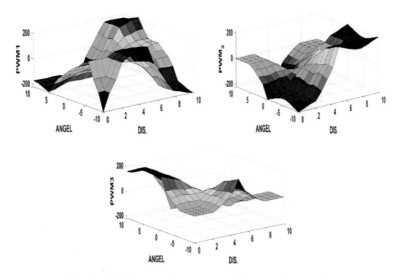

Fig. 9. Rule base surface output of fuzzy logic controller

By applying the rule base on kinematics model of a robot, the output path is shown in Fig. 10 where the target location was (0.5, 0.5) point. It's noted that the robot didn't achieve exactly the target point, but the path is smoother than PID controller.

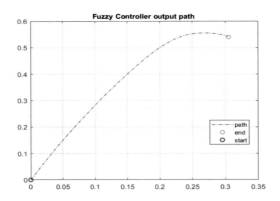

Fig. 10. Fuzzy controller path output

4.3 Self-tuning FUZZY PID

Fuzzy PID self-tuned controller is used where the actual controller is PID and its parameters are updated error using fuzzy logic based on error of robot location and change of error [14]. Mamdani rules are used in building membership functions between input which is error and delta error of robot location and PID parameters (K_p, K_i, K_d) as shown in Fig. 11. The rule base of self-tuning fuzzy PID is shown in Table 2.

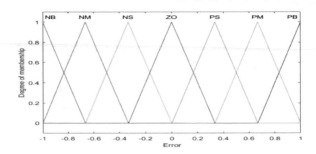

Fig. 11. Self-tuning fuzzy PID membership functions

Table 2. Rule base of self-tuning PID

Kp - Kd - Ki	Error						
	NB	NM	NS	Z	PS	PM	PB
D-Error NB	PB - PS - NB	PB - PS - NB	PM - Z - NB	PM - Z - NM	PS - Z - NM	PS - PB - Z	Z - PB - Z
NM	PB - NS - NB	PB - NS - NB	PM - NS - NM	PM - NS - NM	PS - Z - NS	Z - NS - Z	Z - PM - Z
NS	PM - NB - NM	PM - NB - NM	PM - NM - NS	PS - NS - NS	Z- Z - Z	NS - PS - PS	NM - PM - PS
Z	PM - NB - NM	PS - NM - NS	PS - NM - NS	NS - Z - PS	NM - PS - PS	NM - PM - PM	NM - PM -PM
PS	PS - NB - NS	PS - NM - NS	Z - NS - Z	NS - NS - PS	NS - Z - PS	NM - PS - PM	NM - PS - PM
PM	Z - NM - Z	Z - NS - Z	NS - NS - PS	NM - NS - PM	NM - Z - PM	NM - PS - PB	NB - PS - PB
PB	Z - PS - Z	NS - Z - Z	NS - Z - PS	NM - Z - PM	NM - Z - PB	NB - PB - PB	NB - PB - PB

As a result, the fuzzy controller could tune K_p, K_i, K_d according to the current error and delta error of the running system which could result in more accurate output as shown in Fig. 12.

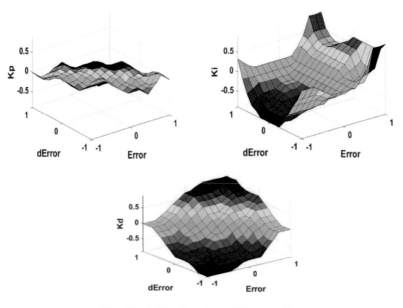

Fig. 12. Self-tuning fuzzy PID controller

The Fuzzy self-tuning path output as shown in Fig. 13 gives the best result as the robot reaches its target point (1,1) with smooth path. This method also allows to simulate more paths like circular and infinity paths as shown in Fig. 14 but it takes more computational time and can't be applied on Arduino controller (as the used prototype is Nexis kit comes with Arduino) but it needs more powerful controller.

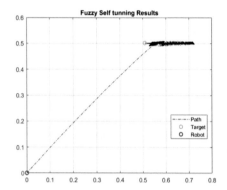

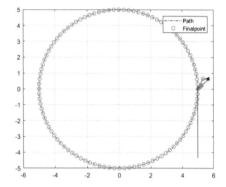

Fig. 13. Fuzzy self-tuning path output   **Fig. 14.** Circular path using fuzzy self-tuning

5   Conclusion

From the simulation and prototype testing, the best path tracking motion is obtained by using self-tuning Fuzzy PID controller with less computational process than Fuzzy logic controller. The results also illustrated that PID controller tuned with loop shaping had some dispersion before reaching target point due to delay in real time and because of non-linearity behavior resulting from the slipping force with the ground, input delay from 3 encoders and IMU even though from non-linearity behavior of DC-motors. The comparing between the fuzzy logic controller and self-tuning fuzzy-PID revealed that the second method is more efficient as fuzzy logic rule-base is not right in all conditions and needed more experiments and trials to reach perfect behavior also it needs more computational time with more processing capability microcontroller.

References

1. Wang, C., Liu, X., Yang, X., Hu, F., Jiang, A., Yang, C.: Trajectory tracking of an omnidirectional wheeled mobile robot using a model predictive control strategy. Appl. Sci. **8**(2), 231 (2018)
2. Zhao, D.B., Qiang, Y.J., Yue, D.X.: Structure and kinematic analysis of omni-directional mobile robots. Robot **25**, 394–398 (2003)
3. Zadeh, L.A.: Fuzzy sets. Inform. Contr. **8**(3), 338–353 (1965)
4. Ahmed, H., Singh, G., Bhardwaj, V., Saurav, S., Agarwal, S.: Controlling of DC motor using fuzzy logic controller. In: Conference on Advances in Communication and Control Systems, pp. 666–670 (2013)

5. Watanabe, K., Shiraishi, Y., Tzafestas, S.G., Tang, J., Fukuda, T.: Feedback control of an omnidirectional autonomous platform for mobile service robots. J. Intell. Robot. Syst. **22**(3–4), 315–330 (1998)

6. Takei, T., lmamura, R., Yuta, S.: Baggage transportation and navigation by a wheel inverted pendulum mobile robot. IEEE Trans. Ind. Electron. **56**, 3985–3994 (2009)

7. Azar, A.T., Ammar, H.H., Barakat, M.H., Saleh, M.A., Abdelwahed, M.A.: Self-balancing robot modeling and control using two degree of freedom PID controller. In: International Conference on Advanced Intelligent Systems and Informatics, pp. 64–76. Springer, Cham (2018)

8. Al-Ammri, A.S., Ahmed, I.: Control of omni-directional mobile robot motion. Al-Khwarizmi Eng. J. **6**(4), 1–9 (2018)

9. Li, X., Zell, A.: Motion control of an omnidirectional mobile robot. In: Filipe, J., Cetto, J.A., Ferrier, J.L. (eds.) Informatics in Control, Automation and Robotics. Lecture Notes in Electrical Engineering, vol. 24. Springer, Heidelberg (2009)

10. Meza, J.L., Santibáñez, V., Soto, R., Llama, M.A.: Fuzzy self-tuning PID semiglobal regulator for robot manipulators. IEEE Trans. Ind. Electron. **59**(6), 2709–2717 (2012)

11. Kessler, C.: Das symmetrische optimum. Regelungstetechnik **6**, 432–436 (1958)

12. Azar, A.T., Ammar, H.H., Mliki, H.: Fuzzy logic controller with color vision system tracking for mobile manipulator robot. In: Hassanien, A., Tolba, M., Elhoseny, M., Mostafa, M. (eds.) The International Conference on Advanced Machine Learning Technologies and Applications (AMLTA 2018). Advances in Intelligent Systems and Computing, vol. 723, pp. 138–146. Springer, Cham (2018)

13. Azar, A.T., Vaidyanathan, S.: Advances in system dynamics and control. Advances in Computational Intelligence and Robotics (ACIR) Book Series. IGI Global, USA (2018). ISBN 9781522540779

14. Zhou, K., Doyle, J.C.: Essentials of robust control, vol. 104. Prentice hall, Upper Saddle River (1998)

Agricultural Service Mobile Robot Modeling and Control Using Artificial Fuzzy Logic and Machine Vision

Mohamed Hesham Barakat[1], Ahmad Taher Azar[1,2(✉)], and Hossam Hassan Ammar[1]

[1] School of Engineering and Applied Sciences, Nile University Campus, Sheikh Zayed District, Egypt
ahmad_t_azar@ieee.org
[2] Faculty of Computers and Information, Benha University, Benha, Egypt
ahmad.azar@fci.bu.edu.eg

Abstract. This paper represents modeling and control of an agricultural service skid steering mobile robot for the purposes of grass cutting using Proportional-Integral-Derivative (PID) controller and Fuzzy Logic techniques and feedback signals from sensors as IMU, encoders, and Machine Vision. The paper deals with the system modeling into two methods: The first is using Fuzzy modeling as a modeling tool for complex nonlinear system, the second is using MATLAB software system Identification Tool. The study Uses PID, Fuzzy logic controller and fuzzy self-tuning of PID controller to control the path tracking of the skid steering mobile robot and improve its response. Utilizing Machine vision provides better results than that of the IMU sensor and the encoders. The fuzzy self-tuning of PID controller shows better dynamic results than the fuzzy logic and PID controllers.

Keywords: Agricultural service robot · Skid steering mobile robot · Modeling · Fuzzy logic control · Fuzzy self-tuning PID · Fuzzy-PID · IMU · Encoders · Vision

1 Introduction

As a result of the shortening of food, individuals and governments are highly concerned due to the rise in the demand of agricultural crops, because of the rapid growing of the population around the world. One of the proposed solutions is to increase the area of the agricultural land. This increase in the cultivated areas led to the rise in the need to the industrialization of agriculture as a necessary response to the challenge of feeding a growing and increasingly urban global population, this is done by introducing autonomous agricultural robots, leading to improve the productivity, efficiency and elimination of the human factor issues in the farming and harvesting of the crops, this produce a great interest in the research field to enhance and introduce new engineering and technological tools in the field of autonomous farming as mobile robots [1]. Mobile robots are simply the robots that can travel or navigate from one point to another autonomously forming a specific path specified by the target points the robot travels.

© Springer Nature Switzerland AG 2020
A. E. Hassanien et al. (Eds.): AMLTA 2019, AISC 921, pp. 453–465, 2020.
https://doi.org/10.1007/978-3-030-14118-9_46

This great features allow the mobile robots to enter and be used in a large scale of applications not only in the field of agriculture but also in another fields for example but not limited to military, rescue, transporting, and exploring applications.

In the agricultural field applications Mobile robots for example can perform the job of inspecting and irrigation, transplanting, harvesting and monitoring the crops [2]. The basics of the skid steering is built on controlling the comparative speeds of both tracks in the same way as differential drive wheeled vehicles. Nevertheless, the control of tracked mobile robots pretenses a problem of the slippage due to the variation of the comparative speed of the two tracks, as well as soil shearing and compacting to achieve steering [3], the matter that makes a control system for trajectory or path tracking is a big challenge. Proportional–Integral–Derivative Controller (PID Controller) is the most common control algorithm used in industry, The attractiveness of PID controllers can be attributed partly to their strong performance in a wide range of operating conditions and partly to their functional simplicity, the PID algorithm main concern is to equal the error value which is the difference between the set-point and the resultant value from the feedback of the plant to be a zero value [4]. Another control method is the Fuzzy logic which is dependent on fuzzy logic, fuzzy logic has the benefits of modeling and control of the nonlinear systems in a verbal variable rather than the traditional techniques of modeling and control as these techniques when working with nonlinear system experiences an intensive computation and complex stability challenges [5].

2   Mobile Robot Modelling

The principles of the skid steering are founded on controlling the relative speeds of both tracks in the same way as differential drive wheeled vehicles. However, the control of tracked mobile robots poses a problem of the slippage due to the difference of the relative velocity of the two tracks, as well as soil shearing and compacting in order to achieve steering [3]. A skid-steered robot can be characterized by two features First, the robot steering depends on controlling the relative speeds of the left and right side tracks. Second, tracks stay parallel to the longitudinal axis of the robot, turning necessitates slippage of the tracks. Many of the difficulties associated with modeling and operating both classes of skid-steered vehicles arise from the track and terrain interaction. Skid steered robots, in general, are modeled by curvilinear motion, the tracks roll and slide at the same time. This makes it difficult to develop kinematic and dynamic models, which accurately describe the motion [6]. Kinematics is not forthright, as it is not possible to forecast the exact motion of the vehicle only from its control inputs. This means motion control methods appropriate for differential wheeled robots cannot be directly used for tracked robots or at least not in a trustworthy way [6]. From this the modelling of the demonstrator is done by two methods, First method using MATLAB^{TM} software identification tool to estimate system transfer function as its approved that system identification is a beneficial tool when its used to build a mathematical model for the dynamic systems [7], Second method using Fuzzy modeling technique to analyze the system's behavior and to design the related controller.

2.1   Robot Modeling and Identification Based on Linear Methods

The transfer function of the demonstrator is identified by the usage MATLAB's system Identification Toolbox, where the system identification is the process of constructing the mathematical model for the dynamic system, this model can be used in modeling and simulation, error detection and controller design. This process involves experimental procedures carried out for variable input signal, followed by generation of the mathematical model regarding to the system response to these input variables [7–9]. The demonstrator is NEXUS tracked mobile tank robot kit 10022 that contains two 12 V DC motors with encoders from which the speed of the robot can be detected, Arduino 328 Microcontroller as the main controller. An IMU (Inertial Measurement Unit) which is an electronic device that can be used to measure body's velocity, acceleration and yaw bitch and roll of a system, is added to the platform to detect speed and angle position of the robot [4]. A vision system is also introduced to the system to detect the inclination angle as a feedback to the system Fig. 1 shows the block diagram of the system, Fig. 2 shows the robot with an object to be detected with the vision system to indicate the orientation angle.

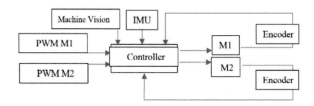

Fig. 1. System block diagram

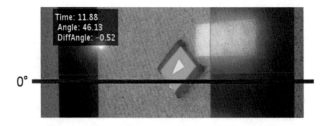

Fig. 2. Vision system to detect angle of inclination

Using MATLAB software for image processing, And Arduino UNO board is used as a data acquisition card to detect and transmit the speed of the two DC motors and the angle of the robot in order for the MATLAB system identification Toolbox to generate the transfer function using the feedback from the encoders and the IMU sensor [4].

In Fig. 3 a comparison between the readings of the orientation angle of the robot using the encoders with kinematics, IMU and the vision system, vision system produces better orientation angle with high accuracy and low noise.

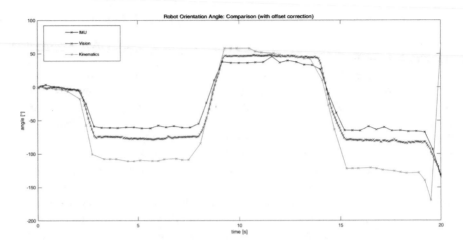

Fig. 3. Robot orientation angle

Table 1. Robot linear speed to a step PWM inputs

Both tracks PWM	Robot linear speed m/s
255	0.05
255	0.15
255	0.2
255	0.25
255	0.3
255	0.35
255	0.38

Table 2. Robot angular speed to a step PWM inputs

Left track PWM	Right track PWM	Robot angular speed rad/s
255	−255	0.15
255	−255	0.2
255	−255	0.25
255	−255	0.3
255	−255	0.35
255	−255	0.38
255	−255	0.4

Table 1 represents the robot speed in response to the step input of the controller as a PWM. Table 2 represents the angular speed of the robot to a step input PWM to the right wheel driving the right track. From these experimental data the software can generate the transfer function of the robot, where the transfer functions of the system are as follows:

$$sys = \frac{926}{s^2 + 79s + 1958} \tag{1}$$

$$sys = \frac{0.3791s + 0.00011}{s^2 + 0.2387s + 0.002604} \qquad (2)$$

Where Eq. 1 is the transfer function of the angular velocity, Eq. 2 is the transfer function of the linear velocity, in Fig. 4 the step response of the generated angular velocity transfer function of the system, while Fig. 5 shows the step response of the generated linear velocity transfer function of the system.

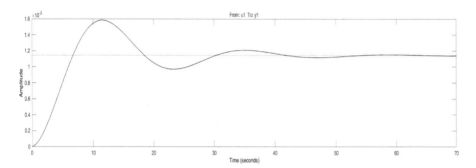

Fig. 4. Step input response of the generated angular velocity transfer function

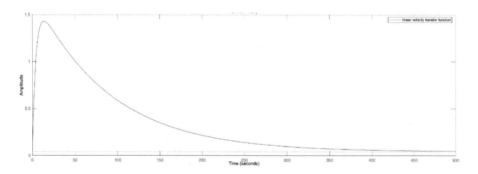

Fig. 5. Step input response to the generated linear velocity transfer function

2.2 Robot Modeling and Identification Based on Non-linear Methods

The second method for generating the transfer function in this paper is done by the usage of fuzzy logic, Zadeh [5] presented fuzzy logic for the first time in 1965. Fuzzy logic is a key development of fuzzy set theory. This is a multivalued logic that allows transitional values to be defined between the conservative evaluations like yes/no, black/white [10]. Fuzzy logic was designed to signify knowledge in linguistic or verbal form [11]. There are key benefits to applying fuzzy logic tools, fuzzy logic tools provide a simplified platform where the development and analysis of models require reduced development time than other approaches. Wherefore, fuzzy logic tools are easy to implement and modify. Notwithstanding, fuzzy tools have shown to perform just as or better than other soft approaches to decision making under uncertainties [12]. The model is based on sending PWM to each motor in a certain Way and measure the speed of the robot using kinematics with the motor encoders to detect the speed as a feedback sensor, Vision and an IMU sensor to detect the orientation angle of the robot and the angular speed, Fig. 1 illustrates the block diagram.

Table 3 shows an experimental sample data got from applying different PWM values for each motor and recording the linear speed and the angular velocity of the mobile robot on the same subjected terrains. Then using the MATLAB Fuzzy Toolbox to analyze the data to generate the mobile robot model.

From the experimental data the rule base is generated regarding to the experimental experience, Tables 4 and 5 shows the rule base of the system.

Table 3. Sample of the experimental data

M1 PWM	M2 PWM	V m/s	W rad/s
100	110	7.67	7.37
100	130	13.25	22.22
100	140	14.99	26.89
100	210	21.24	63.49
100	230	24.15	75.15
100	250	25.76	74.12

The overall rule base as the inputs are a PWM from the two motors and the outputs are the linear and angular velocity, this shows the relation between the two inputs and the two outputs and the overall behavior of the system when applying PWM to a single motor or the two motors and how the linear and angular velocity are related to the inputs. Figures 6 and 7 shows the input membership function for motor 1 and motor 2.

Table 4. Fuzzy rule base with output linear velocity

PWM M2	PWM M1					Very fast counter Clockwise (VFCC)
	FC	Slow C	Dead	Slow CC	FCC	Slow Counter Clockwise (slow CC)
VFCC	HB	Low B	Low B	Low B	Z	Dead zone (Dead)
Slow CC	Low B	Low B	Z	Z	Low F	Slow Clockwise (slow C)
Dead	Low B	Low B	Z	Low F	Low F	Fast clockwise (FC)
						Fast counter clockwise (FCC)
Slow C	Z	Z	Low F	Low F	Low F	High clockwise (HC)
VFC	Z	Z	Z	Low F	HF	Medium clockwise (Mc)
						Zero (Z)
						medium counter clockwise (mCC)
						High counter clockwise (Hcc)

Table 5. Fuzzy rule base with output angular velocity

PWM M2	PWM M1					Very fast counter Clockwise (VFCC)
	FC	Slow C	Dead	Slow CC	FCC	Slow Counter Clockwise (slow CC)
VFCC	Z	Z	mCC	mCC	HCC	Dead zone (Dead)
Slow CC	mC	mC	mCC	mCC	mCC	Slow Clockwise (slow C)
Dead	mC	mC	Z	mCC	mCC	Fast clockwise (FC)
						Fast counter clockwise (FCC)
Slow C	mC	mC	mC	Z	mCC	High clockwise (HC)
VFC	HC	mC	mC	mC	Z	Medium clockwise (Mc)
						Zero (Z)
						medium counter clockwise (mCC)
						High counter clockwise (Hcc)

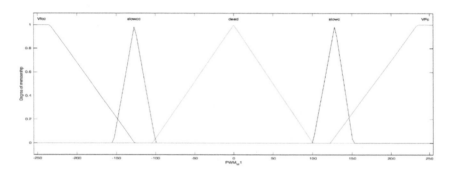

Fig. 6. M1 PWM membership

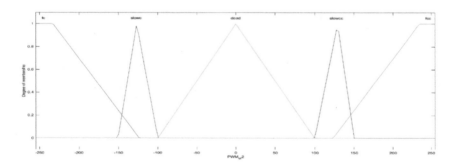

Fig. 7. M2 PWM membership

3 Mobile Robot Control

In this paper three method are used for controlling the mobile robot, the first method is proportional-integral-derivative (PID) controller, the second method is using Fuzzy like PID controller, and the third method is using the Fuzzy self-tuning of PID.

3.1 Proportional-Integral-Derivative (PID) Controller

Using the optimal tuning method to check the performance criterion using the Cohen coon, Integral of Absolute Error (IAE) and Integral of Squared Error (ISE), Figs. 8 and 9 show the performance criterion of the PID controller for the linear and angular velocities transfer function [4].

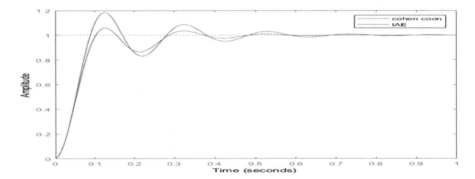

Fig. 8. Step response to the PID controller of the linear velocity

From the figure the Cohen coon optimization technique has a better overshoot and settling time, the ISE optimizer gives an unstable performance.

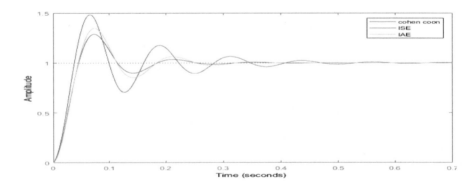

Fig. 9. Step response to the PID controller of the angular velocity

The Cohen coon optimization technique gives the best performance regarding to the overshoot and the settling time. Figure 10 shows the path of the robot while the target coordinates is (4, 3) meters using the point-to-point technique.

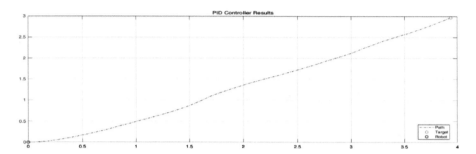

Fig. 10. Robot path to target point with PID controller

3.2 Fuzzy-PID Controller

In this method a fuzzy controller is made to detect the error of orientation angle and distance error, then generating an output PWM to the right and left motor in order to drive the robot to the target point, Fig. 11 shows the orientation error Ø membership functions.

Figure 12 shows the distance error membership, Figs. 13 and 14 shows the rule base of PWM output for the right track and the left track respectively, as for the membership function for PWM of the right and the left tracks are shown in Figs. 15 and 16 respectively.

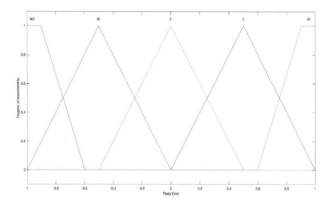

Fig. 11. Orientation angle error membership

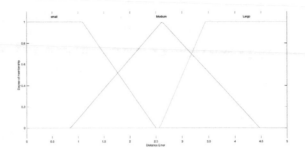

Fig. 12. Distance error membership

Distance error / Ø error	Small	Medium	Large	Very big counter Clockwise (VCC)
				Counter Clockwise (CC)
				Zero (Z)
VCC	HN-R	MR-R	HD-R	Clockwise (C)
				Very beg clockwise (VC)
CC	PR	MP-R	MP-R	High negative right (HN-R)
Z	off	MP-R	HP-R	Medium negative right (MN-R)
C	N-R	P-R	P-R	Negative Right (N-R)
				Positive Right (P-R)
VC	HN-R	P-R	P-R	Medium positive right (MP-R)
				High positive right (HN-R)

Fig. 13. Right track PWM rule base

Distance error / Ø error	Small	Medium	Large	High negative left (HN-L)
				Medium negative left (MN-L)
				Negative left (N-L)
VCC	HN-L	P-L	P-L	Positive left (P-L)
				Medium positive left (MP-L)
CC	N-L	P-L	P-L	High positive left (HN-L)
Z	off	MP-L	HP-L	
C	P-L	MP-L	HP-L	
VC	P-L	MP-L	HP-L	

Fig. 14. Left track PWM rule base

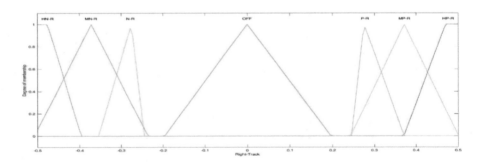

Fig. 15. Membership PWM of the right track

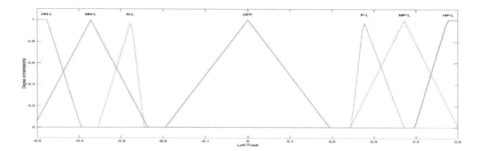

Fig. 16. Membership PWM of the left track

Figure 17 shows the resultant path of the robot while using fuzzy like PID controller against the conventional PID controller.

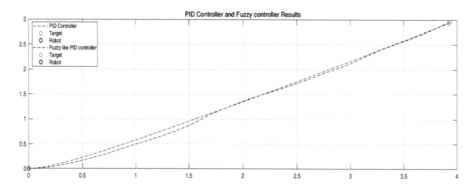

Fig. 17. Fuzzy controller compared to the conventional PID controller resultant path

3.3   Fuzzy Self-tuning of PID

In this method using fuzzy logic to continuously generate K_p, K_i and K_d parameters of the PID controller [13, 14], Fig. 18 shows the technique rule base, Fig. 19 shows the path of the robot with different controller algorithm.

E\EC	NB	NM	NS	ZO	PS	PM	PB
NB	PB	PB	PM	PM	PS	ZO	ZO
NM	PB	PB	PM	PS	PS	ZO	NS
NS	PM	PM	PM	PS	ZO	NS	NS
ZO	PM	PM	PS	ZO	NS	NM	NM
PS	PS	PS	ZO	NS	NS	NM	NM
PM	PS	ZO	NS	NM	NM	NM	NB
PB	ZO	ZO	NM	NM	NM	NB	NB

E\EC	NB	NM	NS	ZO	PS	PM	PB
NB	PS	NS	NB	NB	NB	NM	PS
NM	PS	NS	NB	NM	NM	NS	ZO
NS	ZO	NS	NM	NM	NS	NS	ZO
ZO	ZO	NS	NS	NS	NS	NS	ZO
PS	ZO	ZO	ZO	ZO	ZO	ZO	ZO
PM	PB	NS	PS	PS	PS	PS	PB
PB	PB	PM	PM	PM	PS	PS	PB

E\EC	NB	NM	NS	ZO	PS	PM	PB
NB	NB	NB	NM	NM	NS	ZO	ZO
NM	NB	NB	NM	NS	NS	ZO	ZO
NS	NB	NM	NS	NS	ZO	PS	PS
ZO	NM	NM	NS	ZO	PS	PM	PM
PS	NM	NS	ZO	PS	PS	PM	PB
PM	ZO	ZO	PS	PS	PM	PB	PB
PB	ZO	ZO	PS	PM	PM	PB	PB

Linguistic value
Negative large (NL)
Negative Small (NS)
Negative Medium (NM)
Zero (ZO)
Positive Large (PL)
Positive Small (PM)
Positive Small (PS)
Error (E)
Diffrential error (EC)

Fig. 18. Fuzzy self-tuning of PID rule base

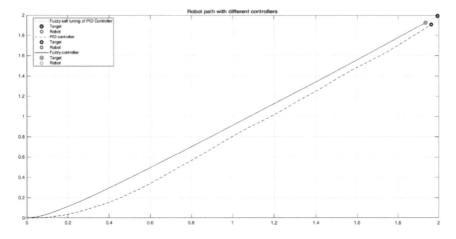

Fig. 19. Path of the robot with different controller algorithm

4 Results and Conclusion

The robot modelling using Fuzzy modeling or system identification and parameter estimation is more reliable and may give a better result than the conventional mathematical model as due slipping and the interaction between the track and terrain.

Also, it is not possible to predict the exact motion of the vehicle only from its control inputs. This means that motion control methods suitable for differential wheeled robots cannot be directly used for tracked vehicles or at least not in a reliable way [6].

However, the robot needs to be modeled each time the track or the type of the terrains changed. Machine vision system gives a better results, low noise and accurate measurements than that of the encoders and the IMU while measuring the orientation angle of the robot also the robot speed, as the encoders gives not accurate values due the system slipping. Controlling the robot using the fuzzy logical controller gives an advantage over the conventional PID controller as it has less overshoot with faster computations than that of the PID controller giving better response to target tracking as shown in Fig. 17. Fuzzy self-tuning PID controller shows a better path tracking with minimal overshoot and a higher accuracy to reach the target than the PID controller or the fuzzy controller, this is shown in Fig. 19.

References

1. Hajjaj, S.S.H., Sahari, K.S.M.: Review of research in the area of agriculture mobile robots. In: Mat Sakim, H., Mustaffa, M. (eds) The 8th International Conference on Robotic, Vision, Signal Processing & Power Applications. Lecture Notes in Electrical Engineering, vol 291. Springer, Singapore (2014)
2. Monta, M.: Robots for bioproduction systems. IFAC Proc. Vol. **33**(29), 1–11 (2000)
3. Martinez, J.L., Mandow, A., Morales, J., Pedraza, S., Garcia-Cerezo, A.: Approximating kinematics for tracked mobile robots. Int. J. Robot. Res. **24**(10), 867–878 (2005)
4. Azar, A.T., Ammar, H.H., Barakat, M.H., Saleh, M.A., Abdelwahed, M.A.: Self-balancing robot modeling and control using two degree of freedom PID controller. In: International Conference on Advanced Intelligent Systems and Informatics. Springer, Cham (2018)
5. Zadeh, L.A.: Outline of a new approach to the analysis of complex systems and decision processes. IEEE Trans. Syst. Man Cybern. **1**, 28–44 (1973)
6. Mandow, A., Martinez, J.L., Morales, J., Blanco, J.L., Garcia-Cerezo, A., Gonzalez, J.: Experimental kinematics for wheeled skid-steer mobile robots. In: IEEE/RSJ International Conference on Intelligent Robots and Systems, 2007. IROS 2007. IEEE (2007)
7. Nelles, O.: Nonlinear System Identification: From Classical Approaches to Neural Networks and Fuzzy Models. Springer, Heidelberg (2013)
8. Jakeman, A.J., Littlewood, I.G., Whitehead, P.G.: Computation of the instantaneous unit hydrograph and identifiable component flows with application to two small upland catchments. J. Hydrol. **117**(1-4), 275–300 (1990)
9. Jelali, M., Kroll, A.: Hydraulic Servo-Systems: Modelling, Identification and Control. Springer, Heidelberg (2012)
10. Klir, G., Yuan, B.: Fuzzy Sets and Fuzzy Logic, vol. 4. Prentice Hall, New Jersey (1995)
11. Salimi, A., Subaşı, M., Buldu, L., Karataş, C.: Prediction of flow length in injection molding for engineering plastics by fuzzy logic under different processing conditions. Iran. Polym. J. **22**(1), 33–41 (2013)
12. Azadegan, A., Porobic, L., Ghazinoory, S., Samouei, P., Kheirkhah, A.S.: Fuzzy logic in manufacturing: a review of literature and a specialized application. Int. J. Prod. Econ. **132**(2), 258–270 (2011)
13. Fu, K. (ed.): Learning Systems and Intelligent Robots. Springer, London (2012)
14. He, S.-Z., et al.: Fuzzy self-tuning of PID controllers. Fuzzy Sets Syst. **56**(1), 37–46 (1993)

Performance Evaluation of Research Reactors Under Different Predictive Controllers

Mina S. Andraws[1,2(✉)], Asmaa A. Abd El-Hamid[1],
Ahmed H. Yousef[2,3], Imbaby I. Mahmoud[1], and Sherief A. Hammad[2]

[1] Nuclear Research Center, Egyptian Atomic Energy Authority, Cairo, Egypt
mina.shoukrey@yahoo.com, asm28aa@yahoo.com,
imbabyisma@yahoo.com
[2] Computer and Systems Department, Ain Shams University, Cairo, Egypt
{ahassan, sherif.hammad}@eng.asu.edu.eg
[3] Communication and Information Technology, Nile University, Cairo, Egypt
ahassan@nu.edu.eg

Abstract. This paper is concerned with the evaluation of nuclear research reactor under two types of predictive controllers. The first one is Receding Horizon Predictive Controller (RHPC) which is considered a simple linear predictive controller. The other one is Neural Network Predictive Controller (NNPC) which is a type of nonlinear predictive controller. These controllers are applied over multi-point reactor core model. This model takes into consideration the nonlinearity of the reactor. It also takes into consideration some important physical phenomena like temperature effect, time variant fuel depletion and nonlinear xenon poisoning concentration. Simulation results showed the superiority of RHPC in both tracking and regulation.

Keywords: Neural Network Predictive Controller ·
Receding Horizon Predictive Controller · Nuclear research reactor

1 Introduction

Research reactors represent a wide range of reactors that are not used for power generation. The main purpose of these reactors is generating electrons. This electron beam has a great value in producing isotopes which are used in many nondestructive fields. One of the most important application is medicine, as these produced isotopes are used in the detection of tumors and radiation therapy of cancer. They also have a use in the art as the radiography analyze the material to detect the authenticity of the art. These isotopes have a wide range of applications in agriculture, industry and scientific research [1].

The production of these isotopes requires different operating points for the reactor to be set. The reactor needs a controller to work at these operating points. That controller must ensure safety and can work under the system constraints. It also has to handle system nonlinearity and time-varying parameters.

Researchers had suggested different controllers for research reactors. The traditional control is still predominant for research reactors such as proportional differential (PD) controller [2].

© Springer Nature Switzerland AG 2020
A. E. Hassanien et al. (Eds.): AMLTA 2019, AISC 921, pp. 466–477, 2020.
https://doi.org/10.1007/978-3-030-14118-9_47

An artificial intelligent such as Fuzzy controller [3] and fractional order fuzzy control are applied over reactor models [4], while other researchers proposed optimum controllers which shows its effectiveness in regulation in nuclear reactors.

The model predictive control (MPC) is one of these optimal controllers which is applied to power reactor [5, 6]. The motive of this paper is to evaluate the performance of different controllers over a research reactor by comparing PD with MPC and NNPC which has a good stability against nonlinear systems [7]. The rest of this study is organized as follows. In Sect. 2, reactor model is presented. In Sect. 3, proposed controllers are formulated. In Sect. 4, the results are showed and discussion about it is presented. In Sect. 5 the conclusion of this paper is provided.

2 Research Reactor Model

There are different ways of modeling the reactors [8], but multipoint reactor core modeling is used in this paper. The reason behind using that model that it provides sufficient details of the reactor without ignoring important factors. Yet it is still simple to use it for designing a controller. The details of model (Fig. 1) block will be discussed in the following sections.

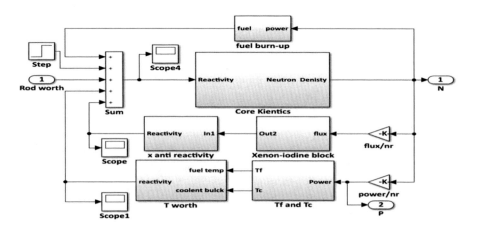

Fig. 1. Simplified block diagram of the reactor model [9]

2.1 Point Reactor Kinetics

Reactor core kinetics evaluate the dynamics of neutron flux, as its main factor is total reactivity. The total reactivity is calculated with enough accuracy by representing it in two terms. The first term evaluates the neutron density, while the second term calculates delayed neutron precursors [10, 11].

$$\frac{dn}{dt} = \frac{\rho - \beta}{n}\bigcap + \sum_{i=1}^{6} \lambda_i C_i \tag{1}$$

$$\frac{dC_i}{dt} = \frac{\beta_i}{n}\bigcap + \lambda_i C_i \tag{2}$$

Where n is neutron density, ρ is reactivity, β is effective delayed neutron fraction, $\bigcap$ neutron generation time and λ_i, C_i, β_i donate the ith group of precursor decay constant, precursor density, effective delayed neutron fraction respectively.

For computational purposes, it is convenient to use the equivalent normalized equations for normalized neutron density $N = \frac{n}{n_o}$ [12]. The equivalent normalized versions of Eqs. (1) and (2) use the delayed neutron relative abundance ai instead of βi, to improve the flexibility of the model. The βi values are changing with the effective delayed neutron fraction β, while the ai values are constant for a given fissile material. So, the ith group of normalized precursor density is $\theta_i = \frac{\lambda_i C_i \bigcap}{\beta_i n_o}$ and $a_i = \frac{\beta_i}{\beta_i}$ is the relative abundance of ith group delayed neutron. It's better to represent the reactivity in dollars $\$ = \frac{\rho}{\beta}$.

$$\frac{dN}{dt} = \frac{\beta}{\bigcap} [(\$ - 1)N + \sum_{i=1}^{6} a_i \theta_i \tag{3}$$

$$\frac{d\theta_i}{dt} = \lambda_i (N - \theta_i) \tag{4}$$

2.2   Temperature Effect Reactivity

The Doppler coefficient affects primarily the resonance escape probability temperature coefficients α_c and α_f which represent coolant temperature coefficient and fuel temperature coefficient respectively. So the reactivity change due to the temperature of the fuel T_f and moderator T_c with the following equations [10].

$$\frac{\partial \rho_t}{\partial T_c} = \alpha_c \tag{5}$$

$$\frac{\partial \rho_t}{\partial T_f} = \alpha_f \tag{6}$$

The temperature of fuel rises due to the operation. The coolant works in a separate closed loop system which its responsibility is to maintain the temperature of the fuel at a point to achieve the highest neutron energy. The temperature can be calculated with simple set of point reactor equations without considering the spatial temperature distribution inside the fuel or the coolant.

$$M_f C_f \left(\frac{dT_f}{dt}\right) = Po - hA(T_f - T_c) \tag{7}$$

$$M_c C_c \left(\frac{dT_c}{dt}\right) = -2F\,C_c(Tc - Ti) + hA(T_f - T_c) \tag{8}$$

Where M_f and M_c are the masses of fuel and coolant respectively. C_f and C_c are the specific heat of fuel and coolant respectively. Po is the reactor power in Watt, F is the coolant mass flow kg/s and Ti is the inlet temperature.

2.3 Fission Product Poisoning

The fission of Uranium (^{235}U) produces many different materials [13]. The overall production process and half-lives involved are summarized below.

$$(^{135}Tel)\frac{(\beta < 0.5\,\text{min})}{fission} \rightarrow (^{135}I)\frac{(\beta(6.7\,\text{hr}))}{fission} \rightarrow (^{135}xe)\frac{(\beta(9.2\,\text{hr}))}{fission}$$

$$\rightarrow (^{135}Cs)\frac{(\beta(9.2\,\text{hr}))}{fission} \rightarrow (^{135}Ba)stable$$

Xenon (^{135}xe) is the most important element of fission as it makes high anti reactivity ρ_x because it has high thermal neutron absorption cross section $\sigma_{ax} = 2.6 * 10^{-6}$ barns over the average Σ_F.

$$\rho_x = \frac{\sigma_{ax}}{\Sigma_F nu} x \tag{9}$$

The Xenon X can be produced directly from Uranium (^{235}U) fission or indirectly from iodine I decay.

$$\frac{dI}{dt} = \Gamma_i \Sigma_F \Phi_T + \lambda_I I \tag{10}$$

$$\frac{dX}{dt} = \Gamma_x \Sigma_F \Phi_T + \lambda_I I - \lambda_x X - \sigma_{ax} \Phi_T X \tag{11}$$

Where Γ_i, Γ_x are the fission yield of iodine and xenon respectively, λ_I, λ_x are the decay constant of iodine and xenon respectively and Φ_T is thermal flux.

2.4 Fuel Density

The fuel reactivity ρ_f is function of Uranium (^{235}U) density which is decreasing by its consumption to produce power P_o. The reactivity dynamic behavior can be explained the next equation [2].

$$\frac{d\rho_f}{dt} = -K_f P_o \tag{12}$$

2.5   Control Rods

The reactor in this paper has six control rods, these rods are made of cadmium to control the reactivity by absorbing neutrons. These six cadmium bars are categorized into two safety rods and four compensating rods which is controlled by stepper motors. The safety rods shut down the reactor in case of emergency and the compensating rod are responsible for regulating the power by changing its length in the reactor Z_{irod}. Where ρ_{rod}, ρ_{ifrod} are reactivity control rod and control rod fuel respectively.

$$\rho_{rod} = \sum_{i=1}^{6} \rho_{ifrod}(1-0.01Z_{irod}) \tag{13}$$

The rods aren't moved simultaneously. Only one moves at a sample time which is determined by this simple control logic.

```
If (CS>0)
  {for (k=1; k<4; k++)
   If (Cr[k] > Cr [pos- 1]) {pos = k+1;}}
Else if (CS<0)
  {for (k=1; k<4; k++)
   If (Cr[k] < Cr [pos- 1]) {pos = k+1;}}
```

From Eqs. (2–13) the state space of the system can be obtained.

$$X^T = \begin{bmatrix} N & \theta_1 & \theta_2 & \theta_3 & \theta_4 & \theta_5 & \theta_6 & T_f & T_C & I & X & \rho_f & \rho_{rod} \end{bmatrix}.$$

$$Am = \begin{bmatrix}
\frac{-\beta}{\Lambda} & \frac{\beta_1}{\Lambda} & \frac{\beta_2}{\Lambda} & \frac{\beta_3}{\Lambda} & \frac{\beta_4}{\Lambda} & \frac{\beta_5}{\Lambda} & \frac{\beta_6}{\Lambda} & \frac{\alpha_c}{\Lambda}N_o & \frac{\alpha_f}{\Lambda}N_o & 0 & \frac{\sigma_{ax}}{\Sigma_p nu}N_o & \frac{K_f}{\Lambda}N_o & \frac{N_o}{\Lambda} \\
\lambda_1 & -\lambda_1 & 0 & 0 & 0 & 0 & 0 & 0 & 0 & 0 & 0 & 0 & 0 \\
\lambda_2 & 0 & -\lambda_2 & 0 & 0 & 0 & 0 & 0 & 0 & 0 & 0 & 0 & 0 \\
\lambda_3 & 0 & 0 & -\lambda_3 & 0 & 0 & 0 & 0 & 0 & 0 & 0 & 0 & 0 \\
\lambda_4 & 0 & 0 & 0 & -\lambda_4 & 0 & 0 & 0 & 0 & 0 & 0 & 0 & 0 \\
\lambda_5 & 0 & 0 & 0 & 0 & -\lambda_5 & 0 & 0 & 0 & 0 & 0 & 0 & 0 \\
\lambda_6 & 0 & 0 & 0 & 0 & 0 & -\lambda_6 & 0 & 0 & 0 & 0 & 0 & 0 \\
\frac{PK}{M_f C_f} & 0 & 0 & 0 & 0 & 0 & 0 & \frac{-hA}{M_f C_f} & \frac{hA}{M_f C_f} & 0 & 0 & 0 & 0 \\
0 & 0 & 0 & 0 & 0 & 0 & 0 & \frac{-2F C_c}{M_c C_c} - \frac{hA}{M_c C_c} & \frac{hA}{M_c C_c} & 0 & 0 & 0 & 0 \\
0 & 0 & 0 & 0 & 0 & 0 & 0 & 0 & 0 & -\lambda I & 0 & 0 & 0 \\
\Gamma_x \Sigma_p K & 0 & 0 & 0 & 0 & 0 & 0 & 0 & 0 & -\lambda I & \lambda_x + kN_o & 0 & 0 \\
K_f & 0 & 0 & 0 & 0 & 0 & 0 & 0 & 0 & 0 & 0 & 0 & 0 \\
K_f & 0 & 0 & 0 & 0 & 0 & 0 & 0 & 0 & 0 & 0 & 0 & 0 \\
0 & 0 & 0 & 0 & 0 & 0 & 0 & 0 & 0 & 0 & 0 & 0 & 0
\end{bmatrix}$$

B_m matrix is $\begin{bmatrix} 0 & 0 & 0 & 0 & 0 & 0 & 0 & 0 & 0 & 0 & 0 & 0 & 1 \end{bmatrix}$

And C_m matrix is $\begin{bmatrix} 1 & 0 & 0 & 0 & 0 & 0 & 0 & 0 & 0 & 0 & 0 & 0 & 0 \end{bmatrix}$ and $D_m = [0]$.

3 Control Methods

There are many methods for controlling nuclear reactor [8], but only three methods will be discussed in this paper the proportional-derivative PD which is currently applied to this reactor [2], model predictive control (MPC) and Neural Network Predictive control (NNPC). There are some constraints which must applied over all this controller. The control signal is between ± 1, the rate change of any rod position must not exceed one step per second. These constraints are applied by rod control logic.

3.1 Proportional-Derivative Controller (PD)

Many research reactors are controlled by PD. The traditional controllers have proven their robustness against many systems. Although due to systems nonlinearity, it fails to track some changes [2].

$$CS = K_p \left(\frac{P_o}{P_{ref}} - 1 \right) + K_d f_r \tag{14}$$

Where CS is control signal, K_p, K_d are proportional and derivative gains, f_r is flux rate and P_{ref} is the reference power.

3.2 Model Predictive Control (MPC)

Model predictive control (MPC) is a type of optimum controller. This controller uses the mathematical model, previous outputs and inputs to predict the future outputs. Then uses this prediction to change the future control signal to minimize the difference between the actual and desired output. N_p and N_c are prediction horizon and control horizon respectively.

$$A = \begin{bmatrix} A_m & 0_m^T \\ C_m A_m & 1 \end{bmatrix} \tag{15}$$

$$B = \begin{bmatrix} B_m \\ C_m B_m \end{bmatrix} \tag{16}$$

$$C = \begin{bmatrix} 0_m & C_m \end{bmatrix} \tag{17}$$

$$\phi = \begin{bmatrix} CB & 0 & 0 & 0 & \ldots & 0 \\ CAB & CB & 0 & 0 & \ldots & 0 \\ CA^2B & CAB & CB & 0 & \ldots & 0 \\ \cdot & \cdot & \cdot & \cdot & \ldots & \cdot \\ \cdot & \cdot & \cdot & \cdot & \ldots & \cdot \\ CA^{N_p-1}B & CA^{N_p-2}B & CA^{N_p-3}B & \cdot & \ldots & CA^{N_p-N_c}B \end{bmatrix} \tag{18}$$

$$F = \begin{bmatrix} CA \\ CA^2 \\ CA^3 \\ . \\ . \\ CA^{N_p} \end{bmatrix} \tag{19}$$

$$R = r_w I_{N_C * N_C} \tag{20}$$

$$R_s = [1 \quad 1 \quad . \quad . \quad 1] * r(K_i) \tag{21}$$

$$\Delta CS(k+1) = (\phi^T \phi + R)^{-1} \phi^T (R_s - Fx(K_i)) \tag{22}$$

$$CS(k+1) = CS(k) + \Delta CS(k+1) \tag{23}$$

The control signal can be obtained from Eq. 23.

3.3 Neural Network Predictive Control (NNPC)

It is a type of nonlinear predictive controller which uses the features in the neural network for modeling of nonlinear system. The use of Artificial Neural Network (ANN) has been carried using Multilayer Feedforward Neural Networks. Which have been successful in identification and control dynamics system [14, 15]. The identified model is used to design controller with generalized predictive control, which updates the weights of ANN online.

This ANN is used to identify plant model off-line by a technique called "Levenberg Marquardt" [16]. The training is done by subjecting the plant model and the ANN to the same input CS(k). The error between output of plant model y(k) and output of neural network $y_n(k)$ is used to modify the weights of ANN. This step is repeated until the error is at its minimum threshold.

The ANN is consisting of the input layer, output layer and Hidden layer. The inputs are time delayed inputs [cs(k) ...cs(k−Ni)] and time delayed outputs [y(k−1) ...y(k−Nj)]. The Network has one hidden layer with 2 1 neurons that have linear activation function. The output node has a tanh activation function (see Fig. 2).

When the network is trained, the predicted output j can be obtained [17]

$$\hat{y}(k+j) = W_2^T * \varphi_2 \left(W_1^T * x(k+j) \right) \tag{24}$$

As the weights of hidden layer W_1 and the output layer W_2 can be represented in the matrix form. Where d is open loop input, output plant delay, n is the size of input vector x(k) = Nj + Ni − d + 1 and φ_2 is the second layer (output) activation vector.

$$W_1^T = \begin{bmatrix} b_{10} & v_{11} & \cdots & v_{1n} \\ \vdots & \vdots & & \vdots \\ & & \cdots & \\ b_{L0} & v_{L1} & \cdots & v_{Ln} \end{bmatrix} \tag{25}$$

$$W_2^T = [b \quad w_{11} \quad \cdots \quad w_{1L}] \tag{26}$$

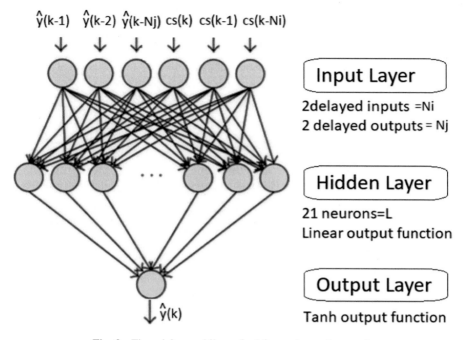

$\widehat{y}(k-1)$ $\widehat{y}(k-2)$ $\widehat{y}(k-Nj)$ cs(k) cs(k-1) cs(k-Ni)

Input Layer

2delayed inputs =Ni
2 delayed outputs = Nj

Hidden Layer

21 neurons=L
Linear output function

Output Layer

Tanh output function

$\widehat{y}(k)$

Fig. 2. Time delay multilayer feed-forward neural network

$$x(k) = [\widehat{y}(k+1) \quad \ldots \quad \widehat{y}(k+Nj) \quad cs(k-d) \quad cs(k-d-1) \quad \ldots \quad cs(k-Ni)] \tag{27}$$

The value of control signal is obtained from Eq. (28) and the cost function $J(N_1, N_2, N_u)$ (30), as N_1, N_2 are minimum and maximum prediction horizon. And N_u are the control horizon.

$$CS(n+1) = CS(n) - \left[\frac{\partial^2 J}{\partial CS^2}(n)\right]^{-1} * \frac{\partial J}{\partial CS}(n) \tag{28}$$

Where

$$CS(n) = \begin{bmatrix} cs(k+1) \\ \vdots \\ cs(k+N_u) \end{bmatrix} \tag{29}$$

$$J(N_1, N_2, N_u) = \mu \sum_{j=N_1}^{N_2} (y_d(k+j) - \widehat{y}(k+j))^2 + \vartheta \sum_{j=1}^{N_u} \Delta cs(k+j-1)^2 \tag{30}$$

The partial differentiation in Eq. (28) can be calculated as follows.

$$\frac{\partial J}{\partial CS}(n) \triangleq \begin{bmatrix} \frac{\partial J}{\partial cs(k+1)} \\ \vdots \\ \frac{\partial J}{\partial cs(k+N_u)} \end{bmatrix} \tag{31}$$

$$\frac{\partial^2 J}{\partial CS^2}(n) = \begin{pmatrix} \frac{\partial^2 J}{\partial cs(k+1)^2} & \cdots & \frac{\partial^2 J}{\partial cs(k+1)\partial cs(k+N_u)} \\ \vdots & \ddots & \vdots \\ \frac{\partial^2 J}{\partial cs(k+N_u)\partial cs(k+1)} & \cdots & \frac{\partial^2 J}{\partial cs(k+N_u)^2} \end{pmatrix} \tag{32}$$

As each element in $\frac{\partial J}{\partial CS}$ and the Hessian matrix $H = \frac{\partial^2 J}{\partial CS^2}$ are calculated as following.

$$\frac{\partial J}{\partial cs(k+h)} = -2\sum_{j=N_1}^{N_2} (y_d(k+j) - \widehat{y}(k+j)) \frac{\partial \widehat{y}(k+j)}{\partial cs(k+h)} + \sum_{j=1}^{N_u} \Delta cs(k+j-1) \frac{\partial cs(k+j)}{\partial cs(k+h)} \tag{33}$$

$$\frac{\partial^2 J}{\partial cs(k+m)\partial cs(k+h)} =$$

$$2\sum_{j=N_1}^{N_2} (y_d(k+j) - \widehat{y}(k+j)) \frac{\partial^2 \widehat{y}(k+j)}{\partial cs(k+m)\partial cs(k+h)} - \frac{\partial \widehat{y}(k+j)}{\partial cs(k+m)} \frac{\partial \widehat{y}(k+j)}{\partial cs(k+h)} +$$

$$2\sum_{j=1}^{N_u} \Delta cs(k+j-1) \frac{\partial^2 cs(k+j)}{\partial cs(k+m)\partial cs(k+h)} + \frac{\partial cs(k+j)}{\partial cs(k+m)} \frac{\partial cs(k+j)}{\partial cs(k+h)} \tag{34}$$

4 Results and Discussion

These controllers have been applied to Simulink® model of 22 MW research reactor. This model takes in consideration the nonlinearity of neutron production and xenon absorption effect. It also takes in consideration the effect of the time-variant parameter fuel depletion.

The controllers that have been tested with time sample of 0.5 s has the following parameters. The PD controller has $K_p = 5.5$ and $K_d = 6.667$. Both MPC and NNPC has prediction horizon of $N_p = N_2 = 10$ and control horizon $N_c = N_u = N_1 = 2$.

The NNPC has a delayed input $N_i = 2$ and delayed output $N_j = 2$. The hidden layer has L = 21 neurons. Three iterations are used to adapt weights online at each sample.

To make a sufficient test over these controllers, it has to detect the time-invariant and nonlinearity in the system by testing the control tracking when the system at high power operation and the effect of fuel depletion on the operation. The starting of reactor is the test case will be used as the compensating rod is placed manually at 65% of its withdrawal length then the controller is engaged with power setpoint of 70%, after that the power stepped up to 100% at time 1000 s as showed in Figs. 3, 4 and 5 are the zoom of the system response at time 1000 to 1100 s to illustrate the tracking of the system and time 1320 to 1400 s to show the steady state error respectively.

Although both MPC and NNPC have no steady state error, the NNPC has a higher overshoot and slower than MPC as showed in Table 1.

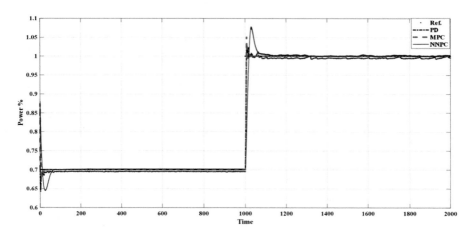

Fig. 3. System response when it stepped the power from 70% to 100% at 1000 s.

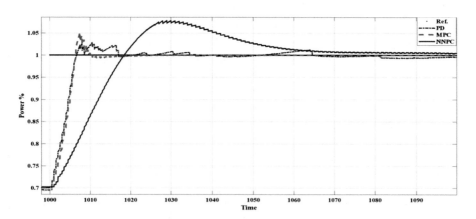

Fig. 4. System response when it stepped up (Zoom in)

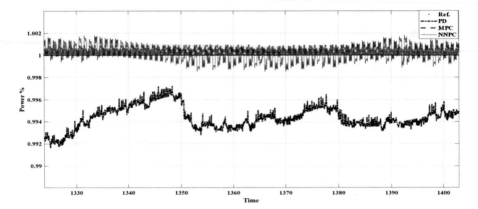

Fig. 5. Zoom in to check steady state error

Table 1. Controllers Performance.

Control parameter	PD	MPC	NNPC
Steady state error	0.7%	No	No
Overshoot	3.2%	5.2%	7.6%
2% Settling time	15 s	9 s	53 s

5   Conclusion

The controllers are implemented on research reactor model. They compensate the changes due to time-variant parameter fuel depletion and deal with nonlinearity at changes of power operation. Both MPC and NNPC show the ability to remove steady-state error. Results showed MPC has a performance better than NNPC. The reason the response of NNPC is slower than MPC is that it identifies the plant at certain point. Thus, when the system changes its parameters some iterations are needed to adapt the neural network weights. These new weights change plant identification which requires design another predictive controller. In the future, the simulation will be verified by hardware implementation of the control.

References

1. IAEA, Applications of Research Reactors. International Atomic Energy Agency, Vienna International Centre (2014)
2. Mahmoud, I.I: Modeling and automatic control of nuclear reactors. In: IEEE Proceedings of the Nineteenth National Radio Science Conference, vol. 1–2. IEEE, New York (2002)
3. Emara, H.M., et al.: Power stabilization of nuclear research reactor via fuzzy controllers. In: Proceedings of the 2002 American Control Conference, vol. 1–6. IEEE, New York (2002)

4. Das, S., Pan, I., Das, S.: Fractional order fuzzy control of nuclear reactor power with thermal-hydraulic effects in the presence of random network induced delay and sensor noise having long range dependence. Energy Convers. Manag. **68**, 200–218 (2013)
5. Kim, J.H., Park, S.H., Na, M.G.: Design of a model predictive load-following controller by discrete optimization of control rod speed for PWRs. Ann. Nucl. Energy **71**, 343–351 (2014)
6. Liu, X.J., Wang, M.Y.: Nonlinear fuzzy model predictive control for a PWR nuclear power plant. In: Mathematical Problem in Engineering, p. 10 (2014)
7. Abdel-Ghaffar, H., et al.: Neural generalized predictive controller stability analysis. In: The 15th International Conference on System Theory, Control and Computing (2011)
8. Li, G., et al.: Modeling and control of nuclear reactor cores for electricity generation: a review of advanced technologies. Renew. Sustain. Energy Rev. **60**, 116–128 (2016)
9. Andraws, M.S., et al.: Performance of receding horizon predictive controller for research reactors. In: 12th International Conference on Computer Engineering and Systems (ICCES) (2017)
10. Cameron, I.R.: Nuclear Fission Reactors. Springer, Boston (1982)
11. Schultz, M.A.: Control of Nuclear Reactors and Power Plants. McGraw-Hill, New York (1961)
12. Keepin, G.R.: Physics of Nuclear Kinetics. Addison-Wesley, Reading, Palo Alto, London (1965)
13. Lamarsh, J.R., Anthony J.B.: Introduction to Nuclear Engineering. Prentice Hall, Upper Saddle River (2001)
14. Gupta, M.M., Jin, L., Homma, N.: Static and Dynamic Neural Networks. Wiley, New Jersey (2005)
15. Abdel-Ghaffar, H., et al.: Neural generalized predictive controller and internal model principle. In: The 17th International Conference on Automation and Computing (2011)
16. Beale, M.H., Hagan, M.T., Demuth, H.B.: Neural Network Toolbox User's Guide: The MathWorks (2018)
17. Abdel-Ghaffar, H.: Neural networked control systems. In: Computer and Systems Engineering Conference, Ain Shams University: Cairo, Egypt (2013)

Enhanced Genetic Algorithm and Chaos Search for Bilevel Programming Problems

Yousria Abo-Elnaga[1], S. M. Nasr[2(⊠)], I. M. El-Desoky[2],
Z. M. Hendawy[2], and A. A. Mousa[2,3]

[1] Department of Basic Science, Higher Technological Institute,
Tenth of Ramadam City, Egypt
[2] Department of Basic Engineering Science, Faculty of Engineering,
Menoufia University, Shebin El-Kom, Egypt
Sarah.nasr.eid@gmail.com
[3] Department of Mathematics and Statistics, Faculty of Sciences,
Taif University, Ta'if, Saudi Arabia

Abstract. In this paper, we propose an enhanced genetic algorithm and chaos search for solving bilevel programming problem (BLPP). Enhanced genetic algorithm based on new effective selection technique. Effective selection technique enables the upper level decision maker to choose an appropriate solution in anticipation of the lower level's decision. Firstly, the upper level problem is solved using genetic algorithm based on new effective selection technique. Secondly, another search based on chaos theory is applied on the obtained solution. The performance of the algorithm has been evaluated on different sets of test problems. Also, comparison between the proposed algorithm results and other best known solutions is introduced to show the effectiveness and efficiency of our proposed algorithm.

Keywords: Bi-level optimization · Evolutionary algorithms ·
Genetic algorithm · Chaos theory

Mathematics Subject Classification: 68T20 · 68T27 · 90C26 · 90C59 ·
90C99

1 Introduction

In the common form of optimization problems, there are just one decision maker seeks to find an optimal solution (Kalashnikov et al. 2015; Ruusk et al. 2012). On the contrary, many optimization problems appeared in real world in engineering design, logistics problems, traffic problems, economic policy and so on consist of a hierarchical decision structure and have many decision makers (Shuang et al. 2016; Gaspar et al. 2015). This hierarchical decision structure can modeled as "k" levels nested optimization problems. The first level is the upper level and its decision maker is the leader. Next levels are lower levels and its decision makers are the followers. Any decision taken by leader to optimize his/her problem is affected by the followers' response. The hierarchical process starts by taking the upper level decision maker his/her decisions.

© Springer Nature Switzerland AG 2020
A. E. Hassanien et al. (Eds.): AMLTA 2019, AISC 921, pp. 478–487, 2020.
https://doi.org/10.1007/978-3-030-14118-9_48

Then, the lower level decision makers take its own decisions upon the leader's decisions (Aihong et al. 2017; Birla et al. 2017; Osman et al. 2018). BLPP solution requirements produce difficulties in solving such as disconnectedness and non-convexity even to the simple problems (Migdalas et al. 1998; Bard et al. 1991).

Since 1960s, the studies of bilevel programming models have been started and since that many strategies have been appeared to solve it. The methods developed to solve BLPP can be classified into two main categories: classical methods and evolutionary methods. The classical methods include the Kth best algorithm, descent direction method, exact penalty function, etc. (Bard et al. 1990; Bard et al. 1982). Classical methods can treat only with differentiable and convex problem. Furthermore, the BLPP solution requirements produce disconnectedness and non-convexity even to the simple problems. Lu et al. proposed a survey on multilevel optimization problems and its solution techniques (Jie et al. 2016). They proposed detailed comparisons between BLPP solution methods. They accentuated that classical methods have been used to solve a specified BLPP and have not been used to solve various multilevel programming problems especially for large-scale problems. The evolutionary methods are conceptually different from classical methods. It can treat with differentiable and convex optimization problem. Evolutionary methods are inspired by natural adaptation as the behavior of biological, molecular, swarm of insects, and neurobiological systems. Evolutionary methods include genetic algorithms, simulated annealing particle swarm optimization ant colony optimization and neural-network-based methods... etc. (Wang et al. 2008; El-Desoky et al. 2016; Hosseini 2017; Carrasqueira et al. 2017).

In this paper, we propose a combination between genetic algorithm based on a new effective selection technique and chaos search denoted by CGA-ES. Genetic algorithm is well suited for solving BLPP and chaos theory is also represented as one of new theories that have attracted much attention applied to many aspects of the optimization (Wang et al. 2001). Using genetic algorithm supported with our new effective selection technique and chaos theory enhances the search performance and improves the search efficiency. The search using chaos theory not only enhances the search characteristics but also it helps in faster convergence of the algorithm and avoids trapping the local optima.

The performance of the algorithm has been evaluated on different sets of test problems. The first set of problems is constrained problems with relatively smaller number of variables named TP problems (Sinha et al. 2014). The second set is SMD test set (Sinha et al. 2012). SMD problems are unconstrained high-dimensional problems that are recently proposed. The proposed algorithm results have been analyzed to show that our proposed algorithm is an effective strategy to solve BLPP. Also, comparison between the proposed algorithm results and best known solutions is introduced to show the effectiveness and efficiency of the proposed algorithm.

This paper is organized as follows. In Sect. 2, formulation of bilevel optimization problem is introduced. Section 3 provides our proposed algorithm to solve BLPP. Numerical experiments are discussed in Sect. 4. Finally, Sect. 5 presents our conclusion and notes for future work.

2 Bilevel Programming Problems

BLPP consists of two nested optimization problems with two decision makers. The two decision makers are constructed in hierarchical structure, leader and follower. The leader problem is to take a decision that make the follower take his/her decision in the interest of leader's problem. Contrariwise, the follower doesn't have a problem. He/She make an optimal decision upon the leader's decision. Thus, the lower-level problem is parameterized by the upper-level decision and appears as a constraint in the upper-level problem. Bilevel Programming Problems can be formulated as follows:

$$\text{BLPP:} \quad \underset{x,y}{\text{Min }} F(x, y) \qquad \qquad \text{(Upperlevel)}$$
$$\text{s.t. } G(x, y) \le 0,$$

Where, for each given x obtained by the upper level, solves y

$$\underset{y}{\text{Min }} f(x, y) \qquad \qquad \text{(Lowerlevel)}$$
$$\text{s.t. } g(x, y) \le 0$$

Where $F(x, y), f(x, y)$: $R^{n1} \times R^{n2} \to R$ are object functions of the upper and lower level problems. $G(x, y)$: $R^{n1} \times R^{n2} \to R^p$, $g(x, y)$: $R^{n1} \times R^{n2} \to R^q$ are the constraint functions of the upper and lower level problems. $x \in R^{n1}$, $y \in R^{n2}$ are the decision variables controlled by the upper and lower level problems, respectively. $n1$, $n2$ are the dimensional decision vectors for the upper and lower level problems. A vector $v^{\bullet} = (x^{\bullet}, y^{\bullet})$ is an feasible solution if it satisfies all upper and lower level constrains $G(x, y)$, $g(x, y)$ and if vector $y^{\bullet}$ is an optimal solution for the lower level objective function $f(x, y)$ solved for the upper level vector solution $x^{\bullet}$ (Zhongping et al. 2013).

3 The Proposed Algorithm (CGA-ES)

In this section, we first introduce basic concepts of genetic algorithm and chaos theory and then we explain the proposed algorithm in details.

3.1 Basic Concepts of Genetic Algorithm

Genetic algorithm (GA) is a powerful search technique inspired by the biological evolution process. GA was proposed originally by Holland (Zhongping et al. 2013) in the early 1970s. Since then it has been widely used for solving different types of optimization problem. GA starts with a set of random individuals (chromosomes) that are generated in feasible boundaries. Then, the individuals move through a number of generations in search of optimal solution. During each generation, genetic search operators such as selection, mutation and crossover are applied one after another to obtain a new generation of chromosomes. New generation of chromosomes are expected to be better than its previous generation. The evaluation of chromosomes is according to the fitness function. This process is repeated until the termination criterion is met, and the best obtained chromosome is the solution.

3.2 Chaos Theory for Optimization Problems

Chaos theory was initially described by Henon and summarized by Lorenz. It is study in mathematics that concerns on the behavior of systems that follow deterministic laws but appear random and unpredictable. Chaos theory distinguishes from other optimization techniques by its inherent ability to search the space of interest efficiently. The chaotic behavior is described using a set of different chaotic maps. Some of chaotic maps are discrete time parameterized or continuous time parameterized. The basic idea of chaos search for solution is to transform the variable of problems from the solution space to chaos space and then perform search for solution. The transformation starts with determining the range of chaos search boundary. Then, chaotic numbers are generated using one of chaotic maps. The transformation finishes by mapping the chaotic numbers into the variance range search (Nasr et al. 2015). Logistic map is known that is more convenient to be used and increase the solution quality rather than other chaotic maps (Nasr et al. 2015; El-Shorbagy 2016).

3.3 CGA-ES for BLPP

The proposed algorithm to solve the BLPP is incorporation between an enhanced genetic algorithm and chaos search. Firstly, the upper level problem using genetic algorithm based on our new effective selection technique. Then, lower level problem is solved for the obtained upper level solution. Secondly, another search based on chaos theory is applied to the obtained solution (Fig. 1).

Effective Selection Technique: The new effective selection technique operates on enabling the upper level decision maker to choose an appropriate solution in anticipation of the lower level's decision. After the individuals are evaluated by upper level objective function and are selected according to its fitness. Then, these selected solutions are evaluated by lower level problem objective function and are selected according to its fitness. The steps of the proposed algorithm are listed in details as follows:

Step 1. Solve upper level problem using genetic algorithm based on effective selection technique with constrain handling

Step 1.1 Constrains handling: Any solution for the upper level is feasible solution for the upper level only if it is feasible and optimal solution for the lower level. The constrains can be handled by adding constrains of the lower level problem to upper level constrains.

Step 1.2 Initial Population: The population vectors are randomly initialized and within the search space bounds (El-Shorbagy et al. 2016).

Step 1.3 Obtaining reference point: At least one feasible reference point is needed to enter the process of repairing infeasible individuals of the population and to complete the algorithm procedure.

Step 1.4 Repairing: If the problem is constrained. Some of generated individuals don't satisfy constrains and become infeasible individuals. The propose of repairing process is to transform the infeasible individuals to be feasible individuals. The reader can refer to (Mousa et al. 2012).

Step 1.5 Evaluation: The individuals are evaluated using both upper level objective function and lower level objective function

Step 1.6 Create a new population: In this step, a new population is generated by applying a new effective selection technique as mentioned above, crossover operator and mutation operator.

- **Crossover:** crossover operation is one of genetic algorithm operators. Crossover combines two individuals to produce new individuals (Mousa et al. 2011).
- **Mutation:** one of problems in using genetic algorithm is premature convergence which occurs when highly fit parent individuals in the population breed many similar new individuals in early time of algorithm (Mousa et al. 2011).

Step 1.7 Migration: In this step, the best individuals of the new generation and old generation is migrated for the new generation (Mousa et al. 2012).

Step 1.8 Termination test for GA: Terminated of algorithm is achieved either when the maximum number of generations is achieved, or when the individuals of the population convergence occur. Otherwise, return to step 1.6.

Step 2. Solve lower level problem using genetic algorithm as upper level steps for the solution of the upper level decision variables.

Step 3. Evaluate the solution obtained from solving lower level problem using upper level objective function

Step 4. Search based on chaos theory for the solution obtained in step3
In this step, chaos search is applied to the solution obtained in step 3 (x_i^*, y_i^*) (El-Shorbagy et al. 2016). The detailed description of chaotic search is described as follows:

Step 4.1 Determine chaotic search range: The range of chaotic search is $[a, b]$ for the upper level variables and $[c, d]$ for the lower level variables. The range is determined by the following equation

$$x_i^* - \varepsilon_1 < a_i, \; x_i^* + \varepsilon_1 > b_i \tag{1}$$

$$y_i^* - \varepsilon_2 < c_i, \; y_i^* + \varepsilon_2 > d_i \tag{2}$$

where $\varepsilon_1, \varepsilon_2$ is specified radiuses of chaos search for the upper level and lower level variables receptively.

Step 4.2 Generate chaotic number using logistic map: we choose the logistic map because it is more convenient to use and increase the solution quality rather than other chaotic maps (El-Shorbagy et al. 2016). Chaotic random numbers z^k is generated by the logistic map by the following equation

$$z^{k+1} = \mu z^k (1 - z^k), \; z^0 \in (0, 1), \; z^0 \notin \{0.0, 0.25, 0.50, 0.75, 1.0\}, k = 1, 2, \ldots \tag{3}$$

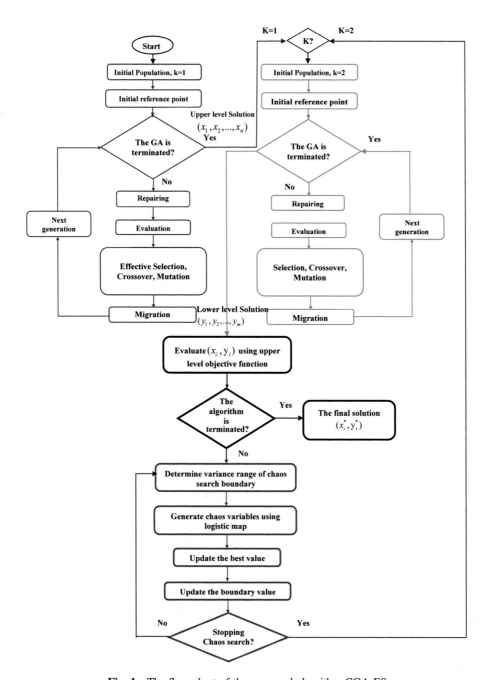

Fig. 1. The flow chart of the proposed algorithm CGA-ES

Step 4.3 Map the chaos variable into the variance range: Chaos variable z^k is mapped into the variance range of optimization valuable $[a, b]$ and $[c, d]$ by:

$$x_i = x_i^* - \varepsilon_1 + 2\varepsilon_1 z^k \quad \forall i = 1, \ldots, n \tag{4}$$

$$y_i = y_i^* - \varepsilon_2 + 2\varepsilon_2 z^k \quad \forall i = 1, \ldots, m \tag{5}$$

Step 4.4 Update the best value: Set chaotic iteration number as $k = 1 \rightarrow$ Do

$$x_i^k = x_i^* - \varepsilon + 2\varepsilon z \quad \forall i = 1, \ldots, n \tag{6}$$

$$y_i^k = y_i^* - \varepsilon_2 + 2\varepsilon_2 z^k \quad \forall i = 1, \ldots, m \tag{7}$$

If $F(x^k, y^k) < F(x^*, y^*)$ then set $(x^*, y^*) = (x^k, y^k)$.
Else if $F(x^k, y^k) \geq F(x^*, y^*)$ then give up the Kth iterated
Result (x^*, y^*).
Loop runs until $F(x^*, y^*)$ is not improved after k searches.

Step 4.5 Update the boundary: Update the boundary value $[a, b]$ and $[c, d]$ of the new optimal point (x^*, y^*) as the new chaos search range. Use Eqs. (4) and (5) to map the chaos variables into the new search range then go to 4.2.

Step 4.6 Stopping Chaos search: Stop chaos search for, K^{th} for the specified iterations and put out (x^*, y^*) as the best solution.

Step 5. Solve lower level problem using genetic algorithm for the solution after chaos search then evaluate the solution obtained from solving lower level problem. In this step, repeat step 2 and 3 for the solution obtained from chaos search (x^*, y^*).

Step 6: Stopping algorithm: Compare the result before chaos search at step3 and after chaos search at step 5. The best result is the algorithm solution.

4   Numerical Experiments

Our proposed algorithm has been evaluated on several sets of well-known multimodal test problems. CGA-ES is coded in matlab 7.8 and the simulations have been executed on an Intel core (TM) i7-4510u cpu @2.00GHZ 2.60 GHz processor. CGA-ES involves a number of parameters that affect the performance of algorithm.

(a) Generation gap	0.9
(b) Crossover rate	0.9
(c) Mutation rate	0.7
(d) Selection operator	Stochastic universal sampling
(e) Crossover operator	Single point
(f) Mutation operator	Real-value
(g) GA generation	200–1000
(h) Chaos generation	10000
(i) Specified neighborhood radius	1E–3

4.1 Results and Analyses

In this section, we first introduce analyses of our results for the chosen test set. We compare our results with best known solutions and other algorithms results solved same chosen test set. The results and comparisons reveal of our proposed algorithm feasibility and efficiency to solve BLPP and that it has better ability and precision than other proposed methods in literature.

(a) Results Analyses for TP Test Set

In Table 1, TP problems result is presented. The results for upper level and lower level objective functions are introduced in second column. The best known solutions of upper level and lower level objective functions according to the reference of the problems are in the third column (Sinha et al. 2014). As indicated in Table 1, our proposed algorithm find solutions better than the best known solutions for four problems TP3, TP4, TP5, TP6 and reach to same solutions as best known solutions for both levels TP7 and reach to same solutions as best known solutions for upper levels for TP2. The remainder problems have also a small difference from best known solutions.

Table 1. CGA-ES results and best known solutions for TP1–TP10.

Problem	CGA-ES Results		Best known Solutions	
	$F(x^{\bullet}, y^{\bullet})$	$f(x^{\bullet}, y^{\bullet})$	$F(x^{*}, y^{*})$	$f(x^{*}, y^{*})$
TP1	225.0001	99.9999	225.0000	100.0000
TP2	0	200.0000	0	100.0000
TP3	−18.9365	−1.1563	−18.6787	−1.0156
TP4	−29.3529	3.008	−29.2000	3.2000
TP5	−3.9014	−2.0300	−3.6000	−2.0000
TP6	−1.2520	8.0708	−1.2091	7.6145
TP7	−1.9600	1.9660	−1.9600	1.9600
TP8	8.7468E–05	199.9989	0	100.0000
TP9	8.6845E–06	2.7183	0	1.0000
TP10	1.23793E–04	2.7183	0	1.0000

(B) Results Analyses for SMD Test Set

Table 2 presents results for SMD problems. In the second column our proposed results for upper level and lower level objective functions. The optimal solution of upper level and lower level objective functions SMD problems are in the third column (Sinha et al. 2012). From Table 2, our proposed algorithm reach to optimal solutions for two problems SMD1, SMD3. The remainder problems have also a small error.

Table 2. CGA-ES results and best known solutions for SMD1–SMD6.

Problem	CGA-ES Results		Optimal Solutions	
	$F(x^\bullet, y^\bullet)$	$f(x^\bullet, y^\bullet)$	$F(x^*, y^*)$	$f(x^*, y^*)$
SMD1	0	0	0	0
SMD2	4.7659e–06	2.2190e–06	0	0
SMD3	0	0	0	0
SMD4	5.5692E–12	3.4094E–11	0	0
SMD5	1.1324e–09	1.1324e–09	0	0
SMD6	9.3428E–11	9.3428e–11	0	0

5   Conclusions

In this paper, we proposed a genetic algorithm based on a new effective selection technique and chaos theory to solve bilevel programming problems. To verify the performance of CGA-ES, extensive numerical experiments based on a suite of multi-modal test have applied. A careful observation will reveal the following benefits of the CGA-ES:

- CGA-ES is presented as a powerful global feasible and efficient technique to solve BLPP.
- CGA-ES overcomes the difficulties associated with BLPP as non-linearity, multi-modality and conflict in upper and lower level objectives.
- The combination between genetic algorithm technique and chaos search offer the advantages of both genetic algorithm as a powerful global searching technique and chaos search as an efficient and fast searching technique.
- The new effective selection technique enables the upper level decision maker to choose an appropriate solution in anticipation of the lower level's decision.
- The new effective selection technique helps in dispensing of solving lower level problem for every generation that fast convergence to the optimal solution.

References

Kalashnikov, V., Dempe, S., Pérez-Valdés, A., Kalashnykova, I., Camacho-Vallejo, J.: Bilevel programming and applications. Math. Probl. Eng. 1–16 (2015)

Ruusk, S., Miettinen, K., Wiecek, M.: Connections between single-level and bilevel multiobjective optimization. J. Optimiz. Theory App. **153**, 60–74 (2012)

Shuang, M.: A nonlinear bilevel programming approach for product portfolio management. Plus **5**(1), 1–18 (2016)

Gaspar, I., Benavente, J., Bordagaray, M., Jose, B., Moura, L., Ibeas, A.: A bilevel mathematical programming model to optimize the design of cycle paths. Transport Res. Procedia **10**, 423–443 (2015)

Aihong, R., Yuping, W., Xingsi, X.: A novel approach based on preference-based index for interval bilevel linear programming problem. J. Inequal. Appl. 1–16 (2017)

Birla, R., Agarwal, V., Khan, I., Mishra, V.: An alternative approach for solving bi-level programming problems. Am. J. Oper. Res. **7**, 239–247 (2017)

Osman, M., Emam, M., Elsayed, M.: Interactive approach for multi-level multi-objective fractional programming problems with fuzzy parameters. J. Basic Appl. Sci. **7**(1), 139–149 (2018)

Migdalas, M., Pardalos, M., Värbrand, P.: Multilevel Optimization: Algorithms and Applications, 1st Edn. pp. 149–164. Kluwer, U.S.A (1998)

Bard, J.: Some properties of the bilevel programming problem. J. Optimiz. Theory App. **68**(2), 371–378 (1991)

Bard, J., Moore, J.: A Branch and Bound Algorithm for the Bi-level Programming Problem Siam. J. Sci. Stat. Comp. **11**(2), 281–292 (1990)

Bard, J., Falk, J.: An explicit solution to the multi-level programming problem. Comput. Oper. Res. **9**(1), 77–100 (1982)

Jie, L., Jialin, H., Yaoguang, H., Guangquan, Z.: Multilevel decision-making: a survey. Inform. Sci. **346–347**, 463–487 (2016)

Wang, G., Wan, Z., Wang, X., Yibing, L.: Genetic algorithm based on simplex method for solving linear-quadratic bilevel programming problem. Comput. Math. Appl. **56**, 2550–2555 (2008)

El-Desoky, I., El-Shorbagy, M., Nasr, S., Hendawy, Z., Mousa, A.: A hybrid genetic algorithm for job shop scheduling problem. Int. J. Adv. Eng. Technol. Comput. Sci. **3**(1), 6–17 (2016)

Hosseini, E.: Solving linear tri-level programming problem using heuristic method based on bi-section algorithm. Asian J. Sci. Res. **10**, 227–235 (2017)

Carrasqueira, P., Alves, M., Antunes, C.: Bi-level particle swarm optimization and evolutionary algorithm approaches for residential demand response with different user profiles. Inform. Sci. **418**, 405–420 (2017)

Wang, L., Zheng, D., Lin, Q.: Survey on chaotic optimization methods. Comput. Technol. Automat. **20**(1), 1–5 (2001)

Sinha, A., Malo, P., Kalyanmoy, D.: Test problem construction for single-objective bilevel optimization. Evol. Comput. **22**(3) (2014)

Sinha, A., Malo, P., Deb, K.: Unconstrained scalable test problems for single-objective bilevel optimization. In: Proceedings of the 2012 IEEE Congress on Evolutionary Computation, Brisbane, Australia, June 2012

Zhongping, W., Guangmin, W., Bin, S.: A hybrid intelligent algorithm by combining particle swarm optimization with chaos searching technique for solving nonlinear bi-level programming problems. Swarm Evol. Comput. **8**, 26–32 (2013)

Nasr, S., El-Shorbagy, M., El-Desoky, I., Hendawy, I., Mousa, A.: Hybrid genetic algorithm for constrained nonlinear optimization problems. Brit. J. Math. Comp. Sci. **7**(6), 466–480 (2015)

El-Shorbagy, M., Mousa, A., Nasr, S.: A chaos-based evolutionary algorithm for general nonlinear programming problems. Chaos Soliton Fract. **85**, 8–21 (2016)

Mousa, A., El-Shorbagy, M., Abd-El-Wahed, W.: Local search based hybrid particle swarm optimization algorithm for multiobjective optimization. Swarm Evol. Comput. **3**, 1–14 (2012)

Mousa, A., Abd-El-Wahed, W., Rizk-Allah, R.: A hybrid ant optimization approach based local search scheme for multiobjective design optimizations. Elec. Power Sys. Res. **81**, 1014–1023 (2011)

The Design of Power Quality Detecting System Based on ADSP-BF606

Yunhua Zhang[1], Fenghou Pan[1(✉)], Qiang Gao[1,2], Feng Yuan[1],
Jiayu Pan[1], Zailin Li[1], and Jicheng Dai[1]

[1] State Grid Liaoning Electric Power Research Institute,
Shenyang 110003, China
panfenghou@dbdky.com
[2] Shanghai Jiao Tong University, Shanghai 200240, China

Abstract. The low cost and the multi-function have been the main contradiction of power quality detecting devices. For it, this paper designs a power quality detecting system based on the latest ADI's ADSP-BF606 dual-core processor. With the high-speed AD7606 chip, the system achieves real-time data acquisition, processing, and analysis. The hardware design makes the system highly integrated, lower cost and easy to be integrated into the enterprise information management system. The fast flourier transformation with window and interpolation is adapted to detect harmonic, and S-Transform is used to detect and locate interferences automatically and quickly, wavelet package transform and binary tree support vector machine are also used to make interference classifications. Experiment results show that this design meets the requirements and different algorithms make the system more automatic and intelligent. The system with low cost and the multi-function is better than others.

Keywords: Power quality detecting system · Dual-core processor ·
Data acquisition processing analysis

1 Introduction

Nowadays, the use of electronic equipment, high-voltage electronic devices, large capacitance, and nonlinear load has brought serious pollution to the power quality [1]. It causes losses to the user. The power quality detecting device is an effective way to detect the disturbing signal timely and accurately. At present, the device adopts various programs, but they all have some problems in the applications. The low cost and the multi-function have been the main contradiction. Therefore, this paper designs a dual-core power quality detecting system based on the latest ADI's dual-core ADSP-BF606 processor [1–3]. With the high-speed AD7606 chip, the system achieves real-time data acquisition, processing, and analysis. The Fast Flourier Transform (FFT) with window and interpolation is adapted to detect the harmonic. S-transform is used to locate disturbances automatically and quickly [4]. And the design of binary tree Support Vector Machine (BT-SVM) makes classifications intelligently.

© Springer Nature Switzerland AG 2020
A. E. Hassanien et al. (Eds.): AMLTA 2019, AISC 921, pp. 488–496, 2020.
https://doi.org/10.1007/978-3-030-14118-9_49

2 Hardware Design

2.1 The System Architecture

Power quality detection device could complete a wide range of tasks. The system should have the functions of online detection, real-time analysis, and network [2]. The design should be suitable for installing in large numbers.

This paper comes up with a new design of power quality detecting system with the latest ADI's dual-core chip. The hardware architecture is divided into three parts: signal acquisition module, signal processing module and a power module. It's shown in Fig. 1.

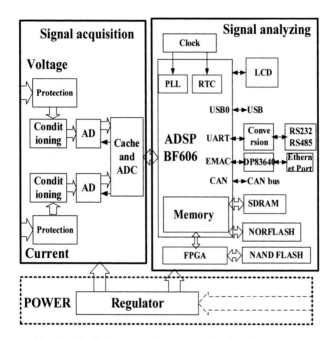

Fig. 1. Block diagram of power quality detection system

Signal acquisition module is used to acquire signal analog and convert analog to digital for the back-end digital signal processing module. Signal to analyze and scheduling module is the main part of the whole system, including the core processing chip ADSP-BF606 and its various peripherals, mainly in charge of the signal harmonic analysis and calculations, storing and uploading calculation data. The power supply module is used as a power supply for the other two parts.

2.2 Design of Modules

In the signal analyzing module, ADSP-BF606 is a new generation of high-speed DSP with dual-core (Core #1 and Core #2). Core #1 can realize embedded kernel of uCLinux 3.3 and support the IEC61850. Core #2 can make different real-time calculations.

It has two ethernet modules to fit the IEC61850. And the port of ePPI can be connected to LCD or other parallel peripherals. It also has two external memory expansion interfaces, which can directly expand the SDRAM, Flash, DDR2 and other external memory [5].

The signal acquisition module can be divided into two parts: protection and condition part, and AD converter part. This module is mainly used to convert the three-phase power signal to ±10 V voltage by the voltage transformer SPT204A, then convert the signal into a digital signal by AD chip. The AD chip is the high-speed six-channel AD7606, its parallel sampling frequencies are up to 200kSPS, and it's better and faster than other AD chips. The current transformer is CKSR 6-NP with high sensitivity, low linear error, and high precision. The AD7606 has eight current and voltage channels, so only one chip is needed.

The power supply module is used to supply power for the other two modules. Since ADSP-BF606 has three power supplies (core is 1.25 V, the I/O is 1.8 V and 3.3 V), the ADP2119, ADP1715, and ADP2384 are used to supply power for ADSP-BF606 and other types of equipment. And another ADP2384 is used to supply 5 V power for USB and CAN modules. The structure of the power supply is shown in Fig. 2.

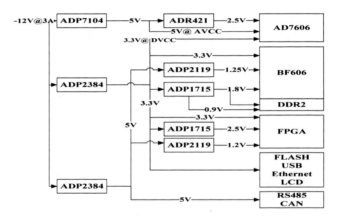

Fig. 2. Diagram of 1.2 V, 1.8 V and 3.3 V power supply circuit

3 Power Quality Detecting Methods

Power quality detecting methods have consisted of harmonic detection, disturbances detection, and disturbances classification. First, the system initializes ADSP-BF606. Second, the system starts collecting signal, analyzing the harmonic, detecting the disturbances and making a classification [6]. Finally, the system calculates various parameters, stores, and uploads data. The detecting flowchart is shown in Fig. 3.

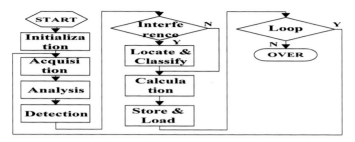

Fig. 3. The detecting flowchart of power quality detecting system

3.1 Harmonic Detection

The harmonic detection method is FFT. FFT has the advantage of fast calculation and high precision. The FFT method also has some problems, such as spectrum aliasing, picket fence effect and spectrum leakage [3]; these cause the results' deviation. But the FFT can be improved by the window function and interpolation algorithm. There are kinds of window functions, such as Hanning, Blackman and Blackman-Harris window function and so on. Hanning window function has a small amount of calculation, it can reduce the leakage of harmonics [7, 8]. This design adopts the Hanning window function [10]. The Hanning window is as followed:

$$w(n) = 0.5 - 0.5\cos(2\pi n/N), \ (n = 0, 1, 2 \dots, N - 1) \tag{1}$$

The interpolation algorithm based on bimodal spectrums means that bimodal spectrums are extracted from the FFT results and then gets the weighted average of them and the correct value. The interpolation algorithm is as follows:

$$m = 1.5q, \ \theta = \arg[\tilde{X}(k_i \Delta f)] + \pi/2 - (m - (-1)^i \times 0.5)\pi, \ i = 1, 2$$
$$A = N^{-1}(y_1 + y_2)(2.3561940 + 1.15543682m^2 + 0.32607873m^4 + 0.07891461m^6) \tag{2}$$

3.2 Harmonic Detection

Disturbances can be divided into steady-state power quality and transient power quality disturbances. Transient disturbances include transient pulse, transient oscillation, voltage dropping, and voltage rising and interrupting. S transform is used as disturbances detection algorithm [4]. It's a reversible time-frequency analysis algorithm. The one-dimensional continuous S-transform of signal x (t) is as followed:

$$S(\tau, f) = \int_{-\infty}^{\infty} x(t) \cdot w(\tau - t, f) \cdot \exp(-i \cdot 2pft) \, dt$$
$$w(\tau - t, f) = |f| / \sqrt{2\pi} \cdot \exp(-(t - \tau)^2 \cdot f^2/2) \tag{3}$$

Where w (τ–t, f) is Gaussian window function, the τ controls the timeline position of Gaussian window function, f means frequency, and i is an imaginary number. The x [k · T] is the discrete time series of signal x (t). T is the sampling interval. N is the total number of sampling points. The discrete form of the formula is shown as followed:

$$S[jT, n/NT] = \sum_{m=0}^{N-1} X[(m+n)/NT] \cdot \exp(-2\pi^2 m^2/n^2) \cdot \exp(i \cdot 2\pi mj/N) \quad (4)$$

After the S transform of x [k · T], get the two-dimensional complex time-frequency matrix [4]. The matrix columns correspond to the discrete frequency, and its row corresponds to the sampling time point and local spectrum.

3.3 Disturbances Classification

In order to make system efficiency, it needs to make a fast and accurate disturbance classification. Artificial Neural Networks, Fuzzy logic, Expert system and Support Vector Machine (SVM) are the main methods [5]. SVM has the advantages of avoiding local minima and dealing with the curse of dimensionality. It has good scalability and classification accuracy and has the advantage to solve the small sample and nonlinear problems. This design combines the Wavelet Package Transform (WPT) with SVM. The wavelet package decomposition energy of the signal is got by WPT as SVM's sample. The formula of the fast Wavelet Package Transform [5] is as follows:

$$\begin{aligned} d_{j+1}^{2p}(k) &= d_j^p(k) * \bar{h}_0(2k) = \sum_{m=-\infty}^{m=\infty} d_j^p(m) * h_0(m - 2k) \\ d_{j+1}^{2p+1}(k) &= d_j^p(k) * \bar{h}_1(2k) = \sum_{m=-\infty}^{m=\infty} d_j^p(m) * h_1(m - 2k) \end{aligned} \quad (5)$$

The formula of signal energy in every node is as follows:

$$e(i) = \sum_{k=1}^{N} d_i^2(k), \ E_{\max} = \sum_{i=1}^{N} e(i), E(i) = e(i)/E_{\max} \quad (6)$$

The h0 and h1 are decomposition filter coefficients, dj + 12p is low-frequency wavelet coefficient, the dj + 12p + 1 is high-frequency wavelet coefficient, dj p(k) is node wavelet decomposition coefficients. E (i) is the energy of i-node wavelet decomposition coefficients. The signal is decomposed with WPT, and the energy of the disturbances is got as the samples of SVM. The energy results are shown in Fig. 4. They are different from each other, so they can be the SVM's samples and classified by SVM.

SVM is put forward to solve a two class problem. While it is used to classify the power quality disturbances, it needs multi-class classification. Multi-class classification methods are one vs. one (OVO), one vs. rest (OVR) and Directed Acyclic Graph (DAG) [6]. However, they all have problems that there is a linear region that can't be divided and sub-classifications influent each other. So this design adopts BT-SVM. K classifiers just need to construct K-1 classifiers with BT-SVM. Classifiers of this method can be less than others.

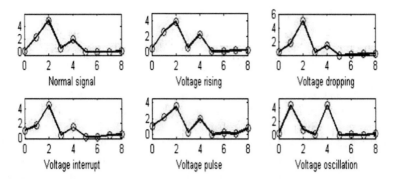

Fig. 4. The energy of different signals

A BT-SVM needs to select an appropriate tree structure and arrange the nodes of the tree reasonably. The sample should be easy to be divided into two groups with the best way in every node. The basic idea of the classification algorithm is the concept of the class distance of the clustering analysis. Calculate the shortest distance between different classes in the eigenvectors space. The shortest distance between classes is d, the $\varphi(x)$ is the x that mapped from the original space into the eigenvectors space. $K(x1, x2) = \varphi(x1) \varphi(x2)$ is the kernel function.

$$d(x_1, x_2) = ||\varphi(x_1) - \varphi(x_2)||_2 = \sqrt{K(x_1, x_1) - 2K(x_1, x_2) + K(x_2, x_2)} \quad (7)$$

After getting the distance, the structure can be designed. There are 5 kinds of disturbances. They are voltage pulse, oscillation, dropping, rising and interrupt. If another multi-class classification method is adopted, the classifiers of the structure are 5 and the number of calculation is 5 at least. But the BT-SVM structure has 4 classifiers, and the number of calculation is just 3 at most. So the number of calculation and classifiers of the structure are less than OVO, OVR and DAG method. The structure of BT-SVM is shown in Fig. 5.

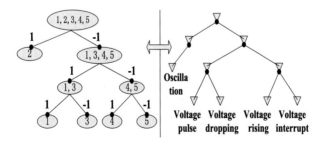

Fig. 5. Structure of binary tree SVM

4   Experimental Results

In this paper, the amplitude and phase are set in different frequencies. The harmonic analysis uses the improved FFT. Results show that the calculated amplitudes and phases in different frequencies are correct. And the errors meet the requirement, and the errors of amplitudes become large in high frequencies. The original value of the harmonic amplitude and phase is shown in Fig. 6(a). The errors are shown in Fig. 6(b).

Disturbance detection uses S-transform to detect signal which is joined in 30 dB Gaussian noise and harmonic noise. If there are mutations in the disturbance detection, it's able to judge that some disturbance has happened. When the disturbance happens, amplitude mutation appears, and when the disturbance stops, another amplitude mutation appears. Then get the duration of the disturbance.

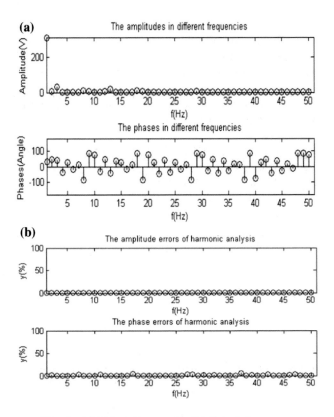

Fig. 6. (a) Calculation results; (b) Calculation errors

Compared the classification results of BT-SVM with OVO's; it shows that the accuracy of BT-SVM is better than OVO's. The results are listed in Tables 1 and 2.

Table 1. The results of classification by OVO

Interference type	Rising	Dropping	Interrupt	Oscillation	Pulse
Voltage rising	194	0	0	4	2
Voltage dropping	0	193	5	2	0
Voltage interrupt	0	0	195	5	0
Oscillation	0	4	0	196	0
Voltage pulse	7	4	0	0	189

Table 2. The results of classification by BT-SVM

Interference type	Rising	Dropping	Interrupt	Oscillation	Pulse
Voltage rising	196	0	0	4	0
Voltage dropping	0	199	1	0	0
Voltage interrupt	0	1	198	0	1
Oscillation	0	0	0	192	8
Voltage pulse	5	0	0	0	195

The results show that the system can identify the disturbances automatically and intelligently by BT-SVM. The classification accuracy meets the requirement. But the error rate of oscillation is a little high; it still needs to be improved.

5 Conclusion

This paper designs a power quality detecting system with ADI's dual-core ADSP-BF606. The hardware design makes the system highly integrated, lower cost and multifunction. The improved FFT makes the harmonic detection more accurate. The disturbances can be located automatically and quickly with the S-transform. And the system can make classifications intelligently with the WPT and BT-SVM. Experiments show that the system meets the requirements and the whole design of detecting methods makes the system more automatic and intelligent. The system is better than others and fit to be installed in large numbers.

References

1. Xiao, X.: Power quality analysis and control. China Electric Power Press Beijing **44**(2), 45–47 (2004). (in Chinese)
2. Zhang, G., Ma, X., Lei, Y.: The design of the power quality monitoring devices based on TMS320C6713. Electron. Measur. Technol. **1**(33), 83–86 (2010). (in Chinese)
3. Haydn, G.T.: Applications of the windowed FFT to electric power quality assessment. IEEE Trans. Power Deliv. **14**(14), 6–11 (2004)
4. Zhu, X., Su, H.: Power quality disturbances analysis based on the S-transform. Yunnan Electric Power **3**(7), 7–15 (2009). (in Chinese)
5. Ma, Z., Li, P., Yang, Y.: Power quality detecting based on wavelet multi-resolution method. J. North China Electric Power Univ. **3**(1), 3–6 (2003). (in Chinese)
6. Qiao, Z., Sun, W.: A multi-class classifier based on the SVM decision tree. Comput. Appl. Softw. **11**, 227–230 (2009)
7. Sun, F., Zhang, S., Wang, T.: Debugging and operation technology for the intelligent substation. China Electric Power Press, **27**(10) (2014). (in Chinese)
8. Wang, F., Pan, S.: Debugging technology of secondary system for intelligent substation. China Electric Power Press, **29**(15) (2013). (in Chinese)
9. Harris, F.J.: On the use of windows for harmonic analysis with the discrete Fourier transform. Proc. IEEE **66**(1), 51–83 (1978)

Role of Robotic Process Automation in Pharmaceutical Industries

Nitu Bhatnagar[(⊠)]

Department of Chemistry, Manipal University Jaipur, Jaipur, India
nitu.bhatnagar@jaipur.manipal.edu

Abstract. Robotic Process Automation (RPA) is a technological revolution in the offing and is aimed at taking up the mundane and repetitive tasks from people's daily workload. It throws up a new vista of research to the research community and lot many types of research are going on in this domain. It is not Robotics but is different technology altogether. RPA is a recent and fast-growing sub-domain of Robotics. The healthcare and pharmaceutics domain generate a lot of data or we may call it medical big data, and it is all the more pertinent to analyze & evaluate such data coming from varied sources. New drug discovery, drug formulation process, drug delivery mechanisms or in-patient and out-patient activities are some of the key processes in the Healthcare and Pharmaceutical industries generating a tremendous amount of data. Therefore, data science and RPA provides handy tools to work with such huge data volumes. In this paper, the authors highlight the key aspects of RPA and review its usage in the all-important healthcare and pharmaceutics domain. RPA is proving to be the technology of future and its goal is to provide a sustainable solution that reduces costs and delivery time, improves quality, speed and operational efficiency of a business process. The application of Machine Learning (ML) technologies in the healthcare domain are proving to be beneficial and effective in gaining new insights. The author also proposes a generic RPA/ML-based framework to ensure the standardization and quality of Bhasma – an end product obtained after multiple activities in the traditional Indian System of Medicine – Ayurveda.

Keywords: Robotic process automation · Machine learning · Healthcare · Pharmaceutics

1 Introduction

Robotic process automation (or RPA) is an emerging form of business process automation technology based on the notion of software robots or artificial intelligence (AI) workers [1]. RPA is a software-based automation approach which automates the business processes by understanding the existing processes & practices. In essence, it is software which mimics the virtual human workforce and performs the tasks & activities which are repetitive in nature reducing the involvement of human in the process. Today the business world is extremely competitive and everyone wants to be ahead of its competitor for gaining the advantage. This type of automation helps the human employees to focus on more critical tasks, be more innovative and devote time to

© Springer Nature Switzerland AG 2020
A. E. Hassanien et al. (Eds.): AMLTA 2019, AISC 921, pp. 497–504, 2020.
https://doi.org/10.1007/978-3-030-14118-9_50

enhance their knowledge & skills in the domain. Hence, the RPA technology is proving to be effective in increasing profitability, the efficiency of the business houses when certain recurrent processes are automated.

RPA is considered to be a significant technological evolution in the sense that new software platforms are emerging which are sufficiently mature, resilient, scalable and reliable to make this approach viable for use in large enterprises [2]. Researchers opine that RPA will usher in a new wave of productivity & efficiency gains in the global workforce once the technology becomes more matured and stable in the coming decade or so [3]. Owing to its tendency to quickly adapt to the situations and high throughput, RPA is now spilling out of this business house and are being frequently being adopted across different sectors such as pharmaceutical, clinical, biotechnology and chemical as represented in Fig. 1.

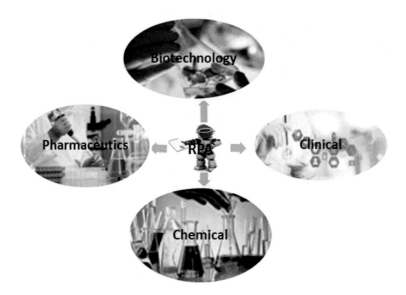

Fig. 1. Broad Areas of applications of RPA

2   RPA in Pharmaceutics

Pharmaceutical industries which require its workers to invest hours in the labs evaluating some new formulated medicine has always been a mundane task. This is because the discovery of an active ingredient in a drug and formulating it into a drug is time-consuming and involved a company spending around $1.6 billion for finally bringing the finished product into the market. This marked the beginning of automation in the pharmaceutical industry which began with the discovery of an active ingredient and has now spread to formulation development and other applications. Robotic process automation has changed the scenario of pharmaceutical business with the use of "robots" which perform a huge amount of work and repeatable tasks which earlier needed human workforce [4]. One can easily understand how this technology has revolutionized the pharmaceutical area by the example of a high content screening where an automated Perkin Elmer system can acquire 200,000 images a day and perform relevant scientific experiments like culturing cells, dispensing compounds and re-arraying them at the same time which would have otherwise required two full-time employees to perform the same task [5].

The application of RPA in the field of pharmaceutics has seen a tremendous revolution with the introduction of the first robot scientist, "Adam" who could discover new scientific knowledge in the field of functional genomics without human intervention [6]. It has been found to work in tandem with another robot scientist, "Eve" for drug screening [7]. Pharmaceutical companies had been employing robots in Chemistry and Biology labs for many purposes such as auto samplers for analytical devices and for synthesizing and testing chemical entities [8, 9]. Many companies have been employing small robots (Andrew) [10] to handle liquid samples. A structural determination is one such area of pharmaceutical research where automation has been employed. For e.g.: The use of robotic arms for sample preparation before NMR (Nuclear Magnetic Resonance) spectroscopy and HPLC-MS (High-Performance Liquid Chromatography-Mass Spectroscopy) analysis [11]. Also, the use of a micropipette machine helps to create a million crystals before analysis through X-ray crystallography [12]. The concept of TLA (Total Laboratory Automation) had come into the picture with the introduction of first fully automated lab set up by Dr. Masahide Sasaki's group [13], but could not become popular due to the high cost and lack of communication between different devices. Recent research in the use of automation language like AutoIt [14] has made it possible to integrate any device. RPA in pharmaceutics has further been applied to the areas of forensic science, food, and agriculture, genomics and proteomics (Fig. 2)

Fig. 2. Application areas of Pharmaceutics

In forensic labs, DNA sample processing is being taken as an important step to solve many criminal cases. With the introduction of robotic liquid handling instrument, DNA sample processing has become fully automated with increased throughput and improved reliability, chain of custody and reduced errors [15]. However, DNA sequencing may involve the risk of sample contamination as the typical automated process lack the capability to open and close the sample tubes [16].

Automation is being frequently adopted in the food industry to eliminate contamination in food processing plants with sterile human-and-bacteria free production line. In beverage production, rotary fillers are being used for soft drink, wine, milk, and water. Drones equipped with infrared cameras are being employed by farmers to monitor irrigation patterns and map soil variation which can help to detect pests and fungal infestations [17].

The field of genomics and proteomics has seen tremendous improvement in the throughput at lower flexibility with the introduction of automation in DNA sequencing. Workstations for functional genomics are useful for replication of clone sets, PCR, sequencing set up and clean up, hit picking, gel loading and nucleic acid purification procedures [18].

2.1 Need for a Smarter Solution

In the past, RPA tools had brought about tremendous development in many sectors by executing specifically defined tasks, but limited in the sense that they could not adjust to changing conditions or learn from experience. Deficiencies in sample handling protocols and lack of an audit trail could prove to be the weak link in a case that relies on forensic evidence [15].

Machine learning, on the other hand, applies artificial intelligence (AI) capabilities to lend business context to tasks executed by RPA systems, enabling the latter to make better decisions and be more productive overall. For example, when extracting field values from unstructured data, RPA can extract values based on the rules set. Machine learning, on the other hand, "learns" the most common labels for fields, while working with a human trainer to confirm what is being learned, which can then be applied to future scenarios. This results in a 10x faster path to automation because explicit programming isn't needed for quick improvement gains. By contrast, RPA without this learning capability requires a human to explicitly program these improvements, defeating the purpose of automation all together [19].

3   RPA/ML-Based Intelligent Framework for Standardization of End Products in Indian Traditional System of Medicine – Ayurveda

This section discusses the intelligent framework as proposed by the author for obtaining the standardized end-products after applying relevant Ayurvedic processes. Implementation of the framework will help to maintain the best quality end products even if different manufacturers are there doing the business of Ayurvedic preparations. The section begins with an introduction to Ayurveda, stages involved in the preparation of *Bhasma* – an Ayurvedic preparation (can be equated to modern-day Nanomedicines). Every stage generates data and the proposed RPA/ML-based framework is meant to analyze and work upon the data being generated in the process.

3.1 The Indian Traditional System of Medicine Ayurveda

Ayurveda, the Indian traditional system of medicine has been in existence since centuries. It is well known for its healing powers for a long time and it has been proved that it helps individuals gain a long and healthy life without using synthetic drugs or undergoing painful surgeries. Recent studies have claimed that the herbal-mineral formulations of Ayurveda constituting bhasmas to be equivalent and in tune with nanotechnology [20]. Also, the nanoparticles in the form of bhasma have been found to have an advantage over other preparations in terms of their stability, lower dose and easy availability [21, 22]. In this way, the use of metals and minerals in the form of bhasma particles became the strength of Ayurveda [23].

3.2 Bhasma Preparation

The process of Bhasma preparation is a very complex process and involves the following steps:

3.2.1 Shodhan

The principal objective of shodhan is to remove the unwanted part from the raw materials and separate out impurities. It is the process of purification and detoxification by which physical and chemical impurities and toxic materials are eliminated. The shodhan process mentioned in the Rasa text is not only a process of chemical purification but also process of addition and separation which causes physical, chemical and biological changes in the metal. These changes depend on the structure, constituents, impurity, and properties of a particular substance [24]. Shodhan can be done by two sub-processes:

a. **Samanya shodhan** can be done by Dhalana (liquefying and pouring) method in Kanji (sour gruel), Takra (butter milk), Kulattha (*Dolichus biflorus*) kwatha (decoction), Go-mutra (Cow's urine) and Tila (*Sesamum indicium*) Taila (oil).
b. **Specific shodhan** can be done by same Dhalana process, but in this process, specific liquid medias are required for specific metal.

3.2.2 Bhavana

It helps to make the metal and mineral particles become finer and change into organometallic compounds. This is done by mixing shodhit metal with specific plant extracts.

3.2.3 Marana

It is the process in which metals and minerals are made into a paste with various drugs and juice. A change is brought about in the chemical form or state of the metal. This makes it lose its metallic characteristics and physical nature. Metals can be converted into powder or another form suitable for administration. It is carried out by heating the metal in presence of mercury, plant extracts, and sulfur.

3.2.4   Jarana

Metals are melted and mixed with some plant powder and are rubbed by an iron ladle on the inner surface of the pot until metal become a complete powder form.

3.3   Data Obtained During Bhasma Preparation

All these steps mentioned above take into account different parameters like pH, temperature, the volume of different liquid media, and physical state and weight of the metals after each step of the process to ascertain whether the desired product has been formed. Even after the end product has been formed, there are different classical and physical characterization techniques as mentioned in the Ayurvedic texts which help us confirm the formation of the end product.

As the preparation and characterization of these bhasmas are done with utmost care when prepared on a small scale by the trained practitioners, there is hardly any chance of toxicity involved in the consumption of these bhasmas. But, some recent kinds of literature have reported of toxicity in these bhasmas. One of the main reason behind this is that with rapid industrialization and commercialization of services and products, Ayurvedic products are also facing the heat. The pharmaceutical companies involved in manufacturing Ayurvedic products on large scale are often neglecting or compromising with the detailed procedure as mentioned in the ancient texts. Thus, the end products delivered by them are of inferior quality containing toxic remains which are not fit for human consumption. Therefore, there is an urgent need to standardize the manufacturing processes and develop standardization methods after they have been formed [25]. Standardization is the need of the hour and all the companies who are in the business of manufacturing Ayurvedic products must abide by these set procedures. There must be some mechanism to check the deviation and/or rectify the same before it is too late.

4   RPA/ML-Based Proposed Framework

With the success of RPA in pharmaceutics, it is expected that if the manufacturing process of the Ayurvedic bhasmas and their characterization techniques are automated by bringing in RPA to work on the same line as it has been used for pharmaceutical applications, it would give a big boost to this ancient science of Indian traditional medicine. Therefore, the author has come up with a novel proposal involving the RPA and ML-based algorithmic framework which when implemented will result in quality Ayurvedic product which will be effective, less costly and suitable for human consumption and wellbeing.

Ayurvedic product preparations generate a lot of data and there are multiple sources of Ayurvedic data e.g. traditional dataset from Ayurvedic texts dating back to 5000 years, knowledgebase of different Ayurvedic drug manufacturing companies because each company has its own set of customized processes based on the technologies in use by them, data as collected from traditional practitioners and some other sources. The processes contain a set of activities which are repetitive in nature and hence RPA can be utilized for them. The proposed ML framework/application would contain various algorithms based on ML to sniff through the data and looking out for unusual patterns which may interfere with getting the quality end product. It will sit somewhere in between and interact with the dataset & RPA tool

Figure 3 below depicts the proposed framework:

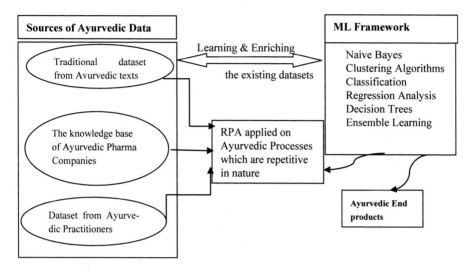

Fig. 3. Proposed RPA and ML-based Framework for Standardized Ayurvedic End Products

5 Conclusion

RPA has a lot of applications in a number of industries including healthcare and pharmaceutics, but just like any other new technology, it suffers from various challenges. With the support from Machine learning to tasks executed by RPA systems, it would help advance the learning capabilities of RPA systems so that they can take better decisions and be more productive. The two can together create such a framework which can help in the standardization of the complex process involved in the preparation of Ayurvedic Bhasma, the well-known Indian traditional system of medicine.

References

1. Hodson, H.: AI interns: software already taking jobs from humans, New Scientist Technology News, 31 March 2015
2. Robotic Automation Emerges as a Threat to Traditional Low-Cost Outsourcing, HfS Research. Accessed 21 Nov 2018
3. Willcocks, L.P., Lacity, M.C.: Nine likely scenarios arising from the growing use of software robots, London School of Economics (2015)
4. http://pharmaceuticscongress.alliedacademies.com/events-list/robotic-process-automations
5. King, A.: Let the robots take the strain (2016). https://www.chemistryworld.com/features/lab-automation/2500121.article

6. King, P., Rowland, J., Aubrey, W., Liakata, M., Markham, M., Soldatova, L.N., Whelan, K.E., Clare, A., Young, M., Sparkes, A., Oliver, S.G., Pir, P.: The Robot Scientist Adam. Computer **42**(7), 46–54 (2009). https://doi.org/10.1109/MC.2009.270

7. Robot scientist' holds key to new drugs, BBC News

8. Mortimer, J.A., Hurst, W.J.: Laboratory robotics: a guide to planning, programming, and applications. VCH Publishers, New York (1987). ISBN 0-89573-322-6

9. Ward, K.B., Perozzo, M.A., Zuk, W.M.: Automatic preparation of protein crystals using laboratory robotics and automated visual inspection". J. Cryst. Growth **90**, 325 (1988). https://doi.org/10.1016/0022-0248(88)90328-4

10. Hands-free use of pipettes, October 2012, Accessed 30 Sept 2012

11. Gary, A., McClusky, B.T.: Automation of Structure Analysis in Pharmaceutical R&D. J. Manag. Inf. Syst. (1996)

12. Heinemann, U., Illing, G., Oschkinat, H.: High-throughput three-dimensional protein structure determination. Curr. Opin. Biotechnol. **12**(4), 348–354 (2001)

13. Felder, R.A.: The Clinical Chemist: Sasaki, M., MD, PhD (August 27, 1933 – September 23, 2005). Clin. Chem. **52** (4), 791–792 (2006)

14. Carvalho, M.C.: Integration of Analytical Instruments with Computer Scripting. J. Lab. Autom. **18**(4), 328–333 (2013)

15. Robotics to the Rescue: Automated Sample Processing Fri, 01/15/2016 by Lois C. Tack and Laurent Baron. https://www.forensicmag.com/article/2016/01/robotics-rescue-automated-sample-processing

16. Comar, C., Duncan, G., Kevin W.P.: MillerFlipTube™ technology promotes clean manipulation of forensic samples on automated robotic workstations (2017). https://doi.org/10.1016/j.fsigss.2017.09.019

17. How Robotic Automation Will Benefit Food and Agriculture Kagan Pittman posted on 27 April 27 2017

18. Lorenz, M.G.O.: Liquid Handling workstations for Functional Genomics, Jala tutorial (2004). https://journals.sagepub.com/doi/pdf/10.1016/j.jala.2004.03.010

19. Kakhandiki, A.: The symbiosis of RPA and machine learning, 29 May 2017. Business. https://www.itproportal.com/features/the-symbiosis-of-rpa-and-machine-learning/

20. Virupaksha, G.K.L., Pallavi, G., Patgiri, B.J., Kodlady, N.: Relevance of Rasa Shastra in the 21st century with special reference to lifestyle disorders (LSDs). Int. J. Res. Ayurveda Pharm. **2**(6), 1628–1632 (2011)

21. Sharma, R.N.: Ayurveda-sarsangrha, 13th edn, pp. 101–102. Shri Baidhyanath Ayurveda Bhavan Ltd., Varanasi (1985)

22. Mishra, L.C.: Scientific bases for Ayurvedic therapies, Handbook of Ayurveda, pp. 83–100. CRC Press, Washington (2004)

23. Pal, S.K.: The ayurvedic bhasma: The ancient science of nanomedicine. Recent Patents on Nanomedicine. Bentham Science Publishers, 5, 12–18 (2015). 1877-9131/15

24. Raveendran, P., Fu, J., Wallen, S.L.: Completely "green" synthesis and stabilization of metal nanoparticles. J. Am. Chem. Soc. **125**, 13940 (2003)

25. Pareek, A., Bhatnagar, N.: Revisiting ancient therapeutic potential of bhasma. Int. J. Pharm. Sci. Res. **9**(8), August 2018

Text Mining, Summarization and Language Identification

Using Related Text Sources to Improve Classification of Transcribed Speech Data

Niraj Shrestha[(✉)], Elias Moons, and Marie-Francine Moens

Department of Computer Science, KU Leuven, Leuven, Belgium
{niraj.shrestha,elias.moons,marie-francine.moens}@cs.kuleuven.be
http://liir.cs.kuleuven.be

Abstract. Today's content including user generated content is increasingly found in multimedia format. It is known that speech data are sometimes incorrectly transcribed especially when they are spoken by voices on which the transcribers have not been trained or when they contain unfamiliar words. A familiar mining tasks that helps in storage, indexing and retrieval is automatic classification with predefined category labels. Although state-of-the-art classifiers like neural networks, support vector machines (SVM) and logistic regression classifiers perform quite satisfactory when categorizing written text, their performance degrades when applied on speech data transcribed by automatic speech recognition (ASR) due to transcription errors like insertion and deletion of words, grammatical errors and words that are just transcribed wrongly. In this paper, we show that by incorporating content from related written sources in the training of the classification model has a benefit. We especially focus on and compare different representations that make this integration possible, such as representations of speech data that embed content from the written text and simple concatenation of speech and written content. In addition, we qualitatively demonstrate that these representations to a certain extent indirectly correct the transcription noise.

Keywords: Speech data · Word embeddings

1 Introduction

The main objectives of this paper are to test the validity of the hypothesis that related written data help in correcting the noise in transcribed speech and to evaluate different types of representations that accomplish this task. These objectives are extrinsically evaluated in a text categorization task.

As multimedia data including user generated multimedia grows every day, it is important to correctly search and mine such data. Many technologies have been developed to deal with written texts, but only very few works deal with transcribed speech data (further in this article called speech data). We do not have yet proper ways to deal with the noise caused by transcription errors like

© Springer Nature Switzerland AG 2020
A. E. Hassanien et al. (Eds.): AMLTA 2019, AISC 921, pp. 507–517, 2020.
https://doi.org/10.1007/978-3-030-14118-9_51

spelling errors and inserted or deleted words. These errors are especially prominent when the speech is spoken by voices on which the transcribers have not been trained or when they contain unfamiliar words such as named entities.

In this paper, we show that by incorporating content from related written sources - when training the classification model - has a benefit. In an encoder model we train a mapping from the input speech data to related written data, which we call an *encoding* model as we encode the speech data guided by its related written texts as output. This model is inspired by common multimodal mappings such as done with language and visual data [1]. By this procedure we hope that the resulting representations of the speech data are enriched by related information from the paired written texts. In an alternative model we just *concatenate* the input speech vectors with the vector that contains related written information. For these models we experiment with different ways of selecting content from the written texts, including the selection of named entities. A straightforward approach here is to use classical word *embeddings* that are pretrained on written and speech data and expand the speech words with their close neighbors that come from the written texts. We assume here that speech words and written words that appear in similar contexts are closely related.

We extrinsically evaluate the learned representations in a speech classification task. More specifically we classify the speech of online news in the predefined categories "Business", "Entertainment", "Health", "International", "Politics" and "Technology", where we empirically show that the multimodal representations trained on speech and written data outperform a baseline model that performs the classification solely by relying on features of the speech data. Finally, we qualitatively demonstrate that the multimodal representations to a certain extent indirectly correct the transcription noise.

2   Related Work

Although speech classification has many applications, the number of works that classify the content of speech data is very limited. Most of the existing work concerns the classification of audio signals. In [2], the authors classify and segment news videos into speech, music, commercials, environmental background noise or other acoustic noise. Similarly, [3] classify videos based on audio signals to segment, classify and cluster broadcast news. [4] propose a method to segment and classify the audio into five different acoustic classes: speech, music, speech with music, speech with noise, and others. Event detection and classification is a common task in multimedia processing. [5] use a hierarchical approach for audio based event detection for surveillance: They first classify the audio frame into vocal and nonvocal events, and then perform further classification into normal and excited events. [6] use visual and audio signals to detect events in multimedia. On the language side, older work [7] classifies transcribed broadcast news in different topics. They use a generative approach to assign probabilistic topics to the transcriptions and use these as features in a classification model. [8] propose a recurrent neural network language modelling (RNNLM) that predicts the probability of an extractive summary based on the likelihood of extracted sentences, but they evaluate their method on speech data that is manually transcribed. [8] propose a two-step method for hate speech detection: They learn distributed representations in a joint space using the continuous BOW (CBOW) neural language model which results in low-dimensional text embeddings, where semantically similar comments and words reside in the same part of the space. These embeddings are used to train a binary classifier to distinguish between hateful and clean comments. None of the above works considers improving the representations of speech data based on related written data and use the enriched features in a classification model. [9] used audio (transcribed speech), visual and text (obtained with optical character recognition in the images) features in an early and late fusion setting to detect events. They show the value of multimodal data, but their methods require the availability of all sources at test time, they do not focus on the problem of noise in the transcribed speech, and do not study embedded multimodal representations, as done in this paper.

3   Methodology

The main objective of this paper is to investigate the impact of using related written data to improve the classification of noisy speech transcripts. We focus on how the related written data can be embedded with the speech data. First we will discuss the two classification models that we have tested: a neural network classifier and a logistic regression classifier[1]. They will be used as baselines to classify the transcribed speech, but will also be used as classification model for the speech data enriched with information from the written text. In a next section we discuss how information from the written data is embedded with the

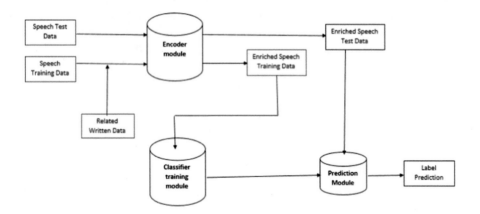

Fig. 1. Figure showing general architecture of classifier using encoder.

---

[1] We have tested several other models such as a support vector machine and a Naive Bayes classifier, but the results did not change substantially.

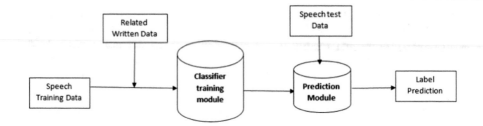

Fig. 2. Figure showing general architecture of classifier without using the encoder.

speech data during training of the classifier. Finally we describe how we select information from the written data.

3.1   Classification Models

Logistic Regression (LR) Classifier. The multiinomial logistic regression classifier learns a log-linear decision function from the training data. As input we use the words of the transcribed speech possibly augmented with words from related written text (where we use a bag-of-words representation computed as the sum of the one-hot vectors that represent the input words) or we use the enriched vectors that represent the encoded speech input (see below).

Neural Network (NN) Classifier. The classifier is a simple feedforward neural network. Again as input we use the words of the transcribed speech possibly augmented with words from related written text (sum of the one-hot vectors that represent the input words) or we use the enriched vectors that represent the encoded speech input (see below). We call the input text representation $\boldsymbol{x}^{enr}$. This vector passes through two fully connected layers with transformation matrices $\boldsymbol{W}^{enr,1}$ and $\boldsymbol{W}^{enr,2}$, obtaining respectively a hidden representation $\boldsymbol{h}^{enr}$ and an output $\boldsymbol{y}^{enr}$. The state of one hidden node $h_i^{enr} \in \boldsymbol{h}^{enr}$ is calculated as:

$$h_i^{enr} = \sum_{j=1}^{o} F(w_{i,j}^{enr,1} \cdot x_j^{enr}) \tag{1}$$

with $o = 4096$. The state of an output node $y_i^{enr} \in \boldsymbol{y}^{enr}$ becomes:

$$y_i^{enr} = \sum_{j=1}^{p} F(w_{i,j}^{enr,2} \cdot h_j^{enr}) \tag{2}$$

with $p = 256$.

Both classifiers predict the membership of the transcribed speech to each of the subject categories used in the classification.

As baselines, we use the logistic regression and neural network classifier for classification of the transcribed speech where no features are enriched with the written data. We call these models **Baseline-LR** and **Baseline-NN** respectively.

For the models that we propose we assume that during training with the transcribed speech we have access to related written texts, but during testing or prediction we only have transcribed speech.

3.2   Use of Related Written Data

Encoding the Speech Data with the Written Data as Output. The input of the encoder is a vector representation x^{enc} of transcribed speech data (in the form of 1-hot vectors) for each word. This vector passes through two fully connected layers with transformation matrices $W^{enc,1}$ and $W^{enc,2}$, obtaining respectively a hidden representation h^{enc} and an output y^{enc}. The state of one hidden node $h_i^{enc} \in h^{enc}$ is calculated as follows (with F being a softmax activation function):

$$h_i^{enc} = \sum_{j=1}^{m} F(w_{i,j}^{enc,1} \cdot x_j^{enc}) \tag{3}$$

with $m = 4096$. The state of an output node $y_i^{enc} \in y^{enc}$ becomes:

$$y_i^{enc} = \sum_{j=1}^{n} F(w_{i,j}^{enc,2} \cdot h_j^{enc}) \tag{4}$$

with $n = 1024$. At training time, the encoder learns to predict the related written text representation for a given transcribed speech representation (both 4096-dimensional vectors). At test time, it produces an enriched speech representation for a given transcribed speech test instance.

The encoder is trained on pairs of transcribed speech and related written text. The encoder enriches the speech data with different data obtained from the written texts (see below).

At test time a speech sample is sent through the encoder to obtain an enriched representation and then through the speech classifier to obtain subject category membership predictions. As mentioned above the classifier is either a NN or LR classifier.

In the experiments below we call this embedding model "With encoder". The architecture of this model is shown in Fig. 1.

Concatenating Speech and Related Written Data During Training. In this simple model, during training of the classifier the transcribed words are expanded with words selected from related written text (see below) and all words used as input. During test time only the speech data will be inputted (without additional information from the written text). We call this model "Without encoder". The architecture of this model is shown in Fig. 2.

3.3 Selection of Related Terms from Written Data

The main idea of this approach is to use important words from the related written data among which are named entities as we assume that the transcribed speech has a tendency to erroneously transcribe named entities which the transcriber has not encountered during its training. We have collected the written data from a news source on the same date span during which we collected the speech data. For each speech document, we found the related written data using cosine similarity using a bag-of-words representation of the transcribed speech and the written documents.

Selection of First n Words from Written Data. We take the first n words from the related written data. n is defined as 50, 100, 200 and 300. This model is called **First-n-words**.

Selection of n Words with Highest $tf-idf$ Value from Written Data. We select the n words with highest $tf-idf$ value from related written data. Again n is defined as 50, 100, 200 and 300. We name this model as **Top-n-words**.

Selection of Named Entities from Written Data. We use the Stanford NER system to recognize named entities in the written texts. We select all recognized named entities from the k most related texts.[2] We choose k as 3, 5, 7, and 10, but only report results on k being 3. We call this model **NER-top-k**.

Selection of Related Terms Using Google Word Embeddings. For each named entity selected in the written text with the above method, we select related terms in the vector space created by Google Word Embeddings.[3] We have selected the ten most similar words from the vector space for the selected named entity using the cosine similarity between the word vectors. We call this model **GWE**.

Selection of Related Terms Using Word Embeddings Created on the Concatenated Speech and Written Data. Here, we train the word2vec algorithm on a corpus which is obtained by merging the speech data and the related written data to build the word embeddings. For each named entity selected in the written text with the above method, we select the 10 most related terms in the vector space created by the obtained word embeddings using the cosine similarity between the word vectors. We name this method **Sp-wrt-WE**.

---

[2] If a named entity consists of more than one token, then we convert it to a single token representation by joining its components with underscore ("_"), for example "new york" is converted to "new_york". Thus "new_york" is a single feature rather than two different features when we use "new" and "york".

[3] https://code.google.com/archive/p/word2vec/.

Table 1. Number of video files in each category.

Category	# Files
Business	79
Entertainment	208
Health	113
International	107
Politics	247
Technology	78
Total	832

4   Data and Experiment

4.1   Preparing Speech Data

We have collected 832 news videos of different categories from the *"abcnews"*[4] and *"newsy"*[5] news site. The news sites categorize the videos into *"Business"*, *"Entertainment"*, *"Health"*, *"International"*, *"Technology"* and *"Politics"*. These categories are used as class labels. We have collected these videos in the period of June–August 2016. The number of news videos per category is shown in Table 1. We have split the speech data into a train and test set with a ratio of 80:20 of each category. We have transcribed the video news using the CMUsphinx [10]. CMUSphinx takes only sound format (mp3), therefore we have used "ffmpeg" tool [11] to convert the video into sound format. The format of the sound for the CMUSpinx tool should be in 16-bit mono 16000 Hz (or 8000 Hz). We have converted the video into sound in 16-bit mono 16000 Hz format. After that we transcribed the mp3 data into text using CMUsphinx, this comprises our speech data[6].

4.2   Written Data

We have crawled 65905 written news from Google news (http://www.news. google.com) for the same period (June–Aug 2016). They constitute our related written data.

5   Evaluation

We use the standard precision, recall and F_1 scores for evaluation of the speech classification. We report micro-average precision and recall.

---

[4] http://abcnews.go.com/.
[5] http://www.newsy.com/.
[6] We will make the speech transcriptions of the collected dataset available after publication including its split in train and test data.

6 Results and Discussion

We evaluate here the two classifiers LF and NN in two different modalities, one using the encoder and the other without encoder as discussed earlier. We have tested the different ways of selecting information from the written data that is used during training as discussed earlier.

Table 2 shows the results of the classifiers using the encoder in which speech data is passed to the encoder that during training was enriched with words from the related written texts. This table presents the precision (P), recall (R) and F_1 scores obtained with the classifiers using different models of selecting words from the written data. We observe that speech data enriched with the top or first n words gives better classification performance than the **Baseline** model. The model **Top-100-words** outperforms the rest. The model improves recall and F_1 scores by 5.1% and 3.3% respectively compared to the results obtained with the **Baseline** model without hurting precision, when using the LR classifier. While applying the NN classifer, the model **Top-300-words** outperforms the rest. The model improves recall and F_1 scores by 5.5% and 4% respectively as compared to the **Baseline** model.

Table 3 shows the results of classifiers without using the encoder, but just expanding the speech input with relevant words from related written texts. In this setting, the model **Top-100-words** outperforms the rest by improving recall and F_1 scores by 4.1% and 4% respectively compared to the **Baseline** model without hurting precision when using the LR classifier. Similarly, while applying

Transcribed speech data	Top 100 words from related written data	True Label	Prediction by Baseline model	Prediction by top-100-words model
the headline i see got a c d c how thistles morning it's very likely the virus will strive for misuse the new concerns forty years will have a seat the virus and the speed of this reddit easy to be very no full you must saying it is hoped very white read for mosquitoes to humans in america from her right around the corner what's the best thing you when your family to do to combat the disease right euro last spring consumer proven actually reduce the incidence of nuts you know borne diseases what else can you do around your whole are now with more michael i think you learn the habit of abc news the medical enter dr richard betts of and about winning streak of matthew question i've wandered yeah some freak out if i get about mosquito well i i think if if if your sunscreen you need to reapply that on top fine without you have to thank you won't forget what you give on the scale you should menthol and got the best movie answer your questions you tweet him and dr richard betts or or go to the ladies facebook page	zika virus workers mosquito health pregnant said birth become women cdc officials may also states us disease experts cases guidelines friday infection united mosquitoes control employers areas risk active season american pregnancy reported protect april new defect bites transmission defects mosquitoborne centers prevention consider reduce prevent travel federal babies confirmed many long people recommend far infections according item nt latin america healthy percent know help exposure transmitted microcephaly partner infected healthday issued urge insect wear standing water male travelers partners businesses danger news known cause abnormally small heads thought limited sexual even debate government country expect specialists avoid dr frieden	Health	Entertainment	Health
we have put nuclear disaster and survival on the american agenda tonight it has been ten years almost to the bay since the chernobyl explosion in ukraine by far the worst of it in nuclear accident in history the extent of the destruction and the lingering effect on all kinds of life are still unfolding and will continue to do so the margin reported father has filed the animals ah not only living in the midst of this on imaginable pollution but are actually in some cases tried long never in the history of the album durbin that accident quite like this one two hundred times is much radiation of hiroshima they the land around chernobyl is considered a bit so what is that you provided the largest now and yet would want to asserting with this research thing he was surprised to find a good zone was a threat to us and wildlife prisoner this area is deceptively normal til you turn on the geiger counter and spoil the mood you really are struck by the beauty of desert the real surprise came when the scientists collected field mice stayed in the area of the spiky accident they were apparently thriving despite	chernobyl plant nuclear explosion ukraine world radiation reactor april could worst accident numbers even fallout people days workers power fire may look amount begin area eventual death toll deaths cancer anything aerial view f sent large radioactive billion square soviet reuters wednesday arrow pointing location morning one yet tell meltdown thensoviet poisoning air much deadly radioactivity would become reutersfile disaster email telling story years later involves dauntingly figures others vexing still unknown hint scope money spent internationally funded project build longterm shelter building containing exploded structure place work remove lavalike waste miles land around abandoned heavy half located rest belarus	International	International	Health

Fig. 3. Figure showing predication by **Baseline** and **Top-100-words** on the sample speech data and its top 100 words from related written data.

Table 2. Classification results using two classifiers (*LR* and *NN*) **with Encoder** compared to the results of the Baseline. Note that the test data is without any enrichment, i.e., only speech data.

	NN			LR		
	Precision	Recall	F_1	Precision	Recall	F_1
Baseline	0.638	0.511	0.525	0.623	0.526	0.54
First-50-words	0.521	0.524	0.513	0.542	0.563	0.546
First-100-words	0.589	0.543	0.551	**0.592**	**0.577**	**0.573**
First-200-words	0.597	0.526	0.526	0.56	0.521	0.522
First-300-words	0.551	0.509	0.518	0.48	0.466	0.463
NER-top-3	0.241	0.297	0.264	0.351	0.348	0.333
Top-50-words	0.601	0.523	0.526	0.571	0.584	0.565
Top-100-words	0.521	0.478	0.486	0.543	0.545	0.533
Top-200-words	0.508	0.499	0.5	0.559	0.54	0.545
Top-300-words	**0.567**	**0.566**	**0.565**	0.568	0.563	0.559
GWE	0.334	0.342	0.297	0.325	0.327	0.317
Sp-wrt-WE	0.348	0.328	0.304	0.4	0.409	0.397

the NN classifier, **Top-50-words** improves recall and F_1 scores by 2.1% and 2% respectively compared to the **Baseline** model.

Table 3. Classification results using two classifiers (*LR* and *NN*) **without Encoder** compared to the results of the Baseline. Note that the test data is without any enrichment, i.e., only speech data.

	NN			LR		
	Precision	Recall	F_1	Precision	Recall	F_1
Baseline	0.638	0.511	0.525	0.623	0.526	0.54
First-50-words	0.556	0.521	0.526	0.601	0.548	0.56
First-100-words	0.553	0.506	0.512	0.589	0.534	0.546
First-200-words	0.601	0.528	0.538	0.571	0.511	0.525
First-300-words	0.51	0.442	0.438	0.559	0.492	0.507
NER-top-3	0.53	0.481	0.483	0.563	0.511	0.521
Top-50-words	**0.601**	**0.532**	**0.545**	0.614	0.557	0.572
Top-100-words	0.582	0.538	0.544	**0.613**	**0.567**	**0.58**
Top-200-words	0.547	0.516	0.518	0.604	0.551	0.563
Top-300-words	0.564	0.514	0.522	0.607	0.541	0.558
GWE	0.407	0.396	0.358	0.528	0.493	0.496
Sp-wrt-WE	0.495	0.450	0.436	0.572	0.536	0.541

From the experiments, we observe that incorporating features like the top n words or first n words from external related texts through encoding or expansion improves the performance compared to the baseline classification models that are not trained with such information.

Figure 3 shows the predictions of **Baseline** and **Top-100-words** on the sample speech data and its top 100 words from related written data. In the first example, i.e., first row, the baseline model predicts transcript speech data as *Entertainment* while its true label is *Health*. But when we incorporate the top 100 words from related written data, the classifier classifies the document correctly as *Health*. But in some cases, incorporating features from related written data does not help as shown in second sample of the figure where our model **Top-100-words** wrongly predicts the document's class as *Health* while its true label is *International*. But overall, we observe that embedding features from related written text helps to improve the prediction of the classifiers.

7   Conclusion and Future Work

Transcribed speech is known to be noisy and to contain transcription errors. We have proposed several models that embed information from related written texts during training of a classifier that operates on the transcribed speech. From the experiments, we have observed that embedding selected words from related written data during the training improves the performance of a logistic regression and neural network classifier. One of the achievements of our approach is that we don't need related written text during classification. This research opens several avenues on how to enrich content representations with related information that might have applications beyond dealing with noisy texts.

References

1. Collell, G., Zhang, T., Moens, M.: Imagined visual representations as multimodal embeddings. In: Proceedings of the Thirty-First AAAI Conference on Artificial Intelligence, pp. 4378–4384 (2017)
2. Huang, R., Hansen, J.H.: Advances in unsupervised audio classification and segmentation for the broadcast news and NGSW corpora. In: IEEE Transactions on Audio, Speech, and Language Processing, pp. 907–919 (2006)
3. Siegler, M.A., Jain, U., Raj, B., Stern, R.M.: Automatic segmentation, classification and clustering of broadcast news audio. In: Proceedings DARPA Speech Recognition Workshop, pp. 97–99 (1997)
4. Castán, D., Ortega, A., Miguel, A., Lleida, E.: Audio segmentation-by-classification approach based on factor analysis in broadcast news domain. EURASIP J. Audio Speech Music Process. **2014**, 34 (2014)
5. Atrey, P.K., Maddage, N.C., Kankanhalli, M.S.: Audio based event detection for multimedia surveillance. In: 2006 IEEE International Conference on Acoustics Speech and Signal Processing Proceedings, vol. 5 (2006)

6. Jiang, Y., Zeng, X., Ye, G., Ellis, D., Chang, S., Bhattacharya, S., Shah, M.: Columbia-UCF TRECVID2010 multimedia event detection: combining multiple modalities, contextual concepts, and temporal matching. In: TRECVID, National Institute of Standards and Technology (NIST) (2010)

7. Schwartz, R.M., Imai, T., Kubala, F., Nguyen, L., Makhoul, J.: A maximum likelihood model for topic classification of broadcast news. In: Kokkinakis, G., Fakotakis, N., Dermatas, E. (eds.) EUROSPEECH, ISCA (1997)

8. Chen, K., Liu, S., Chen, B., Wang, H., Jan, E., Hsu, W., Chen, H.: Extractive broadcast news summarization leveraging recurrent neural network language modeling techniques. IEEE/ACM Trans. Audio Speech Lang. Process. **23**, 1322–1334 (2015)

9. Djuric, N., Zhou, J., Morris, R., Grbovic, M., Radosavljevic, V., Bhamidipati, N.: Hate speech detection with comment embeddings. In: Proceedings of the 24th International Conference on World Wide Web, pp. 29–30. ACM (2015)

10. CMU: CMU sphinx toolbox. "CMU" (2016). https://cmusphinx.github.io/wiki/download/

11. FFmpeg: FFmpeg tool (2016). http://ffmpeg.org/

A Semantic Text Summarization Model for Arabic Topic-Oriented

Rasha M. Badry[1]([⊠]) and Ibrahim F. Moawad[2]

[1] Department of IS, Faculty of Computers and Information,
Fayoum University, Fayoum, Egypt
rmb01@fayoum.edu.eg
[2] Department of IS, FCI, Ain Shams University, Cairo, Egypt
ibrahim_moawad@cis.asu.edu.eg

Abstract. In the era of data overloading, Text Summarization systems (TSs) is one of the important Natural Language processing applications. These systems provide a concise form for the input document(s). According to the type of output summary, Text Summarization can be classified into extractive and abstractive. While the extractive text summarization is the process of identifying the important sections of the input text and producing them verbatim, the abstractive text summarization produces a new material in a generalized form. To facilitate the topic-oriented summarization, current research efforts focus on query-based text summarization, which summarizes the input document according to the user query. Although, the Arabic language is one of the Semitic languages and is spoken by 422 million people, there are very limited research efforts in Arabic query-based text summarization. In this paper, we propose a new Arabic query-based text summarization model. The model accepts both user query and Arabic document and then generates the extractive summary. The proposed model generates the extractive summary for the input document semantically by applying the Latent Semantic Analysis technique and exploiting the Arabic WordNet (AWN) ontology. Finally, to show the importance of the proposed model, a case study is presented.

Keywords: Semantic text summarization ·
Arabic query-based summarization · Topic-oriented ·
Extractive summary · AWN

1 Introduction

Text summarization (TS) research is an effective and ongoing field of research. Text summarization is defined as the process of creating a short version of a given text (Radev and McKeown 2002). It can be used in different fields. For example, TS can be useful in search engine to provide the user with summarized information of each page. There are different types of summarization systems, which are categorized according to many features (Gholamrezazadeh et al. 2009) like output, language, no of input documents, and the used approach. The summary can be extractive or abstractive summary (Gupta and Lehal 2010; Das and Martins 2007). In extractive summary, the important sentences are extracted from the original document. In extractive summary, no new sentences are

© Springer Nature Switzerland AG 2020
A. E. Hassanien et al. (Eds.): AMLTA 2019, AISC 921, pp. 518–528, 2020.
https://doi.org/10.1007/978-3-030-14118-9_52

generated from the original document (Moawad and Aref 2012). For the summary which is based on the purpose feature, the summary can be generic or query-based summary. In generic summarization, the summary contains the important points of the whole document (Hovy and Lin 1999). In query-based summarization systems, the summary contains the most important sentences for the user's query (El-Haj and Homma 2008). The generated summary can be generated from single or multi-document. Furthermore, the summary can be from single language document or multi-lingual document. Also, the summary can be based on supervised technique using specific data sets or unsupervised technique which is based on linguistics and statistical information.

There are hundreds of millions of people speaking and interested in the Arabic language, and the research in automatic Arabic summarization is still very active. Current research efforts focus on query-based text summarization to facilitate the topic-oriented summarization, which summarizes the input document according to the user query. There are very limited research efforts in Arabic query-based text summarization. In this paper, a semantic approach for Arabic query-based extractive text summarization is proposed. The Arabic WordNet (AWN) is used to semantically expand the user query. The proposed approach exploits the latent semantic technique, which provides the similarity of sentences and words using the Singular Value Decomposition (SVD). Then, the sentences are scored according to their semantic similarity to the users and the expanded query. Finally, the sentences are ranked and the top N sentences are selected to generate the summary.

The rest of the paper is organized as follows. Section 2 presents the related work in Arabic query-based summarization models. While Sect. 3 discusses the proposed approach in details, Sect. 4 describes a detailed case study scenario. Finally, Sect. 5 concludes the paper.

2 Related Work

Krishna et al. (2013) have done a hybrid method for query based summarization system. The proposed method is based on discovering the relationship between document sentences and the user query. It is mainly based on two techniques, which are the statistical and linguistic to find the relationship between the document sentences and the given query to calculate the similarity score between them.

Leuva (2016) have done a survey on a hybrid method for query based summarization system. First, the proposed system is based on graph method. The system is divided into two stages, which are offline and online stages. In the offline stage, it is a preprocessing step. Each paragraph is represented by a graph node. The semantic relationship is represented by an edge between two nodes. Then, TF-IDF is used to calculate the similarity score. The AHA approach and the neighbor algorithm is used to cluster the nodes of the graph. In the online stage, Okapi equation is used to calculate the similarity measure between each cluster and query. Okapi is based on TF-IDF. Another author proposed system which is based on extractive and abstractive summarization technique. Extractive technique is applied then the summary is enhanced by applying extractive technique. In addition to the use of WordNet for generated the abstracted summary.

Yang (2014) proposed a novel Contextual Topic Model for Query-focused Multi-document Summarization. It is based on Bayesian topic models. The proposed contextual topic model (CTM) is effectively used to determine the relevance of the sentences and to recognize the latent topics. Then, they are hierarchically arranged.

El-Haj and Hammo (2008) proposed two Arabic text summarization systems. The first is the Arabic Query-Based Text Summarization System (AQBTSS). The second is the Arabic Concept-Based Text Summarization System (ACBTSS) (Rahman and Borah 2015; El-Haj et al. 2011). AQBTSS is based on standard retrieval methods. AQBTSS and ACBTSS consist of two modules which are the document selection and the document summarization module. In document selection module, finding the documents that satisfy the user query. Then, a selected document is summarized. In document summarization, it starts by document segmentation phase into sentences. Then a sentence matcher phase, AQBTSS is used to compare the sentences with the user query. But in ACBTSS, the sentences are matched with a given concept. The cosine measure is one of the matching techniques. Followed by a sentence selector phase, in this phase a weighting scheme which is based on vector space model is applied. In addition, the top ranked sentences are selected. Finally, the fusion phase is used to merge the top ranked sentences according to their position in the source document (El-Haj et al. 2011).

Ikhtasir is an automatic Arabic text summarization system. The system is based on RST method (Azmi and Al-thanyyan 2009; Almani et al. 2015). The importance of the sentences is dependent on some features which are used to calculate the sentence score. The root and the use of a rhetorical analysis is used to calculate the word frequency. Then, the rhetorical tree of the text is generated to form the primary summary of the text.

As noted, all the previous summarization approaches described above are based on the statistical, linguistic, and RST approaches. Besides, there are other two limitations. Firstly, the lack of using an approach to semantically identify the relationships between the query terms and the document. Secondly, the lack of applying Arabic topic-oriented with the Latent Semantic Analysis.

Many summarization systems have been done for English language. On the other side, there are hundreds millions of people speaking and interested in the Arabic language. Although the research in Arabic summarization is still very active, few works have been done for Arabic language. In addition, there are very limited research efforts in Arabic query-based summarization. In this paper, we propose a new Arabic query-based text summarization model. The proposed model semantically generates the extractive summary for the input document by applying the Latent Semantic Analysis technique and exploiting the Arabic WordNet (AWN) ontology.

3   Model Architecture

The main goal of the proposed model is to produce a short summary of a lengthy Arabic document. The model is based on Latent Semantic Analysis (LSA) technique and exploits the Arabic WordNet (AWN) ontology. As shown in Fig. 1, it consists of five modules: query expansion, preprocessing, input matrix creation, Singular Value Decomposition (SVD), Singular Value Decomposition with reduction, sentence scoring and ranking.

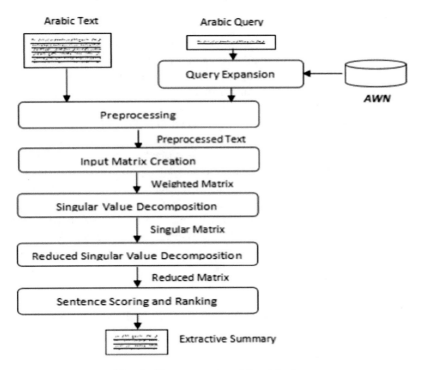

Fig. 1. The proposed model architecture

3.1 Query Expansion (QE)

The input query is expanded by supplementing additional terms to improve the retrieval performance. In query expansion, thesaurus is an important issue to help in the selection of the additional terms and identify the synonymous. The input user's query is expanded by using the Arabic WordNet, which is a lexical resource for Arabic language. AWN follows the Euro WordNet methodology. Moreover, AWN is based on the contents of the Princeton WordNet (PWN) (Black et al. 2006; Imam et al. 2013). For Example, the term "مصر" (Egypt) has many equivalent additional terms like ["الكنانة ارض" ,"الدنيا ام"]. Figure 2 shows the algorithm of the query expansion module.

Input: Query
Output: Expanded query terms
1. Read the query
2. Decompose the query into set of terms Q={$t_1, t_2, \ldots, t_n$}
3. For each query term (t_i) in Q
    Get the synonyms of term (t_i) and append the Q
  Repeat read
Return Q.

Fig. 2. The query expansion algorithm

3.2 Preprocessing

There are major preprocessing steps performed to enhance the performance of the created matrix. Figure 3 shows the algorithm of the preprocessing module. It includes six steps. In Sentence detection step, segmentation is an essential process where the document is divided into meaningful units. These units can be a paragraph, a sentence, or word (Hovy and Lin 1999). In Arabic, the white spaces and punctuations are used as a segmentation point. In tokenization step, after dividing the input document into sentences, the sentences are divided into words using the white spaces and punctuation marks (Imam et al. 2013). In remove diacritics step, there are some special notations in Arabic language which are called diacritics. Diacritics are used for Arabic grammar. These notations need to be removed. In remove punctuations step, there are many punctuations like " = + -× ÷ ':° >< | \ @ #ء". These punctuations do not have any value, so they should be removed. In Stop words removal step, there some words are high-frequency, and less meaningful like "فى" (in), "الى" (to), "على" (on). There is a need to remove these words. In stemming step, many words in the text document have different morphological variants. It is useless to treat such various forms of word as different words. There is a need to map these forms to the same term using a stemmer algorithm like Khoja's stemmer (Khoja and Garside 2010; Badry et al. 2013a). For example the root for "يعلم" (educate) is the basic block for many derivations like "تعليم" (education), "تعلم" (educated), "تعليم" (educating).

Input: Arabic Document, Query, Synonyms Query Terms
Output: Document terms, Query terms, Expanded query terms
1. Read the text from input file
2. Split the text document (D) into sentences $(S_0,S_1,S_2,....,S_n)$
3. For each sentence (S_i) in D
// Split the sentence into tokens
       For each sentence (S_j) Check segmentation marks
          Case (Not exist) continue reading
          Case (Exist) return sequence of characters as a token
       Repeat read
// To remove diacritics
       For each Token (T_j) Check diacritic marks
          Case (Exist) return remove token diacritics
       Repeat read
//To remove punctuations
       For each Sentence (S_j) Check punctuation
          Case (Exist) return remove punctuation
       Repeat read
// To remove stop words
       For each Token (T_j) in (S_i) Check stop words existence
          Case (Not exist in stop word list) return Token (T_j)
          Case (Exist) return delete word
       Repeat read
// To apply stemming rules
       For each Token (T_j) in (S_j)
          Check Token weight (T_j) , then apply stem rule
       Repeat read
Return Document terms, Query terms, Expanded query terms.

Fig. 3. The preprocessing algorithm

3.3   Input Matrix Creation

Input matrix creation is the first task in LSA (Ozsoy 2011; Ozsoy et al. 2010; Badry et al. 2013b). The input document is represented as a matrix, where the matrix columns represent the sentences of the source document, and the rows represent the document words. Furthermore, the cells represent the term frequency in each sentence. The input matrix creation implies the term weighting for document sentences. Suppose we have a document with n sentences and m total number of words, Fig. 4 represents the document matrix, where m represents the document words, n represents the sentences, and $w_{i,j}$ represents the term i weight in sentence j.

	S_0	S_1	S_2	....	S_n
T_1	W_{10}	W_{11}	W_{12}	....	W_{1n}
T_2	W_{20}	W_{21}	W_{22}	....	W_{2n}
....	....	....	....	....	....
T_m	W_{m0}	W_{m1}	0	....	W_{mn}

Fig. 4. The weighted matrix

It is important for the SVD calculations to calculate the term weights. There are many possible weighting schemes (Ozsoy et al. 2010) like frequency weight (FQ), Binary weight (BI), and Tf-Idf (Term Frequency–Inverse Document Frequency). In the proposed model, the term weighted is calculated using the Term Frequency (TF) method, where TF is the number of times the term appears in the sentence.

3.4   Singular Value Decomposition

The second task is to apply factor analysis, which is named singular value decomposition (SVD). SVD is a mathematical calculation, which applied to the created input matrix. Its main aim is to model and identify the relationships between terms and sentences, SVD is calculated according to the main formula (1) (Ozsoy 2011; Ozsoy et al. 2010; Badry et al. 2013a):

$$A = U \sum V^T \tag{1}$$

Where A is the input matrix (m × n); U is words × extracted concepts (m × n) which represents; $\sum$ represents scaling values, diagonal descending matrix (n × n); and V is sentences × extracted concepts (n × n).

The importance of the concept is determined by the magnitude of singular vectors. Then, the related sentences to the concept are projected along the singular vectors (concepts). Finally, the highest index sentences which are related to that singular vector with are represented.

3.5 Reduced Singular Value Decomposition

Reduced SVD is a mathematical technique used for breaking the term x document matrix into linearly independent components. The noisy correlations of the original data are removed. These components are mapped to sets of values that independently represent the structure of the dataset along each dimension. This means that the word vectors contain the elements of the most important correlations among words. The dimension reduction is important for enhancing the performance of the computation (Ozsoy 2011; Ozsoy et al. 2010; Badry et al. 2013b), see formula (2). Figure 5 shows the algorithm of weighted matrix, singular, and reduced singular matrix module.

$$M = U_n \sum\nolimits_n V^* \tag{2}$$

Where U_n is (m × n), $\sum n$ is (n × n) diagonal and V^* is (n × n)

Input: Document Terms, Query Terms, Expanded Query Terms
Output: SVD document matrix, SVD query matrix
1. Read the document terms, Query Terms, Expanded Query Terms
2. Create Weighted Matrix: Fill the matrix cell values
    For each term (t_j) in a sentence (S_i)
        Get term frequency (tf_{ji})➔ W (t_{ji}) = tf_{ji}
    Repeat reading
 3. Create SVD matrix (weighted matrix)
    Calculate the SVD calculation A= $U\sum V^T$,
    Get SVD calculations, Weighted matrix is decomposed into three matrices (U, $\sum$,V^T)
    Compute U, $\sum$,V^T, then compute A (SVD matrix)
4. Compute a reduced version of the SVD
    Thin SVD : M= $U_n \sum_n V^*$, Get a Reduced version of SVD
Return SVD document matrix, SVD query matrix.

Fig. 5. The algorithm of weighted and singular matrix creation

3.6 Sentence Scoring and Ranking

In sentence scoring, the semantic score for each sentence is calculated. The algorithm uses the results of the SVD to determine the most related sentences to the user query. The algorithm is based on measuring the similarity between the user query, the expanded query, and the sentences of the original document. The algorithm can be summarized as follow

1. Calculate sentence semantic score: The semantic scores are calculated by applying cosine similarity function. The cosine similarity is one of the most popular similarity measures and it is used to measure the similarity between the query and the document, see formula (3).

$$\cos\_sim(Q, D) = \frac{\sum_i Q_i.D_i}{\sqrt{\sum_i (Q_i)^2}.\sqrt{\sum_i (D_i)^2}} \qquad (3)$$

Where cos_sim (Q, D) is the similarity between query and document, Q_i is the weight of term i in query Q_i, and D_i is the weight of term i in document D_i

2. Rank the sentences: the sentences are ranked according to their semantic scores
3. Select the highest scores sentences: Selecting the top N sentences with the highest semantic scores to generate the summary.

4   Case Study

A case study is presented to illustrate how the proposed approach works. Table 1 shows an example for Arabic document. To semantically generate the extractive summary for the input document by applying the Latent Semantic Analysis technique and exploiting the Arabic WordNet (AWN) ontology, the presented approach phases should be followed step-by-step.

Table 1. Example for Arabic document

Arabic Document	Translation
الرياضة هي مجهود جسدي يمارسه الشخص لتطوير المهارات، أو المتعة، أو تقوية الثقة بالنفس، أو المنافسة . ويتساءل كثيرٌ من النّاس عن أهميّة الرّياضة في حياة الإنسان؛ إذ تزيد الثقة بالنفس، وتقوي الذاكرة، وتحافظ على الوزن المثالي للجسم . و تحافظ على المفاصل، وتعمل على تقوية عضلة القلب .ومن فوائدها أيضا أنها تقلل من نسبة إصابة الجسم بأمراض القلب .تتحكم بالتوتر والضغوطات بشكل أفضل .وتساعد على النوم بشكل أفضل.	A sport is a physical effort by a person to develop skills, joy, strengthen self-confidence or competition. Many people wonder about the importance of sports in human life; it increases self-confidence, strengthens memory, and maintains the ideal body weight. It maintains joints and works to strengthen the heart muscle. It also benefits from re-ducing the incidence of heart diseases. It controls stress and pressures better. And helps sleep better.

Firstly, consider the user query "فوائد الرياضة لقلب الانسان" (Benefits of sports for the human heart). The terms of input user query (UQ) is expanded by using AWN as in Table 2.

Table 2. User query with synonym query terms

Query term	Synonym terms
فوائد (Benefits)	هدف-عائد - أهمية - ثمرة - نفع - ربح -جدوى
الرياضة (Sport)	_____
قلب (Heart)	فؤاد
الإنسان (Human)	الناس - آدمي - شخص -بشر

In preprocessing step, the Arabic text document D is decomposed into set of sentences{S0, S1, S2, S3, S4, S5}, which are divided into tokens. Next, the diacritics, punctuation marks, and the stop words are removed. Finally, the stemming algorithm is applied to the document tokens. In input matrix creation step, the term incidence matrix A for the document D is created and weighting the cell value using the term frequency as shown as in Table 3.

Table 3. Weighted matrix for the Arabic document

Term	S_0	S_1	S_2	S_3	S_4	S_5
الرياضة (Sport)	1	1	0	0	0	0
مجهود (Effort)	1	0	0	0	0	0
جسدي (Physical)	1	0	0	0	0	0
يمارس (Perform)	1	0	0	0	0	0
الشخص (Person)	1	0	0	0	0	0
....	....	....	....	....	....	....
النوم (Sleep)	0	0	0	0	0	1

In SVD and its reduction calculations, they are applied on the created incidence matrix A to calculate the sentence semantic score. In Sentence Scoring and Ranking, the cosine similarity is used to calculate the semantic similarity score. For each sentence, the sentence semantic similarity is calculated according to the user's query UQ and the expanded query terms EQ. The semantic scores for the sentences (S_0, S_1, S_2, S_3, S_4, S_5) are (0.041, 0.051, 0.065, 0.064, 0.061, 0.041) respectively. Then, the sentences are ranked according to their score values. Finally, select the top N weighted sentences, which have the highest score values. Assuming that the highest two weighted sentences are considered, therefore, S_2 and S_3 are selected to compose the summary. Therefore, the output summary for the case study input user query is: "وتحافظ على المفاصل، وتعمل على تقوية عضلة القلب. ومن فوائدها أيضا أنها تقلل من نسبة اصابة الجسم بأمراض القلب" (It maintains joints and works to strengthen the heart muscle. It also benefits from reducing the incidence of heart diseases).

5   Conclusions

In this paper, a new Arabic query-based text summarization model was proposed. The model accepts both user query and the Arabic document, and then generates the extractive summary. Focusing on the user-query summarization using semantic oriented analysis and Arabic WordNet to select the most important sentences, the model passes through two phases. First, the user-query is expanded using the Arabic Word-Net. Then, the weighted matrix is constructed from the source document. Before creating the matrix, the text document is preprocessed. Then, the SVD calculations are applied on the created matrix. Finally, all the sentences are semantically scored and ranked to select the top N ranked sentences.

References

Alamin, N., Meknassi, M., Rais, N.: Automatic texts summarization: current state of the art. J. Asian Sci. Res. **5**(1), 1–15 (2015)

Azmi, A., Al-thanyyan, S.: Ikhtasir—a user selected compression ratio Arabic text summarization system. In: International Conference on Natural Language Processing and Knowledge Engineering, Dalian, China, 24–27 September 2009, pp. 1–7 (2009)

Badry, R., Sharaf Eldin, A., Elzanfally, D.: Text summarization within the latent semantic analysis framework: comparative study. Int. J. Comput. Appl. (IJCA) **81**(11), 40–45 (2013a)

Badry, R., Sharaf Eldin, A., Elzanfally, D.: Text summarization of arabic and english texts. In: Proceedings of the Sixth International Conference on Intelligent Computing and Information Systems (ICICIS), Cairo, Egypt, 14–16 December 2013, pp. 123–129 (2013b)

Black, W., Elkateb, S., Vossen, P.: Introducing the Arabic WordNet project. In: Third International WordNet Conference (GWC-06), Korea (2006)

Das, D., Martins, A.: A survey on automatic text summarization. Technical report, Carnegie Mellon University, US, November 2007

El-Haj, M., Hammo, B.: Evaluation of query-based arabic text summarization system. In IEEE Proceeding of the NLP-KE 2008, Beijing, China, 19–22 October 2008 (2008)

El-Haj, M., Kruschwitzand, U., Fox, C.: Experimenting with automatic text summarisation for arabic. In: Human Language Technology: LTC 2009. Lecture Notes in Computer Science, vol. 6562, pp. 490–499. Springer, Heidelberg (2011)

Gholamrezazadeh, S., Salehi, M., Gholamzadeh, B.: Comprehensive survey on text summarization systems. In: 2nd International Conference on Computer Science and its Applications, Jeju, Korea, pp. 1–6 (2009)

Gupta, V., Lehal, G.S.: A survey of text summarization extractive techniques. J. Emerg. Technol. Web Intell. **2**(3), 258–268 (2010)

Hovy, E., Lin, C.: Automated text summarization in SUMMARIST. In: Mani, I., Maybury, M. (eds.) Advances in Automated Text Summarization, pp. 81–94. MIT Press, Cambridge (1999)

Imam, I., Hamouda, A., Abdul Khalek, H.: An ontology-based summarization system for arabic documents (OSSAD). Int. Conf. Comput. Appl. **74**(17), 38–43 (2013)

Khoja, S., Garside, R.: Stemming arabic text. Computing Department, Lancaster University, Lancaster (1999). http://www.comp.lanc.ac.uk/computing/users/khoja/stemmer.ps

Krishna, R.V.V.M., Kumar, S.V.P., Reddy, C.S.: A hybrid method for query based automatic summarization system. Int. J. Comput. Appl. **68**(6), 39–43 (2013)

Leuva, S.: A Survey on A hybrid method for query based automatic summarization system for text. In: International Institution for Technological Research and Development, vol. 1, no. 3 (2016)

Moawad, I., Aref, M.: Semantic graph reduction approach for abstractive text summarization. In: 2012 Seventh International Conference on Computer Engineering & Systems (ICCES), 2012, pp. 132–138 (2012)

Ozsoy, M.G.: Text summarization using latent semantic analysis. Masters thesis, Department of Computer Engineering, Middle East Technical University, Turkey (2011)

Ozsoy, M.G., Cicekliand, I., Alpaslan, F.: Text summarization of Turkish texts using latent semantic analysis. In: Proceedings of the 23rd International Conference on Computational Linguistics (Coling 2010), pp. 869–876 (2010)

Radev, D.R., McKeown, K.: Introduction to the special issue on summarization. Comput. Linguist. – Summarization **28**(4), 399–408 (2002)

Rahman, N., Borah, B.: A survey on existing extractive techniques for query-based text summarization. In: International Symposium on Advanced Computing and Communication (ISACC) (2015)

Yang, G.: A novel contextual topic model for query-focused multi-document summarization. In: Proceedings of IEEE 26th International Conference on Tools with Artificial Intelligence, pp. 576–583 (2014)

A Language Identification System Based on Voxforge Speech Corpus

Khaled Lounnas[1]([✉]), Mourad Abbas[2], Hocine Teffahi[1],
and Mohamed Lichouri[1,2]

[1] University of Sciences and Technology Houari Boumediene, Bab Ezzouar, Algeria
`lounnas-khaled@outlook.com`
[2] Computational Linguistics Department, CRSTDLA, Bouzaréah, Algeria

Abstract. In this work, we address the problem of identifying languages based on Voxforge speech corpus. We downloaded corpora for three languages: English, German and Persian from Voxforge. In addition, we recorded two additional corpora, the first one for Modern Standard Arabic (MSA) and the other one for Kabyl, one of the Algerian Berber dialects. To tackle this task, we used three classifiers, namely: k-Nearest Neighbors (kNN), Support Vector Machines (SVM) and Extra Trees Classifier. We obtained an average precision of 87.45% for binary classification compared to 44% for the multi-class one.

Keywords: kNN · SVM · Extratrees · Language identification

1 Introduction

Nowadays, language identification becomes an essential solution, especially in cases where several languages are present, for example international trade such as stock exchange, checking into a hotel, arranging a meeting or making travel arrangement comfortable for non-native speakers [1].

Language identification is the key of success of many other tasks. Indeed, a Correct identification of speaker's language can improve the performance of speech recognition systems as well as speaker identification, audio stream tagging in spoken document retrieval, channel monitoring, or voice conversion.

The main objective of language identification is to detect automatically the language either in speech format or in a textual document. Language identification can be achieved in many levels: syntactic, morphological, acoustic, phonetic, prosodic, etc.

It is worth mentioning that although the existence of several works on identification of several languages, only few works on Modern Standard Arabic (MSA) had been achieved and no works at all had been carried out on Kabyl, which is one of the dialects in the Arabs' area.

The objective of this preliminary work is to achieve experiments on language identification using a set of data related to four languages in addition to one

© Springer Nature Switzerland AG 2020
A. E. Hassanien et al. (Eds.): AMLTA 2019, AISC 921, pp. 529–534, 2020.
https://doi.org/10.1007/978-3-030-14118-9_53

dialect. We used speech corpora downloaded mainly from Voxforge (corpora related to Persian, English and German), and we made recordings for Arabic and Kabyl dialect. This paper is organized as follows. In the second section, we present an overview of the related work. In third section, we briefly present the used methods of classification. A description of the dataset and the features is presented in Sects. 4 and 5. In Sect. 6, we discuss the conducted experiments. Finally, we conclude in Sect. 7.

2   Related Work

Biadsy et al. [2] designed a system dealing with Arabic dialect identification (Gulf, Iraqi, Levantine, Egyptian and MSA) by using a phonotactics approach, representing better the dialects and provided an accuracy of 81.6% using 30 s utterances.

Bhattacharjee et al. presented in [1] a Language identification system using Mel Frequency Cepstral Coefficients (MFCC) and prosodic features using a recently collected multilingual speech database namely Arunachali Language Speech Database (ALS-DB) which includes three Indian languages. They used a Gaussian mixture model (GMM) and obtained an encouraging result in term of EER (EER= 10.02%).

In [3], Laguna et al. carried out an experiment to identify Philippine Languages by training a Gaussian Mixture Model using Expectation Maximization algorithm, with acoustic parameters like: MFCC, Perceptual Linear Prediction (PLP), Shifted Delta Cepstra (SDC) and Linear Predictive Cepstral Coefficients (LPCC). They found that using PLP with 16 Mixture GMM-EM yielded the best performance among the four feature vectors in discriminating seven languages with an accuracy of 48.07%.

In [4], Hanani et al. classified transcriptions of automatic speech recognition according to five Arabic dialects by an SVM and using character tri-gram as features and obtained an accuracy of 57%.

Lopez-Moreno et al. [5] proposed two different DNN based approaches. In the first one, the DNN acts as an end-to-end probabilities of the target languages. In the second approach, the DNN is used to extract bottleneck features that are then used as inputs for a state-of-the-art i-vector system which give them an accuracy of 45% when applied to twenty-three languages.

Sarma and Sarma presented in [6] a dialect identification system for Assamese Speech using prosodic features. Indeed, they identified dialects from speech utterances by using Neuro fuzzy classifier and two sets of pertinent information: vowel sounds and their four formants. Experiments show that both of tow sets have given an encouraging identification rate especially the second which yielded an accuracy of 98.02%.

Itrat et al. developed [7] a method for identifying two languages of Pakistan via a vector quantization for phoneme codebook generation and multi-class classification using SVM. They obtained an accuracy of 82%.

An Arabic Dialect Identification based on Motif Discovery using GMM-UBM with different motif lengths was introduced by Moftah et al. in [8]. A new technique for extracting the characteristics of different Arabic dialects from the speech signal has been presented. The idea is to discover the repeated sequences (motifs) that characterize each dialect that was the input to train GMM-UBM classifier. The obtained accuracy was 62.5%.

3   Methods of Classification

3.1   Support Vector Machines (SVM)

It is a binary classifier, i.e. in the case where there are two classes It projects an input vector x into a scalar value f(x) in [11], as follows:

$$f(x) = \sum_{i=1}^{I} \alpha_i k(x; x_i) + \beta \tag{1}$$

where vector x_i are support vectors, I is the number of support vectors, α_i are the weights, and β is a bias, however it can be extended to the multi-class case, it enters into the supervised classification because it requires the prior knowledge of data and modelling of each language is done while learning SVM, it is a classifier of the discriminative [7] type because it seeks to find the optimal hyper plane of separation while maximizing the margin, it tries to distinguish data of different classes as best as possible.

In addition even in the case where the data is non-linearly separable SVM have proven its capacity and solve this problem through the so-called kernel functions.

The following equation shows the kernel function:

3.2   k-Nearest Neighbors (kNN)

kNN is a non-parametric and lazy learning algorithm, its use began in the 1970s in the field of artificial intelligence and pattern recognition, it is used for classification and regression. It assigns the unknown sample to the corresponding class according to a decision criterion that is generally a distance like Euclidean, Manhattan, and Minkowski. For more details, reader may refer to [9,10].

3.3   Extremely Randomized Trees (Extratrees)

It is a supervised learning method similar to Random Decision Forests but differs in the way of introducing randomness during training phase where The training of an extratrees involves the training of several trees, this implies that every tree is learned on all whole training data. For more details, reader may refer to [12].

4   Datasets

The quality of any system depends essentially on the quantity and the size of data. It depends also on the way the data is processed. The better the data is processed, the more accurate the system's performance is. In this work we built our system based on five languages, namely: Arabic (MSA), English, German, Persian and Kabyl dialect. The corpus is made of 635 utterances sampled at 16 kHz and coded on 16 bits. The number of speakers involved in the recording of this data is 220. 65% of this dataset had been used for training, and the remaining for the test phase.

Table 1 presents a description of the used dataset showing the number of utterances, the duration and the number of speakers for each language.

Table 1. Data set description

Language	Utterances	Duration(s)	Speakers
Persian	127	788	3
English	127	436	16
German	127	655	180
Arabic	127	170	15
Kabyle	127	123	6

5   Short Term Feature Extraction

In order to represent the speech signal, we used short-term feature extraction. The signal is split into short-term windows, then a number of features are computed for each window. This results in a sequence of feature vectors for the overall signal. The vector is made of the following components: 13 MFCC coefficients, energy, energy entropy, Zero Crossing Rate, Spectral Centroid, Spectral Spread, Spectral Entropy, Spectral Flux, Spectral Rolloff, Chroma Vector and Chroma Deviation. The total number of elements of the feature vector is 34.

6   Experiments and Results

In this section we show the results of the experiments that we carried out. We performed binary classification for each couple of language as well as multi-class classification. The metrics we used to evaluate the performance of the system are: Precision, Recall, and F1. Table 2 shows the obtained performance for binary classification for each couple of languages.

We can clearly see that the performance is high for some languages. For instance, the couples (Persian-Arabic), (German-Arabic), (English-Arabic) and

Table 2. Performance of binary classification of the five languages.

Data/Models	kNN			SVM			Extra		
	Pre	Recal	F1	Pre	Recal	F1	Pre	Recal	F1
Per Vs Ar	97.8	97.8	97.8	97.8	100	98.8	97.8	100	98.8
Per Vs Ka	89.3	68.8	77.7	80.8	69	74.4	63.8	62.5	63.1
Per Vs Ger	78.7	100	88	95.7	100	97.8	100	100	100
Per Vs Eng	87.2	61.1	71.8	87.2	61.1	71.8	85.1	60.6	70.8
Ger Vs Ar	100	100	100	100	100	100	97.8	100	98.8
Ger Vs Ka	100	66.1	79.5	100	73.4	84.6	100	72.3	83.9
Eng VS Ar	100	97.9	98.9	97.8	97.8	97.8	97.8	97.8	97.8
Eng VS Ka	93.6	70.9	80.6	48.6	56.4	51.1	53.1	59.5	56.1
Eng VS Ger	97.8	100	98.8	97.8	100	98.8	97.8	100	98.8
Ar Vs Ka	100	69.1	81.7	100	73.4	84.6	100	73.4	84.6
Average	94. 44	83.17	87.4	90.57	83.11	86	89.32	82.61	85.2

(Persian-German) are well identified by the three classifiers, with an F1 ranging from 97.8% to 100%. However, some other couple of languages as (Arabic-Kabyl), (English-Kabyl), (German-Kabyl) and (Persian-Kabyl) are less identified with an F1 measure ranging from of 51% to 84.6%. We should note that this difference in performance was expected, because for the first case, the languages are different, consequently, the identification was easier. Nevertheless, in the second case, Kabyl dialect is less identified with respect to Arabic, Persian, German and English. This can be interpreted by its closeness to these languages at phonetic and lexical levels. In fact, Kabyl shares a set of phonemes with German, Persian and English like /v/ and /p/, while it intersects with Arabic at lexical level.

Moreover, we should not that kNN and SVM have shown their efficiency compared to Extra trees classifier. In addition, we conducted experiments on multi-class identification, however the complexity of this operation yielded a lower performance than binary classification. This can be noticed through the results shown in Table 3, where the performance has been reduced dramatically to 40% and 44% and 35.8% obtained by kNN, SVM and Extratrees respectively.

Table 3. Multi-class classification performance

kNN			SVM			Extra		
Pre	Recall	F1	Pre	Recall	F1	Pre	Recall	F1
44.33	36.04	40	41.44	44.67	44	37.23	34.5	35.81

7 Conclusion

In this preliminary work we have designed a spoken language identification system based on Voxforge seech corpus. For the binary classification case, the five languages used in our experiments have been identified successfully using kNN, SVM and Extratrees. Contrary to the binary classification that yielded an encouraging average F1 score 87.45%, the multi-class case showed a degraded performance 44% because of the complexity of processing a higher number of classes at the same time. For further works, our aim will be the enrichment of our corpora as well as the use of other reliable classifiers.

References

1. Bhattacharjee, U., Sarmah, K.: Language identification system using MFCC and prosodic features. In: 2013 International Conference on Intelligent Systems and Signal Processing (ISSP), pp. 194–197. IEEE, March 2013
2. Biadsy, F., Hirschberg, J., Habash, N.: Spoken Arabic dialect identification using phonotactic modeling. In: Proceedings of the EACL 2009 Workshop on Computational Approaches to Semitic Languages, Association for Computational Linguistics, pp. 53–61, March 2009
3. Laguna, A.F., Guevara, R.C.: Experiments on automatic language identification for philippine languages using acoustic Gaussian mixture models. In: 2014 IEEE Region 10 Symposium, pp. 657–662. IEEE, April 2014
4. Hanani, A., Qaroush, A., Taylor, S.: Classifying ASR transcriptions according to Arabic dialect. In: Proceedings of the Third Workshop on NLP for Similar Languages, Varieties and Dialects (VarDial3), pp. 126–134 (2016)
5. Lopez-Moreno, I., Gonzalez-Dominguez, J., Martinez, D., Plchot, O., Gonzalez-Rodriguez, J., Moreno, P.J.: On the use of deep feedforward neural networks for automatic language identification. Comput. Speech Lang. **40**, 46–59 (2016)
6. Sarma, M., Sarma, K.K.: Dialect identification from assamese speech using prosodic features and a neuro fuzzy classifier. In: 3rd IEEE International Conference on Signal Processing and Integrated Networks (SPIN), pp. 127–132, February 2016
7. Itrat, M., Ali, S.A., Asif, R., Khanzada, K., Rathi, M.K.: Automatic language identification for languages of Pakistan. Int. J. Comput. Sci. Netw. Secur. (IJCSNS) **17**(2), 161 (2017)
8. Moftah, M., Fakhr, M.W., El Ramly, S.: Arabic dialect identification based on motif discovery using GMM-UBM with different motif lengths. In: The 2nd IEEE International Conference on Natural Language and Speech Processing, (ICNLS2018), pp. 1–6, April 2018
9. Dasarathy, B.V. (ed.): Nearest Neighbor (NN) Norms: NN Pattern Classification Techniques. IEEE Computer Society Press, Washington D.C. (1991). ISBN 0-8186-8930-7
10. Shakhnarovish, G., Darrell, T., Indyk, P. (eds.): Nearest-Neighbor Methods in Learning and Vision. MIT Press, Cambridge (2005). ISBN 0-262-19547-X
11. Li, H., Ma, B., Lee, K.A.: Spoken language recognition: from fundamentals to practice. Proc. IEEE **101**(5), 1136–1159 (2013)
12. Geurts, P., Ernst, D., Wehenkel, L.: Extremely randomized trees. Mach. Learn. **63**(1), 3–42 (2006)

A Novel Automated Financial Transaction System Using Natural Language Processing

Sachin Agarwal$^{(\boxtimes)}$, Prasenjit Mukherjee, Baisakhi Chakraborty,
and Debashis Nandi

Department of Computer Science and Engineering,
National Institute of Technology, Durgapur, India
sachinl3agarwal@gmail.com, prasen.mscit09@gmail.com,
baisakhichak@yahoo.co.in,
debashis.nandi@it.nitdgp.ac.in

Abstract. This paper proposes an automated financial transaction system (AFTS) that accepts a natural language transaction from a user in a query-response model that will be automatically converted to corresponding journal and ledger entries. This model uses the POS tags assigned to each token in a transaction to determine the name of account associated with the transaction and insert them in semantic table. The Journal and ledger entries will be produced from the semantic table. The type of transaction means debit or credit detection is dependent on relationship attributes in the semantic table. The proposed system generates journal and ledger entries from natural language transaction text in automated way. The proposed model uses a well-organized database to store keywords that helps to determine the account name and the type of transaction in time of semantic analysis.

Keywords: AFTS · Financial Transaction System · Natural language query · Query-Response Model · Automated accounting

1 Introduction

Natural language processing (NLP) is used in several applications like Query Response Systems, automatic text summarization, sentiment analysis, topic extraction, text mining, named entity recognition, parts-of-speech tagging, machine translation, relationship extraction, automated question answering and many more. The authors of the paper have proposed an account maintaining application of natural language text inputs which are posted as Queries and Responses generated to individual queries are the financial transactions (debit and credit entries) that are automatically created.

In the proposed system, the code parses the inputs and records the respective journal and ledger entries. The process begins with tokenization of the inputs which are then passed to the Parts of Speech (POS) tagger. The tags for each of the list of tokens are stored in a separate file named tags.txt. After extracting some specific information on the transactions like the date of transaction, the verbs and nouns are associated with the tags. The verbs are used to determine the nature of transaction i.e. whether it is a debit or credit. The nouns are used to determine the account name. The amount of transaction can be extracted if the corresponding tag is numeral. The verbs are compared to file

© Springer Nature Switzerland AG 2020
A. E. Hassanien et al. (Eds.): AMLTA 2019, AISC 921, pp. 535–545, 2020.
https://doi.org/10.1007/978-3-030-14118-9_54

name credit.txt and the nouns are looked up in the list of words of the 'accnts' folder to determine the account. The proposed system is an automated financial transaction system that is capable of handling general natural language transactions. The proposed system is not dependent on user's accounting knowledge. User may be unaware about the accounting knowledge but user can post total transaction in natural language into the proposed AFTS which will process and create the journal and ledger accounts dynamically with the ongoing postings in automated way. Section 2 discusses the previous work, Sect. 3 elaborates on the AFTS architecture followed by the algorithm in Sect. 4 describes methodology and tools used in proposed system. Application of proposed system refers in Sect. 5. Section 6 points out limitations and future work of the system and Sect. 7 winds up with conclusion.

2 Related Works

Several NLP applications are based on Query-Response models. Paper [1] discusses an Automated SQL Query Generator system that converts natural language statements to MySQL query that extracts information from particular database. This system is able to handle queries with aggregate functions, having group by clause. The natural language input statement will pass through tokenization, part of speech tagging, stemming and lemmatization to get the desired form for further processing. The system extracts the type of query and basic clause that specifies the required entities from the database and conditional clause. Finally the SQL query will be generated by the conversion of basic and conditional clause as in [1]. Paper [2] elaborates on utilization of database by organizations where insertion and extraction from database is difficult for a user who is not familiar with SQL. Natural language processing removes the barrier between novice user and database. NLP converts English statement to SQL. Novice user can insert or extract information from database without any knowledge of SQL using NLP. An Automated Knowledge Provider System (AKPS) has been proposed in [3] where natural language queries are converted to the conceptual form of database. AKPS is based on grammatical rules that can able to extract knowledge data from knowledge database without any manual intervention. Several NLP work focuses on using the Natural Language Toolkit (NLTK) [4, 5] library in the Python environment. Yumusak, Dogd and Kodaz have focused on the analysis of the corpora and the respective POS tagging methods which are critically compared by using the Python language. They have analyzed Brown, Penn Treebank and NPS Chat corpuses, and the taggers have been used for this analysis are namely; the default, the regex and the n-gram taggers. The paper specifies applying all these taggers to these three corpuses, resultantly they have shown that whereas Unigram tagger does the best tagging in all corpora, the combination of taggers performs a tad bit better if it is correctly ordered. Additionally, the paper gives us critical insight that NPS Chat Corpus gives different accuracy results than the other two corpuses as in [6]. Natural Language Systems and their respective methods for automated natural language translation take natural language as user-based input and translate it into a target natural language which can be used by the system as elaborated in [7]. The language translation requires to proceed further into the query based system that is facilitated through the strategic use of domain specific dictionaries, grammar rules, and probability values that are associated with some dictionary entries. To determine the likelihood of a particular multi-word combination, probability values are used, which in turn help us to determine a particular part of speech.

An approach to support programmers during program maintenance is established where they have presented a technique to extract concepts and relations from the source code, through the application of natural language parsing to sentences constructed from the terms that appear in program element identifiers. A dependency tree has been used to display the result of the parsing that has been computed by the system as in [8]. The foremost challenge in today's world is the appropriate recording of various accounting transactions carried out by the evidence of the transactions, and then formally preparing a system to automate such a query based response system in natural language. Each of the accounting transactions can be of the form of various invoices, receipts, letters of intent, electricity bill, telephone bill, etc. Iswandi, Suwardi and Maulidevi, had proposed a design [9] of rules to identify the entities located on the sales invoice. There are some entities identified in a sales invoice, namely: invoice date, company name, invoice number, product id, product name, quantity and total price. The rapid identification of all these entities is rigorously done using the named entity recognition method as elaborate in [15]. The entities generated from the rules used as a basis for automation process of data input into the accounting system.

In this paper, we have proposed to build a fully functional Automated Accounting System, to which a query (in the form of an accounting transaction, written in natural language) would be given by the user and the system would generate a double-entry journal. Along with the creation of a double-entry journal, the transactions would also be generated in a ledger. One of the salient features of this Automated Accounting System is that each transaction would also generate totally separate cash, debit and credit accounts to allow the user to keep track of each and every transaction that has taken place via the automated system as designed. This system proposes to reduce human labour time which is spent in making these accounts.

3   Architecture of the System

The AFTS is an automated query-response model that accepts a natural language transaction from a user in a query-response model that will be automatically converted to corresponding journal and ledger entries. The proposed system handles cash transactions only. The client feeds the day-to-day transactions to the system in natural language as query and the system prepares the corresponding journal and ledger entries as response. In the system's architecture, there is a default database consisting of two tables accounting based synonyms table and transaction type detection table. The accounting-based synonyms table contains synonymous words associated with each account name. On the other hand, the transaction type detection database contains detection keywords that determine whether the 'cash a/c' is debited or credited. The system uses the default data-base to prepare the journal and ledger entries and posts them to the journal and ledger databases respectively. Figure 1 shows the context diagram corresponding to the architecture of Automated Book Keeping System. A valid client makes entry to the system through login. A natural language query is posted by the client to the system. The query after segmentation and tokenization is tagged by the POS tagger. The system searches the default database for the tokens present in the query. The POS tags help the system to choose which of the two tables in the default database to be traversed for a particular token. Once a match is found, account name and the type of transaction (debit or credit) are identified. The system then prepares the journal and ledger entry ready to be posted to their respective databases. This system has been proposed because the proposed system is capable of handling accounting based natural language transactions using NLTK library that is most popular tool in natural language processing. The accounting based natural language transactions are different type natural language sentences. Most NLP systems are information extraction, question-answer type but they are unable to process accounting based natural language sentences. This necessitates designing the proposed Automated Financial Transaction Systems.

3.1   Algorithmic Steps

The step-wise description of the system is as follows.

1. Post Transaction in Natural Language
The client stores day-to-day transactions, in natural language, in a file. The file is then fed to the system as input. Our system assumes the transactions being posted are simple

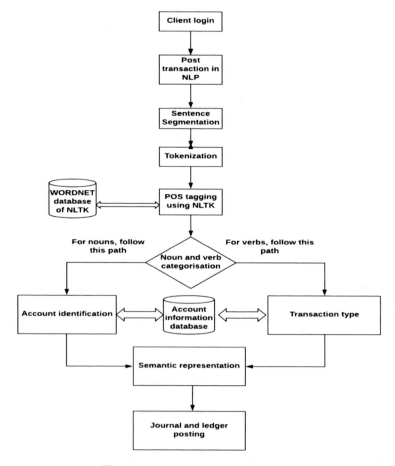

Fig. 1. Modular representation of AFTS

assertive sentences. The format of the input is Date: Transaction. For example, 'Jan 2: Paid electricity bill of Rs. 5,000'.

2. Sentence Segmentation

The system uses the separator ':' and splits the input into a tuple whose first element is the date and the second is the raw transaction in natural language (Fig. 2).

The above transaction is split into:

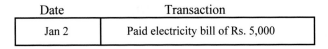

Date	Transaction
Jan 2	Paid electricity bill of Rs. 5,000

Fig. 2. Sentence segmentation

3. Tokenization

Each natural language transaction is then divided into constituent parts (words or phrases), called tokens. The result of this step is an array of tokens (Fig. 3), i.e.

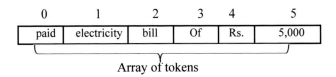

Fig. 3. Sentence tokenization

4. POS Tagging

The array of tokens for each natural language transaction is then fed to a POS-tagger, which tags each token with its part of speech using a built-in tagger available in the NLTK library. This results in a sequence of key-value pairs where key is the token and value is its part of speech.

For the above transaction, Fig. 4 shows the sequence is [('paid', 'VBN'), ('electricity', 'NN'), ('bill', 'NN'), ('of', 'IN'), ('Rs', 'NN'), ('5,000', 'CD')].

VBN - verb, past participle;
NN - noun, common, singular or mass
IN - preposition or conjunction;
CD numeral, cardinal

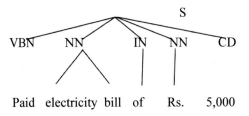

Fig. 4. POS tagging

5. Noun and Verb Categorization

The POS-tagged tokens of each transaction are then classified into two broad categories, namely noun and verb. The noun array contains all the tokens having POS tags as nouns (NN, NNS, NNP, NNPS), and the verb array contains the tokens having POS tags as verbs (VB, VBD, VBG, VBP, VBN, VBZ). The 'CD' tagged tokens give us the amount being mentioned in the transaction.

In our example, Amount - 5,000;
Verb array - [('paid', 'VBN')];
Noun array - [('electricity', 'NN'), ('bill', 'NN')]

6. Account Identification

The set of nouns help us identify the account name. We have accounting-based synonyms table in the default database. Each row consists of account name and a synonymous keyword that can uniquely determine the account name. The table is traversed to match the tokens in the noun array, if a match is found the account name is returned else the search continues. The 'Amount' variable will be fixed because it detects as a value. This amount variable has been used in semantic table.

In our example, the match is found for the account name – 'electricity a/c'.

7. Account Type Identification

The set of verbs identify whether the above mentioned account is being credited or debited. The transaction type detection table in the default database is used to identify the type of transaction. Each row consists of the type of transaction (debit or credit) and detection keywords. The table is traversed to match the tokens in the verb array.

Here, the verb 'paid' is found in the 'credit' file. It implies that 'cash' goes out, so 'electricity a/c' is debited while the 'cash a/c' is credited.

8. Semantic Representation

The system reads the nouns, verbs and Amount and inserts them in the semantic table (Table 1) as an entity and relationship.

Table 1. Semantic table

Attribute	POS	Relationship	Amount
Paid	Verb	Cash a/c credit	5000
Electricity	Noun	Electricity a/c	5000

9. Journal and Ledger Posting

The final double-entry journal and ledger is prepared and posted to the accounting database. Tables 2, 3 and 4 show the journal and ledger entry of the above natural language transaction.

Table 2. Journal entry of a transaction

Date	Particulars	Amount (Dr.)	Amount (Cr)
Jan. 2	Electricity A/c	5000	
	Cash A/c		5000

Table 3. The ledger entry of electricity A/c

Date	Particulars	Amount (Dr.)	Amount (Cr)
Jan. 2	Cash A/c		5000

Table 4. The ledger entry of cash A/c

Date	Particulars	Amount (Dr.)	Amount (Cr)
Jan. 2	Electricity A/c	5000	

3.2   Structure of the Default Database

The Automated Financial Transaction System processes the natural language transactions and generates the semantic table with entities and their relationship in the default database. The part of speech tags are used to make relationship between the entities that generates the conceptual form of database. The semantic table has been used to create the journal and ledger entries from the conceptual form.

The default database consists of two tables-Accounting-based synonyms table and transaction type detection table. The accounting-based synonyms table contains synonymous words associated with each account name. This table has been used to identify the account name with the help of nouns in the transaction. On the other hand, the transaction type detection database contains detection keywords that determine whether the 'cash a/c' is debited or credited. This table uses the verbs in the transactions to identify the type of transaction.

The attributes in the accounting-based synonyms table (Table 5) has been given below.

(1) ID - The primary key that uniquely identifies a row in the table.
(2) Account name - The name of the account to be used in the journal and ledger entries.
(3) Synonyms - The words that are synonymous to the account name, i.e., the words with the help of which one can identify the account name.

Table 5. Accounting-based synonyms table

ID	Account name	Synonyms
1	Postage	POSTAL
2	Communication	Telegram
3	Communication	Telegrams
4	Communication	Fax
5	Postage	Postage
6	Communication	Telephone
7	Electricity	Electricity
8	Conveyance	Travel

The attributes in the Transaction type Detection table (Table 6) has been given below.

(1) ID - The primary key that uniquely identifies a row in the table.
(2) Type of transaction - The type of transaction, i.e., debit or credit.
(3) Detection keywords - The words (verbs) with the help of which one can identify the type of transaction.

Table 6. Transaction type detection table

ID	Type of transaction	Detection keywords
1	Credit	Pay
2	Credit	Pays
3	Credit	Paid
4	Credit	Purchased
5	Credit	Purchased
6	Debit	Received
7	Debit	Received
8	Debit	Received

These tables should be regularly updated because increase in the number of synonymous words and the detection keywords will help in more efficient conversion of the natural language transactions to their respective journal and ledger entries.

4 Methodology and Tools Used

The Automated Financial Transaction System accepts a set of transactions in natural language and processes these transactions using NLTK library and prepares the corresponding journal and ledger entries.

1. The input file is the list of transactions in natural language (Fig. 5). This file is fed to the system as input.
2. The system processes these transactions one by one. Firstly, the date and the text are separated using a ':' separator. The text is then tokenized which results in a list of tokens present in the text.
3. The list of tokens is fed to the POS tagger of the NLTK library where each token is tagged with their respective part of speech.
4. For each transaction, the set of nouns and verbs are used to identify the account name and the type of transaction (Table 7) respectively.
5. Using the information obtained in the last step, i.e., the account name and the type of transaction (debit or credit), the corresponding journal entry is prepared and stored in the database.

6. The entries are then posted to the respective ledger accounts like assets, cash, communication, dividends, electricity, furniture, garlands, prepaid rent, purchase, sales etc.

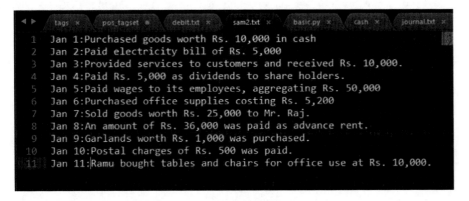

Fig. 5. List of transaction in natural language

Table 7. Types of transaction after processing

Date	Particulars	Amount (Dr.)	Amount (Cr)
Jan 1	Purchase A/c	10000	
	Cash A/c		10000
Jan 2	Electricity A/c	5000	
	Cash A/c		5000
Jan 3	Cash A/c	10000	
	Service revenue A/c		10000
Jan 4	Dividends A/c	5000	
	Cash A/c		5000
Jan 5	Salary & wages A/c	50000	
	Cash A/c		50000
Jan 6	Assets A/c	5200	
	Cash A/c		5200
Jan 7	Cash A/c	25000	
	Sales A/c		25000
Jan 8	Prepaid rent A/c	36000	
	Cash A/c		36000
Jan 9	Garlands A/c	1000	
	Cash A/c		1000
Jan 10	Postage A/c	500	
	Cash A/c		500
Jan 11	Furniture A/c	10000	
	Cash A/c		10000

AFTS has been implemented using Python. The database interface has been implemented in MySQL, which is a very popular open source database. The NLTK (Natural Language Toolkit) is a python library which has been used to process the transactions in natural language as it provides easy-to-use interfaces along with a suite of text processing libraries for classification, tokenization, stemming, POS tagging, parsing, semantic reasoning, wrappers, etc. for NLP libraries.

5   Time Complexity

Let there are n numbers of token in a query (Fig. 6).

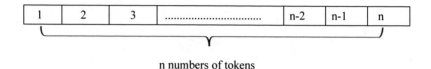

n numbers of tokens

Fig. 6. Token array

  i. Time taken for date head is k unit time and time taken for transaction is (n−k) unit time in sentence segmentation step.
 ii. Time taken = (n−k) unit time [the first token has already been used in step i].
iii. Let there are h number of tags in NLTK library. Time taken for one token out of (n−1) tokens to be tagged in k number of tags in NLTK = h unit time. There for total time taken for (n−1) tokens to be tagged in this process = (n−1)h unit time.
 iv. Let there are x number of noun tokens and y number of verb tokens out of (n−1) token. Time taken for all noun tokens, verb tokens and CD tagged tokens = (x + y + 1).
  v. Let, there are m numbers of rows and two fixed column in Accounting-based Synonyms Table. There for time taken for all noun tokens to be searched = 2xm unit time.
 vi. Let, there are b numbers of rows and two fixed column in Transaction type Detection Table. There for time taken for all noun tokens to be searched = 2by unit time.
vii. Time taken to prepare semantic table will take (m + b) unit time.

Total time complexity of the proposed system:
$f(n, k, h, x, y, m, b) = k + (n − k) + (n − k) + (n − 1)h + (x + y + 1) + 2xm + 2by + (m + b)$
[$\cong$ - congruence, Here, $k \cong h \cong x \cong y \cong b \cong m \cong n$.]
$\cong n + (n − n) + (n − n) + (n − 1)n + (n + n + 1) + 2n.n + 2n.n + n + n = n + n^2 − n + 2n + 1 + 2n^2 + 2n^2 + 2n$
$= 5n^2 + 4n + 1 = O(n^2)$

6 Limitations

The input statement has to follow the prototype strictly; a mismatch may populate the system with undesirable results.

The inputs statements are simple and system will not accept complex sentences.

The AFTS handles cash transactions only where debit and credit transactions are created.

7 Conclusion

The proposed AFTS model successfully reduces the barrier of a high-level accounting knowledge required by the user. It tokenises the input statement (Query), tags it, identifies the accounts involved in the transaction and extracts the verbs to determine the nature of the transaction associated with the accounts. The information extracted from the input statements are used to add the journal and ledger entries. The database is queried for the account names and in case, if no relevant account is found, a new entry for the particular account (the queried noun) is added to the database for future retrieval. From a user's perspective the above details are abstracted and the final result involves the ledger and journal entry written using file stream. The proposed system however, requires a very structured input statement and ill-formed queries might lead to wrong database entries being created and hence wrong ledger and journal entries.

References

1. Pagrut, A., Pakmode, I., Kariya, S., Kamble, V.: Haribhakta, Y: Automated SQL query generator by understanding a natural language statement. Int. J. Nat. Lang. Comput. (IJNLC) **7**(3), 1–11 (2018)
2. Kunte, A.S., Hasbe, A., Chavan, A., Patil, K.: Natural language query processing. Int. Res. J. Eng. Technol. (IRJET) **5**(3), 2731–2733 (2018)
3. Mukherjee, P., Chakraborty, B.: Automated knowledge provider system with natural language query processing. IETE Tech. Rev. **33**(5), 525–538 (2016)
4. NLTK 3.0 Documentation. http://www.nltk.org/. Accessed 06 May 2018
5. Loper, E., Bird, S.: NLTK: The Natural Language Toolkit. In: ETMTNLP '02 Proceedings of the ACL-02 Workshop on Effective tools and Methodologies for Teaching Natural Language Processing and Computational Linguistics, Philadelphia, Pennsylvania, vol. 1, pp. 63–70 (2002)
6. Yumusak, S., Dogd, E., Kodaz, H.: Tagging accuracy analysis on part-of-speech taggers. J. Comput. Commun. **2**, 157–162 (2014)
7. Akers, G.A., Kuno, S.: Automated system for generating natural language translations that are domain-specific, grammar rule-based and/or based on part-of speech. US6278967B1. https://patents.google.com/patent/US6278967B1/en# cited by, Logovista Corp, 31 August 1992
8. Abebe, S.L., Tonella, P.: Natural language parsing of program element names for concept extraction. In: IEEE 18th International Conference on Program Comprehension (ICPC), pp. 156–159. University of Minho, Braga, Minho, Portugal (2010)
9. Iswandi, I., Suwardi, I.S., Maulidevi, N.U.,: Designing rules for accounting trans-action identification based on Indonesian NLP. In: Annual Applied Science and Engineering Conference (AASEC), Bandung, Indonesia, vol. 180, no. 1 (2016)

Exploring Different Approaches
for Parsing Telugu

B. Venkata Seshu Kumari[1][✉], A. Giri Prasaad[1], M. Susmitha[1],
Vikram Raju R.[1], and Roheet Bhatnagar[2]

[1] Department of IT, VNR VJIET, Hyderabad, India
venkataseshukumari@gmail.com,
giridhar9ambati@gmail.com, sushmitha_m@vnrvjiet.in,
vikramraju.r@gmail.com
[2] Department of Computer Science and Engineering,
Manipal University Jaipur, Jaipur, India
roheet.bhatnagar@jaipur.manipal.edu

Abstract. In this paper we explore different approaches for parsing Telugu. We consider three popular dependency parsers namely, MaltParser, MSTParser and TurboParser. We first experiment with different parser and feature settings and show the impact of different settings. We then explore different ways of ensembling these parsers. We also provide a detailed analysis of the performance of all the approaches on major dependency labels and different distance ranges. We report our results on test data of Telugu dependency treebank provided in the ICON 2010 tools contest on Indian languages dependency parsing. We obtain state-of-the art performance of 91.8% in unlabelled attachment score and 70.0% in labelled attachment score.

Keywords: Dependency parsing · Telugu · MSTParser · MaltParser · TurboParser

1 Introduction

Dependency parsing is the task of uncovering the dependency tree of a sentence, which consists of labeled links representing dependency relationships between words. Parsing is useful in major NLP applications like Machine Translation, Dialogue Systems, Question Answering etc. This led to the development of grammar-driven, data-driven and hybrid parsers. Due to the availability of annotated corpora in recent years, data driven parsing has achieved considerable success. The availability of phrase structure treebank for English (Marcus et al. 1993) has seen the development of many efficient parsers. Unlike English, many Indian (Hindi, Bangla, Telugu, etc.) languages are free-word-order and are also morphologically rich. It has been suggested that free-word-order languages can be handled better using the dependency based framework than the constituency based one (Bharati et al. 1995). Due to the availability of dependency treebanks, there are several recent attempts at building dependency parsers. Two CoNLL shared tasks (Buchholz and Marsi 2006; Nivre et al. 2007a) were held aiming at building state-of-the-art dependency parsers for different languages. Recently in two

© Springer Nature Switzerland AG 2020
A. E. Hassanien et al. (Eds.): AMLTA 2019, AISC 921, pp. 546–555, 2020.
https://doi.org/10.1007/978-3-030-14118-9_55

ICON tools contests (Husain 2009; Husain et al. 2010), and Coling 2012 Hindi parsing shared task (Bharati et al. 2012), rule-based, constraint based, statistical and hybrid approaches were explored towards building dependency parsers for three Indian languages namely, Telugu, Hindi and Bangla.

In all these efforts, state-of-the art accuracies are obtained by the popular data-driven parsers namely, MaltParser (Nivre et al. 2007b), MSTParser (McDonald 2006) and TurboParser (Martins et al. 2009). Among Indian languages, though there has been significant amount of work on dependency parsing Hindi, there is very little work on parsing Telugu. Most of the work in ICON 2010 tools contest for parsing Telugu used MaltParser. In this paper, we consider all the three popular dependency parsers, MaltParser, MSTParser and TurboParser. We provide related work in Sect. 2 and details of dependency parsing and the Telugu dependency treebank in Sect. 3. In Sect. 4, we first experiment with different parser and feature settings and show the impact of different settings. We then explore different ways of ensembling these parsers. Section 5 provides a detailed analysis of the performance of all the approaches on major dependency labels and different distance ranges. We conclude with possible future directions in Sect. 6. We obtain state-of-the art performance of 91.8% in unlabelled attachment score and 70.0% in labelled attachment score. To the best of our knowledge ours is the only work which explored different features, parsers and parser combinations for Telugu.

2 Related Work

In two ICON Tools Contests (Husain 2009; Husain et al. 2010), different rule-based, constraint based, statistical and hybrid approaches were explored towards building dependency parsers for Indian languages. Kesidi et al. (2010) used a constraint based approach. The scoring function for ranking the base parses is inspired by a graph based parsing model and labeling. Ambati et al. (2009), Kosaraju et al. (2010), Nivre (2009) used MaltParser and explored the effectiveness of local morphosyntactic features, chunk features and automatic semantic information. Parser settings in terms of different algorithms and features were also explored. Ambati et al. (2009) explored the usefulness of MSTParser for parsing Indian languages. Zeman (2009) combined various well known dependency parsers forming a super parser by using a voting method. Recently in Coling 2012 workshop on Machine Translation and Parsing in Indian Languages, Hindi parsing shared task was held with the latest Hindi dependency treebank (Bharati et al. 2012). In this shared task, Kukkadapu et al. (2012) explored voting and blending techniques for parsing Hindi using MaltParser, MSTParser and TurboParser. Kumari and Rao (2012) combined the output of MaltParser, and MSTParser in an intuitive manner to extract pros of both the parsers. In our work, by tuning the features and parser parameters we develop the best parsing models for Telugu using MaltParser, MSTParser and TurboParser. We then combine the output of these parsers using the techniques of Kumari and Rao (2012) and Kukkadapu et al. (2012).

3   Concept of Dependency

Grammar Dependency Grammar (DG) describes the syntactic structure of a sentence through dependency graphs. A dependency graph represents words and their relationship to syntactic modifiers using directed edges. These edges can be labelled with grammatical relations like Subject, Object etc. Dependency trees can either be projective or non-projective. As English is fixed word order language, most English sentences can be analyzed though projective trees. But, in free word order languages, like Czech, Hindi, Telugu etc. non-projective dependencies are more frequent. Rich inflection systems reduce the demands on word order, leading to non-projective dependencies (McDonald 2006). Figure 1 shows the dependency tree for an example/sample Telugu sentence.

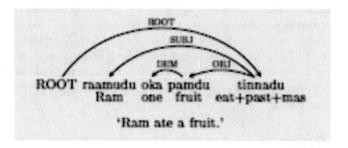

Fig. 1. Dependency tree for an example Telugu sentence

Telugu dependency treebank released in ICON 2010 Tools contest (Husain et al. 2010) is used in our work. The data is annotated using the part-of-speech (POS) tagging, chunking and dependency annotation guidelines (Bharati et al. 2006, 2009). The treebank consists of morphological information (root, coarse pos-tag, gender, number, person, case marking, suffix and TAM (tense, Aspect and Modality marker)), POS, chunk and dependency information. The dependency annotation follows a scheme that can be traced back to Paninian grammar (Bharati et al. 2009), known to be well-suited to modern Indian languages. The dependency labels are syntactico-semantic in nature (Bharati et al. 1995, 2009). For example, 'k1' usually corresponds to subject syntactic role and agent semantic role. Similarly, 'k2' corresponds to the syntactic role of object and the semantic role of patient. For the purposes of readability, instead of original treebank dependency labels, their corresponding English labels are used in this paper (SUBJ, OBJ, DEM for k1, k2, nmod adj respectively).

The treebank is available in SSF (Bharati et al. 2007) and CoNLL1 formats. We work with the CoNLL format in this paper. In this format, word, root, pos tag, chunk tag and morphological features are available in the FORM, LEMMA, POSTAG, and CPOSTAG and FEATS columns respectively. Data released has both fine-grained and coarse-grained versions of dependency labels. We used fine-grained version for our experiments. Table 1 shows the details of the training, development and the testing data sets of the Telugu dependency treebank. Statistics of sentence count, word count and average sentence length are provided in the table, (http://nextens.uvt.nl/depparse-wiki/DataFormat).

Table 1. Telugu treebank statistics

Type	Sent count	Word count	Avg.sent_length
Train	1400	7602	5.43
Devel	150	839	5.59
Test	150	836	5.57

4   Experiments and Results

We first explored MaltParser, MSTParser and TurboParser for parsing Telugu. Exploring different feature and parser settings, we built best models for each parser. We then experimented with different ways of combining the outputs of these parsers. As the training data size is small, we merged training and development data and did 10-fold cross validation for tuning the parameters of the parsers and for feature selection. Best settings obtained using cross-validated data are applied on test set. We used standard Unlabelled Attachment Score (USA), Labelled Attachment Score (LAS) and Labeled Score (LS) metrices for our evaluation.

4.1   MaltParser

MaltParser is a freely available implementation of the parsing models described in Nivre et al. (2007b). It is a classifier-based shift reduce parser. With MaltParser, parsing can be performed in linear time for projective dependency trees and quadratic time for arbitrary (possibly non-projective) trees. MaltParser provides options for nine deterministic parsing algorithms: Nivre arceager, Nivre arc-standard, Covington projective, Covington non-projective, Stack projective, Stack swap-eager, Stack swap-lazy, Planar and 2-planar. It also provides options for libsvm and liblinear learner algorithms. For Telugu dependency parsing liblinear learner and arc-eager parsing algorithm consistently gave better performance. We did a step-by-step analysis of the impact of different features on parsing Telugu. Table 2 provides results of these experiments. In Exp1, we provided word FORM and POSTAG of current word as features which gave an accuracy of 74.1% in UAS and 48.1% in LAS. Adding FORM and POSTAG of context words (Exp2) improved both UAS and LAS by around 12% which shows the importance of context in parsing. Adding LEMMA and CPOSTAG features (Exp3 and Exp4) gave a slight improvement of 1.6% in UAS and 2% in LAS. In Exp5, we added FEATS which contain morphological information and this gave 1% improvement in UAS and boosted LAS by 5.4%. As Telugu is a morphologically rich language, it is expected that morphological information plays a crucial in parsing, especially in identifying correct dependency labels. In Exp6, we provided dependency relations (DEPREL) of the partially formed trees which gave an improvement of 1.8% in UAS and 1.5% in LAS. Adding partial tree (Exp7) and bi-gram (Exp8) features gave further improvements of 1.3% in UAS and 1.7% in LAS. After all these experiments, we achieved a performance of 91.8% in UAS and 70.0% in LAS.

Table 2. Impact of different features on parsing Telugu using MaltParser.

Features	UAS (%)	LAS (%)	LS (%)
Exp1: current FORM, POSTAG	74.1	48.1	51.1
Exp2: Exp1+context FORM, POSTAG	86.1	59.4	61.1
Exp3: Exp2+LEMMA	86.3	61.3	63.3
Exp4: Exp3+CPOSTAG	87.7	61.4	62.8
Exp5: Exp4+FEATS	88.7	66.8	69.1
Exp6: Exp5+DEPREL	90.5	68.3	70.5
Exp7: Exp6+Partial Tree features	90.7	69.6	71.8
Exp8: Exp7+Bi-gram features	91.8	70.0	72.3

4.2   MSTParser

MSTParser is a freely available implementation of the parsing models described in McDonald (2006). It is a graph-based parsing system in which parsing algorithm is equated to finding directed maximum spanning trees from a dense graph of the sentence. MSTParser uses Chu-Liu-Edmonds maximum spanning tree algorithm for non-projective parsing and Eisner's algorithm for projective parsing. It uses online large margin learning as the learning algorithm (McDonald et al. 2005). It also provides options of 1st order and 2nd order features. For Telugu, 2nd order features and non-projective algorithm gave the best results of 90.0% UAS and 62,6% LAS (Table 3, MSTParser: Baseline). It is difficult to do feature tuning with MSTParser as it doesn't provide nice options similar to MaltParser. Labelling module of the MSTParser is not using FEATS column. Exp5 in Table 2 clearly showed that morphological features in FEATS are very important for labelling in case of Telugu. We explored different features using FEATS columns in the labelling module of the MSTParser and selected the settings which gave best results on 10-fold cross-validation. This gave a huge boost of 4.5% improvement in LAS over the baseline model (MSTParser: Extended in Table 3). With this tuning, we achieved a performance of 90.0% in UAS and 67.1% in LAS.

Table 3. Impact of different features on parsing Telugu using MSTParser

Features	UAS (%)	LAS (%)	LS (%)
MSTParser: Baseline	90.0	62.6	63.9
MSTParser: Extended	90.0	67.1	68.6

4.3   Turbo Parser

TurboParser is a freely available implementation of the parsing models described in (Martins et al. 2009). It uses integer linear programming technique for parsing. With default settings, we got an UAS of 90.5% and LAS of 67.5%. As the data is low and as the average sentence length is small, using standard model and considering only first order features gave better results. Final best results we could obtain are 91.2% UAS and 68.8% LAS (Table 4).

Table 4. Impact of different features on parsing Telugu using TurboParser

Features	UAS (%)	LAS (%)	LS (%)
TurboParser: Baseline	90.5	67.5	69.0
TurboParser: Extended	91.2	68.8	70.1

4.4   Combining MaltParser and MSTParser

McDonald and Nivre (2007) compared MSTParser and MaltParser and observed that MSTParser outperformed MaltParser on longer dependencies and dependencies closer to the root of the tree. Whereas MaltParser performs better on short dependencies and those further from the root. Similar observations were made by Ambati et al. (2010) for Hindi. Taking pros of both these parsers, Kumari and Rao (2012) combined the outputs of MaltParser and MSTParser in an intuitive way for a better parser for Hindi. Observing the output on development data, they made a list of dependency labels for which MSTParser gave better performance than the MaltParser. Whenever there is a mismatch between the outputs of both the parsers, they checked the dependency label given by the parsers. If MSTParser marked it with a label in the list, then they considered MSTparsers output. Otherwise Maltparser's output is considered. In this way, they gave more weightage to MSTparser in case of labels for which it gave better performance and gave more weightage to Maltparser for the labels for which it is good at like short distance labels.

We did similar experiment for Telugu as well. With this approach, we got 0.2% improvement in UAS which is not significant but a significant decrease of 0.5% in LAS (Table 5). Unlike Kumari and Rao (2012), we couldn't get better results with this approach as the accuracy of MSTParser is much lower than the MaltParser.

4.5   Voting

Voting is a simple method of combining the output of different parser as in Zeman (2009). In this method, we first obtain the outputs of different parsers. Then for each word, dependency relation that has the maximum number of votes from different parsers is included in the output. In case of a tie, the dependency relation predicted by the high accurate parser is picked in the final output. We explored this approach for Telugu combining the outputs of MaltParser, MSTParser and TurboParser. As the performance of MaltParser is much better than the other parsers, voting didn't help as the output of MaltParser is chosen all the times. Table 5 shows that the performance of voting is same as MaltParser.

4.6   Blending

Blending is another but a better method of combining the outputs of different parsers introduced by Sagae and Lavie (2006). The drawback of voting approach is that the final dependency tree for each sentence may not be fully connected as we are including each dependency relation at a time. Blending approach overcomes this drawback. In this approach, a graph is built for the dependency relations obtained from the various outputs and then the maximum spanning tree is extracted. We experimented with this approach for parsing Telugu using the outputs of MaltParser, MSTParser and Turbo-Parser. Table 5 shows that the performance of the blending approaches. Though the performance is better than both MSTParser and TurboParser, this approach didn't give better results than MaltParser.

Table 5. Performance of different systems on Telugu dependency treebank test data

Parser	UAS (%)	LAS (%)	LS (%)
MaltParser	91.8	70.0	72.3
MSTParser	90.0	67.1	68.6
TurboParser	91.2	68.8	70.1
Malt+MST Parser	92.0	69.5	71.8
Voting	91.8	70.0	72.3
Blending	91.7	69.5	71.0

Table 6. Performance of different approaches on top eight dependency labels

Labels	MaltParser	MSTParser	TurboParser	Malt+MST Parser	Blending
MAIN	**97.0**	95.3	96.3	97.0	96.0
SUBJ	63.0	59.4	**64.1**	60.3	62.7
OBJ	58.8	**62.7**	59.9	**62.8**	59.1
COORD	**83.1**	74.4	77.6	83.8	81.5
TIME	**61.2**	60.5	57.8	58.3	**64.0**

5   Discussions and Result Analysis

We could achieve state-of-the-art performance of 91.8% in UAS and 70.0% in LAS using MaltParser. As the majority of dependencies are short distance, MaltParser outperformed other parsers as it is good at short distance dependencies. Performance of MSTParser and TurboParser are much lower than the MaltParser. Hence the techniques of combining the outputs of different parsers didn't perform better than MaltParser. As the training data is very low, and also as Telugu is agglutinative language, LAS for the all the systems is very low. With more training data and specialized techniques for handling agglutinative languages like Telugu, we can achieve better results in LAS. Table 6, gives an overview of the performance of the individual parsers for the top five dependencies in Telugu dependency treebank. MAIN, SUBJ, OBJ, COORD, TIME, are the dependency labels for sentence root, subject, object, co-ordination, and time expression. MaltParser performed better for MAIN, COORD and TIME dependency labels. For SUBJ label TurboParser performed better and for OBJ label MSTParser performed better. TurboParser can handle linguistic constraints better which could be the reason for better handling of SUBJ label. As MaltParser is a greedy transition based parser, it is subject to error propagation. But MSTParser doesn't have this problem which could be the reason for better handling of OBJ label. Malt+MST Parser and Blending gave better results than MSTParser and TurboParser but are not better than MaltParser in most of the cases.

Performance of different approaches on different distance ranges is provided in Table 7. When distance between words is less than 5, MaltParser gave better results. But when the distance is greater than 5, MSTParser performed well. This observation is similar to previous results of McDonald and Nivre (2007), Ambati et al. (2010).

Table 7. Performance of different approaches on different distance ranges

Labels	MaltParser	MSTParser	TurboParser	Malt+MST Parser	Blending
<5	**98.7**	98.0	97.9	98.7	98.5
>5	36.4	**40.0**	36.4	**36.5**	39.4

6   Conclusions and Future Work

Experimenting with different settings, we built best models of MaltParser, MSTParser and TurboParser for Telugu. We explored different ways of ensembling these parsers. We also provided a detailed analysis of the performance of all the approaches on major dependency labels and different distance ranges. We obtained state-of-the art performance of 91.8% in unlabelled attachment score and 70.0% in labelled attachment score. In future, we would like to see the usefulness of external resources like WordNet. We also plan to explore the usefulness of large un-annotated data using self-training and co-training techniques to improve the performance of the Telugu dependency parsers.

Acknowledgements. We would also like to thank Language Technologies Research Centre (LTRC), International Institute of Information Technology, Hyderabad (IIIT-H) for providing the Telugu dependency treebank.

References

Ambati, B.R., Gadde, P., Jindal, K.: Experiments in Indian language dependency parsing. In: Proceedings of the ICON09 NLP Tools Contest: Indian Language Dependency Parsing, pp. 32–37 (2009)

Ambati, B.R., Husain, S., Jain, S., Sharma, D.M., Sangal, R.: Two methods to incorporate 'Local Morphosyntactic' features in Hindi dependency parsing. In: Proceedings of the NAACL HLT 2010 First Workshop on Statistical Parsing of Morphologically-Rich Languages, Los Angeles, USA, pp. 22–30 (2010)

Bharati, A., Chaitanya, V., Sangal, R.: Natural Language Processing: A Paninian Perspective, pp. 65–106. Prentice-Hall of India, New Delhi (1995)

Bharati, A., Sangal, R., Sharma, D.M., Bai, L.: AnnCorra: Annotating Corpora Guidelines for POS and Chunk Annotation for Indian Languages. Technical Report (TR- LTRC-31), LTRC, IIIT-Hyderabad (2006)

Bharati, A., Sangal, R., Sharma, D.M.: SSF: Shakti Standard Format Guide. Technical Report (TR-LTRC-33), LTRC, IIIT-Hyderabad (2007)

Bharati, A., Sharma, D.M., Husain, S., Bai, L., Begum, R., Sangal, R.: AnnCorra: TreeBanks for Indian Languages, Guidelines for Annotating Hindi TreeBank (version 2.0) (2009). http://ltrc.iiit.ac.in/MachineTrans/research/tb/DS-guidelines/DSguidelines-ver2-28-05-09.pdf

Bharati, A., Mannem, P., Sharma, D.M.: Hindi parsing shared task. In: Proceedings of Coling Workshop on Machine Translation and Parsing in Indian Languages, Kharagpur, India (2012)

Buchholz, S., Marsi, E.: CoNLL-X shared task on multilingual dependency parsing. In: Proceedings of the Tenth Conference on Computational Natural Language Learning, New York City, New York, pp. 149–164 (2006)

Husain, S.: Dependency parsers for Indian languages. In: Proceedings of the ICON09 NLP Tools Contest: Indian Language Dependency Parsing, India (2009)

Husain, S., Mannem, P., Ambati, B.R., Gadde, P.: The ICON-2010 tools contest on indian language dependency parsing. In: Proceedings of ICON- 2010 Tools Contest on Indian Language Dependency Parsing, Kharagpur, India (2010)

Kesidi, S.R., Kosaraju, P., Vijay, M., Husain, S.: A two stage constraint based hybrid dependency parser for Telugu. In: Proceedings of the ICON- 2010 Tools Contest on Indian Language Dependency Parsing (2010)

Kosaraju, P., Kesidi, S.R., Ainavolu, V.B.R., Kukkadapu, P.: Experiments on Indian language dependency parsing. In: Proceedings of the ICON-2010 Tools Contest on Indian Language Dependency Parsing (2010)

Kukkadapu, P., Malladi, D., Dara, A.: Ensembling various dependency parsers: adopting turbo parser for Indian languages. In: Proceeding of Coling 2012 Workshop on MT and Parsing in Indian Languages (2012)

Kumari, B.V.S., Rao, R.R.: Hindi dependency parsing using a combined model of Malt and MST. In: Proceeding of Coling 2012 Workshop on MT and Parsing in Indian Languages (2012)

Marcus, M.P., Santorini, B., Marcinkiewicz, M.A.: Building a large annotated corpus of English: the penn treebank. Comput. Linguist. 19(2), 313–330 (1993)

Martins, A., Smith, N., Xing, E.: Concise integer linear programming formulations for dependency parsing. In: Proceedings of the Joint Conference of the 47th Annual Meeting of the ACL and the 4th International Joint Conference on Natural Language Processing of the AFNLP, Suntec, Singapore, pp. 342–350 (2009)

McDonald, R.: Discriminative learning and spanning tree algorithms for dependency parsing. PhD thesis, Philadelphia, PA, USA (2006)

McDonald, R., Nivre, J.: Characterizing the errors of data-driven dependency parsing models. In: Proceedings of the Conference on Empirical Methods in Natural Language Processing and Natural Language Learning (2007)

McDonald, R., Crammer, K., Pereira, F.: Online large-margin training of dependency parsers. In: Proceedings of the 43rd Annual Meeting on Association for Computational Linguistics, Ann Arbor, Michigan, pp. 91–98 (2005)

Nivre, J.: Parsing Indian languages with MaltParser. In: Proceedings of the ICON09 NLP Tools Contest: Indian Language Dependency Parsing (2009)

Nivre, J., Hall, J., K"ubler, S., McDonald, R., Nilsson, J., Riedel, S., Yuret, D.: The CoNLL 2007 shared task on dependency parsing. In: Proceedings of the CoNLL Shared Task Session of EMNLP-CoNLL 2007, Prague, Czech Republic, pp. 915–932 (2007a)

Nivre, J., Hall, J., Nilsson, J., Chanev, A., Eryigit, G., Kübler, S., Marinov, S., Marsi, E.: Maltparser: a language-independent system for data-driven dependency parsing. Nat. Lang. Eng. **13**(2), 95–135 (2007b)

Sagae, K., Lavie, A.: Parser combination by reparsing. In: Proceedings of the Human Language Technology Conference of the NAACL, Companion Volume: Short Papers, pp. 129–132 (2006)

Zeman, D.: Maximum spanning malt: hiring world's leading dependency parsers to plant Indian trees. In: Proceedings of the ICON09 NLP Tools Contest: Indian Language Dependency Parsing (2009)

A Proposed Approach for Arabic Semantic Annotation

Ghada Khairy[1]([✉]), A. A. Ewees[1], and Mohamed Eisa[2]

[1] Computer Department, Damietta University, Damietta, Egypt
ghadakhairy89@yahoo.com
[2] Computer Science Department, Port Said University, Port Said, Egypt

Abstract. Semantic annotation refers to the process of annotating documents using the ontology in order to data becomes meaningful. Most of the techniques and methods of the field of semantic annotation and retrieval are used for dealing broadly in the English language. This paper aims to enhance the process of information retrieval for Arabic language that depends on the ontology in the process of document annotation. To achieve this aim, it is determined and processed the problems of the Arabic language through the proposed approach. This paper depends on semantic annotation based on ontology and Resource Description Framework (RDF). The results achieved high precision and high recall for the semantic annotation based on the proposed approach.

Keywords: Arabic semantic annotation · Ontology ·
Resource Description Framework · Stemming

1 Introduction

The conception of the semantic web has become the Web of data instead of the Web of documents in a model that can be prepared by computers. This approach could be implemented in the current Web adopting semantic annotation. Due to exponential extension and the enormous dimension of the Web references, there is a requirement to become a rapid and automatic semantic annotation of Web documents. The Arabic language acquired a significant research study from academia due to the complexity and challenges in semantic web study compared to Latin languages especially in the field of semantic annotation [1]. Semantic Annotation (SA) is the process of including metadata, which is thoughts of ontology (i.e. classes, instances, properties, and relations), in order to specify semantics, and knows what precisely the concepts annotated expect in the context [2]. Arabic language is a complicated language that may limit the development of the mechanisms for semantic web in that language. It has several particularities as compare to English like low vowels, the lack of uppercase letters and complicated morphology. Arabic language is comprised of nouns, verbs, and particles. Arabic is also greatly inflectional and derivational, which makes morphological interpretation a highly difficult task [3]. Arabic words could have further than one affix and can be represented as a sequence of affix such as prefixes lemma and suffixes, which make it more complicated for stemming. Furthermore, they have various types of ambiguities correlated with typographic patterns and spelling [1]. These challenges increase the difficulty in

© Springer Nature Switzerland AG 2020
A. E. Hassanien et al. (Eds.): AMLTA 2019, AISC 921, pp. 556–565, 2020.
https://doi.org/10.1007/978-3-030-14118-9_56

dealing automatically with Arabic documents. The computer science techniques are widely used to solve many general complicated problems such as image processing [4–7], forecasting problems [8], prediction [9–12], and features reduction [13]. Moreover, there are many studies work to solve the Arabic semantic annotation challenges; the authors of [14] suggested an approach for improving the process of information retrieval for Arabic language based on the ontology in the manner of document annotation. The results of this approach showed significant improvement in the process of documents retrieval based on the two standard evaluation measures accuracy and recall. Add to that it was presented as an automatic annotation tool that assisted the semantic annotation of Arabic web documents. The results had proved that an encouraging achievement remained in leveraging semantic web technologies to assist the Arabic and providing semantically annotated web documents for several fields in an automatic way [1]. Further, it had been reported the design, implementation, and evaluation of a lexical ontology for Arabic semantic relationships. The principal objective of the ontology is to aid the task of semantic annotation of the Arabic textual content. The results of the evaluation registered that the ontology was fit for the direction of semantic annotation of Arabic text with lexical relations [15]. Besides, it had suggested a framework to promote a semantic annotation tool for encouraging Arabic contents and matching it with other tools in a similar field [16].

In addition, it was presented as an automatic annotation of the Arabic web resources related to food, nutrition, and health fields to improve Arabic OWL ontologies associated to those domains. The results confirmed that encouraging accuracy and recall [17]. Furthermore, it had offered a successful way to perform clustering with semantic similarities by combining k-means document clustering with semantic feature extraction and document vectorization to group the Arabic web pages according to semantic similarities and then explained the semantic annotation [18]. Besides, it was recommended a technique to obtain taxonomic relationships to build ontology automatically from original Arabic text on political News domain. The results reached 92% and recall 91% [19].

The main motivation for this paper is to propose an approach for enhancing the process of information retrieval for Arabic language that depends on the ontology in the process of document annotation. The sections of this paper are organized as follow: Sect. 2 provides materials and methods based on semantic annotation tools. Section 3 presents the details of the proposed Arabic semantic annotation approach. Section 4 describes the experimental setup and discuss results. Finally, Sect. 5 concludes the paper.

2   Materials and Methods

The semantic web, as described by the W3C, is a web of data - the combination of semantic web technologies, e.g. RDF, OWL, provides an environment where a software application can query the data, make conclusions utilizing vocabularies (ontology). The achievement of the semantic web needs the broad extent availability of semantic annotation for current and new documents on the web [20]. In this section, RDF and Ontology techniques had explained for semantic annotation in details.

2.1 Resource Description Framework (RDF)

Resource Description Framework (RDF) is a language for describing information about support resources in the World Wide Web [21]; it is a basic ontology language. RDF is written in XML. By using XML, RDF information can simply be replaced among various types of computers utilizing several types of operating systems and application languages. RDF was intended to present a general way to represent information so it is device readable. RDF representations are not designed to be presented on the web [22]. RDF is a data model for describing objects and relationships between them [23]. RDF identifies resources with Uniform Resource Identifiers (URI) [24].

For example, "Machine Learning part of Artificial Intelligence", for the triple "Machine Learning" is subject, "Part of" is a predicate, "Artificial Intelligence" is an object. A set of RDF statements is called an RDF graph [25]. RDF-Schema is a language for defining vocabulary for describing properties and classes of RDF resources. RDFS is used to define graphs of trio RDF, with semantics of generalization/prioritization of such properties and classes [23]. In RDFS, predefined web resources rdfs: Class, rdfs: Resource, and rdf: Property can be used to declare classes, resources, and properties respectively. From the first view, RDFS is a simple ontology language that supports only class and property hierarchies, as well as domain and range restrictions for properties [25].

2.2 Ontology

Ontology is a formal, explicit specification of a given conceptualization in the pattern of theories and relationships [26]. The ontology includes individuals, classes, properties, and relationships. Individuals are the essential elements of ontology. Individuals represent objects in the domain in which we are involved such as people, animals, and plants, as well as abstract individuals such as numbers and words. Individuals are also identified as examples. Individuals can be pointed to as existing 'examples of classes'. Classes are the collections or sets of objects represent by the collection of properties. Classes may classify individuals with aid of these properties. Properties are attributes and features that classes can have. For example, a person class or object has the properties; name, age, height, etc. Relations between objects in ontology define how objects are associated with other objects [27]. The Web Ontology Language (OWL) helps larger machine know the ability of web resources than that recommended by RDFS by combining further constructors for producing class and property declarations (vocabulary) and new axioms (constraints), along with a formal semantics [25].

In the top-down approach, the theories in the ontology are inferred from an investigation and study of appropriate information sources about the field. A top-down development method begins with the definition of the most common concepts in the field and subsequent specialization of the theories [28].

3   The Proposed Approach

The proposed approach consists of three main phases, the first one is pre-processing consists of fifth steps: prepare the Arabic documents, remove diacritics, remove stop-words, stem and part of speech. The second phase is Arabic documents annotation consists of two steps: get ontology, RDF and correct text. The third phase is text retrieval. Figure 1 presents the stages of the proposed approach for Arabic semantic annotation.

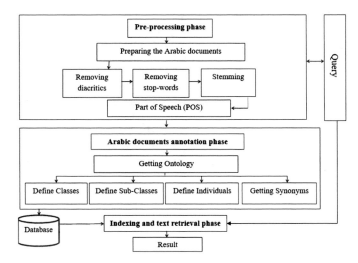

Fig. 1. Proposed approach for Arabic semantic annotation

3.1   Pre-processing Phase

This phase consists of fifth steps: prepare the Arabic documents, remove diacritics, remove stop-words, stem and part of speech.

Prepare the Arabic documents: it is one of the most important steps. This step uses to collect documents manually in (Computer science) domain. Then it uses these to annotate and retrieval Arabic documents.

Removing diacritics: it removes diacritics from the Arabic documents.

Removing stop-words: it removes all stop-words from the Arabic documents because stop-words are terms that are too frequent in the text. These terms are insignificant. The stop-words collected by the proposed approach consists of 13,019 stop-words.

Stemming: it returns Arabic words into their stems. Stemming removes all prefixes and suffixes letters from word to get the stem of the word. This is an important process because stem word leads to reduce the space of words in order to improve the effectiveness of the retrieval process. This step is divided into three stage. The first stage removes all prefixes from word. It depends on the light stemmer technique.

Part of speech (POS): it is used to find POS tags for the stem and affixes to overcome determining the POS tags of a word in a particular context, primarily because the same word may be spelled in different ways. Further, detecting the difference between Arabic derivatives represents a very challenging issue for the majority of POS taggers. Hence, the task of tagging the correct POS tags requires advanced processing and the use of considerable resources.

3.2   Arabic Documents Annotation Phase

This phase consists of the step of getting the ontology. In this step uses a top-down approach [29] in building the ontology where the ontology consists of the following stages:

Define concepts (i.e. classes and sub-classes): it is used in our ontology domain.

Define instances (i.e. real elements in the chosen domain): Creating instances (individuals) is a very important step to enrich the ontology with direct relation with classes and sub-classes.

Getting synonyms: that are very similar words to the basic word, it can be useful and also deploying thresholds to eliminate poor scoring words that unfairly punish a particular set.

3.3   Indexing and Text Retrieval Phase

In this phase, the purpose of indexing is to optimize speed and performance in finding relevant documents for a search query. In our work, we used term frequency–inverse document frequency (TFIDF) which is a numerical statistic that is intended to reflect how important a word is to a document in our (Computer science) domain as shown in Eq. (1) calculated (TF), Eq. (2) calculated (IDF) and Eq. (3) calculated (TFIDF).

$$TF_{i,j} = \sum_{k} n_{i,j} \tag{1}$$

$$IDF(w) = \log\left(\frac{N}{df_t}\right) \tag{2}$$

$$w_{i,j} = TF_{i,j} * \log\left(\frac{N}{df_t}\right) \tag{3}$$

where $TF_{i,j}$ (number of occurrences of I in j), DF_i (number of documents containing i) and N (total number of documents). This phase is performed in order to retrieve the relevant documents by entering the query over the user by linking the research word, synonyms, and documents and arranging them from the closest to the farthest links. Also, the text is corrected at this stage which helps to find misspellings in the research word and automatically corrects it in order to get the result of the research word. For example, the word "تعلم الآه" is corrected to the word "تعلم الآله".

4 Experiments and Results

This section shows the description of the dataset, performance measures, parameters settings, experiment description and the results.

4.1 Dataset

The dataset consists of 63 Arabic documents related to (Computer science) Domain. The smallest Arabic document consists of 260 words, while the largest Arabic document consists of 2570 words. The longest sentence in the dataset consists of 43 words while the shorter sentence consists of 6 words. These documents were collected manually from Wikipedia website "ar.wikipedia.org" where these documents have been included in classes ((programming), (artificial intelligence), (databases), (operating systems), (computer security), (multimedia)). programming class is divided into sub-classes ((programming languages), (application programs)). Artificial intelligence class is divided into sub-classes ((machine learning), (image processing), (natural language processing)). These classes are divided into subclasses. Unicode has been used as a universal character encoding standard, it defines the way individual characters are represented in text files, web pages, and other types of documents. In the dataset, a site "thesaurus.com" was used to obtain synonyms in English for the difficulty of obtaining synonyms in Arabic in this field, to solve this problem, we translate these synonyms from English to Arabic, and so we have synonyms in Arabic. DBpedia Ontology "wiki.dbpedia.org" is a simple, cross-domain ontology, which has been manually designed based on the most commonly used infoboxes within Wikipedia. DBpedia Ontology was used in the dataset to link among classes as a requirement to come up with the ontology through RDF and annotating Arabic documents. The dataset contains 6 classes, 17 subclasses, 13 individuals, and 30 synonymous.

4.2 Performance Measures

To evaluate the performance of the proposed approach accuracy, precision and recall measures [21] are used. Equation (4) show Accuracy that is a measure of the overall correctness of the approach, it's the number of documents that are correctly classified divided by the sum of the total documents. Equation (5) illustrates the precision, it measures the number of retrieved positive documents divided by the total positive retrieved documents. Equation (6) illustrates the recall, it measures the number of retrieved positive documents divided by the number of existing relevant documents.

$$Accuracy = \frac{Tp + Tn}{Tp + Tn + Fp + Fn} \qquad (4)$$

$$Precision = \frac{Tp}{Tp + Fp} \qquad (5)$$

$$Recall = \frac{Tp}{Tp + Fn} \qquad (6)$$

where TP is true positive, FP is false positive, FN is false negative, and TN is true negative.

4.3   Experiment Description

This part describes the experiment environment which used to perform the proposed approach for Arabic semantic annotation. All experiments are performed using "Intel Corei5 2.5 GHz CPU", 8 GB RAM, and Windows 10 46 bit. And the software descriptions are DBpedia ontology, Protégé 5, and Python 3.

4.4   Results and Discussion

All results are tabulated in the following parts. Table 1 shows the calculated results of TP (True Positive), TN (True Negative), FP (False Positive) and FN (False Negative) for each class of computer science domain (Computer, Programming, Artificial Intelligence, Machine Learning, Databases, Operating Systems, Natural Language Processing, Image Processing, Computer Security, and Multimedia). The results obtained by (a) the proposed approach (b) searching using traditional method.

Table 1. TR, TN, FP and FN results for each class

Classes	(a) The proposed approach results				(b) The traditional method			
	TP	TN	FP	FN	TP	TN	FP	FN
Computer	58	2	2	4	32	1	7	25
Programming	55	3	3	5	31	3	6	26
Artificial Intelligence	57	1	3	5	24	1	6	25
Machine Learning	51	5	4	6	22	5	5	32
Databases	55	2	1	3	26	5	6	28
Operating Systems	55	6	1	4	19	6	5	33
Natural Language Processing	57	2	2	5	27	1	6	32
Image Processing	59	6	0	1	23	3	5	35
Computer Security	52	10	0	4	20	10	3	37
Multimedia	55	8	1	0	28	8	4	30

The results in Table 2 show differences in the evaluation measures of the ten mentioned classes. This is due to the following reasons: The Machine Learning class has achieved less efficient precision (92.7%) because of the nature of some documents

Table 2. Results of precision, recall and accuracy

Annotation types	(a) The proposed approach results			(b) The traditional method		
	Recall	Precision	Accuracy	Recall	Precision	Accuracy
Computer	93.5%	96.6%	90.77%	56.14%	82.05%	50.77%
Programming	91.6%	94.8%	87.88%	54.39%	83.78%	51.52%
Artificial Intelligence	91.9%	95%	87.87%	48.98%	80.00%	44.64%
Machine Learning	87.9%	92.7%	83.58%	40.74%	81.48%	42.19%
Databases	94.8%	98.2%	93.75%	48.15%	81.25%	47.69%
Operating Systems	92.9%	98%	92.19%	36.54%	79.17%	39.68%
Natural Language Pro.	91.9%	96.6%	89.23%	45.76%	81.82%	42.42%
Image Processing	98.3%	100%	98.41%	39.66%	82.14%	39.39%
Computer Security	94.3%	100%	95.21%	35.09%	86.96%	42.86%
Multimedia	91.6%	98.2%	98.44%	48.28%	87.50%	51.43%
Average	**92.87%**	**97%**	**91.73%**	**45.37%**	**82.62%**	**45.26%**

content which have mixed content, and thus exists in more than one class. It also achieved less efficient recall value (87.9%) since there is some documents classified to other classes due to the same reason of mixed content. The Databases class has achieved high efficient recall (94.8%) and precision (98.2%) because it is a specific category in the domain computer science and it is not mixed content in other classes.

Table 2 shows the calculated values of precision, recall and accuracy for the ontology concepts, we listed some of the results due to the space limitation. The results obtained by (a) the proposed approach (b) searching using the traditional method. The results are calculated based on Eqs. (4), (5) and (6). We can conclude that, the results of the proposed approach are better than the traditional (manual) method, it achieved the high accuracy equals 91.7% whereas, the traditional method obtained only 45.2%.

On one hand, the studies [14, 19] depended on GATE software to perform document annotations. They used Onto Root Gazetteer to produce ontology-based annotations and produce a part-of-speech tag as an annotation on each Arabic word.

In this research, ASA approach was implemented online through the site https://pos-project.herokuapp.com and used python code to execute pre-processing phase which divided into fifth sub-phases: prepare the Arabic documents, remove diacritics, remove stop-words, stem, and part of speech. ASA approach used part of DBpedia machine learning ontology and it was placed and translated in the python code. Consequently, our result improved the accuracy compared with previous researches. On the other hand, [1, 15] researches depended on a local GUI (Graphical User Interface). In this research, we used GUI online. Thus, our result improved the recall rate than the previous works.

5   Conclusion

This paper has presented a proposed approach based on Arabic semantic annotation for Arabic information retrieval that facilitates information retrieval with high precision and high recall. Our approach consists of several stages: Preprocessing stage (prepares the Arabic documents, remove diacritics, remove stop-words, stem and part of speech), Arabic documents annotation stage (get ontology, RDF and correct text), Indexing and text retrieval stage. Using our approach, we overcome the problem of difficulty in Arabic and the traditional way used in the process of documents search and retrieval. The results achieved high Precision and high Recall for all the annotation types as in Experiments section. For future work, we intend to increase our corpus of documents to retrieve more documents in the domain "computer science" and obtain more accurate results. In the future, we intend to improve our proposed approach by helping people with special needs (blind) by turning it into a dynamic interactive application to facilitate their search while developing the Arabic semantic annotation mechanism in which the approach works.

References

1. Al-Bukhitan, S., Helmy, T., Al-Mulhem, M.: Semantic annotation tool for annotating Arabic web documents. Procedia Comput. Sci. **32**, 429–436 (2014)
2. Oliveira, P., Rocha, J.: Semantic annotation tools survey. In: 2013 IEEE Symposium on Computational Intelligence and Data Mining (CIDM), pp. 301–307. IEEE, April 2013
3. Beseiso, M., Ahmad, A.R., Ismail, R.: A Survey of Arabic language support in semantic web. Int. J. Comput. Appl. **9**(1), 35–40 (2010)
4. Kaloub, A.: Automatic ontology-based document annotation for Arabic information retrieval. Unpublished master's thesis, Islamic University-Gaza, Deanery of Graduate Studies, Faculty of Information Technology (2013)
5. El Aziz, M.A., Ewees, A.A., Hassanien, A.E.: Hybrid swarms optimization based image segmentation. In: Hybrid Soft Computing for Image Segmentation, pp. 1–21 (2016)
6. El Aziz, M.A., Ewees, A.A., Hassanien, A.E., Mudhsh, M., Xiong, S.: Multi-objective whale optimization algorithm for multilevel thresholding segmentation. In: Advances in Soft Computing and Machine Learning in Image Processing, pp. 23–39 (2018)
7. El Aziz, M.A., Ewees, A.A., Hassanien, A.E.: Multi-objective whale optimization algorithm for content-based image retrieval. Multimed. Tools Appl. **77**(19), 26135–26172 (2018)
8. El Aziz, M.A., Ewees, A.A., Hassanien, A.E.: Whale optimization algorithm and Moth-Flame optimization for multilevel thresholding image segmentation. Expert Syst. Appl. **83**, 242–256 (2017)
9. Sahlol, A.T., Moemen, Y.S., Ewees, A.A., Hassanien, A.E.: Evaluation of cisplatin efficiency as a chemotherapeutic drug based on neural networks optimized by genetic algorithm. In: 2017 12th International Conference on Computer Engineering and Systems (ICCES), pp. 682–685. IEEE, December 2017
10. Ahmed, K., Ewees, A.A., Hassanien, A.E.: Prediction and management system for forest fires based on hybrid flower pollination optimization algorithm and adaptive neuro-fuzzy inference system. In: 2017 Eighth International Conference on Intelligent Computing and Information Systems (ICICIS), pp. 299–304, December 2017

11. Sahlol, A.T., Ewees, A.A., Hemdan, A.M., Hassanien, A.E.: Training feedforward neural networks using Sine-Cosine algorithm to improve the prediction of liver enzymes on fish farmed on nano-selenite. In: 2016 12th International Computer Engineering Conference (ICENCO), pp. 35–40. IEEE, December 2016

12. Oliva, D., Ewees, A.A., Aziz, M.A., Hassanien, A., Peréz-Cisneros, M.: A chaotic improved artificial bee colony for parameter estimation of photovoltaic cells. Energies **10**(7), 865 (2017)

13. Ahmed, K., Ewees, A.A., El Aziz, M.A., Hassanien, A.E., Gaber, T., Tsai, P.W., Pan, J.S.: A hybrid krill-ANFIS model for wind speed forecasting. In: International Conference on Advanced Intelligent Systems and Informatics, pp. 365–372, October 2016

14. Ewees, A.A., El Aziz, M.A., Hassanien, A.E.: Chaotic multi-verse optimizer-based feature selection. Neural Comput. Appl., 1–16 (2017)

15. Al-Yahya, M., Al-Shaman, M., Al-Otaiby, N., Al-Sultan, W., Al-Zahrani, A., Al-Dalbahie, M.: Ontology-based semantic annotation of Arabic language text. Int. J. Mod. Educ. Comput. Sci. **7**(7), 53 (2015)

16. El-ghobashy, A.N., Attiya, G.M., Kelash, H.M.: A proposed framework for Arabic semantic annotation tool. Int. J. Com. Dig. Syst. **3**(1), 47–53 (2014)

17. Albukhitan, S., Helmy, T.: Automatic ontology-based annotation of food, nutrition and health Arabic web content. Procedia Comput. Sci. **19**, 461–469 (2013)

18. Alghamdi, H.M., Selamat, A., Karim, N.S.A.: Arabic web pages clustering and annotation using semantic class features. J. King Saud Univ. Comput. Inf. Sci. **26**(4), 388–397 (2014)

19. El Zraie, B.: Extraction of Taxonomic Relations from Arabic Text for Ontology Construction. Unpublished master's thesis, Islamic University-Gaza, Deanery of Graduate Studies, Faculty of Information Technology (2016)

20. Yang, C.Y., Lin, H.Y.: Semantic annotation for the web of data - an ontology and RDF based automated approach. J. Converg. Inf. Technol. (JCIT), Special Issue Soc. Netw. Appl. Decis. Support **6**(4), 318–327 (2011)

21. Manola, F., Miller, E., McBride, B.: RDF primer. W3C recommendation, 10(1–107), 6 (2004)

22. Champin, P.-A.: RDF Tutorial. Pierre-Antoine Champin, 1–9, 5 April 2001

23. Alatrash, E.: Using Web Tools for Constructing an Ontology of Different Natural Languages, Doctoral dissertation, University of Belgrade (2013)

24. Corcho, O., Fernández-López, M., Gómez-Pérez, A.: Methodologies, tools and languages for building ontologies. Where is their meeting point? Data Knowl. Eng. **46**(1), 41–64 (2003)

25. Pan, J., Horrocks, I.: RDFS(FA): connecting RDF(S) and OWL DL. IEEE Trans. Knowl. Data Eng. **19**, 192–206 (2007)

26. Gruber, T.R.: A translation approach to portable ontology specifications. Knowl. Acquis. **5**(2), 199–220 (1993)

27. Ahmed, Z.: Domain Specific Information Extraction for Semantic Annotation. (Unpublished Master Thesis), Charles University (2009)

28. Al Tayyar, M.S.: Arabic information retrieval system based on orphological analysis (AIRSMA). Ph.D. Thesis DeMonfort University, July 2000

29. López-Pellicer, F.J., Vilches-Blázquez, L.M., Nogueras-Iso, J., Corcho, Ó., Bernabé, M.A., Rodríguez, A.F.: Using a hybrid approach for the development of an ontology in the hydrographical domain (2008)

Automated Essay Evaluation Based on Fusion of Fuzzy Ontology and Latent Semantic Analysis

Saad M. Darwish$^{(\boxtimes)}$ and Sherine Kh. Mohamed$^{(\boxtimes)}$

Department of Information Technology,
Institute of Graduate Studies and Research, University of Alexandria,
163 Horyya Avenue Elshatby, P.O. Box: 832, Alexandria 21526, Egypt
saad.darwish@alex-igsr.edu.eg,
sherinekhamis@alexu.edu.eg

Abstract. New learning researches proved that creativity is an essential concern in the arena of education. The best means to evaluate learning outcomes and students' creativity is essay questions. However, to evaluate these questions is a time-consuming task and subjectivity in scoring assessments remains inevitable. Automated essay evaluation systems (AEE) provide a cost-effective and consistent alternative to human marking. Therefore, numerous automatic essay-grading systems have been developed to lessen the demands of manual essay grading. However, these systems concentrate on syntax and vocabulary, and no consideration is paid to the semantic and coherence of the essay. Moreover, few of the existing systems are able to give informative feedback that is based on extensive domain knowledge to students. In this paper, a system is evolved that uses latent semantic analysis (LSA) and fuzzy ontology to evaluate essays, where LSA will be responsible for checking the semantic. Fuzzy ontology is used to check the consistency and coherence of the essay as it is the best way to overcome the vagueness of the language, and the system will also provide a score with feedback to the student. Experimental results were good in evaluating the essay syntactically and semantically.

Keywords: Natural language processing · Automated essay evaluation ·
Fuzzy ontology · Latent Semantic Analysis · Information retrieval

1 Introduction

Creativity is considered one of the main literacy of contemporary education. Educators report the importance of creativity in learning because the unforeseen problems that continually rise in life will need lifelong learners to produce innovative solutions [1]. Essays are considered by many researchers the most powerful tool to evaluate creativity. However, evaluating essays and open end questions is a time consuming and tiring process. It is assumed that machine learning can provide help to teachers in this field by using automated essay evaluation systems [2]. Educators and educational institutions consider AEE not only a tool to assess learning outcomes, but also helps consume time, effort, and money without reducing the quality.

© Springer Nature Switzerland AG 2020
A. E. Hassanien et al. (Eds.): AMLTA 2019, AISC 921, pp. 566–575, 2020.
https://doi.org/10.1007/978-3-030-14118-9_57

These systems usually face some challenges [3–5] which are: first, language ambiguities and lack of one "correct" answer to any given essay question. Second, communication infrastructures are different between e-learning content objects and e-learning platforms. Third, to be a successful essay evaluation system, it has to get information about a learner's knowledge, many word-based and statistical approaches have supported information retrieval, data mining, and natural language processing systems, and a profounder understanding of text is still a crucial challenge. Fourth, concepts, semantic relationships among them, contextual information needed for the concept disambiguation require further progress in the textual information management. Fifth, the system has to be reliable as teachers and more usable.

All the existing AEE systems have a main drawback [6] which is the concentration on syntax and ignoring semantics, consistence and coherence. Since the AEE is considered an important field of eLearning, a system is suggested that uses Latent Semantic Analysis to measure coherence (semantic development) and consistency of facts (compared to common sense knowledge and other facts in essays). LSA is used in this paper as is has some benefits [7]. Noise is abridged through the dimensionality-reduction step. LSA discourses synonymy as synonyms are frequently used in the similar context and as a result LSA concepts are expected to mirror them. Polysemy is dealt with through the noise reduction. To overcome the drawbacks of LSA, it is integrated with cosine similarity. The consistency attributes calculate the number of semantic errors in a student essay using information extraction and logical reasoning. Fuzzy ontology is integrated in the system to provide the students with feedback about his errors. It is used [8] because the query development is determined by the membership value, customization is modest, constructed on alteration of membership values, intermediate locations for grouping is needless and Knowledge demonstration is associated with the use.

The rest of this paper is organized as follows: Sect. 2 describes some of the recent related works. The detailed description of the proposed system has been shown in Sect. 3. In Sect. 4, the results and discussions on the dataset are given. Finally, the conclusion is annotated in Sect. 5.

2 Literature Review

The existing systems look at the essay from different points of view. Some of them scores the essay according to the attributes describing its quality which can approximately be alienated into three groups [3, 9]: style, content and semantic attributes. Style attributes pays attention to lexical complexity, grammar and mechanics (spelling, capitalization, and punctuation). Style attributes scoring usually relied on statistical methods which gave good results but they didn't succeed in dealing with words' morphology. Content attributes imprecisely define semantics of an essay and are grounded on associating an essay with source text and other already graded essays. To evaluate content, systems use pattern matching techniques (PMT) and extensions to LSA such as Generalized Latent Semantic Analysis (GLSA) (which uses an n-gram-by-document matrix instead of a word-by-document matrix) and improvement that considers semantics by means of the syntactic and shallow semantic and overcoming the drawback of the LSA.

For validating the accuracy and consistency of essay content, approaches such as (Open) Information Extraction (OIE), Semantic Networks (SN), Ontologies are used. Semantic attributes are built on validating the accuracy of content meaning. Different systems use numerous approaches to extract attributes from essays. The supreme extensively used methodology is based on NLP. Systems focusing on content are frequently using Latent Semantic Analysis (LSA) - a machine learning method that analyses correlated conceptions between some documents and the contained terms. It is good in making relations between documents but it doesn't take into consideration word order. These techniques related the essay to its domain correctly but failed to handle the vagueness of the language. Most of the systems use machine learning algorithms (usually regression modelling) to calculate the final grade. To know more about the previous techniques used in essay evaluation see [5, 10–12].

Although automated essay scoring has been studied for few decades, there is still a room to make it extra efficient and practical in real applications. According to the aforementioned review, it can be found that past studies were primarily devoted to: (1) score essay grammar and lexical sophistication, (2) not addressing essay coherence (3) doesn't provide feedback to students about their mistakes. However, to the best of our knowledge, little devotion has been given to advising new technique to evaluate essay syntactically, semantically, check its coherence and give feedback to students.

3   Proposed Methodology

This paper proposes a new method that evaluates all the attributes of the essay (Lexical, semantic and content) so it does not concentrate on syntax only. It focuses on automatic semantic evaluation and delivers semantic feedback to students. It involves semantic attributes that checks coherence as a function of semantic relatedness through the overall essay and not only between neighboring sentences. The system also examines text consistency by detecting entities in an essay, their relations and considering co-references of concepts. It also endeavors common sense knowledge fuzzy ontologies, taxonomies, and can therefore work on different domains. The main diagram of the suggested system is shown in Fig. 1. The following paragraphs describe in details the steps of the system.

3.1   Preprocessing of the Essay

In the preprocessing phase, the system first goes through [13] an essay and divides it into sentences. Then, it generates a duplicate of each sentence and does a number of preprocessing steps: tokenization; part-of-speech tagging; finding and labeling stop-words, punctuation marks, determiners and prepositions; transformation to lower-case; and stemming. This step is important as the system cannot deal with the essay as a whole.

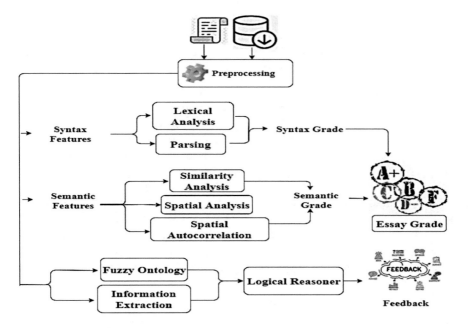

Fig. 1. The proposed essay evaluation system

3.2   Calculate Syntax Grade

To obtain the syntax grade the system goes throw two steps after which the grammar, spelling and lexical richness is scored.

Lexical Analysis: The best extensively used measure of lexical richness, as used in [14], is mainly a type/token ratio, stated as the ratio of the number of the different lexical items(tokens) to the whole number of lexical items in the essay. In defining the number of diverse lexical items (i.e., types), however, changed forms counted only once. See [15] to calculate lexical richness of an essay.

Parsing: In this step grammar is checked. It analyzes essay constituent words grounded on the underlying grammar. The product of the parsing process is a parse tree. If the parse tree is broken, it is considered a syntax error. Parse tree is used as it can make several passes over the data without having to re-parse the input [16]. The result of these two steps is the syntax grade.

3.3   Calculate Semantic Grade

Coherence attributes are grounded on the hypothesis that the semantic content of a coherent essay changes gradually through its text. The system starts by first dividing essays into many consecutive intersecting parts, attained by stirring a window over an essay by steps of 10 words. Window's size is defined so that it contains 25% of the average number of words per essay. For each essay corpus (dataset) the term frequency - inverse document frequency (*TF-IDF*) depiction is calculated [17]. *TF-IDF* vectors of essay parts characterize points in high dimensional semantic space, as the dataset is considered the sematic space and each vector is a point, which ought to be close to each other in coherent essays, according to our assumption. Then these points are used to check the following in the essay.

Similarity Analysis: These are important coherence measures to calculate the detachment between parts of the essay, which are characterized as points in the semantic space. These measures [18] are: (1) average distance between neighboring point that is used to measure the cosine similarity between sentences in an essay that extremely coherent dissertations have trivial movements in semantic space and vice versa. (2) Average distance between any two points which is important to check how well an idea perseveres within the essay. (3) Maximum difference between any two points that is used to calculate the diameter of area that is enclosed with points and thus the range of the discussed concept in the space. (4) Clark and Evans' distance to the nearest neighbor of each point in the semantic space which is important for measuring spatial relationships. (5) Cumulative frequency distribution of the nearest neighbors 'distances. It is used to check the percentage of content deviancies from the main idea. If the result of these measures is below 0.5 it is considered a semantic error.

Spatial Data Analysis: It targets to extract implicit knowledge such as spatial statistics and patterns. The proposed attributes measure the central spatial leaning and the spatial dispersion and are defined by using (1) average Euclidean distance between the centroid and each point. This measures an amount of dispersion in a point pattern. It is measured with the same way as average distance between two points in the previous section except the distance is not between two neighboring points but from the centroid. (2) Standard distance (a spatial equivalent of standard deviation). It is used to calculate an amount of absolute dispersion in a point pattern. Similar to the standard deviation, the standard distance is also toughly subjective to extreme values. Because distances to the mean center are squared, the atypical points have a dominant influence on the magnitude of this metric, which permits identifying deviating (incoherent) essay parts. (3) Relative distance which is used as a descriptive ration of the relative spatial dispersion. For more details, see [19].

Spatial Autocorrelation: Measures of spatial autocorrelation define how data inclines to be clustered together in space or discrete. They allow us to identify global and local semantic coherence of the essays' content. Distinctive measures of spatial autocorrelation are Moran's I [20] that is vital to evaluates the overall clustering pattern and calculated by (1), Geary's C [21] shown in Eq. (2), and Getis's G [22] which enables us to examine point patterns at a more local scale and calculated by (3). These three measures were adjusted so it can be used high-dimensional semantic space.

$$I = \frac{N}{S} \cdot \frac{1}{n} \sum_{k=1}^{n} \left[\frac{\sum_{i=1}^{N} \sum_{j=1}^{N} w_{ij} \left(D_i^k - \overline{D_c^k} \right) \left(D_j^k - \overline{D_c^k} \right)}{\sum_{i=1}^{N} (D_i^k - \overline{D_c^k})^2} \right] \tag{1}$$

where $D_i^k = 1.......n; i = 1.......N$ is a kth coordinate component of point i, $\overline{D_c^k}$ is a kth coordinate component of a mean center, n is the number of dimensions, N is the number of points, and S is a sum of all weights w_{ij}. Weights w_{ij} are given to every pair of points, with value $w_{ij} = 1$, if i and j are neighbors, and value $w_{ij} = 0$ otherwise. The range of I varies from -1 to $+1$. A positive sign of I designates positive spatial autocorrelation and means that neighboring points cluster together, while the opposite is true for the negative sign. Values nearby to zero designate complete spatial randomness.

$$C = \frac{(N-1)}{2} \cdot \frac{1}{n} \sum_{k=1}^{n} \left[\frac{\sum_{i=1}^{N} \sum_{j=1}^{N} w_{ij}(D_i^k - D_j^k)^2}{\sum_{i=1}^{N} \sum_{j=1}^{N} w_{ij}(D_i^k - \overline{D_c^k})^2} \right] \tag{2}$$

where D_j^k, $k = 1......n; i = 1$ is a kth coordinate component of point i, $\overline{D_c^k}$ is a kth coordinate component of a mean center, n is the number of dimensions, N is number of points, and w_{ij} are point weights as mentioned before.

$$G(d) = \frac{1}{n} \sum_{k=1}^{n} \left[\frac{\sum_{i=1}^{N} \sum_{j=1}^{N} w_{ij}(d)D_i^k D_j^k}{\sum_{i=1}^{N} \sum_{j=1}^{N} D_i^k D_j^k} \right] \tag{3}$$

where D_j^k, $k = 1......n; i = 1$ is a kth coordinate component of point i, $\overline{D_c^k}$ is a kth coordinate component of a mean center, n is the number of dimensions, N is number of points, and d is the average distance between two points in the semantic space. A weighting function $wij(d)$ is used to describe binary weights to every pair of points, where $wij(d) = 1$, if i and j are within distance d and $wij(d) = 0$, otherwise.

3.4   Calculating Essay Grade

The majority of systems use machine learning technique to forecast the final score. It is used for its ease of interpretability. In this system Multiple Linear Regression (MLR) is going to be used as there are a lot of features. Formal definition of MLR can be found on [23].

3.5   Automated Error Detection System

The system begins by building a fuzzy ontology grounded on common sense knowledge and enhances it by means of a source text, domain and target knowledge. After building the fuzzy ontology, the system uses entity recognition, co-reference resolution and open information extraction so that it links these constituents to the fuzzy ontology. Then, the system goes on as it adds the extractions in an iterative way into the fuzzy ontology and uses the logical reasoner to decide whether a fuzzy ontology is consistent after each extraction is added. If the logicl reasoner discovers a conflict in the fuzzy ontology, it accounts a revealed consistency error and embraces it in the ultimate feedback (Fig. 2).

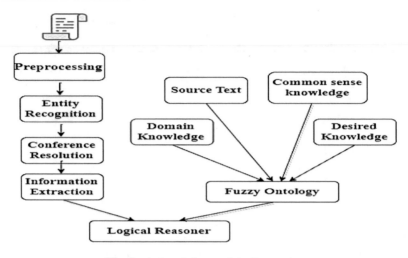

Fig. 2. Automated error detection system

4 Experimental Results

To analyze the potential benefits of the proposed system, the functions of the system are written in Python with a spellchecking library PyEnchant, and it was run on a HP Pavilion g6-1304tx -15.6″ - 2.5 GHz Intel Core i5-2450M - 4 GB RAM running windows 10. We select benchmark datasets provided within the Automated Essay Scoring competition on the Kaggle website to validate the accuracy of suggested system. The proposed system utilizes 72 non-redundant attributes (42 lexical attributes, 29 coherence attributes and 3 consistency attributes) as described in [20].

In the first experiments, we compared our system with some pioneer systems to evaluate if semantic attributes yield to better model per performance: AGE which is a system with only linguistic and content attributes. AGE+ that is a system AGE, augmented with additional coherence attributes. SAGE which is a system AGE+ improved with additional consistency attributes. As the system SAGE needs source-based essay to build a fuzzy ontology for the logical reasoner, we had the chance to evaluate it only on dataset 4. Table 1 shows the quadric weighted Kappas and precise agreement for AEG and AGE+. The outcome that the forecast accuracy significantly increases when coherence attributes are used in the system. Table 2 shows that consistency attributes assisted in attaining higher Kappa values on all four observed datasets.

Table 1. Comparison of AEG (syntactic attributes only), AGE+ (syntactic and coherence attributes) using quadric weighted kappa (1st row) and exact agreement (2nd row).

System	DS1	DS2	DS3	DS4
AGE QW kappa	0.9045	0.7473	0.6619	0.8096
Exact agg.	0.7224	0.7716	0.7379	0.7886
AGE+ QW kappa	0.9251	0.7924	0.6714	0.8272
Exact agg.	0.7807	0.8057	0.7481	0.8236

Table 2. Comparison of AEG (syntactic attributes only), AGE+ (syntactic and coherence attributes) and the proposed system (syntactic, coherence and consistency attributes) on source-based datasets using quadric weighted kappa (1st row) and exact agreement (2nd row).

System	DS1	DS2	DS3	DS4
AGE QW kappa	0.8096	0.840	0.8701	0.7737
Exact agg.	0.7886	0.7273	0.7847	0.7314
AGE+ QW kappa	0.8272	0.8109	0.8729	0.7817
Exact agg.	0.8036	0.7375	0.7805	0.7400
Proposed system	0.8520	0.8482	0.9095	0.8207
	0.8277	0.7714	0.8132	0.7770

The superiority in our system relies on the fact that the first system checks syntax attributes only. The second system check syntax and coherence attributes and the third system check syntax and consistence attribute. Whereas the proposed system integrates all the three attributes (syntax, coherent and consistence) which led to better results. Furthermore, these systems don't pay attention to the word order or the coherence and consistence of the essay as these systems use statistical methods that deals with words as numbers only. Whereas in this system the use of the spatial autocorrelation measures indicating the existence of redundant information. When both the high and low values cluster, they tend to cancel each other out and the use of spatial auto-correction overcomes this draw back.

In the third experimental result, we tried to prefatory assess the proposed automated error detection system; we built an artificial dataset containing of 60 sentences talking about tourism. We wrote sentences as correct and incorrect by hand to represent the base truth, having 41 correct and 19 wrong sentences. As an input to the automated error detection system, we used only a common sense fuzzy ontology and were targeting to check how successfully the system identifies wrong sentences. The sensitivity and specificity of our system were tested, where the sensitivity means the ratio of wrong sentences that are successfully detected as such, and the specificity checks the ratio of correct sentences that are correctly detected as such. By doing the experiment, we achieved 100% specificity and 80.2% sensitivity. The 100% specificity was expected, because the system deals with each sentence as correct unless it identifies an error in the sentence. The use of the fuzzy ontology led to 75.2% sensitivity as it is successful in conducting vague linguistic concepts and get rid of the loss of precision.

5 Conclusion and Future Work

This paper proposes an automated approach is used in essay evaluation. It scores the syntax of the essay and measures semantic coherence and consistence of the essay content. This is done by dividing attributes into three groups describing distance, spatial patterns, and spatial autocorrelation between parts of the essay. It also gives students feedback about their mistakes which was done by the use of fuzzy ontology.

It treats uncertain, incomplete, vague, or imprecise information, therefore giving the ontology more malleable models as well as permitting incomplete but real-life queries. As a result, it gave 80.2% sensitivity in error detecting. Experiments using different types of data sets accomplished superior results on 3 out of 4 data sets, scored with other state-of-the art AEE systems. To set a plan for future works, the system may be upgraded to be able to evaluate handwritten essays also. There is also further need to develop diverse semantic attributes and enhancement of feedback algorithm. Other approaches than TF-IDF for transforming text into attribute space have to be tested to find out how the alternatives influence the grades.

References

1. Chuang, T., Liu, E., Shiu, W.: Game-based creativity assessment system: the application of fuzzy theory. Multimed. Tools Appl. **74**(21), 9141–9155 (2015)
2. Wilson, J.: Universal screening with automated essay scoring: evaluating classification accuracy in grades 3 and 4. J. School Psychol. **68**(2), 19–37 (2018)
3. Zupanc, K., Bosnić, Z.: Automated essay evaluation with semantic analysis. Knowl. Based Syst. **120**(9), 118–132 (2017)
4. Ghosh, S., Fatima, S.: Design of an automated essay grading (AEG) system in Indian context. Int. J. Comput. Appl. **1**(11), 1–6 (2010)
5. Elsayed, E., Eldahshan, K., Tawfeek, S.: Automatic evaluation technique for certain types of open questions in semantic learning systems. Hum. Centric Comput. Inf. Sci. **3**(1), 1–15 (2013)
6. Hofmann, T.: Probabilistic latent semantic indexing. In: Proceedings of ACM SIGIR Forum, USA, pp. 211–218. ACM (2017)
7. Vrana, S.R., Vrana, D.T., Penner, L.A., Eggly, S., Slatcher, R.B., Hagiwara, N.: Latent Semantic Analysis: a new measure of patient-physician communication. Soc. Sci. Med. **198** (3), 22–26 (2018)
8. Devadoss, N., Ramakrishnan, S.: Knowledge representation using fuzzy ontologies–a review. Int. J. Comput. Sci. Inf. Technol. **6**(5), 4304–4308 (2015)
9. Chali, Y., Hasan, S.: On the effectiveness of using syntactic and shallow semantic tree kernels for automatic assessment of essays. In: Proceedings of the 6th International Conference on Natural Language Processing, Japan, pp. 767–773. Asian Federation of Natural Language Processing (2013)
10. Cutrone, L., Chang, M.: Automarking: automatic assessment of open questions. In: Proceedings of the 10th International Conference on Advanced Learning Technologies, Tunisia, pp. 143–147. IEEE (2010)
11. McNamara, D., Crossley, S., Roscoe, R., Allen, L., Dai, J.: A hierarchical classification approach to automated essay scoring. Assess. Writ. **23**(3), 35–59 (2015)
12. Ruseti, S., Dascalu, M., Johnson, A., McNamara, D., Balyan, R., McCarthy, K., Trausan-Matu, S.: Scoring summaries using recurrent neural networks. In: Proceedings of the International Conference on Intelligent Tutoring Systems, Canada, pp. 191–201. Springer, Cham (2018)
13. Thomas, N., Kumar, A., Bijlani, K.: Automatic answer assessment in LMS using latent semantic analysis. Procedia Comput. Sci. **58**(1), 257–264 (2015)
14. Wang, X.: The relationship between lexical diversity and EFL writing proficiency. Univ. Syd. Pap. TESOL **1**(1), 1–9 (2014)

15. Bestgen, Y.: Beyond single-word measures: L2 writing assessment, lexical richness and formulaic competence. System **69**(6), 65–78 (2017)
16. Schuster, S., Manning, Ch.: Enhanced English universal dependencies: an improved representation for natural language understanding tasks. In: Proceedings of the 10th International Conference on Language Resources and Evaluation, Slovenia, pp. 23–28. European Language Resources Association (2016)
17. Kusner, M., Sun, Y., Kolkin, N., Weinberger, K.: From word embeddings to document distances. In: Proceedings of the 32nd International Conference on Machine Learning, France, pp. 957–966 (2015). Journal of Machine Learning
18. Han, S., Zhao, C., Meng, W., Li, C.: Cosine similarity based fingerprinting algorithm in WLAN indoor positioning against device diversity. In: Proceedings of the International Conference on Communications, UK, pp. 2710–2714. IEEE (2015)
19. Zupanc, K., Bosnic, Z.: Automated essay evaluation augmented with semantic coherence measures. In: Proceedings of the International Conference on Data Mining, China, pp. 1133–1138. IEEE (2014)
20. Diniz-Filho, J., Barbosa, A., Collevatti, R., Chaves, L., Terribile, L., Lima-Ribeiro, M., Telles, M.: Spatial autocorrelation analysis and ecological niche modelling allows inference of range dynamics driving the population genetic structure of a Neotropical savanna tree. J. Biogeogr. **43**(1), 167–177 (2016)
21. Viney, N., Van Capelleveen, J., Geary, R., Xia, S.: Antisense oligonucleotides targeting apolipoprotein (a) in people with raised lipoprotein (a): two randomised, double-blind, placebo-controlled, dose-ranging trials. Lancet **388**(10057), 2239–2253 (2016)
22. Amiri, S., Lutz, R., Socías, E., McDonell, M., Roll, J., Amram, O.: Increased distance was associated with lower daily attendance to an opioid treatment program in Spokane County Washington. J. Subst. Abuse Treatm. **93**(3), 26–30 (2018)
23. Phandi, P., Chai, K., Ng, H.: Flexible domain adaptation for automated essay scoring using correlated linear regression. In: Proceedings of the International Conference on Empirical Methods in Natural Language Processing, Portugal, pp. 431–439. Association for Computational Linguistics (2015)

Towards a Portable SLU System Applied to MSA and Low-resourced Algerian Dialects

Mohamed Lichouri[1,2]([⊠]), Rachida Djeradi[2], Amar Djeradi[2], and Mourad Abbas[1]

[1] Computational Linguistics Department, CRSTDLA, Bouzaréah, Algeria
licvol@gmail.com, m.lichouri@crstdla.dz, mlichouri@usthb.dz
[2] University of Science and Technology Houari Boumediene, Bab Ezzouar, Algeria

Abstract. As the most used approach to extend a Spoken language Understanding (SLU) from a language to another, Machine translation achieves high performance for English domains, which is not the case for other languages, especially low-resourced ones as Arabic and its dialects. To avoid Machine Translation approach which requires huge parallel corpora, we will investigate, in this paper, the problem of user's intent interpretation from natural language queries to a system's semantic representation format across the languages and dialects, namely: English, Modern Standard Arabic (MSA) and four vernacular Algerian dialects from different regions: Blida, Djelfa, Tenes and Tizi-Ouzou. We should note that the domain we have chosen to run our experiments is a special application of school management. For this, We use three classifiers: kNN, Gaussian Naive Bayes and Bernoulli Naive Bayes which led to an average accuracy of 90%.

Keywords: Spoken Language Understanding · Multilingual · Dialects · Portability · Human-machine dialog · Utterance · Thematic approach

1 Introduction

The tremendous growing number of applications and the variety of available technologies on the net and mobile, have allowed users to get multiple access to information via human-machine interfaces (touching, gesture, speech, ...). But that still not enough so far to easily access the information because of the human nature (non-mastery of technology and unavailable time made people lazy). To remedy to this problem, the implementation of a multimodal, multilingual and multi-domain human-machine dialogue system is necessary.

In the present paper, we are interested in the multilingual aspect in human-machine dialog system. Strictly speaking, we first look to the problem of SLU portability from a source language to a target language. Second, we initiate a

© Springer Nature Switzerland AG 2020
A. E. Hassanien et al. (Eds.): AMLTA 2019, AISC 921, pp. 576–585, 2020.
https://doi.org/10.1007/978-3-030-14118-9_58

study to the problem of portability of SLU System from languages to dialects, which is considered as a problem not fully exploited yet.

Our goal is to provide a portable SLU system, with minimal intervention from human experts, (i) across languages, where two target languages (Arabic and English) will be extended from the original version (French) and (ii) across dialects, where four Algerian dialects (Blida, Djelfa, Tenes and Tizi-Ouzou) will be considered as target languages.

In this regard, we first presented a Spoken Language Understanding system based on thematic approaches [11], followed by a description of the LSA (Latent Semantic Analysis) used for information search that was considered. Then, we showed the minimal changes to perform on our systems to adapt to the different languages and dialects. After that, we carried out six different experiments, where we tested our application on two languages and four dialects. To conduct this experiment, we consider the Scikit-learn [1] machine library of python. For this first version of our SLU system the input is textual and the output answer is under two forms: text and speech.

This paper is organized as follows. In the second section, we present an overview of the related work. In third section, we introduce the architecture of an SLU System based on the proposed approach; Fourth section discusses the conducted experiments and we conclude in the last section.

2   Related Work

So far there are many works on SLU portability across languages from the early 2000. Where the first multilingual SLU is the VOYAGER System which was developed by Glass et al. [2] by extending the first English version to the Japanese, Italian, French and German. Then there is the work done by Lefevre et al. [3], where the authors have achieved an overall accuracy of 92%. They used the Semantic Tuple Classifiers (STCs) in conjunction with statistical machine translation (SMT) models applied on a small amount of data in the target language (French) bootstrapped from English data. The following work of Lefevre et al. [4] was in the framework of the PORTMEDIA project, they tried to evaluate their Tourist Information system (French) in terms of robustness and portability across languages and domains. They considered Italian as the new language and ticket reservation as the new domain. Stepanov et al. [6] evaluated end-to-end SLU porting on close and distant language pairs: Spanish-Italian and Turkish-Italian respectively; and achieved a significant drop in Character Error Rate (CER) to about 25%. In addition, they addressed, in [7], the problem of creating multilingual aligned corpora and their evaluation in the context of an SLU porting task, which was concluded by developing the Multilingual LUNA Corpus, a translation of Italian LUNA Corpus to Spanish, Turkish, and Greek. Another effort to solve the problem of SLU portability was carried out by Misu et al. [5]. Indeed, they propose a bootstrapping method of building a new SLU system in a target language using SMT, applying a back-translation to check whether the translation result maintains the semantic meaning of the original sentence, i.e. to solve erroneous translation results. The best results achieved by this approach is around 86% for Concept detection and 84% for Intent detection. As opposed to most of researchers that used Machine Translation, Upadhyay et al. [8] proposed an approach for extending SLU models and grammars to two new languages, Hindi and Turkish, based on bilingual model (i.e., jointly with English) with little supervision. This model enables faster learning, in that the model requires fewer labeled instances in the target language to generalize and also has attained an intent classification of 80%.

In another conjecture, for Arabic Dialect Understanding, Graja et al. [9] presented a method to understand spoken Tunisian dialect based on lexical semantic that takes into account the specificity of the Tunisian dialect which has no linguistic processing tools. The proposed method allows exploiting the ontological concepts for semantic annotation and ontological relations for speech interpretation. This allows increasing the rate of comprehension and limits the dependence on linguistic resources. This model has achieved an F-measure of 66% on a corpus that contains 175 user utterances. An other work has been achieved for Egyptian dialect by Elmadany et al. [10] in which they proposed a novel approach to user's utterances labeling for Egyptian spontaneous dialogues and Instant Messages using Machine Learning approach without relying any special lexicons, cues, or rules. The system was evaluated by a multi-genre corpus which includes 4725 utterances for three domains, and collected and annotated manually from Egyptian call-centers. Their method achieved an F1 score of 70.36% overall domains.

3   Spoken Language Understanding

We assume that the cognitive properties of the human being are his tendency to understand an utterance in two different ways:

- By identifying significant terms of this utterance followed by the identification of relationships between these terms, which leads him to understand the meaning of the utterance [12] (i.e Slot Filling).
- By identifying the subject of the utterance without understanding the words one by one [11] (i.e Intent Identification or Classification). The thematic approach for Spoken Language Understanding is based on this property.

3.1   Thematic Approach

The architecture of the SLU System in a Human-Machine Dialog (HMD) that is proposed in this paper is based on the Thematic Approach [11] (see Fig. 1). It is considered as a two-stage system where in the first stage we will take a corpus to generate a model that will be used in the second stage to identify or classify the Intent of a given request. We describe this approach in the following subsection.

First Stage: Semantic Representation. It consists in the extraction of semantic concepts in the corpus. This is why we relied on three types of analyzers which are **Structural**, **Lexical** and **Semantic**.

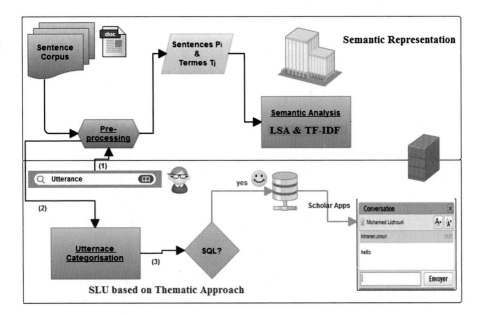

Fig. 1. Spoken Language Understanding System in HMD based on the Thematic approach

Structural Analyzer. It is in charge of extracting sentences contained in our corpus of study [13]. This corpus is made of 146 sentences (query or requests and not dialogues) of school type (marks, diploma or certificate requests). To extract these requests, we applied a detection of the boundary of the sentences (Sentence boundary disambiguation) [14], using "**PUNKT**" of NLTK (Natural Language Toolkit) [15]. This tool, detects the borders by considering capital letter and punctuation [16]. We should note that capital letters cannot be used for Arabic and its dialects because of the intrinsic characteristics of this language.

Lexical Analyzer. Let us consider the set of sentences:

$S = \{p_1, p_2, ..., p_m\}$ as an input of the analyzer, where m represents the number of sentences contained in the corpus. The output is the set of significant terms $T = \{t_1, t_2, ..., t_n\}$, where n represents the number of terms. Two Stages are necessary, the first one is to split (tokenize) the sentence into substrings using a regular expression and white space criterion followed by the second stage which is filtration [17] of unnecessary words to the understanding, by taking either the words with a length superior to three characters or those which do not belong to the list of empty words (Stop List).

Thematic Semantic Analyzer. During this analysis, we can get the presentation of sentence by the mean of LSA analysis. The objective of the LSA [18] is the representation of terms' ideas, by considering the context of these terms. This model considers that two contexts (Sentences) are similar if they contain similar terms. To achieve this goal, we built the occurrence matrix A which is composed of elements representing the occurrences of each term t_i in every sentence p_j and weighted by TF-IDF (Term Frequency-Inverse Document Frequency). Then, the singular value decomposition "SVD" of A is calculated. The SVD [19] is given by the formula:

$$A_{(n \times m)} = U_{(n \times r)} \times S_{(r \times r)} \times V^T_{(r \times m)} \tag{1}$$

U stands for the matrix of similarity term-concept, while the diagonal elements of S represent the "strength" of every concept, and finally V^T represents the matrix of similarity sentence-theme.

After applying a dimension reduction by selecting the most informative elements of S (respectively V^T). The similarity between sentences is given by calculating the distance between these elements (vectors representing the sentences). By using the appropriate classifier, we can determine the themes of these sentences.

Second Stage: Thematic Approach Based SLU. In this stage, three types of requests for Marks, Diploma and Certificate are considered as the three themes used in our experiments that we carried out using three classifiers, namely Naive Bayes (Gaussian and Bernoulli) and kNN (The choice of these classifiers is justified by the fact that they do not require a large corpus during learning phase). In regard to the School Management application, the query (sentence) of the user is classified into one of the three themes representing the tasks covered by this application. The final result is either the Diploma, Certificate or the Marks of the student.

4   Experiments

The main purpose of this work is to study the possibility to generalize an SLU system across languages and dialects. In order to evaluate the proposed approach for Spoken Language Understanding, we compared the performance of the three aforementioned classifiers using two sets of vector features: TF-IDF and LSA. If the theme of the sentence (request) is correctly identified then the sentence understanding is achieved. Otherwise it is counted as error.

In our work, we are interested in School Management Application which belongs to the domain of Database Information Retrieval. The used corpus has been collected from around 300 students which formulated their requests to access to their information from the education office. After discarding the repeated requests we obtained a corpus made of 127 different requests expressed in French. The collected corpus, which was initially in French, was translated manually by experts to English and Arabic, and also to the four Algerian dialects. To train the different Classifiers we used 75% of the data and 25% for test. Some examples of the requests are given in Table 1.

Table 1. Corpus sample of some request expressed in the three languages and four dialects.

Languages	**French**	*Marks Requests*	**Je veux ma note de Math**
		Diploma Requests	**Donner moi mon Diplome**
		Certificate Requests	**Pourrais-je avoir mon Certificat**
	English	*Marks Requests*	**I want my Math marks**
		Diploma Requests	**Give me my Diploma**
		Certificate Requests	**Could I have my Certificate**
	Arabic	*Marks Requests*	اريد نقطة مادة الرياضيات الخاصة بي
		Diploma Requests	اعطيني الديلوم الخاص بي
		Certificate Requests	هل استطيع الحصول على شهادتي المدرسية
Algerian Arabic Dialects	**Blida**	*Marks Requests*	راني حاب ندي لا نوت تاع المات نتاعي
		Diploma Requests	اطيني الديبلوم نتاعي
		Certificate Requests	نقدر ندي السارتيفيكا نتاعي
	Djelfa	*Marks Requests*	اطيني النقطة المات
		Diploma Requests	أطيني ديبلومي
		Certificate Requests	نقدر ندي السارتيفيكا نتاعي
	Tenes	*Marks Requests*	راني باغي النقطة تاعي
		Diploma Requests	أطيني الديبلوم تاعي
		Certificate Requests	ننجم ندي سارتيفيكا نتاعي
	Tizi-Ouzou	*Marks Requests*	أبغيغ لا نوت ايو
		Diploma Requests	افكير الدبلوم ايو
		Certificate Requests	مازمراغ اذويغ سيرتيفيكا ايو

4.1 Results and Discussion

For the proposed approach, we achieved a comparison between the three afore-mentioned classifiers using TF-IDF and n-truncated LSA features by SVD (different values of n SVD parameters from the original features are tested). n ranges from 10 to 30. We should note that greater values of n degrades the performance. The obtained results are presented in detail in Figs. 2, 3 and 4.

We note that the performance of the kNN classifier is almost the same for both TF-IDF and LSA. However, Gaussian and Bernoulli using TF-IDF features outperforms LSA. In Table 2, we summarized the best results of the conducted experiments, where we extracted different LSA vectors with $n = 10, 20, 30$. The results showed that the kNN classifier gives the best scores for all the languages and dialects where the best results are 91.30% for English and French and 93.48% for Tenes Dialect.

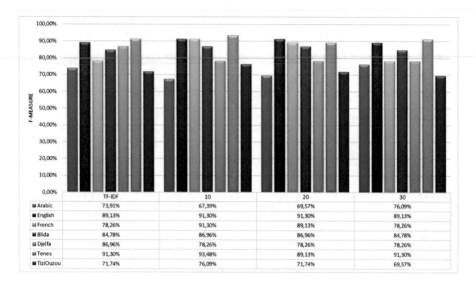

Fig. 2. kNN (TF-IDF vs LSA).

	TF-IDF	10	20	30
▣ Arabic	73,91%	67,39%	69,57%	76,09%
▪ English	89,13%	91,30%	91,30%	89,13%
▣ French	78,26%	91,30%	89,13%	78,26%
▣ Blida	84,78%	86,96%	86,96%	84,78%
▣ Djelfa	86,96%	78,26%	78,26%	78,26%
▣ Tenes	91,30%	93,48%	89,13%	91,30%
▪ TiziOuzou	71,74%	76,09%	71,74%	69,57%

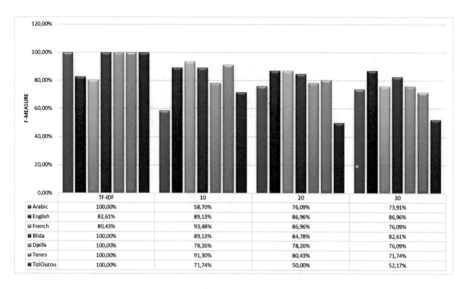

Fig. 3. Gaussian Naive Bayes (TF-IDF vs LSA).

	TF-IDF	10	20	30
▣ Arabic	100,00%	58,70%	76,09%	73,91%
▪ English	82,61%	89,13%	86,96%	86,96%
▣ French	80,43%	93,48%	86,96%	76,09%
▣ Blida	100,00%	89,13%	84,78%	82,61%
▣ Djelfa	100,00%	78,26%	78,26%	76,09%
▣ Tenes	100,00%	91,30%	80,43%	71,74%
▪ TiziOuzou	100,00%	71,74%	50,00%	52,17%

Whereas by using the TF-IDF vectors as features for the different classifiers (see Table 3), kNN yielded a performance similar to that obtained by using LSA vectors. The best results have been achieved by the Bernoulli Naive Bayes Classifier, (100%) for English and the four Dialects and over 93% for Arabic and French.

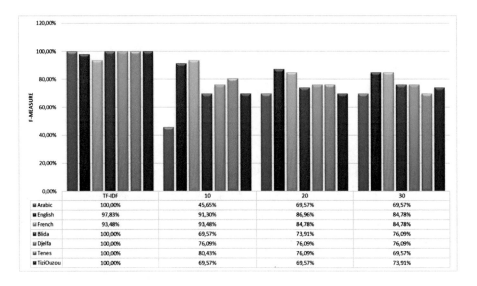

Fig. 4. Bernoulli Naive Bayes (TF-IDF vs LSA).

Table 2. Classification using LSA: Best results of $n = 10, 20, 30$

	Languages			Algerian Arabic Dialects			
	Arabic	English	French	Blida	Djelfa	Tenes	Tizi-Ouzou
K-Nearest Neighbors	76.09	91.30	91.30	86.96	78.26	93.48	76.09
Gaussian Naive Bayes	76.09	89.13	93.48	89.13	78.26	91.30	71.74
Bernoulli Naive Bayes	69.57	91.30	93.48	76.09	76.09	80.43	73.91

Table 3. Classification using TF-IDF

	Languages			Algerian Arabic Dialects			
	Arabic	English	French	Blida	Djelfa	Tenes	Tizi-Ouzou
K-Nearest Neighbors	73.91	89.13	78.26	84.78	86.96	91.30	71.74
Gaussian Naive Bayes	100.00	82.61	80.43	100.00	100.00	100.00	100.00
Bernoulli Naive Bayes	100.00	97.83	93.48	100.00	100.00	100.00	100.00

5 Conclusion

This paper is a contribution to the ongoing researches about the generalization of a Spoken Language Understanding (SLU) System in a multi-languages and multi-dialects Human-Machine Dialog. To our knowledge, this is the first study to investigate the possibility of a portable SLU system across low-resourced Algerian dialects (Blida, Djelfa, Tenes and Tizi-Ouzou) and across languages

(French, Arabic and English). The findings were quite encouraging since the minimum accuracy to adapt the School Management Application to other languages is about 93%.

Acknowledgment. Special thanks to Dhia El Hak Megtouf, Amel Elbachir and Karima Mahdjane for their contribution in corpus enrichment.

References

1. Pedregosa, F., Varoquaux, G., Gramfort, A., Michel, V., Thirion, B., Grisel, O., Blondel, M., Prettenhofer, P., Weiss, R., Dubourg, V., Vanderplas, J.: Scikit-learn: machine learning in Python. J. Mach. Learn. Res. **12**(Oct), 2825–2830 (2011)
2. Glass, J., Flammia, G., Goodine, D., Phillips, M., Polifroni, J., Sakai, S., Seneff, S., Zue, V.: Multilingual spoken-language understanding in the MIT Voyager system. Speech Commun. **17**(1–2), 1–18 (1995)
3. Lefevre, F., Mairesse, F., Young, S.: Cross-lingual spoken language understanding from unaligned data using discriminative classification models and machine translation. In: Eleventh Annual Conference of the International Speech Communication Association (2010)
4. Lefevre, F., Mostefa, D., Besacier, L., Esteve, Y., Quignard, M., Camelin, N., Favre, B., Jabaian, B., Barahona, L.M.R.: Leveraging study of robustness and portability of spoken language understanding systems across languages and domains: the PORTMEDIA corpora. In: The International Conference on Language Resources and Evaluation, May 2012
5. Misu, T., Mizukami, E., Kashioka, H., Nakamura, S., Li, H.: A bootstrapping approach for SLU portability to a new language by inducting unannotated user queries. In: 2012 IEEE International Conference on Acoustics, Speech and Signal Processing (ICASSP), pp. 4961–4964. IEEE, March 2012
6. Stepanov, E.A., Kashkarev, I., Bayer, A.O., Riccardi, G., Ghosh, A.: Language style and domain adaptation for cross-language SLU porting. In: 2013 IEEE Workshop on Automatic Speech Recognition and Understanding (ASRU), pp. 144–149. IEEE, December 2013
7. Stepanov, E.A., Riccardi, G., Bayer, A.O.: The development of the multilingual LUNA corpus for spoken language system porting. In: LREC pp. 2675–2678, May 2014
8. Upadhyay, S., Faruqui, M., Tur, G., Hakkani-Tur, D., Heck, L.: (Almost) Zero-Shot Cross-Lingual Spoken Language Understanding (2018)
9. Graja, M., Jaoua, M., Belguith, L.H.: Building ontologies to understand spoken tunisian dialect. arXiv preprint arXiv:1109.0624 (2011)
10. Elmadany, A.A., Abdou, S.M., Gheith, M.: Towards understanding Egyptian Arabic dialogues. arXiv preprint arXiv:1509.03208 (2015)
11. Lichouri, M., Djeradi, A., Djeradi, R.: A new automatic approach for understanding the spontaneous utterance in human-machine dialogue based on automatic text categorization. In: Proceedings of the International Conference on Intelligent Information Processing, Security and Advanced Communication, p. 50. ACM, November 2015
12. Lichouri, M., Djeradi, A., Djeradi, R.: Une approche Statistico-Linguistique pour l'extraction de concepts sémantiques: Une première étape vers un système générique de dialogue Homme-Machine

13. Indurkhya, N., Damerau, F.J. (eds.): Handbook of Natural Language Processing, vol. 2. CRC Press, Boca Raton (2010)
14. Palmer, D.D., Hearst, M.A.: Adaptive multilingual sentence boundary disambiguation. Comput. Linguist. **23**(2), 241–267 (1997)
15. Bird, S., Loper, E.: NLTK: the natural language toolkit. In: Proceedings of the ACL 2004 on Interactive Poster and Demonstration Sessions, p. 31. Association for Computational Linguistics, July 2004
16. Kiss, T., Strunk, J.: Unsupervised multilingual sentence boundary detection. Comput. Linguist. **32**(4), 485–525 (2006)
17. Ramshaw, L.A., Marcus, M.P.: Text chunking using transformation-based learning. In: Armstrong, S., Church, K., Isabelle, P., Manzi, S., Tzoukermann, E., Yarowsky, D. (eds.) Natural Language Processing Using Very Large Corpora, pp. 157–176. Springer, Dordrecht (1999)
18. Steinberger, J., Jezek, K.: Using latent semantic analysis in text summarization and summary evaluation. In: Proceedings of ISIM, vol. 4, pp. 93–100 (2004)
19. Leskovec, J.: Dimensionality reduction PCA, SVD, MDS, ICA, and friends. Machine Learning recitation, 27 April 2006
20. Yang, Y.: An evaluation of statistical approaches to text categorization. Inf. Retrieval **1**(1–2), 69–90 (1999)
21. Schütze, H., Manning, C.D., Raghavan, P.: Introduction to Information Retrieval, vol. 39. Cambridge University Press, New York (2008)

Ans2vec: A Scoring System for Short Answers

Wael Hassan Gomaa[1(✉)] and Aly Aly Fahmy[2]

[1] Beni-Suef University, Beni-Suef, Egypt
Wael.goma@gmail.com
[2] Cairo University, Cairo, Egypt
aly.fahmy@cu.edu.eg

Abstract. Automatic scoring is a complex task in computational linguistics, particularly in an educational context. Sentences vectors (sent2vec) approaches affirmed their prosperity recently as favorable models for sentence representation. In this research, we propose an efficient and uncomplicated short answer grading model named Ans2vec. Skip-thought vector approach is used to convert both model and student's answers into meaningful vectors to measure the similarity between them. Ans2vec model achieves promising results on three different benchmarking data sets. For Texas data set; Ans2vec achieves the best Pearson correlation value (0.63) compared to all related systems.

Keywords: Automatic scoring · Short answer grading · Sentence embeddings

1 Introduction

Automatic scoring for educational systems has attracted much attention during these past years. Automatic scoring guarantees fairness and saves time. Automatic assessment for True/False questions and multiple choice questions may be easy to handle while Automatic Essay Scoring (AES) and Short Answer Grading (SAG) are a real challenge to researchers [1, 2]. AES scores essay questions that are characterized by long answers where no model answer is provided. They consider spelling analysis, grammar, sentence coherency and relatedness to main topic in answers grading. Scoring for short-answer questions helps evaluating student understanding for specific concepts which empower the importance of SAG that handle short answers ranging from 1 to 3 sentences considering a model answer in the scoring process. In SAG students answer is graded in the light of model answer where grammar and coherency are not of interest in many approaches [3].

Word embeddings or Word2vec models are widely permeating on Natural Language Processing (NLP) applications. These word representations improved downstream tasks in many domains such as machine translation, Paraphrase detection, text classification, and sentiment analysis, among others. Word embedding techniques tackle words as low dimensional continuous vectors which represent semantic and syntactic relations of words [4, 5]. Sentence embeddings (Sent2vec) try to achieve something similar: use a fixed-dimensional vector to represent a sentence. According to their purposes, sentence embeddings generally fall into two categories: task-specific sentence embeddings and general-purpose sentence embeddings [6]. The first category focuses on training sentence

© Springer Nature Switzerland AG 2020
A. E. Hassanien et al. (Eds.): AMLTA 2019, AISC 921, pp. 586–595, 2020.
https://doi.org/10.1007/978-3-030-14118-9_59

embeddings for a particular task using supervised learning methods. Researchers have proposed many models along this line, and they typically use Recursive Neural Networks (RNN) [7, 8], Convolutional Neural Networks (CNN) [9–11] or Recurrent Neural Networks with Long Short-Term Memory (LSTM) [12] as an intermediate task in building sentence embeddings to implement different NLP tasks including paraphrase identification, question answer matching and sentiment classification [13–15]. The other category focuses on global sentence embeddings, which are usually built using unsupervised or semi-supervised learning and can be served as features for different NLP tasks like text classification and semantic textual similarity. This include recursive auto-encoders [16], Paragraph Vector [17], Skip-thought vectors [18], Siamese CBOW [19] FastSent [20], Sent2Vec [21], GRAN [22], etc.

In this research, based on Skip-thought vectors [18], we introduce a simple SAG model named Ans2vec that depends on converting both student's and model answers into two vectors and then measuring the similarity between them. Skip-thought vectors are – in a way – word2vec for sentences. Where word2vec attempts to predict surrounding words from certain words in a sentence, skip-thought vector extends this idea to sentences: it predicts surrounding sentences from a given sentence. Skip-thought vectors use the encoder-decoder model to first encode a sentence into a vector and then decode that representation into the surrounding sentences. Skip-thought vectors not take the ordering of both words and sentences into account. This allows it to encode rich information into the embedding. The skip-thought model has been proven to be effective at learning sentence representations and capturing sentence semantics. Skip-thought vectors have achieved outstanding results in many complicated tasks including image-sentence ranking, question-type classification, paraphrase detection, semantic relatedness and sentiment analysis [18].

In Sect. 2 related work on SAG is presented. Section 3 dives into the proposed model (Ans2vec) and the used SAG benchmarking data sets. Section 4 discusses the experiment results, and finally, Sect. 5 sums up the conclusion.

2   Related Work

A comprehensive review of automatic short answer grading can be found in [23, 24]. Authors in [23] provided a review of 12 systems and sketched their characteristics, such as C-Rater [25], CarmelTC [26], Intelligent Assessment Technologies (IAT) [27], Oxford-UCLES [28] and Texas [29]. Authors in [24] traced the deepening interest in SAG task over recent years. They identified five eras represented by clusters of 35 systems and two competitions [30, 31] that share a common theme. The five ears were classified into two groups: method-based and evaluation-based. Method-based eras include concept mapping, information extraction, corpus-based methods and machine learning. Sample of SAG systems that covers concept mapping era are Burstein [32], ATM [33], C-rater [25] and Wang [34]. Sample systems that covers information extraction era are WebLas [35], eMax [36], Auto-Assessor [37] and IndusMarker [38]. Sample systems that covers corpus-based methods era are Atenea [39], Willow [40] and SAMText [41]. Sample systems that covers machine learning era are e-Examiner [42], CAM [43] and CoMiC-EN [44].

We will zoom into five works [3, 29, 45–47] as we will compare our proposed system to them. Authors in Texas system [29] enhanced lexical semantic similarity by applying dependency graph alignments generated by machine learning. They experimented many WordNet-based measures in addition to two corpus-based measures which are Semantic Analysis (LSA) and Explicit Semantic Analysis (ESA) [48]. The data set used in their experiments was a computer science data set that will be described in next section. The Texas system achieved Pearson correlation (r) = 0.518 and a Root Mean Square Error (RMSE) of 0.978 as its best result.

The computer science data sets in [45] were subjected to three stages of examination. Firstly similarity between model answer and student answer was measured by using fourteen String-based algorithms. Seven algorithms were Character-based while the rest were Term-Based. Secondly similarity was measured by using DICSO1 and DISCO2 Corpus-based similarity [48]. Finally a combination between String-based and Corpus based measures was applied where the best correlation value 0.504 was obtained from mixing N-gram with Disco1 similarity values. The work in [46] utilized the English version of Cairo University data set with a total of 536 different runs such that 256 of the runs used string-based similarity, 64 used corpus-based similarity, while the remaining 216 used knowledge-based similarity measures. A combination task was performed through examining three supervised models: simple linear regression, linear regression and SMOreg models. The best r and RMSE values were 0.83 and 0.75 respectively. Authors in [47] tested two data sets: computer science and SemEval-2013. They implemented a fast system that depended on both text similarity and semantic vector similarity features. The semantic vector feature employed off-the-shelf word embeddings (Word2Vec). A sentence level semantic vector was computed for each input sentence as the sum of its content word embeddings. The best r and RMSE values of computer science data set were 0.63 and 0.85 respectively. The best F1 score on the SemEval-2013 dataset was 0.55 on average. Authors in [3] presented a vector summation model, for each type of vector representation Word2Vec, GloVe and Sense aware vectors, a sentence is represented as a vector by adding up the vectorized representation of its words. For each type of vector representation cosine similarity is then calculated between the vectors representing the two sentences. They examined three data sets: computer science, SemEval-2013 and the Arabic version of Cairo university data set. The best r and RMSE values of computer science data set were 0.55 and 0.91 respectively. The best F1 score on the SemEval-2013 dataset was 0.51 on average. The best r and RMSE values of the Arabic version of Cairo university data set were 0.84 and 0.89 respectively.

To conclude this section, much current published scoring systems are implemented using complex models. These models mainly depend on different text similarity approaches and machine learning algorithms, and include many preprocessing, postprocessing and combining modules. Some systems are focused on Word2vec models and sentence embeddings that are produced by adding up the vectors of its words.

3 Proposed Model and Data Sets

The proposed model (Ans2vec) is characterized by simplicity; it employs an unsupervised pre-trained sentence embedding model that is not so complex that you really don't know what is happening behind the scenes. There is no need to apply any natural language preprocessing or post-processing modules such as stop word removing, stemming, part of speech (POS) tagging, normalization or lemmatization. Furthermore human tagged data or lexical databases such as WordNet are not required. As shown in Fig. 1, Ans2vec model simply depends on converting the whole answers into semantic vectors using skip-thought approach. Vectors is an array with as many rows as the length of X, and each row is 4800 dimensional (combine-skip model). The first 2400 dimensions is the uni-skip model, and the last 2400 is the bi-skip model. Bi-skip model contains two encoders with different parameters: one encoder is given the answer in correct order, while the other is given the answer in reverse. The combine-skip vectors are highly recommended, as they are almost universally the best performing in previous research [18]. To represent student's answer (SA) and model answer (MA), two features

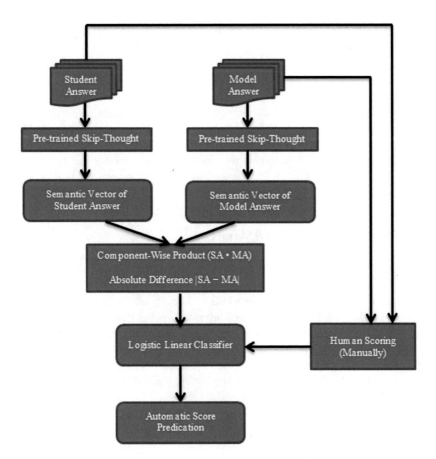

Fig. 1. Proposed model (Ans2vec) architecture

are used. Given two skip-thought vectors SA and MA, Their component-wise product (SA • MA) and their absolute difference |SA − MA| are computed and concatenated together. To predict a score, a logistic linear classifier is trained on top of these extracted features with the same setup steps explained in [49], with no additional fine-tuning or backpropagation through the skip-thoughts model [18].

Proposed system is experimented on three benchmark SAG data sets, Texas, Cairo University and SCIENTSBANK. Firstly, Texas data set consists of 80 undergraduate Data Structures questions that are scattered across ten different assignments and two tests, each on a related set of topics along with 2,273 student responses. Students Answers are graded by two human judges who considers the model answer provided for each question. Average of the two human scores is used as the final gold score for each student answer [29].

Secondly, Cairo University data set consists of 61 questions, 10 answers for each, with a total number of 610 answers for only one chapter of the official Egyptian curriculum for Environmental Science course. The average length of a student's answer is 2.2 sentences, 20 words or 103characters. The data set contains a collection of students' answers along with their grades that vary between 0 and 5 according to two evaluators' assessment. Average of the two human scores is used as the final gold score for each student answer [2]. Also an English version of Cairo University data set is used in this research too.

Finally, SCIENTSBANK data set is a part of SemEval-2013 task #7 [50]. It doesn't depend on numeric-valued scores instead the 5-way task depends on assigning one of the following five labels to a student response: correct, partially correct/incomplete, contradictory, irrelevant, and non-domain (answers that are clear from domain content). The SCIENTSBANK data contains training data and 3 types of test data; the Unseen Answers (UA) dataset consists of responses to questions that are present in the training set. Unseen Questions (UQ) contains responses to in-domain but previously unseen questions. Unseen Domains (UD) test set, containing completely out of domain question-response pairs.

4   Experiments Results and Analysis

The evaluation metrics Pearson's correlation (r) and root mean squared error (RMSE) are used in the experiments of Texas and Cairo University data sets. F1 measure is used to evaluate the experiments of SCIENTSBANK data set. As we mentioned in the related work section, we will compare our proposed system (Ans2vec) to five works [3, 29, 45–47]. Texas and Cairo University data sets are partitioned across 100 random runs into 60% for training, 20% for development and 20% for testing. The training, development and testing answers in Texas data set are 1363, 455 and 455, while in Cairo university are 366, 122 and 122 respectively. Tables 1 and 2 show the results of Texas and Cairo University data sets respectively.

Table 1. Results of Texas data set

System	Pearson's (r)	RMSE
Mohler et al. [29]	0.52	0.98
Gomaa et al. [45]	0.50	NA
Sultan et al. [47]	0.59	**0.89**
Magooda et al. [3]	0.55	0.91
Ans2vec	**0.63**	0.91

Table 2. Results of Cairo University data set

System	Pearson's (r)	RMSE
Gomaa et al. [46]	0.83	**0.75**
Magooda et al. [3]	**0.84**	0.98
Ans2vec	0.79	0.92

Table 3. F1 results of SCIENTSBANK data set

System	UA	UQ	UD
Sultan et al. [47]	**0.58**	**0.55**	**0.54**
Magooda et al. [3]	0.47	0.51	0.46
Ans2vec	**0.58**	0.47	0.46

The training and development answers in SCIENTSBANK data set are 3976 and 993 respectively. The testing answers of UA, UQ and UD are 733, 540 and 4562 respectively. Table 3 shows the F1 results of SCIENTSBANK data set.

Given the simplicity of our approach; it achieves better or competitive results in experimental evaluation on the three different benchmarking data sets. For Texas data set; Ans2vec achieves the best Pearson correlation value (0.63) compared to all the systems. The RMSE value (0.91) is very promising compared to the best RMSE value (0.89) achieved in [47]. For Cairo University; Ans2vec achieves promising results if we take into consideration the complex work performed in [3, 46] described in details in related work section. For SCIENTSBANK data set; Ans2vec achieves the best F1 score (0.58) in Unseen Answers (UA) that is the same score reported in [47]. The comparison between the results of UA, UQ and UD indicates that our approach is performing well if the test questions and domains present in the training set.

5   Conclusions

In this research, based on Skip-thought vectors, we introduce a simple SAG model named Ans2vec that depends on converting both student's and model answers into two semantic vectors and then measuring the similarity between them. Ans2vec model achieves better or competitive results in experimental evaluation on three different benchmarking data sets: Texas, Cairo University and SCIENTSBANK. Future work will be concerned with three main points. First is testing the other Sent2vec approaches such as Paragraph Vector, Siamese CBOW and FastSent. Second is merging our approach with the classical machine learning and text similarity techniques. The final point is testing data set with other natural languages like Arabic.

References

1. Gomaa, W.H., Fahmy, A.A.: Tapping into the power of automatic scoring. In: The Eleventh International Conference on Language Engineering, Egyptian Society of Language Engineering (ESOLEC) (2011)
2. Gomaa, W.H., Fahmy, A.A.: Arabic short answer scoring with effective feedback for students. Int. J. Comput. Appl. **86**(2), 35–41 (2014)
3. Magooda, A.E., Zahran, M.A., Rashwan, M., Raafat, H.M., Fayek, M.B.: Vector based techniques for short answer grading. In: FLAIRS Conference, pp. 238–243, March 2016
4. Mikolov, T., Chen, K., Corrado, G., Dean, J.: Efficient estimation of word representations in vector space (2013). arXiv preprint: arXiv:1301.3781
5. Mikolov, T., Sutskever, I., Chen, K., Corrado, G.S., Dean, J.: Distributed representations of words and phrases and their compositionality. In: Advances in Neural Information Processing Systems, pp. 3111–3119 (2013)
6. Jiao, X., Wang, F., Feng, D.: Convolutional neural network for universal sentence embeddings. In: Proceedings of the 27th International Conference on Computational Linguistics, pp. 2470–2481 (2018)
7. Socher, R., Huval, B., Manning, C.D., Ng, A.Y.: Semantic compositionality through recursive matrix-vector spaces. In: Proceedings of the 2012 Joint Conference on Empirical Methods in Natural Language Processing and Computational Natural Language Learning, pp. 1201–1211. Association for Computational Linguistics, July 2012
8. Socher, R., Perelygin, A., Wu, J., Chuang, J., Manning, C.D., Ng, A., Potts, C.: Recursive deep models for semantic compositionality over a sentiment treebank. In: Proceedings of the 2013 Conference on Empirical Methods in Natural Language Processing, pp. 1631–1642 (2013)
9. Kalchbrenner, N., Grefenstette, E., Blunsom, P.: A convolutional neural network for modelling sentences (2014). arXiv preprint: arXiv:1404.2188
10. dos Santos, C., Gatti, M.: Deep convolutional neural networks for sentiment analysis of short texts. In: Proceedings of COLING 2014, the 25th International Conference on Computational Linguistics: Technical Papers, pp. 69–78 (2014)
11. Kim, Y.: Convolutional neural networks for sentence classification (2014). arXiv preprint: arXiv:1408.5882
12. Chung, J., Gulcehre, C., Cho, K., Bengio, Y.: Empirical evaluation of gated recurrent neural networks on sequence modeling (2014). arXiv preprint: arXiv:1412.3555
13. Yin, W., Schütze, H.: Convolutional neural network for paraphrase identification. In: Proceedings of the 2015 Conference of the North American Chapter of the Association for Computational Linguistics: Human Language Technologies, pp. 901–911 (2015)
14. Tan, M., Dos Santos, C., Xiang, B., Zhou, B.: Improved representation learning for question answer matching. In: Proceedings of the 54th Annual Meeting of the Association for Computational Linguistics. Long Papers, vol. 1, pp. 464–473 (2016)

15. Lin, Z., Feng, M., Santos, C.N.D., Yu, M., Xiang, B., Zhou, B., Bengio, Y.: A structured self-attentive sentence embedding (2017). arXiv preprint: arXiv:1703.03130

16. Socher, R., Huang, E.H., Pennin, J., Manning, C.D., Ng, A.Y.: Dynamic pooling and unfolding recursive autoencoders for paraphrase detection. In: Advances in Neural Information Processing Systems, pp. 801–809 (2011)

17. Le, Q., Mikolov, T.: Distributed representations of sentences and documents. In: International Conference on Machine Learning, pp. 1188–1196, January 2014

18. Kiros, R., Zhu, Y., Salakhutdinov, R.R., Zemel, R., Urtasun, R., Torralba, A., Fidler, S.: Skip-thought vectors. In: Advances in Neural Information Processing Systems, pp. 3294–3302 (2015)

19. Kenter, T., Borisov, A., de Rijke, M.: Siamese cbow: Optimizing word embeddings for sentence representations (2016). arXiv preprint: arXiv:1606.04640

20. Hill, F., Cho, K., Korhonen, A.: Learning distributed representations of sentences from unlabelled data (2016). arXiv preprint: arXiv:1602.03483

21. Pagliardini, M., Gupta, P., Jaggi, M.: Unsupervised learning of sentence embeddings using compositional n-gram features (2017). arXiv preprint: arXiv:1703.02507

22. Wieting, J., Gimpel, K.: Revisiting recurrent networks for paraphrastic sentence embeddings (2017). arXiv preprint: arXiv:1705.00364

23. Ziai, R., Ott, N., Meurers, D.: Short answer assessment: establishing links between research strands. In: Proceedings of the Seventh Workshop on Building Educational Applications Using NLP, pp. 190–200. Association for Computational Linguistics, June 2012

24. Burrows, S., Gurevych, I., Stein, B.: The eras and trends of automatic short answer grading. Int. J. Artif. Intell. Educ. 25(1), 60–117 (2015)

25. Leacock, C., Chodorow, M.: C-rater: automated scoring of short-answer questions. Comput. Humanit. 37(4), 389–405 (2003)

26. Rosé, C.P., Roque, A., Bhembe, D., VanLehn, K.: A hybrid approach to content analysis for automatic essay grading. In: Proceedings of the 2003 Conference of the North American Chapter of the Association for Computational Linguistics on Human Language Technology: Companion Volume of the Proceedings of HLT-NAACL 2003–Short papers, vol. 2, pp. 88–90. Association for Computational Linguistics, May 2003

27. Mitchell, T., Russell, T., Broomhead, P., Aldridge, N.: Towards robust computerised marking of free-text responses. In: Proceedings of the Sixth International Computer Assisted Assessment Conference. Loughborough University, Loughborough, UK (2002)

28. Pulman, S.G., Sukkarieh, J.Z.: Automatic short answer marking. In: Proceedings of the Second Workshop on Building Educational Applications Using NLP, pp. 9–16. Association for Computational Linguistics, June 2005

29. Mohler, M., Bunescu, R., Mihalcea, R.: Learning to grade short answer questions using semantic similarity measures and dependency graph alignments. In: Proceedings of the 49th Annual Meeting of the Association for Computational Linguistics: Human Language Technologies, vol. 1, pp. 752–762. Association for Computational Linguistics, June 2011

30. Hewlett Foundation: Automated student assessment prize: phase two – short answer scoring. Kaggle Competition (2012)

31. Dzikovska, M.O., Nielsen, R.D., Brew, C., Leacock, C., Giampiccolo, D., Bentivogli, L., Clark, P., Dagan, I., Dang, H.T.: SemEval-2013 task 7: the joint student response analysis and eighth recognizing textual entailment challenge. In: Diab, M., Baldwin, T., Baroni, M. (eds.) Proceedings of the 2nd Joint Conference on Lexical and Computational Semantics, Atlanta, pp. 1–12 (2013)

32. Burstein, J., Kaplan, R., Wolff, S., Lu, C.: Using lexical semantic techniques to classify free responses. In: Viegas, E. (ed.) Proceedings of the ACL SIGLEX Workshop on Breadth and Depth of Semantic Lexicons, Santa Cruz, pp. 20–29. Association for Computational Linguistics (1996)

33. Callear, D., Jerrams-Smith, J., Soh, V.: CAA of short non-MCQ answers. In: Danson, M., Eabry, C. (eds.) Proceedings of the 5th Computer Assisted Assessment Conference, pp. 1–14. Loughborough University, Loughborough (2001)

34. Wang, H.-C., Chang, C.-Y., Li, T.-Y.: Assessing creative problem-solving with automated text grading. Comput. Educ. 51(4), 1450–1466 (2008)

35. Bachman, L.F., Carr, N., Kamei, G., Kim, M., Pan, M.J., Salvador, C., Sawaki, Y.: A reliable approach to automatic assessment of short answer free responses. In: Tseng, S.C., Chen, T.E., Liu, Y.F. (eds.) Proceedings of the 19th International Conference on Computational Linguistics, COLING 2002, Taipei, vol. 2, pp. 1–4. Association for Computational Linguistics (2002)

36. Sima, D., Schmuck, B., Szöll, S., Miklós, A.: Intelligent short text assessment in eMax. In: Rudas, I.J., Fodor, J., Kacprzyk, J. (eds.) Towards Intelligent Engineering and Information Technology. Studies in Computational Intelligence, vol. 243, pp. 435–445. Springer, Heidelberg (2009)

37. Cutrone, L., Chang, M., Kinshuk: Auto-assessor: computerized assessment system for marking student's short-answers automatically. In: Narayanaswamy, N.S., Krishnan, M.S., Kinshuk, Srinivasan, R. (eds.) Proceedings of the 3rd IEEE International Conference on Technology for Education, Chennai, pp. 81–88. IEEE (2011)

38. Siddiqi, R., Harrison, C.J.: A systematic approach to the automated marking of short-answer questions. In: Anis, M.K., Khan, M.K., Zaidi, S.J.H. (eds.) Proceedings of the 12th International Multitopic Conference, Karachi, pp. 329–332. IEEE (2008)

39. Alfonseca, E., Pérez, D.: Automatic assessment of open ended questions with a BLEU-inspired algorithm and shallow NLP. In: Vicedo, J., Martínez-Barco, P., Muńoz, R., Saiz Noeda, M. (eds.) Advances in Natural Language Processing. Lecture Notes in Computer Science, vol. 3230, pp. 25–35. Springer, Berlin (2004)

40. Pérez-Marín, D., Pascual-Nieto, I.: Willow: a system to automatically assess students' freetext answers by using a combination of shallow NLP techniques. Int. J. Contin. Eng. Educ. Life Long Learn. 21(2), 155–169 (2011)

41. Bukai, O., Pokorny, R., Haynes, J.: An automated short-free-text scoring system: development and assessment. In: Proceedings of the 20th Interservice/Industry Training, Simulation, and Education Conference, pp. 1–11. National Training and Simulation Association (2006)

42. Gütl, C.: e-Examiner: towards a fully-automatic knowledge assessment tool applicable in adaptive e-learning systems. In: Ghassib, P.H. (ed.) Proceedings of the 2nd International Conference on Interactive Mobile and Computer Aided Learning, pp. 1–10. Amman (2007)

43. Bailey, S., Meurers, D.: Diagnosing meaning errors in short answers to reading comprehension questions. In: Tetreault, J., Burstein, J., De Felice, R. (eds.) Proceedings of the 3rd ACL Workshop on Innovative Use of NLP for Building Educational Applications, Columbus, pp. 107–115. Association for Computational Linguistics (2008)

44. Meurers, D., Ziai, R., Ott, N., Bailey, S.M.: Integrating parallel analysis modules to evaluate the meaning of answers to reading comprehension questions. Int. J. Contin. Eng. Educ. Life-Long Learn. 21(4), 355–369 (2011)

45. Gomaa, W.H., Fahmy, A.A.: Short answer grading using string similarity and corpus-based similarity. Int. J. Adv. Comput. Sci. and Appl. (IJACSA) 3(11), 115–121 (2012)

46. Gomaa, W.H., Fahmy, A.A.: Automatic scoring for answers to Arabic test questions. Comput. Speech Lang. **28**(4), 833–857 (2014)
47. Sultan, M.A., Salazar, C., Sumner, T.: Fast and easy short answer grading with high accuracy. In: Proceedings of the 2016 Conference of the North American Chapter of the Association for Computational Linguistics: Human Language Technologies, pp. 1070–1075 (2016)
48. Gomaa, W.H., Fahmy, A.A.: A survey of text similarity approaches. Int. J. Comput. Appl. **68**(13), 13–18 (2013)
49. Tai, K.S., Socher, R., Manning, C.D.: Improved semantic representations from tree-structured long short-term memory networks (2015). arXiv preprint: arXiv:1503.00075
50. Dzikovska, M.O., Nielsen, R.D., Brew, C., Leacock, C., Giampiccolo, D., Bentivogli, L., Dang, H.T.: Semeval-2013 task 7: the joint student response analysis and 8th recognizing textual entailment challenge. North Texas State Univ Denton (2013)

Machine Learning in Cyber Security

Applications of Machine Learning in Cyber Security - A Review and a Conceptual Framework for a University Setup

Rishabh Jain[1] and Roheet Bhatnagar[2(✉)]

[1] Department of IT, Manipal University Jaipur, Jaipur, India
[2] Department of CSE, Manipal University Jaipur, Jaipur, India
`roheet.bhatnagar@jaipur.manipal.edu`

Abstract. Machine learning is a growing technical field due to its versatility and stability for the ever increasing data flow from heterogeneous sources and computational demands. Machine learning techniques are deployed nearly in every aspect of computing today because of its highly adaptive and scalable characteristics. It has the potential to adapt to new and unknown challenges. Cyber-Security is a field which is rapidly developing these days because of the attention that is required to secure the net-works and applications with the growth in social networks, internet and mobile banking, cloud computing, web technologies, smart grid etc. The domain of Cyber Security owing to its diversity and applications generate lot of data which is voluminous and coming from different sources. Such data provides great scope to the data scientists, cyber security specialists and machine learning enthusiasts, since they have the potential to provide insights which could help in curbing Cyber Crimes using Machine Learning Algorithms. This paper reviews literature in the field of machine learning and cyber security and highlighting the key developments and improvements in these fields. The applications of machine learning algorithms in cyber security have been discussed in detail in the paper. This paper focuses on the critical and the technical aspects of the previous work carried out by other researchers in these fields culminating with a comprehensive conclusion about the state-of-art in the fields. It also discusses as to what should be done to further improvise the situation currently faced by data scientists and cyber security researchers around the world. The current paper also discusses a conceptual framework for a typical university setup to tackle the Cyber Security issues and the implementation work on the proposed solution strategies is in progress and the authors are working on it.

Keywords: Machine learning · Cyber security · Social networks ·
Cloud computing

© Springer Nature Switzerland AG 2020
A. E. Hassanien et al. (Eds.): AMLTA 2019, AISC 921, pp. 599–608, 2020.
https://doi.org/10.1007/978-3-030-14118-9_60

1   Introduction

Machine learning is the branch of computer science aimed at enabling computers to learn new behaviours based on training set (a subset of data is provided to algorithm). The goal is to design algorithms that allow a computer to display behaviours learned from past experiences, rather than human interactions.

Machine Learning Applications can be found in almost all the possible domains and are being used for gaining new insights which cannot be visualised otherwise. Image recognition, speech recognition, medical diagnosis, statistical arbitrage, learning associations, classification, prediction, extraction and regression are some of the key areas where ML algorithms & techniques are being applied extensively. Thus they are helping in improving our lives with the use of technology.

Companies like Apple, Microsoft Palo Alto networks etc. are also taking advantages of machine learning algorithms and investing millions of dollars in their development. New Generation companies and start-ups in various fields are quickly adopting to machine learning algorithms to protect their data. Specifically, AI encompasses any case where a machine is designed to complete tasks which, if done by a human, would require intelligence.

Data has posed perhaps the single greatest challenge in cyber security over the past decade. For a human, or even a large team of humans, the amount of data produced daily on a global scale is unimaginable. So, in order to protect the world against cyber crimes there is a need to develop the cyber security industry through using machine learning and artificial techniques.

2   Related Literature

Some of the key reviews carried out as part of our study are discussed in the following sections:

2.1   Applications of Artificial Intelligence (AI) to Network Security

Originally written by Alberto Perez Veiga in University of Maryland, University College in March 2018 [1]. This paper very beautifully starts with the basics of machine learning. What it actually is and how it can be deployed in the industries so as to defend the networks. Here, the difference between artificial intelligence and machine learning is also explained. The two main parts of machine learning that are supervised and unsupervised machine learning are explained here in detail with proper algorithmic and functional proofs. The author tries to explain that there is a problem in developing algorithms for Intrusion Detection Systems (IDS) and Intrusion Prevention System (IPS) due to lack of sample data. Despite this, supervised ML has delivered the best results for Cyber Security in recent experiments conducted. The author talks about Self Organizing Maps (SOMs) in unsupervised Learning part, it's basically a neural network in which a neuron is connected to its neighbour and represents a point in multidimensional space. SOMs together with other clustering techniques will help in identifying malicious IP traffic. This paper finishes with perfection and as such no error can be detected. Future prospects of machine learning applications in cyber security were discussed in the paper.

2.2   Applications of Machine Learning in Cyber Security

Written by Vitaly Ford and Ambareen Siraj, Computer Science Department, Tennessee Tech University, published in Dec 2015 [2]. In this paper the authors have tried to discuss all the possible applications across all the domains of cyber security. They have very intelligently covered domains such as phishing detection, network intrusion detection, authentication with keystroke dynamics etc. The paper was remarkably written with actual mathematical proofs and the information provided was very crisp and scientific in nature. The authors tells us about how they have derived a way to classify phishing sites using a feature selection algorithm for extracting various phishing email traits by means of: Hierarchical Clustering (HC) Algorithm that adopted cosine similarity (using the TF-IDF metric) for measuring the similarity between two points, and K-Medoids (KM) Clustering approach. The proposed methods for phishing website and malware categorization have about 85% performance. The author also discusses here various algorithms for Human Interaction Proofs (CAPTCHAs) as well and talks about Cryptography too. Also the algorithm (Probabilistic Neural Net-work) discussed in this paper for authentication with keystroke dynamics is a bit old, there have been better and improvised versions of that algorithm (e.g. Recurrent Neural Network) which is discussed in [4].

2.3   Characterization of TOR Traffic using Time based Features

Written by Arash Habibi Lashkari, Gerard Draper Gil, Mohammad Saiful Islam Mamun and Ali A. Ghorbani in Canadian Institute for Cybersecurity (CIC), University of New Brunswick (UNB) [3]. This paper was published in Sep 2017. This paper focuses on the encryption aspect of cyber security. Encryption has various protocols or methods or algorithms. Here TOR is discussed. Current approaches used for TOR traffic detection depends on blocking known entry nodes of the TOR network. This is not a scalable approach and can be easily bypassed. So, the approach used in this paper which is time based featuring using deep learning has proven to be effective. Over the past few years there have been some significant advances in characterizing the TOR traffic. This paper is successful in its attempt to explain the concept it has picked.

2.4   Application of Recurrent Neural Networks for User Verification based on Key-Stroke Dynamics

Written by Pawel Kobojek and Khalid Saeed, Faculty of Mathematics and Information Sciences, Warsaw University of Technology. This paper was published in

Jan 2016 [4]. This paper mainly focuses on verification using keystroke dynamics which is basically a biometric authentication. This algorithm uses Recurrent Neural Network (LSTM and GRU). The main goals of the authors were high accuracy without false positive errors and high scalability in terms of user count. High accuracy without false positive errors as well as high scalability in terms of user count. Some attempts were made to mitigate natural problems of the algorithm (e.g. generating artificial data). The author talks about false positive error minimization at length. The scale of explanation and information provided in this paper is just at par and is so remarkably understandable that a novice can also get a gist of the topic. The authors conduct the experiment using their own data set and produced some productive results, but due to small size of the data set used, a non-standard evaluation has been carried out but the information and mathematics that is already supplied here is adequate and apt. This research may prove to be of importance in near future if some better and large data sets are used for computation.

2.5   Convolutional Neural Networks for Malware Classification

Written by Daniel Gibert of Universitat de Barcelona (UB). In the recent past malwares have become so stable and smart that the traditional techniques of identifying them using signatures is obsolete and ineffective [5]. In this paper CNN (Convolutional Neural Network) is used to detect the malware and stop it. Here CNN is used to learn a feature hierarchy to discriminate among samples of malware represented as grayscale images. The paper is written in a systematic fashion with an eye for details and errors. Overall the paper was very good but then there was no practical explanation of the topic or any given result from any kind of research conducted.

2.6   A Hybrid Spectral Clustering and Deep Neural Network Ensemble Algorithm for Intrusion Detection in Sensor Networks

Written by Tao Ma, Fen Wang, Jianjun Cheng, Yang Yu and Xiaoyun Chen from Lan-zhou University and Published in October, 2016 [6]. The main focus of this paper is IDS (Intrusion Detection Systems) that are adapted to allow routers and network defence systems to detect and report malicious network traffic signal. This paper presents an innovative approach known as SCDNN, which combines spectral clustering (SC) and deep neural network (DNN) algorithms. This SCDNN approach for IDS is explained in great detail with usage of proper data sets and obtaining relevant experimental results. The authors use Six KDD-Cup99 and NSL-KDD datasets and a sensor network dataset to test the performance of the model. Here, the authors successfully prove the fact that SCDNN classifier performs better than traditional backpropagation neural network (BPNN), support vector machine (SVM), random forest (RF) and Bayes tree models in detection accuracy of types of abnormal attacks found and also provides an effective tool of study and analysis of intrusion detection in large systems. The working of the whole SCDNN model is explained in great detail with a lot of experiments and proofs. Also to make the SCDNN network more efficient, this model adapts a SAE and denoising auto-encoders (DAEs) for the DNN so as to overcome the limitations of traditional auto-encoders. At each step of the research, proper algorithms are given for the reference of the reader. Therefore, the research conducted in this paper is very original and it can be said with great confidence that this model can prove to be very beneficial for future smart ML based IDS systems.

2.7 An Analysis of Recurrent Neural Networks for Botnet Detection Behavior

Written by Pablo Torres, Carlos Catania, Sebastian Garcia and Carlos Garcia Garino, published in June 2016 [7]. A botnet is basically a group of compromised or hacked computers which can be controlled remotely by hackers to execute coordinated attacks or to commit fraudulent acts. The fact that botnets keep continuously evolving has made traditional detection approaches useless. So as to overcome the issue there has been research on behaviour analysis of network traffic. The behaviour analysis approach aims to look at the common patterns that botnets follow across their life cycle. With the current research that has been going on, RNN has proved to be effective for large sequences which makes it a viable candidate on the task of behaviour analysis. This paper provides an insight into how Recurrent Neural Network (RNN) can be used to detect behaviour of network traffic by modelling it as a sequence of states that change over time. In particular, an analysis of the application of Large Short Term Memory (LSTM) network for recognizing the different sequences of states that change over time. The idea proposed in this work is to use LSTM network for building detection models based on the behaviour of connections using classical supervised machine learning then train the LSTM and to finally obtain a detection model capable of recognizing connections behaviours. The authors used datasets from Czech Technical University in Prague (CTU) network from Malware Capture Facility Project (MCFP). The research conducted revealed that RNN is capable of classifying the traffic with high values of ADR (Attack Detection Rate) and very small values of FAR (False Alarm Rate) making RNN a very potential candidate for future implementation and deployment in real-world scenarios. The research con-fronted the traditional problems in detection of botnet behaviour and helped mitigate them but still more experiments have to be conducted in the future so as to make the LSTM network more trained, adaptive and effective which will help it in differentiating traffic behaviours.

2.8 A Novel LSTM-RNN decoding algorithm in CAPTCHA recognition

Written by Chen Rui, Yang Jing, Hu Rong-gui, Huang Shu Guang of Institute of Hefei, China [8]. This paper was published in September 2013. Completely Automated Public Turning Test to tell Computers and Humans apart (CAPTCHA) provides a way for automatically distinguishing a human from a computer program. Currently, the research on CAPTCHA is focused on the design and recognition technology. This paper mainly discusses the recognition technology of text-based CAPTCHA. Traditionally researchers mainly use SVM, Neural Network, HMM etc for CAPTCHA pattern recognition but none of them could produce satisfactory results because they are all based on segmentation, however once a character in the CAPTCHA touching or merged are encountered, the segmentation becomes difficult. Therefore, a segmentation-free strategy based on two dimensional Long-Short Term Memory Recurrent neural network (2DRNN) is developed which uses merged-type CAPTCHA recognition without segmentation in this paper. Basically 2DRNN can learn not only horizontal text, but also the surrounding context automatically, thus making 2DRNN more adaptable to CAPTCHA recognition. The authors conducted several experiments to evaluate the performance of CAPTCHA recognition method based on 2DRNN and the new decoding algorithm based on GA (Genetic Algorithm) using custom generated CAPTCHA values for data set as there are no public data sets available. The experiments revealed some fascinating results, 2DRNN method when applied to CAPTCHA recognition can learn vertical context automatically, which is more suitable and the proposed decoding algorithm based on GA is better than DA-NS (Decoding Algorithm based on neighbourhood search) in improving the recognition rate further. It can be inferred that though the research in this paper produced some useful results but further research in this field will accelerate the development of CAPTCHA security technology. This field needs some more research to improve the technology further so as to improve the reliability of CAPTCHA recognition.

2.9   Malware Detection with Deep Neural Network Using Process Behaviour

Written by Shun Tobiyama, Yukiko Yamaguchi, Hajime Shimada, Tomonori Ikuse and Takeshi Yagi of Nagoya University. This paper was published in June 2016 [9]. Increase of malware and cyber-attacks are now becoming a serious problem. Every time a better version of malware is seen creating a mess for computer and web applications of many people and companies. Traditional methods of determining malware using signatures and by analysing traffic data are becoming obsolete and ineffective as these malwares are becoming more and more adaptive. So, in order to face the issue, this paper discusses a technique of malware identification through the application of DNN (Deep Neural Network) in which the RNN (Recurrent Neural Network) is first trained to ex-tract features of process behaviour and then CNN (Convolution Neural Network) is trained to classify feature images which are generated by extracted features from trained RNN. The authors used custom trained data sets as they trained the RNN by feeding process behaviour like API calls in the data set, the feature vector thus extracted are converted into an image and CNN is applied to classify by training local features thus creating a classifier. The experiment conducted by the authors produced very good results, AUC (Area Under the Curve) = 0.96 in best case.

Even though the authors conducted some serious and innovative research work here by using two stage DNN and by producing some very good experimental results but the data sets used here were pretty small and thus the work in large data sets is still something that is to be done in future (Table 1).

Table 1. Comparative summarization of key work done in the field of ML & AI in Cyber Security

Name of paper	Authors	Date of publish	Cyber security aspect touched	ML concept used	Data set used
Applications of Artificial Intelligence (AI) to Network Security	Alberto Perez Veiga	March, 2018	Network Security	RNN, SOMs, HMM, Clustering, ARL	MIT Dataset
Applications of Machine Learning in Cyber Security	Vitaly Ford and Ambareen Siraj	December, 2015	Phishing, Cryptography, CAPTCHA, IDS, Spam Detection	SVM, PNN, LR, CART, BART, RF	KDDcup99, Custom Dataset, Phishing emails
Characterization of Tor Traffic using Time based Features	Arash Habibi Lashkari, Gerard Draper Gil, Mohammad Saiful Islam Mamun and Ali A. Ghorbani	September, 2017	TOR encryption	(SE+BF), (IG+RK), KNN, C4.5, Random Forest	Custom Generated TOR Traffic Dataset
Application of Recurrent Neural Networks for User Verification based on Keystroke Dynamics	Paweł Kobojek and Khalid Saeed	January, 2016	Biometric Verification	RNN	Custom Generated Dataset
Convolutional Neural Networks for Malware Classification	Daniel Gibert	February, 2018	Malware analysis and Prediction	CNN	Malfease, VXheavens
A Hybrid Spectral Clustering and Deep Neural Network Ensemble Algorithm for Intrusion Detection in Sensor Networks	Tao Ma, Fen Wang, Jianjun Cheng, Yang Yu and Xiaoyun Chen	October, 2016	IDS	SCDNN (SC+DNN)	KDD-CUP-99, NLS-KDD
An Analysis of Recurrent Neural Networks for Botnet Detection Behavior	Pablo Torres, Carlos Catania, Sebastian Garcia and Carlos Garcia Garino	June, 2016	Botnet Behavioural Analysis	RNN, LSTM	MCFP IDs 1. CTU13-42 2. CTU13-47
A Novel LSTM-RNN decoding algorithm in CAPTCHA recognition	Chen Rui, Yang Jing, Hu Rong-gui, Huang Shu Guang	September, 2013	CAPTCHA	2DRNN	Custom Generated Experimental Data due to lack of availability of public Datasets
Malware Detection with Deep Neural Network Using Process Behaviour	Shun Tobiyama, Yukiko Yamaguchi, Hajime Shimada, Tomonori Ikuse and Takeshi Yagi	June, 2016	Malware analysis and Prediction	DNN	Custom Trained Datasets

3   A Proposed Conceptual Framework for a Typical University Setup

Upon examining in detail the different works that is been going on in the field of machine learning for the development of existing cyber security tools and services, we are proposing two security system solution strategies for our University so that the breakthrough research in the field of machine can also be applied for the benefits of our network infrastructure. The first implementation shown in Fig. 1 is based on De-militarised zone (DMZ) and the second implementation is a Cloud based Security Infrastructure (CBSI).

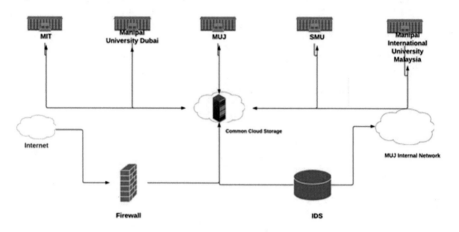

Fig. 1. Proposed machine learning based security system for Manipal University Jaipur

Figure 1 is the proposed system, where the IDS (Intrusion Detection System) and the firewalls will be secured using machine learning algorithms, making them smart enough to automatically learn and mitigate emerging threats to our network infrastructure. Also securing the DMZ with the help of machine learning algorithms can be thought of and planned. The authors are working to realize the model as discussed above.

The Fig. 2 is yet another proposition for cyber security system for our university. In this system, we are basically using cloud based solution for our security system as cloud based systems are more secure and robust, also cloud infrastructure is protected by some strict rules and regulations which are to be followed mandatorily by all the companies. Also, we get the added advantage, we can get security reports and information from all the other institutions which comes under Manipal Group. It can further help us train our machine learning algorithms with more sound and advanced datasets for each kind of cyber-attack.

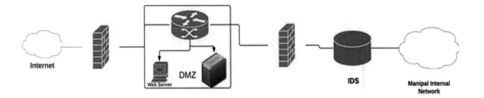

Fig. 2. Another proposition for the ML based Cyber Security systems

4   Conclusion

The IT market has a tendency to quickly assimilate buzz words pushed by marketing departments. During the last years, technologies such as Big Data, Cloud Computing, Artificial Intelligence, machine learning, etc., have been repeated again and again in multiple forums and in many cases without a clear understanding of their significance or their application to solving real problems effectively. It is a known fact that when humans do not completely understand a technology, two kinds of effects usually occur: either the technology is irrationally rejected (e.g. new operating systems) or, if it is properly marketed, it is assumed to be the silver bullet capable of solving every problem (Artificial Intelligence). It usually takes some time, even years, for the dust to settle down and for the market to realize the true potential of it. The eruption of ML in Cyber Security is forcing a paradigm shift from proactive rule-based prevention, to reactive real-time detection. Security threats have become so varied, different and smart, that traditional techniques, based on rules inferred from known attacks, that stop the attack before it happens, do not seem to be a viable approach anymore. Many attacks escape these mechanisms and cause tremendous damage that can not be stopped once it has started. ML aims at identifying attacks in real-time, with little to no human interaction and stopping them before they provoke serious harm. We can also conclude that Artificial Intelligence is not being used nowadays, as it is expected to be, to solve Network or Cyber Security problems in general. For now, only Machine Learning, a branch of AI, is being successfully applied to solve a small part of the problems. There is still a lot of development that has to happen in the future so as to make artificial intelligence and machine learning fully capable to deal with the threats without any human interference. For now it can be fairly assumed that machine learning is being deployed and used by organizations around the world but it is actual true potential has still not been exploited as we still lack efficient algorithms to make that happen. Machine learning clearly is an innovative and creative field with a lot of research work already going on and we sure do hope for some magical or significant developments in the field that are going to help us solve problems that we currently are not able to tackle and solve. Machine learning and AI are somewhat new fields and as any other field they still need time and research to be able to solve real world challenges efficiently.

References

1. Veiga, A.P.: Applications of artificial intelligence to network security. arXiv preprint arXiv:1803.09992 (2018)
2. Ford, V., Siraj, A.: Applications of machine learning in cyber security. In: Proceedings of the 27th International Conference on Computer Applications in Industry and Engineering (2014)
3. Lashkari, A.H., Draper-Gil, G., Mamun, M.S.I., Ghorbani, A.A.: Characterization of TOR traffic using time based features. In: ICISSP, pp. 253–262 (2017)
4. Kobojek, P., Saeed, K.: Application of recurrent neural networks for user verification based on keystroke dynamics. J. Telecommun. Inf. Technol. **3**, 80–90 (2016)
5. Gibert, D.: Convolutional neural networks for malware classification. Ph.D. thesis, MS Thesis, Department of Computer Science, UPC (2016)
6. Ma, T., Wang, F., Cheng, J., Yu, Y., Chen, X.: A hybrid spectral clustering and deep neural network ensemble algorithm for intrusion detection in sensor networks. Sensors **16**(10), 1701 (2016)
7. Torres, P., Catania, C., Garcia, S., Garino, C.G.: An analysis of recurrent neural networks for botnet detection behavior. In: 2016 IEEE Biennial Congress of Argentina (ARGENCON), pp. 1–6. IEEE (2016)
8. Rui, C., Jing, Y., Rong-gui, H., Shu-guang, H.: A novel LSTM-RNN decoding algorithm in captcha recognition. In: 2013 Third International Conference on Instrumentation, Measurement, Computer, Communication and Control (IMCCC), pp. 766–771. IEEE (2013)
9. Tobiyama, S., Yamaguchi, Y., Shimada, H., Ikuse, T., Yagi, T.: Malware detection with deep neural network using process behavior. In: 2016 IEEE 40th Annual Computer Software and Applications Conference (COMPSAC), vol. 2, pp. 577–582. IEEE (2016)

Applying Cryptographic Techniques for Securing the Client Data Signals on the Egyptian Optical Transport Network

Kamel H. Rahoma[1(✉)] and Ayman A. Elsayed[2]

[1] Electrical Engineering Department, Faculty of Engineering,
Minia University, Minia, Egypt
kamel_rahouma@yahoo.com
[2] Telecom Egypt, Cairo, Egypt
aymanelmelm@gmail.com

Abstract. The physical layer of the optical transport network (OTN) is the weakest layer in the network, as anyone can access the optical cables from any unauthorized location of the network and stat his attack by using any type of the vulnerabilities. The paper discusses the security threats and the practical challenges in the Egyptian optical network and presents a new technique to protect the client's data on the physical layer. A new security layer is added to the OTN frames for the first time since the network infrastructure has been installed. The design of the proposed security layer is done by using a structure of XOR, a linear feedback shift register (LFSR), and random number generator in a synchronous model. We propose the security model for different rates in the OTN and wavelength division multiplexing (WDM) system. The proposed model is implemented on the basis of protecting the important client signals only over the optical layers by passing these signals into extra layer called security layer, and before forming the final frame of the OTN system, this done by adding a new card in the Network Element (NE) to perform this job. The results show that the proposed model of the OTN encryption scheme is providing a high security against any wiretapping attack. If the attacker has the ability to access the fiber cables from any unauthorized location, he will find encrypted data and many years will be needed to find one right key to perform the decryption process.

Keywords: Linear feedback shift registers (LFSR) ·
Random number generators · XOR · Optical transport network (OTN) ·
Wiretapping · Client data signals

1 Introduction

The huge use of the mobile applications nowadays forced the telecom operators to depend on the core optical network with high capacities to transfer huge amount of data from one location to another. The Optical Transport Network (OTN) Technology was originally designed to transfer different types of clients' signals at different capacity rates through the optical transmission network. This takes place when using the optical wavelengths associated with wavelength-division multiplexing (WDM) system.

© Springer Nature Switzerland AG 2020
A. E. Hassanien et al. (Eds.): AMLTA 2019, AISC 921, pp. 609–622, 2020.
https://doi.org/10.1007/978-3-030-14118-9_61

The OTN is defined as a set of Network Elements connected by optical links. It has the ability of multiplexing, switching, transporting, managing, and supervising the optical channels carrying client signals through the core optical network. Many mobile operators use standard encryption algorithms to secure the client signals at the application layers only. These signals travel through different layers in the backbone core transmission networks which could be owned by another operator as the case in Egypt. Similarly, optical transmission network operators rely mostly on security algorithms of the client signals at the applications layers only to secure the clients data during its travel journey in the optical networks without making any extra effort. However; the problem with the majority of the optical transmission networks is that any attacker can access the physical layer or the optical fibers of the existing optical transmission network and split the optical signal by wiretapping the signals from the optical cables, and taking a live copy of the optical signals, then, by trying different types of reverse engineering for this live copy of data and de-mapping this data to the standard structure of the OTN frame or any other alternatives systems such as the synchronize digital hierarchy (SDH), he can reach the original layer of the client signals. On the other hand, if the attacker tries different types of decryption algorithms in the service layer, the probability that he understands the contents of the client data will be very high, and he will be 100% able to break the security system in the optical transmission network at the very first try for the whole network [1].

Despite the fact that the core transmission network in Egypt is extended all over its area through huge amounts of optical cables, the necessity to secure the important clients data as it travels through this network remains crucial especially those optical cables which pass through long routes in deserts and unpopulated areas (The operators of these networks have no ability to make any physical protection for the majority of optical network locations), and are vulnerable to being split by attackers at any unsecured location with the aim of copying live optical signals from any transmission network route by wiretapping it [2]. At the same time, how to secure customer data while it travels through a wide array of kilometers in the optical transmission network until it reaches its destination every time is the difficult matters in the most of any transmission network. Most of the available vendors and the owners of the optical transmission networks in many countries do not pay any attention to securing clients' data over their optical transmission networks. Nevertheless, most of previous studies conducted to solve this problem discuss how to use the XOR and optical LFSR to secure optical data despite the difficulties one may face distributing the encryption keys between different NE's in the optical network [3, 4]. Other studies discuss new algorithms for optical data encryption that uses quantum noise inherent in laser light and modulations algorithms that use two different cycles of the M-ary phase-shift-keyed (PSK) signal, a technique which allows the receiver who owns the short secret key to transform M-ary signals to standard modulated signals. The attacker, who does not own the secret-key, will be forced to try different M-ary measurements for several times, which is very difficult [2, 5]. Despite the huge number of studies proposed many years ago on optical security, the optical transmission networks are still the weakest loop in communication networks from security perspective overview.

In this paper, we study a new proposed model for securing transmission optical network. This model provides new service related to securing clients' signals over the whole optical network to the customers by adding a new security layer through an extra separate card in the NE's that makes the required functions of the encryption algorithms for the selected clients' signals as per customer choices. On adding this new layer of the security for the different client frames in the OTN structure, the wavelengths over optical cables will be more secured from any unauthorized access and wiretaps hacking. On the other hand, implementing security algorithms in a separate stage in the OTN system for certain client signals only as an option will reduce the complexity of transmission systems, especially in the optical networks with huge capacity and at the same time the number of the faults which will arise from using these security algorithms in the transmission system will be fewer, in addition to the troubleshooting of the network routes faults will be a lot easier when separating the security layer in standalone cards. All encryption algorithms in the proposed security layer will be built with the usage of XOR gates and LFSR considering that the client signal is 100 Gb/s or 10 Gb/s and the wavelength rates of the DWDM transmission systems are the same as that of the client signal.

The paper is structured as following: Sect. 2 presents the current structure of OTN frames in the optical transmission network Sect. 3 presents the proposed security model in the OTN system Sect. 4 presents proposed model analysis, Sect. 5 presents the numerical analysis and the results of the proposed model, and finally Sect. 6 presents the conclusion of the paper.

2 The Structure of OTN Frames in the Optical Network

2.1 The Construction of the OTN Frame

The different mapping stages of the client signals inside the NE's of the transmission network consist of:

Stage1. The transponder card receives the client signals with rates 10/100 Gb/s and transforms it to electrical domain by using small form-factor pluggable (SFP) module with a suitable laser frequency for the client signal. The next step in this stage is the mapping of the client signals to be part of the OTN frames structure which is done by putting the client data in containers with fixed size as per client data rates. These containers are used to form the optical payload units (OPU_k) (where k represents the capacity of the OTN frame with k = 2 represents a frame with data rates equal to 10 Gb/s and k = 4, OTN frame with data rates equal to 100 Gb/s) by adding its overheads, the next step is to form the optical data unit ODU_k. Finally, on adding the final overheads and alignments words with forward errors corrections (FEC), the optical transport unit (OTU_k) will be formed (see Fig. 1), and by converting the OTU_k to the optical domain it will ready it be multiplexed as a wavelength in the WDM transmission system [6].

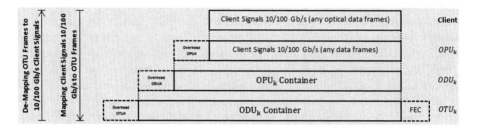

Fig. 1. The construction of the OTN frame

The final frame of the OTU_k in the OTN system is shaped in 4 different rows of 4080 bytes and repeated every certain period (12.191 usec for OTU_2 frame and 1.167 usec for OTU_4 frame). This frame consists of bytes that stand for frame alignment word (FA), different layered overheads bytes (OH), payload data and finally bytes that represent forward error correction algorithms. After that, this frame is converted to standard optical signals with standard wavelengths (see Fig. 2) [4, 7–13].

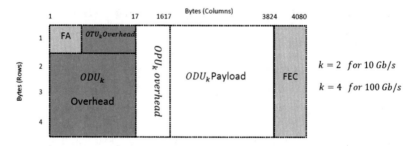

Fig. 2. The structure of the OTU_k frame

Stage2. Giving that the network consists of two network elements with two directions in the optical network, the 2'nd step in the optical transmission system processes multiplexes many standard optical signals to form one beam of laser that consists of many wavelengths and different standard frequencies in the c band.

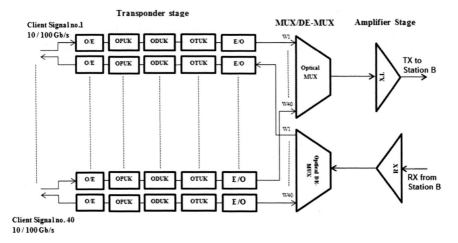

Fig. 3. The transmission model of two network elements

Stage3. The last stage in the transmission network inside the NE's is the amplification process that uses any type of the optical amplifiers such erbium doped fiber amplifiers (EDFAs) with the suitable gains according to different factors on the network such as the travelling distance between the two network elements and the attenuations on the available fiber cable.

The transmission model in this paper consists of two network elements connected by a fiber cable with 40 wavelengths (W1 to W40) in the DWDM system and client signals with different bit (see Fig. 3).

2.2 The Security Problem in the Current Transmission Model

The problem in the current model from security perspective can be explained as follows:

Assume that Alice, in location A wants to send a message to Bob, in another location B Where A&B are connected through the pervious optical transmission network. Then, the attacker Eve in location C can split the laser signals of the fiber cable from any point of the transmission network into two paths by using optical splitters where the first path of the laser signals will be returned back to the original route of the network to transfer original message of Alice to Bob and the 2'nd will be used by Eve's equipment where Eve can have a live copy of the OTU frames without affecting the original routes of the network and in case Eve's hacking system made a little effort in the de-mapping process according to the standards of the OTN structures, Eve will be able to break the optical network structure and will know the contents of the messages between Alice and Bob while they are not realizing that, all their communications messages are known by third person who is Eve.

As shown in the previous optical transmission network, there aren't any security algorithms that are implemented to prevent the interested attacker such as Eve from understanding the contents of the optical signals even though he can access the fiber cable and keep a live copy of the client signal.

3   The Proposed System Model

The proposed model to secure the client signals over the OTN system in the DWDM transmission network from wiretapping is to add a new security layer in the mapping process of the client signal before forming the final frame of the OTN system, and it is suggested that this security layer to be implemented after forming the ODU_k bytes of the frame and before the final OTU_k stage (see Fig. 4).

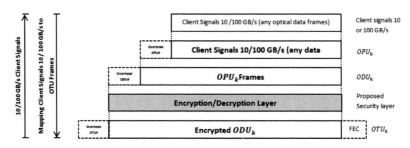

Fig. 4. The proposed security layer in the OTN frame structure

To achieve this security layer, there are two types of the pseudo random generators (RNG) that could be used to generate the keys of the encryption/decryption processes. The 1'st one synchronizes RNG's while the 2'nd is a non-synchronies RNG's, and provided that both types of the RNG's systems are working with the same linear feedback shift register (LFSR) and XOR operations to perform the encryption/decryption processes and the only difference is that the first type is the RNG in the source and destination stations are synchronized with the same clock and it generates the same keys in both stations as result of using the same clock for synchronization in both stations, Where as in the 2'nd type every RNG in the source and destination stations is working separately in every station with different keys generation.

The proposed security layer will be performed according to the traditional techniques of the encryption/decryption processes as the following concept: If we considered the plaintext of the data as m, the output sequence of the security layer as y and by using secret key for the encryption operations as k, then the processes of securing the plaintext m at the source station will be according to $y_i = m_i \oplus k_i$ where i is the bit number in the plaintext frame and $\oplus$ is the XOR operation, to retrieve the original plaintext at the destination station, the decryption process will be implemented using the same secret key according to $m_i = y_i \oplus k_i$ [4, 8].

The transmission over the optical networks takes place by transmitting the bits within periodic frames with different frame rates (the synchronize digital hierarchy system (SDH) is 125 us, and the OTN is varying from 12.191 usec for OTU_2 frame to 1.167 usec for OTU_4 frame) which means that for every 1 s there are 8000 frames in the SDH system, 82027 frames for OTU_2 and 598802 frames for OTU_4 will be transmitted from source to destination [6].

The proposed model for the new security layer of the OTN frame is to use dual synchronize pseudo random number generator (SRNG) to generate the same secret key and the polynomial of the LFSR with the trigger of every clock, by considering both SRNG's as synchronized by the same distributed clock on the optical transmission network, then both sides will have the same generator polynomials P_n and the same seeds of initializing key words k_n for the LFSR's [9].

We assumes that the synchronize random number generation (SRNG) generates random polynomial with degree n and random initializing key word for the LFSR of length n bits with every clock pulse. The proposed model for the encryption system is shown in (see Fig. 5).

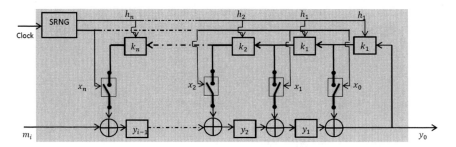

Fig. 5. The proposed model of the security layer

The proposed model consists of:

- Dual synchronized random number generator (SRNG) which generates random polynomial $P_n = f(x_n)$ with degree n, which used to enable or disable the switches in the proposed model to implement different possibilities of the random number generator and at same time to generate random key k_n with stream of bits h_1 to h_n as well as to initialize the LFSR with every reset cycle of the encryption period.
- Flip-flops to keep the encryption bits with every bit shift encryptions.
- XOR to implement the encryptions/decryptions algorithms.
- Plaintext which is equal the data of the ODU_k and equal to $M = N^* m$ where N is the number of the reset cycles and m is the length of the part of the plain text which will be encrypted/decrypted by the same initialized key and the generated polynomial.
- Encrypted plain text y_i.

As assumed, the synchronized Random number generator (SRNG) randomly selects one prime polynomial according to the following equation [10]:

$$P \leq \psi(n-1)/n \tag{1}$$

Where $\psi(.)$ is Euler function, n is the degree of the polynomial which is limited by the range of $n_{min} \leq n \leq n_{max}$, and n_{min} is equal to the key length which is used to encrypt data message m. Due to the high speed of the OTN system the length of the message of the ODU_k becomes very long and encrypting the current length with the same length as that of the encryption key becomes more difficult. Therefore, the original message M which is equal to ODU_k in every frame with length L_M will be divided to many equal parts of message with length $L_m = L_M/N$, where N is the number of the reset cycles to encrypt/decrypt the messages with length l_m as per the following condition [11]:

$$n_{min} \geq (\frac{L_M}{N}-1) \tag{2}$$

The uncertainty of the generated polynomial and initialized secret key from the SRNG is measured by the entropy of probability of true polynomial which may be detected according to the following equation [12]:

$$H_1 = \sum\nolimits_{i=0}^{n} P_i(\log 1/P_i) \tag{3}$$

Where P_i is the probability to guess one polynomial of degree n out of 2^{n-1} polynomials, and the entropy of the probability to guess the initialized secret key of the LFSR as per the following equation:

$$H_2 = \sum\nolimits_{i=0}^{n} k_i(\log 1/k_i) \tag{4}$$

Where k_i is the probability of detecting one secret keyword of length l_k out of 2^n keys. If we considered the generations of the polynomials and the generations of the initialized keys are independent, then the joint entropy of the polynomial and the secret keys of the whole system will be $H = H_1 + H_2$. Afterwards, the generated polynomial can be selected randomly with every clock from 2^{n-1} polynomials, which can be stored in buffers with the required size. The implementation of the proposed model in the OTN transmission system in our case is done between 2 NE's source and destination (see Fig. 6), where client signals are mapped into the OTN frames according to its bit rates, then, after the ODU_k level the data is encrypted according to the proposed security model of Fig. 5 and the encrypted data (Encrypted ODU_k) is returned back to fill the frame of the OTU_k in the OTN system, after that it will be multiplexed optically with other wavelengths, and the amplifier stage will be done over the optical fiber cables to reach the destination station. In the receive side, the same processes will be done but in reverse actions. After the preamplifier stage in the receive direction, the de-multiplexing stage will isolate the OTU_k and forward it to the de-mapped operations until it reaches the encrypted ODU_k, then, the decryption processes will start searching for the original ODU_k by using the same model of the encryption system, as shown in Fig. 5 whereas the dual SRNG polynomial with the same secret key in both sides of source and destination stations is used.

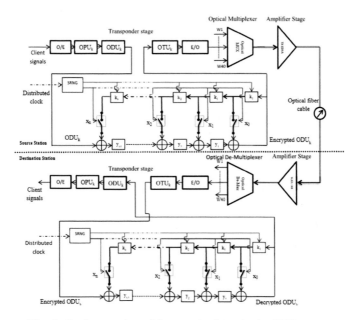

Fig. 6. Implementation of the security layer in the OTN system

4 The Analysis of the Proposed Security Model

With every clock, the dual synchronized Random number generator (RNG) will generate two independent outputs. The 1'st output is the generation of one Random polynomial with degree n out of 2^{n-1} polynomials to enable or disable the switches of the LFSR security model. The 2'nd output is the generation of one initialized secret key with length $l_k = n$ out of 2^n keys to initialize and rest the key of the LFSR model with every frame in the OTN transmission system. Afterwards, the reset cycle starts with the trigger of every clock in both sides of source and destinations and the system makes the encryptions with the same polynomial function and the produced keys from the combinations of the initialized key and the shifted data bits of the original data message with length $l_m = l_n$.

In case the attacker has the ability to manage and access the fiber cable by splitting the optical laser signals and take a live copy of the OTN frames to start the de-mapping process, he will only get the encrypted ODU_k instead of the original ODU_k which included in the OTN frame (see Fig. 7). Then, the attacker will need to be aware of 4 variables to understand the encrypted ODU_k to be able to retrieve the original ODU_k of the OTN frames. The 1'st variable is to know the algorithms of the dual SRNG with its synchronization clock. The 2'nd variable is to know the secret keyword to which used in the reset cycle with every trigger of the clock with probability $p_k = 1/2^n$ and entropy equal $\sum_{i=0}^{n} k_i(\log 1/k_i)$ as listed in Eq. (3) to initialize the original message. The 3'rd variable is to know the polynomial degree, with the probability $p_P = 1/2^{n-1}$ and entropy $\sum_{i=0}^{n-1} P_i(\log 1/P_i)$ as listed in Eq. (4) and finally the 4th variable is to know clock rate of the dual SRNG.

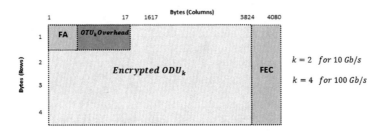

Fig. 7. The structure of the OTN 10/100 Gb/s frame with encrypted ODU

The time required to break the proposed system and to guess the right key to decrypt the data can be estimated by T^E where [11]:

$$T^E \geq N. \; F.\tau. \; 2^{n-1} \tag{5}$$

Where: τ is the time required for the attacker system to guess the right key for the decryption process, 2^{n-1} is the number of the right polynomials functions with degree n and the right number of initialized secret key with length $l_k = n$ with probability to find right polynomial function with degree n and right initialized secret key with length l_k is equal to $1/2^{n-1}$, F is the number of the initialized cycles in 1 s, and finally N is the number of the secret keys which is used in the same encryption cycle.

5 The Numerical Analysis of the Proposed Model

In order to be able to test the impact of the proposed model in protecting the client signals over the OTN system in the optical network, we will calculate the time required to break the security model for one hypothetical case which is considered as the clock rate that will be used to trigger the dual SRNG is equal to the same rate of the frames in the optical transmission system and estimated by the same rate of the frames in the SDH system as equal to 125 usec, 12.191 usec for OTU_2 and 1.167 usec for OTU_4 frame. Then the OTN system will transmit 82027 frames every 1 s for 10 Gb/s, and 598802 frames every 1 s for 100 Gb/s as a result, the dual SRNG will generate 82027 polynomials and initialized secret kays every 1 s in case of OTN system is 10 Gb/s and 598802 polynomials and initialized secret kays every 1 s in case of OTN system is 100 Gb/s as well. Table 1 gives the results from Eq. (1) 31 of the primitive polynomials of the SRNG for the polynomial degree with range $10 \leq n \leq 31$ (see Fig. 8).

Table 1. The number of the generated primitive polynomial for $10 \leq n \leq 31$

n	2^n	ψ (n-1)	ψ (n-1)/n	n	2^n	ψ (n-1)	ψ (n-1)/n
10	1024	600	60	21	2097152	1778112	84672
11	2048	1936	176	22	4194304	2640704	120032
12	4096	1728	144	23	8388608	8210080	356960
13	8192	8190	630	24	16777216	6635520	276480
14	16384	10584	756	25	33554432	32400000	1296000
15	32768	27000	1800	26	67108864	44717400	1719900
16	65536	32768	2048	27	134000000	113000000	4202496
17	131072	131070	7710	28	268000000	133000000	4741632
18	262144	139968	7776	29	537000000	534000000	18407808
19	524288	524286	27594	30	1070000000	535000000	17820000
20	1048576	480000	24000	31	2150000000	2150000000	69273666

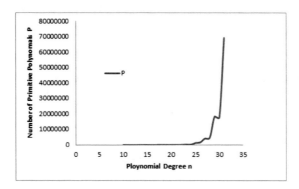

Fig. 8. The relation between the polynomial degree and generated primitive polynomials for $10 \leq n \leq 31$

The generated polynomials increased in huge rate with every little step increase in the n variable. From Table 1 for 10 Gb/s bit rate to achieve 82027 polynomials and initialized secret keys every 1 s, n is estimated to be equal to $n \geq 21$, and for 100 Gb/s bit rate to achieve 598802 polynomials and initialized secret keys every 1 s, n is estimated to be equal to $n \geq 29$, then for $l_m = l_M/N$ to measure the time required to break the proposed security system, we will consider n = 21 for 10 Gb/s and n = 29 for 100 Gb/s [13].

– For the client signals 10 Gb/s and n = 21 the length of the ODU$_2$ equal to $10.037 \frac{Gb}{s}$, then

$$N = 10.037 \text{ Gbits/s} /n = 4779 * 10^5 \quad (6)$$

- For client signals 100 Gb/s and n = 29 the length of the ODU_4 is equal to 104.7944 Gb/s, then

$$N = 104.7944 \text{ Gbits/s}/n = 3742 * 10^6 \tag{7}$$

From Eq. 6 the time required to break the security system and guess the correct key for the encryption process of the 10 Gbit/s client signals in the proposed model with n = 21 and for proposed $\tau = 10^{-12}$ s, F = 82027 $\frac{frames}{s}$ (the number of the OTN frames in 1 s) will be:

$$T^E \geq F \cdot N \cdot \tau \cdot 2^{n-1} = 82027 * 4779 * 10^5 * 10^{-12} * 2^{20} = 475.75 \text{ days} = 1.3 \text{ years} \tag{8}$$

From Eq. 7 the time required to break the security system and guess the correct key for the encryption process of the 100 Gbit/s client signals in the proposed model with n = 29 and for proposed $\tau = 10^{-12}$ s, F = 598802 $\frac{frames}{s}$ will be:

$$T^E \geq F \cdot N \cdot \tau \cdot 2^{n-1} = 598802 * 3742 * 10^6 * 10^{-12} * 2^{28} = 6961665 \text{ days} = 19073 \text{ years} \tag{9}$$

From Eqs. 8 and 9 we found that to guessing one of the keys that will be used in the encryption process of the ODU_k in the new proposed security layer in the OTN frames it will require 1.3 years for 10 Gb/s client signals and 24170 years for 100 Gb/s which is not possible in practical life to do that, and this indicates the our proposed system makes it very difficult to break the structure of the OTN frames by any interested hacker in normal conditions and understand the contents of the client data in the OTN frames.

The limitations of the proposed model are its very difficult to implement the proposed model over all the wavelengths of the OTN in the DWDM system as the transmission network will be very complex and the probability of the faults in routes of the services will be high, also the quality of synchronized clock which will be used in the RNG's of the proposed model should be very high such the Primary Reference Clock (PRC).

6   Conclusion and Future Work

The future work of this paper is to use the machine learning technology such the Intrusion Detection and Response System with the optical cryptographic system to drive the automation of the security system in the optical network, and this will be done by using neural network to monitor and trace any changes in the values of the Optical Signal to Noise Ratio (OSNR) over all the sections of the optical network and enables automatic optical cryptographic system in the optical network section which may has any strange change in its OSNR value and is likely to be wiretapping attack. By using machine learning technology in the optical cryptographic system the detection of the attack and the response will be done automatically in the affected sections only and this technique will provide robust network security system and at the same time will reduce the complexity and the cost of using cryptographic system over all optical network without usefulness in the sections which have no attacks.

In this paper, we addressed, for the first time, one of the practical challenges that face the security of the optical physical layer associated with the OTN transmission network and this was done by adding new layer for the security system in the stages of mapping client signals into the OTN frame. The results were very good where in case of the attacker has the ability to access the fiber cables from any unauthorized place he will only be able to copy the encrypted data signals without understanding what is the contents of this data as the result of using the encryption algorithm which was implemented before on this data by our proposed security model, also the attacker will need many years to find out one right key to perform the decryption processes. The proposed security model is based on the idea of protecting the important client signals only on the optical physical layers according to the requests of the customers. This system can be achieved by passing the selected client signals only through extra layer, called the security layer and before forming the final frame of the OTN system. This system can be executed on many sections in the transmission network of Egypt to protect the important client signals for certain people, also the system can be used in the military services over the public transmission network, and finally the system will be very useful in the near future especially with the needs to transfer a huge amount of the clients data through the optical network as the result of the tremendous progress in using the internet of things (IOT) technology and machine to machine communication.

References

1. Skorin-Kapov, N., et al.: Physical-layer security in evolving optical networks. IEEE Commun. Mag. **54**(8), 110–117 (2016)
2. Fok, M.P., et al.: Optical layer security in fiber-optic networks. IEEE Trans. Inf. Forensics Secur. **6**(3), 725–736 (2011)
3. Han, M., Kim, Y.: Unpredictable 16 bits LFSR-based true random number generator. In: SoC Design International Conference (ISOCC17((2017)
4. Liu, X.-B., et al.: A study on reconstruction of linear scrambler using dual words of channel encoder. IEEE Trans. Inf. Forensics Secur. **8**(3), 542–552 (2013)
5. Dimitriadou, E., Zoiros, K.E.: All-optical XOR gate using single quantum-dot SOA and optical filter. J. Lightwave Technol. **31**(23), 3813–3821 (2013)
6. Mobilen, E., Bernardo, R., Monte, L.R.: 100 Gbit/s optical transport network 40 nm test chip design and prototyping. In: Microwave and Optoelectronics Conference (IMOC), 2017 SBMO/IEEE MTT-S International. IEEE (2017)
7. Loprieno, G., Losio, G.: Timeslot encryption in an optical transport network. U.S. Patent No. 8,942,379, 27 January 2015
8. Liu, X.-B., et al.: Investigation on scrambler reconstruction with minimum a priori knowledge. In: Global Telecommunications Conference (GLOBECOM 2011), 2011 IEEE. IEEE (2011)

9. Kumar, M.C., Praveen Kumar, Y.G., Kurian, M.Z.: Design and implementation of logical scrambler architecture for OTN protocol. IJARCET **3**(4), 1260–1262 (2014)
10. Engelmann, A., Jukan, A.: Computationally Secure Optical Transmission Systems with Optical Encryption at Line Rate. arXiv preprint arXiv:1610.01315 (2016)
11. Engelmann, A., Jukan, A.: Computationally Secure Optical Transmission Systems with Optical Encryption at Line Rate. arXiv preprint arXiv:1610.01315 (2016)
12. Carter, T.: An Introduction to Information Theory and Entropy. Complex systems summer school, Santa Fe (2007)
13. Barlow, G.: A G. 709 Optical Transport Network Tutorial. Innocor Ltd. Capturado em (2003). http://www.innocor.com/pdf_files/g709_tutorial.pdf

A Blind Fragile Based Medical Image Authentication Using Schur Decomposition

Abdallah Soualmi[1], Adel Alti[1(✉)], Lamri Laouamer[2],
and Morad Benyoucef[3]

[1] Department of Computer Science, LRSD Laboratory University of Sétif-1,
P.O. Box 19000, Sétif, Algeria
{sabdallah,adel.alti}@univ-setif.dz
[2] Lab-STICC (UMR CNRS 6285), University of Bretagne Occidentale,
29238 Brest Cedex, France
laoamr@qu.edu.sa
[3] Telfer School of Management, University of Ottawa, Ottawa, Canada
benyoucef@telfer.uottawa.ca

Abstract. Image watermarking is an effective and powerful solution in multimedia security. Imperceptibility and less computational complexity are the most desirable properties for any watermarking approach. For that purpose, several fragile watermarking based methods have been proposed in the last decade. Many of those methods suffer from low ratios of imperceptibility. In order to authenticate medical images, this paper proposes a new blind fragile watermarking method using Schur decomposition. Indeed, perturbation due to watermark embedding is reduced using Schur decomposition, which can be considered in designing a watermarking approach to enhance imperceptibility and computational complexity. The main idea is to embed the watermark in the Schur decomposition coefficients of the host image using a new embedding technique. The experiments results show interesting values of imperceptibility with low computational complexity.

Keywords: Fragile watermarking · Authentication · Integrity ·
Blind watermarking · Schur decomposition · Medical image

1 Introduction

With the increasing adoption of new technologies in everyday life in general, and in medicine in particular, the exchange of huge amounts of patient and medical data by telemedicine applications through the Internet has become a great concern. Indeed, exchanging digital images through the Internet has become easier and faster. But ensuring the integrity and authenticity of patients' information remains a challenge. Many researchers have undertaken the task to develop new security mechanisms to protect digital patients' information by targeting special features of medical data.

Cryptography is the most common solution used in data security. It involves making data unreadable, hence useless, by an unauthorized party [1]. Cryptography protects data only during transmission, in such a way that once data is decrypted, we cannot protect its ownership rights or prevent its illegal reproduction [2]. One solution

© Springer Nature Switzerland AG 2020
A. E. Hassanien et al. (Eds.): AMLTA 2019, AISC 921, pp. 623–632, 2020.
https://doi.org/10.1007/978-3-030-14118-9_62

to this problem is to use watermarking techniques, which consist of embedding secret data (i.e., a watermark) into a cover image document in order to improve data security and prevent illegal reproduction of the cover image and preserving ownership [3]. A watermarking method can be *robust, semi-fragile* or *fragile* [6]. Briefly speaking, for the first method, the watermark must resist any intentional attacks or operations targeting the watermarked image. For the second method, the watermark must resist any minor alteration applied on the watermarked image, while for last method the watermark collapses if it undergoes any kind of alteration. These three watermarking methods have different applications and they can be employed according to the purpose of the watermarking system. For example, the robust technique is used for proof of ownership and copyright protection, while fragile and semi-fragile are employed for image authentication and data integrity checking.

Based on the watermark extracting techniques, digital watermarking approaches can be classified into three categories: *blind, semi-blind* or *non-blind* [7]. Blind watermarking approaches do not require the original image or the watermark. Semi-blind watermarking approaches require the original watermark, while for the non-blind watermarking approaches, the original image is required. Compared to other images, medical images watermarking presents an additional challenge since the watermark must be embedded in the medical image without affecting its quality. Indeed, the slightest distortion in the medical image could provoke the wrong physician's interpretation and consequently result in the wrong diagnosis. Accordingly, proposing a new blind watermarking method that provides a high level of imperceptibility is necessary to authenticate transmitted medical images across unsecured networks. This requirement can be met by reducing perturbations due to embedding a watermark into a digital image. This paper proposes a new blind fragile-based medical image watermarking using Schur decomposition [4, 5] with high imperceptibility and low computational complexity.

The rest of paper is organized as follows. Section 2 describes recent related work. Section 3 describes the proposed approach in detail. Section 4 presents and discusses the experimental results. The conclusion and perspectives are presented in Sect. 5.

2   Related Work

Several fragile image watermarking algorithms have been proposed in the literature, but each one has a number of associated advantages as well as drawbacks. This section discusses some of those algorithms [8–10, 13–15].

The authors in [9] present a blind watermarking method for image authentication of the holy Quran. The main idea is to use two layers of embedding to enhance the sensitivity of fragile watermarking: on wavelet and on spatial domains, the Discrete Wavelet Transform (DWT) is performed on the cover image to embed the watermark on DWT coefficients. These coefficients are inverted to spatial domain then the Least Significant Bits (LSB) is used to embed another watermark. The proposed method gives good results in terms of data embedding capacity, imperceptibility and capacity of tamper detection, as demonstrated by the experimental results. However, the algorithm requires an important computational complexity for embedding, extracting and tamper detection processes.

Lin et al. [10] presented a reversible watermarking method for image authentication. The main idea is to generate watermark and authentication code from the cover image, and then embed it into the two rebuilt components of each color pixel by using an authentication table. The proposed method ensures tamper detection performance while keeping good imperceptibility. However, it require an important time for embedding, extracting and tamper detection processing (generate the authentication code, generate/read the authentication table, etc.). Moreover, the embedding capacity is rather mediocre.

In [13], the authors present a new watermarking method for image authentication using Local Binary Pattern (LBP). The watermarking process is achieved in the spatial domain by embedding the watermark using two LSB. Then, to secure the watermark, the pixel information is used by the LBP to generate a data key. The proposed approach presents moderate imperceptibility and low computational complexity.

In [14], the authors propose a hybrid watermarking method for image authentication and tamper detection based on Singular Value Decomposition (SVD) and LSB. First, a watermark is generated from the cover image texture information, where the texture information is extracted from the singular value's matrix. Then the generated watermark is embedded in the LSB of the cover image. The proposed method has a high capacity for detecting and locating the altered area while keeping good imperceptibility. However, the embedding capacity is mediocre. Also, an attacker could easily modify the watermarked image, extract the texture information from the modified watermarked image, generate a new watermark, and embed it in LSB without the receiver noticing it.

In [15], the authors propose a new watermarking method based on both Slant Transform (ST) and Lower Upper Decomposition (LUD). Firstly, the watermark image is scrambled, and then the host image is separated into red, green, and blue components. The red channel is divided into 8 × 8 non-overlapping blocks. Secondly, the ST method is applied to convert each block in ST coefficients. Finally, LUD is applied to obtain the lower and upper triangular matrix from the obtained ST coefficients. The watermark data are embedded into the upper triangular matrix of ST coefficients. The proposed technique offers good imperceptibility but it requires high computational complexity.

From the abovementioned discussion we see that the majority of methods suffer from low imperceptibility and/or high computational complexity. In this paper, we propose a fast watermarking method for medical images that ensures watermark security without affecting the quality of the watermarked image.

3 Proposed Watermarking Technique

The proposed approach includes two main phases: embedding and extracting. The general structure of the proposed approach is illustrated in Fig. 1.

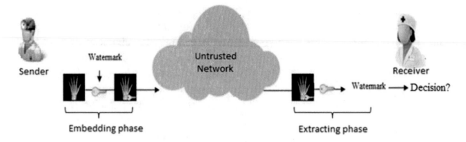

Fig. 1. Generic model for the proposed method.

3.1 Watermark Embedding Phase

The watermark embedding phase is illustrated in Fig. 2 and detailed below.

Step 1. Apply Schur decomposition [4, 5] on the cover image I:

$$\text{Schur (I)} = [U, V] \tag{1}$$

The reason behind choosing Schur decomposition for watermark embedding is that the perturbation which may result from data embedding in the host image could be reduced when data is embedded in Matrix V. This could improve the imperceptibility significantly [5].

Step 2. Decompose the V matrix into 2×2 non-overlapping blocks.

Step 3. Select a block's coefficients (BC_i) and compute its weight (Weight (BC_i)) using Eq. (2) where I is the block number.

$$\text{Weight } (BC_i) = \sum_{k=1}^{2} \sum_{m=1}^{2} \left(BCi(k,m) * (-1)^{(k+m+i)mod2} \right) \tag{2}$$

Step 4. Select a watermark Bit (WB_i) from the scrambled watermark bits after performing Arnold chaotic map [16] and embed it in BC_i using the following case:
If $WB \sim = \text{Weight } (BC_i) \bmod 2$

$$\text{Min}(BC_i(1,2), BC_i(2,1)) = \text{Min}(BC_i(1,2), BC_i(2,1)) + 1 \bmod S.$$

Where S is 2^8 for the JPEG format and 2^{16} for the DICOM format.

Note that adding 1 to minimum ($BC_i(1,2)$, $BC_i(2,1)$) modifies(increments or decrements) one pixel intensity from the 2×2 block pixel intensities (see Fig. 3), which could improve the imperceptibility.

Else:
Go to Step 5.

Step 5. If not all the watermark bits are embedded go back to step 3 with i + 1 Else go to step 6.

Step 6. Apply Schur^{-1} to obtain the watermarked image.

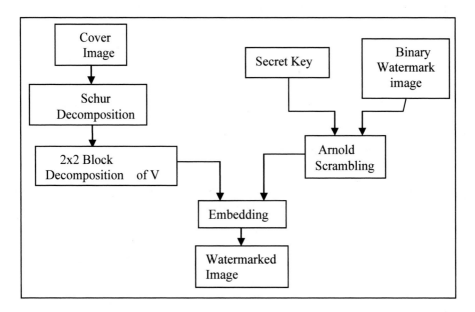

Fig. 2. Watermark embedding main steps.

Blocks Values		V matrix		Watermark Bits	Block Weights	New Blocks Values	
261	132	21	13	0	5	261	132
354	220	3	0			353	220
186	228	5	2	0	1	185	228
245	298	4	2			245	298
4080	4080	580	580	1	-2	4080	4079
4080	4082	580	578			4080	4080

Fig. 3. Example of new blocks pixel intensities after embedding of bits.

3.2   Watermark Extraction Phase

The watermark extraction phase is illustrated in Fig. 4 and detailed below.

Step 1. Apply Schur decomposition on the cover image using Eq. (1).

Step 2. Decompose the V matrix into 2×2 non-overlapping blocks.

Step 3. Select a block's coefficients (BC_i) and compute its weight using Eq. (2).

Step 4. Extract a watermark Bit (WB_i) using the following case:
If Weight (BC_i) mod 2 == 1

$$WB_i = 1$$

Else

$$WB_i = 0.$$

Step 5. If not all the watermark bits are extracted go back to Step 3 with i + 1 Else go to Step 6.

Step 6. Decrypt the extracted encrypted watermark bits, and compare them with the original watermark bits, then decide if the image is authorized.

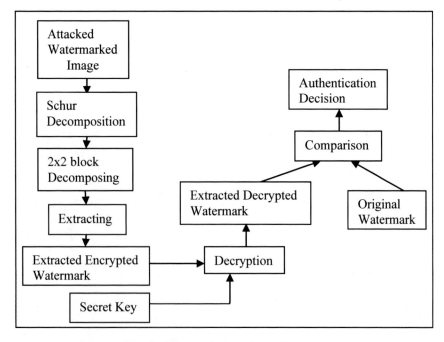

Fig. 4. Watermark extracting main steps.

4   Experimental Results and Discussion

In order to measure the performance of the proposed technique in terms of imperceptibility and computational complexity, a data set of DICOM images of size 256×256 are used (from [11, 12]), while the watermark image is of size 32×32.

Figure 5 shows the binary watermark image used in the experimentation, while Fig. 6 show a sample of medical images used in the experimentation.

Fig. 5. Watermark image used in the experimentation.

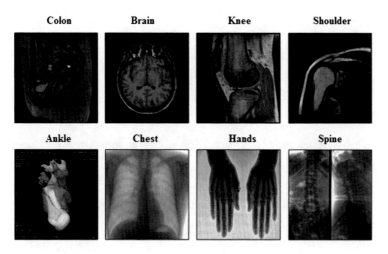

Fig. 6. Sample of image used in the experimentation.

Table 1. Imperceptibility evaluation of the proposed watermarking method.

Images	Colon	Knee	Hands	Spine	Brain	Shoulder	Ankle	Chest
PSNR(dB)	53.67	54.96	57.01	57.03	54.02	53.09	54.12	57.03

In order to estimate the imperceptibility degree of the proposed technique, we employ technical similarity measures used in the watermarking literature such as Mean Square Error (MSE) [13] and Peak Signal to Noise Ratio (PSNR) [13]. Table 1 shows the imperceptibility degree in terms of PSNR. The PSNR average value of all dataset images is around 54 dB.

Figure 7 show a comparison of the results obtained with the approaches described in [9–10, 13–15] and our proposed approach in terms of PSNR (dB).

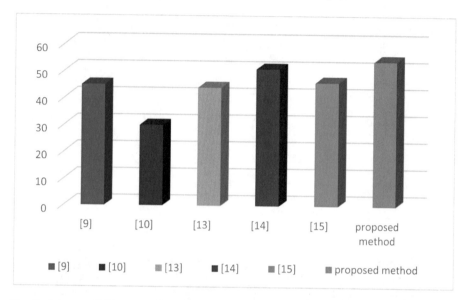

Fig. 7. Imperceptibility comparison between our proposed technique and methods described in [10–15].

Table 2 show the execution time needed to embed and extract the watermark in different medical images.

Table 2. Execution time measurement.

Images	Colon	Knee	Hands	Spine	Brain	Shoulder	Ankle	Chest
Embedding time (seconds)	2.99	2.99	3.48	3.22	3.04	2.96	3.08	3.49
Extracting time (seconds)	2.04	2.05	2.56	2.22	2.19	2.11	2.14	2.59

Clearly, our proposed method yields better imperceptibility. The reason is that the embedding of a watermark DCT coefficient necessitates a small change in the Schur DCT coefficients of the cover image blocks (Fig. 3). Consequently, the water-marked images keep a good imperceptibility compared to the original ones. This is proved by results described in Table 1 and Fig. 7.

It is also noticed that the computational complexity for the proposed method is acceptable (i.e., about 3 s for embedding and 2 s for extracting) [18]. Such results are obtained on a workstation with the following specifications: DELL LATITUDE E5410 Laptop/Intel core i5 2.67 GHz, 4 GB using MATLAB.

However, the major drawback of the proposed method is the mediocre embedding capacity (a watermark bit is embedded in a block of size 2 × 2, which means a binary watermark of size N × N necessitates at least a cover medical image of size 2N × 2N). This must be enhanced in future research.

5 Conclusion and Perspectives

Preserving the security and authenticity of medical data is becoming a fundamental and necessary requirement. In this paper, we presented a new blind fragile watermarking approach using Schur decomposition for medical image authentication. The proposed approach chooses Schur decomposition in order to reduce perturbation due to the embedding watermark process. Experimental results show that the proposed method provides good performance in terms of imperceptibility while preserving medical image quality. A comparison with some recent methods reveals that the proposed method is more imperceptible. However, the embedding capacity must be improved in future research efforts. Possible future enhancements to this work include proposing robust blind watermarking methods for real-time e-Health applications.

References

1. Al-shaikh, M.: Protection des contenue des images médicales par camouflage d'information secrète pour l'aide à la télémédecine, Ph.D thesis (2016)
2. Saini, S.: A survey on watermarking web contents for protecting copyright. In: The IEEE Second International Conference on Innovations in Information Embedded and Communication Systems ICIIECS, pp. 1–4. IEEE (2015)
3. Su, Q., Niu, Y., Wang, Q., Sheng, G.: A blind color image watermarking based on DC component in the spatial domain. Optik Int. J. Light Electron Opt. **124**(23), 6255–6260 (2013)
4. Mohammad, A.: A new digital image watermarking scheme based on Schur decomposition. Multimed. Tools Appl. **59**, 851–883 (2012)
5. Golub, G.H., Van Loan, C.F.: Matrix Computations. Johns Hopkins Univ Press, Baltimore (1989)
6. Kamran, A., Khan, A., Malik, S.: A high capacity reversible watermarking approach for authenticating images: Exploiting down-sampling, histogram processing, and block selection. Inf. Sci. **256**, 162–183 (2014)
7. Agarwal, H., Raman, B., Venkat, I.: Blind reliable invisible watermarking method in wavelet domain for face image watermark. Multimed. Tools Appl. **74**(17), 6897–6935 (2014)
8. Kurniawan, F., Khalil, M.S., Khan, M.K., Alginahi, Y.M.: DWT + LSB-based fragile watermarking method for digital Quran images. In: International Symposium on Biometrics and Security Technologies (ISBAST), pp. 290–297. IEEE (2014)
9. Khalil, M.S., Kurniawan, F., Khan, M.K., Alginahi, Y.M.: Two-layer fragile watermarking method secured with chaotic map for authentication of digital holy Quran. Sci. World J. 1–29 (2014)
10. Lin, C., Liu, X.L., Lin, C.H., Yuan, S.M.: Fragile watermarking-based authentication scheme for demosaicked images. In: International Conference on Intelligent Information Hiding and Multimedia Signal Processing, pp. 97–100. IEEE (2015)

11. http://dicom.nema.org. Accessed 15 Oct 2018
12. http://deanvaughan.org/wordpress/2013/07/dicom-sample-images/. Accessed 15 Oct 2018
13. Pinjari, S.A., Patil, N.N.: A pixel based fragile watermarking technique using LBP (Local Binary Pattern). In: International Conference on Global Trends in Signal Processing, Information Computing and Communication, pp. 194–196. IEEE (2016)
14. Zhang, H., Wang, C., Zhou, X.: Fragile watermarking for image authentication using the characteristic of SVD. Algorithms **10**(27), 1–12 (2017)
15. Sikder, I., Kumar Dhar, P., Shimamura, T.: A semi-fragile watermarking method using slant transform and LU decomposition for image authentication. In: International Conference on Electrical, Computer and Communication Engineering (ECCE), IEEE, Cox's Bazar, Bangladesh, 16–18 February (2017)
16. Soualmi, A., Alti, A., Laouamer, L.: A new blind medical image watermarking based on weber descriptors and Arnold chaotic map. Arab. J. Sci. Eng. 1–13 (2018)
17. AL-Nabhani, Y., Jalab, H., Wahid, A., Noor, R.: Robust watermarking algorithm for digital images using discrete wavelet and probabilistic neural network. J. King Saud Univ. Comput. Inf. Sci. **27**(4), 393–401 (2015)
18. Ghadi, M., Laouamer, L., Nana, L., Pascu, A.C.: A novel zero-watermarking approach of medical images based on Jacobian matrix model. Secur. Commun. Netw. **16**(1), 62–71 (2016)

A New Image Watermarking Technique in Spatial Domain Using DC Coefficients and Graph Representation

Lamri Laouamer[✉]

Department of Management Information Systems,
College of Economics and Administration, Qassim University,
P.O. Box 6633, Buraidah 51452, Kingdom of Saudi Arabia
laoamr@qu.edu.sa

Abstract. In this paper, we propose a new watermarking technique that ensures a suitable compromise between the authentication level of host images and the computational complexity with maintaining a good visual quality of the watermarked image and a good robustness against common image processing attacks. The DC components of host image and image-to-graph representation are used to select particular positions of the original image as inputs of the embedding process in spatial domain. Knowing that the DC value provides a measure of the texture or smooth nature. The proposed model is based on building a directed graph from the original image after dividing it into $N \times N$ blocks. Each block is represented as a vertex and the DC value obtained from Discrete Cosine Transform (DCT) process to the block is used as an edge cost. A simply connected path algorithm is suggested to process the directed graph and to find a simply connected path whose blocks will be used to hide secret data (watermark). One main contribution of this paper is achieving high robustness and low degradation of images in comparison to other existing approaches in spatial domain, by embedding watermark in high textured image blocks. Another main contribution is providing the possibility to recover the watermark even in case of cropping or rotation attacks, due to the embedding of the watermark in several blocks. The proposed model has been tested on gray scale images under several attacks scenario and the experiments results show significant ratios of robustness and bit error rates against the attacks.

Keywords: Watermarking · Spatial domain · Robustness · DC components · Image-to-graph · Textured block

1 Introduction

The amount of images transmitted over wireless sensor networks is increasing exponentially, and this is the case in application domains requiring high security (medicine, industrial systems, etc.). Therefore, it is of capital importance to design efficient models for the security, integrity, and authentication of images transmitted over wireless sensor networks. This requires considering their main constraints including computational complexity, robustness, and fault tolerance against various attacks.

© Springer Nature Switzerland AG 2020
A. E. Hassanien et al. (Eds.): AMLTA 2019, AISC 921, pp. 633–644, 2020.
https://doi.org/10.1007/978-3-030-14118-9_63

Generally, there are two domains based-image watermarking: the frequency domain [1–4] and the spatial domain [5–8]. In frequency domain-based image watermarking, the processed image is segmented into multiple frequency bands, and the secret data called watermark is embedded by modifying the frequency coefficients. The frequency coefficients result from the application of a transform algorithm such as DCT [3, 9], Discrete Wavelet Transform (DWT) [4], Singular Value Decomposition (SVD) [2], Fourier Transform (FT) [9], etc. In spatial domain based-image water-marking, the watermark is embedded directly by modifying the pixels values of the processed image [5, 7]. The Least Significant Bit (LSB) is one of the popular spatial domain based-image watermarking systems [6].

By evaluating the performance of the watermark robustness against different image processing attacks and the computational complexity, the frequency domain-based image watermarking is more robust comparing to spatial domain based-image water-marking [6]. Nevertheless, the computational complexity and capacity factors in spatial domain approaches are better than in frequency domain [10, 11].

This paper presents a new spatial domain based-image watermarking system based on DC components of DCT process. The proposed model represents an host image as a directed graph by processing it as $N \times N$ blocks. Then, each block is represented as a vertex, and its DC value as an edge cost. The resulted graph is used to build an adjacency matrix which is processed by simply connected path algorithm to find set of blocks that are used in the watermark embedding process. The reasons for using DC components and representing image as a graph in designing this model come from two points of literature. (i) The watermarking approaches should ensures the robustness by embedding watermark in largely capacity and without distortion to the visual quality of the host image. (ii) Embedding watermark in high textured image zones make it possible to achieve high robustness and low error rate against common image processing attacks.

2 Proposed Technique

The proposed watermarking technique depends on representing an image as a graph utilizing the DC components to select particular positions of the host image. These positions represent the high textured regions of host image, which are more appropriate to embed the watermark with preserving high visual quality of embedded image and high robustness against various attacks. The proposed model consist of many processes including computing DC coefficients, building an adjacency matrix and applying simply connected path algorithm, besides the embedding and extraction techniques.

The host image I of size $M \times M$ in spatial domain is embedded with watermark of size $N \times N$ directly by modifying its pixels values. The host image is partitioned into $(M/N \times M/N)$ blocks $B_{i,j}(i = 1, \ldots, M/N; j = 1, \ldots, M/N)$, in order to find several locations in the image that will be used in the embedding process. Such approach enforces the robustness of the embedded watermark against the most influential geometric attacks (i.e. cropping, rotation, translation). Indeed, the malicious attack that may destroy the watermark if the watermark is embedded in a single location will have less chance to destroy the watermark if it is dispatched in many locations.

The proposed model considers a weighted graph G, which results by transforming the host image I into an adjacency matrix R, defined through an adjacency relation. Figure 1 presents an example of image-to-graph representation of grayscale Lena image sized 32×32 that partitioned into 8×8 blocks. Based on Fig. 1, the distance between two adjacent blocks B_{ij} and B_{kl} is less than 2 as follows.

$$\sqrt{(i-k)^2 + (j-l)^2} < 2$$

The adjacent relation between the blocks B_{ij} and B_{kl} is defined based on Lemma 1.

Lemma 1

$$adjacent(B_{ij}, B_{kl}) \text{ is exist iff}$$
$$(i = k \text{ and } l \neq M/N \text{ and } l = j+1)$$

or

$$\left(i \neq \frac{M}{N} \text{ and } k = i+1\right) \text{ and } ((j = 1 \text{ and } (l = j \text{ or } l = j+1)) \text{ or}$$
$$\left(j = \frac{M}{N} \text{ and } (l = j-1 \text{ or } l = j)\right) \text{ or } \left(1 < j < \frac{M}{N} \text{ and } (l = j-1 \text{ or } l = j \text{ or } l = j+1)\right))$$

The weight value of the vertices between blocks B_{ij} and its adjacent blocks B_{kl} is the absolute value of DC of block B_{ij}. These values will be used to identify the textured blocks in the host image which will be used to embed the watermark. The simply connected path algorithm is used to find these elements. The structure of the proposed

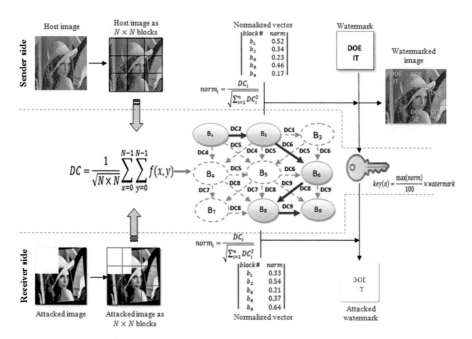

Fig. 1. Structure of the proposed spatial domain-based watermarking approach

model is presented in Fig. 1. First of all, we start with partitioning host image into $N \times N$ blocks and computing the DC coefficients of the partitioned blocks. In the second step, we transforming the host image into graph and building the adjacency matrix. In the third step, simply connected path based on maximum DC values is defined and the DC values for all blocks that are located on simply path are normalized into values between (0–1). In the fourth step, the public key is computed with means to the maximum normalized value and the original watermark, and then the embedding process takes place and results with watermarked image. After that, the key which built before is used to extract the attacked watermarks with means to the normalized values of the DC coefficients of the watermarked image. The details what follows the main steps of the proposed approach are illustrated below.

The pseudo-code of building an adjacency matrix is illustrated in Algorithm 1.

Algorithm 1: Builds an adjacency matrix	
Input:	(I) is $M \times M$ host image.
Initialization:	Defining a DC_Matrix of size $M/8 \times M/8$. Defining an adjacency matrix R sized $(M/N) \times (M/N)$. Defining an cost matrix C sized $(M/N) \times (M/N)$. Let $L \leftarrow M/N$.
Step 1:	Partitioning I as 8×8 blocks
Step 2:	Computing the DC coefficient for each block using equation (1), and set it in corresponding location in DC_Matrix
Step 3:	Partitioning DC_Matrix as $N/8 \times N/8$ blocks
Step 4:	Computing the mean value of each $N/8 \times N/8$ block in DC_Matrix
Initial output:	DC_Matrix sized $(M/N) \times (M/N)$
Step 5:	Queuing all DC values in DC_Matrix into vector V

Step 6:	**For** $x \leftarrow 1$ *to size of* R
	    If $index(x)$ *located in row* M/N *and* $index(x) \neq (M/N) \times (M/N)$ **Then**
	       $R[x][x+1] \leftarrow 1$
	    else if $index(x) == (M/N) \times (M/N)$ **Then**
	       doing nothing // it is the destination vertex
	    else if $index(x)$ *located in column* 1 **Then**
	       $R[x][x+1] \leftarrow 1$
	       $R[x][x+L] \leftarrow 1$
	       $R[x][x+L+1] \leftarrow 1$
	    else if $index(x)$ *located in column* M/N **Then**
	       $R[x][x+L] \leftarrow 1$
	       $R[x][x+L-1] \leftarrow 1$
	    Else
	       $R[x][x+1] \leftarrow 1$
	       $R[x][x+L] \leftarrow 1$
	       $R[x][x+L+1] \leftarrow 1$
	       $R[x][x+L-1] \leftarrow 1$
	 Loop

Step 7:	Building a cost matrix C by replacing any 1 in matrix R with corresponding DC value from vector V, otherwise input -1.
Output:	- An adjacency matrix R between all blocks of original image I. - A cost matrix C between all blocks of host image I.

Once the adjacency matrix R has been built by applying Algorithm 1, the proposed model suggests Algorithm 2 to find simply connected path based on maximum DC value (which represents the edge cost). The selected path from the source vertex B_1 to the destination vertex $B_{(M/N \times M/N)}$ is not the optimum path, where finding the optimum path is NP-hard problem and it requires complex calculations in terms of time and memory. The pseudo-code of the simply connected path algorithm is illustrated in Algorithm 2.

Algorithm 2: Finds the simply connected path	
Input:	- (R) is $(M/N) \times (M/N)$ adjacent matrix, where $R(i,j)$ is nonzero (*equal* 1) iff an edge connects block i to block j. - (C) is $(M/N) \times (M/N)$ cost matrix, where $C(i,j)$ contains the value of the cost (DC value) to move from block i to block j. - (idS) is the index of the start point. - (idD) is the index of the destination point.
Step 1:	$E \leftarrow edgeList\ (R)$, returns a vector contains the linear indices of each nonzero element in matrix R. $L \leftarrow M/N$.

Step 2:	$iVector \leftarrow NaN(L)$, returns a vector of L elements and all entries are $\emptyset$.
	$maxCost \leftarrow Inf(L)$, returns a vector of L elements and all entries are ∞.
	$IsSettled \leftarrow False(L)$, returns a vector of L elements and all entries are 0.
	$Path \leftarrow NaN(L)$, returns a vector of L elements and all entries are $\emptyset$.
	$S \leftarrow idS$.
	$maxCost(S) \leftarrow 0$.
	$iVector(S) \leftarrow 0$.
	$isSettled(S) \leftarrow True$.
	$Path(S) \leftarrow \{S\}$.
	
	While $any(\sim isSettled(idD))$ // while exist any elements in isSettled vector is not false
	    $jVector \leftarrow iVector$.
	    $iVector(S) \leftarrow \emptyset$.
	    $nodeIndex \leftarrow find((E) == S)$. // returns all linear indices has value I (returns all neighbors of node I)
	    for $p \leftarrow 1: length(nodeIndex)$
	        $T \leftarrow E(nodeIndex(p))$
	        If $\sim isSettled(T)$
	            $cost \leftarrow C(S,T)$
	            $empty \leftarrow isnan(jVector(T))$. // check if it is $\emptyset$ or no
	            if $empty$ or $(jVector(T) < jVector(S) + cost)$
	                $iVector(T) \leftarrow jVector(S) + cost$
	                $path(T) \leftarrow [path\{S\}\ T]$
	            else
	                $iVector(T) \leftarrow jVector(T)$
	            endif
	        endif
	    endfor
	    $Q \leftarrow find(\sim isnan(iVector))$,
	                    // returns a vector contains indices of each nonzero element in $iVector$
	    if $isEmpty(Q)$
	        $break$
	    else
	        $B \leftarrow max(iVector(Q))$
	                    // returns the index of neighbor which has the max cost among all neighbors, if there
	                    are two or more neighbors have same cost value, select the first one and proceed
	        $S \leftarrow Q(B)$
	        $maxCost(S) \leftarrow iVector(S)$
	        $isSettled(S) \leftarrow True$
	    endif
	endwhile

Output:	$[Path] \leftarrow Path(idD)$, returns the simply connected path $[Cost] \leftarrow maxCost(idD)$, returns the cost of the simply connected path

2.1 Watermark Embedding Process

The pseudo-code of embedding watermark process is illustrated in Algorithm 3, where equation is used for embedding watermark in host image.

Algorithm 3: Embedding watermark	
Input:	The host image I sized $M \times M$, the blocks of simply connected path, and the watermark w sized $N \times N$.
Initialization:	Dividing I to $M/N \times M/N$ blocks, results with n-blocks
Step 1:	**For** each $block_i$, $i = 1 : n$ 　If $block_i$ is located in simply connected path　**Then** $$block_{iw} = block_i + \frac{norm(block_i)}{100} \times w$$ **Loop**
Output:	watermarked image I_w

2.2 Watermark Extraction Process

The pseudo-code of extraction process is illustrated in Algorithm 4.

Algorithm 4: Attacked watermark extraction	
Input:	The attacked watermarked image I_{wa} sized $M \times M$, the indexes of blocks of simply connected path, and the public key alpha (σ).
Initialization:	Dividing I_{wa} as N×N blocks to results with n-blocks
Step 1:	**For** each $block_j$, $j = 1 : n$ 　If $block_j$ is located in simply connected path　**Then** $$wa_j = \sigma \times \frac{100}{norm(block_j)}$$ **Loop**
Output:	Set of attacked watermarks wa

3 Experiments Result

The proposed watermarking system is tested for its performance against common image processing attacks. These attacks are of the three categories: geometric, non-geometric and hybrid. The experiments result has been analyzed on grayscale images of size 512×512 and using 64×64 watermark image. The capacity, perceptual quality and robustness against different attacks has been evaluated using four well-known image quality metrics including: SSIM [12], PSNR, CC, and BER [12]. The presentation of these metrics are presented below.

The PSNR: measures the peak signal-to-noise ratio between two images. This ratio often expresses the perceptual quality of the watermarked image with respect to the original image. Higher PSNR proves that the embedded watermark is imperceptible and did not degrade the quality of the original image. The PSNR computes according the following equation.

$$PSNR(X,Y) = 10log_{10}\left(\frac{255^2}{\frac{1}{M \times N}\sum_{i=1}^{M}\sum_{j=1}^{N}(X_{ij} - Y_{ij})^2}\right) dB$$

Where X_{ij} is the pixel (i, j) in the original image X and Y_{ij} is the pixel (i, j) in the watermarked image Y, $M \times N$ is the size of image.

The SSIM: measures the perceptual quality of embedded image based on the original image. Using SSIM metric to measure the image quality is more useful than other traditional metric like PSNR and Mean Square Error (MSE), where using PSNR in some cases presents an inconsistent with the principles of HVS. The SSIM computes according to the following equation.

$$SSIM(X,Y) = \frac{(2\mu_X\mu_Y + C_1)(2\sigma_{XY} + C_2)}{(\mu_X^2 + \mu_Y^2 + C_1)(\sigma_X^2 + \sigma_Y^2 + C_2)}$$

Where μ_X is the average of original image X; μ_Y is the average of watermarked image Y; σ_{XY} is the covariance of X and Y; σ_X^2 is the variance of X; σ_Y^2 is the variance of Y; $C_1 = (K_1L)^2, C_2 = (K_2L)^2$ are two variables to stabilize the division with weak denominator; L the dynamic range of the pixel-values (typically is $2^{\# bits\,per\,pixel} - 1$), $K_1 = 0.01$ and $K_2 = 0.03$ by default.

The CC: measures the similarity between the original watermark image and the attacked watermark image. The CC ranges [1, −1]; if CC = 1 this means that two images are absolutely identical, if CC = 0 this means that two images are completely dissimilar, if CC = −1 this means that two images are completely anti-similar. The CC computes according to the following equation.

$$CC(w,wa) = \frac{\sum_i\sum_j(w_{ij} - \overline{w})(w_{ij} - \overline{wa})}{\sqrt{\left(\sum_i\sum_j(w_{ij} - \overline{w})^2\right)\left(\sum_i\sum_j(wa_{ij} - \overline{wa})^2\right)}}$$

Where w_{ij} is the intensity of (i,j) pixel in original watermark image w, wa_{ij} is the intensity of (i,j) pixel in extracted watermark image wa, $\overline{w}$ is the mean intensity of watermark image w, and $\overline{wa}$ is the mean intensity of extracted watermark image wa.

The BER: measures the percentage of erroneous extracted watermark bits to the total number of original watermark bits. Lower BER expresses high robustness of watermark against different attacks. The BER computes according to the following equation.

$$BER(w,wa) = 1/n\left[\sum_{i=1}^{n}w(i) \oplus wa(i)\right] \times 100$$

Where $w(i)$ represents the ith original watermark bit, $wa(i)$ represents the ith extracted watermark bit and n is the total number of original watermark bits.

3.1 Performance of the Proposed Model on Grayscale Images

The performance of the proposed model on grayscale images is analyzed in terms of capacity, perceptual quality and robustness against various attacks. As well as, to ensure the performance of the proposed model the comparison studies with other related models are introduced.

Processed image			
Number of concerned blocks	11 out of 64	13 out of 64	13 out of 64
Total number of changed pixels	$11 \times 4096 = 45056$ out of 262144 pixels	$13 \times 4096 = 53248$ out of 262144 pixels	$13 \times 4096 = 53248$ out of 262144 pixels

Fig. 2. Locations and number of embedded blocks (in case of grayscale images)

Figure 2 shows that the proposed model provides high availability to embed the watermark bits in more than 32768 pixels, where the minimum number of blocks that are located in simply connected path for any grayscale image is 8 blocks.

3.2 Perceptual Quality Analysis

To evaluate the perceptual quality of the watermarked image with respect to the host image. The PSNR and SSIM are computed without any image processing attacks on the watermarked image. Figure 3 presents the PSNR and SSIM values in case of Lena, Pepper and Airplane images.

Figure 3 shows an interesting perceptual quality of watermarked images, where the PSNR exceeds 55 dB and the SSIM reaches 0.99 for the three images. This reveals that embedding watermark did not degrade the visual quality of host image. On the other hand, Fig. 4 illustrates a comparison between the PSNR obtained for watermarked grayscale Lena image in this model and the PSNR obtained in the other related works in [1, 2, 4, 9, 10].

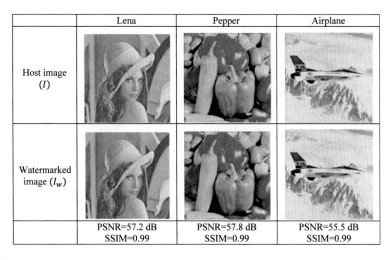

	Lena	Pepper	Airplane
Host image (I)			
Watermarked image (I_w)			
	PSNR=57.2 dB SSIM=0.99	PSNR=57.8 dB SSIM=0.99	PSNR=55.5 dB SSIM=0.99

Fig. 3. The perceptual quality analysis in case of Lena, Pepper and Airplane images

Figure 4 shows that the PSNR of the proposed model outperforms the PSNR obtained in other models proposed in [1, 2, 4, 9, 10]. Although some models based on frequency domain.

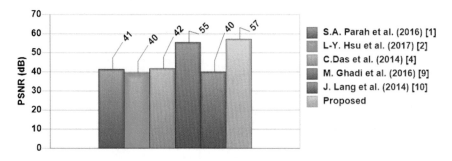

Fig. 4. PSNR comparisons (in case of grayscale Lena image)

3.3 Robustness Analysis

To show the robustness of embedded watermark, the watermarked images are exposed to common geometric attacks. The geometric attacks that are tested including: rotation, translation, cropping, affine transformation, ReMove Line (RML), Latestrnddist, and scaling. The presentation of these attacks are illustrated in [14]. The robustness results of grayscale Lena image in terms of CC and BER are presented below, then a comparison study with other related models is introduced to evaluate the performance of the proposed model.

Attack	Affine transformation (2)	RML (10)	Rotation (5°)	Translation vertically (10%)	Latestrnddist (1)
iw_a					
w_a1	CC=0.57 BER=9.4	CC=0.93 BER=11.2	CC=0.93 BER=10.3	CC=0.99 BER=41	CC=0.45 BER=9.6
w_a2	CC=0.87 BER=11	CC=0.68 BER=9.7	CC=0.95 BER=10.9	CC=076 BER=9.3	CC=0.53 BER=9.5
w_a3	CC=0.74 BER=10	CC=0.70 BER=9.8	CC=0.87 BER=10.5	CC=0.50 BER=9.5	CC=0.48 BER=9.6
w_a4	CC=0.79 BER=10.5	CC=0.73 BER=9.3	CC=0.75 BER=9.2	CC=0.56 BER=9.4	CC=0.56 BER=9.4
w_a5	CC=0.22 BER=9.4	CC=0.22 BER=9.4	CC=0.19 BER=9.4	CC=0.41 BER=9.6	CC=0.21 BER=9.4
w_a6	CC=0.62 BER=9.6	CC=0.59 BER=9.7	CC=0.60 BER=9.7	CC=0.48 BER=9.5	CC=0.64 BER=9.5
w_a7	CC=0.56 BER=9.4	CC=0.48 BER=9.6	CC=0.57 BER=9.4	CC=0.27 BER=9.4	CC=0.55 BER=9.4

Fig. 5. Attacked watermarked Lena image and the extracted watermarks after some geometric attacks

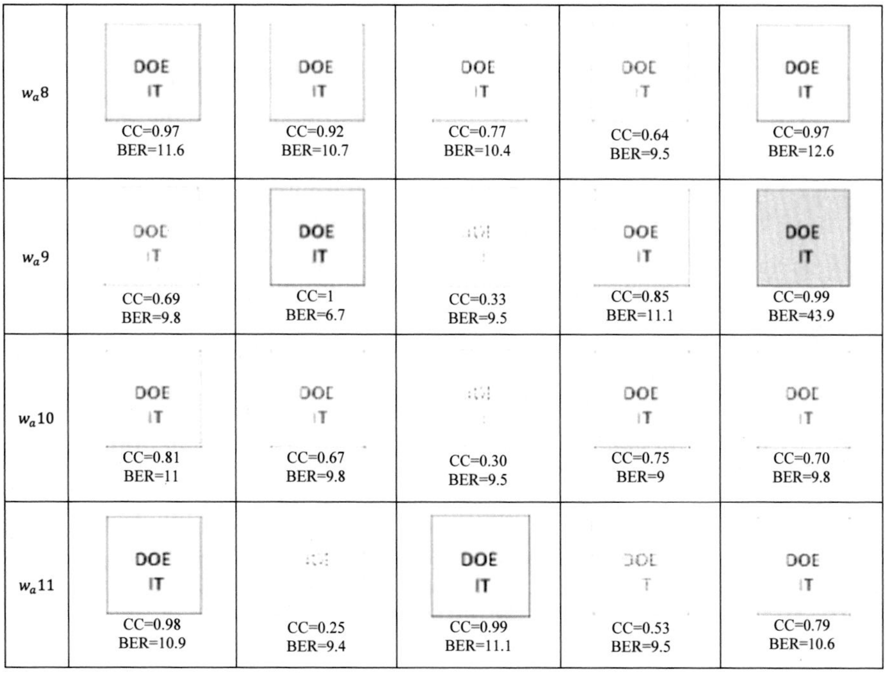

w_a8	CC=0.97 BER=11.6	CC=0.92 BER=10.7	CC=0.77 BER=10.4	CC=0.64 BER=9.5	CC=0.97 BER=12.6
w_a9	CC=0.69 BER=9.8	CC=1 BER=6.7	CC=0.33 BER=9.5	CC=0.85 BER=11.1	CC=0.99 BER=43.9
w_a10	CC=0.81 BER=11	CC=0.67 BER=9.8	CC=0.30 BER=9.5	CC=0.75 BER=9	CC=0.70 BER=9.8
w_a11	CC=0.98 BER=10.9	CC=0.25 BER=9.4	CC=0.99 BER=11.1	CC=0.53 BER=9.5	CC=0.79 BER=10.6

Fig. 5. (*continued*)

From the results in Fig. 5, it is clear that against different geometric attacks we have a good chance to extract the watermark with high CC and low BER. In case of affine_2 transformation attack the w_a11 is extracted with CC equal 0.98 and BER equal 10.9%. In case of RML_10 attack the w_a9 is extracted with CC equal 1 and BER equal 6.7%. Against Rotation 5° attack the w_a11 is extracted with CC equal 0.99 and BER equal 11.1%, and against translation vertically _10% attack the w_a9 is extracted with CC equal 0.85 and BER equal 11.1%. Lastly, in case of Latestrnddist_1 attack the w_a8 is extracted with CC equal 0.97 and BER equal 12.6%.

4 Conclusion

A new spatial domain-based image watermarking system based on DC components and graph representation has been proposed in this paper. The proposed watermarking system utilized image-to-graph representation to select some connected blocks with maximum DC to embed the watermark data. The proposed algorithm consist of many processes: computing the DC coefficients, image-to-graph representation and finding set of blocks that are located within a simply connected path. The normalized values of DC coefficients of simply connected path blocks are computed to build the public key and to complete watermark embedding and extraction processes. The proposed approach achieved high level of watermark capacity and visual quality of watermarked image, as well as an interesting robustness against several attacks. The results of perceptual quality and robustness for grayscale, color and medical images are analyzed in terms of PSNR, SSIM, CC and BER. Where the PSNR exceeding 57 dB, the SSIM and CC reaching 1 and the BER ranged 3.6–12.9%.

References

1. Parah, S.A., Sheikh, J.A., Loan, N.A., Bhat, G.M.: Robust and blind watermarking technique in DCT domain using inter-block coefficient differencing. Digit. Signal Proc. **53**, 11–24 (2016)
2. Hsu, L.-Y., Hu, H.-T.: Robust blind image watermarking using crisscross inter-block prediction in the DCT domain. J. Vis. Commun. Image R. **46**, 33–47 (2017)
3. Han, J., Zhao, X., Qiu, C.: A digital image watermarking method based on host image analysis and genetic algorithm. J. Ambient Intell. Hum. Comput. **7**, 37–45 (2016)
4. Das, C., Panigrahi, S., Sharma, V.K., Mahapatra, K.K.: A novel blind robust image watermarking in DCT domain using inter-block coefficient correlation. Int. J. Electron. Commun. **68**, 244–253 (2014)
5. Benoraira, A., Benmahammed, K., Boucenna, N.: Blind image watermarking technique based on differential embedding in DWT and DCT domains. EURASIP J. Adv. Sign. Process. **55**, 1–11 (2015)
6. Su, Q., Niu, Y., Wang, Q., Sheng, G.: A blind color image watermarking based on DC component in the spatial domain. Int. J. Light Electron. Opt. **124**(23), 6255–6260 (2013)
7. Singh, R.K., Shaw, D.K., Alam, M.J.: Experimental studies of LSB watermarking with different noise. Procedia Comput. Sci. **54**, 612–620 (2015)
8. Ghadi, M., Laouamer, L., Pascu, N.A.: Fuzzy rough set based image watermarking approach. In: Proceedings of the 2nd International Conference on Advanced Intelligent Systems and Informatics, vol. 533, pp. 234–245 (2016)
9. Ghadi, M., Laouamer, L., Nana, L., Pascu, A.: A robust associative watermarking technique based on frequent pattern mining and texture analysis. In: Proceedings of the 8th International ACM Conference on Management of computational and collective IntElligence in Digital EcoSystems, pp. 73–81 (2016)
10. Lang, J., Zhang, Z.: Blind digital watermarking method in the fractional Fourier transform domain. Opt. Lasers Eng. **53**, 112–121 (2014)
11. Bansal, N., Bansal, A., Deolia, V.K., Pathak, P.: Comparative analysis of LSB, DCT and DWT for digital watermarking. In: International Conference on Computing for Sustainable Global Development, pp. 40–45. IEEE (2015)
12. Su, B., Chen, B.: Robust color image watermarking technique in the spatial domain. Soft Comput. **22**, 1–16 (2017)
13. Moosazadeha, M., Ekbatanifard, G.: An improved robust image watermarking method using DCT and YCoCg-R color space. Optik **140**, 975–988 (2017)
14. Ghadi, M., Laouamer, L., Nana, L., Pascu, A.: A novel zero-watermarking approach of medical images based on Jacobian matrix model. Secur. Comm. Netw **9**(18), 5203–5218 (2016)

Image Encryption Algorithm Methodology Based on Multi-mapping Image Pixel

W. M. Abd-Elhafiez[1,2]([⊠]), Omar Reyad[1], M. A. Mofaddel[1], and Mohamed Fathy[1]

[1] Faculty of Science, Sohag University, Sohag, Egypt
w_a_led@yahoo.com, ormak4@yahoo.com
[2] College of Computer Science and Information Systems, Jazan University, Jazan, Kingdom of Saudi Arabia

Abstract. Image encryption and decryption using a key sequence based on the multi-mapping method of an image pixel is proposed in this paper. The generated key sequences are based on the logistic map, elliptic curves (EC) and Henon map. In the beginning, the one-dimensional sequence is generated using logistic map for the bifurcation parameter and initial value. The generated sequences are XORed with image pixels as a first mapping step. Then the resulted pixels are XORed with key sequence based on elliptic curve random number generator (ECRNG) as a second mapping step. Finally, the resulted pixels XORed with key sequence generated by Henon map. The obtained cipher image introduces good cryptographic properties in the case of multi-mapping. The performance of the proposed scheme is analyzed by computing Histogram, Entropy, and Correlation between the original and encrypted images. The proposed method provides an immense improvement in the encrypted image.

Keywords: Encryption · Gray image · Henon map · Logistic map · Elliptic curves

1 Introduction

Information security's primarily focused on the balanced protection of confidentiality, integrity, and availability of cipher image. To standardize this discipline academic and professionals collaborate and seek to set basic guidance, policies, and industry standards on passwords, antivirus software, firewall, and encryption software. Various type of image encryption methods had been developed to protect secret information against unauthorized users, or against falling in wrong hands [1–3]. The cryptographic image encryption schemes play the main role in securing the images sent or received via insecure channels. Cryptography utilizes algorithms in the process of ciphering. The oldest and best-known technique is the secret key which is called symmetric encryption, as long as the sender and recipient know the secret key they can encrypt and decrypt all messages [4, 5]. As well as in asymmetric encryption utilizes two related keys (a key pair). The first one is public and available free to anyone who wants to send someone massage. The Second key is a private and kept secret, the only authorized person knows it, that is encrypted by using public key can be decrypted only by

© Springer Nature Switzerland AG 2020
A. E. Hassanien et al. (Eds.): AMLTA 2019, AISC 921, pp. 645–655, 2020.
https://doi.org/10.1007/978-3-030-14118-9_64

applying the same algorithm using that matching private key [6]. In this paper, A new encryption scheme based on three mapping levels of an image pixel is presented. The logistic map, elliptic curves (EC) and Henon map used to generate key sequences for the multi-mapping method in order to raise the level of the security and resistance of cipher images versus chosen-plaintext attacks.

The paper is ordered as follows. In Sect. 2, the literature review is presented. In Sect. 3, the preliminaries of Logistic map, EC mode, and Henon map are discussed. Section 4 presents a description of the proposed image encryption scheme. In Sect. 5, the experimental results and discussion are presented. Conclusions are given in Sect. 6.

2 Literature Review

Cryptography is the most standard field of encryption that could be helpful for secure data. The field of new cryptography provides a theoretical foundation based on which one can understand what these problems are, this encrypted connection provides secure access to personal and business information which should not be available to third parties. It includes many problems such as encryption, authentication, and key distribution to name a few. In recent years, there is a significant increase in chaos/based cryptography research area. Shah and Dhobi [7] have proposed a gray image cryptosystem using the Chaos system of various methods. Dhiman and Singh [8] have provided better security, to choose a fast and secure algorithm which provides superior security. Lagmiri and Elalami [9] proposed Chaotic and Hyperchaotic systems, that donate an algorithm for color image and gray image encryption was utilized. Zhang, Han and Niu [10] presented a hyperchaos digital image encryption technique that is based on bit permutation and dynamic DNA encoding. Zang, Li and Hou [11] proposed a study image cryptosystem based on AES to testify the viewpoint of AES not suitable for image encryption. Chen and Hu [12] proposed improved chaotic mapping image encryption algorithm. Huang and Yang [13] presented a new image compression encryption algorithm combining compressive sensing with double random-phase encoding. Wang, Zhu and Zhang [14] introduced a novel chaotic image encryption scheme based on Josephus traversing and mixed chaotic map.

3 Preliminaries

3.1 The Logistic Map

The logistic map is a very straightforward chaotic map and its mathematical expression formula illustrated in Eq. (1):

$$x_{n+1} = rx_n(1 - x_n) \tag{1}$$

which used in biology to model how a population x_n changes with the seasons (time is represented by the integer numbers n); here x1 is the initial condition, $3.99465 \leq r$ 4 and r is a control parameter [15].

3.2   Elliptic Curve Over a Binary Finite Field

The field F_2^m is called a binary finite field and it can be displayed as a vector space of dimension m over the field F_2 which formed of two binary elements $\{0, 1\}$. A non-supersingular elliptic curve E over a binary finite field F_{2m} is defined by Eq. (2) as follows:

$$y^2 + xy = x^3 + ax^2 + b \qquad (2)$$

where the parameters $a, b \in F_2^m$ with $b \neq 0$. The obtained set of points $E\ (F_2^m)$ consists of all the points (x, y), $x \in F_2^m$, $y \in F_2^m$, which satisfy the defining equation given in (5), together with a special point O called the point at infinity [16–18].

3.3   The Henon Map

Henon map is a two-dimensional dynamic system proposed to simplify the Lorenz map with the same properties [19, 20]. Consider the following equation:

$$x_{i+1} = y_{i+1} + 1 - \alpha x_{i^2}, y_{i+1} = \beta x_i \qquad (3)$$

The initial parameters are α, β and the initial point is (x_0, y_0). Each point (x_n, y_n) is mapped to a new point (x_{n+1}, y_{n+1}) through the Henon map. For $\alpha = 1.4$ and $\beta = 0.3$, the Henon function has chaotic behavior and the iterations have a boomerang-shaped chaotic attractor. Minute variations in the initial point will lead to major changes and different behavior.

4   The Proposed Image Encryption Schemes

In this section, two encryption schemes for grayscale are proposed. *The first scheme* is described in (Algorithm 1) it starts with reading the grayscale image and use a logistic map to generate a logistic key, then XORed the grayscale image with the logistic key. The result is XORed with the Elliptic Curve (EC) key, again the result is XORed with the Henon key which generated by Henon map. It eventually produces the final encrypted image.

The second scheme is described in (Algorithm 2) it starts with reading the grayscale image and use the logistic map to generate a logistic key, which XORed with the plain image. Then use the permutation method [10] with the encryption result to obtain the cipher image. The cipher image is XORed with the EC then the permutation method is used to obtain the second encryption image. This encryption image is XORed with the Henon key which generated by Henon map, then the permutation method is used with the encryption result to obtain the cipher image.

Algorithm 1: First Proposed Scheme (FPS)

1) Read gray image.
2) Use a logistic map to generate the logistic key.
3) Apply XOR operation between gray image and the logistic key to generate an encrypted image (this step called First encryption (FE)).
4) Use an encrypted image of FE as a key then XORed with EC (this step called Second encryption(SE)).
5) Generate a Henon key using Henon map.
6) Apply XOR operation between the encrypted image of SE and Henon key to generate a final encrypted image (this step called Third encryption (TE)).

Algorithm 2: Second Proposed Scheme (SPS)

1) Read gray image.
2) Use a logistic map to generate the key.
3) Apply XOR operation between grayscale image and the logistic key to generate an encrypted image.
4) Convert the encrypted image of the previous step to binary then use a permutation method to obtain an encrypted image (this step called First encryption (FE)).
5) Use the encrypted image of FE as a key and XORed with EC.
6) Convert the encrypted image of step 5 to binary and then use a permutation method to obtain an encrypted image (this step called Second encryption (SE)).
7) Use Henon map to generate a key.
8) Apply XOR operation between the encrypted image of SE and Henon key to generate an encrypted image.
9) Convert the encrypted image of step 8 to binary and then use a permutation method to obtain a final encrypted image (this step called Third encryption (TE)).

5 Experimental Results

In this section, we analyzed the efficiency of the proposed schemes by using various security test measures. These measures are taken as follows: statistical analysis including histogram analysis and calculus the peak signal-to-noise ratio (PSNR) for the image, information entropy analysis, and test security against differential attack including calculating the number of pixel change rate (NPCR) and unified average changing intensity (UACI). Six grayscale images (Lena, Baboon, clock, plane, tank, and earth) are used with size 256×256.

5.1 Histogram Analysis

It is important to ensure that encrypted image does not have any statistical similarity with its original image to prohibit the leakage of information. A robust scheme for image encryption should generate a cipher image of the unique histogram for any original image all the time. The histogram is plotted for original and ciphered images. Figure 1(a–d) show that the histograms of the ciphered image are fairly unique and significantly different about those of the original image. The statistical characteristic of the original images gets better in such a way that the cipher images had good balance property and a unique level distribution.

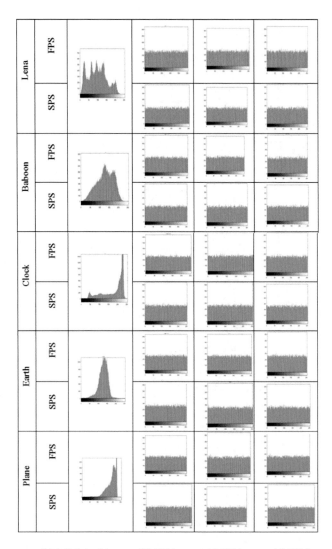

Fig. 1. Histogram of (a) Original image, (b) FE image, (c) SE image, (d) TE image for different images.

5.2 Entropy Test

The degree of uncertainties in the system is measured using entropy. It is proved that the entropy $H(m)$ of a message source m can be computed as:

$$H(m) = - \sum_{i=0}^{255} P(m_i) log_2 P(m_i) \tag{4}$$

where $P(m_i)$ represents the probability of m_i symbol. For all the used cipher images shown in Table 1, the number of occurrence of each grayscale recorded and the probability of occurrence is computed for grayscale images, respectively. This table indicates the different values of the entropies for the encrypted images by the presented schemes. We can be noticed that the entropy of the cipher images are closed to the ideal value 8, which indicate that all the pixels in the encrypted images occur with nearly equal probability. So, the information leakage in the introduced encryption schemes is negligible, and it is safe against the entropy attack. Also, it is compared with the entropy values presented by recent references in [14, 21, 22] as shown in Table 2.

Table 1. Entropy values for different images.

	Gray image	FE	SE	TE
FPS	Lena	7.9973	7.9972	7.9973
	Baboon	7.9973	7.9972	7.9970
	Clock	7.9973	7.9972	7.9977
	Earth	7.9974	7.9972	7.9970
	Plane	7.9972	7.9973	7.9975
	Tank	7.9975	7.9976	7.9971
SPS	Lena	7.9974	7.9962	7.9970
	Baboon	7.9973	7.9968	7.9970
	Clock	7.9973	7.9974	7.9970
	Earth	7.9974	7.9970	7.9968
	Plane	7.9972	7.9971	7.9975
	Tank	7.9975	7.9974	7.9971

Table 2. Comparison of entropy value for lena image.

Scheme	Entropy
The proposed scheme (FPS)	7.9973
The proposed scheme (SPS)	7.9974
Ref. [21]	7.9891
Ref [22]	7.9972
Ref. [14]	7.9971

5.3 PSNR Test

The measure of the reconstruction of the encrypted image PSNR is used. This metric is utilized for discriminating between the original and encrypted image. The advantage of this measure is an easy computation [23]. It is calculated as:

$$PSNR = \frac{20 \log 255^2}{MSE} \tag{5}$$

The MSE is the average of the squares of the difference between the intensities of the encrypted image and the original image. Table 3 explains the PSNR value for various cipher images.

Table 3. PSNR value for different images.

	Gray image	FE	SE	TE
FPS	Lena	28.70	28.68	28.65
	Baboon	27.38	27.43	27.38
	Clock	25.68	25.71	25.69
	Earth	27.45	27.42	27.43
	Plane	25.50	25.52	25.51
	Tank	27.46	27.46	27.45
SPS	Lena	28.68	28.67	28.65
	Baboon	27.37	27.40	27.39
	Clock	25.69	25.68	25.68
	Earth	27.45	27.43	27.43
	Plane	25.51	25.52	25.52
	Tank	27.64	27.40	27.48

5.4 UACI and NPCR

The attacker tries to derive a relationship between the plain image and the cipher image, by studying the effect of differences in input on the resultant difference at the output to find out the key. Two common measures are used here, the number of pixels changes rate (NPCR) and unified average changing intensity (UACI) [24]. Trying to make a slight change such as modifying one pixel of the cipher image, the change of the plain image is noticed by the attackers. To test the effect of one-pixel change on the whole encrypted image by the proposed algorithm, the two tests are defined as follows:

$$NPCR = \frac{\sum_{i,j} D(i,j)}{W \times H} \times 100 \tag{6}$$

$$UACI = \frac{100}{W \times H} \sum_{i,j} \frac{|C_1(i,j) - C_2(i,j)|}{255} \times 100 \tag{7}$$

Table 4. NPCR and UACI values for different images.

	Gray image	FE		SE		TE	
		NPCR	UACI	NPCR	UACI	NPCR	UACI
FPS	Lena	0.9960	0.1019	0.9962	0.0964	0.9960	0.0971
	Baboon	0.9963	0.1446	0.9961	0.1419	0.9960	0.1443
	Clock	0.9963	0.2934	0.9960	0.2907	0.9960	0.2917
	Earth	0.9963	0.1319	0.9960	0.1322	0.9960	0.1327
	Plane	0.9963	0.2976	0.9960	0.2955	0.9960	0.2963
	Tank	0.9963	0.1283	0.9960	0.1290	0.9960	0.1299
SPS	Lena	0.9963	0.0960	0.9960	0.0958	0.9960	0.0971
	Baboon	0.9960	0.1437	0.9961	0.1441	0.9960	0.1431
	Clock	0.9962	0.2924	0.9962	0.2921	0.9963	0.2919
	Earth	0.9962	0.1318	0.9964	0.1327	0.9963	0.1322
	Plane	0.9963	0.2968	0.9962	0.2958	0.9962	0.2956
	Tank	0.9961	0.1291	0.9962	0.1301	0.9961	0.1287

Table 5. Comparison of the NPCR value for lena image with recent methods.

Scheme	NPCR
The proposed scheme (FPS)	99.60
The proposed scheme (SPS)	99.60
The method in Ref. [21]	99.59
The method in Ref. [22]	99.63
The method in Ref. [13]	99.59

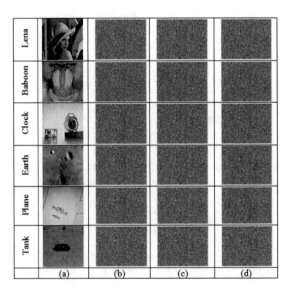

Fig. 2. (a) Original image, (b) FE image, (c) SE image, (d) TE image for grayscale images of the first scheme (FPS).

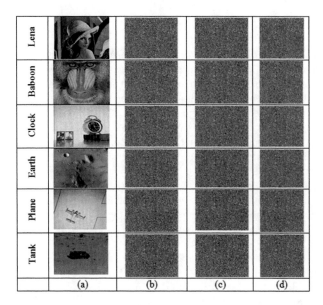

Fig. 3. (a) Original image, (b) FE image, (c) SE image, (d) TE image for grayscale images of the second scheme (SPS).

where C_1 and C_2 are two ciphered images with the same size WxH, whose corresponding original images have an only one-pixel difference. $C_1(i, j)$ and $C_2(i, j)$ are grey-scale values of the pixels at grid (i, j). $D(i, j)$ is determined by $C_1(i, j)$ and $C_2(i, j)$, if $C_1(i, j) = C_2(i, j)$, then, $D(i, j) = 1$; otherwise, $D(i, j) = 0$. From the results shown in Table 4 the proposed schemes are very sensitive with respect to less change in the plain image. Also the values of NPCR for image lena is compared with recent methods as shown in Table 5.

The performance of the two proposed schemes is shown in Figs. 2 and 3 by using different images.

6   Conclusion

Image encryption schemes play an important role in the security of digital images and are considered as a common method to protect the image information. In this paper, image encryption using key sequence produce by a sequence of logistic map, EC and Henon map is proposed. The Performance of the proposed method is analyzed using various measurements, like histogram, entropy. After comparing the proposed method with the related work it was observed that the presented method is more reliable due to the long secret key and its randomness and complexity.

References

1. Auyporn, W., Vongpradhip, S.: A robust image encryption method based on bit plane decomposition and multiple chaotic maps. Int. J. Signal Process. Syst. **3**(1), 8–13 (2015)
2. Reyad, O., Mofaddel, M.A., Abdelhafiez, W.M., Fathy, M.: A novel image encryption scheme based on different block size for grayscale and color images. In: 12th International Conference on Computer Engineering and Systems (ICCES), pp. 455–461. IEEE (2017)
3. Reyad, O., Kotulski, Z., Abdelhafiez, W.M.: Image encryption using chaos-driven elliptic curve pseudo-random number generators. Appl. Math. Inf. Sci. **10**(4), 1283–1292 (2016)
4. Fridrich, J.: Symmetric ciphers based on two-dimensional chaotic maps. Int. J. Bifurcat. Chaos **8**, 1259–1284 (1998)
5. Ahmad, S., Alam, K., Rahman, H., Tamura, S.: A comparison between symmetric and asymmetric key encryption algorithm based decryption mixnets. In: International Conference in Networking Systems and Security (NSysS), pp. 1–5. IEEE (2015)
6. Reyad, O.: Text message encoding based on elliptic curve cryptography and a mapping methodology. Inf. Sci. Lett. **7**(1), 7–11 (2018)
7. Shah, J., Dohbi, J.S.: Review of image encryption and decryption techniques for 2D images. Int. J. Eng. Technol. Manage. Res. **5**(1), 81–84 (2018)
8. Dhiman, R., Singh, B.: Image encryption techniques: a literature Review. Int. J. Adv. Res. Comput. Sci. **8**(7), 239–244 (2017)
9. Lagmiri, S.N., Elalami, N., Elalami, J.: Color and gray images encryption algorithm using chaotic systems of different dimensions. Int. J. Comput. Sci. Network Secur. **18**(1), 79–86 (2018)
10. Zhang, X., Han, F., Niu, Y.: Chaotic image encryption algorithm based on bit permutation and dynamic DNA encoding. Hindawi Comput. Intell. Neurosci. 1–11 (2017)
11. Zang, Y., Li, X., Hou, W.: A fast image encryption scheme based on AES. In: 2nd International Conference on Image, Vision and Computing, pp. 624–628. IEEE (2017)
12. Chen, X., Hu, C.: Adaptive medical image encryption algorithm based on multiple chaotic mapping. Saudi J. Biol. Sci. **24**(8), 1821–1827 (2017)
13. Huang, H., Yang, S.: Image encryption technique combining compressive sensing with double random-phase encoding. Hindawi Math. Probl. Eng. (2018)
14. Wang, X., Zhu, X., Zhang, Y.: An image encryption algorithm based on Josephus traversing and mixed chaotic map. IEEE Access **6**, 23733–23746 (2018)
15. Bresten, C.L., Jung, J.H.: A study on the numerical convergence of the discrete logistic map. Commun. Nonlinear Sci. Numer. Simul. **14**, 3076–3088 (2009)
16. Silverman, J.H.: The Arithmetic of Elliptic Curves. Springer, New York (2009)
17. Reyad, O., Kotulski, Z.: On pseudo-random number generators using elliptic curves and chaotic systems. Appl. Math. Inf. Sci. **9**(1), 31–38 (2015)
18. Reyad, O., Kotulski, Z.: Statistical analysis of the chaos-driven elliptic curve pseudo-random number generators. In: CCIS, vol. 448, pp. 38–48. Springer, Heidelberg (2014)
19. Szczepanski, J., Kotulski, Z.: Pseudorandom number generators based on chaotic dynamical systems. Open. Syst. Inf. Dyn. **8**, 137–146 (2001)
20. Sarmah, H.K., Paul, R.: Period doubling route to chaos in a two parameter invertible map with constant Jacobian. Int. J. Res. Rev. Appl. Sci. (IJRRAS) **3**, 72–82 (2010)
21. Liu, H., Wang, X.: Color image encryption based on one-time keys and robust chaotic maps. J. Comput. Appl. Math. **59**, 3320–3327 (2010)
22. Rhouma, R., Meherzi, S., Belghith, S.: OCML-based colour image encryption. Chaos Soliton Fract. **40**(1), 309–318 (2009)

23. Gonzalez, R.C., Woods, R.E.: Digital Image Processing, 3rd edn. Prentice-Hall Inc., Upper Saddle River (2006)
24. Wu, Y., Noonan, J.P., Agaian, S.: NPCR and UACI randomness tests for image encryption. IEEE Transl. J. Sel. Areas Telecommun. (JSAT) 31–38 (2011)

City Crime Mapping Using Machine Learning Techniques

Nitish Yadav[1], Ashish Kumar[1(✉)], Roheet Bhatnagar[1],
and Vivek Kumar Verma[2]

[1] Department of Computer Science and Engineering, Manipal University Jaipur,
Jaipur, Rajasthan, India
yadav.nitish007@yahoo.in, aishshub@gmail.com,
roheet.bhatnagar@jaipur.manipal.edu
[2] Department of Information and Technology, Manipal University Jaipur,
Jaipur, Rajasthan, India
vermavivek123@gmail.com

Abstract. In order to prevent a crime it is very important to analyze and understand the patterns of criminal activity of that place. Police Department can work effectively and efficiently if the crime pattern is known to them. In this work, we attempted an exploratory analysis of a standard dataset in order to predict the resolution that was given for the crimes that occurred from 2003 to 2015. The dataset is obtained from San Francisco Police Department Crime Incident Reporting System. We used Machine Learning Algorithms like CART, K-NN, Gaussian Naive Bayes, and Multilayer Perceptron (MLP). Validation and cross validation were used to test the results of each technique. The experiment shows that we can obtains higher accuracy by using CART algorithm.

Keywords: Crime classification · CART · K-NN ·
Gaussian Naive Bayes

1 Introduction About the City of Our Study - San Francisco

Right after California Gold Rush in 1849, the city of San Francisco first boomed and started expanding in terms of population and land area. After the Internet boom of 1990s, San Francisco became the hub for technological driven economy and still holds an important position in the world city network today. The technology firms and Internet content production created over 50,000 jobs and the job growth rate was at 4.9% between the years of 1999 and 2000. In the second technological boom driven by social media in the mid 2000's, San Francisco became a popular location for companies such as Apple, Google, Facebook and Twitter to base their tech offices and for their employees to live. Since then, tech employment has continued to increase. In 2014, San Francisco's tech employment grew nearly 90% between 2010 and 2014, beating out Silicon Valley's 30% growth rate over the

© Springer Nature Switzerland AG 2020
A. E. Hassanien et al. (Eds.): AMLTA 2019, AISC 921, pp. 656–668, 2020.
https://doi.org/10.1007/978-3-030-14118-9_65

same period. The rapid population increase led to social problems and high crime rates fueled in part by the presence of red light districts [1].

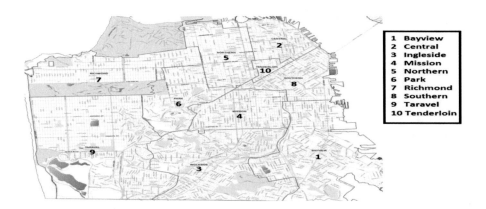

Fig. 1. Police districts in San Francisco

Figure 1 shows the area controlled by a Police Districts. A machine learning agent can analyze the crime dataset and find patterns in them, but finding these patterns for a crime will not prevent it 100% from happening, but to some extent it will help the police department to effectively utilize it as a resource. The above technique of finding patterns based on time, type or any other factor is known as classification and it allows us to predict labels The crime dataset of San Francisco contains the reported criminal activities in the neighborhoods of the city for a duration of 12 years. So, to analyze the dataset we follow the underlying phases:

Phase 1: Data is collected from various sources. Data can be either numerical or categorical or both.

Phase 2: After determining the type of data we can select any of the data mining methods like classification, clustering or regression.

Phase 3: In this phase with the help of algorithms we try to find a pattern in the crime dataset.

Phase 4: In this phase we use the dataset to predict certain variables, evaluate certain results and also, visualize these results.

We have discussed various classifiers like CART, K-NN, Gaussian Naive Bayes, MLP and according to evaluation metric (i.e. Accuracy and ROC Curve) we have shown which algorithm gives better prediction accuracy and best ROC curve. This paper contains following further sections: Literature Review, Dataset Analysis, Pre-Processing of data, Evaluation Metrics, Classification, Conclusion & Future Work.

2   Literature Review

For all governments across the world, combating criminal activity has been their top priority. Many researches have been done effectively to find indicators and analyze the pattern of crime in order to avoid them in future.

Research has been done in order to find a relation between the criminal activities and socio economic variables [6,8,9,12].

Some researchers were able to analyze mobile network infrastructure and concluded that it can be used to predict crime hot spots in London [5].

Sojayee et al. [14] employed a number of algorithms where k-Nearest Neighbor performed better than other algorithms and gave the prediction accuracy by 89.50%.

Wang et al. [16] proposed an algorithm that can find the patterns of crime committed by an offender or a group of offenders.

Sadhana and Sangareddy [13] proposed that social network can also be used as possible criminal activity indicator. They have used twitter data along and applied sentiment analysis to predict crime in real time.

3   Dataset Analysis

The San Francisco Crime Classification dataset [4] contains the following set of features:

1. X Coordinate - X coordinate on the map where the crime occurred.
2. Y Coordinate - Y coordinate on the map where the crime occurred.
3. Date - The entered date is in the format YYYY-MM-DD HH:MM:SS. Thus we can easily conclude time and date for a particular crime.
4. Address - The complete address of crime incident.
5. Category - Type of Crime that has occurred. There are in total 39 distinct category of crimes.
6. Resolution - The outcome derived for the crime. There are 17 distinct Resolutions. This is the target variable that we are predicting.
7. District - Every crime is assigned a police district. There are 10 distinct police Districts.
8. Day of week - The day of the week (like Monday).
9. Descript - Description of the crime.

On analyzing the column named 'District' of the crime dataset we, concluded that there are 10 distinct Police districts in San Francisco. After identifying the districts, we were able to derive the Table 1.

From the Table 1, we can infer that Southern Police District got maximum number of cases which is almost four times that of Richmond and Park Police District.

On analyzing the column named 'Resolution' of the crime dataset, we found that there are 17 distinct Resolutions. We have plotted a graph shown in Fig. 2 between total number of cases and 'Resolutions'.

Table 1. Police district wise number of cases

Sl. No	Police districts	Number of cases
1	Northern	105296
2	Park	49313
3	Ingleside	78845
4	Bayview	89431
5	Richmond	45209
6	Central	85460
7	Tarval	65595
8	Tenderloin	81809
9	Mission	119908
10	Southern	157182

In Fig. 3, using the results of Table 1 and Fig. 2, a histogram graph is plotted between total number of cases and number of police districts. Figure 3 shows as to how many cases were handled by each district and every histogram bar shows what percentage of resolution was given for the total number of cases handled by that particular police district.

Figure 4 shows that crime occurs throughout the year and it shows that from June to September the crime rate is comparatively lower than rest of the year. Table 2 shows each crime frequency given in the dataset.

Table 2. Category wise crime frequency

Category	Frequency
Larceny/Theft	174900
Other offenses	126182
Non-Criminal	92304
Assault	76876
Drug/Narcotic	53971
Vehicle theft	53781
Vandalism	44725
Warrants	42214
Burglary	36755
Suspicious occ	31414

4   Pre-processing of the Data

The San Francisco crime dataset contains both numeric and nominal data. In order to apply certain classification techniques we had to convert the entire nominal part of the dataset into numeric dataset. The algorithm identifies the nominal data and assigns a numeric value to it. The Fig. 5 shows the original dataset's top 10 values and its equivalent converted dataset.

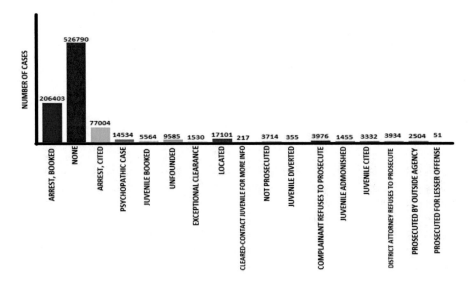

Fig. 2. Histogram plot between distinct resolution and number of cases

The dataset contains a total of 878049 instances, and after checking the dataset for redundancies or missing data, we concluded that there were a total of 57 instances that were redundant or contained missing data. These redundancies were removed from the dataset. The time stamp (i.e. the column 'date' is out dataset) contained the date, year and time of occurrence of each crime. We were able to break it into five features: date, month, year, hour, minute.

5   Evaluation Metrics

There are two metrics that have been used to evaluate the quality of the classifier: (a) Accuracy percentage (b) ROC Curve.

Accuracy Percentage: In binary classification, accuracy is a statistical measure which tells about how well a binary classifier correctly classifies the instances. According to ISO 5725-1, (Reference BS ISO 5725-1) the overall term "accuracy" is used to define the nearness of a quantity to the true value. It is the number of correct predictions made divided by the total number of predictions made, multiplied by 100 to turn it into a percentage i.e.

$$AccuracyPercentage = ((TP + TN)/All) * 100 \qquad (1)$$

where, TP = True Positive; TN = True Negative and All = total number of instances.

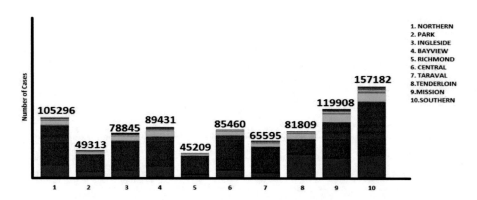

Fig. 3. Histogram plot between police district and number of cases.

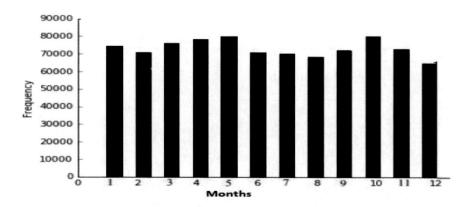

Fig. 4. Crimes occurring in different months of the year.

ROC Curve: It is a graphical plot that illustrates the diagnostic ability of a binary classifier system as its discrimination threshold is varied. The ROC curve is created by plotting the true positive rate (TPR) against the false positive rate (FPR) at various threshold settings [10].

A ROC plot in Fig. 6 shows the following features:

1. Relationship between specificity and sensitivity which have an inverse relationship.
2. Accuracy: The closer the graph is to the top left hand side of the plot border, the more accurate the test is.

6   Classification

In this section, we explored four classification models along with one clustering model: Gaussian Naive Bayes, Multilayer Perceptron, K-Nearest Neighbors, Classification and Regression tree and k-means. We compared their accuracy for predicting Resolution for the dataset.

	Dates	Category	Descript	DayOfWeek	PdDistrict	Address	X	Y	Resolution
0	5/13/2015 23:53	WARRANTS	WARRANT ARREST	Wednesday	NORTHERN	OAK ST / LAGUNA ST	-122.425892	37.774599	ARREST, BOOKED
1	5/13/2015 23:53	OTHER OFFENSES	TRAFFIC VIOLATION ARREST	Wednesday	NORTHERN	OAK ST / LAGUNA ST	-122.425892	37.774599	ARREST, BOOKED
2	5/13/2015 23:33	OTHER OFFENSES	TRAFFIC VIOLATION ARREST	Wednesday	NORTHERN	VANNESS AV / GREENWICH ST	-122.424363	37.800414	ARREST, BOOKED
3	5/13/2015 23:30	LARCENY/THEFT	GRAND THEFT FROM LOCKED AUTO	Wednesday	NORTHERN	1500 Block of LOMBARD ST	-122.426995	37.800873	NONE
4	5/13/2015 23:30	LARCENY/THEFT	GRAND THEFT FROM LOCKED AUTO	Wednesday	PARK	100 Block of BRODERICK ST	-122.438738	37.771541	NONE
5	5/13/2015 23:30	LARCENY/THEFT	GRAND THEFT FROM UNLOCKED AUTO	Wednesday	INGLESIDE	0 Block of TEDDY AV	-122.403252	37.713431	NONE
6	5/13/2015 23:30	VEHICLE THEFT	STOLEN AUTOMOBILE	Wednesday	INGLESIDE	AVALON AV / PERU AV	-122.423327	37.725138	NONE
7	5/13/2015 23:30	VEHICLE THEFT	STOLEN AUTOMOBILE	Wednesday	BAYVIEW	KIRKWOOD AV / DONAHUE ST	-122.371274	37.727564	NONE
8	5/13/2015 23:00	LARCENY/THEFT	GRAND THEFT FROM LOCKED AUTO	Wednesday	RICHMOND	600 Block of 47TH AV	-122.508194	37.776601	NONE
9	5/13/2015 23:00	LARCENY/THEFT	GRAND THEFT FROM LOCKED AUTO	Wednesday	CENTRAL	JEFFERSON ST / LEAVENWORTH ST	-122.419088	37.807802	NONE

	Dates	Category	Descript	DayOfWeek	PdDistrict	Address	X	Y	Resolution
0	193133	24	493	3	2	21943	-122.425892	37.774599	7
1	193133	31	539	3	2	21943	-122.425892	37.774599	7
2	349110	31	539	3	2	19362	-122.424363	37.800414	7
3	103921	21	761	3	2	6505	-122.426995	37.800873	16
4	103921	21	761	3	6	20222	-122.438738	37.771541	16
5	103921	21	151	3	1	18437	-122.403252	37.713431	16
6	103921	38	863	3	1	8349	-122.423327	37.725138	16
7	103921	38	863	3	5	11952	-122.371274	37.727564	16
8	125248	21	761	3	0	19968	-122.508194	37.776601	16
9	125248	21	761	3	8	19758	-122.419088	37.807802	16

Fig. 5. Original dataset's top 10 values and its equivalent converted dataset.

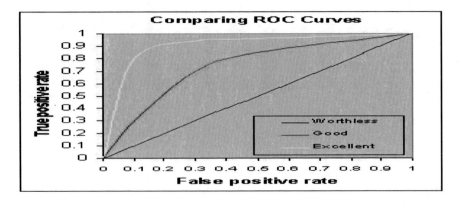

Fig. 6. Type of ROC curve and their accuracy.

6.1   Gaussian Naive Bayes Classifier

Gaussian Naive Bayes algorithm is a special type of Naive Bayes algorithm and is specifically used when the features have continuous values. It's also assumed that all the features are following a Gaussian distribution i.e., normal distribution. It is a supervised classifier with an assumption that the two features have no dependency between them. According to the Naive Bayes theorem, if there is a class A and a set of dependent vectors (i.e. $a_0, a_1, ..., a_n$) then the relationship will be

$$P(A|a_0, a_1, ..., a_n) = P(a_0, a_1, ..., a_n|A)/P(a_0, a_1, ..., a_n) \qquad (2)$$

The above model along with a decision model constitutes a Naive Bayes Classifier [11].

Using the naive independence assumption for all i, the relationship can be simplified to

$$P(y|x_1, ..., x_n) = (P(y)\Pi_{i=1}^{n}P(x_i|y))/P(x_1, ..., x_n) \qquad (3)$$

There are three popular Naive Bayes Algorithms namely Gaussian Naive Bayes, Multinomial Naive Bayes and Bernoulli Naive Bayes.

Gaussian Naive Bayes

$$P(x_i|y) = 1/\sqrt{2\pi\sigma_y^2}\exp(-((x_i - \mu)^2/2\sigma_y^2)) \qquad (4)$$

where σ_y and μ_y are estimated using maximum likelihood.

Multinomial Naive Bayes. It is used when the data is distributed in a multinomial manner and is generally used in text classification. For each class y, where n is the number of features (in text classification, the size of the vocabulary) and θ_{yi} is the probability $P(x_i|y)$ of feature i appearing in a sample belonging to class y. The parameters θ_y is estimated by a smoothed version of maximum likelihood:

$$\hat{\theta}_{yi} = N_{yi} + \alpha/N_y + \alpha n \qquad (5)$$

where $N_{yi} = \sum_{x \in T} x_i$ is number of times feature i appears and $N_y = \sum_{i=1}^{|T|} N_y i$ is total count of all features for class y.

Bernoulli Naive Bayes. It follows multivariate Bernoulli distribution where a dataset may contain multiple features and every one of them is binary valued. It is given by

$$P(x_i|y) = P(i|y)x_i + (1 - P(i|y))(1 - x_i) \qquad (6)$$

In this paper we have used Gaussian Naive Bayes. On applying the Gaussian Naive Bayes Classifier we have achieved the prediction accuracy by 65.132801%. We have also plotted the ROC Curve (Receiver Operating Characteristic curve) for the same with Area under the Curve (AUC) equal to 0.78 which is shown in Fig. 7.

6.2   Multilayer Perceptron (MLP)

Rosenblatt, in 1958 introduced a concept of single perceptron. MLP is a supervised learning algorithm. If 'i' is the number of dimensions and for input and 'o' is the number of dimension for output then the function is defined as:

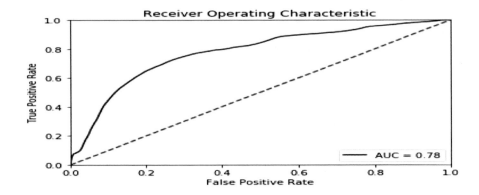

Fig. 7. ROC Curve for Gaussian Naive Bayes classifier.

$$f(x) : R^i \longrightarrow R^o \tag{7}$$

A perceptron consist of multiple real valued inputs and computes a single output by using its input weights and forming a linear combination out of it and then passing the output through a nonlinear activation function. It can be written as:

$$f = \varphi(\sum_{i=1}^{n} x_i y_i + a) = \varphi(x^T y + a) \tag{8}$$

where x denotes the weight's vector; y is the input vector; a is the bias and φ is the activation function.

Nowadays, the activation function is selected to be either the logistic sigmoid $1/(1 + e^{-x})$ or the hyperbolic tangent $\tanh(x)$. They are related by $(\tanh(x) + 1)/2 = 1/(1 + e^{-2x})$ [17]. With the help of MLP, we can learn non-linear models as well as we can learn them in real time.

The left most layer is called the input layer which contains a set of neurons $(x_1, x_2, ..., x_n)$. The layer in between the output and the input is called the hidden layer and every neuron in that layer using the weighted summation equation $w_1 x_1 + w_2 x_2 + ... + w_n x_n$ converts the previous layers value. After this, an activation function like the tan function is applied.

On applying the Multilayer Perceptron Classifier we have achieved the prediction accuracy by 60.169383%. We have also plotted the ROC Curve (Receiver Operating Characteristic curve) for the same with Area under the Curve (AUC) equal to 0.70, which is shown in Fig. 8.

6.3   k-Nearest Neighbours

The k-Nearest Neighbor method is commonly used for both supervised and unsupervised learning. k indicates the number of neighbors that are needed to classify a data point. k-NN can be measured by distance function (if the data is numeric) or Hamming distance (if the dataset is categorical) [2]. Distance function can include Euclidean, Manhattan, Minkowski. In our Experiment we have used Euclidean Distance to calculate distances.

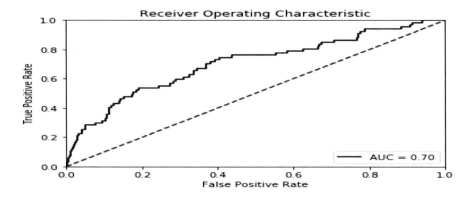

Fig. 8. ROC curve for multilayer perceptron.

On applying the K-NN classifier at various values of k, we have achieved the prediction accuracy by 60.169383% at very large values of k like 10 and so on. We have also plotted the ROC Curve (Receiver Operating Characteristic curve) for the same with Area under the Curve (AUC) equal to 0.63 which is shown in Fig. 9.

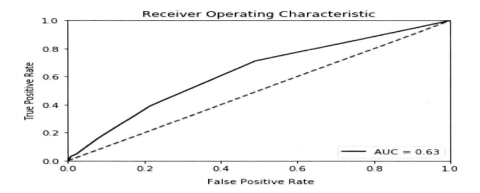

Fig. 9. ROC curve for k-NN classifier.

6.4   Classification and Regression Trees (CART)

It was given by Breiman et al. in 1984 and is based on Classification and Regression Trees. A CART is a binary tree that is constructed by splitting a node into 2 child nodes recursively.

It was an umbrella term for 2 types of trees, Classification trees and Regression Trees. In classification trees the target variable is of categorical type and the tree is used to identify the "class". In Regression Trees the target variable is of continuous nature and the tree is used for prediction of the target variable

There are 3 main elements in CART algorithm [7]:

1. Splitting rules for the node must be determined by using the value of one variable.
2. Stopping rule for determining when a branch can be split no more.
3. A prediction for the target variable in each node.

The measure used to build a CART is Gini Index and is given by:

$$Gini = \sum_{i \neq j} p(i)p(j) \tag{9}$$

where i and j are levels of the target variable.

For a nominal variable with k level the maximum value of Gini index is given by $1 - 1/k$ and the minimum value is 0 when all the observation belong to one label [7].

On applying the CART classifier, we have achieved the prediction accuracy by 80.86211%. We have also plotted the ROC Curve (Receiver Operating Characteristic curve) for the same with Area under the Curve (AUC) equal to 0.81, which is shown in Fig. 10. Among the five classifiers the best result is given by CART algorithm.

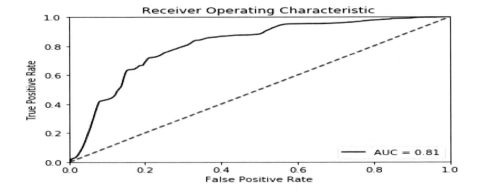

Fig. 10. ROC Curve for CART classifier.

The Table 3 provides the overall experimental results.

Table 3. Experimental results

Technique	Accuracy %	AUC
Gaussian Naive Bayes	65.132	0.78
Multilayer perceptron	60.169	0.70
k-NN	60.169	0.63
CART	**80.862**	**0.81**

6.5 k-Means

The k-means algorithm is a clustering algorithm that separates samples in n groups of equal variance, minimizing within cluster sum of square. It divides N samples into k disjoint clusters, each defined by the mean of the samples [3]. It chooses centroids that can minimize the sum of square within a cluster. The equation is:

$$\sum_{i=0}^{n} min_{\mu_i \in C}(||x_j - \mu_i||^2) \tag{10}$$

We have used k-means algorithm to make 10 clusters which are actually 10 police districts and plotted the centroid for the crime coordinates (i.e. X and Y coordinates present in dataset) for that district and shown in Fig. 11.

Fig. 11. Centroids for all the crimes in each police district.

7   Conclusion and Future Work

In this paper, we have used many classifiers to best predict the Resolution for San Francisco crime dataset and we can conclude that out of the following (Naive Bayes, k-NN, MLP, CART), CART algorithm is giving the best prediction accuracy (i.e. 80.86211%) and the best ROC with AUC = 0.81.

We still think that a higher accuracy can be achieved if we employ more analysis on the Address field. Some researchers suggested that we can use Learning by Counting, explained in [15], in order to generate log odds feature that might be useful as claimed. Someone can also use a Neural Network with large numbers of hidden layer like 512, to train on the data.

References

1. San Francisco - crime rates and statistics. https://www.neighborhoodscout.com/ca/san-francisco/crime
2. Abdulrahman, N., Abedalkhader, W.: KNN classifier and naïve bayse classifier for crime prediction in San Francisco context (2017)
3. Abouelnaga, Y.: San Francisco crime classification. arXiv preprint arXiv:1607.03626 (2016)
4. Ang, S.T., Wang, W., Chyou, S.: San Francisco crime classification. University of California San Diego (2015)
5. Bogomolov, A., Lepri, B., Staiano, J., Oliver, N., Pianesi, F., Pentland, A.: Once upon a crime: towards crime prediction from demographics and mobile data. In: Proceedings of the 16th International Conference on Multimodal Interaction, pp. 427–434. ACM (2014)
6. Braithwaite, J.: Crime Shame and Reintegration. Cambridge University Press, Cambridge (1989)
7. Breiman, L.: Classification and Regression Trees. Routledge, New York (2017)
8. Ehrlich, I.: On the relation between education and crime. In: Education, Income, and Human Behavior, pp. 313–338. NBER (1975)
9. Freeman, R.B.: The Economics of Crime. Handbook of Labor Economics, vol. 3, pp. 3529–3571 (1999)
10. Hanley, J.A., McNeil, B.J.: The meaning and use of the area under a receiver operating characteristic (ROC) curve. Radiology 143(1), 29–36 (1982)
11. Michalski, R.S., Carbonell, J.G., Mitchell, T.M.: Machine Learning: An Artificial Intelligence Approach. Springer, Heidelberg (2013)
12. Patterson, E.B.: Poverty, income inequality, and community crime rates. Criminology 29(4), 755–776 (1991)
13. Sadhana, C.: Survey on Predicting Crime Using Twitter Sentiment and Weather Data (2015)
14. Shojaee, S., Mustapha, A., Sidi, F., Jabar, M.A.: A study on classification learning algorithms to predict crime status. Int. J. Digital Content Technol. Appl. 7(9), 361 (2013)
15. Data Transformation: Learning with Counts. Microsoft Azure Machine (2015)
16. Wang, X., Gerber, M.S., Brown, D.E.: Automatic crime prediction using events extracted from twitter posts. In: International Conference on Social Computing, Behavioral-cultural Modeling, and Prediction, pp. 231–238. Springer (2012)
17. Witten, I.H., Frank, E., Hall, M.A., Pal, C.J.: Data Mining: Practical Machine Learning Tools and Techniques. Morgan Kaufmann (2016)

Fragile Watermarking Techniques
for 3D Model Authentication: Review

Kariman M. Mabrouk[1(✉)], Noura A. Semary[1,2],
and Hatem Abdul-Kader[1]

[1] Faculty of Computers and Information, Menoufia University, Menoufia, Egypt
kariman.mamdouh@ci.menofia.edu.eg
[2] Faculty of Computers and Information Technology, Jeddah University,
Khulais, Kingdom of Saudi Arabia

Abstract. The need for multimedia data in many fields such as industry, medicine, and entertainment has become a common need nowadays. 3D graphics, as a type of multimedia, attracts the attention of many researchers. The spread of programs that are used to modify and duplicate this data easily leads to increase the necessity to develop a variety of watermark technologies for various protection purposes such as copyrights protection and authentication. In this study, we provide a review study of fragile watermarking techniques for 3D models over the last decade and clarify what are the criteria and requirements for design such systems. We briefly show the different types of 3D models representation and what are the most common kinds of attacks could be applied to them, and finally discuss the challenges encountered at designing such techniques.

Keywords: 3D model · Fragile watermarking · Content authentication · Attacks

1 Introduction

Information hiding (IH) and Cryptography are two classes of information security where their goal is to protect information from unauthorized access, use, modification, or destruction to achieve confidentiality, integrity, and availability of the data. Encryption means converting data into a form that is non-recognizable by its attackers (changes the message itself). Information hiding (IH) embeds the message (watermark) over a cover signal such that its presence cannot be detected during transmission. IH can be classified into two categories: steganography and watermarking that could be used for many applications like copyright protection, content authentication, and cover communication. While the goal of steganography is to protect message itself and hide as much data as possible in the cover signal, the goal of the watermarking is to protect the cover signal by hiding data (watermark) in it. Therefore, digital watermarking is considered as an efficient solution for multimedia security as it does not change the content but preserves the original media as it is.

Watermarking methods are classified into three categories; robust watermarking, fragile watermarking and semi-fragile watermarking. Whereas, the aim of the robust watermark is to protect the ownership of the digital media, so the embedded watermark

A. E. Hassanien et al. (Eds.): AMLTA 2019, AISC 921, pp. 669–679, 2020.
https://doi.org/10.1007/978-3-030-14118-9_66

should remain detectable after being attacked. While the goal of the fragile watermark is to be sensitive to any attacks and identifies tamper localization and possibly what the model was before modification. The semi-fragile watermark has the advantages of both the robust and fragile watermark so that it is more robust than fragile watermark and less sensitive to classical user modifications. It is used to discriminate between malicious and non-malicious attack [1].

In this paper, we concerned with 3D models watermarking. This paper presents a review study of the fragile watermarking techniques, clarifies the challenges that should be taken into consideration and determines the requirements must be available for design. This paper is organized as follows, Sect. 2 discusses the different 3D model format and explains the various attacks could be applied to it. Sections 3 and 4 explains the fragile watermarking algorithm (FWA) life cycle and various FWA from the state of the art respectively and shows the evaluation measures used in the watermarking. Section 5 discusses the challenges at designing FWA. Finally, Sect. 6 shows the conclusion and sets of the future direction of the research domain.

2   3D Object Representation and Manipulations

3D model representation methods are classified into: voxels, polygonal mesh, constructed solid geometry (CSG), an implicit set of parameterized equations, such as non-uniform relational B-splines (NURBS), or point cloud 3D models as shown in (Fig. 1). Most researchers use the polygonal mesh method due to their simple structure. While common types of polygon mesh are triangle strip and quadrilateral mesh. In this paper, we concerned with fragile watermarking with the triangular 3D polygonal model representation formats where the 3D triangular model M is formed by a set of 3D points (vertices), connected between them with a set of polygons (faces) [2].

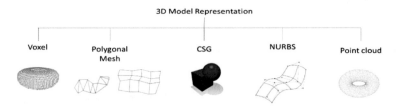

Fig. 1. 3D model representation.

The method of designing the watermark algorithm depends on which attacks should be robust against. For designing a robust watermarking algorithm, the watermark must be robust for any attacks, and able to extract the watermark after any modification.

For designing FWA, the watermark must be sensitive to any attack and detect and localize any modification in the model (tamper localization). Generally, these attacks can be classified as follows:

Geometric Attacks. Modifies the geometric part of the watermarked model. It doesn't remove the embedded watermark, but it intends to distort the watermark detector synchronization with the embedded information. Such as *similarity Transformation* (translation, rotation, uniform scaling or non-uniform scaling) and *Signal Processing Attacks* (random additional noise, smoothing, enhancement, and lossless compression). Therefore, to immune similarity transformation attack, one of these strategies should be considered: The *first solution* is to use the primitives that are invariant to similarity transformations. The *second solution* is to embed the watermark in an invariant space [3]. The spectral watermarking techniques that modify the low and median frequency parts are more robust to Signal Processing attacks than the spatial techniques.

Connectivity Attacks. Like cropping, remeshing, subdivision, and simplification. This type of attacks is considered quite difficult to handle as it tampers the connectivity of the model. Cropping is a special attack while some researchers regard it as a geometric attack. Watermark repetition in different patches and the indexed method seem the most efficient ways to resist cropping.

Attacks Affecting the Geometric Representation. Like file attack, format attack, and representation attack. File attack refers to the reordering of the vertices and/or the facets in the mesh description file. The mesh file format conversion attack alters the underlying mesh data structure, so the processing order of the vertices and facets can also be changed. So, there is a need to turn the synchronization scheme independent of these intrinsic orders, to be invariant to these two attacks. The representation conversion may be the most destructive attack to 3D mesh watermarks as the mesh itself will no longer exist.

3 Digital Watermarking Life Cycle

The general stages of 3D FWA consist of two stages; watermark embedding stage and watermark extraction stage. In the embedding stage, watermark W was embedded in the original model (cover model) M using specific key which producing a watermarked model (stego model) M' that transmitted over the network, then a specific attack can be applied to it which produces distorted watermarked model that transmitted to the receiver. The watermark W' is extracted from the distorted watermarked model M' and compared against the watermark W, to verify the integrity of the stego model as shown in (Fig. 2). To design a watermarking algorithm, there is a need to determine what is the most suitable primitives to add the watermark, then to determine the method of producing the watermark and finally, to determine the watermark embedding method.

Embedding Primitives (EP). Selection of EP to design the FWA, decides a few other factors like robustness towards content preserving attacks, attack localization capability, the complexity of the procedure and resulting distortion to the surface. Generally, the common EP are classified as Geometrical Embedding Primitives, Topological embedding primitives and Non-geometrical quantity. In the *Geometrical Embedding*

Primitives, the coordinates of points and vertices are modified to embed data as shown in Table 1. While in *Topological embedding primitives*, the watermark is embedded by changing the topology of a model, which involve a change in geometry as a side effect (e.g., inserting or displacing vertices). An example of a topological embedding primitive is the connectivity of triangles in a triangle strip [3]. Other examples of topological embedding primitives include encoding of a binary symbol by using two alternative ways of triangulating a quadrilateral, or two different mesh sizes, like in (Fig. 3).

The *Non-geometrical quantity* is associated with the vertex, line, face, or volume include color, 2D and 3D texture coordinates, normal vector and refractive index. All these attributes, essentially a set of numerical values, can be modified to embed a watermark. Frequently, geometry embedding primitives are preferred due to their higher attack localization capability and resulting with less distortion.

Arrangements of EP can be established for 3-D polygonal models by the arrangement of both geometrical and non-geometrical EP which can be performed by using one of the following; topology, geometrical quantity or non-geometrical quantity of geometrical primitives. For a proper arrangement of EP, it is often necessary to find an initial condition. For example, to determine a unique topological arrangement of vertices by using a vertex tree, an initial vertex and an initial traverse direction must be found. So, both the arrangement and initial condition must be robust against expected disturbances.

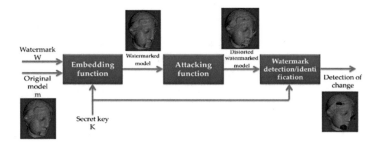

Fig. 2. 3D watermarking algorithm life cycle

Table 1. Geometrical EP and their invariance towards certain operation

Embedding primitives	Invariance towards
–Coordinates of a point	Altered by all the transformations
–Length of a line –Area of a polygon –The volume of a polyhedron –Length of mesh centroid to face centroid –Length of mesh centroid to vertex –The distance between one ring vertex neighbor and centroid to vertex –Angles of a face and Area of a face	Invariant to translation and rotation

(continued)

Table 1. (*continued*)

Embedding primitives	Invariance towards
–Two quantities that define a set of similar triangles –The ratio of the areas of two polygons –The ratio of face edges –The ratio between the two angles –Distribution of vertex norms –First order moments –Second order moments	Invariant to rotation, uniform-scaling, and translation
–The ratio of the lengths of two segments of a straight line –The ratio of the volumes of two polyhedrons	Invariant to affine transformation
–Cross-ratio of four points on a straight line	Invariant to projection transformation

Watermark Generation Pattern. Watermark generation relays on the application type, the watermark may be an external information specific to the model and must be kept secured or may be information that is not related to the model. Generally, there are two ways of watermark generation pattern: *Self-embedding* which means the watermark embedded in the cover model is a compressed version (the hash of the cover model or error correction code) of the same model by some embedding strategy. And *External information embedding* which means the watermark is an external information related or not related to the cover model. This external information could be text data, image data or pseudo-random bit sequence. And it is a need to transform the embedded data to binary bit sequence before embedding.

Fig. 3. Topological embedding primitives

4 Overview of 3D Fragile Watermarking Algorithms

Embedding method (style) primarily is divided into two classes; additive and substitutive [4]. In the case of *additive embedding* method, the watermark is considered as a random noise pattern which is added to the mesh surface which leads to increase the interference between the watermark signal and the watermarked model. But in the case of *substitutive embedding* method, the watermark is embedded in the numerical values of the mesh elements by a selective bit substitution. In the extraction stage, the watermark restoration is performed without any interference. Compared to the additive scheme, substitutive embedding capacity is enhanced. Based on this embedding style, the watermark may be added in different embedding primitives as follows:

Data File Organization. This category utilizes the redundancy of polygon models to carry information. The first FWA was proposed by Ichikawa et al. [5] where they modified the order of the triangles or the order of the triplet of vertices forming a given triangle. This method only uses the redundancy of description, it can be combined with other methods that use geometrical and topological redundancy. This method isn't blind as the need for the order of vertices in both embedding and extraction process. Under this circumstance, Wu et al. [6] used the mesh partitioning to divide the mesh into patches with a fixed number of vertices. While the geometrical and topological information of each patch, as well as other properties (color, texture, and material), are used to produce the hash value which represents the signature embedded in the model. Therefore, embedding and retrieval of the signature are performed within each patch individually, so that the tamper can be localized into the specific patches.

According to the visual presentations of a 3D model, Bennour et al. [7, 8] proposed an alternative framework for the watermarking of 3D objects. Whereas, his goal is to protect the visual presentations of a 3D object in images or videos after it has been marked. He also proposed an extension of 2D contour watermarking algorithm to a 3D silhouette. Sales et al. [9] presented a method based on the protection of the intellectual rights of 3D objects through their 2D projections. Since the most common use of 3D models is done through their visual projections in 2D or stereovision, they developed 3D watermarking schemes able to resist the visualization process undergone by the object during the 3D to 2D transformation to protect the 3D models in its 2D representations. Werghi et al. [10] addressed the problem of representation of the mesh data with an ordered structure of the matrix and the array by encoding the mesh data into a novel ordered structure named the Ordered Rings Facets (ORF). While this structure composed of concentric rings in which the triangles are arranged in a circular fashion. This representation exhibits several interesting features that include a systematic traversing of the whole mesh model, simple mechanisms for avoiding the causality problem, and an efficient computation of the embedding distortion.

Topological Data. These algorithms use the topology (connectivity) of the 3D object to insert the watermark which leads to change in the triangulation of the mesh. Ohbuchi et al. [3] presented two visible algorithms where the local triangulation density is changed to insert a visible watermark depending on the triangle similarity quadruple (TSQ) algorithm. Whereas the second is to embed a blind watermark by topological ordering TVR (Tetrahedral Volume Ratio) method. Mao et al. [11] proposed a new watermarking technique that triangulated a part of a triangle mesh to embed the watermark into the newly positions of the vertices, that lead to achieving high capacity as up to 8 bytes data can be invisibly embedded into an edge of the triangle mesh without causing any distortion. This algorism is considered reversible because it allows full extraction of the embedded information and complete restoration of the cover signal.

Geometrical Data. Most of the 3D FWA embed the watermark by modifying the geometry of the 3D object. In this category, some of them operate in the spatial domain by modifying vertices, normal (direction or length) and geometrical invariants (i.e. the length of a line, area of a polygon, etc.). While, others embed information in a transform domain (spectral decomposition, wavelet transform, or spherical wavelet transform).

Yeo and Yeung [12] proposed the first 3D FWA where each of the vertex information is modified by slightly perturbing the vertex based on a pre-defined hash function to make all vertices valid for authentication. At the extraction phase, one simply examines the validity of each vertex and locates the possible attacks on the invalid vertices. Whereas Lin et al. [13], and Chou et al. [14] solved the causality problem raised in Yeo's method by setting both hash functions depending only on the coordinates of the current vertex [13] and proposed a multi-function vertex embedding method and an adjusting-vertex method [14], whereas this method does not need the original model and watermarks for authentication.

With considering high-capacity watermarks that often fragile and precise attack localization capability, Cayre and Macq [15] proposed a high-capacity blind data-hiding algorithm for triangular meshes. They considered a triangle as a two-state geometrical object and classify the triangle edges based on the traversal into entry edge and exit edge, where the entry edge is modulated using Quantization index modulation (QIM) to embed watermark bits. By quantizing the distance from a facet to the mesh center. Due to the tradeoff between embedding capacity and achieving authentication, Recently Tsai et al. [16] proposed two strategies to achieve higher embedding capacity by modifying the message grabbing method to raise the embedding capacity exceeding 11%. Further, they also modify the embedding ratio generation method, to increase the embedding capacity to 50%.

To immune similarity transformation attacks, Chou et al. [17] proposed a transformation invariant public FWA for 3D model authentication. Where they embedded watermarks in a subset of the model's faces so that any changes will ruin the relationship between the mark faces and neighboring vertices. Huang et al. [18] proposed a new spherical coordinate based FWA, where the 3D model is translated to the spherical coordinate system, then they used the QIM technique to embed the watermark into the r coordinate for authentication and verification. Another QIM based FWA uses the spherical coordinate system introduced by Molaei et al. [19], where the data was embedded into the middle of three sides of the marked triangle in the spherical coordinates. They also adopted the concept of multi-function vertex embedding strategy to embed the mark vertex identification, the watermark and the hash bit to x, y, and z coordinates. Xu and Zhao-Quan [20] used the Principal Component Analysis PCA to generate a parameterized spherical coordinates mapping square-matrix to embed a binary image (watermark).

According to the substitutive embedding style, Wang et al. [21] proposed a numerically stable fragile watermarking scheme by considering the mantissa part of the floating-point number as an unsigned integer and operates on it by the bit XOR operator, where the number of ones in the mantissa bit sequence used to check the integrity of the content at the extraction process. Wang et al. [22] modified the integral invariant of the vertices to embed the watermark image. Although this method is robust against normal use modifications, the capacity is very less, and could not achieve attack localization. Chen et al. [23] used the adaptive authentication scheme (AAS), where the position and the connectivity relationship of a vertex, the vertex degree, coordinate x and y, and watermark are combined with a hash function and embedded into the vertex z coordinate. They used the vertex digest scheme (VDS) which follows the same rationale, but the watermark is replaced by a key.

Wang et al. [24] used the hamming code to calculate the parity bits that embedded in each vertex coordinate with the LSB substitution to achieve verification during the extraction stage, whereas this lead to increase the data hiding capacity but the embedding distortion to the model was uncontrollable. Authors claimed the method to be immune to the causality, convergence and embedding holes problems. Due to the problem of high collision characteristic of the hash function used for generating the watermark from the mesh model. Wang et al. [25] depended on the concept of error detecting code to achieve the verification task. Where they use spherical coordinates to embed the watermark into the angular vertex coordinates by simple LSB substitution. Another error correcting code based semi-fragile watermarking is proposed by Chang et al. [26] in a spherical coordinate system that enables the authentication of 3D models for detecting unauthorized alterations. An adaptive watermark is generated from each cover model by using Hamming code technique, which is embedded by employing a simple LSB substitution technique. According to external-embedding watermark generation, and the problem of high collision characteristic of hash function used for generating the watermark from the mesh model Wang et al. [27] employed a chaotic sequence generator to generate the embedded watermark to achieve both the authentication and verification.

Performance Evaluation. To measures the performance of the mesh watermarking algorithm there is an evaluation metrics could be used as shown in Table 2. Generally, there are three types of measures; (1) *Imperceptibility measures* used to check the model quality after the watermark is embedded. There is a set of imperceptibility measures proposed to be sensitive to mesh perceptual degradation to detect any change in mesh geometry. Many of the FWA have used the perceptual correlation measure [13], where M and M' identifying the original mesh and the embedded mesh, v_i are the vertices, x_i, y_i, z_i are the coordinate details. (2) *Robustness measures* used to measure if the method withstands to any change operation. Where most researchers use the Correlation coefficient to determine the robustness of the watermarking scheme at the watermark extraction stage. (3) In *tamper detection measures,* the aim is to determine the rate of the correctness of the method using the Bit Error Rate (BER). While False Positive Rate (FPR) shows the rate of correctly classified attacked mesh $N_{correct}$ out of N_{total} number of meshes, the False Negative Rate (FNR) shows the rate of falsely rejected tampered mesh $N_{incorrect}$, where the actual number of the tampered mesh is $N_{tampered}$ [28].

Table 2. Performance assessment measures used in mesh watermarking

Assessment type	Assessment measure	Formula
Imperceptibility measures	Hausdorff distance (HD)	$d(v, M') = min_{v' \in M'} \|v - v'\|$
	Modified Hausdorff distance (MHD)	$D(M, M') = \max(d_H(M, M'), D_H(M', M))$
	Signal to noise ratio (SNR)	$SNR_{mesh} = 10log_{10} \frac{\sum_{i=1}^{n_v}(x_i^2, y_i^2, z_i^2)}{\sum_{i=1}^{n_v}(x_i' - x_i)^2 + (y_i' - y_i)^2 + (z_i' - z_i)^2}$
	Root mean square error (RMSE)	$d_{rms}(M, M') = \sqrt{d_{v \in M}(v, M')^2 dM}$
	Geometrical Laplacian (GL)	$GL(v_i) = v_i - \frac{\sum_{j \in N(i)} l_{ij}^{-1} v_j}{\sum_{j \in N(i)} l_{ij}^{-1}}$
Robustness measures	Correlation coefficient	$corr(w^d, w) = \frac{\sum_{i=0}^{N-1}(w_i^d - \overline{w^d})(w_i - \overline{w})}{\sqrt{\sum_{i=0}^{N-1}(w_i^d - \overline{w^d})(w_i - \overline{w})}}$
Tamper detection measures	Bit Error rate	$BER(M, M') = 1 - \frac{1}{n_b}\sum_{n=1}^{n_b} \delta_{(m_i, m_i')}$
	False Positive Rate (FPR) False Negative Rate (FNR)	$FP = \frac{N_{correct}}{N_{tempered}}, \quad FN = \frac{N_{incorrect}}{N_{total} - N_{tempered}}$

5 Watermarking Requirements and Challenges

The main requirements to provide an effective watermark are imperceptibility, robustness against intended or non-intended attacks and capacity. Imperceptibility indicates that the watermarked model should look like the original model (perceptually equivalent). Robustness means that the watermark should be able to resist different attacks and manipulation operations. Capacity corresponds to the amount of information embedded in the model without producing any perceptible distortions on the content.

There is a clear trade-off between these requirements. If the capacity increased, this might lead to visible deformations in the content. If there is a need to achieve robustness of the watermark, this might also lead to degradation of imperceptibility. So, the optimum compromise between these requirements is dependent on the application. While the number of hidden bits, in some applications, such as broadcast monitoring, should be sufficient to differentiate all broadcasts from each other. Moreover, in copyright protection, it might require a fewer bit of hidden information that indicates the owner of the content.

6 Conclusion and Future Work

In this study, we presented a review study of the most cited 3D fragile watermarking algorithms by classifying major algorithms into three classes: Data file organization, topological data, and geometrical data. To find out subclasses we highlighted a few other attributes and the requirements on which the algorithms are highly dependent on. While the major function of fragile watermarking is for content authentication. The design goal of fragile watermarking is to make the embedded watermarks sensitive to any attack. So, the researcher must take the criteria and requirements of the watermarking algorithm as we mentioned in mind at the designing process.

A future work could focus on the tamper recovery for 3D meshes, which need a good amount of prior information. Moreover, another work is to increase the tamper detection accuracy and minimize the false positive detections.

References

1. Chou, C.M., Tseng, D.C.: Technologies for 3D model watermarking: a survey. Int. J. Comput. Sci. Netw. Secur. **7**(2), 328–334 (2007)
2. Bors, A.G.: Watermarking mesh-based representations of 3-D objects using local moments. IEEE Trans. Image Process. **15**(3), 687–701 (2006)
3. Ohbuchi, R., Masuda, H., Aono, M.: Watermarking three-dimensional polygonal models through geometric and topological modifications. IEEE J. Sel. Areas Commun. **16**(4), 551–560 (1998)
4. Dugelay, J.L., Baskurt, A., Daoudi, M. (eds.): 3D Object Processing: Compression Indexing and Watermarking. Wiley, Hoboken (2008)
5. Ichikawa, S., Chiyama, H., Akabane, K.: Redundancy in 3D polygon models and its application to digital signature. J. WSCG **10**(1), 225–232 (2002)
6. Wu, H.T., Cheung, Y.M.: Public authentication of 3D mesh models. In: 2006 IEEE/WIC/ACM International Conference on Web Intelligence (WI 2006 Main Conference Proceedings) (WI 2006), Hong Kong, pp. 940–948, December 2006
7. Bennour, J., Dugelay, J.L.: Protection of 3D object visual representations. In: 2006 IEEE International Conference on Multimedia and Expo, Toronto, Ontario, pp. 1113–1116, July 2006
8. Bennour, J., Dugelay, J.L.: Protection of 3D object through silhouette watermarking. In: 2006 IEEE International Conference on Acoustics Speech and Signal Processing Proceedings, Toulouse, pp. 221–224, May 2006
9. Sales, M.M., Rondao Alface, P., Macq, B.: 3D objects watermarking and tracking of their visual representations. In: The Third International Conferences on Advances in Multimedia (2011)
10. Werghi, N., Medimegh, N., Gazzah, S.: Watermarking of 3D triangular mesh models using ordered ring facets. In: 10th International Multi-Conference on Systems, Signals and Devices, SSD, pp. 1–6 (2013)
11. Mao, X., Shiba, M., Imamiya, A.: Watermarking 3D geometric models through triangle subdivision. In: Security and Watermarking of Multimedia Contents III, vol. 4314, pp. 253–260, August 2001
12. Yeo, B.L., Yeung, M.M.: Watermarking 3D objects for verification. IEEE Comput. Graph. Appl. **19**(1), 36–45 (1999)

13. Lin, H.Y., Liao, H.Y., Lu, C.S., Lin, J.C.: Fragile watermarking for authenticating 3-D polygonal meshes. IEEE Trans. Multimedia **7**(6), 997–1006 (2005)
14. Chou, C.M., Tseng, D.C.: A public fragile watermarking scheme for 3D model authentication. Comput. Aided Des. **38**(11), 1154–1165 (2006)
15. Cayre, F., Macq, B.: Data hiding on 3-D triangle meshes. IEEE Trans. Signal Process. **51**(4), 939–949 (2003)
16. Tsai, Y.Y., Tsai, Y.S., Chang, C.C.: An improved region-based tampered detection algorithm for 3D polygonal models. In: 2018 IEEE International Conference on Applied System Invention (ICASI), Chiba, pp. 1163–1166 (2018)
17. Chou, C.M., Tseng, D.C.: Affine-transformation-invariant public fragile watermarking for 3D model authentication. IEEE Comput. Graph. Appl. **29**(2), 72–79 (2009)
18. Huang, C.C., Yang, Y.W., et al.: Spherical coordinate based fragile watermarking scheme for 3D models. In: International Conference on Industrial, Engineering and Other Applications of Applied Intelligent Systems, pp. 566–571. Springer, Heidelberg, June 2013
19. Molaei, A.M., Ebrahimnezhad, H., Sedaaghi, M.H.: A blind fragile watermarking method for 3D models based on geometric properties of triangles. 3D Res. **4**(4), 4 (2013)
20. Tao, X., Zhao-Quan, C.: A novel semi-fragile watermarking algorithm for 3D mesh models. In: 2012 International Conference on Control Engineering and Communication Technology, Liaoning, pp. 782–785 (2012)
21. Wang, W.B., Zheng, G.Q., Yong, J.H., Gu, H.J.: A numerically stable fragile watermarking scheme for authenticating 3D models. Comput. Aided Des. **40**(5), 634–645 (2008)
22. Wang, Y.P., Hu, S.M.: A new watermarking method for 3D models based on integral invariants. IEEE Trans. Vis. Comput. Graph. **15**(2), 285–294 (2009)
23. Chen, T.Y., Hwang, M.S., Jan, J.K.: Adaptive authentication schemes for 3D mesh models. Int. J. Innovative Comput. Inf. Control **5**(12), 4561–4572 (2009)
24. Wang, J.T., et al: Hamming code based watermarking scheme for 3D model verification. In: 2014 International Symposium on Computer, Consumer and Control (I3C), Taichung, pp. 1095–1098 (2014)
25. Wang, J.T., et al.: Error detecting code based fragile watermarking scheme for 3D models. In: 2014 International Symposium on Computer, Consumer and Control (I3C), Taichung, pp. 1099–1102, June 2014
26. Chang, Y.C., et al.: An error-detecting code based fragile watermarking scheme in spherical coordinate system. In: 2016 International Symposium on Computer, Consumer and Control (IS3C), Xi'an, pp. 287–290, July 2016
27. Wang, J.T., et al.: A novel chaos sequence based 3D fragile watermarking scheme. In: 2014 International Symposium on Computer, Consumer and Control (IS3C), Taichung, pp. 745–748, June 2014
28. Lavoué, G.: A roughness measure for 3D mesh visual masking. In: Proceedings of the 4th Symposium on Applied perception in Graphics and Visualization, pp. 57–60. ACM (2007)

Supervised Performance Anomaly Detection in HPC Data Centers

Mohamed Soliman Halawa[1]($\boxtimes$), Rebeca P. Díaz Redondo[2],
and Ana Fernández Vilas[2]

[1] Information System Department,
Arab Academy for Science Technology and Maritime Transport, Cairo, Egypt
halawamohamed@hotmail.com
[2] Information & Computing Lab., AtlantTIC Research Center
School of Telecommunications Engineering, University of Vigo, Vigo, Spain
{rebeca, avilas}@det.uvigo.es

Abstract. High Performance Computing (HPC) systems play an important role in advancing scientific research due to a significant demand for processing power and speed grows. In practice, HPC systems are in the spot of interest of different businesses which account on this growing technology. The growing complexity of the HPC systems made it exposed to a great range of performance anomalies. Permanent management of such systems health has a huge impact financially and operationally. Several machine learning techniques can be used to identify these performance anomalies in such complex systems. This study compares the most commonly used three supervised machine learning algorithms for anomaly detection. We had applied these algorithms on the Fundación Pública Galega Centro Tecnolóxico de Supercomputación de Galicia (CESGA) memcpy metrics which is a benchmark used to measure memory performance for each CPU socket. Our study shows that Neural Network algorithm had the highest accuracy (93%), KNN algorithm had the highest value of precision (0.97), Gaussian Anomaly Detection algorithm had the highest value of recall (0. 99), and Neural Network algorithm had the highest value of F-measure (0.96).

Keywords: Cloud computing · High Performance Computing ·
Anomaly detection · Machine learning

1 Introduction

In recent years, the demand for High-Performance Computing (HPC) data centers has increased due to high computing technology [1]. HPC data centers attract different businesses and institutions of different fields and capacities to rely on HPC computing power in their complex applications [2]. HPC often consists of thousands of computing services. As HPC systems and applications continue to increase in complexity, they become more vulnerable to face more performance problems like resource contention, software and firmware-related problems, etc. This probably leads to premature job termination, reduced performance, and/or wasted computing platform resources.

© Springer Nature Switzerland AG 2020
A. E. Hassanien et al. (Eds.): AMLTA 2019, AISC 921, pp. 680–688, 2020.
https://doi.org/10.1007/978-3-030-14118-9_67

HPC systems have high costs for the setup and maintenance of such systems. Therefore, it is essential for the HPC control and operators to monitor and analyze the performance of such a complex system environment [2]. HPC system monitoring may help in early detection of the performance data that may contain evidence of anomalies. HPC systems generate a huge amount of data per day as metrics of resource usage data and other Key Performance Indicators (KPI) form thousands of computational nodes. Therefore, manually monitoring systems in this size and complexity is an impossible task.

The KPIs are subcategories of the Service Measurement Index (SMI) [3] which were introduced by Cloud Services Measurement Initiative Consortium (CSMIC) [4] as shown in Fig. 1. The KPIs are a set of metrics that provide a standardized method for measuring and comparing business services such as the cloud service. The KPIs metrics are organized in groups of limited system resources, such as memory, CPU, disk, network, storage and so on [5]. These metrics help the system providers identify the source of any performance problems (anomalies).

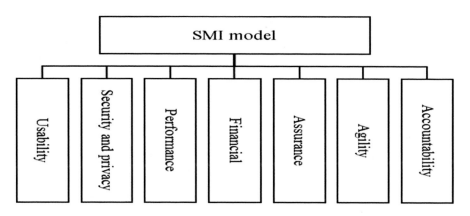

Fig. 1. SMI model

HPC performance anomalies is a major challenging topic today as HPC systems grow in capacity and complexity. These challenges directed the research on computational intelligence into a new era to develop the tools and algorithms to identify these anomalies and their root causes. These tools use different data analytics techniques such as statistical techniques, machine learning, etc. These tools are able to capture data on a large number of the time-varying system performances metrics and analyze the relationships among system components and applications.

In this study, we compare some supervised machine learning algorithms that can detect compute nodes that have shown sign performance anomalies. The rest of the paper is organized as follows: Sect. 2 gives a brief related work of how supervised learning approach is being used to detect anomalies. The methodology is described in Sect. 3. Experiments and Results are given in Sect. 4. Finally, a conclusion is given in Sect. 5.

2 Related Works

Detecting anomalies in HPC systems has experienced extremely valuable research in various domains over recent years, in which classification techniques were widely and commonly used by researchers to detect anomalies. However, such classification techniques require a dataset with a sufficient and diverse set of annotated anomalies. Hence, authors in [6] specified the pros and cons of the most common classification techniques, namely, clustering analysis, isolations forests, and neural networks classifiers. They specified that both the clustering analysis and isolations forests techniques do not only require annotations for training, but also are effective only when there are many features. Therefore, such classifiers are non-useful on time-series data, especially when there are no other available features. On the other hand, they mentioned that artificial neural networks have the ability to predict periodic time-series data. However, each problem potentially required its own model because whether a point is considered an anomaly or not might differ based on the data. Accordingly, they trained a separate model for each data stream, in which no significant correlation was found between each data stream and another. Moreover, those models considered a current point anomalous by returning predictions to the actual data stream values using the anomaly detection rules.

Since their proposed models only predicted a point through regression, the authors mentioned the necessity for a defined rule to determine whether a point is anomalous or not. They initially suggested to employ the Euclidean distance approach along with a predefined threshold, in order to compare the predicted and actual values. However, the main drawback of such approach is that it could lead to many false positives especially with noisy data. Accordingly, they employed two other different rules to detect anomalies; the first one utilized an accumulator to detect continuous outages, whereas the other employed a statistical method presented by Ahmad and Purdy [7]. It had been found that the accumulator approach was easy to be modified and controlled, whereas the probabilistic approach gave a percentage chance of being an outlier, which could be multiplied between models to potentially give a more robust score.

Tuncer et al. [1] proposed a model that detected performance anomalies in HPC systems via machine learning. Throughout several application runs, time series data was collected from the resource usage and performance counters. Consequently, this collected data went through a dimensionality reduction technique to speed-up computation of their machine learning algorithms. Then these algorithms were applied to classify the types of anomalies detected during running applications. Their framework was tested on HPC cluster and on a public cloud, achieving an outstanding F-score that exceeded 0.97.

In order to detect performance anomalies in distributed multi-server systems, Peiris et al. [8] developed a Performance Anomaly Detector (PAD). That PAD tool aimed to collect performance data which provided developers of distributed systems with insights about the system performance, which in turn helped in troubleshooting performance problems and finding root causes. Besides, PAD tool reduced the developers' time spent in analyzing the numerous amounts of performance data generated across multi-server systems. The authors tested the PAD tool on the Orleans framework. It had significantly proven to assist developers in detecting the performance anomalies found in distributed multi-server systems. Such assistance enabled the distributed system developers to perform a deeper analysis of such performance behaviors.

While the approach showed potential benefits related to the developers, the approach also faced some challenges which are common in other classification techniques used to detect the anomalies. The challenges were in analyzing performance counters like huge data sizes, insufficient training data, and time correlation. Huge data sizes generated from such systems performance data is unbearable to examine manually. In addition, it is difficult to select a group of performance counters to be considered. Secondly, insufficient annotated training data needed by classification techniques to train and build the predictive model, are rarely found in such systems. Thirdly, time correlation as the physical clock are different across machines, which make it hard sometimes to associating data collected from different machines across time. Sometimes time approximation may possibly give false results [8].

For the mentioned reasons, our study intends to evaluate some classification techniques on well-known defined labeled dataset to identify anomaly in memory performance for CPU sockets.

3 Methodology

The overall framework for detecting performance anomalies in HPC data centers is illustrated in Fig. 2. Supervised machine learning algorithms were used to classify and detect anomalies in HPC system KPI's. Consequently, the algorithms were compared to identify which had the highest accuracy.

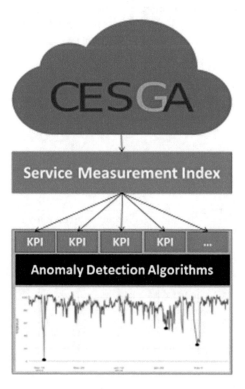

Fig. 2. Proposed anomaly detection framework

3.1   Dataset Collection

The dataset was acquired from the Fundación Pública Galega Centro Tecnolóxico de Supercomputación de Galicia (CESGA) [9] in Spain. The dataset contains 85 annotated jobs with more than 30 parallel nodes from June 1, 2018 to July 31, 2018 in a CSV format. The study used time series memcpy datasets which are a manufactured metric ex-process to detect the anomalies correspond to a change point in the metric, with a peak and a quite pronounced level change in the CPU. Also, it is a benchmark used to measure memory performance for each CPU socket. The values of memcpy performance metrics were recorded every 15 min.

3.2   Modeling

We used the software Python 3.7 for modeling our data. Python 3.7 is rich with many machine learning and data analytic libraries' such as pandas [10], numpy [11] and scikit-learn [12]. These libraries are helpful in the preprocessing phase, visualizing the time series and applying the machine learning algorithms to estimate the accuracy of our predictive models.

3.3   Performance Measures

In order to measure the model performance, the dataset was spited into two sets: training set and testing set. The training was used to train the model, and the testing set was used to appraise the model in order to compute its accuracy, precision, recall, and F-value. These performance measures are defined by Eqs. (1–4), whose terms are explained in the confusion matrix shown in Table 1 [13, 14]:

$$\text{Accuracy} = \frac{TP + TN}{TP + FN + FP + TN} \tag{1}$$

$$\text{Recall} = \frac{TP}{TP + FN} \tag{2}$$

$$\text{Precision} = \frac{TP}{TP + FP} \tag{3}$$

$$\text{F--Measure} = \frac{2 * \text{Recall} * \text{Precision}}{\text{Recall} + \text{Precision}} \tag{4}$$

Table 1. Confusion matrix for anomaly detection

	Predicted anomaly	
	Anomaly	Normal
Anomaly	Correctly detected (TP)	False negative (FN)
Normal	False detected (FP)	False negative (FN)

4   Experiment

In this study, we compared the current commonly used three supervised machine learning algorithms to detect and classify anomalies in our memcpy dataset to determine the algorithm of the highest accuracy. The algorithms were Neural network, KNN, and Gaussian anomaly detection. Each one has its own characteristic and setup to detect and classify anomalies.

4.1   Experiments Using Neural Networks

Our neural network shown in Fig. 3, has 3 layers: an input layer, a hidden layer and an output layer. The input layer reflects the number of nodes in each job. We had 32 input nodes, which is estimated as the number nodes in each job plus the extra bias unit which always outputs +1.

The neural network algorithm was loaded with two data files, the first one contained a training data file, and the second file contained a set of network parameters $\Theta 1$ and $\Theta 2$ that were randomly initialized. The parameters $\Theta 1$ and $\Theta 2$ had a dimension size based on the hidden layer and the output layer. The hidden layer size was determined using the following equation [15]:

$$\text{Hidden layer size} = (\text{Number of inputs} + \text{Number of outputs}) * (2/3) \quad (5)$$

So, based on this rule our hidden layer was 23 nodes plus the extra bias unit which always outputs +1 and the output layer was 1 node which is either 0 or 1.

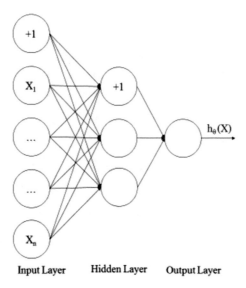

Fig. 3. Neural Network

We trained the Neural Network algorithm using randomly initialized weights to the get output value of the hypothesis hθ $(X^{(I)})$ for any $(X^{(I)})$ with forward propagation, then calculated the cost function, where $(X^{(I)})$ is the training examples. Then, the algorithm computed the partial derivatives by applying backpropagation. Consequently, the algorithm used an optimization function to minimize the cost function with the weights in theta. The algorithm looped on every training instance during forward and back propagation [16].

4.2   Experiment Using KNN

KNN algorithm was loaded with one data file as the training dataset which contains the nodes performance data and its corresponding classes, either 0 for normal or 1 for an anomaly.

KNN algorithm used k = 3, where the dataset was split into 60% for training data and 40% for testing. KNN algorithm used the 60% training data to calculate the Euclidean Distance for all the points training set, and then compare the distance of each point in the testing set to determine which were the closest points based on K. The algorithm chooses the predicted class based on the closest point [17].

4.3   Experiment Using Gaussian Anomaly Detection

In this study, the Gaussian Anomaly Detection algorithm was loaded with one dataset split in two sets; (1) a training set which contains a majority of normal performance to fit the data distribution model. And (2) a testing set which contains a majority of anomalies' performance to test and validate the model.

We trained the Gaussian anomaly detection algorithm with the training set to estimate the Gaussian distribution using μ and σ^2 for each point in the time series. Where μ is the mean and σ^2 is the variance. The algorithm then selected a threshold using the F1 score on a cross validation set. The testing set was used to compare the probability of each point with the threshold. The points with a probability lower than the threshold were considered anomalies, whereas the points with a probability higher than the threshold were considered normal.

4.4   Experimental Results

Our study shows that Neural Network algorithm had the highest accuracy (93%), followed by the KNN algorithm (90%) and Gaussian Anomaly Detection algorithm (79%) for detecting anomaly in the memcpy metrics. The relatively low accuracy of Gaussian Anomaly Detection algorithm may be due to the training data having a higher distribution of normal instances and may need further annotated data to train. Figure 4, shows the accuracy of these three algorithms.

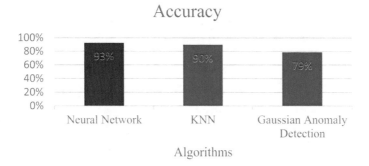

Fig. 4. Anomaly detection algorithms accuracy

Our study shows that KNN algorithm had the highest value of Precision (0.97) followed by the Neural Network algorithm (0.94), and Gaussian Anomaly Detection algorithm (0.51) for detecting anomaly in the memcpy metrics.

On the other hand, our study shows that Gaussian Anomaly Detection algorithm had the highest value of Recall (0. 99) followed by the Neural Network algorithm (0.98), and KNN algorithm (0.89) for detecting anomaly in the memcpy metrics.

Furthermore, our study shows that Neural Network algorithm had the highest value of F-Measure (0.96) followed by the KNN algorithm (0.93), and Gaussian Anomaly Detection algorithm (0.68) for detecting anomaly in the memcpy metrics. Table 2 presents the results of the algorithms regarding Precision, Recall and F-Measure for detecting anomaly in the memcpy metrics.

Table 2. Algorithms results.

Machine learning algorithms	Precision	Recall	F-Measure
Neural Network	0.94	0.98	0.96
KNN	0.97	0.89	0.93
Gaussian Anomaly Detection	0.51	0.99	0.68

5 Conclusion

This paper gives a comparative study of supervised machine learning algorithms: Neural Network, KNN and Gaussian Anomaly Detection for anomaly detection in HPC systems. The performance of the algorithms was tested on CESGA memcpy metrics which is a benchmark used to measure memory performance for each CPU socket. Each algorithm has a different result for detecting anomaly. But the overall accuracy of the Neural Network is the highest among the other two algorithms. The supervised learning was the best option in this study, since we were supported with annotated data samples, which is rarely the case in such system. So in the future, we need to overcome these types of challenges that face the HPC anomaly detection like large data volumes, insufficient training data, and time correlation. In addition, we need to explore unsupervised machine learning approaches that are more compatible with the nature of data generated from these complex systems.

Acknowledgments. The European Regional Development Fund (ERDF) and the Galician Regional Government under the agreement for funding the Atlantic Research Center for Information and Communication Technologies (AtlantTIC), the Spanish Ministry of Economy and Competitiveness under the National Science Program (TEC2014-54335-C4-3-R and TEC2017-84197-C4-2-R). Finally, the authors would like to thank the Supercomputing Center of Galicia (CESGA) for their support and resources in this research.

References

1. Tuncer, O., et al.: Diagnosing performance variations in HPC applications using machine learning. In: International Supercomputing Conference. Springer (2017)
2. Sorkunlu, N., Chandola, V., Patra, A.: Tracking system behavior from resource usage data. In: IEEE International Conference on Cluster Computing (CLUSTER). IEEE (2017)
3. Garg, S.K., Versteeg, S., Buyya, R.: SMICloud: a framework for comparing and ranking cloud services. In: Fourth IEEE International Conference on Utility and Cloud Computing (UCC). IEEE (2011)
4. Kumar, N., Agarwal, S.: QoS based cloud service provider selection framework. Res. J. Recent Sci. (2014). ISSN 2277-2502
5. Ibidunmoye, O., Hernández-Rodriguez, F., Elmroth, E.: Performance anomaly detection and bottleneck identification. ACM Comput. Surv. (CSUR) **48**(1), 4 (2015)
6. Shipmon, D.T., et al.: Time series anomaly detection; detection of anomalous drops with limited features and sparse examples in noisy highly periodic data. arXiv preprint arXiv: 1708.03665 (2017)
7. Ahmad, S., Purdy, S.: Real-time anomaly detection for streaming analytics. arXiv preprint arXiv:1607.02480 (2016)
8. Peiris, M., et al.: PAD: performance anomaly detection in multi-server distributed systems. In: IEEE 7th International Conference on Cloud Computing (CLOUD). IEEE (2014)
9. Cesga. http://www.cesga.es/en/. Accessed 16 Oct 2018
10. Seabold, S., Perktold, J.: Proceedings of the 9th Python in Science Conference (2010)
11. Walt, S.v.d., Colbert, S.C., Varoquaux, G.: The NumPy array: a structure for efficient numerical computation. Comput. Sci. Eng. **13**(2), 22–30 (2011)
12. Pedregosa, F., et al.: Scikit-learn: machine learning in Python. J. Mach. Learn. Res. **12**, 2825–28309 (2011)
13. Shyu, M.-L., et al.: A novel anomaly detection scheme based on principal component classifier. Miami Univ Coral Gables Fl Dept of Electrical and Computer Engineering (2003)
14. Thill, M., Konen, W., Bäck, T.: Online anomaly detection on the webscope S5 dataset: a comparative study. In: Evolving and Adaptive Intelligent Systems (EAIS). IEEE (2017)
15. Panchal, G., Panchal, M.: Review on methods of selecting number of hidden nodes in artificial neural network. Int. J. Comput. Sci. Mob. Comput. **3**(11), 455–464 (2014)
16. Agrawal, S., Agrawal, J.: Survey on anomaly detection using data mining techniques. Procedia Comput. Sci. **60**, 708–713 (2015)
17. Hodo, E., et al.: Shallow and deep networks intrusion detection system: a taxonomy and survey. arXiv preprint arXiv:1701.02145 (2017)

Online Signature Verification Using Deep Learning and Feature Representation Using Legendre Polynomial Coefficients

Amr Hefny[1](✉) and Mohamed Moustafa[2](✉)

[1] Mathematics Department, Faculty of Science, Cairo University, Giza 12613, Egypt
amrhefny@sci.cu.edu.eg
[2] Computer and Systems Engineering Department, Faculty of Engineering,
Ain Shams University, Cairo, Egypt
m.moustafa@aucegypt.edu

Abstract. Handwritten signing are one of the most popular behavioral biometrics. They are widely accepted for verification purposes, such as authenticating legal documents and financial contracts. In this paper, Legendre polynomials coefficients are used as features to model the signatures. The classifier used in this paper is deep feedforward neural network and the deep learning algorithm is stochastic gradient descent with momentum. The experimental results show better Equal Error Rate reduction and accuracy enhancement on SigComp2011 Dataset presented within ICDAR 2011 in comparison with state-of-the-art methods.

Keywords: Online signature · Deep learning · Machine learning · Legendre polynomials

1 Introduction

Bio-metrics refers to the automatic recognition of individuals based on their physiological and/or behavioral traits, rather than the recognition of individuals based on knowledge and based on some tokens that relies on what the person possesses that can be easily stolen or relies on what the person remembers that can be easily forgotten, respectively Plamondon et al. (2014). So, biometrics is divided into two categories: physiological and behavioral biometrics. Physiological biometrics is based on the measurement of biological traits of users, like hand geometry, fingerprint, face and iris, while behavioral biometrics consider behavioral traits of users, such as handwritten signature or voice Khalil et al. (2009). Bio-metric traits are supposed to be universal (each person should have the trait), unique (there is no two persons should have the same trait), invariable (the trait should neither change nor be alterable) and obtainable (the trait can be obtained easily) Jain et al. (2011). However, there is no ideal bio-metric trait completely satisfies all supposed characteristics required for a bio-metric system Plamondon et al. (2014). Although signatures are not good traits with

© Springer Nature Switzerland AG 2020
A. E. Hassanien et al. (Eds.): AMLTA 2019, AISC 921, pp. 689–697, 2020.
https://doi.org/10.1007/978-3-030-14118-9_68

respect to universality, uniqueness, and permanence, but they are collectable and have been accepted in government, legal, and commercial transactions as a method of verification Khalil et al. (2009). The task of determining whether or not the signature is that of a given person is called Signature verification. Hafemann et al. (2017) Unfortunately, signature verification is usually done by visual inspection. Moreover, in the majority of situations no verification takes place at all where a signature is required, due to the amount of time and effort that would be required for manual signature verification. Automating the signature verification process tries to improve the current situation and eliminate fraud Jain et al. (2011). Automating signature verification is divided into offline and online signature verification. In the online case, the two-dimensional coordinates of successive points of the writing as a function of time are stored in order, i.e., the order of strokes made by the writer is readily available. In the offline case, only the completed signature is available as an image. The online case deals with a spatio-temporal representation of the input, whereas the offline case involves analysis of the spatio-luminance of an image Parodi and Gómez (2014). In despite of online signature is far from being natural, in online signature the dynamics of writing, which is not present in the 2-D representation of the signature, is captured in addition to shape, and hence online signature is difficult to forge. And that gets more attention to online signature over offline counterpart. The main methods have been used for verification techniques are: probabilistic classifiers, time warping or dynamic matching, neural networks, hidden Markov models, Euclidean or other distance measure Jain et al. (2011), Plötz and Fink (2009).

2   Related Work

Any online signature verification method involves the following stages: enrollment and data acquisition, pre-processing, feature extraction, classification and verification. In enrollment and data acquisition stage, there are two approaches reference-based and model-based, which depend on how the signatures are represented in the database, a set of templates to each signer or statistical model is derived for each signer respectively. In pre-processing stage, most of the approaches do some pre-processing before extracting features from the signature such as re-sampling, smoothing, stroke concatenation and normalization. In feature extraction stage, there are two approaches: feature-based extraction which depend on global features, as in Zalasiński et al. (2014), and function-based extraction which depend on local features as in Liu et al. (2015). Some methods suggest a combination of both approaches. Combination of both approaches have some of the best results obtained because online signature is a complex signal of several dimensions and one method may concentrate on one aspect of the signal such as shape, while another method may focus on another such as timing Yanikoglu and Kholmatov (2009). In classification and verification stage, researchers propose the methods used to compute a similarity score between signatures comparing it to a global or user-depended threshold to produce the verification result. There are three matching approaches here: global Zalasiński et al. (2014), local Parodi and Gómez (2014), Iranmanesh et al. (2014), Liu et al. (2015) and combination of local and global Nanni et al. (2010), Zalasiński et al. (2016), Fierrez-aguilar et al. (2005), depending on what features used in previous stage. Dynamic Time Warping (DTW) Yu et al. (2011), Fernandes and Bhandarkar (2014), Sharma and Sundaram (2016) is the most popular and successful local method used in verification stage when signatures are represented as a set of templates to each signer, and the winner in the first signature verification competition Yeung et al. (2004). Yu et al. in Yu et al. (2011) replace Euclidean distance by Mahalanobis distance to improve the performance of DTW-based signature verification. Compared Euclidean distance, Mahalanobis distance can take account of the correlations among different features and can put weights on different features. While HMM Kaur and Choudhary (2015), Rúa and Castro (2012) is the most popular and successful local method used in this stage when statistical model is derived for each signer. Many techniques have been proposed in the literature to select the best feature combination to minimize the verification error Khalil et al. (2009), Parodi and Gómez (2014). In Khalil et al. (2009), authors experimentally showed that the change of signature curvature and speed are the most productive features, and pressure doesn't give significant enhancement for accuracy. On the other hand, authors in Parodi and Gómez (2014) experimentally showed that the pen pressure improves performance when combined with the other local features. Many techniques have been proposed in the literature to approximate the time functions associated with the signing process. In Yanikoglu and Kholmatov (2009), the Fourier Transform was used while the Wavelet Transform is proposed in Chang et al. (2012), Thumwarin et al. (2013), Bharadi and Singh (2014) Nagbhidkar and Bagdi (2015). In Parodi and Gómez (2014)Legendre orthogonal polynomials was used. In Yanikoglu and Kholmatov (2009), a fixed-length representation of the signatures is proposed based on the Fast Fourier Transform (FFT). Authors exploited the ability of the Fourier domain to compactly represent an online signature using a fixed number of coefficients, regardless of the signature length, which leads to fast matching algorithms. Authors used the pen-up duration information in identifying forgeries. Moreover, they showed that most of the timing information of signatures are removed by sub-sampling which is commonly done to normalize the length of a signature significantly that reduces verification performance. Because this approach based on global features (Fourier descriptors), its error rates are higher than function-based approaches for the same databases. Authors experimentally showed that fusion of this approach with Dynamic Time Warping (DTW) system lowers the error rate of the DTW. In Parodi and Gómez (2014) Parodi et al. approximated the time functions using orthogonal polynomials series and then estimated and used the series coefficients as features to model signatures. In verification stage, they used two classifiers: Support Vector Machines (SVMs) and Random Forests (RFs), which showed more promising. Authors focused on Western signatures and Chinese signatures styles. They used the cost of the log-likelihood ratios $\hat{C}_{llr}$ besides the EER (Equal Error Rate) to quantify the verification performance. In the last few years, Deep Learning has strongly re-emerged into scientific world. In Hafemann et al. (2017), Authors tried to overcome losing dynamic information in offline signature verification using deep convolutional neural networks. Saffar et al. in Saffar et al. (2018) introduce a deep learning approach to verify signatures, the features, they used to verify the signatures, have been learned from dataset by using a sparse auto-encoder with one hidden layer. In this paper, we'll introduce new approach to verify online signatures using deep feedforward neural network and the features, are used to model signatures, approximated by Legendre polynomials. The experimental results show better Equal Error Rate reduction and accuracy enhancement in comparison with state-of-the-art methods.

3   Proposed Work

In this study, Legendre polynomials is used to approximate time functions of signatures, Legendre polynomials approximation is very convergent to original signature time functions. Legendre polynomials coefficients in these series are used as features to model the signatures. These coefficients gives static length for features which make signature matching easy.

3.1   Feature Extraction and Selection

Typically, we used the three measured discrete time functions: pen coordinates x and y, and pen pressure p. In addition to the raw data, we used x and y velocities. Depending The procedural steps are simplified as follows.

1. Let $i = 1, \ldots, L_{sign}$ be the discrete time index of the measured functions and L_{sign} the time duration of the signature in sampling units, then used x and y velocities are computed as
   - x velocity: $(v_x)_i$ is the first derivative of x_i.
   - y velocity: $(v_y)_i$ is the first derivative of y_i.
2. perform size normalization on the x and y trajectories of the signature.

$$x_{normalized} = \frac{x_i - x_{min}}{x_{max} - x_{min}}$$

$$y_{normalized} = \frac{y_i - y_{min}}{y_{max} - y_{min}}$$

$$p_{normalized} = \frac{p_i - p_{min}}{p_{max} - p_{min}}$$

$$(v_x)_{normalized} = \frac{(v_x)_i - (v_x)_{min}}{(v_x)_{max} - (v_x)_{min}}$$

$$(v_y)_{normalized} = \frac{(v_y)_i - (v_y)_{min}}{(v_y)_{max} - (v_y)_{min}}$$

3. Find the mean value of each feature variable (x, y, p, v_x, v_y) as follows:

$$mean\ f = \sum_{i=1}^{n} \frac{f_i}{N}$$

4. Subtract the mean value from each feature variable as shown in the following equation to have a new matrix (data adjust) with the same dimension

$$f_{normalized} = f_{normalized} - mean\ f$$

5. Approximate all selected functions using Legendre orthogonal polynomials

$$f(t) \approx \sum_{i=0}^{N} \alpha_i L_i(t)$$

$$\begin{pmatrix} f(t_1) \\ f(t_2) \\ \vdots \\ f(t_M) \end{pmatrix} = \begin{pmatrix} L_0(t_1) & L_1(t_1) & \dots & L_N(t_1) \\ L_0(t_2) & L_1(t_2) & \dots & L_N(t_2) \\ \vdots & \vdots & \ddots & \vdots \\ L_0(t_M) & L_1(t_M) & \dots & L_N(t_M) \end{pmatrix} \begin{pmatrix} \alpha_0 \\ \alpha_1 \\ \vdots \\ \alpha_N \end{pmatrix}$$

6. Use the series expansion coefficients α's as features.
7. Calculate the best fit between the measured and the approximated time functions, to determine the best Legendre polynomial orders using the following equation:

$$best\ fit = 100 \times \frac{\| f_{measured} - f_{approx.} \|}{\| f_{mean} - f_{approx.} \|}$$

3.2 Verification

The classifier used in this paper is deep feedforward neural network. The goal of the deep feedforward network classifier is to approximate the feature functions. The classifier, y = f (x) maps an input x (as a feature vector for a signature) to a class y (genuine or forged signature). A deep feedforward network defines a mapping

$$y = f(x; \theta)$$

and learns the value of the parameters θ that result in the best function approximation.

The hyper-parameters to adjust are the order of the Legendre polynomials and the internal parameters of the deep feedforward neural network.

The network is composed of an input layer, three hidden layers, and an output layer. The features vector of signatures feed to the input layer and the activation function of the first layer is ReLU (rectified linear unit) function, the activation function of the hidden layers are tanh (hyperbolic tangent) function and the activation function of the output layer is sigmoid function dropout regulation is used to overcome over-fitting, the deep learning algorithm is stochastic gradient descent with momentum.

We perform a 5-fold cross-validation (5-fold CV) over the Training Set to estimate the validation losses.

4   Experimental Results

4.1   Handwritten Signature Database

The publicly available SigComp2011 Dataset presented within ICDAR 2011 is used. It has two separate datasets, one containing genuine and forged Western signatures (Dutch ones) and the other one containing genuine and forged Chinese signatures. The available forgeries are skilled forgeries, which are simulated signatures in which forgers (different signers than the reference one) are allowed to practice the reference signature for as long as they deem it necessary. The data was collected from realistic and forensically relevant scenarios. The signatures were acquired using a ballpoint pen on paper (WACOMIntuos3 A3 Wide USB Pen Tablet), which is the natural writing process. This is in contrast to the approach of other researchers who tested signatures produced on a PDA or with a Wacom-stylus on a glass or plastic surface. Each of the datasets in the SigComp2011 Database is divided into two sets, namely, the Training Set and the Testing Set (Fig. 1 and Table 1).

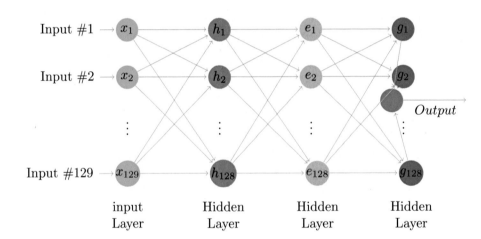

Fig. 1. Architecture of the deep feedforward network classifier

Table 1. Online Dutch signature dataset

Authors	Genuines	Forgeries
Training set	240	119
Testing set	1296	611

4.2 Results

We've used some performance tools as AUC, Area Under the ROC (Receiver Operating Characteristic Curve), Fig. 2 include ROC curve for the proposed classifier. AUC is a way to summarize an ROC curve in a single number. Equal Error rate (EER) has been used, EER can be defined such that false positive and false negative fractions are equal. Table 2 indicates the best verification results obtained for feature combination x (horizontal trajectory) and y (vertical trajectory) and p (pressure) v_x (horizontal velocity) and v_y (vertical velocity). Table 3 compare between the proposed results and Parodi's results Parodi and Gómez (2014) and Saffar's results Saffar et al. (2018).

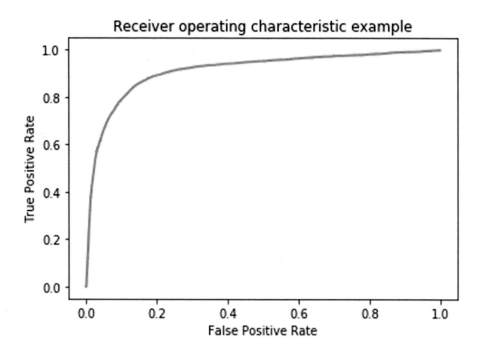

Fig. 2. ROC (Receiver Operating Characteristic Curve)

Table 2. Best verification results for the Dutch dataset, for the proposed classifier.

Features	Equal Error Rate (EER)	Log-likelihood ratio $\hat{C}_{llr}$	Area Under the ROC
$xypv_xv_y$	0.49	0.186	0.92

Table 3. Comparison between the proposed Classifier' results and Parodi's and Saffar's classifiers results.

Classifier	Equal Error Rate (EER)	$\hat{C}_{llr}$	Dataset
Proposed	0.49	0.186	SigComp2011
Parodi and Gómez (2014)	5.5	0.2039	SigComp2011
Saffar et al. (2018)	0.83	–	SVC2004

5   Conclusion

In this paper, a new approach has been introduced based on deep learning and Legendre polynomials. Using Legendre polynomials to model signatures yields good signature representation because Legendre polynomial approximation is very convergent to original signature time functions and gives fixed length representation for features which make signature matching easy. Using the stochastic gradient descent with momentum deep learning algorithm yields better Equal Error Rate reduction.

References

Bharadi, V.A., Singh, V.I.: Hybrid wavelets based feature vector generation from multidimensional data set for on-line handwritten signature recognition. In: 5th International Conference - Confluence The Next Generation Information Technology Summit (Confluence), pp. 561–568 (2014)

Chang, H., Dai, D., Wang, P., Xu, Y.: Online signature verification using wavelet transform of feature function architecture of an online signature verification system. **11**, 3135–3142 (2012)

Fernandes, J., Bhandarkar, N.: Enhanced online signature verification system. Int. J. Emerg. Trends Technol. Comput. Sci. (IJETTCS) **3**(6), 205–209 (2014). ISSN 2278-6856

Fierrez-Aguilar, J., Krawczyk, S., Ortega-Garcia, J., Jain, A.K.: Fusion of local and regional approaches for on-line signature verification. In: IWBRS 2005. LNCS, vol. 3781, pp. 188–196 (2005)

Hafemann, L.G., Sabourin, R., Oliveira, L.S.: Learning features for offline handwritten signature verification using deep convolutional neural networks. Pattern Recogn. **70**, 163–176 (2017)

Iranmanesh, V., Ahmad, S.M.S., Adnan, W.A.W., Yussof, S., Arigbabu, O.A., Malallah, F.L.: Online handwritten signature verification using neural network classifier based on principal component analysis. Sci. World J. **2014**, 1–9 (2014)

Jain, A.K., Ross, A.A., Nandakumar, K.: Introduction. In: Introduction to Biometrics, pp. 1–49. Springer (2011)

Kaur, M.R., Choudhary, M.P.: Handwritten signature verification based on surf features using HMM. Int. J. Comput. Sci. Trends Technol. **3**(1), 187–195 (2015)

Khalil, M., Moustafa, M., Abbas, H.: Enhanced DTW based on-line signature verification. In: 16th IEEE International Conference on Image Processing (ICIP), pp. 2713–2716 (2009)

Liu, Y., Yang, Z., Yang, L.: Online signature verification based on DCT and sparse representation. IEEE Trans. Cybern. **45**(11), 2498–2511 (2015)

Nagbhidkar, K.P., Bagdi, P.V.: Online signature verification on smart phone using discrete wavelet transforms. IORD J. Sci. Technol. **2**(2), 1–6 (2015)

Nanni, L., Maiorana, E., Lumini, A., Campisi, P.: Combining local, regional and global matchers for a template protected on-line signature verification system. Expert Syst. Appl. **37**(5), 3676–3684 (2010). https://doi.org/10.1016/j.eswa.2009.10.023

Parodi, M., Gómez, J.C.: Legendre polynomials based feature extraction for online signature verification. Consistency analysis of feature combinations. Pattern Recogn. **47**(1), 128–140 (2014). https://doi.org/10.1016/j.patcog.2013.06.026

Plamondon, R., Pirlo, G., Impedovo, D.: Online signature verification. In: Handbook of Document Image Processing and Recognition, pp. 917–947. Springer (2014)

Plötz, T., Fink, G.A.: Markov models for offline handwriting recognition: a survey. Int. J. Doc. Anal. Recogn. **12**, 269–298 (2009)

Rúa, E.A., Castro, J.L.A.: Online signature verification based on generative models. IEEE Trans. Syst. Man Cybern. B Cybern. **42**(4), 1231–1242 (2012)

Saffar, M.H., Fayyaz, M., Sabokrou, M., Fathy, M.: Online signature verification using deep representation: a new descriptor. arXiv preprint arXiv:1806.09986 (2018)

Sharma, A., Sundaram, S.: An enhanced contextual DTW based system for online signature verification using vector quantization. Pattern Recogn. Lett. **84**, 22–28 (2016)

Thumwarin, P., Pernwong, J., Matsuura, T.: FIR signature verification system characterizing dynamics of handwriting features (2013). http://asp.eurasipjournals.com/content/2013/1/183

Yanikoglu, B., Kholmatov, A.: Online signature verification using Fourier descriptors. EURASIP J. Adv. Signal Process. (2009)

Yeung, D.-y., Chang, H., Xiong, Y., George, S., Kashi, R., Matsumoto, T., Rigoll, G.: This is the Pre-published Version SVC2004: First International Signature Verification Competition, pp. 1–7 (2004)

Yu, Q., XingXing, W., Chunjing, X.: Learning Mahalanobis distance for DTW based online signature verification. In: IEEE International Conference on Information and Automation (ICIA), June 2011, pp. 333–338 (2011)

Zalasiński, M., Cpałka, K., Hayashi, Y.: New method for dynamic signature verification based on global features. In: International Conference on Artificial Intelligence and Soft Computing, pp. 231–245. Springer (2014)

Zalasiński, M., Cpałka, K., Rakus-Andersson, E.: An idea of the dynamic signature verification based on a hybrid approach. In: International Conference on Artificial Intelligence and Soft Computing, pp. 232–246. Springer (2016)

Machine Learning in Image and Signal Processing

Non-invasive Calibration-Free Blood Pressure Estimation Based on Artificial Neural Network

Nashat Maher[1](✉), G. A. Elsheikh[2](✉), Wagdy R. Anis[1](✉),
and Tamer Emara[3](✉)

[1] ASU Faculty of Engineering, Cairo, Egypt
nmr277@gmail.com, wagdy_anis@eng.asu.edu.eg
[2] PHI Institute, Giza, Egypt
gaelsheikh@gmail.com
[3] ASU Faculty of Medicine, Cairo, Egypt
thmfe@yahoo.com

Abstract. This paper presents a non-invasive method for Blood Pressure (BP) estimation based on extracted features from photoplethysmogram (PPG) and Electrocardiogram (ECG) signals. The proposed method depends on a machine learning technique, namely Artificial Neural Networks (ANN), to estimate blood pressure. The training is conducted on a real data set (more than 2000 BP, ECG and PPG signals) recorded by patients' monitoring at various hospitals between 2001 and 2008. In addition to the ten features that are usually used in literature, the proposed method uses the cross validation technique between features to provide more robust estimation of the blood pressure. Furthermore, the proposed method provides accurate and reliable blood pressure estimation while it is calibration-free. Compared to previous works, we used half of the data and the results clarified that we achieved more accuracy in the systolic pressure measurements. These results are expected to improve more by increasing the training samples, which is planned in future work.

Keywords: Pulse wave velocity · BP monitoring · Machine learning

1 Introduction

World Health Organization (WHO) in 2014's world health statistics reported that the hypertension causes 9.4 million people death annually [1]. According to 2008's survey, 29.2% of men and 24.8% of women suffer from the high blood pressure problem [2]. The hypertension has been recognized as the second factor of cardiovascular disease after diabetes. It is also called the silent killer, as many people are not aware of their hypertension and the way to control it. The Blood Pressure (BP) is a periodic signal with the heart rate frequency. The upper bound of the blood pressure is called the Systolic Pressure (SP) while its lower bound is called the Diastolic Pressure (DP). The mean arterial pressure (MAP) is defined as the average of the blood pressure in a cardiac cycle. If SP is above 140 mmHg or DP is above 90 mmHg, it is called hypertension that can damage internal body organs. The normal range of MAP is between 70 mmHg and 110 mmHg. Patients with hypertension usually measure their

A. E. Hassanien et al. (Eds.): AMLTA 2019, AISC 921, pp. 701–711, 2020.
https://doi.org/10.1007/978-3-030-14118-9_69

blood pressure occasionally. However, their blood pressure varies over time due to many factors such as food taking, mental situations or stress. Therefore, continues blood pressure monitoring seems necessary for accurate diagnosis and treatment of such patients. On the other hand continuous 'beat-to-beat' blood pressure monitoring is very useful in patients who are likely to display sudden changes in blood pressure (e.g. vascular surgery), in whom close control of blood pressure is required (e.g. head injured patients), or in patients receiving drugs to maintain the blood pressure. It also relies in the improvement of patient comfort, especially for those who are likely to need close blood pressure monitoring for a long period of time e.g. ICU patients. The most accurate and common blood pressure measurement devices are sphygmomanometers, which must inflate a cuff around the arm so that BP can be measured with the height of a column of mercury [3]. This method requires inflatable cuff, which is inconvenient and prevents continues measurements due to physiological limitations. Invasive (intra-arterial) blood pressure (IBP) monitoring methods are commonly used to achieve this purpose but these techniques generally used in the Intensive Care Unit (ICU) and are also often used in the operating theatre. This technique involves direct measurement of arterial pressure by inserting a cannula needle in a suitable artery. The cannula must be connected to a sterile, fluid-filled system, which is connected to an electronic patient monitor, which is known to carry a risk, albeit a small one and its measurement is complex and time consuming. Risks of these kinds could be avoided, however, if there was a noninvasive method offering a high degree of accuracy and real time operation in a continuous, beat-to-beat, mode. Further, the method should be insensitive to the patient's movement (artifacts) and respond rapidly to cardiovascular changes, such as a sudden drop in blood pressure. Recent researches suggest new cuff-less blood pressure estimation methods. Although, there have been some attempts on estimating SP and DP based on the Photoplethysmograph (PPG) signal shape, no clear relation between PPG and BP has been found yet [4]. Another cuff-less method is based on the wave propagation theory for fluids, which is founded on the natural relationship between the fluid pressure and wave propagation velocity. The theory implies that the blood pressure can be calculated from the heart pulse wave velocity PWV. One of much simpler way to get PWV is by looking for arterial properties such as arterial stiffness.

There are a number of disadvantages associated with this method, such as the need for calibration for each person and the expiration of this calibration in short time intervals [5]. A novel method is proposed for calibration free and accurate estimation of the blood pressure. The proposed scheme is accomplished by the extraction of a number of physiological parameters from Electrocardiogram (ECG) and PPG signals along with machine learning theories. The main features of PPG signal are shown in Fig. 1.

2   Background

The main idea that motivate this work is that the velocity of the pressure pulse, which is initiated by the heart beat and propagates through arteries, is highly correlated with the elastic properties of arteries, similar to a pipe with elastic walls. The velocity of

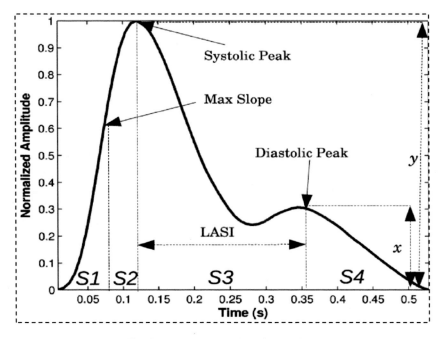

Fig. 1. Features extraction of PPG signal

displaced fluid in a pipe is a function of its tension and elasticity. The relation between the Pulse Wave Velocity (PWV), vessel parameters and blood properties can be represented as [6]:

$$PWV = \sqrt{\frac{Et}{2R\rho}} \qquad (1)$$

Where R is the inner radius of vessels, ρ represents the blood density, **t** is the vessel thickness and **E** is Young's modulus, which is related to the vessels elasticity. For an elastic vessel, the relation between the blood pressure and E is given by:

$$E = E_0 + e^{\alpha(P-P_0)} \qquad (2)$$

Where E_0 and P_0 are some constants and P can be interpreted as the blood pressure in arteries. In fact (2) indicates that there exists an exponential relation between E and the blood pressure. The formulations in (1) and (2) show that there is a relation between the pulse wave velocity and BP [7]. There are several methods for calculation of the pulse wave velocity, among which one of the most well-known methods is the Pulse Transit Time (PTT) [8]. PTT is defined as the time it takes for the heart beat pulse to propagate from heart to the body peripherals. The pulse wave velocity can be estimated by dividing the distance from heart to a specific peripheral $(d_{h,p})$ by the measured PTT through the following equation:

$$PWV = \frac{d_{h,p}}{PTT} \tag{3}$$

The calculation of the blood pressure from Eq. (3) incurs several challenges. One of them is that arterial properties differ from person to another and are highly dependent to an individual's age. Moreover, according to Eq. (3), it is required to have the distance between the peripheral and the heart, which is related to the person's height. PTT can be estimated as the time interval between the R peak of the ECG signal as shown in Fig. 2, which indicates the electrical activity of the heart, and certain points in the finger PPG waveform.

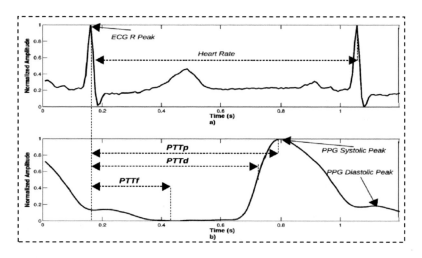

Fig. 2. Calculation of PTT from the heart beat pulse. (a) ECG signal. (b) PPG signal

3   Analysis Stepwise Procedure

The proposed method estimates the BP by extracting features from ECG and PPG signals by building a regression model as shown in Fig. 3.

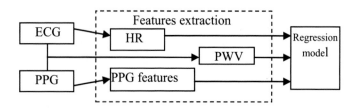

Fig. 3. Overview of the method to estimate BP

The work consists of the following steps:

(1) Database: collection of a database with adequate sample size.
(2) Preprocessing: smooth and remove invalid signals.
(3) Feature extraction: extract useful features from signals.
(4) Partition the samples into three subsets i.e. train, validation and test samples.
(5) train the regression models using Machine Learning (ANN algorithm).
(6) Evaluate the trained models' performance.

3.1 Database

Multi-parameter Intelligent Monitoring in Intensive Care (MIMIC) II online waveform database [9] provided by PhysioNet organization is used in this paper as the reference database. It consists of thousands signals recorded by patient monitors at various hospitals between 2001 and 2008. Waveform signals were sampled at the frequency of 125 Hz with at least 8 bit accuracy. We extracted PPG, ECG and arterial blood pressure waveform signals from this database. Figure 4 shows one sample of a measured data from a patient, the first is the ECG signal, second is the PPG and the last is continuous Arterial Blood Pressure (ABP). All of these three signals are measured simultaneously.

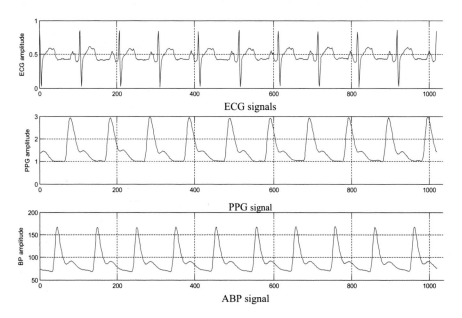

Fig. 4. Sample of ECG, PPG and ABP signals

3.2 Preprocessing

In order to make data ready for feature extraction, removing distorted and unreliable signals is a vital task. Preprocessing is performed by dividing samples into fixed size signal blocks as follows:

- Step-1: Smoothing all signals with a simple averaging filter.
- Step-2: Removing signal blocks with irregular and unacceptable human blood pressure values.
- Step-3: Removing signal blocks with unacceptable heart rates.
- Step-4: Removing signal blocks with severe discontinuities, which was not resolved with the help of smoothing filter in step 1.

3.3 Feature Extraction

Some useful features of the PPG signal are added to PTT features to improve the BP estimation. These features are shown as follow:

1. PTT features: obtained by calculating the time distance between the ECG peak (R-peak) and three points on the PPG signal. These features including (a) The maximum PPG peak (PTTp), (b) The PPG minimum (PTTf), and (c) The point of maximum slope of the PPG waveform (PTTd) (Refer to Fig. 2).
2. Heart rate: The heart rate is obtained by calculation of the peak-to-peak time interval of the PPG or ECG signals.
3. The PPG features: in order to achieve a calibration free method various features related to the blood pressure are selected as follow:

   - Augmentation Index (AI): it is a measure of the wave reflection on the arteries [4], which is calculated as the ratio between the diastolic peak and the systolic peak as clarified in Fig. 1.
   - Large Artery Stiffness Index (LASI): it is a measure of the arterial stiffness and is related to the time interval between the systolic peak and the diastolic peak as shown in Fig. 1.
   - Inflection Point Area ratio (IPA): it is defined as a function of the areas under the PPG curve between selected points, denoted by S1, S2, S3 and S4 as shown in Fig. 1.

4. Feature crosses: In this work cross validation between features is used to enhance the regression model as shown in Fig. 5.

3.4 Data Partitioning

After processing of the database, more than 2000 records were obtained. The BP distribution histograms are shown in Fig. 6. The new database is then randomly partitioned into three sets: 70% as training, 15% as validation and the remaining as test samples.

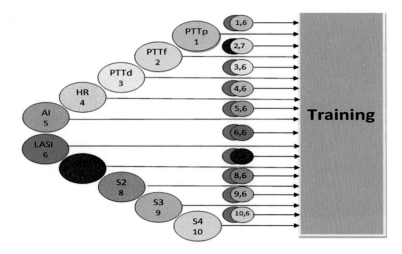

Fig. 5. Features cross validation

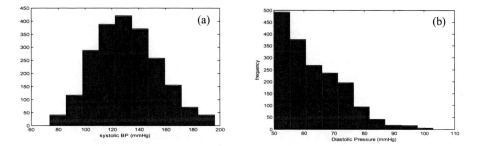

Fig. 6. Samples histogram (a) Systolic BP (b) Diastolic BP

3.5 Model Based on Machine Learning

As mentioned before, there is an exponential relationship between the blood pressure and the PWV, which causes a huge non-linearity near the high blood pressure values [10]. Consequently, the estimation of the high BP values becomes erroneous using the selected features. To overcome this problem, two approaches can be used. First approach, polynomial terms of the extracted features are added as extra features and then the training is performed via the linear regression algorithm [11]. In this approach, Regularized Linear Regression (RLR) with the Mean Squared Error (MSE) cost function used as a basic regression algorithm with a low computational cost and fast training [11]. Second approach, non-linear regression algorithms, which are proven to have a better capability to handle this non-linearity issue compared to the linear regression approach at the cost of an extra computational complexity. ANN, which is one set of algorithms used in machine learning for modeling data, is among these non-linear methods, described in the following:

- ANN: Levenberg-Marquardt back propagation algorithm is used as the learning function and MSE is used as the cost function. The network topology, trained with this algorithm, consists of one input layer with the size of the feature vector, one hidden layer with sizes between 5 to 15 and one neuron in the output layer as shown in Fig. 7.
- Choosing more than 2000 samples from the same data used in literature work to make the comparison more reliable.
- The correlation between the crossed features and the outputs is checked such that highest relation maintained. This work proved that crosses feature between LASI and all ten features used before in literature are enhanced as shown in Table 1 within next section.

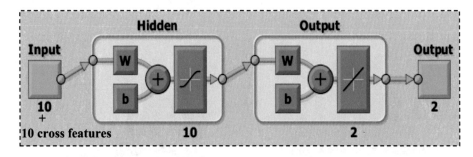

Fig. 7. Simplified block diagram for Artificial Neural Networks

3.6 Model Evaluation

The Mean Absolute Error (MAE) and Standard Deviation (STD) of estimation errors are used for the model evaluation, which are calculated as,

$$STD = \sqrt{\frac{\sum d^2}{n}} \tag{4}$$

$$MAE = \frac{\sum |d|}{n} \tag{5}$$

Where d is the error, n is the number of samples and σ is the standard deviation. These estimation errors are calculated and shown in (Fig. 8) via the histogram of systolic and diastolic error.

4   Experimental Results

Figure 8 shows the histogram of the BP estimation error using the ANN regression, which is more accurate and reliable method in the model evaluation. Most of estimated error is ±20. the results of ANN algorithm are presented in Table 1. It is clear that MAE in both DP and SP enhanced in proposed work although the number of samples used in literature work is approximately double. In proposed work MAE of DP reduced to 6.8132 while it was 6.86 i.e. enhanced by 0.6822%, and of SP reduced to 13.2493 while it was 13.78 in the previous work, i.e. enhanced by 3.85%. STD of DP enhanced by 1.943% from 8.96 to 8.7859 while STD of SP increase by 0.369% from 17.46 to 17.5247.

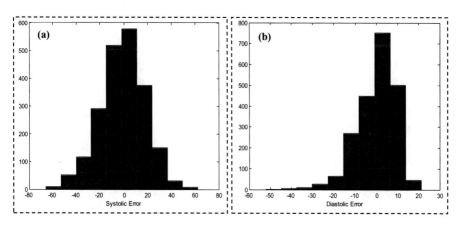

Fig. 8. The proposed work results (a) Systolic BP error histogram (b) Diastolic BP error histogram

To evaluate the accuracy, cumulative percentage of readings falling within 5, 10 and 15 mmHg, respectively is calculated for diastolic and systolic blood pressure. The proposed work enhanced result in systolic BP but still the literature result better in DP as shown in Table 2.

Table 1. Performance of proposed & literature work

Algorithm	DP		SP		Number of samples
	MAE	STD	MAE	STD	
Literature	6.86	8.96	13.78	17.46	4254
Work	6.8132	8.7859	13.2493	17.5247	2126

Table 2. Cumulative percentage of readings comparison

Absolute difference		≤ 5 mmHg		≤ 10 mmHg		≤ 15 mmHg	
		Literature	Work	Literature	Work	Literature	Work
Results	Diastolic	51.2%	27.33%	78.9%	52.59%	93.6%	74.13%
	Systolic	28.8%	34.01%	51.5%	59.92%	69.5%	75.96%

Although number of samples is one half of that used in literature, the proposed model results are more accurate. Table 2 shows that in systolic pressure, the number of samples that has absolute error less than 5 mmHg is increased by 5.21%, the number of samples that has absolute error less than 10 mmHg is increased by 8.42% and the number of samples that has absolute error less than 15 mmHg is increased by 6.46%.

Other cuff-less BP estimation designs in literature suffer from serious drawbacks such as the need of calibration per each patient [5] or utilizing relatively small sample size database [10]. The proposed method in this work resolves all these issues while providing enough capability for reliable and calibration-free, blood pressure estimation.

5   Conclusions

Continuous BP monitoring is vital; nevertheless it is unreachable through conventional cuff-based BP measurement devices which require cuff inflation and deflation that and prevent continuous measurement due to physiological limitation, moreover it is slow and uncomfortable for many patients. The proposed method establishes a cuff-less BP estimation system without calibration using ANN and cross validation of input features. The used cross feature validation method enhances the results of the output model compared to results achieved in Literature work. Although number of samples is one-half of that used in the previous work, the results obtained from the proposed model are more accurate in systolic pressure measurements. This model clarified that the MAE in both systolic and diastolic are decreased. In addition, the results are promising and expected to increase accuracy of the model via increasing the number of samples, which is planned for further work.

References

1. World Health Organization: World Health Statistics 2014 (2014)
2. Sanuki, H., Fukui, R., Inajima, T., Warisawa, S.: Cuff-less calibration-free blood pressure estimation under ambulatory environment using pulse wave velocity and photoplethysmogram signals. In: Proceedings of the 10th International Joint Conference on Biomedical Engineering Systems and Technologies (BIOSTEC 2017), pp. 42–48 (2017)
3. Van Montfrans, G.: Oscillometric blood pressure measurement: progress and problems. Blood Press. Monit. 6(6), 287–290 (2001)
4. Elgendi, M.: On the analysis of fingertip photoplethysmogram signals. Curr. Cardiol. Rev. 8(1), 14 (2012)
5. Wong, M., Poon, C., Zhang, Y.: An evaluation of the cuffless blood pressure estimation based on pulse transit time technique: a half year study on normotensive subjects. Cardiovasc. Eng. 9(1), 32–38 (2009)
6. Obrist, P., Light, K., McCubbin, J., Hutcheson, J., Hoffer, J.: Pulse transit time: relationship to blood pressure. Behav. Res. Methods Instrum. 10(5), 623–626 (1978)
7. Hughes, D.J., Geddes, L.A., Babbs, C.F., Bourland, J.D.: Measurements of Young's modulus of the canine aorta in-vivo with 10 MHz ultrasound. In: Proceedings of Ultrasonics Symposium, p. 326, September 1978

8. Peter, L., Noury, N., Cerny, M.: A review of methods for non-invasive and continuous blood pressure monitoring: pulse transit time method is promising? IRBM **35**(5), 271–282 (2014)
9. Goldberger, A., Amaral, L., Glass, L., Hausdorff, J., Ivanov, P., Mark, R., Mietus, J., Moody, G., Peng, C., Stanley, H.: Physiobank, physiotoolkit, and physionet components of a new research resource for complex physiologic signals. Circulation **101**(23), 215–220 (2000)
10. Gesche, H., Grosskurth, D., Kuchler, G., Patzak, A.: Continuous blood pressure measurement by using the pulse transit time: comparison to a cuff-based method. Eur. J. Appl. Physiol. **112**(1), 309–315 (2012)
11. Kachuee, M., Kiani, M.M., Mohammadzade, H., Shabany, M.: Cuff-less high-accuracy calibration-free blood pressure estimation using pulse transit time. In: IEEE International Symposium on Circuits and Systems (ISCAS) (2015)

Data Augmentation and Feature Fusion for Melanoma Detection with Content Based Image Classification

Rik Das[1], Sourav De[2], Siddhartha Bhattacharyya[3(✉)], Jan Platos[3],
Vaclav Snasel[3], and Aboul Ella Hassanien[4]

[1] Department of Information Technology, Xavier Institute of Social Service,
Ranchi, Jharkhand, India
rikdas78@gmail.com
[2] Department of Computer Science and Engineering,
Cooch Behar Government Engineering College, Cooch Behar, India
dr.sourav.de79@gmail.com
[3] Faculty of Electrical Engineering and Computer Science,
VSB Technical University of Ostrava, Ostrava, Czechia
{siddhartha.bhattacharyya, jan.platos,
vaclav.snasel}@vsb.cz
[4] Cairo University, Giza, Egypt
aboitcairo@gmail.com

Abstract. Computer aided diagnosis has leveraged a new horizon for accurate diagnosis of numerous fatal diseases. Melanoma is considered as one of the most lethal form of skin cancer which is increasingly affecting the population in recent times. The disease can be completely healed if diagnosed and addressed at an early stage. However, in most of the cases patients receive delayed care which results in fatal consequences. The authors have attempted to design an automated melanoma detection system in this work by means of content based image classification. Extraction of content based descriptors can nullify the requirement for manual annotation of the dermoscopic images which consumes considerable time and effort. The work has also undertaken a fusion based approach for feature combination for evaluating classification performances of hybrid architecture. The results have outclassed the state-of-the-art outcomes and have established significant performance improvement.

Keywords: Computer aided diagnosis · Melanoma ·
Content based image classification · Feature fusion · Colour histogram · HOG

1 Introduction

Content based image classification is instrumental in recent times for computer aided diagnosis leading to data driven decision making in medical science [1]. It has assisted medical practitioners and patients choosing correct line of treatment by providing accurate analysis of the genre of the disease. Skin cancer has created a major concern in contemporary times and has largely affected the well-being of the society. Melanoma is reportedly the deadliest among all forms of skin cancer which has a capricious

© Springer Nature Switzerland AG 2020
A. E. Hassanien et al. (Eds.): AMLTA 2019, AISC 921, pp. 712–721, 2020.
https://doi.org/10.1007/978-3-030-14118-9_70

proliferation [2]. One of the vital reasons of high mortality rate for cancer is delayed detection. It eventually leads to postponed medical care and premature demise of the victim due to shortage of time in determining the course of treatment. In case of melanoma, it is observed that identification at an early stage has promising likelihood for full cure of the ailment [3]. Melanoma is widespread in Europe, North America and Australia and is seen to have an endurance rate of five years for survival for late detection. However, this form of cancer is recorded to have a survival rate of almost 99% if detected at its inception [4]. Usually, skin disease is treated by the dermatologists with a non-invasive technique of skin imaging, named, Dermoscopy. This method of imaging reduces the interference for skin reflection using a high factor of magnification, thus making it possible to have a superior observation of skin lesion in contrast to the naked eyes [5]. Nevertheless, appointment with a dermatologist is seldom possible promptly and consume considerable amount of valuable time before addressing the aggressiveness of the disease. Therefore, it is extremely advantageous to devise an automated arrangement for detection of melanoma by means of computer aided diagnosis (CAD). Dermoscopic images can act as a perceptive resource to design a content based image identification system for binary classification of melanoma into benign or malignant cases. However, the design of such a system is quite exigent since it is challenging to segment the skin lesion due to lack of perceptible boundary between normal and lesion skin. Moreover, the intra-class variations (types of melanoma) are quite apparent compared to the inter-class variations (malignant or benign). Consequently, designing of effective techniques for robust feature vector extraction becomes evident for content based categorizing of malignancy from the dermoscopic images of melanoma. In recent times, skin cancer classification by supervised learning with Convolutional Neural Networks has asserted accurate performance to the level of Dermatologists' expertise. However, concerns are expressed by the authors related to scarcity of data which has restricted the application to limited visual evidences.

In this paper, the authors have attempted to create enhanced visual evidences of melanoma images by means of data augmentation using image rotation technique. The process is endeavored to create increased training examples for the classification process.

The motivations for this work are as follows:

- Data augmentation for increased training samples of dermoscopic images
- Feature extraction using automatic learning with deep neural networks and handcrafted techniques
- Classification evaluation using three different classifiers
- Early fusion of automatically learned features and handcrafted features for enhanced classification results

A public dataset named PH^2 [6] is considered for the experiment which comprises of 200 dermoscopic images. These images are augmented with rotation to form a dataset of 2,400 images on the whole. The classification outcomes achieved in this work are compared with respect to different feature extraction techniques used in this paper and a fusion based architecture. The analysis of the comparisons has revealed encouraging inferences.

2 Related Work

The procedure of designing an automated architecture for diagnosis of melanoma is envisioned by means of content based image classification. Efficiency of the technique is principally dependent on the robustness of the extracted descriptors from the dermoscopic images. Researchers have mostly grouped the feature extraction methodologies into two different categories, namely, hand-crafted descriptor extraction and automatically learned descriptors. Recent literature has a broad coverage for both of these categories and has shown the versatility of the techniques under different classification environments.

Hand-crafted descriptors are usually extracted by exploiting the low level characteristics of the image data, namely, shape, colour, texture and so on. Aforesaid properties of the skin lesion have greatly influenced the classification performance for melanoma detection. A number of customary literature of computer science have utilized the discussed descriptors to address traditional problems of computer vision.

Hand-crafter features using colour and texture in classification of lesion is studied in [6] where both the global and local methods for extracting features are utilized. The classification results have shown high accuracy for both specificity and sensitivity. Three different colour spaces based feature extraction is carried out from the melanoma images using colour histogram and texture based feature is extracted using Gray level co-occurrence matrix (GLCM) [7]. Combination of diverse texture and colour features has shown encouraging results in melanoma classification [8]. Relative colour descriptors are extracted from morphologically filtered images of melanoma for the purpose of classification [9]. Uni-dimensional colour descriptors have shown 100% sensitivity and 93% specificity in melanoma classification [10]. Colour-SIFT descriptors are used for melanoma detection in [11]. Colour feature extraction using clustering, estimating probability of density and by use of dissimilarity measure is observed in [12]. Shape and symmetry features are used in a recent approach [13] for successful melanoma classification.

Very deep residual networks are successfully introduced for identifying the malignancy in dermoscopic images [14]. A combined approach including deep learning, sparse coding and support vector machine (SVM) has carried out effective melanoma recognition. Transfer of features by means of unsupervised learning has eradicated the requirement for annotation [15]. Classification results of melanoma images are positively influenced with transfer learning in [16]. Ensemble of deep learning techniques has revealed improved analysis of melanoma images [17]. Amalgamation of deep learning technique and hand-crafted features have resulted in improved results for melanoma classification [18]. Melanoma classification using extreme learning machine (ELM) has addressed the generalization problem in [19].

All This work has attempted to fetch useful image descriptors by augmenting the melanoma images by means of image rotation. Further, it has evaluated the features individually for classification performance and also attempted to elevate the classification performance of features using early fusion technique. It has combined transfer learning approach of feature extraction to that of the hand-crafted techniques which is hardly noticed in the approaches of existing literature for elevated classification performance.

3 Our Approach

The authors have evaluated the classification performance of three different feature extraction techniques in this work out of which two are hand-crafted descriptors and one is automated feature learning. The process is illustrated with block diagram in Fig. 1.

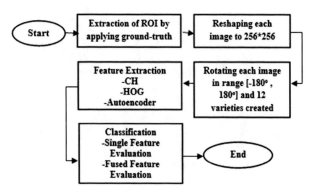

Fig. 1. Block diagram for feature extraction and classification

The feature extraction techniques are carried out on a widely used dermoscopic image dataset named PH2. The dataset has 200 images including 80 common nevi, 80 atypical nevi and 40 melanomas. This work has attempted for a binary classification of the images into benign and malignant categories respectively. The images are supplied with ground truth segmentation mask which is applied on each of them to extract the region of interest (ROI) containing the lesion as shown in Fig. 2. These images are further applied with bilinear interpolation to convert each image to a dimension of 256 * 256. Henceforth, a rotation range of [−180°, 180°] is applied on each of the images to create 12 different varieties of each. Thus the training data now comprises of (12 * 200) = 2400 images as in Fig. 2.

The hand-crafted feature extraction techniques, namely, colour histogram (CH) and Histogram of Oriented Gradients (HOG) are applied consecutively on each of the 2400 images in the dataset followed by the automated feature learning which is performed with autoencoder. The learned weights with autoencoder are shown in Fig. 3.

The extracted features are evaluated with three different classifiers, namely, Random Forest (RF), K-nearest neighbour (KNN) and Support Vector Machine (SVM) and the results are compared for performance measures. The metric used to judge classification performances are Specificity and Sensitivity and is expressed as in Eqs. 1 and 2 respectively.

$$Specificity = \frac{correctly..identified..benign..cases}{all..benign..cases} \tag{1}$$

$$Sensivity = \frac{correctly..identified..malignant..cases}{all..malignant..cases} \tag{2}$$

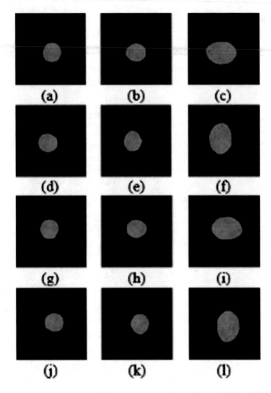

Fig. 2. Image varieties with rotation of [−180°, 180°]

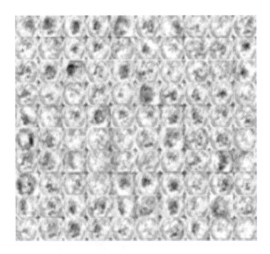

Fig. 3. Learned weights with autoencoder used as features

4 Results and Discussions

The experiments have been carried out with a core i5 system installed with Matlab 2015b. The classification results are evaluated using a popular data mining tool named Weka. The results for classification with three different classifiers, namely, RF, KNN and SVM for three different feature extraction techniques are given in Figs. 4 and 5 respectively. The results are evaluated using 80:20 train-test split where 80% of data is used for training and the rest 20% is used for testing in Fig. 4 and using 10-fold cross validation in Fig. 5.

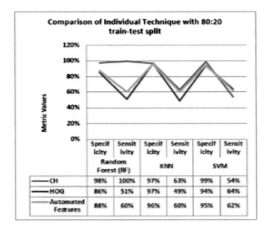

Fig. 4. Comparison of individual techniques with 80:20 train-test split

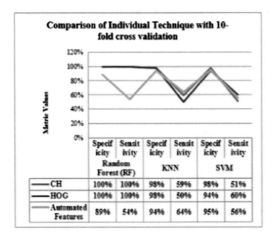

Fig. 5. Comparison of individual techniques with 10-fold cross validation

The results illustrated in Figs. 4 and 5 have revealed that hand-crafted features, namely, CH and HOG has performed better with respect to the automatically learned features using autoencoder. CH has shown the highest performance for specificity with all the three classifiers for both 80:20 train test split and 10-fold cross validation. On, the other hand, HOG has shown highest sensitivity with SVM in case of 80:20 train-test split and 10 fold cross validation. It has shown 100% sensitivity with RF classifier for 10 fold cross validation. However, automatically learned features have shown poor classification performances with respect to CH and HOG. Therefore, feature fusion technique is applied to combine the automatically learned features to CH and HOG to elevate the classification performance. The fusion is carried out using feature scaling technique, where, individual set of feature vectors are normalized using min-max normalization and then concatenated horizontally.

Since RF classifier has given the highest classification performance, the fused features are tested with RF classifier for both 80:20 train-test split and 10 fold cross validation. The results are shown in Figs. 6 and 7.

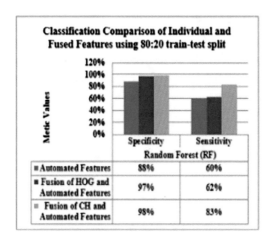

Fig. 6. Comparison of individual and fused features for 80:20 train-test split

Thus it is inferred that feature fusion can lead to elevated classification performance with respect to individual features which can yield higher specificity and sensitivity values for detection of malignancy of dermoscopic images. The performance comparisons in Figs. 6 and 7 have clearly shown elevated specificity and sensitivity when automatically learned features are fused with HOG and CH. Further, the results are compared to state-of-the art results in Table 1.

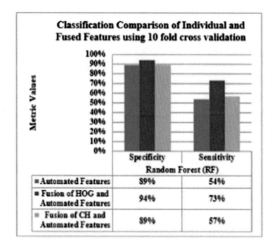

Fig. 7. Comparison of individual and fused features for 80:20 train-test split

Table 1. Comparison of the classification performance of the proposed method and existing methods

Types of techniques	Techniques	Specificity	Sensitivity
Proposed	CH (80:20 train test split)	98%	100%
Proposed	HOG (80:20 train test split)	94%	64%
Proposed	Automated Featured (80:20 train test split)	95%	62%
Proposed	CH (10 fold cross validation)	100%	100%
Proposed	HOG (10 fold cross validation)	100%	100%
Proposed	Automated Featured (10 fold cross validation)	94%	64%
Proposed	Fusion of HOG and Automated Features (80:20 train test split)	97%	62%
Proposed	Fusion of CH and Automated Features (80:20 train test split)	98%	83%
Proposed	Fusion of HOG and Automated Features (10 fold cross validation)	94%	73%
Proposed	Fusion of CH and Automated Features (10 fold cross validation)	89%	57%
State-of-the-art	Color Features (Global Method) [4]	89%	90%
State-of-the-art	Texture Features (Global Method) [4]	78%	93%
State-of-the-art	Color Features (Local Method) [4]	84%	93%
State-of-the-art	Texture Features (Local Method) [4]	76%	88%

The comparison reveals that the proposed techniques in this work have shown higher specificity in most of the cases compared to the state-of-the-art. Higher sensitivity is observed only in four cases.

5 Conclusion

Melanoma is a major concern in present time and a significant health hazard which requires immediate attention. This paper has discussed a model for a computer aided diagnosis system for malignancy detection by means of content based image classification. It has also increased the training data from 200 to 2400 by means of image rotation which has helped in potential learning of the classifier for distinguishing the malignant and benign varieties of the affected region.

The comparison of hand-crafted features in contrast to the automatically learned features has revealed better classification performances for the prior one. However, there are benchmarked results with automatically learned features in recent literature which our system is unable to achieve due to lack of training evidences required to extract meaningful automatically learned descriptors. Nevertheless, the performance of the automatically learned features is enhanced with feature fusion technique and the results have shown interesting improvements.

Acknowledgement. This work was supported by the ESF in "Science without borders" project, reg. nr. CZ.02.2.69/0.0/0.0/16_027/0008463 within the Operational Programme Research, Development and Education.

Dr. Rik Das would like to acknowledge Calcutta University Data Science group for continuous brainstorming and support towards innovative research ideas relevant to this work.

References

1. Esteva, A., Kuprel, B., Novoa, R.A., Ko, J., Swetter, S.M., Blau, H.M., Thrun, S.: Dermatologist-level classification of skin cancer with deep neural networks. Nature **542**, 115–118 (2017)
2. Amelard, R., Glaister, J., Wong, A., Clausi, D.A.: High-level intuitive features (HLIFs) for intuitive skin lesion description. IEEE Trans. Biomed. Eng. **62**(3), 820–831 (2015)
3. Mayer, J.: Systematic review of the diagnostic accuracy of dermatoscopy in detecting malignant melanoma. Med. J. Aust. **167**(4), 206–210 (1997)
4. Ali, A.-R.A., Deserno, T.M.: A systematic review of automated melanoma detection in dermatoscopic images and its ground truth data. In: Proceedings of SPIE 8318(8318-8318-11) (2012). https://doi.org/10.1117/12.912389
5. Silveira, M., et al.: Comparison of segmentation methods for melanoma diagnosis in dermoscopy images. IEEE J. Sel. Top. Sign. Proces. **3**(1), 35–45 (2009)
6. Barata, C., et al.: Two systems for the detection of melanomas in dermoscopy images using texture and color features. IEEE Syst. J. **8**(3), 965–979 (2014)
7. Kavitha, J.C., Suruliandi, A.: Texture and color feature extraction for classification of melanoma using SVM. In: International Conference on Computing Technologies and Intelligent Data Engineering (ICCTIDE), pp. 1–6 (2016)
8. Alfed, N., Fouad, K.: Bagged textural and color features for melanoma skin cancer detection in dermoscopic and standard images. Expert Syst. Appl. **90**, 101–110 (2017)
9. Cheng, Y., Swamisai, R., Umbaugh, S.E., Moss, R.H., Stoecker, W.V., Teegala, S., Srinivasan, S.K.: Skin lesion classification using relative color features. Skin Res. Technol. **14**(1), 53–64 (2008)

10. Ruela, M., Barata, C., Mendonça, T., Marques, J.S.: What is the role of color in dermoscopy analysis? In: Sanches, J.M., Micó, L., Cardoso, J.S. (eds.) Pattern Recognition and Image Analysis. IbPRIA 2013. Lecture Notes in Computer Science, vol. 7887. Springer, Berlin (2013)
11. Barata, C., Jorge, S.M., Rozeira, J.: Evaluation of color based keypoints and features for the classification of melanomas using the bag-of-features model. In: International Symposium on Visual Computing, pp. 40–49. Springer, Heidelberg (2013)
12. Møllersen, K.: Melanoma Detection Colour. Clustering and Classification (2016)
13. Ruela, M., Barata, C., Marques, J.S., Rozeira, J.: A system for the detection of melanomas in dermoscopy images using shape and symmetry features. Comput. Methods Biomech. Biomed. Eng. Imaging Vis. 5(2), 127–137 (2017)
14. Yu, L., Chen, H., Dou, Q., Qin, J., Heng, P.-A.: Automated melanoma recognition in dermoscopy images via very deep residual networks. IEEE Trans. Med. Imaging 36(4), 994–1004 (2017)
15. Codella, N., Cai, J., Abedini, M., Garnavi, R., Halpern, A., Smith, J.R.: Deep learning, sparse coding and SVM for melanoma recognition in dermoscopy images. In: Zhou, L., Wang, L., Wang, Q., Shi, Y. (eds.) Machine Learning in Medical Imaging. MICCAI 2015. Lecture Notes in Computer Science, vol. 9352. Springer, Cham (2015)
16. Menegola, A., Fornaciali, M., Pires, R., Avila, S., Valle, E.: Towards Automated Melanoma Screening: Exploring Transfer Learning Schemes. Computer Vision and Pattern Recognition, abs/1609.01228 (2016)
17. Codella, N.C.F., Nguyen, Q.-B., Pankanti, S., Gutman, D.A., Helba, B., Halpern, A.C., Smith, J.R.: Deep learning ensembles for melanoma recognition in dermoscopy images. IBM J. Res. Dev. 61(4/5), 1–15 (2017)
18. Majtner, T., Yildirim-Yayilgan, S., Hardeberg, J.Y.: Combining deep learning and hand-crafted features for skin lesion classification. In: 2016 6th International Conference on Image Processing Theory Tools and Applications (IPTA), pp. 1–6 (2016)
19. Rahman, Md.M., Alpaslan, N.: Automated melanoma recognition in dermoscopic images based on extreme learning machine (ELM). In: Medical Imaging 2017: Computer-Aided Diagnosis, vol. 10134. International Society for Optics and Photonics (2017)

Dental Age Estimation: A Machine Learning Perspective

Jiang Tao[1], Jian Wang[1], Andrew Wang[2], Zhangqian Xie[3], Ziheng Wang[4], Shaozhi Wu[5], Aboul Ella Hassanien[6], and Kai Xiao[7(✉)]

[1] Department oft of General Dentistry, Ninth People's Hospital, Shanghai Jiao Tong University School of Medicine, No. 639 Zhi Zao Ju Road, Shanghai 200011, China
taojiang_ng_doctor@hotmail.com, doc_oc_wangj@126.com
[2] Department of Electrical Engineering and Computer Sciences, University of California, Berkeley Soda Hall, Berkeley, CA 94720-1776, USA
andrewawang@berkeley.edu
[3] School of Mathematics, Shandong University, 27 Shanda Nanlu, Jinan 250100, China
xzq_zq_201609@mail.sdu.edu.cn
[4] The School of Aerospace Engineering and Applied Mechanics, Tongji University, No. 1239 Siping Road, Shanghai 200000, China
wangzh1364@hotmail.com
[5] School of Computer Science and Engineering, University of Electronic Science and Technology of China, No. 4, Section 2, North Jianshe Road, Chengdu 610054, China
wszfrank@126.com
[6] Information Technology Department, Cairo University, 5 Ahmed Zewal Street, Orman, Giza, Cairo, Egypt
aboitcairo@gmail.com
[7] School of Electronic Information and Electrical Engineering, Shanghai Jiao Tong University, 800 Dongchuan Road, Shanghai 200240, China
showkey@sjtu.edu.cn

Abstract. Dental age estimation is important for determining the actual age of an individual. In this paper, for the purpose of improving the accuracy of dental age estimation, we present several machine learning algorithms. We apply Demirjian's method, Willem's method, and our methods to a dataset with 1636 cases; 787 males and 849 females. The Multi-layer Perceptron algorithm is used to predict dental age in our experiments. In order to avoid overfitting, we use Leave-one-out cross-validation when training the model. Meanwhile, we employ root-mean-square error, mean-square-error and mean-absolute-error to measure the error of the results. Through experiments, we verify that this algorithm is more accurate than traditional dental methods. In addition, we try to use a new set of features that are converted by traditional dental methods. Specifically, we find that using Demirjian's method converted data for males and using Willem's method converted data for females can improve the accuracy of the dental age predictions.

Keywords: Dental age estimation · Demirjian · Willem · Multi-layer Perceptron

© Springer Nature Switzerland AG 2020
A. E. Hassanien et al. (Eds.): AMLTA 2019, AISC 921, pp. 722–733, 2020.
https://doi.org/10.1007/978-3-030-14118-9_71

1   Introduction

Age estimation, which unveils its integral aspect of biological profile, is a critical procedure in the field of forensic science [1–4]. It is of vital importance under several scenarios such as identifying the victims of accidents or disasters and confirming the imputability of teenagers of whether has reached the age of criminal responsibility [5,6]. The authorities would also seek forensic age examination in order to obtain an asylum-seeker's age information with undocumented confirmation of their birth date [7]. Moreover, precise age estimation provides crucial clues for resolving dispute over the qualification of participation of events with strict age limits such as gymnastics.

Dental age method, compared with other general physical examinations such as left-hand radiographs, is fundamentally controlled by genetics and is less affected by nutritional and environment factors [8]. Methods for estimating human dental age can be roughly divided into three categories: Morphological observation, biochemical analysis and radiological methods. Radiological methods play an indispensable role in the human age determination in which radiological images are utilized in the process of age estimation. In this category, chronological age was calculated based on the time from date of birth to the day of the panoramic X-ray study. Nolla [15] proposed the pioneering method for quantitative dental age estimation through evaluating development of each tooth by the radiographs which were graded on a scale from 0 to 10. Demirjian method, which was proposed in 1973 by Demirjian et al. [9], has been widely applied as a radiographic method for dental age assessment. It is based on eight calcification stages which represented the stages of tooth development from calcification of the crown cusps to the closure of the apex of the seven left permanent mandibular teeth. Maturity scores are calculated by gender, then converted to dental age according to the Demirjian table. In 2001, Willems [10] put forward a new age estimation standard which is a modified version of Demirjian method [10]. A different transformation table helped dental maturity score directly convert into dental age. Not only simplifying transformation steps, several reports also voiced Willems method was more accurate when compared with Demirjian method [11–14].

In order to create a conversion table, Demirjian et al. [9] firstly studied and constructed the typical features in dental maturity ranged from the beginning of the calcification to the final apical closure [10]. With the objective of transforming the special development points into the desired dental age, an advocated approach proposed by Tanner et al. [16] was adopted. The approach assigns a score to each tooth of study interest with respect to individual development stage. The scores of all the studied teeth are then summed to form a total maturity score which can be converted into the final dental or chronological age. The method was created and applied on the population of French-Canadian and desirable results were obtained. Based on this idea, a similar but more sophistically tuned method was later proposed by Willems et al. [10] which was based on the population of Belgian children.

Although the two popular methods have been proven effective for the tasks of dental age estimation, two critical issues can be concluded:

1. In both Demirjian and Willems methods, stage scores of all the studied tooth are summed for creating the final dental age. The summation operation suggests that the two methods assume maturity of each tooth contributes equally to the derivation of the final dental age. However, no theory validates this assumption of equal weights. For example, the first molar completes its apical closure earlier than the incisor. As a result, the first molar of a 13-year-old child has a less impact on estimating dental age in comparison with his/her incisor or other tooth. However, models in both Demirjian and Willems methods exclude the possibility of varied tooth maturity speed thus deteriorating the estimation accuracy.

2. Even origins of French-Canadian (studied by Demirjian method [9]) and Belgian (studied by Willems method [10]) are both European, reports of unmodified uses of Demirjian and Willems methods on other races are not available. It suggests that, the conversion table created in the methods of this kind can only be confined in certain populations or races or relevant modifications are necessary in order to achieve desirable results. Since it is obvious that bone structure of Asian population is significantly different with that of Europeans, the conversion tables within these methods needs to be re-designed.

One approach to resolve these issues is to apply machine learning techniques on the radiological methods, such as Demirjian's and Willems', where the statistical model is created through optimization of the model deducted from the 'known' dataset. In this approach, weights of features are normally derived dynamically and iteratively from the model optimization processes, with the constraints created by treating actual chronological age for optimizing the model. Furthermore, in machine learning techniques, a general framework dynamically creates models which are derived from the dataset which eliminate the uses and modifications of the conversion tables in the Demirjian and Willems methods.

Our previous work tries to test the hypothesis by replacing the conversion tables with the use of the simplest machine learning regression method - least square regression (LSR) method [17]. In this work, experiments on east Asian population shows that, LSR methods creates higher dental age estimation accuracy. This suggests the practicality and effectiveness of the use of machine learning methods. However, since LSR is by nature a linear model which seems not being capable of precisely associating features and chronological ages. In the present work, 1636 cases of 787 boys and 849 girls of East Asian population are treated as dataset. The traditional Radiological methods (Demirjian's, Willem's) and the machine learning regression method Multi-Layer Perceptron Neural Networks (MLP-NN) are implemented to discriminatively evaluate the performances of all methods studied. Through Experimental results have been analyzed through statistical measurement of errors of the prediction to validate the used machine learning methods for dental age estimation.

2   Method

2.1   Overview

2.1.1   Introduction to Regression in Machine Learning

Machine learning is a subset of artificial intelligence in the field of computer science, which typically uses statistical techniques to enable computers to "learn" from data [18]. Regression analysis is a way to find the relationship between data features and continuous target values. Applying machine learning to regression is an effective and practical technique used today that yields more accurate and universal results. That is to say, regression analysis using machine learning can be adapted to different samples and may result in satisfactory accuracy. In addition, there are many successful examples of applications of machine learning in real problems today, whether it be in the natural sciences, sociology or medicine.

The core idea of machine learning is to generalize the rules of existing data so that when new and unknown data is encountered, the target value of the sample can be obtained through said data features. Compared to traditional regression methods, machine learning pursues a more general model. When using machine learning, we want to be able to achieve high accuracy, and prevent overfitting to specific data. Hence, we have decided to use machine learning algorithms to predict dental age. We hope that by using machine learning algorithms, we can improve the accuracy of dental age predictions, and be able to cope with any data.

2.1.2   The Intuition Between Dental Age Estimation and Regression in Machine Learning

We refer to the traditional dental methods and collect data including seven scores of maturity of tooth and chronological age [9]. Then we decide to use the data as training data for machine learning algorithms. Specifically, for each case, seven scores of maturity of tooth are taken as features, and chronological age is used as the target value. We obtain the model between the features and the target by using machine learning. The resulting model can be used for dental age estimation on unknown data, as long as the features of each sample in the data are known. Meanwhile, since the chronological age of each sample is continuous, this process is doing a regression analysis. In this way, dental age estimation becomes a regression in machine learning.

2.2   Estimating Dental Age by Machine Learning

At first, we extracted the features from the teeth in the cases using Demirjian's method [9]. These features will be used as data features in the experiment. Meanwhile, we record the actual age of each case. Then we use machine learning algorithms to find the relationship between these features and the actual age of the case, which is often called the target or the label. The above process is called training a model. Finally, we use this trained model on a different data set and get the target values of the different data based on the features. This process is called prediction.

The process of training the model is actually finding the values of parameters in the machine learning algorithm. The process of finding parameters requires mathematics and some optimization programs that are already well developed. Machine learning also includes some hyperparameters, which are parameters that are set before the training starts. Optimizing these hyperparameters can improve the efficiency of model training and the accuracy of the model. The process of optimizing hyperparameters, known as tuning, requires some inspiration and trying different values.

It is noted that we use a special cross-validation, called Leave-one-out cross-validation, to avoid overfitting when training the model [19]. The main idea of cross-validation is to divide the training process into multiple parts, using only one part of the data at a time, until all the data is traversed. While Leave-one-out cross-validation is to use data removed from one sample during one training. Specifically, for a data with n samples, n-1 sample points are used each time to train the model. And then the trained model is used to predict the target value of the remaining sample point. By performing these steps n times, the model is completely trained. Using such a model training method can eliminate certain overfitting.

Our experiment is divided into two parts: using the same data and features as Demirjian's method and the Willem's method for comparative experiments and using Demirjian's method and the Willem's method converted features for exploratory experiments. On the one hand, we compare machine learning algorithms with traditional dental methods. On the other hand, we use the same machine learning algorithm for both types of data to study how the differences in data characteristics affect the predictions.

2.3   Description of the Used Machine Learning Algorithms

When using machine learning for regression, a suitable algorithm is needed. Since there are many machine learning algorithms, we have chosen one machine learning algorithm to experiment with. In the experiment, we will comprehensively compare the effects of the algorithm. The algorithm may have not been designed for dental age estimation, but it can be applied to dental age estimation.

2.3.1   Multi-layer Perceptron

A Multi-layer Perceptron is a class of feedforward artificial neural networks. A Multi-layer Perceptron consists of at least three layers of nodes. Except for the input nodes, each node is a neuron that uses a nonlinear activation function. MLP utilizes a supervised learning technique called backpropagation for training [21,22]. Multi-layer Perceptrons are useful in research for their ability to solve problems stochastically, which we think can be used in prediction of dental age. Because MLPs are also universal function approximators as showed by Cybenko's theorem [23], we have selected it for our experiments.

3   Data and Setting of Experiments

3.1   Data Description

Our experimental data consists of 1636 cases, including 849 girls and 787 boys, recorded by Shanghai Ninth People's Hospital. The data includes the dental age of each case, the seven features extracted from the seven teeth on the left side, the features transformed by several dental medical methods, and the dental age of each case predicted by these dental methods. The dental ages of the cases in the data range from 11 to 19 years old and is recorded as two decimal points. Considering that the difference between the seven teeth on the left side and all fourteen teeth is small [22], the raw seven features are extracted from the seven teeth on the left side. These seven features are determined by the shape of the corresponding seven teeth, as shown in Fig. 1. For the convenience of our later experiments, we have mapped the letters A to H to the numbers 1 to 8. The processed features are obtained by the dentist's expert using Demirjian's method (D-method) and Willem's method (W-method) to convert the raw features. The dental age estimated by the D-method and the W-method in each case is also obtained by dental professionals [9,10].

In addition, we have divided the data set into three parts: the raw feature data, the D-method processed data, and the W-method processed data. We have also experimented separately for males and females. The raw feature data is used to compare the effects of our proposed machine learning algorithms with traditional D-methods and W-methods. The last two processed data are used for exploratory research, and we think that using these data may result in higher prediction accuracy.

3.2   Setting of Experiments

We have decided to use the Multi-layer Perceptron (MLP) machine learning algorithm. This machine learning algorithm has been chosen mainly because of its better accuracy in dealing with regression problems [23,24]. On the other hand, we have adopted a special cross-validation method to prevent issues from overfitting [19]. Specifically, for a data set with n sample points, we use n-1 sample points each time to train the model. The trained model is then used to predict the target value of the remaining sample point. By performing such steps n times, we have obtained the predicted values for each sample point and have eliminated any overfitting to an extent.

For MLP, we have used a logistic activation function [25], as shown in Eq. (1). For each training value, connection weights are changed based on the mean-square-error of the predicted value and the expected value. The error in node j of the nth training value is given by Eq. (2), where e is the error, d is the expected value, and y is the predicted value. The weights of the nodes are changed to minimize the entire model error given by Eq. (3). With the use of gradient descent [26], the delta for each weight can be calculated by Eq. (4). The model is finished training when the neural net weights are updated with each training value.

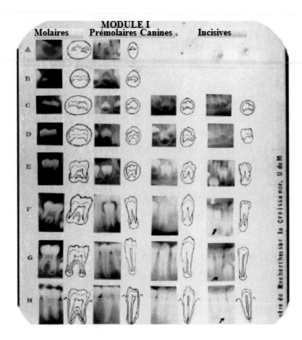

Fig. 1. Illustration of how to determine the value of feature based on the shape of tooth.

$$y(x_i) = \frac{1}{1 + e^{-x_i}} \tag{1}$$

$$e_j = d_j(n) - y_j(n) \tag{2}$$

$$\epsilon(n) = \frac{1}{2} \sum_j e_j^2(n) \tag{3}$$

$$\Delta\omega_{ji}(n) = \eta y_i(n) \frac{\partial \epsilon(n)}{\partial v_j(n)} \tag{4}$$

To compare the predictions of our three algorithms and traditional medical methods, we have selected three indicators to measure prediction accuracy: root-mean-square error (RMSE), mean-square-error (MSE), and mean-absolute-error (MAE) [28]. The values of these three indicators can be obtained by Eq. (5). In Eq. (5), p_i represents the predicted value, e_i represents the expected value, and n is the total number of sample points. RMSE, MSE, and MAE are often used to assess the accuracy of model predictions [29], and the smaller their values are, the higher the accuracy of the predictions is.

$$RMSE = \sqrt{\frac{1}{n}\sum_{i=1}^{n}(p_i - e_i)^2}$$

$$MSE = \frac{1}{n}\sum_{i=1}^{n}(p_i - e_i)^2$$

$$MAE = \frac{1}{n}\sum_{i=1}^{n}|p_i - e_i| \tag{5}$$

4   Experimental Results

We have applied traditional medical methods and machine learning algorithms to estimate the dental age for 787 males and 849 females using directly the seven input feature's data (I_1, I_2, C, PM_1, PM_2, M_1, and M_2). The traditional methods we used include Demirjian's method (D-method) and Willem's method (W-method), while the machine learning algorithm is Multi-layer Perceptron (MLP). In order to improve the accuracy of machine learning algorithms, we have employed an improved cross-validation, specifically for each set of n sample points using n-1 sample point's features to train the model and predict the age of the remaining one sample point. We have used three indicators to score the predictions of these methods in order to obtain a comprehensive and objective evaluation. These three indicators are root-mean-square error (RMSE), mean-squared-error (MSE) and mean-absolute-error (MAE). The smaller the value of these indicators are, the higher the accuracy of the prediction is. The results of these methods are briefly illustrated in Fig. 2. It can be seen that, for males, regardless which of the three machine learning algorithms is used, the values of RMSE, MSE, and MAE are significantly reduced. That is, the accuracy of the prediction is improved. Compared to the D-method and the W-method, MLP can significantly improve the accuracy of the prediction of male dental age. From this figure, the changes in RMSE, MSE, and MAE are not significant for females. Therefore, further numerical comparisons are needed.

Experimental Results of RMSE, MSE, and MAE are listed in Table 1. It can be observed that the values of RMSE, MSE, and MAE for males are significantly reduced by using a machine learning method. For females, the use of a machine learning method also results in some improvement, but the effect is not as obvious compared to predictions for males. Meanwhile, we have found that for both the traditional methods and the method proposed by us, when predicting the female dental age, the method will produce greater errors than the prediction for males. Taking MSE as an example, when we use the traditional method, the MSE value predicted for females is up to 25% higher than for males. When we use machine learning algorithms, the predicted MSE value for females is up to 47.4% higher than the value for males. It can be thought that the prediction of female dental age is likely to be more difficult. The values of the main parameters used in our machine learning method and average time to train the model are tabulated in Table 2.

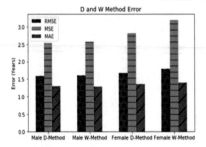 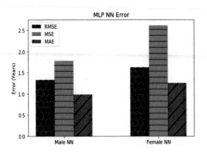

Fig. 2. The error of traditional medical methods and machine learning algorithm. The subgraph on the left is the result of traditional medical methods, and the subgraph on the right is the result of the machine learning algorithm. Each subgraph distinguishes between gender and uses three indicators.

Table 1. RMSE, MSE, MAE values of experimental results using D-method, W-method, and MLP

Male	RMSE	MSE	MAE	Female	RMSE	MSE	MAE
D-method	1.596	2.548	1.307	D-method	1.677	2.812	1.364
W-method	1.602	2.556	1.291	W-method	1.788	3.196	1.407
MLP	1.332	1.775	0.990	MLP	1.617	2.616	1.261

The values in the table are rounded and kept to three digits after the decimal point.

We have noticed that for the data set we processed, the error of the D-method was generally smaller than that of the W-method. Therefore, we calculated the rate of error reduction after using the machine learning algorithm versus the error value of the D-method. The calculation results of the reduction rate are listed in Table 3. It shows that the error reduction of males is obviously greater than that of females in all the indicators we used.

As can be seen from the above results, the machine learning algorithm we used does improve the accuracy of predictions for children's dental age. Furthermore, we have switched to a data set with dental experts using D-methods and W-methods to transform features [9,10] and have trained it with the machine learning algorithm. Similarly, we have calculated the RMSE, MSE, and MAE values for these methods and the results are presented in Table 4. We have observed that, after using the converted data set, the accuracy of the predictions of the machine learning algorithm is improved to some extent. However, which one is better, using D-method converted data or W-method converted data, has no definitive result. As far as the current results are concerned, we suggest using D-method converted data for male predictions and W-method converted data for female predictions.

Table 2. Values of main parameters used in MLP and Average time to train the model on all n points and predict one age of each method

MLP	
Hidden layers	88
Activation function	Logistic
Weight optimization solver	Limited-memory BFGS
Average time (Seconds)	0.962

Table 3. The Reduction rate of MLP machine learning algorithm compared to D-Method

Male	RMSE	MSE	MAE	Female	RMSE	MSE	MAE
MLP	16.5%	30.3%	24.3%	MLP	3.6%	7.0%	7.6%

The values in the table are rounded and kept to one digit after the decimal point.

Table 4. RMSE, MSE, MAE values of experimental results using MLP with D- and W-Method weights pre-applied to input data

Male	RMSE	MSE	MAE	Female	RMSE	MSE	MAE
Raw	1.332	1.775	0.990	Raw	1.617	2.616	1.261
D-method	1.214	1.475	0.864	D-method	1.464	2.144	1.116
W-method	1.230	1.514	0.884	W-method	1.446	2.090	1.090

The values in the table are rounded and kept to three digits after the decimal point.

5   Conclusion

In this paper, we have introduced a machine learning algorithm to improve the accuracy of dental age prediction. Through experiments, we have verified that this algorithm is indeed more accurate than the two commonly used traditional dental methods for predicting dental age. In addition, we have tried to use a new set of features that are converted by the traditional dental methods. We have experimentally verified that this new feature set does make the prediction of dental age more accurate. However, we are only experimenting with a specific population. In future work, we will test our methods on other different populations.

Acknowledgments. The data in this paper is provided by the Ninth People's Hospital affiliated to Shanghai Jiao Tong University School of Medicine. We also sincerely thank 1636 volunteers who have supplied the collected dental data for research.

References

1. Williams, G.: A review of the most commonly used dental age estimation techniques. J. Forensic Odontostomatol. **19**(1), 9–17 (2001)
2. Olze, A., Geserick, G., Schmeling, A.: Age estimation of unidentified corpses by measurement of root translucency. J. Forensic Odontostomatol. **22**(2), 28–33 (2004)
3. Kvaal, S.I.: Collection of post mortem data: DVI protocols and quality assurance. Forensic Sci. Int. **159**, S12–S14 (2006)
4. Karkhanis, S., Mack, P., Franklin, D.: Dental age estimation standards for a Western Australian population. Forensic Sci. Int. **257**, 509-e1 (2015)
5. Ritz-Timme, S., Cattaneo, C., Collins, M.J., Waite, E.R., Schütz, H.W., Kaatsch, H.J., Borrman, H.I.M.: Age estimation: the state of the art in relation to the specific demands of forensic practise. Int. J. Legal Med. **113**(3), 129–136 (2000)
6. Lopez, T.T., Arruda, C.P., Rocha, M., de Oliveira Rosin, A.S.A., Michel-Crosato, E., Biazevic, M.G.H.: Estimating ages by third molars: stages of development in Brazilian young adults. J. Forensic Legal Med. **20**(5), 412–418 (2013)
7. Melo, M., Ata-Ali, J.: Accuracy of the estimation of dental age in comparison with chronological age in a Spanish sample of 2641 living subjects using the Demirjian and Nolla methods. Forensic Sci. Int. **270**, 276-e1 (2017)
8. Garn, S.M., Lewis, A.B., Kerewsky, R.S.: Genetic, nutritional, and maturational correlates of dental development. J. Dent. Res. **44**(1), 228–242 (1965)
9. Demirjian, A., Goldstein, H., Tanner, J.M.: A new system of dental age assessment. Hum. Biol. **45**(2), 211–227 (1973)
10. Willems, G., Van Olmen, A., Spiessens, B., Carels, C.: Dental age estimation in Belgian children: Demirjian's technique revisited. J. Forensic Sci. **46**(4), 893–895 (2001)
11. Ye, X., Jiang, F., Sheng, X., Huang, H., Shen, X.: Dental age assessment in 7–14-year-old Chinese children: Comparison of Demirjian and Willems methods. Forensic Sci. Int. **244**, 36–41 (2014)
12. Kumaresan, R., Cugati, N., Chandrasekaran, B., Karthikeyan, P.: Reliability and validity of five radiographic dental-age estimation methods in a population of Malaysian children. J. Invest. Clin. Dent. **7**(1), 102–109 (2016)
13. Djukic, K., Zelic, K., Milenkovic, P., Nedeljkovic, N., Djuric, M.: Dental age assessment validity of radiographic methods on Serbian children population. Forensic Sci. Int. **231**(1–3), 398-e1 (2013)
14. Urzel, V., Bruzek, J.: Dental age assessment in children: a comparison of four methods in a recent French population. J. Forensic Sci. **58**(5), 1341–1347 (2013)
15. Nolla, C.M.: The development of the permanent teeth. J. Dent. Child. **27**, 254–266 (1952)
16. Tanner, J.M.: Growth at Adolescence. Blackwell Scientific Publications, Oxford (1962)
17. Tao, J., Chen, M., Wang, J., Liu, L., Hassanien, A.E., Xiao, K.: Dental age estimation in East Asian population with least squares regression. In: International Conference on Advanced Machine Learning Technologies and Applications, pp. 653–660. Springer, Cham (2018)
18. Samuel, A.L.: Some studies in machine learning using the game of checkers. IBM J. Res. Dev. **3**(3), 210–229 (1959)
19. Kohavi, R.: A study of cross-validation and bootstrap for accuracy estimation and model selection. In: Ijcai, vol. 14, no. 2, pp. 1137–1145 (1995)

20. Altman, N.S.: An introduction to kernel and nearest-neighbor nonparametric regression. Am. Stat. **46**(3), 175–185 (1992)
21. Rosenblatt, F.: Principles of neurodynamics: perceptrons and the theory of brain mechanisms (No. VG-1196-G-8). Cornell Aeronautical Lab Inc., Buffalo (1961)
22. Rummelhart, D.E.: Learning internal representations by error propagation. In: Parallel Distributed Processing: I. Foundations, pp. 318–362 (1986)
23. Cybenko, G.: Approximation by superpositions of a sigmoidal function. Math. Control Signals Systems **2**(4), 303–314 (1989)
24. Hastie, T., Tibshirani, R., Friedman, J.: The Elements of Statistical Learning, 2nd edn, pp. 587–588. Springer, New York (2008)
25. Pinkus, M.L.V.L.A., Schocken, S.: Multilayer feedforward networks with non-polynomial activation functions can approximate any continuous function. Neural Netw. **6**, 861–867 (1993)
26. Govan, A.: Introduction to optimization. In North Carolina State University, SAMSI NDHS, Undergraduate workshop (2006)
27. Breiman, L.: Bagging predictors. Mach. Learn. **24**(2), 123–140 (1996)
28. Panik, M.J.: Advanced Statistics from an Elementary Point of View, vol. 9. Academic Press, Amsterdam (2005)
29. Lehmann, E.L., Casella, G.: Theory of Point Estimation. Springer, Heidelberg (2006)
30. Mood, A.M., Graybill, F.A., Boes, D.C.: Introduction to the Theory of Statistics, pp. 540–541. McGraw-Hill, New York (1974)
31. Nasrabadi, N.M.: Pattern recognition and machine learning. J. Electron. Imaging **16**(4), 049901 (2007)

A Survey on Human Activity Recognition Based on Temporal Signals of Portable Inertial Sensors

Reda Elbasiony[1(✉)] and Walid Gomaa[2,3]

[1] Faculty of Engineering, Tanta University, Tanta 31527, Egypt
`reda@f-eng.tanta.edu.eg`
[2] Egypt Japan University of Science and Technology, Alexandria 21934, Egypt
`walid.gomaa@ejust.edu.eg`
[3] Faculty of Engineering, Alexandria University, Alexandria 11432, Egypt

Abstract. In recent years, automatic human activity recognition has drawn much attention. On one hand, this is due to the rapid proliferation and cost degradation of a wide variety of sensing hardware, which resulted in the tremendous explosion of activity data. On the other hand there are urgent growing and pressing demands from many application areas such as: in-home health monitoring especially for the elderly, smart cities, safe driving by monitoring and predicting driver's behavior, healthcare applications, entertainment, assessment of therapy, performance evaluation in sports, etc. In this paper, we introduce a detailed survey on multiple human activity recognition (HAR) systems which use portable inertial sensors (Accelerometer, Magnetometer, and Gyro), where the sensor's produced temporal signals are used for modeling and recognition of different human activities based on various machine learning techniques.

Keywords: Human activity recognition · Machine learning ·
Inertial measurement unit · Accelerometer · Gyroscope

1 Introduction

Automatic recognition of human activities has become a very substantial research topic. A wide range of computer applications depend mainly on *human activity recognition (HAR)* in their work such as monitoring patients and elderly people, surveillance systems, robots learning and cooperating, and military applications. The idea of automatic HAR system depends on collecting measurements from some appropriate sensors which are affected by selected human motion attributes. Then, depending on these measurements, some features are extracted to be used in the process of training activity models, which in turn will be used to recognize these activities later.

© Springer Nature Switzerland AG 2020
A. E. Hassanien et al. (Eds.): AMLTA 2019, AISC 921, pp. 734–745, 2020.
https://doi.org/10.1007/978-3-030-14118-9_72

Based on the data acquisition paradigm, HAR systems can be divided into two categories: surrounding fixed-sensor systems and wearable mobile-sensor systems. In the first category, the required data are collected from distributed sensors attached to fixed locations in the activity environment (where users activities are monitored) such as surveillance cameras, microphones, motion sensors, etc. Alternatively, the sensors are attached to interactive objects in order to detect the type of interaction with them, such as a motion sensor attached to the cupboard doors or microwave ovens (to detect opening and/or closing), or on water tap to feel turning it on or off, and so on. Although this method can detect complex actions efficiently, it has many limitations due to its fixed nature.

In the wearable based systems, the measurements are taken from mobile sensors mounted to human body parts like wrists, legs, waist, and chest. Many sensors can be used to measure selected attributes of human body motion, such as accelerometers, gyroscopes, compasses, and GPSs. Or to measure phenomena around the user, such as barometers, magnetometers, light sensors, temperature sensors, humidity sensors, microphones, and mobile cameras. On the contrary of the fixed-sensor based systems, wearables are able to measure data from the user everywhere, while sleeping, working, or even traveling anywhere since it is not bounded by a specific place where the sensors are installed. Also, it is very easy to concentrate on directly measuring data of particular body parts efficiently without a lot of preprocessing that are needed, for example, in fixed depth cameras. However, carrying and installing a lot of sensors, wires, and power supplies mounted to the user may be uncomfortable and annoying. A comprehensive review on the use of wearables in the medical sector can be found in [34].

During the recent few years, there has been a tremendous evolution in the manufacturing of mobile devices. Particularly mobile phones, tablet PCs, and smart watches. All such devices contain various types of sensors, such as accelerometers, barometers, gyroscopes, microphones, GPS, etc. The evolution of the Internet of Things and ubiquitous sensing have encouraged mobile device manufacturers to provide more types of sensors and improve the accuracy and efficiency of the existing ones. Smart phones have also become more and more popular. Recent statistics show that the total number of smartphone subscribers reached 3.9 billion in 2016 and is expected to reach 6.8 billion by 2022 [1]. Other statistics show that in some countries, the percentage of smart phone subscribers reaches 88% of the population [35]. Therefore, the disadvantages of wearable mobile-sensors of being intrusive, uncomfortable, and annoying have vanished to a great extent, making this method of on-board sensing from smart devices very suitable for HAR data acquisition.

Many kinds of attributes can be measured using wearable sensors [26]. These include: (i) environmental data such as barometric pressure, temperature, humidity, light intensity, magnetic fields, and noise level, (ii) vital health signs such as pulse, body temperature, blood pressure, and respiratory rate, (iii) location data which are typically identified by longitudes and latitudes using GPS sensors, and (iv) body limbs motion such as acceleration and rotation of body parts like arms, legs, and torso using accelerometers and gyros. For the purpose of activity recognition, the latter type of attributes have proven to represent human motion accurately, where some methods depending only on acceleration measurements have achieved very high accuracy [8,20]. However, it is very difficult to extract the pattern of motion from acceleration raw data directly because of high frequency oscillation and noise. So, feature extraction techniques should be applied on the raw data before using it. Many types of features can be extracted from acceleration data: (1) time-domain features like mean and variance [21], (2) frequency-domain features like Fourier transform, discrete cosine transform, and wavelet transform, and (3) applying dimensionality reduction techniques like PCA and LDA [8].

An essential function of HAR systems is to provide a general model for each activity. The activity models are mostly generated based on the features extracted from training data using supervised machine learning techniques. Thus, the role of machine learning techniques in HAR systems is to build general models to describe each activity and use these models to detect or classify the target activities later. Many classification techniques are used such as support vector machine (SVM), random forests, C4.5 decision trees, hidden Markov models (HMM), k-nearest neighbor (KNN), Bayesian networks, and artificial neural networks (ANN) [12,44].

Among all wearable sensors, accelerometers are considered as the most used sensors in HAR systems [12]. Being small-sized, inexpensive, and embedded in most of smart mobile devices has encouraged many researchers to use acceleration in their work. Compared to cameras, accelerometers are more suitable for HAR systems, it is very difficult to fix a camera to monitor a user everywhere; also mounting the camera to the user's body is very annoying and uncomfortable. From the privacy point of view, it is not acceptable nor convenient by many people to be monitored all the time. As well, the videos or images collected using cameras are very sensitive to many environmental conditions like lighting and surrounding barriers. However, accelerometers can be easily mounted to users or embedded into many devices such as smart phones and/or smart watches which are naturally carried by many users everywhere most of the time. Also acceleration data preserve user privacy and are not affected by any outside conditions.

2   Related Work

Sensors locations and count are very important issues that have to be taken into consideration while designing an accelerometer-based HAR system. As seen in Table 1, many settings have been studied through the previous work. Regarding the locations of wearable sensors, different body locations have been used from feet to chest. However, the relevance of the selection to activities plays an important role in specifying the sensor's location. For example, ambulation activities (such as walking, running, jumping, etc.) can be detected efficiently using a chest or waist mounted sensor [16,20,37]. Whereas, non-ambulation activities (such as brushing teeth, combing hair, eating, etc.) can be classified more efficiently using a wrist-worn sensor [8,22]. This makes using more than one sensor in different

Table 1. Related work classified by sensors (type, count, location), sampling rate, learning method, number of activities, and average accuracy (A: Accelerometer, M: Magnetometer, G: Gyroscope).

Reference	Used sensors			No. of sensors	Sensors location	Sampling rate (HZ)	Learning method/algorithm	No. of activities	Accuracy (%)
	A	M	G						
[29]	X			2	Hip left and right sides	256	Multilayer perceptron	4	90
[27]	X		X	2	Waist, front trouser pocket	5	Threshold-based	5	86.7
[36]	X			1	Waist	50	Multiple classifiers	8	84
[19]	X			1	Waist	45	Threshold-based	12	90.8
[31]	X			6	Wrist, waist, necklace, trouser pocket, shirt pocket, bag	50	Decision Trees Naive-Bayes	6	88
[18]	X			12	Ankles, knees, elbows, shoulders, wrists, hip left and right sides	92	Multiple Eigenspaces combined with SVM	8	88.3
[17]	X			1	Different locations (clothes pocket, waist belt, and trouser pocket)	100	SVM	4	97.21
[37]	X			1	Waist	14	Multilayer perceptron	9	95.5
[11]	X			1	Chest	50	Threshold-based	3	81.25
[23]	X			3	Hip, dominant ankle, non-dominant thigh	50	AdaBoost, SVM, RLogReg	7	88.2
[16]	X			1	Trouser pocket	100	SVM	4	97.5
[47]	X			2	Foot, waist	150	Feed forwarn NN, HMM	4	89.7
[22]	X			1	Wrist	100	KNN	5	92.2
[46]	X			1	Waist	1	SVM	6	82.8
[20]	X			1	Chest	20	Feed-forwarn NN	15	97.9
[24]	X			1	Trouser pocket	20	Decision trees (J48), logistic regression, neural network	6	91.7
[28]	X			1	Thigh	250	SVM	5	99
[9]	X			1	Wrist	33	C4.5, neural network	5	94.1
[13]	X		X	9	Wrists, arms, thighs, ankles, waist	50	Hierarchical clustering, K-means, decision trees	25	93.3

(*continued*)

Table 1. (*continued*)

Reference	Used sensors			No. of sensors	Sensors location	Sampling rate (HZ)	Learning method/algorithm	No. of activities	Accuracy (%)
	A	M	G						
[48]	X			1	Right thigh	20	Feed Forwarn NN, HMM	8	85
[25]	X			1	Chest	50	C4.5	3	92.6
[43]	X			1	Waist	100,50, 16,5	J48 adaptive decision tree	10	87
[33]	X			1	Torso	40	Threshold-based	6	100
[10]	X			2	Dominant wrist, ankle	30	K-NN probabilistic neural network	7	96
[42]	X	'	X	7	Waist, wrists, right arm, left thigh, ankles	30	Bayesian Sparse representation-Based	14	87.7
[3]	X			1	Waist	50	MC-HF-SVM	6	89
[45]	X		X	1	Waist	100	K-NN, naive Bayes, SVM	9	95.2
[2]	X			1	Waist	80	Stochastic Approximation classifier	2	94.5
[38]	X			3	Chest, right thigh, left ankle	25	HMM	12	91.4
[8]	X			1	Right wrist	32	GMM, GMR	8	68.7
[39]	X			1	Trouser pocket	1	Hierarchical-SVM	4	98.5
[4]	X			1	Trouser pocket	20	KNN	6	99.4
[14]	X			2	Abdomen,right thigh	50	Threshold-based, Random Forest	4	98.85
[32]	X	X	X	1	Right shoe	100	KNN	20	77
[6]	X			3	Wrist, chest, foot	100	Neural network	14	89.7
[7]	X		X	3	Chest, left crus, right thigh	50	Neuro-fuzzy classifier	7	97.2
[40]	X			1	Back mounted	50	KNN	5	95.6
[21]	X			1	Wrist	32	Neural network	7	91
[30]	X			1	Wrist	20	Template-matching-based	22	80
[41]	X			3	Chest,dominant wrist, dominant ankle	100	Back propagation neural network	12	93.7
[15]	X		X	1	Dominant wrist	50	Random forests	14	80
[5]	X		X	1	Dominant wrist	50	LSTM	31	97

body locations a good idea for improving accuracy for both ambulation and non-ambulation activities [13].

Accelerometer sampling frequency is also an important parameter. Maurer et al. [31] and Yan et al. [43] studied the effect of different sampling frequencies on the classification accuracy. Maurer et al. [31] found that the accuracy stabilizes between 15 to 20 Hz and not considerably improved on higher sampling rate. Yan et al. [43] studied sampling frequency from the energy saving point of view. They found that there is a an activity specific tradeoff between consumed energy and classification accuracy based on sampling frequency. So, they introduced a smart adaptive method which changes the sampling frequency in real-time based on the type of activity.

There are mainly two approaches to HAR: threshold-based and machine learning-based. HAR systems which are threshold-based don't require any training processes, however, they can be used to classify relatively small number of activities compared to systems which are based on machine learning techniques. The majority of HAR systems use supervised classification algorithms in order to classify the relevant activities as shown in Table 1 making use of the ability of learning algorithms to detect and discriminate between different hidden patterns of activities.

Several kinds of activities have been considered by HAR systems. These include: ambulation activities, daily activities, fitness activities, and industrial activities. However, most of the studied HAR systems concentrate on ambulation activities as shown in Table 2, because such activities are regularly performed by almost all people. Ambulation activities are also easy to recognize because of having many repetitive motions and having the same patterns everywhere for any subject. The effectiveness of any HAR system depends not only on the recognition accuracy but also on the number and types of activities which can it recognize. According to Table 1, the number of activities for developed HAR systems widely vary, starting from only 2 or 3 activities to more than 20 activities. However, the average recognition accuracy is affected by the count and the diversity of the types of considered activities.

Raw acceleration data can not be used directly to recognize human activities. It is generally hard getting the same acceleration values for the same activity twice even if the activity is performed by the same person. Thus, to measure the similarity between acceleration time series, preprocessing have to be applied first to extract more informative features which can substitute the raw data. Table 3 summarizes the selected features for our related work. Two common types of features can be extracted from acceleration data (and generally from any time series): time domain features and frequency domain features. Frequency domain features are avoided in many studies, especially those assuming limited computing resources or requiring real-time recognition, to avoid the costly computation overhead required to transform the signals from the time domain to the frequency domain using fast Fourier transform (FFT). Time domain feature represent more the values of the acceleration, whereas the frequency domain features usually reflect the periodicity in the signal. As seen in Table 3, the most popular

Table 2. Related work classified by tested activities.

Ref.	Walking	Stairs (Up/Down)	Sitting	Standing	Running	Lying	Jumbing	Other
[29]	X	X						X
[27]	X	X	X	X				
[36]	X	X		X	X			X
[19]	X		X	X		X		X
[31]	X	X	X	X	X			
[18]	X	X	X	X				X
[17]	X				X		X	X
[37]	X		X	X	X	X		X
[11]	X				X			X
[23]	X	X	X	X	X	X		X
[16]	X				X		X	X
[47]	X	X			X			
[22]								X
[46]	X		X	X		X		X
[20]	X	X	X	X	X	X		X
[24]	X	X	X	X	X			
[28]	X		X	X	X			X
[9]	X		X	X	X	X		
[13]	X	X	X	X		X	X	X
[48]	X		X	X		X		X
[25]	X		X		X			
[43]	X	X	X	X				X
[33]	X	X				X		
[10]	X	X	X	X	X			X
[42]	X		X	X		X	X	X
[3]	X	X	X	X		X		
[45]	X	X	X	X	X		X	
[2]	X				X			
[38]	X	X	X	X		X		
[8]	X	X	X	X				X
[39]	X		X	X	X			
[4]	X	X	X	X	X			
[14]			X			X		X
[32]	X				X			X
[6]	X	X	X	X	X	X	X	X
[7]	X	X	X	X	X	X		
[40]	X				X		X	X
[21]	X	X	X	X				X
[30]	X				X			X
[41]	X	X	X	X	X	X		X
[15]	X	X	X	X		X		X
[5]	X	X	X	X	X			X

Table 3. Accelerometer data classified by extracted features.

Ref.	Time Domain												Frequency Domain		Other
	Mean	Standard Deviation	Correlation	Variance	RMS	Mean Crossing Rate	Zero Crossing Rate	Normalized Signal Magnitude Area (SMA)	Interquartile Range	Minimum	Maximum	Median	Energy	Entropy	
[29]															X
[27]		X													X
[36]	X	X	X								X				
[19]								X							X
[31]	X	X		X	X	X	X								X
[18]	X			X											
[17]															X
[37]															Raw data
[11]	X							X							
[23]	X		X	X									X	X	
[16]															X
[47]	X														
[22]	X	X	X							X	X	X			X
[46]										X	X				
[20]	X	X	X					X							
[24]	X	X													X
[28]	X			X					X			X			X
[9]	X	X	X	X						X			X	X	X
[13]	X	X			X										X
[48]	X			X											
[25]	X	X	X	X	X								X		X
[43]	X			X									X	X	X
[33]	X											X			X
[10]		X											X	X	X
[42]	X	X	X	X	X	X	X		X			X			X
[3]	X	X	X					X						X	
[45]	X	X	X	X	X	X	X	X				X	X	X	X
[2]	X	X	X	X	X	X	X		X			X			X
[38]															Raw data
[8]															Raw data
[39]				X											X
[4]	X	X	X		X			X	X	X	X	X		X	X
[14]															X
[32]	X														
[6]	X	X									X	X	X		X
[7]	X	X					X				X	X	X		X
[40]															X
[21]	X				X	X									X
[30]	X				X	X				X	X		X	X	
[41]															X
[15]			X												
[5]													X		X

time domain features are mean, standard deviation, correlation, and variance. While the most popular frequency domain features are energy and entropy.

3   Conclusion

A review of HAR systems which uses temporal signals generated from portable inertial sensors has been introduced. Types of HAR systems according to data acquisition paradigms, types of attributes, and sensors' types, counts and locations have been presented. Moreover, various machine learning algorithms which are used with HAR systems have been stated. Finally, some important related proposed systems are illustrated.

References

1. Ericsson Mobility Report on The Pulse of The Networked Society. Technical report, Ericsson, November 2016. https://www.ericsson.com/assets/local/mobility-report/documents/2016/ericsson-mobility-report-november-2016.pdf
2. Alshurafa, N., Xu, W., Liu, J.J., Huang, M.C., Mortazavi, B., Sarrafzadeh, M., Roberts, C.: Robust human intensity-varying activity recognition using stochastic approximation in wearable sensors. In: IEEE International Conference on Body Sensor Networks (BSN), pp. 1–6. IEEE (2013)
3. Anguita, D., Ghio, A., Oneto, L., Parra, X., Reyes-Ortiz, J.L.: Energy efficient smartphone-based activity recognition using fixed-point arithmetic. J. UCS **19**(9), 1295–1314 (2013)
4. Arif, M., Bilal, M., Kattan, A., Ahamed, S.I.: Better physical activity classification using smartphone acceleration sensor. J. Med. Syst. **38**(9), 95 (2014)
5. Ashry, S., Elbasiony, R., Gomaa, W.: An lstm-based descriptor for human activities recognition using imu sensors. In: Proceedings of the 15th International Conference on Informatics in Control, Automation and Robotics, ICINCO, Vol. 1, pp. 494–501. INSTICC, SciTePress (2018)
6. Basterretxea, K., Echanobe, J., del Campo, I.: A wearable human activity recognition system on a chip. In: 2014 Conference on Design and Architectures for Signal and Image Processing (DASIP), pp. 1–8. IEEE (2014)
7. Braojos, R., Beretta, I., Constantin, J., Burg, A., Atienza, D.: A wireless body sensor network for activity monitoring with low transmission overhead. In: 12th IEEE International Conference on Embedded and Ubiquitous Computing (EUC), pp. 265–272. IEEE (2014)
8. Bruno, B., Mastrogiovanni, F., Sgorbissa, A., Vernazza, T., Zaccaria, R.: Analysis of human behavior recognition algorithms based on acceleration data. In: IEEE International Conference on Robotics and Automation (ICRA), pp. 1602–1607. IEEE (2013)
9. Chernbumroong, S., Atkins, A.S., Yu, H.: Activity classification using a single wrist-worn accelerometer. In: 5th International Conference on Software, Knowledge Information, Industrial Management and Applications (SKIMA), pp. 1–6. IEEE (2011)
10. Chuang, F.C., Wang, J.S., Yang, Y.T., Kao, T.P.: A wearable activity sensor system and its physical activity classification scheme. In: International Joint Conference on Neural Networks (IJCNN), pp. 1–6. IEEE (2012)

11. Chung, W.Y., Purwar, A., Sharma, A.: Frequency domain approach for activity classification using accelerometer. In: 30th Annual International Conference of the IEEE Engineering in Medicine and Biology Society, EMBS 2008, pp. 1120–1123. IEEE (2008)
12. Cornacchia, M., Ozcan, K., Zheng, Y., Velipasalar, S.: A survey on activity detection and classification using wearable sensors. IEEE Sens. J. **17**(2), 386–403 (2017)
13. Ghasemzadeh, H., Jafari, R.: Physical movement monitoring using body sensor networks: a phonological approach to construct spatial decision trees. IEEE Trans. Industr. Inf. **7**(1), 66–77 (2011)
14. Gjoreski, H., Kozina, S., Gams, M., Lustrek, M.: Rarefall–real-time activity recognition and fall detection system. In: IEEE International Conference on Pervasive Computing and Communications Workshops (PERCOM Workshops), pp. 145–147. IEEE (2014)
15. Gomaa, W., Elbasiony, R., Ashry, S.: Adl classification based on autocorrelation function of inertial signals. In: 16th IEEE International Conference on Machine Learning and Applications (ICMLA), pp. 833–837. IEEE (2017)
16. He, Z., Jin, L.: Activity recognition from acceleration data based on discrete consine transform and svm. In: IEEE International Conference on Systems, Man and Cybernetics, SMC 2009, pp. 5041–5044. IEEE (2009)
17. He, Z., Liu, Z., Jin, L., Zhen, L.X., Huang, J.C.: Weightlessness feature–a novel feature for single tri-axial accelerometer based activity recognition. In: 19th International Conference on Pattern Recognition, ICPR 2008, pp. 1–4. IEEE (2008)
18. Huynh, T., Schiele, B.: Towards less supervision in activity recognition from wearable sensors. In: 10th IEEE International Symposium on Wearable Computers, pp. 3–10. IEEE (2006)
19. Karantonis, D.M., Narayanan, M.R., Mathie, M., Lovell, N.H., Celler, B.G.: Implementation of a real-time human movement classifier using a triaxial accelerometer for ambulatory monitoring. IEEE Trans. Inf. Technol. Biomed. **10**(1), 156–167 (2006)
20. Khan, A.M., Lee, Y.K., Lee, S.Y., Kim, T.S.: A triaxial accelerometer-based physical-activity recognition via augmented-signal features and a hierarchical recognizer. IEEE Trans. Inf. Technol. Biomed. **14**(5), 1166–1172 (2010)
21. Kilinc, O., Dalzell, A., Uluturk, I., Uysal, I.: Inertia based recognition of daily activities with anns and spectrotemporal features. In: IEEE 14th International Conference on Machine Learning and Applications (ICMLA), pp. 733–738. IEEE (2015)
22. Koskimaki, H., Huikari, V., Siirtola, P., Laurinen, P., Roning, J.: Activity recognition using a wrist-worn inertial measurement unit: a case study for industrial assembly lines. In: 17th Mediterranean Conference on Control and Automation, MED 2009, pp. 401–405. IEEE (2009)
23. Krishnan, N.C., Panchanathan, S.: Analysis of low resolution accelerometer data for continuous human activity recognition. In: IEEE International Conference on Acoustics, Speech and Signal Processing, ICASSP 2008, pp. 3337–3340. IEEE (2008)
24. Kwapisz, J.R., Weiss, G.M., Moore, S.A.: Activity recognition using cell phone accelerometers. ACM SigKDD Explor. Newsl. **12**(2), 74–82 (2011)
25. Lara, O.D., Labrador, M.A.: A mobile platform for real-time human activity recognition. In: IEEE Consumer Communications and Networking Conference (CCNC), pp. 667–671. IEEE (2012)
26. Lara, O.D., Labrador, M.A.: A survey on human activity recognition using wearable sensors. IEEE Commun. Surv. Tutor. **15**(3), 1192–1209 (2013)

27. Lee, S.W., Mase, K.: Activity and location recognition using wearable sensors. IEEE Pervasive Comput. **1**(3), 24–32 (2002)
28. Mannini, A., Sabatini, A.M.: On-line classification of human activity and estimation of walk-run speed from acceleration data using support vector machines. In: Annual International Conference of the IEEE Engineering in Medicine and Biology Society, EMBC, pp. 3302–3305. IEEE (2011)
29. Mantyjarvi, J., Himberg, J., Seppanen, T.: Recognizing human motion with multiple acceleration sensors. In: IEEE International Conference on Systems, Man, and Cybernetics, vol. 2, pp. 747–752. IEEE (2001)
30. Margarito, J., Helaoui, R., Bianchi, A.M., Sartor, F., Bonomi, A.G.: User-independent recognition of sports activities from a single wrist-worn accelerometer: a template-matching-based approach. IEEE Trans. Biomed. Eng. **63**(4), 788–796 (2016)
31. Maurer, U., Smailagic, A., Siewiorek, D.P., Deisher, M.: Activity recognition and monitoring using multiple sensors on different body positions. In: International Workshop on Wearable and Implantable Body Sensor Networks, BSN 2006, pp. 4–pp. IEEE (2006)
32. Mortazavi, B., Nyamathi, S., Lee, S.I., Wilkerson, T., Ghasemzadeh, H., Sarrafzadeh, M.: Near-realistic mobile exergames with wireless wearable sensors. IEEE J. Biomed. Health Inf. **18**(2), 449–456 (2014)
33. Naranjo-Hernández, D., Roa, L.M., Reina-Tosina, J., Estudillo-Valderrama, M.A.: Som: a smart sensor for human activity monitoring and assisted healthy ageing. IEEE Trans. Biomed. Eng. **59**(11), 3177–3184 (2012)
34. Patel, S., Park, H., Bonato, P., Chan, L., Rodgers, M.: A review of wearable sensors and systems with application in rehabilitation. J. Neuroeng. Rehabil. **9**(1), 21 (2012)
35. Poushter, J.: Smartphone ownership and internet usage continues to climb in emerging economies (2016). http://www.pewglobal.org/2016/02/22/smartphone-ownership-and-internet-usage-continues-to-climb-in-emerging-economies
36. Ravi, N., Dandekar, N., Mysore, P., Littman, M.L.: Activity recognition from accelerometer data. Aaai **5**, 1541–1546 (2005)
37. Song, S.k., Jang, J., Park, S.: A phone for human activity recognition using triaxial acceleration sensor. In: International Conference on Consumer Electronics, ICCE 2008. Digest of Technical Papers, pp. 1–2. IEEE (2008)
38. Trabelsi, D., Mohammed, S., Chamroukhi, F., Oukhellou, L., Amirat, Y.: An unsupervised approach for automatic activity recognition based on hidden markov model regression. IEEE Trans. Autom. Sci. Eng. **10**(3), 829–835 (2013)
39. Weng, S., Xiang, L., Tang, W., Yang, H., Zheng, L., Lu, H., Zheng, H.: A low power and high accuracy mems sensor based activity recognition algorithm. In: IEEE International Conference on Bioinformatics and Biomedicine (BIBM), pp. 33–38. IEEE (2014)
40. Wilson, J., Najjar, N., Hare, J., Gupta, S.: Human activity recognition using lzw-coded probabilistic finite state automata. In: IEEE International Conference on Robotics and Automation (ICRA), pp. 3018–3023. IEEE (2015)
41. Xu, H., Liu, J., Hu, H., Zhang, Y.: Wearable sensor-based human activity recognition method with multi-features extracted from hilbert-huang transform. Sensors **16**(12), 2048 (2016)
42. Xu, W., Zhang, M., Sawchuk, A.A., Sarrafzadeh, M.: Co-recognition of human activity and sensor location via compressed sensing in wearable body sensor networks. In: Ninth International Conference on Wearable and Implantable Body Sensor Networks (BSN), pp. 124–129. IEEE (2012)

43. Yan, Z., Subbaraju, V., Chakraborty, D., Misra, A., Aberer, K.: Energy-efficient continuous activity recognition on mobile phones: an activity-adaptive approach. In: 16th International Symposium on Wearable Computers (ISWC), pp. 17–24. IEEE (2012)

44. Ye, L., Ferdinando, H., Seppänen, T., Huuki, T., Alasaarela, E.: An instance-based physical violence detection algorithm for school bullying prevention. In: 2015 International Wireless Communications and Mobile Computing Conference (IWCMC), pp. 1384–1388. IEEE (2015)

45. Zhang, M., Sawchuk, A.A.: Human daily activity recognition with sparse representation using wearable sensors. IEEE J. Biomed. Health Inf. **17**(3), 553–560 (2013)

46. Zhang, S., McCullagh, P., Nugent, C., Zheng, H.: Activity monitoring using a smart phone's accelerometer with hierarchical classification. In: 2010 Sixth International Conference on Intelligent Environments (IE), pp. 158–163. IEEE (2010)

47. Zhu, C., Sheng, W.: Human daily activity recognition in robot-assisted living using multi-sensor fusion. In: IEEE International Conference on Robotics and Automation, ICRA 2009, pp. 2154–2159. IEEE (2009)

48. Zhu, C., Sheng, W.: Motion-and location-based online human daily activity recognition. Pervasive Mob. Comput. **7**(2), 256–269 (2011)

3D Geolocation Approach for Moving RF Emitting Source Using Two Moving RF Sensors

Kamel H. Rahouma$^{(\boxtimes)}$ and Aya S. A. Mostafa

Department of Electrical Engineering,
Faculty of Engineering, Minia University, Minia, Egypt
kamel_rahouma@yahoo.com, ayasami89@yahoo.com

Abstract. The three-dimensional geolocation of a radio frequency RF emitting source is commonly determined using two RF sensors. Most researchers work on one of three emitter-sensors motion platforms. These are: (a) stationary sensors - stationary emitter, (b) moving sensors - stationary emitter, (c) stationary sensors - moving emitter. The present work aims to investigate a fourth scenario of moving RF sensors and emitter to determine the emitter location. A proposed algorithm is designed to deal with this case as well as the three formal ones. We consider the straight line and maneuvering motions of the emitter and sensors. The presented algorithm uses a hybrid situation of angle of arrival (AOA) and time of arrival (TOA) of the emitter RF signal to estimate the 3D moving emitter geolocation. We test the algorithm for long and short distances and it is found be reliable. The algorithm is also tested for different values of AOAs, and TOAs with different standard deviations. Compared with the previous works, relatively small resulting emitter position error has been detected. A MATLAB programming environment is utilized to build up the algorithm.

Keywords: 3D geolocation · Moving sensors moving emitter platforms ·
AOA estimation · TOA estimation · Hybrid AOA and TOA estimation

1 Introduction

Geolocation refers to the methods used for locating an object or set of objects in the space. In general, space means the most widely used coordinate system, latitude, longitude, and elevation. This means: x, y, and z. In this paper, we are interested in the geolocation of an RF emitting source using two RF sensors in the 2D or 3D coordination systems [1, 2]. Emitter and sensors movements mostly obey four possible platforms: (a) Stationary sensors - stationary emitter, (b) Moving sensors - stationary emitter, (c) Stationary sensors - moving emitter, and (d) Moving sensors - moving emitter. The first three platforms are studied in many previous research works [3–10]. The last platform is rarely discussed. Creating such a dynamic platform is not simple and it is considered challenging. This platform is important for military and space applications. Nowadays, having high speed computers and powerful programing environments make it possible to simulate and study this platform. Depending on the

© Springer Nature Switzerland AG 2020
A. E. Hassanien et al. (Eds.): AMLTA 2019, AISC 921, pp. 746–757, 2020.
https://doi.org/10.1007/978-3-030-14118-9_73

features of the RF signal, there are many techniques used for the geolocating the RF emitting source. Received signal strength (RSS), time difference of arrival (TDOA), frequency difference of arrival (FDOA), and angle of arrival (AOA) may be used individually or combined with each other to carry out the geolocation process. The present work uses a hybrid application of TDOA and AOA methods. A generalized algorithm, named **Mo**ving sensors **Mo**ving emitter (MoMo) is designed to solve such a problem and carry out the following tasks: (1) Simulate the emitter- sensors motion platforms, (2) Calculate AOAs and TOAs for the sensors, and (3) Apply a hybrid AOA/TOA method to locate the emitter in the 2D and 3D coordinates. This paper is divided into six sections. Section 1 is an introduction. Section 2 presents a literature review. Section 3 explains the four signal techniques of geolocating the emitter source. Section 4 presents the proposed system and algorithm. Results of the system are discussed in Sect. 5. Section 6 shows the resulting MoMo position error measurements for different platforms. Also, a comparison with earlier work is discussed in this section. Section 7 depicts the conclusions and at the end a list of utilized references is given.

2　Literature Review

Most researchers use different the platforms described in Sect. 1 to study the geolocation problem of an RF emitter. Kaune et al. [3] used a hybrid method of TDOA and FDOA with two scenarios, fixed emitter moving sensor and fixed sensors moving emitter. Fisher [4] investigated multiple moving sensors platforms to locate a stationary emitter using AOA, TDOA, and FDOA methods. Bamberger et al. [5] built a moving sensor emitter platform using unmanned small aircraft system (UAS), but they assumed a horizontal flat platform. This means that no altitude is considered leaving the system acting as a 2D system. Thoresen et al. [6], used UAS system applying the power difference of arrival PDOA assuming a ground fixed RF emitter. Bailey [7], carried out the geolocation process for a single stationary RF emitter using a joint AOA/FOA method. The used platform is a single moving sensor platform. Kim et al. [8] presented an approach for estimating the 2D position of a moving emitter using a combined TDOA and FDOA method utilizing three sensors. The process is carried out for stationary and moving sensors. Li et al. [9] carried out an emitter location estimation using TOA and TDOA method utilizing three sensors for 2D location and 4 sensors for 3D location estimation. The tested platforms are either all fixed, or fixed sensors moving emitter. Lui et al. [10] used an array of 6 moving sensors to find a moving emitting source. They did the estimation process using the combined TDOA and FDOA method.

3　Geolocation Estimation Methods

The Emitter signal received by the sensors is investigated based on the signal strength (RSS), time of arrival (TOA), frequency of arrival (FOA), and angle of arrival (AOA) at each sensor. It is known that, the received signal strength (RSS) method depends on the signal amplitude. Factors due to physical environment like, attenuation, fading, shadowing and multipath reflection of the received emitter signal make geolocation process unreliable [11, 12]. So, RSS method is not commonly used for geolocation applications. Thus, the RSS method will not be discussed here. In the following subsections, we will introduce the other three methods.

3.1 Time Difference of Arrival TDOA

Estimation of the emitter position using TDOA method is based on the time difference of arrival measurement between the two sensor locations [13]. TDOA of the signal between sensor1 and sensor2 is calculated as:

$$\Delta t_{2,1} = \frac{D_{s2} - D_{s1}}{C} \tag{1}$$

Where: D_{s1}, and D_{s2} are the distances between emitter and sensors1 and sensor2 respectively, C is the speed of light, and $\Delta t_{1,2}$ is the time difference of signal arrival between sensor1 and sensor2 respectively, assuming that the signal arrived at sensor1 first. The distance between sensors and emitter in space is calculated as:

$$D_{s1(2)} = \sqrt{\left(x_e - x_{s1(2)}\right)^2 + \left(y_e - y_{s1(2)}\right)^2 + \left(z_e - z_{s1(2)}\right)^2} \tag{2}$$

Where e is denoting the emitter, s for sensors, and 1(2) refers to sensor1 or sensor2.

3.2 Frequency Difference of Arrival FDOA

The Doppler frequency shift fd in Hz is given by [14, 15]:

$$f_d = \frac{f_0}{C} V_s \cos(\alpha) \tag{3}$$

Where Vs = sensor speed (m/s), C = speed of radio waves (m/s), f0 = frequency of radio transmission (Hz), and α = angle between the sensor speed vector and a line to the emitter (deg.). Then, the FDOA is calculated from (3) as follows:

$$FDOA = f_{ds1} - f_{ds2} \tag{4}$$

Equation (4) is rewritten as:

$$FDOA = \frac{f_0}{C}(V\_s1\cos(\alpha 1) - V\_s2\cos(\alpha 2)) \tag{5}$$

Where Vs1(2) = sensor1 or sensor2 speed (m/s), C = speed of radio waves (m/s), f0 = frequency of radio transmission (Hz), and $\alpha 1(2)$ = angle between sensor1 or sensor2 speed vector and the line from the sensor to the emitter (rad).

3.3 Angle of Arrival AOA

The angle of arrival method is estimating the RF emitter source position based on the direction that the signal came from. Estimating the AOA in space is based on measuring the angle of arrival from the emitter to the sensor of the received signal in azimuth, θ, and the angle of arrival in elevation ϕ. Thus, neglecting the noise interference, the two angles of arrival are given by [16]:

$$\theta = \arctan\left(\frac{y_e - y_s}{(x_e - x_s)}\right) \tag{6}$$

$$\emptyset = \arctan\left(\frac{z_e - z_s}{\sqrt{(y_e - y_s)^2 + (x_e - x_s)^2}}\right) \tag{7}$$

In the following section, we will discuss the motion of emitter and sensors as point objects. Stationary means zero speed, i.e., fixed object. The trajectory of a moving object is chosen arbitrarily. We select linear motion of sensors and emitter for simplicity, and most of the flying objects like aircrafts and rockets are moving approximately linearly. Emitter and sensors maneuverability motion is also tested.

4 The Proposed Geolocation System

The proposed algorithm aims to determine the location of the emitter source in the space, i.e. determining the emitter position coordinates x_e, y_e, and z_e. The algorithm is designed using equations of calculating AOA, and TOA. To estimate the emitter coordinates, we have to solve three equations for the three variables x_e, y_e, and z_e. Equation (2) is rearranged as follows:

$$\left(TOA_{1(2)}.c\right)^2 = \left(x_e - x_{s1(2)}\right)^2 + \left(y_e - y_{s1(2)}\right)^2 + \left(z_e - z_{s1(2)}\right)^2 \tag{8}$$

Equation (6) can be rewritten as:

$$\tan\theta_{1(2)} = \frac{y_e - y_{s1(2)}}{x_e - x_{s1(2)}} \tag{9}$$

Equation (7) is rearranged to be:

$$(\tan\emptyset_{1(2)})^2 = \frac{\left(z_e - z_{s1(2)}\right)^2}{\left(x_e - x_{s1(2)}\right)^2 + \left(y_e - y_{s1(2)}\right)^2} \tag{10}$$

Sensors coordinates $x_{s1(2)}$, $y_{s1(2)}$, and $z_{s1(2)}$ are assumed to be known. The values of TOA and AOAs (θ, $\emptyset$) are continuously measured during the process. Thus, solving the three Eqs. (8, 9, and 10) for the three unknown emitter coordinates $x_e, y_e,$ and z_e we get:

$$x_e = \frac{TOA_{1(2)}.C}{\sqrt{\left(1 + (\tan \theta_{1(2)})^2\right)\left(1 + (\tan \emptyset_{1(2)})^2\right)}} + x_{s1(2)} \qquad (11)$$

Then, y_e is calculated using Eqs. (9, and 11):

$$y_e = \left(x_e - x_{s1(2)}\right) \tan \theta_{1(2)} + y_{s1(2)} \qquad (12)$$

z_e coordinate is calculated using Eqs. (8, 11, and 12):

$$ze = (TOA1(2).C)2 - (xe - xs1(2))2 + (ye - ys1(2))2 + Z_{s1}(2) \qquad (13)$$

Emitter coordinate z_e is calculated using Eqs. (10, 11, and 12):

$$z_e = \tan \emptyset_{1(2)} \sqrt{\left(x_e - x_{s1(2)}\right)^2 + \left(y_e - y_{s1(2)}\right)^2} \qquad (14)$$

The emitter 3D coordinates x_e, y_e, and z_e are now determined. Equations (11, 12, and 13) and/or Eq. (14) are the calculation steps of the proposed algorithm.

5 Simulation Results

The algorithm simulates the four platforms previously mentioned in Sect. 1. This section discusses the results of applying the algorithm to an arbitrary example of each platform. We will assume that all coordinates are in meters, and all angles are in degrees. A stationary object is located in a fixed point in the 2D or 3D system while the moving one is represented in the 2D or 3D system by continuously changing coordinates. For example, the 3D coordinates of a stationary emitter are [100, 300, 200] which mean that xe = 100, ye = 300, and ze = 200. The linear motion of an object, as an example, is xe = 1000:100:2000, ye = 500:200:2500, ze = 100:50:600. It means that the emitter is moving linearly in the x coordinate starting from xe = 1000, up to 2000 with a step equals 100. The same applies for the motion in the y and z coordinates.

5.1 Platform (a): Stationary Sensors and Stationary Emitter

We suppose that the emitter and sensors are located at: emitter [800; 1000; 2000], sensor1 [1000; 3000; 3000], and sensor2 [3000; 2000; 1000]. The 2D x-y and x-z locations are shown in Fig. 1 and the 3D locations are shown in Fig. 2.

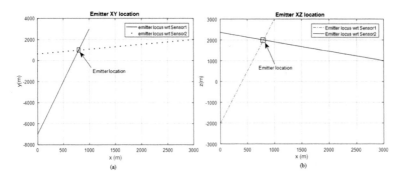

Fig. 1. Fixed sensors and fixed emitter: (a) xy emitter location, (b) emitter xz location.

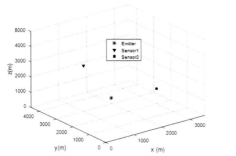

Fig. 2. Fixed sensors and fixed emitter in 3D locations

Fig. 3. The 3D moving sensors and stationary emitter platform

5.2 Platform (b): Moving Sensors and Stationary Emitter

The stationary emitter coordinates are [xe = 1000, ye = 2000, ze = 4000]. Sensor1 and sensor2 are moving such that sensor1 motion trajectory is [xs1 = 3000:200:5000, ys1 = 2000:200:4000, zs1 = 1500:100:2500], and sensor2 movement trajectory is [xs2 = 2000:200:4000, ys2 = 1000:200:3000, zs2 = 5000:100:6000]. The 3D locations are shown in Fig. 3 and the 2D x-y and x-z locations are shown in Fig. 4.

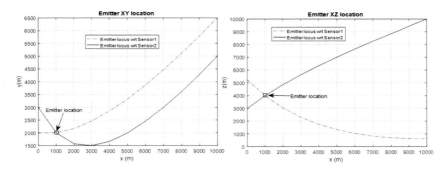

Fig. 4. (a) Emitter locus with respect to sensor1, (b) Emitter locus with respect to sensor2

5.3   Platform (c): Stationary Sensors and a Moving Emitter

Sensors are stationary and emitter is moving such that, sensor1 coordinates are [xs1 = 2000, ys1 = 2500, zs1 = 1000], sensor2 coordinates are [xs2 = 4500, ys2 = 2000, zs2 = 2000]. The emitter movement trajectory is [xe = 2100:100:3100, ye = 1000:200:3000, ze = 2000:100:3000]. The 2D x-y and x-z locations are shown in Fig. 5 and the 3D locations are shown in Fig. 6.

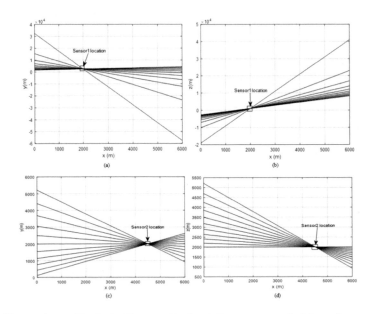

Fig. 5. The moving emitter loci with respect to the stationary sensors in (a) xy of sensor 1, (b) xz of sensor 1, (c) xy of sensor 2, (d) xz of sensor2

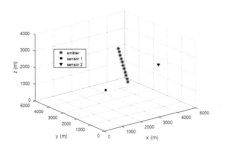

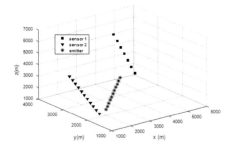

Fig. 6. The 3D locations of the moving emitter with respect to the stationary sensors

Fig. 7. The 3D locations of a linearly moving emitter and linearly moving sensors

5.4   Platform (d): Moving Sensors and a Moving Emitter

d1. Linear Motion

Sensors and emitter are all moving such that: sensor1 movement trajectory is [xs1 = 3500:200:5500, ys1 = 2000:300:5000, zs1 = 4000:200:6000], sensor2 movement trajectory is [xs2 = 1000:100:2000, ys2 = 1500:200:3500, zs2 = 2000:100:3000], and the emitter movement trajectory is [xe = 2000:200:4000, ye = 2000:100:3000, ze = 1500:100:2500]. The 3D trajectories of emitter and sensors are shown in Fig. 7.

d2. Maneuvering Motion

The case of maneuvering objects (emitter and sensors) is illustrated in Fig. 8.

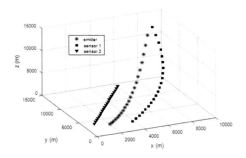

Fig. 8. The 3D locations of a maneuvering emitter and maneuvering sensors.

6   Resulting Position Error and Comparative Analysis

6.1   MoMo Resulting Emitter Position Error

The emitter position error, in meters, is a measure of the algorithm accuracy. It is calculated as shown below:

$$\text{position estimation error } (R_e) = \sqrt{(\hat{x}_e - x_e)^2 + (\hat{y}_e - y_e)^2 + (\hat{z}_e - z_e)^2} \quad (15)$$

From Eq. (15):

$$\Delta R_e = \sqrt{\Delta x_e^2 + \Delta y_e^2 + \Delta z_e^2} \quad (16)$$

Where $\hat{x}_e$, $\hat{y}_e$, and $\hat{z}_e$ are the emitter location estimated values, x_e, y_e, and z_e are the emitter position true values, and ΔR_e, Δx_e, Δy_e, and Δz_e, are the emitter position error, and errors in x, y, and z. The mean values of the emitter 3D coordinates $\hat{x}_e$, $\hat{y}_e$, and $\hat{z}_e$ are calculated by the MoMo algorithm. Different values of standard deviations of input parameters azimuth θ, elevation $\emptyset$, and time of arrival TOA are tried to evaluate the MoMo performance. For all platform examples stated in Sect. (5) standard deviations are selected to be 0.05, 0.1, 0.15, and 0.2 for all parameters. For each platform example, the emitter position error is computed by averaging 1000 simulation iterations. The emitter position error against standard deviation values for the emitter sensors platform examples a, b, c, d1, and d2 are illustrated in Fig. 9.

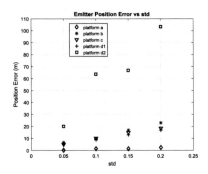

Fig. 9. Position error vs. std for emitter sensors platforms.

From Fig. 9, the algorithm achieves precise geolocation estimation for stationary platforms. The accuracy decreases when start moving but the error values are considered good. For linear motion it is shown from Fig. 9 that the output errors are approximately the same. As for nonlinear motion platforms, maneuvering sensors and emitter, the error is higher due to the complex motion platforms of sensors and emitter. Considering platform d2 example, the emitter and sensors motion ranges reached about 12000 m. For such positions, the output error is accepted. From previously introduced results analysis, it is found that the algorithm realized remarkable emitter geolocation estimation results. To enhance the emitter position error of d2 platform, we suggest considering the frequency of arrival FOA with respect to the sensors. That means, redesign the algorithm to utilize the data offered by FOA beside AOA, and TOA to reduce the emitter position resulting error and enhance the geolocation process.

6.2   Comparison with Previous Work

We compare between the results of the MoMo algorithm and the previous research ones using similar platform conditions. Huai and Lee [17], gave geolocation platforms as shown in Fig. 10. The location estimation error in distance according to different standard deviations of TDOA and AOA are summarized in Table 1. Standard deviations of AOAs θ, and $\emptyset$, are assumed to be equal and have the value of $0.2°$. Two standard deviation values of TDOA (10 ns and 30 ns) are considered during the estimation process. In our work, considering the same sensors-emitter platforms, the location estimation error in distance is calculated as in Table 1. Gudrun [18] presented 2D sensors-emitter platforms as shown in Fig. 11. The system accuracy for 10 km distance could reach 10 m. Momentary error for 10 km distance is 500 m. Applying the MoMo algorithm for the Gudrun case, the accuracy reaches 6.4 m, and the momentary error is 60 m. Tufan and Tuncer [19] introduced sensors-emitter platforms as shown in

Fig. 12. The position error is calculated for different TDOA and AOA standard deviations. The best result is obtained using Cramer-Rao Lower Bound (CRLB) method. The used TDOA standard deviation is 1 ns. The resulting positio\n error is varying from 25 up to 55 m if the AOA standard deviation is changed from 1 up to 5° and the AOA standard deviation is fixed to be 5°. The position error varies from 50 up to 83 m if the TDOA standard deviation is changed from 0.5 up to 1.5 ns. The technical report published by ITU-R [20], introduced 2D measurements of maximum position estimation error of 50 m. Applying MoMo algorithm for the same sensors and emitter platforms, the maximum position error is found to be 5 m.

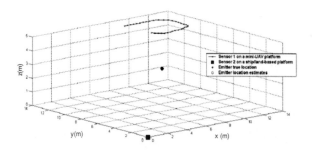

Fig. 10. Huai sensors-emitter platforms

Table 1. Location estimation error in distance

		Location estimation error in distance (m)	
		Huai, and Lee (2004) research results	MoMo algorithm results
Std of TDOA in ns	10	15	10
	30	25	18

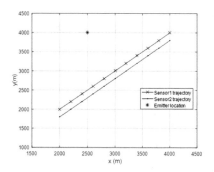

Fig. 11. Gudrun sensors and emitter platforms.

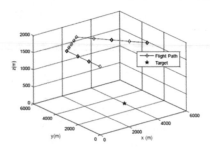

Fig. 12. Tufan and Tuncer sensors and emitter platforms.

7   Conclusions and Future Work

The MoMo algorithm is designed to determine the geolocation of an RF emitter using two RF sensors. This algorithm is found to be dynamic and reliable. It achieved good results. The geolocation of the RF emitting source is done in both the 2D and 3D coordination systems. Different platforms including stationary emitter - stationary sensors, stationary emitter – moving sensors, moving emitter – stationary sensors, and moving emitter – moving sensors are considered. The distance position error is much reduced compared with other previous techniques. The topic is still open to study different moving and maneuvering platforms.

References

1. Okello, N.: Emitter geolocation with multiple UAVs. In: 9th International Conference on Information Fusion, Florence (2006)
2. Scerri, P., Glinton, R., Owens, S., Scerri, D., Sycara, K.: Geolocation of RF Emitters by Many UAVs. Carnegie Mellon University, Pittsburch (2007)
3. Kaune, R., Musicki, D., Koch, W.: On Passive emitter tracking in sensor networks, sensor fusion and its applications. In: Ciza, T., (Ed.) (2010). ISBN: 978- 953-307-101-5
4. Fisher, G.W.: Robust Geolocation Techniques for Multiple Receiver Systems. MSc. Thesis, Department of Electrical and Computer Engineering, Graduate Faculty, Baylor University, USA (2011)
5. Bamberger, R.J., Moore, J.G., Goonasekeram, R.P., Scheidt, D.H.: Autonomous geolocation of RF emitters using small, unmanned platforms. Johns Hopkins APL Technical Digest **32** (3), 636–646 (2013)
6. Thoresen, T., Moen, J., Engebråten, S.A., Kristiansen, L.B., Nordmoen, J.H., Olafsen, H.K., Gullbekk, H., Hoelster, I.T., Bakstad, L.H.: Distribuerte COTS UAS for PDOA WiFi geolokalisering med Android smarttelefoner. Technical report, Forsvarets forskningsinstitutt, FFI-rapport 14/00958 (2014)
7. Bailey, E.J.: Single Platform Geolocation of Radio Frequency Emitters. MSc. thesis, Air Force Institute of Technology, Ohio, USA (2015)
8. Kim, Y.-H., Kim, D.-G., Kim, H.-N.: Two-step estimator for moving-emitter geolocation using time difference of arrival/frequency-difference of arrival measurements. IET Radar Sonar Navig. **9**(7), 881–87 (2015). The Institution of Engineering and Technology, UK

9. Li, X., Deng, Z.D., Rauchenstein, L.T., Carlson, T.J.: Source localization algorithms and applications using time of arrival and time difference of arrival measurements. Rev. Sci. Instrum. **87**, 041502, 1–12 (2016)
10. Liu, Z., Zhao, Y., Hu, D., Liu, C.: A moving source localization method for distributed passive sensor using TDOA and FDOA measurements. Int. J. Antennas Propag., 8625039, 12 (2016). Accessed 18 May 2018. https://doi.org/10.1155/2016/8625039
11. Boukerche, A., Oliveira, H.A., Nakamura, E.F., Loureiro, A.A.: Localization systems for wireless sensor networks. IEEE Wirel. Commun. **14**, 6–12 (2007)
12. Alfandi, O., Bochem, A., Bulert, K., Maier, A., Hogrefe, D.: Received signal strength indication for movement detection. In: Eighth International Conference on Mobile Computing and Ubiquitous Networking (ICMU), Hakodate, Japan, pp. 82–83 (2015)
13. Fowler, M.L., Hu, X.: Signal models for TDOA/FDOA estimation. IEEE Trans. Aerosp. Electron. Syst. **44**(4), 1543–1550 (2008)
14. Mušicki, D., Koch, W.: Geolocation using TDOA and FDOA measurements. In: 11th International Conference on Information Fusion, pp. 1987–1994, Germany (2008)
15. Lee, B.H., Chan, Y.T., Chan, F., Du, H., Dilkes, F.A.: Doppler frequency geolocation of uncooperative radars. In: MILCOM 2007 - IEEE Military Communications Conference, USA (2007)
16. Lee, J., Liu, J.: Passive emitter AOA determination and geolocation using a digital interferometer. In: RTO SET Symposium on Passive and LPI Radio Frequency Sensor, Poland, p. 23–25 (2001)
17. Du, H.-J., Lee, J.P.Y.: Simulation of Multi-Platform Geolocation using a Hybrid TDOA/AOA Method, Technical Report, Ministry of Defence R&D Canada, Ottawa (2004)
18. Høye, G.: Analyses of the geolocation accuracy that can be obtained from shipborne sensors by use of time difference of arrival (TDOA), scanphase, and angle of arrival (AOA) measurements. Forsvarets forsknings institutt Norwegian Defence Research Establishment (FFI), Norway (2010)
19. Tufan, B., Tuncer, T.E.: Combination of emitter localization techniques with angle, frequency and time difference of arrival. In: IEEE 21st Signal Processing and Communications Applications Conference (SIU), Turkey (2013)
20. International Telecommunication Union, Radio Sector Comparison of time difference-of-arrival and angle-of-arrival methods of signal geolocation, Report ITU-R SM.2211-2, Switzerland (2018)

Optimized Feed Forward Neural Network for Microscopic White Blood Cell Images Classification

Shahd T. Mohamed[1,3(✉)], Hala M. Ebeid[1(✉)], Aboul Ella Hassanien[2,3(✉)], and Mohamed F. Tolba[1(✉)]

[1] Faculty of Computer and Information Sciences, Ain Shams University, Cairo, Egypt
shahd.tmohamed@gmail.com, {halam,fahmytolba}@cis.asu.edu.eg
[2] Faculty of Computers and Information, Cairo University, Giza, Egypt
srge1964@gmail.com
[3] Scientific Research Group in Egypt (SRGE), Giza, Egypt
http://www.egyptscience.net

Abstract. Solving the slow convergence of the traditional neural network and searching for weights in Feed Forward Neural Network (FNN) is important to achieve the minimum training error. This paper presents an Optimized Feed Forward Neural Network (OFNN) for Microscopic white blood cell (WBC) images classification. Particle swarm optimization (PSO) and Gravitational Search Algorithm (GSA) are used to train feed forward neural network and to search for the weights of the FFN to achieve minimum error and high classification rate. The OFNN is used to classify the white blood cells into Agranulocytes that contains lymphocytes and monocytes cells and Granulocytes that contains neutrophils, eosinophils and basophils accurately. The OFNNs is trained using cells shape features of the segmented cells. The experimental results show that the obtained results are promising with classification accuracy being greater than 93% for all types.

Keywords: Blood microscopic image · Neural network · Particle swarm Optimization · Gravitational Search Algorithm

1 Introduction

Blood is composed of plasma, red cells, white cells and platelets [1]. White Blood Cells (WBC) defend the body against disease and foreign materials by produce protective antibodies that will overpower the germ or surround the bacteria, Normal white Blood cells are classified into Agranulocytes contains lymphocytes and monocytes cells and Granulocytes contains neutrophils, eosinophils and basophils cells [2].

Each type of (WBC) is an important indicator of human health as every type has standard size, shape and count using to identify the kind of diseases. Researchers tried to develop techniques to extract the unique features of every

© Springer Nature Switzerland AG 2020
A. E. Hassanien et al. (Eds.): AMLTA 2019, AISC 921, pp. 758–767, 2020.
https://doi.org/10.1007/978-3-030-14118-9_74

type and use many classifiers to classify them into their five category to analyze different form of diseases according to each type of (WBC) and help in differential count of cells.

Othman and Ali [3] proposed algorithm to recognition white blood cells contains of 3 steps: segmentation by thresholding technique, scanning algorithm to search for candidates (WBC) location and count and feature extraction using contour based shape and region based shape the roundness. These features prove its efficiency in recognizing lymphocyte cells.

Grigoriev et al. [4] presented automated WBC counting system by localizing and extracting white blood cells (WBC) automatically from color images based on images analysis and recognition. The authors used pixels intensity of image plane for (WBC) localization and segmentation and extract geometric and color features for every (WBC) type. The full processing on toke 0.5 s per image.

Akramifard et al. [5] extracted the color and size features of each type of (WBC) from images using RGB color system with accuracy 98.8% then used complex-valued back-propagation neural network (CVNNs) for classifying and recognize each type of (WBC) because of its speed and stable convergence. The method recognize the (WBC) types accurately and can be used on different resolution and color microscopic blood images.

Su et al. [6] compared between Multilayer perceptrons (MLPs), support vector machine (SVM) and Hyperrectangular Composite Neural Networks (HCNN) classifiers to classify the (WBC) types. First, they segmented the (WBC) by distinct white blood cells pixels on the HSI color space, then they extracted geometrical features, color features, and LDP-based texture features from segmented cells to train the three classifiers. The experiment showed that MLP is the most accurate classifier compared with SVM and HCNN with accuracy 99.11%.

Manik et al. [7] proposed new framework to enhance and reduce the execution time of the detection and classification of (WBC). They use intensity maxima and HSV color space for segmentation then extracted some texture features for each type of (WBC) to train the Artificial Neural Networks (ANN) to classify the Eosinophil, Lymphocyte and Neutrophil with accuracy 98.9%.

Gautam and Bhadauria [8] proposed automatic segmentation using Mathematical Operations, Otsu's thresholding technique and Mathematical Morphing. They classified by extracting shape features from (WBC) to classify them in their five category. The classification rate of this algorithm is 73%.

Hiremath et al. [9] developed system to detect and classify the lymphocytes, monocytes and neutrophi white blood cells after segmenting the images by color based method to extract the geometric features to identify and classify these three types of (WBC). This method compared with manual methods and proves its efficiency and educing time.

Garro and Vazquez [10] tried to enhance the performance of Artificial Neural Network (ANN) by design its architecture, transfer function and learning algorithm automatically using Particle Swarm Optimization (PSO), Second Generation of Particle Swarm Optimization (SGPSO), and a New Model of PSO called NMPSO. Evaluation of each solution done by eight different fitness functions based on the mean square error (MSE) and the classification error (CER). The experiment showed that NMPSO algorithm reached the best performance then the basic PSO and SGPSO algorithm.

AKSU and COBAN [11] trained the Multi-feedback Layer Neural Network (MFLNN) by the Particle Swarm Optimization (PSO) to enhance the weight selection. MFLNN structure evaluated by chaotic time series prediction and identification problems and proved its accuracy and high prediction. They also used MFLNN neural network controller design.

Bullinaria and AlYahya [12] optimized the connection weights of feed-forward neural networks by Artificial Bee Colony (ABC) for classification and they compared it with the Back-Propagation (BP). The performance of neural networks that modified by Artificial Bee Colony was better than BP but need more computational cost.

Roy et al. [13] improved the performance of Artificial Neural Network by optimize the network weights using Particle Swarm Optimization (PSO) combined with back propagation learning algorithm. They classified the IRIS data into three category using its features with classified rate of 97.3%. They also proved the efficiency of the optimized neural network in solving optimization, pattern classification, associative memories and function approximations problems.

Rezatofigh and Soltanian-Zadeh [14] applied Artificial Neural Network (ANN) and Support Vector Machine (SVM) to recognize the types of (WBC) after extracted some morphological and texture features. They showed that the SVM was the best with accuracy rate 93% and reduced the processing time.

This paper proposes Optimized Feed Forward Neural Network (OFNN) using Swarm Optimization algorithms to classify the (WBC) into its 2 categories: Granulocytes and Agranulocytes after extracting the unique features for each cell in the two groups.

The rest of the paper is organized as Sect. 2 explains the Particle swarm Optimization (PSO), Gravitational Search Algorithm (GSA), Statistical Moments and discusses the proposed algorithm. Section 3 shows the experiment results. Finally, Sect. 4 contains the conclusion and future work.

2   Preliminary

This section illustrates the main concepts of Feed Forward Neural Network (FNN), Particle swarm Optimization (PSO) and Gravitational Search Algorithm (GSA), and shape features.

2.1   Feed Forward Neural Network (FNN)

The Artificial Neural Networks (ANN) system based on biological neural network that helps in solving many classification and approximation problems [15,16]. The creation of an accurate neural network for the better performance depends on the architecture of Neural Network include number of layers and activation function and the training of the network depends on the values of the weights and learning algorithm [10]. This paper focuses on optimizing the weights to reduce the errors in FNN.

2.2   Particle Swarm Optimization (PSO)

In 1995, Kennedy and Eberhart [17,18] proposed optimization algorithm inspired from social behavior of particles constitute a swarm searching for best solution. PSO can solve difficult and impossible minimization and maximization problems, faster convergence and less parameters to tune. Equations 1 and 2 show how the particles update their velocities (v_i) and positions (p_i).

$$v_i(t+1) = v_i(t) + c1 * rand * (PBest - p_i(t)) + c2 * rand * (GBest - p_i(t)) \quad (1)$$

where *rand* is random number between 0 and 1.

$$p_i(t+1) = p_i(t) + v_i(t+1) \quad (2)$$

2.3   Gravitational Search Algorithm (GSA)

Gravitational search algorithm (GSA) is search algorithm proposed by Rashedi et al. [19], based on the Newtonian gravity: Every particle in the universe attracts every other particle with a force that is directly proportional to the product of their masses and inversely proportional to the square of the distance between them the low of gravity and mass interactions. GSA population are called agents which interact with each other through the gravity force and the performance of the agents is measured by its mass [19,20]. Equations 3 and 4 show how the agents update their velocities (v_i) and positions (x_i).

$$v_i(t+1) = rand * v_i(t) + a_i(t) \quad (3)$$

$$x_i(t+1) = rand * v_i(t) + a_i(t) \quad (4)$$

GSA solves various engineering optimization problems and gives high performance in solving various nonlinear functions, but its limitation is trapped into local optima [19,21].

2.4   Statistical Moments

Statistical moments can provide statistical feature of the objects using their level histogram in the image. This paper will use mean and standard deviation as they provide some information on the appearance of the distribution and skewness and Kurtosis provide information on the shape of the distribution and area and permitter for the measurements of the shape [22].

1- **Mean:** gives the average gray level of each region.

$$\overline{x} = \frac{1}{N} \sum_{i=1}^{N} (x_i) \quad (5)$$

2- **Standard deviation:** gives the amount of gray level fluctuations from the mean gray level value.

$$\sigma = \sqrt{\frac{1}{N-1}\sum_{i=1}^{N}(x_i - \overline{x})^2} \tag{6}$$

3- **Skewness:** measures of the asymmetry of the gray levels around the sample mean.

$$\text{SK} = \frac{1}{N}\sum_{i=1}^{N}\frac{(x_i - \overline{x})^3}{\sigma^3} \tag{7}$$

4- **Kurtosis:** describes the shape of the tail of the histogram.

$$\text{K} = \frac{1}{N}\sum_{i=1}^{N}\frac{(x_i - \overline{x})^4}{\sigma^4} \tag{8}$$

2.5 The Proposed Optimized Feed Forward Neural Networks

The proposed framework as shown in Fig. 1 uses the extracted statistical features: mean, standard deviation, skewness, kurtosis, area and perimeter for each cell in granulocytes and agranulocytes types. These features are used as input to OFNNs to classify the two categories of (WBC).

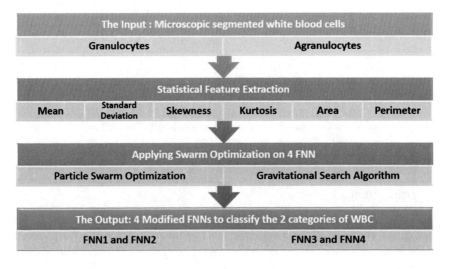

Fig. 1. The proposed optimized feed forward neural networks framework

Feed Forward Neural Network (FNN) uses the extracted features from segmented white blood cells to classify the white blood cells into granulocytes and agranulocytes types as shown in Fig. 2.

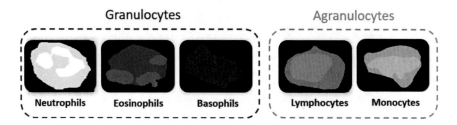

Fig. 2. The white blood cells morphology

The particle Swarm Optimization (PSO) and Gravitational Search Algorithm (GSA) used to train OFNN to minimize the classification error. Sigmoid function is used as an activation function for the OFNN [23]. Its defined using the following Equation.

$$F(\sum_{i=1}^{n} w_{ij}x_i - \theta) = \frac{1}{1 + exp(-\sum_{i=1}^{n} w_{ij}x_i - \theta)} \tag{9}$$

where w is the weight and θ is the bias value.

The extracted statistical features including mean, standard deviation, skewness, kurtosis, area and perimeter for each cell in granulocytes and agranulocytes types. These features are used as input to OFNNs to classify the two categories of WBC as follows:

- **Category (1) classify Granulocytes type:** OFNN is trained using PSO and GSA to classify Granulocytes type and they consist of 6 neurons in the input layer represent the extracted features, three neurons in the output layer and 25 neurons in the hidden layer.
- **Category (2) classify Agranulocytes type:** OFNN is trained using PSO and GSA to classify Agranulocytes type and they consists of six neurons in the input layer, two neurons in the output layer and 25 neurons in the hidden layer.

Figure 3 describe the main steps of the proposed training of the OFNN for white blood cell classification using Gravitational Search Algorithm.

3 Experimental Results

The framework is tested using white blood cells images from LISC (Leukocyte Images for Segmentation and Classification) database, the areas of the nucleus and cytoplasm were manually segmented by an expert and were classified by a hematologist into its morphology: basophil, eosinophil, neutrophil, lymphocyte and monocyte [14].

The experiment is done using 200 segmented images, 40 images for each kind of white blood cells in the training and testing approaches.

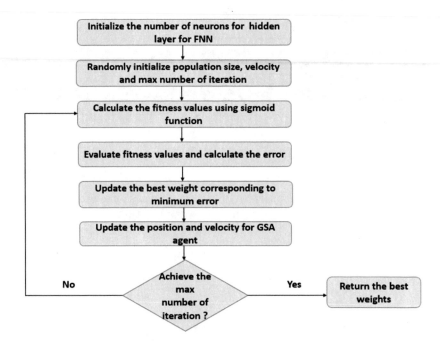

Fig. 3. Training the OFNN by gravitational search algorithm

The OFNNs evaluated by testing the effect of iteration number on the training phase so 50, 100, 150 and 200 number of iterations is tried and testing the classification rate in the aspect of changing the particle of PSO and agents of GSA for applying optimization swarms for reduce the classification error.

As shown in Table 1, the result of classification of Granulocytes type and Table 2 the result of classification of Agranulocytes.

The experiment shows that the number of particles, number of agents and the iteration numbers affect the classification results. The statistical features improved classification rate for Granulocytes type and showed promise results in classifying Agranulocytes type. The OFFNs trained by PSO is close to the OFNNs trained with GSA optimization algorithm in classification rates but PSO achieved better results with less number of iterations than GSA.

Figure 4 at the top left; shows the Converge curve for classifying Granulocytes with PSO and GSA with different numbers of particles and agents for 50 iteration, while the Fig. 4 at top right, Converge curve for classifying Granulocytes with PSO and GSA with different numbers of particles and agents for 200 iteration. Figure 4 bottom left, shows the converge curve for classifying Agranulocytes with PSO and GSA with different numbers of particles and agents for 50 iteration, while Fig. 4 the bottom right shows Converge curve for classifying Agranulocytes with PSO and GSA with different numbers of particles and agents for 200 iteration.

Table 1. The classification accuracy to classify the Granulocytes type

	Number of particles	20	50	80
OFNN Using PSO	50 iteration	81.67%	83.33%	85.33
	100 iteration	87.5%	91.67%	96%
	150 iteration	90.8%	94.2%	96.3%
	200 iteration	93%	95.83%	96.67%
	Number of agents	20	50	80
OFNN Using GSA	50 iteration	80%	82	83.5
	100 iteration	87.16%	92%	95%
	150 iteration	90%	94.16%	95.83%
	200 iteration	92.5%	95%	96%

Table 2. The classification accuracy to classify the Agranulocytes type.

	Number of particles	20	50	80
OFNN Using PSO	50 iteration	97.5%	100%	100%
	100 iteration	98.75%	100%	100%
	150 iteration	99%	100%	100%
	200 iteration	99.3%	100%	100%
	Number of agents	20	50	80
OFNN Using GSA	50 iteration	97%	98.5	98.75
	100 iteration	97.5%	100%	100%
	150 iteration	98.3%	100%	100%
	200 iteration	98.5%	100%	100%

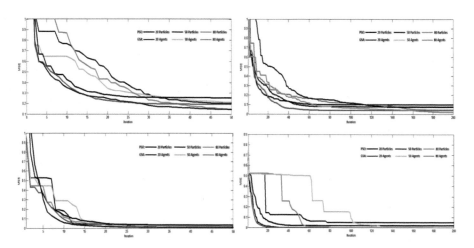

Fig. 4. Converge curve for classifying Granulocytes and Agranulocytes with different iteration

4   Conclusion and Future Work

This paper compares between PSO and GSA to train the Optimized Feed Forward Neural Network (OFNN) to classify the Agranulocytes that contains lymphocytes and monocytes cells and Granulocytes that contains neutrophils, eosinophils and basophils cells. We used the statistical features mean, standard deviation, skewnes, kurtosis, area and perimeter to provide unique information of the white blood cells. The OFNNs evaluated different iteration number and different number of particle and agents for applying PSO and GSA. The experimental results show that the number of Particles and Agents of swarm algorithms enhance the classification accuracy. OFNN obtained classification accuracy up to 100% for Agranulocytes type and 96.67% for Granulocytes type.

In future, we will use more features for classification and another swarm optimization algorithms.

References

1. Mohamed, S.T., Ebeid, H.M., Hassanien, A.E., Tolba, M.F.: Automatic white blood cell counting approach based on flower pollination optimization multilevel thresholoding algorithm. In: International Conference on Advanced Intelligent Systems and Informatics, pp. 313–323. Springer (2018)
2. Prinyakupt, J., Pluempitiwiriyawej, C.: Segmentation of white blood cells and comparison of cell morphology by linear and naïve bayes classifiers. Biomed. Eng. online 14(1), 63 (2015)
3. Othman, M., Ali, A.: Segmentation and feature extraction of lymphocytes WBC using microscopic images. Int. J. Eng. Res. Technol. 3(12), 696–701 (2014)
4. Kovalev, V.A., Grigoriev, A.Y., Ahn, H.S., Myshkin, N.K.: Automatic localization and feature extraction of white blood cells. In: Medical Imaging 1995: Image Processing, vol. 2434, pp. 754–766. International Society for Optics and Photonics (1995)
5. Akramifard, H., Firouzmand, M., Moghadam, R.A.: Extracting, recognizing, and counting white blood cells from microscopic images by using complex-valued neural networks. J. Med. Signals Sens. 2(3), 169 (2012)
6. Su, M.C., Cheng, C.Y., Wang, P.C.: A neural-network-based approach to white blood cell classification. Sci. World J. (2014)
7. Manik, S., Saini, L.M., Vadera, N.: Counting and classification of white blood cell using artificial neural network (ANN). In: IEEE International Conference on Power Electronics, Intelligent Control and Energy Systems (ICPEICES), pp. 1–5. IEEE (2016)
8. Gautam, A., Bhadauria, H.: Classification of white blood cells based on morphological features. In: 2014 International Conference on Advances in Computing, Communications and Informatics (ICACCI), pp. 2363–2368. IEEE (2014)
9. Hiremath, P., Bannigidad, P., Geeta, S.: Automated identification and classification of white blood cells (leukocytes) in digital microscopic images. IJCA special issue on "recent trends in image processing and pattern recognition" RTIPPR, pp. 59–63 (2010)
10. Garro, B.A., Vázquez, R.A.: Designing artificial neural networks using particle swarm optimization algorithms. Comput. Intell. Neurosci. 61 (2015)

11. Aksu, I.O., Coban, R.: Training the multifeedback-layer neural network using the particle swarm optimization algorithm. In: 2013 International Conference on Electronics, Computer and Computation (ICECCO), pp. 172–175. IEEE (2013)

12. Bullinaria, J.A., AlYahya, K.: Artificial bee colony training of neural networks. In: Nature Inspired Cooperative Strategies for Optimization (NICSO 2013), pp. 191–201. Springer (2014)

13. Roy, A., Dutta, D., Choudhury, K.: Training artificial neural network using particle swarm optimization algorithm. Int. J. Adv. Res. Comput. Sci. Softw. Eng. **3**(3) (2013)

14. Rezatofighi, S.H., Soltanian-Zadeh, H.: Automatic recognition of five types of white blood cells in peripheral blood. Comput. Med. Imaging Graph. **35**(4), 333–343 (2011)

15. Yamany, W., Tharwat, A., Hassanin, M.F., Gaber, T., Hassanien, A.E., Kim, T.H.: A new multi-layer perceptrons trainer based on ant lion optimization algorithm. In: 2015 Fourth International Conference on Information Science and Industrial Applications (ISI), pp. 40–45. IEEE (2015)

16. Yamany, W., Fawzy, M., Tharwat, A., Hassanien, A.E.: Moth-flame optimization for training multi-layer perceptrons. In: 2015 11th International Computer Engineering Conference (ICENCO), pp. 267–272. IEEE (2015)

17. Hassanien, A.E., Alamry, E.: Swarm Intelligence: Principles, Advances, and Applications. CRC Press, Inc. (2015)

18. Du, K.L., Swamy, M.: Particle swarm optimization. In: Search and Optimization by Metaheuristics, pp. 153–173. Springer (2016)

19. Rashedi, E., Nezamabadi-Pour, H., Saryazdi, S.: GSA: a gravitational search algorithm. Inf. Sci. **179**(13), 2232–2248 (2009)

20. Chatterjee, R.: Newtonian Law Inspired Optimization Techniques Based on Gravitational Search Algorithm. Ph.D. thesis, KIIT University Bhubaneswar (2011)

21. Rashedi, E., Rashedi, E., Nezamabadi-pour, H.: A comprehensive survey on gravitational search algorithm. Swarm Evol. Comput. (2018)

22. Tiwari, S., Singh, A.K., Shukla, V.: Statistical moments based noise classification using feed forward back propagation neural network. Int. J. Comput. Appl. **18**(2), 36–40 (2011)

23. Mirjalili, S., Hashim, S.Z.M., Sardroudi, H.M.: Training feedforward neural networks using hybrid particle swarm optimization and gravitational search algorithm. Appl. Math. Comput. **218**(22), 11125–11137 (2012)

Renewable Energy

Applying Polynomial Learning for Soil Detection Based on Gabor Wavelet and Teager Kaiser Energy Operator

Kamel H. Rahouma[1](✉) and Rabab Hamed M. Aly[2]

[1] Electrical Engineering Department, Faculty of Engineering,
Minia University, Minia, Egypt
kamel_rahouma@yahoo.com
[2] The Higher Institute for Management Technology and Information,
Minia, Egypt

Abstract. Soil detection is playing an important role in the environmental research. It helps the farmers to determine what kind of plants they can have. Also, it may help to mix plants in certain areas or farm new types. The main target of this paper is to classify the different types of soil. On the other hand, there are many researches which focus on the classification and detection process based on different applications of image processing and computer vision. The paper has two main goals. The first goal is to improve the extraction of soil features based on Gabor wavelet transform but followed by the Teager-Kaiser Operator. The second goal is to classify the types of soil based on group method data handling (polynomial neural networks). We applied these methods using different data sets of soil. Compared with previous work and research, we achieved accuracy limits of (98%–100%) while the previous algorithms were accurate to the limits of (95.1%–98.8%). Behind this improvement in accuracy, there are the methods we used here including the Teager Kaiser operator with Gabor wavelet and polynomial neural networks which have been proved to be more accurate than the methods used before.

Keywords: Soil detection · Gabor wavelet ·
Polynomial neural network (PNN) · Teager-Kaiser

1 Introduction

Nowadays, machine learning and computer vision play an important role in environment image analysis, especially in the detection of features processing (Bhattacharya and Solomatine 2006). Furthermore, the effective role of computer vision and image processing is to classify the type of different soils. The researchers try to make the classification very easier using the modern research technologies. Researchers try to improve the methods of extraction or classification. On the other hand, some researches have introduced some other methods for the features detection of soils or diseases of leaves of trees. In the following, we will show some of the recent researches and how the authors try to improve methods to classify and detect the medical images and environments images such as soils (Bhattacharya and Solomatine 2006; Lu et al. 2018; Odgers and McBratney 2018; Pham et al. 2017).

© Springer Nature Switzerland AG 2020
A. E. Hassanien et al. (Eds.): AMLTA 2019, AISC 921, pp. 771–783, 2020.
https://doi.org/10.1007/978-3-030-14118-9_75

On the other hand, in medical applications, some researchers focused on prediction methods for image analysis (Tekin and Akbas 2017). (Tekin and Akbas 2017) described how to use the Adaptive Neuro Fuzzy Inference System (ANFIS) to predict soil features and comparing to another type of soil. They tried groutability of granular soils with a piece of cement. It is one of the papers which helped us in the work of this paper. We will introduce some of another previous researches in the next section. The remainder of this paper is organized as follows: Section 2 concerns with a background and a literature review of previous work of prediction techniques. Section 3 illustrates the main methodology and it gives the main algorithm used in this paper. Section 4 discusses the results of the applied technique and compares it with the previous works. Section 5 presents a brief conclusion of our work and suggests some future work that can be accomplished.

2 A Literature Review

There are a lot of techniques that achieved accurate results in detection and classification. In the following, we will introduce the important cases of the previous researches about detection methods. Actually, the hybrid technique plays an important role in detection processing. (Sweilam et al. 2010) used a hybrid method based on some of information and support vector machine (SVM) to classify the types of tumors. The accuracy of this method is 90.3% for some specified types of cancer. On another hand, (Cheng and Han 2016) introduced a survey about object detections for optical remote and how to apply machine learning for this detection. Further, (Ford and Land 2014) applied a new model based on latent support vector machine as a model for cancer prognosis. The operations of this model are based on microarray operations and some of the gene expression and they improved the algorithm and technique of this model. The results showed that the increase of quality of curve receiver operating characteristic (ROC) when replacing least regression to SVM. Some authors invested new approaches of cancer features in soil detection and classification. (Khare et al. 2017) applied ANFIS (Neuro fuzzy) as a classification method. They used fuzzy system to detect the severity of the lung nodules depending on IF-Then rules method. They also applied 150 images in computer aided diagnosis (CAD) system. They achieved (sensitivity of 97.27%, specificity of 95% with accuracy of 96.66%). (Potter and Weigand 2018) introduced a study about the image analysis of soil crusts to get the properties of surface heating. Actually, the study approved moderate skewness toward negative tails and the other results can help to improve future mapping for any place having biocrust surfaces such as in the Mojave Desert.

Recently, some authors presented a survey about the most recent image segmentation processing especially for medical images (Dallali et al. 2011). They improved the classification techniques to increase the classification rate accuracy by 99.9%. (Dallali et al. 2011) introduced a new classification algorithm based on fuzzy clustering method and improved the performance by neural networks to classify heart rate (HR) and RR intervals of the ECG signal and they called this method fuzzy clustering method neural network (FCMNN). (Nabizadeh and Kubat 2015) applied Gabor wavelet to extract features of MRI images. They also compared results with statistical features methods. The comparing technique for the features extraction method was based on some classifiers. Authors have evaluated the methods and Gabor wavelet achieved 95.3% based on some of the classification techniques.

Some authors introduced new classification methods for environment images. (Wang et al. 2017) applied deep learning for hyper spectral remote sensing images. They applied classification methods to achieve multi features learning and the accuracy is 99.7%. Furthermore, (Perez et al. 2017) introduced deep learning classification for soil related to illegal tunnel activities. They proposed a new method in handling imbalance learning. The result showed that the method improved the performance significant of soil detection. (Boudraa and Salzenstein 2018) proposed review about Teager Kaiser Operator (TKO) for image enhancement and improved the enhanced technique by following it by an energy operator of TKO. In the next section, based on the previous researches, we will introduce the Gabor wavelet transform followed by TKO in feature extraction of soil image datasets (Bhattacharya and Solomatine 2006) and after that we will classify the result based on polynomial neural network (PNN).

3 Methodologies and Algorithms

The system of this paper consists of two main parts:-

(a) First part is for enhancement and features extraction based on Gabor wavelet and Teager Kaiser.
(b) Second part is for classification based on the polynomial neural network (PNN). We, also, will compare the system methods with the other previous work of the soil classification and detection.

3.1 Enhancement Using Gabor Wavelet Followed by Teager-Kaiser Operators

In this paper, we used soil datasets taken from a set of online recorded images from different places and also based on the database images which have used by (Bhattacharya and Solomatine 2006). In this part, the practical work consists of two main steps:-

(1) The enhancement process of the soil datasets (see Fig. 1).
    (a) Find limits to contrast stretch an image (to increase the contrast of image).
    (b) Convert image to gray level.

Fig. 1. Enhancement Process strategy

(2) Gabor wavelet features extraction followed by Teager Kaiser Operator:

Gabor wavelet is considered one of the most practical methods to extract the optimal features from images after the enhancement process. (Bhattacharya and Solomatine 2006) extracted features based on boundary energy. Furthermore, (Boudraa and Salzenstein 2018) approved that 2D Teager Kaiser Operator reflects better local activity than the amplitude of classical detection operators. The quadratic filter also is used to enhance the high frequency and combined with image gray values to estimate the edge strength value and all of that is used in the enhancement process.

The main function of this method is to generate energy pixels based on 2D Teager Kaiser Operator (TKO). The Teager Kaiser Energy operator is defined by the energy of the signal x(t) as follows (Boudraa and Salzenstein 2018; Rahouma et al. 2017):

$$\Psi_c[x(t)] = [x(\dot{t})]^2 - x(t)x(\ddot{t}) \tag{1}$$

Where x (t) is the signal, $x(\dot{t})$ is the first derivative and $x(\ddot{t})$ is the second derivative.

Actually, we applied Teager Kaiser Energy operator in the discrete as follows Eq. (2):

$$\Psi_d[x(t)] = x_n^2 - x_{n+1}x_{n-1} \tag{2}$$

On the other hand, the Gabor wavelet equations which we applied in our system are based on the Gabor transform employed in one dimension as a Gaussian window shape as shown in Eq. (3) but in the two dimension case, it provides the spectral energy density concentrated around a given position and frequency in a certain direction (Boudraa and Salzenstein 2018).

$$g_{\alpha\mathcal{E}}(x) = \sqrt{\frac{\alpha}{\pi e^{-\alpha x^2}}} e^{-i\mathcal{E}x} \tag{3}$$

Where $\mathcal{E}, x \in R$, $\alpha \in R^+$ and $\alpha = (2\sigma^2)^{-1}$, σ^2 is the variance and $\mathcal{E}$ is the frequency.

Then, the mother wavelet of the Gabor wavelet as follows:

$$g_{\alpha,\mathcal{E},a,b}(x) = |a|^{-0.5} g_{\alpha,\mathcal{E}}\left(\frac{x-b}{a}\right) \tag{4}$$

Where a $\in R^+$ (scale), and b $\in R$ (shift).

Note that, the Gabor wavelet doesn't form orthonormal bases. Actually, the Gabor wavelet can detect edge corner and blob of an image (Boudraa and Salzenstein 2018). We applied blob detection and used the main function of energy operator of Teager Kaiser in the calculation of Gabor energy. Gabor wavelet followed by Teager Kaiser achieved a set of features from soils datasets images which helped us to classify the types of soil. Actually, we extracted 98 as a general but the optimal calculated features from 98 features are Correlation, Energy and Kurtosis based on Eqs. (5–7) (Potter and Weigand 2018). We will discuss the details in Sect. 4.

$$Energy = \sum_{i=1}^{m} \sum_{j=1}^{n} (GLCM(i,j))^2 \tag{5}$$

$$Correlation = \sum_{i=1}^{m} \sum_{j=1}^{n} \frac{\{ij\}GLCM(i,j) - \{\mu_x\mu_y\}}{\sigma_x\sigma_y} \tag{6}$$

Where i, j index instant $\mu_x\mu_y$ and $\sigma_x\sigma_y$ are the mean and standard deviations of probability matrix Gray level coherence Matrix (GLCM) along row wise x and column wise y.

$$\text{Kurtosis} = \frac{1}{\sigma^4} \sum_{i=0}^{m-1} (i - \mu)^4 x(i) - 3 \tag{7}$$

In the following section, we introduce the classification method to classify the types of the soils based on the previous features.

3.2 Polynomial Neural Network Classification Method (PNN)

There are many types of ANN based on a mathematical classification equation such as PNN (Polynomial neural network) which will discussed in this section. On the other hand, ANN is used in classification in data mining and also to predict future data. GMDH is a multilayer network which used quadratic neurons offering an effective solution to modeling non-linear systems. The PNN is one of the most popular types of neural networks based on polynomial equation. It is used for classification and regression. It is more practical and accurate in prediction of behavior of the system model (Rahouma et al. 2017). A class of polynomials (linear, modified quadratic, cubic, etc.) is utilized. We can obtain the best description of the class by choosing the most significant input variables and polynomial according to the number of nodes and layers. Ivakhnenko used a polynomial (Ivakhnenko Polynomial) with the grouping method of data handling (GMDH) to obtain a more complex PNN. Layers connections were simplified and an automatic algorithms was developed to design and adjust the structure of PNN neuron (see Fig. 2).

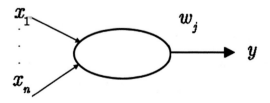

Fig. 2. The neuron inputs and output

To obtain the nonlinear characteristic relationship between the inputs and outputs of the PNN a structure of a multilayer network of second order polynomials is used. Each quadratic neuron has two inputs (x_1, x_2) and the output is calculated as described in Eq. (8) where "Fig. 3" shows the structure of PNN.

$$g = w_0 + w_1 x_1 + w_2 x_2 + w_3 x_1 x_2 + w_4 x_1^2 + w_5 x_2^2 \tag{8}$$

Where $w_i; i = 0, \ldots, 5$ are weights of the quadratic neuron to be learnt. The main equations for PNN structure which is the basic of GMDH-PNN are:

$$(X_i, y_i) = x_{1i}, x_{2i}, \ldots \ldots, x_{Ni}, y_i \tag{9}$$

Where X_i, y_i **are data variables and** $i = 1; 2; 3; \ldots; n$
The input and output relationship of PNN- structure of "Fig. 3" is:

$$Y = F(x_1, x_2, \ldots \ldots \ldots \ldots \ldots x_N) \tag{10}$$

The estimated output is:

$$\begin{aligned} \grave{y} &= \grave{f}(x_1, x_2, \ldots \ldots x_N) \\ &= c_0 + \sum_i c_i x_i + \sum_i \sum_j c_{ij} x_i x_j + \sum_i \sum_j \sum_k c_i c_j c_k x_i x_j x_k + \ldots \end{aligned} \tag{11}$$

The PNN is the best and fastest solution in classification data techniques of data.

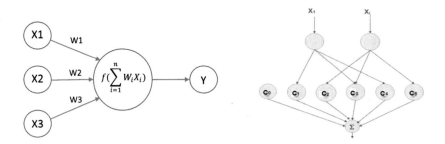

Fig. 3. The polynomial network structure

The main steps to apply the GMDH-PNN algorithm for classification based on polynomial Eq. (8) are:

(a) Determine the system input variables according to Eq. (9).
(b) Formulate the training and testing data according to Eqs. (10, 11).
(c) Select the structure of the PNN.

(d) Estimate the coefficients of the polynomial of nodes to estimate the error between y_i, $\acute{y}_i$ then:

$$E = \frac{1}{n_{tr}} \sum_i^{n_{tr}} \left(y_i - \acute{y}_i \right)^2 \tag{12}$$

Where n_{tr} is the number of training data subsets, i is the node number, k is the data number, n is the number of the selected input variables, m is the maximum order and n_0 is the number of estimated coefficients(Rahouma et al. 2017). By using the training data, the output is given by a linear equation as:

$$Y = X_i C_i \tag{13}$$

$$C_i = (X_i^T X_i)^{-1} X_i^T Y \tag{14}$$

Where
$$Y = [y_1, y_2, y_3, \ldots, y_{ntr}]^T, \quad X_i = [X_{1i}, X_{2i}, X_{3i}, \ldots, X_{ntri}]^T,$$
$$X_{ki}^T = [X_{ki1}, X_{ki2}, X_{kin}, \ldots, X_{ki1}^m, X_{ki2}^m, \ldots, X_{kin}^m]^T \text{ and}$$

$C_i = [C_{0i}, C_{1i}, \ldots, C_{n'i}]^T$, and after that, check the stopping criterion.

(e) Determine the new input variables for the next layer.

Note That: The database which have applied in this paper for classification process based on the online datasets which applied in (Bhattacharya and Solomatine 2006). Furthermore, the Fig. 4 shows the general flow chart of the system.

The general algorithm of the detection and the classification PNN for soil datasets

Start

1) Enter number of images n, Maximum Number of Neurons (Nu) =50 and the Maximum of Layer (NL) =10, Alpha (AL)=0.6,The train ratio (TR)=0.7.

2) Enter n datasets of soil images.

3) Loop i =1: n

  a) Enhance the soil image using Low pass filter

  b) Apply the gray level of the image.

4) Extract the image features (nf =98 features) based on Gabor wavelet and TKO energy operator.

5) Calculate Correlation, Energy, and Kurtosis from discussed equations (5-7).

Start a loop

  Use 80% of the datasets for training using PNN structure to obtain the system coefficient.

  Use the trained system to estimate the classification the rest of 20% of datasets.

End Loop

4) Compute the error of the accuracy of features as follows:

  Err = (actual value-Estimated value)/actual value*100%

5) Calculate the accuracy = 100 – Err.

6) Print the results.

End

Table 1. The percentages values of GW- MSVM classification (Energy Feature)

Datasets	The percentages values of Classification (GW-MSVM) result %																	
Clay	95.1	95.1	95.1	95.1	95.1	96	96	96	96	96.6	96.7	96.7	97	97	97.1	97.1	97.1	98
Clayey_Peat	95.1	95.1	95.1	96.7	96.7	96.77	96.77	96.7	96	96.6	96.77	96.77	98	98	98	98	98	98
Clayey Sand	96	96.6	96.7	96.7	96.7	96.7	98	98.1	98.1	98.1	98	98	98	98	98.7	98.7	98.7	98.7
Peat	96.7	96.7	96.7	96.7	96.7	96.7	96.7	96.7	98.3	98.3	98.3	98.3	98.8	98.8	98.8	98.8	98.8	98.8
Sandy Clay	96	96.8	96.77	96.77	96.77	96.77	98.2	98.2	98.2	98.2	98.37	98.37	98.37	98.37	98.4	98.4	98.4	98.4
Silty Sand	96.77	96.77	96.77	96.77	98	98	98.3	98.3	98.2	98.2	98.2	98.2	98.3	98.3	98.3	98.3	98.3	98.3

4   Results and Discussions

In this paper, we applied our system based on image processing toolbox of MATLAB2017a. Actually, the paper operated two cases, the first case uses the Gabor Wavelet for features extraction and multi support vector machine (MSVM) for classification. The second case uses the Gabor Wavelet followed by Teager Kaiser Energy Operator for features extraction and the PNN for classification. Actually, the accuracy of

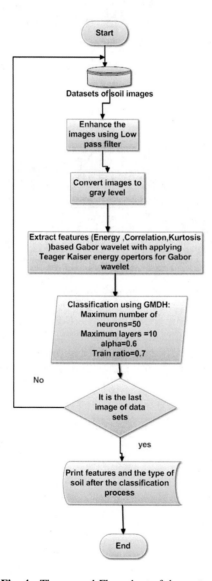

Fig. 4. The general Flow chart of the system

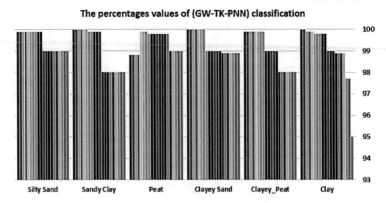

Fig. 5. The result of GW-Multi-SVM of the previous research(Energy)

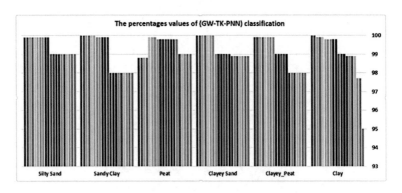

Fig. 6. The result of applying-Gabor wavelet–TK-PNN(Energy)

second case consists of the three optimal features (Correlation, Energy, and Kurtosis) is from 95 to 98%. The three optimal features came from 98 features which are extracted from soil images based GW-TK operators as discussed in pervious section. In Figs. 5 and 6 show the accuracy of the energy feature for all datasets and also Tables 1 and 2.

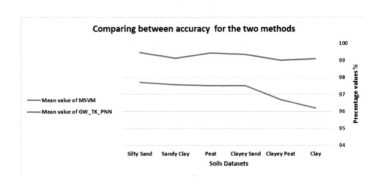

Fig. 7. The comparing between our method with pervious work

Table 2. The percentages values of GW-TK-PNN classification (Energy Feature)

Datasets	The percentages values of Classification (GW-TK-PNN) result %																			
Clay	95	97.7	97.7	98.9	98.9	98.9	99	99	99	99.8	99.8	99.8	99.8	99.8	99.9	99.9	99.9	99.9	100	100
Clayey_Peat	98	98	98	98	98	99	99	99	99	99	99.9	99.9	99.9	99.9	99.9	99.9	99.9	99.9	99.9	99.9
Clayey Sand	98.9	98.9	98.9	98.9	99	99	99	99	99	100	100	100	100	100	100	100	100	100	100	100
Peat	99	99	99	99.8	99.8	99.8	99.8	99.8	99.8	99.8	99.9	99.9	99.8	99.9	99.9	99.9	98.8	98.8	98.8	98.8
Sandy Clay	98	98	98	98	98	99.9	99.9	99.9	99.9	99.9	99.9	100	100	100	100	100	100	100	100	100
Silty Sand	99	99	99	99	99	99	99.9	99.9	99.9	99.9	99.9	99.9	99.9	99.9	99.9	99.9	99.9	99.9	99.9	99.9

The datasets of this work consist of 6 types of soils and each type has 20 different images. Actually, the CPU time approved the operations of GW-Tk-PNN in 13 s for each image in datasets so, applying Gabor wavelet followed by Teager- Kaiser Energy operator improved the accuracy of the extraction which is found to be 98.8% or higher. We compared between our results and the results from previous work (Bhattacharya and Solomatine 2006) and show that in Fig. 7. Actually, the most of pervious work based on SVM or MSVM so, we tried to improve the accuracy as discussed before. The applied techniques of this paper can be utilized in the detection and diagnosis of plants' problems, defects, and diseases. A future work may be done to study images of the plants leaves, roots, and stalks and then extract their features and classify their problems, defects and diseases.

5   Conclusions

This paper aimed to extract features based on Gabor wavelet followed by Teager Kaiser Energy operator and to apply the polynomial learning technique to classify the different types of the soil datasets images. We obtained our results and compared them with the results of previous research. The previous algorithms achieved accuracy limits of (95.1% - 98.8%) while our results achieved accuracy limits of (98% - 100%). However, the applied techniques can be utilized in the analysis, feature extraction, and classification of the plants' problems, defects, and diseases.

References

Bhattacharya, B., Solomatine, D.P.: Machine learning in soil classification. Neural Networks **19**, 186–195 (2006)

Boudraa, A.-O., Salzenstein, F.: Teager-Kaiser energy methods for signal and image analysis: a review. Digit. Signal Process. **78**, 338–375 (2018)

Cheng, G., Han, J.: A survey on object detection in optical remote sensing images. ISPRS J. Photogramm. Remote Sens. **117**, 11–28 (2016)

Dallali, A., Kachouri, A., Samet, M.: Fuzzy C-means clustering, Neural Network, WT, and HRV for classification of cardiac arrhythmia. ARPN J. Eng. Appl. Sci. **6**, 2011 (2011)

Ford, W., Land, W.: A latent space support vector machine (LSSVM) model for cancer prognosis. Procedia Comput. Sci. **36**, 470–475 (2014)

Khare, R.K., Sinha, G., Kumar, S.: Mass segmentation techniques for lung cancer CT images. Int. J. Recent Innovation Trends Comput. Commun. **5**, 184–187 (2017)

Lu, Y., Perez, D., Dao, M., Kwan, C., Li, J.: Deep learning with synthetic hyperspectral images for improved soil detection in multispectral imagery. In: Proceedings of the IEEE Ubiquitous Computing, Electronics & Mobile Communication Conference, New York, NY, USA, pp. 8–10 (2018)

Nabizadeh, N., Kubat, M.: Brain tumors detection and segmentation in MR images: gabor wavelet vs. statistical features. Comput. Electr. Eng. **45**, 286–301 (2015)

Odgers, N.P., McBratney, A.B.: Soil material classes. Pedometrics, pp. 223–264 (2018)

Perez, D., Banerjee, D., Kwan, C., Dao, M., Shen, Y., Koperski, K., Marchisio, G., Li, J.: Deep learning for effective detection of excavated soil related to illegal tunnel activities. In: 2017 IEEE 8th Annual Ubiquitous Computing, Electronics and Mobile Communication Conference (UEMCON) 2017, pp. 626–632. IEEE

Pham, B.T., Bui, D.T., Prakash, I., Dholakia, M.: Hybrid integration of Multilayer Perceptron Neural Networks and machine learning ensembles for landslide susceptibility assessment at Himalayan area (India) using GIS. Catena **149**, 52–63 (2017)

Potter, C., Weigand, J.: Imaging analysis of biological soil crusts to understand surface heating properties in the Mojave Desert of California. CATENA **170**, 1–9 (2018)

Rahouma, K.H., Muhammad, R.H., Hamed, H.F., Eldahab, M.A.A.: Analysis of electrocardiogram for heart performance diagnosis based on wavelet transform and prediction of future complications. Egypt. Comput. Sci. J. **41** (2017). ISSN-1110-2586

Sweilam, N.H., Tharwat, A., Moniem, N.A.: Support vector machine for diagnosis cancer disease: a comparative study. Egypt. Inf. J. **11**, 81–92 (2010)

Tekin, E., Akbas, S.O.: Predicting groutability of granular soils using adaptive neuro-fuzzy inference system. Neural Comput. Appl. (2017)

Wang, L., Zhang, J., Liu, P., Choo, K.-K.R., Huang, F.: Spectral–spatial multi-feature-based deep learning for hyperspectral remote sensing image classification. Soft. Comput. **21**, 213–221 (2017)

Effect of Photo Voltaic Panel on Power Generation by Manual Adjustment with Panel Angle

Xingyu Liu$^{(\boxtimes)}$

State Grid Liaoning Electric Power Supply Co. Ltd., Shenyang, China
xingyu.liuly@163.com

Abstract. The method that dividing a year into multiple time periods and calculating the optimum tilt angle of every period has been proposed. Calculation software is developed using Labview language. Taking some region in Liaoning province as an example, the yearly radiation under fixed angle, two angles each year and three angles each year are compared. The capital investment and increased revenue in the case of fixed angle, adjusting twice a year and adjusting three times a year are compared. The calculation results show that adjusting the angle twice a year and adjusting three times a year can increase the overall yield 4.03% and 4.39% respectively. The manual adjustment shelf is designed which the experiment platform is built up. The experiment results show that the generated energy of the eighth unit after angle adjustment increases significantly comparing with that of the fifth unit under fixed angle.

Keywords: Solar energy · PV generation · Angle adjustment

1 Introduction

In the designing of the PV power system, the inclination of PV array places a great influence on the received solar radiation. Only under the condition of the vertical irradiation of sunlight, the output electricity of PV array can achieve the maximum value [1]. For grid-connected PV array under fixed angle, the design target is to get the maximum amount of solar radiation in a whole year [2]. The fixed shelf is adopted in most of PV station and the inclination cannot be changed. Sunlight tracking can improve the power generation efficiency of the PV panel and the cost of station can be reduced [3–5].

The methods of four type with photo voltaic automatic tracking system have been simulated and compared [6], which is to be automatic tracking system of the simulation, and the results show that the uniaxial pour latitude angle tracking system the most simple and easy to control law project implementation, and through the study compared the different inclination toward and under the power **pv** modules [7]. There is a reported [8] to ZHUHAI region fixed with single and double axes tracking installation flat solar cell module output performance comparison research, single axis and double axes tracking increase output power respectively 18.2% and 23.9%. But high precision tracking system need to be higher cost, and high failure, maintenance, the appropriate

© Springer Nature Switzerland AG 2020
A. E. Hassanien et al. (Eds.): AMLTA 2019, AISC 921, pp. 784–791, 2020.
https://doi.org/10.1007/978-3-030-14118-9_76

precision, low cost, high reliability of the sun tracking system is the key to the development of future [9].

In this study, we will divide a year into multiple time periods and the tilt angle have been optimum for each time period, which were calculated and tracked separately control strategy, respectively. We can manually adjust the PV array inclination to track the sun elevation angle under the condition of designing manual adjustment of the bracket and setting up the experimental platform. Through calculation of the software and experiment have been prove that the manually adjust of inclination can significantly increase the generating capacity of photovoltaic power plants, it is of high economic and practicality.

2 Results and Discussion

2.1 Mathematical Models and Software Designing for the Calculation of the Optimum Tilt Angle

In this study, the calculation of the amount of radiation received by the inclined plane with the instructions of the European Solar Energy Research Center Solar Electricity to calculate the theoretical radiation on inclined surfaces. The basic idea is to calculate the tilt surface solar irradiance value through the horizontal plane of the solar irradiance data. For the net photovoltaic power generation system, in order to ensure the generating capacity of the weakest season, the tilt angle of the solar modules should be large, and then the annual generation capacity will be reduce. And to be an intensive solar power system, the annual maximum generating capacity has been considered. Based on the amount of radiation when the northern hemisphere towards the south, the photovoltaic arrays are discussed in this article which are south toward the computational and the model are as following.

$$\delta = \pi \frac{23.45}{180} \sin\left(2\pi \frac{284+n}{365}\right) \tag{1}$$

δ is the solar declination angle, and n is the number of days from 1th January;

$$\omega_s = \cos^{-1}(-\tan\varphi\tan\delta) \tag{2}$$

φ is the calculating of the latitude, and ωs is the sunrise time angle;

$$B_0 = \frac{24}{\pi} S(1 + 0.033\cos(2\pi d/365)) \\ (\cos\varphi\cos\delta\sin\omega_s + \omega_s\sin\varphi\sin\delta) \tag{3}$$

B0 is the atmosphere outside the irradiance, and S is the solar constant;

$$K_T = G/B_0 \tag{4}$$

KT is sky transmission coefficient, and G is the irradiance of the horizontal surface per unit area;

$$D = G(1 - 1.13K_T) \tag{5}$$

$$B = G - D \tag{6}$$

D is the horizontal plane per unit area of the scattering amount, and B is the horizontal plane per unit area direct of the amount;

$$D(\beta) = \frac{1}{2}(1 + \cos\beta)D \tag{7}$$

D (β) is the tilt of the surface per unit area of scattering radiation exposure, and β is the tilt angle;

$$R(\beta) = \frac{1}{2}(1 - \cos\beta)\rho G \tag{8}$$

R (β) is the tilted surface per unit area reflected radiation exposure, and ρ is the ground reflection coefficient;

$$B(\beta) = B\frac{\cos(\varphi - \beta)\cos\delta\sin\omega_0 + \omega_0\sin(\varphi - \beta)\sin\delta}{\cos\varphi\cos\delta\sin\omega_s + \omega_s\sin\varphi\sin\delta} \tag{9}$$

$$\omega_0 = \min\{\omega_s, \omega_s'\} \tag{10}$$

$$\omega_s' = \cos^{-1}\{-\tan(\varphi - \beta)\tan\delta\} \tag{11}$$

B (β) is the amount of day directly south to the different tilt angle per unit area;

$$G(\beta) = B(\beta) + D(\beta) + R(\beta) \tag{12}$$

G (β) is a total irradiance of the tilted surface;

Radiation under the same angle to accumulate the amount of radiation for every day from 0 to 90° obliquity statistics was to get the optimum tilt angle of the fixed inclination, the inclination corresponding to the horizontal comparison of the maximum amount of radiation is annual optimum tilt angle. The amount of radiation statistical calculations for fixed angle of next year was shown in formula 13, after the maximum value of comparisons of the Gn corresponding angle is from the region of the optimum tilt angle.

$$G_n(m) = \max\left[\sum_{n=a}^{b} G_n(0), \cdots, \sum_{n=a}^{b} G_n(90)\right] \tag{13}$$

Here a, b is the start and end dates, respectively. And m is the optimum tilt angle of the region from the time of $a \sim b$, G (m) in the region from the time of $a \sim b$ in the best angle of the received radiation;

Under the condition of two inclinations each year, the radiation calculation formula (14) was shown, we divide into two time periods for a year, and we were calculated $d \sim e$ and $e \sim f$ period of time, the best angle of m1 and m2, and the maximum amount of radiation Gn1 and Gn2 of the two time periods, were calculated Gn for time-phased manner, the maximum amount of radiation, respectively. Each time segmentation of Gn horizontal comparison, the largest Gn corresponding section and the way two of the optimum tilt angle are the optimum results. The annual three angle was calculated in accordance with the method.

$$G_n = \sum_{n1=d}^{e} G_{n1}(m_1) + \sum_{n2=e}^{f} G_{n2}(m_2) \tag{14}$$

m_1, m_2 is the $0 \sim 90°$, d, e, and f is the time;

In accordance with (1)–(14) calculation of thinking of using the lab view language computing software.

2.2 Calculation of Adjust the Inclination to Increase the Generating Capacity and Economic Analysis

We take the some area in Liaoning as an example for calculation, and the Table 1 shows the data of average horizontal solar radiation exposure from 1971 to 2005 of the provincial meteorological stations. We make the data, the local latitude and other natural conditions enter into software designed in this study, the calculated results shown in Table 2 and the fixed inclination annual peak sunshine time is 1700.4 kwh/m2. And comparing to the fixed angle, the annual two inclination peak sunshine time is 1787.1 kwh/m2 each year, the generating capacity increased 5.1%. Annual three dip peak sunshine time 1795.6 kwh/m2 each year, the generating capacity increased 5.6%. Annually fixed angle inclination and inclination tilted surface of each month received a total amount of radiation are shown in Fig. 1, as can be seen that the fixed inclination in addition to the fixed inclination of the month higher amount of radiation in March and September, other months for the radiation volume is less than polygonal inclination, two inclination of the year for the monthly amount of radiation is very close to the annual three inclination.

Table 1. The average solar radiation data on horizontal surface in local region

Month	January	February	March	April	May
Radiation exposure data $(kw \cdot h/m^2 \cdot year)$	2.15	3.03	4.15	5.00	5.62

Table 2. The calculation results of radiation on tilt surface under fixed angle and multiple angle

	Month	Inclination (degrees)
Fixed angle	From January to December	38
Annual twice year	From April to August	11
	From September to next year of March	55
Annual three year	From April to August	11
	September	35
	From October to next year of March	58

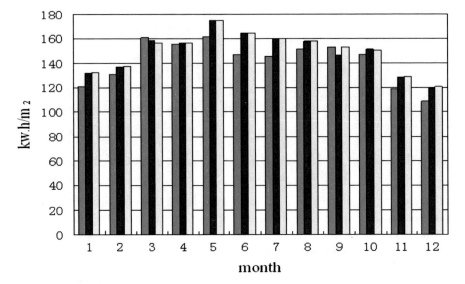

Fig. 1. The monthly radiation on tilt surface under fixed angle and multiple angle

We take the 10 MW power plant as an example to analysis the title adjustments, which can improve the economic benefits. There are 43200 solar panels for each solar panel 235 Wp, and 10 panels need for a stent, it is a total of 4320 stent. We make the two human each group to adjust the bracket, and each bracket adjustment is 3 min, and a group of people can adjust the 160 stents for eight hours, for example, 10 group (20 persons) can complete all bracket angle adjustment in three days. In accordance with the 200 yuan each day for each human costs, which adjusting the inclination can be required a total expenditure of manpower costs 12,000 yuan. Two inclination of each year the expenses are 24,000 yuan and the annual three inclination expenses are 36,000 yuan.

The calculation of the annual peak sunshine time was shown in Table 2, the condition of fixed inclination for generating capacity is calculated to 12 million degrees each year after considering various loss reduction. However, the condition of annual two inclinations for generating capacity was 12.612 million degrees each year, and the condition of annual three inclination for generating capacity was 12.672 million degrees each year. The benefit analysis was shown in the table, the tariff is calculated in accordance with 1 yuan/kWh. The increasing cost in stent designing and material were calculated in accordance with 0.2 yuan/watt, the 10 MW power station was totaling 2 million yuan to increase spending, which was in accordance with the 25-year lifespan of the power station, which was converted to an annual increase of 80,000 yuan. The economic benefits calculation results were shown in Table 3, from the each of the two inclination and three inclinations, the income increased by 4.03% and 4.39%, respectively.

Table 3. The comparison of yearly economy benefit under fixed angle and multiple angles

	Electricity generation (ten thousand kW·h/year)	Inclination to regulate human costs (ten thousand/year)	Increase the proportion of total revenue
Fixed angle	1200	0	0
Annual twice year	1261.2	2.4	4.03%
Annual three year	1267.2	3.6	4.39%

2.3   Manual Adjustment of the Stent Design and Experimental Study

The manual adjustment bracket which was designed is shown in Fig. 2, and the A region is worm regulatory agencies; B region is the sliding rod support. This stent can be installed to double total of 10 panels, the inclination angle ranges of 0° to 75° angle adjustment worm drive, which is of high efficiency and adjust accurate effort. The regulatory agencies are non-stationary; the adjustment is complete after adjusting the worm removed. The stent support columns for the sliding rod type can be fixed individually.

The experimental platform is located in a building roof; it is optional building of the east side of the rooftop on the 5th and the 8th unit comparative experiment. These two units are located in the east of the rooftops of central location; they are basically the same light condition. The panels installed capacity of the two units are 4140 Wp, which use the same model of the solar panels and photovoltaic grid-connected invert power generation, and the length of the cable into the DC power distribution cabinet is basically the same. The 5th unit in accordance with Table 2 to adjust the annual optimum tilt angle of 38° in the experiment. And the 8th Unit was in accordance with Table 2 to adjust to 11° in April and May.

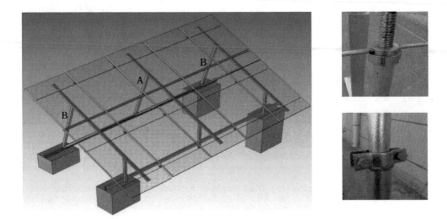

Fig. 2. The structure of manual adjustment bracket

3   Conclusion

At present, the inclination of the fixed angle system and the generation of automatic tracking system are main component for photovoltaic power plant. And the fixed angle system cannot solve problem, which is the drawbacks of the best point of view in summer and winter, so it can be lower power generation efficiency under this condition. However, automatic tracking system can solve the problem of tracking control for the best angle, but there are many shortcomings, for example large investment, large area, high failure rate and so on, and the actual economic benefits are not obvious. In this study, we will divide into multiple time periods for each year, and we have been calculated the optimum tilt angle and optimization the maximum amount of radiation for each time. We prepare calculation software in this method. And we take the some area in Liaoning for an example to calculate the results, the experiment results show that compared to the condition of fixed angle system, the generating capacity were increased by 5.1%, 5.6% and income increased by 4.26%, 4.51% for the condition of annual two inclination and three inclination, respectively. The experimental results show that the title adjustments on the 8th unit with the generating capacity increased significantly comparing to the fixed inclination of the 5th unit under the condition of building artificial adjust the inclination of photovoltaic power station experimental platform. And the added value is 2.27% in April and 6.21% in May, respectively.

References

1. Zhao, C., Liu, J., Sun, X., et al.: Hybrid controller topology for large solar PV installations in high-voltage DC grid-connected applications. Electr. Eng. **100**(4), 2537–2552 (2018)
2. Yang, J., Mao, G., Chen, Z.: Calculation of solar radiation and optimum tilt angle on different azimuth slopes. J. Shanghai Jiao Tong Univ. **36**(7), 1032–1036 (2019)

3. Abdallah, S.: The effect of using sun tracking systems on the voltage-current characteristics and power generation of flat plate photovoltaics. Energy Convers. Manag. **45**, 1671–1679 (2016)
4. Karimov, S., Saqib, M., Akhter, P.: A simple photovoltaic tracking system. Sol. Energy Mater. Sol. Cells **87**, 49–59 (2014)
5. Alata, M., Al Nimr, M.A., Qaroush, Y.: Developing a multipurpose sun tracking system using fuzzy control. Energy Convers. Manage. **46**, 1229–1245 (2019)
6. Chen, W., Shen, H., Shu, F.: Tracking effect analysis of photovoltaic system. J. Univ. Sci. Technol. China **36**(4), 354–359 (2015)
7. Chen, W., Shen, H., Liu, Y.: Experimental research on the influence of photovoltaic array tilt angle on performance. J. Sol. Energy **30**(11), 1519–1522 (2016)
8. Zhang, Q.: To improve the utilization of solar energy and sea. Rate Agric. Mech. Method Res. **10**, 241–244 (2013)
9. Dou, W., Xu, H., Li, J.: Study of tracking photovoltaic system. J. Sol. Energy **28**(2), 169–173 (2017)

The Impact of Large Scale Distributed Generation Grid-Connection on Structure of Electric Power Network

Aihua Wang[⊠], Deming Qi, and Hong Gang

Jinzhou Power Supply Branch,
State Grid Liaoning Electric Power Supply Co. Ltd., Jinzhou, China
jzwangaihua@163.com

Abstract. At present, the research of distributed power grid has made break-throughs, and the proportion of power production is also increasing. The wide application of distributed grid will have a great impact on the traditional power system, including power quality, system reliability, and relay protection and so on. Therefore, by studying the influence of distributed grid on power quality of distribution network, we can better guide how to give full play to the advantages of the distributed grid, such as distribution network loss, grid dispatch, system protection, reliability, power market and so on. In view of the disadvantageous influence of the parallel operation of distributed generators on the local power system, appropriate measures can be taken to better develop distributed generators, so as to adjust the energy structure of our country and realize sustainable development. Therefore, it is of great significance to study the influence of distributed power network on the power quality of the distribution network.

Keywords: Distributed generators · Energy structure · Measures

1 Introduction

In recent years, energy-saving and emission reduction, green energy, low-carbon environmental protection, and sustainable development become the focus of attention [1, 2]. Distributed power generation system with the large power grid increasingly joint operation due to the characteristics that such as investment clean and green, power generation flexible and compatible with the environment [3–5]. On the one hand, it can meet the specific requirements of the power system with the user, such as clipping. On the other hand, it can enhance the reliability and economy of the traditional power system. The access of the distributed power generation system can play the role of automatic voltage regulation, voltage stability and the thermal start of electrical equipment. Driven by the electricity market economy, the distributed power generation system can not only serve as a powerful supplement and effective support for the traditional power supply mode but also occupies a very important position in the comprehensive utilization of energy, which is an important development direction in the future energy field [6, 7].

© Springer Nature Switzerland AG 2020
A. E. Hassanien et al. (Eds.): AMLTA 2019, AISC 921, pp. 792–800, 2020.
https://doi.org/10.1007/978-3-030-14118-9_77

2 Distributed Power Generation Technology

At present, the actual application in the power grid distributed power generation system are small hydropower, wind power, solar photovoltaic power generation, micro-gas turbine, fuel cell power generation [1–3].

2.1 Small Hydroelectric Power Generation

Hydroelectric power is a technology that it could change the water potential energy into electrical energy. The capacity of the small hydropower is generally as 10–100 MW. Water utilization is not high. But the low cost of small hydropower and small hydropower resources almost all over the world, and it is the ideal power supply using distributed power generation. Our country is rich in water resources, water resources can be developed is in the first place in the world, and distribution of water resources is very distinctive. At present, China's hydropower technology development is relatively mature, and the hydropower engineering design, installation, and manufacturing technology have been gradually catching up and reach a world-class level. The development of small hydropower in several decades, we solved the problems that vast rural areas in China scattered electricity because using the nearest water resources, the nearest power supply.

2.2 Wind Power

Wind power [8] is the fastest growing energy power generation technology in the word but also is the most mature in addition to hydroelectric power technology. It is the most development scale and prospects of renewable energy power generation technology. Wind power technology is the power generation technology that converts wind energy into electricity. China's wind power started late, In recent years, in the national policy of encouragement and support, China's wind power progress faster. The wind turbine power generation model is shown in the Fig. 1.

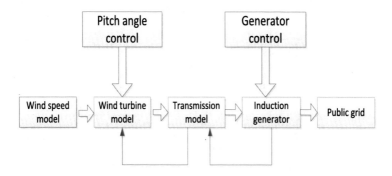

Fig. 1. Wind turbine power generation model

Although the wind power operation with random strong, intermittent obvious, poor control and other shortcomings, but because of wind power to reduce environmental pollution and reduce fuel costs of power systems, and the current wind power project cost is lower than nuclear power, wind power generation will bring considerable economic benefits, to adapt to the commercial scale of production (Refer to Fig. 1).

$$W = 1/2PV3 \tag{1}$$

This formula is the wind energy density formula, the air flow is called wind, it is a kind of energy called wind energy. The energy of the wind should be related to the speed V of the wind and the local air density P. The physical meaning of the above formula is that in air with density p, the wind energy per unit cross-section in unit time is W.

2.3   Solar Photovoltaic Power Generation

Solar energy is the ideal sustainable green energy, with the global depletion of primary energy and the economic and social development of energy demand, solar photovoltaic grid-connected commercialization is inevitable. Solar photovoltaic power generation changes the photoelectric effect directly into solar energy into electricity using semiconductor materials. China's solar energy resources are very rich, with good conditions for the use of solar energy compared to other countries. At present, and China's photovoltaic industry has developed rapidly, the annual output of photovoltaic cells in the world. But the cost of equipment and cost of power generation are high due to technical conditions and photovoltaic power generation market is seriously lagging behind the reasons. Photovoltaic power generation has the advantage of without fuel consumption, and the scale of flexible, non-polluting, safe and reliable, simple maintenance. With the development of technology, policy support and learn from foreign advanced experience, China's photovoltaic power generation prospects are still widely optimistic as shown in Fig. 2.

$$PRT = ET/PehT \tag{2}$$

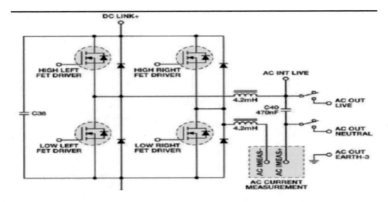

Fig. 2. Solar photovoltaic power generation diagram

The system efficiency of photovoltaic power station is the ratio of the power generation of solar power generation system to the irradiated energy of solar panels, and is the most important index in the quality evaluation of photovoltaic power station. PT represents the average system efficiency of photovoltaic power plants in the T-period, ET represents the input power of photovoltaic power plants into the grid in the T-period, Pe represents the nominal capacity of photovoltaic power plant components installed, and hT is the peak sunshine hours on the square array in the T-period.

2.4   Micro-gas Turbine Power Generation

Micro-gas turbine is a new type of heat engine. In recent years, with changes of the global energy and power demand structure, especially the power system to relax the control and environmental protection requirements, it has been highly concerned and rapid development. Micro-gas turbine power generation is the most mature and most commercially competitive distributed power generation equipment, with the advantage of its small size, lightweight, construction costs and its operating costs are very competitive. China's micro-gas turbine market demand is also great. Therefore, the micro-gas turbine in China has a wide range of applications. Figure 3 shows Micro-gas turbine power generation mechanism.

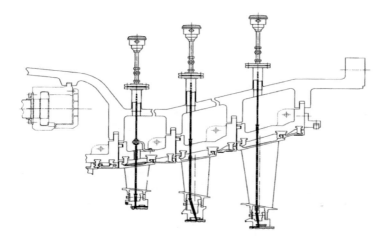

Fig. 3. Microturbine mechanism diagram

$$\int_{T1}^{T2} Cp \frac{dT}{T} = Rln\left(\frac{P2}{P1}\right) \tag{3}$$

P represents the absolute pressure of the gas, T represents the thermodynamic temperature of the gas, R represents the gas constant. This formula indicates that when the gas changes from state 1 to state 2, the upper integral obtains an entropy change of 0.

2.5   Fuel Cell Power Generation

Fuel cell power generation is considered to be one of the new efficient, energy-efficient and environmentally friendly power generations in the 21st century. Fuel cell power generation is an electrochemical reaction under certain conditions to hydrogen, natural gas or gas and oxidant, convert chemical energy directly into electrical energy. Unlike conventional batteries: As long as there are fuel and oxidant supply, the fuel cell will have a continuous power output. Fuel cell power generation has the advantages of high energy conversion efficiency and no restriction from the Carnot cycle, adaptable to the load change, clean and pollution-free, small footprint, fast construction, easy maintenance, and fuel adaptability. Fuel cells can be combined with hydropower, wind power, and solar power, etc., in the high output, the use of electrolytic water hydrogen, low output when the fuel cell power generation, to achieve both energy storage, and efficient power generation purposes.

3   Analysis of the Influence of Distributed Operation System on Grid-Connected Operation

Distributed power generation system and power supply to the regional load, to achieve a variety of complementary forms of energy. With the increasing use of distributed generation systems, the performance of distributed generation systems varies widely, and a wide range of capacity and multiple network connections, which will have multiple effects on regional grid operation and control.

3.1   Frequency

The frequency of the distributed power generation system should be consistent with the frequency of the regional grid. Under normal circumstances, the distributed power generation system capacity is small, its start and stop on the regional power grid frequency impact is small. But some measures should be taken for the active power of random and volatile nature of the large-scale distributed power generation system of the regional power grid. At the same time, the frequency conversion of the power system will affect the power generation system of the distributed generation system. This requires adjusting the speed of the generator of the distributed generation system so that its frequency is consistent with the grid. The power system is a real-time dynamic balance system, the power generation, transmission, electricity must always be balanced. The conventional power supply is adjustable and controllable and the prediction accuracy of the electricity load is high. However, in the case of a fluctuating and intermittent distributed power generation system, the active effort is difficult to accurately predict, this brings a certain degree of impact to the grid frequency modulation.

3.2 Voltage

After the distributed power generation system is connected, the steady-state voltage distribution of the regional power grid changes. The access location and capacity of the distributed generation system have a great influence on the line voltage distribution. The same capacity of the distributed power generation system access in different positions causes a very different consequence. The access point of the distributed power generation system is closer to the end node, the greater the voltage change rate of the node, the greater the influence of the voltage distribution on the line, and the total output of distributed power generation system is bigger, so the overall voltage level is higher. Therefore, the voltage distribution of the regional grid after access to the distributed power generation system is evaluated.

The distributed power generation system is not only affected by the voltage distribution of the regional power grid but even the voltage stability problem. Reactive power shortage is the root cause of voltage instability. In general, access to a distributed generation system contributes to the improvement of voltage stability. However, the interface used in the distributed power generation system needs to absorb the reactive power from the grid, the voltage stability of the system has a negative impact. Such as when the wind power station access to regional power grid, and wind power plants need to absorb a large number of reactive power from the regional power grid, due to wind power plant input of wind energy changes in the random, intermittent, and wind power often use asynchronous generators. This is likely to cause the entire grid voltage stability, and even lead to the entire power grid voltage collapse. From this point of view, it is necessary to study the reactive power and voltage scheduling strategy of the regional power grid with distributed power generation system to ensure that the entire region of the grid reactive power balance. At the same time, we must also study the influence of the distributed power generation system and the voltage stability of the regional power grid.

3.3 Power Quality

Distributed power generation system will bring a variety of disturbances, and affecting the power quality of regional power grids. Mainly embodies two aspects: voltage flicker and harmonics. Distributed power generation voltage flicker caused by the main factors are the stochasticity of the distributed power generation system and the effects of seasonal or climate. The interaction of the distributed power generation system with the voltage feedback control device in the system can cause voltage flicker. At present, mainly by reducing the number of the distributed power supply, or the use of certain equipment such as inverter to reduce the impact on the voltage of the output of the distributed power generation system.

As a large number of power electronic devices used in distributed power generation, so inevitably bring a lot of harmonics to the system, the magnitude and order of harmonics are related to the mode of operation and the mode of operation of the converter. For harmonic problems, special filters or new inverters can be installed on the bus with higher harmonic voltage levels.

Distributed power generation system can both deteriorate and improve the quality of the user's power, usually connected to the distributed power generation system to increase the capacity of the regional power grid, the power quality is beneficial. However, if the capacity of the distributed generation system is too large, such as wind power and solar photovoltaic power plants, the impact of the weather, maybe in the start and close an output waveform case and the power quality will decline.

3.4   Trend

After the distributed power generation system is connected, the structure and operation mode of the regional power grid will change fundamentally. As the output curve of the power supply is different, in particular, some power supply has great volatility and randomness, it leading to changes in the direction and size of the trend. The size and flow of the regional power grid with distributed power generation systems are related to the location of the distributed power generation system, the relative magnitude of the load, and the topology of the network.

The control characteristics of each power supply in the regional power grid with the distributed power generation system have a great influence on the convergence of the power flow calculation method. It is necessary to establish an appropriate interface model of distributed generation systems. Only the effective power flow calculation method can be used to analyze and control the regional power grid operation process of the distributed power generation system. If the regional power grid contains wind power or photovoltaic power generation system or runoff small hydropower and other power, commonly used power flow calculation methods may fail, because their output is affected by the weather, with randomness and volatility, and the regional power grid trend is also random, the trend of the flow or size of a random or volatility, so that new random flow analysis and control methods must be developed.

3.5   Power Loss

In general, the greater the power delivered on the network, the greater the network loss. After the distributed power generation system is integrated into the regional power grid, the relative size, the access position, the operation mode, the network topology and the power factor of the capacity and load will have different effects on the net loss.

3.6   Grid Dispatching

Before the distributed power generation system is connected, the real-time monitoring, control, and coordination of the regional power grid shall be carried out uniformly by the power supply department. After the distributed power generation system is connected, the process of information collection, switching operation, energy dispatching and management of the regional power grid becomes complicated. The information on monitoring and control needs to be re-determined according to the protocol of grid connection. The regional power grid voltage fluctuations and reactive power flow are unreasonable after a large number of distributed power generation systems are connected.

3.7 Relay Protection

The structure and short circuit current of the regional power grid is changed after the distributed power generation system is connected. This fundamental change makes the various settings and mechanisms of the grid change profoundly. The size of the trend on the line and frequent changes in the direction will give the power system relay protection devices and actions to bring some difficulty, especially for a regional power grid with a distributed or intermittent distributed generation system. In most instances, Distributed power generation system access area Power grid side with reverse power relay, the operation will not be injected into the power grid in normal. There will be a momentary flow of distributed power generation system into the power grid, an increase of the regional grid switch short-circuit current level, and the regional power grid switch may be short-circuited current exceeded when the area grid fails. With the increase in the proportion of distributed power generation systems in regional power grids, the regional power grid protection system must be set and coordinate, At the same time, the influence of the access capacity and location of the distributed power generation system on the protection of the power grid need to be studied to make the distributed power generation system must cooperate with and adapt.

3.8 Reliability

Distributed power generation system can partially offset the grid load, reduce the actual transmission power into the line and improve the transmission and distribution power transmission capacity in a regional grid with distributed power generation systems. At the same time, distributed power system on the regional grid voltage support can enhance the system voltage regulation performance, and it plays an important role in improving the reliability of the regional power grid. The grid-connected power generation system will also have a negative impact on the reliability of the regional power grid. Such as if the distributed installation site, capacity, the wiring is not appropriate, it will reduce the reliability of the regional power grid. Due to maintenance or fault circuit breaker tripping to form an unconscious island. There may be an imbalance in power supply and demand, reducing the power supply reliability of the grid.

4 Conclusions

At present, some distributed power generation technology has been matured, and commercial-scale production is formed. Large-scale distributed generation grid-connected, because of the frequent occurrence of distributed generation without reactive power or reactive power, the lack of reactive power in the distribution network will lead to the reduction of the voltage stability of the entire distribution network, which is easy to affect the voltage stability of the grid, and will cause system voltage collapse in serious cases. In the previous distribution network planning and design, the access of distributed generators is not generally considered. The generation grid-connection of distributed generators will change the power flow direction and power flow distribution of the distribution network. With the increase of the power of distributed generators, the voltage and power flow of the local distribution network near the power source will exceed the limit, which will affect the stability of the system. With the increase of the power of distributed generators, the stability of the system will be affected. With the increasing power of power supply, the influence of the unstable output power on the power grid is also increasing. In the most serious case, the power system will lose its dynamic stability, resulting in the collapse of the power grid. At the same time, the access capacity of the wind farm and photovoltaic power station is increasing, which puts forward new requirements for the stability and safety of the power system. From our country's national conditions, we must study the distributed power generation system and the impact of running the grid, and we also should analysis and eliminate the adverse effects of grid operation. This is the key to ensuring that distributed power generation technology is widely used and fully exploited.

References

1. Qian, A., Wang, X.H., Xing, H.: The impact of large-scale distributed generation on power grid and microgrids. Renew. Energy **62**(2), 417–423 (2014)
2. Chen, J.: Spatial analysis of almond leaf scorch disease in the san joaquin valley of california: factors affecting pathogen distribution and spread. Plant Dis. **89**(6), 34–40 (2005)
3. Brown, R.E., Freeman, L.A.: Analyzing the reliability impact of distributed generation. In: Rajan, A., Fleetwood, Z.E. (eds.) Power Engineering Society Summer Meeting 2001, vol. 2, pp. 1013–1018, Vancouver, Canada (2001)
4. Hernandez, J.C., Ogayar, B.: Electrical protection for the grid-interconnection of photovoltaic-distributed generation. Electr. Power Syst. Res. **89**(2), 85–99 (2012)
5. Barker, P.P., De Mello, R.W.: Determining the impact of distributed generation on power systems. I. Radial distribution systems. In: Thippeswamy K. (eds.) Power Engineering Society Summer Meeting 2000, vol. 3, pp. 1645–1656, Vancouver, Canada (2000)
6. Liu, H.: Cable fault diagnosis theory and key technology research. Huazhong Univ. Sci. Technol. **42**(2), 32–35 (2012). (in Chinese)
7. Zhang, C.: Partial discharge monitoring of power cables and insulation fault diagnosis. Hubei Univ. Technol. **37**(2), 44–45 (2013). (in Chinese)
8. Yuan, M.X., Yuan, S.J., Cheng, J., Wen, Y.: Analysis on the application prospects of energy storage technology in solving the problems of large-scale wind power grid integration. Chin. Soc. Electr. Eng. **37**(1), 14–18 (2013)

Wind Power Curtailment Scheme
Based on Wind Tower Method

Aoran Xu[1,2(✉)], Yang Gao[1], Xuemin Leng[1], Weiqing Wu[1,2],
and Hongyu Zhong[2]

[1] Shenyang Institute of Engineering, Shenyang, China
13918678@qq.com
[2] State Grid Tonghua Electric Power Supply Co. Ltd., Jilin, China

Abstract. In order to predict the wind speed at the hub, this paper established
the physical model of the wind speed, wind profile method and analytic method
based on potential flow theory, the wind tower wind speed extrapolated each fan
at hub height wind speed. In order to get the accurate wind power relationship,
each fan speed data obtained by extrapolation of wind power output and the
corresponding data, the measured power curve model of wind turbine generator
by using the maximum likelihood method, the model for the launch of fan power
theory should be made. According to the measured power curve of each unit
tower data and wind power generation capacity, theoretical calculation of wind
farm in the corresponding period, and then obtain the abandoned wind power,
and through the simulation experiment, which lays a theoretical foundation for
the realization of the tower was abandoned wind power assessment model.

Keywords: Wind speed · Wind power generation · Power curve model

1 Introduction

In recent years, with the increasing installed scale of wind power, the problem of peak
shaving and grid constraints, wind power participation in peak shaving and power flow
shaving have emerged. Because there are many wind farms operating in the grid, how
to arrange each wind farm to participate in peak shaving, power flow shaving and
power generation in an open, fair and just way? Reasonable network planning will be a
new problem facing dispatching specialty at present, and the statistics and calculation
method of abandoned wind power will be an important index of this work; abandoned
wind power is obtained by calculating the difference between predicted incoming wind
and actual wind power after restriction, and according to the current development of
related technology, the accuracy of wind power prediction is generally low, and the fair
boundary [1]. Determining the size of abandoned wind power in each wind farm has
important guiding significance for realizing the reuse of abandoned wind, rational
planning of power grid and optimizing the "three public" dispatch of wind power.

The tower is according to the different height of wind tower measuring wind speed
and direction, combined with the wind farm location, wind turbine layout, and
roughness changes, using wind profile method and analytic method based on potential
flow theory, the physical model of wind speed wind tower wind speed, by extrapolated

each fan at hub height wind speed. The wind turbine wind speed data and the corresponding wind power generator output data are obtained from the extrapolation method, and the measured power curve model of each wind power generator is determined by the wind turbine measured power curve model modeling method. Finally, according to the measured power curve of each unit tower data and wind power generation capacity model, theoretical calculation of wind farm in the corresponding period, and then obtain the abandoned wind power. The wind speed model is based on the physical quantity of wind tower wind data, wind power data, considering the wind farm area, terrain roughness changes, combined with the wind farm layout, the development of micrometeorology theory change model wind tower based on wind tower wind speed will be extrapolated to each fan hub height.

2   Related Work

Wind speed is affected by many factors such as temperature, air pressure, and topography. It is difficult to predict directly. Wind speed prediction is a complicated non-linear process involving many factors and has high uncertainty. Therefore, if the original sequence of wind speed can be extracted, properly processed, ill-conditioned data can be eliminated, and the rule can be found, and the prediction model can be established step by step, it will greatly simplify the modeling. Difficulty and improve prediction accuracy [2]. In order to reflect the changes of meteorological conditions in the forecasting time, the wind speed, wind direction and wind tower data of numerical weather prediction are used as input data in the power forecasting system for power grid dispatching. However, the data of numerical weather prediction can only represent the space of each computing grid corresponding to the uniform underlying surface. Mean value, the actual wind farm surface has obvious non-uniform characteristics, and the wind speed and direction of each wind turbine location may be quite different. Obviously, neither the wind tower data nor the numerical weather forecast data can be directly used as wind speed and direction of a wind turbine for power prediction. In addition, wind turbine energy from the wind will form a wake area in the downwind direction at the same time [3]. The wake area develops downstream along the wind direction. If the wind turbine is located in the wake area, its output power will be significantly reduced. According to wind tower data and wind power data of wind farm, the wind farm station will be considered comprehensively. Based on the topographic and roughness changes of the region and the wind farm layout, a wind speed change model based on the wind tower is developed by using the micro-meteorological theory [4]. The wind speed of the wind tower is extrapolated to the hub height of each fan.

3   The Physical Model of Wind Speed

According to the wind tower wind data, wind power data, considering the wind farm area, terrain roughness changes, combined with the wind farm layout, the development of micrometeorology theory change model wind tower based on wind tower wind speed will be extrapolated to each fan hub height. The physical model mainly includes the roughness model, the terrain change model and the wake model, in which the roughness variation model and the terrain change model reflect the local effect of the wind farm [5]. Study on the effect of Wind Farm Bureau in response to the category field belongs to the inhomogeneous underlying surface of the atmospheric boundary layer, in order to reduce the complexity of the model, highlighting the main object of study, to improve the practicality, using roughness change model based on the experimental observations and terrain model based on potential flow theory to simulate the local wind effect. Taking into account the complete theory of the neutral atmospheric boundary layer and another boundary layer, the boundary layer is considered as the neutral boundary layer.

In the atmospheric boundary layer, the average wind speed changes with altitude which is called wind shear or wind velocity profile. When the neutral state is neutral, the wind velocity profile can be expressed as a logarithmic or exponential distribution.

According to the logarithmic rate, the average wind speed $v(z)$ from the ground height z is expressed as:

$$v(z) = \frac{v_*}{k} \ln\left(\frac{z}{Z_0}\right) \tag{1}$$

Where v_* is for the friction speed ($v_* = \sqrt{\tau_W/\rho}$, τ_W for the shear stress); κ is for the Carmen constant, the general approximation to take 0.4; Z_0 is for the surface roughness length.

According to the index, the average wind speed $v(z)$ from the ground height is expressed as:

$$\frac{v(z)}{v(z_0)} = \left(\frac{z}{Z_0}\right)^\gamma \tag{2}$$

In Eq. (2), the mean wind velocity at $v(z_0)$ is the height of z_0, and γ is the wind velocity profile index and roughness index, which is related to the surface roughness length. The log rate distribution of wind velocity profile is proved by theory, but the calculation is more complex; the exponential distribution is empirical relation, but relatively simple. In the wind field engineering, it is used to describe the wind speed profile. The underlying surface of the actual wind farm is not uniform, and the change of roughness will form a new inner boundary layer. At this wind turbine, wind profile mosaic consists of three parts, respectively, corresponding to the roughness of the z_{01}, friction velocity of $u_1(z)$, corresponding to the roughness of the z_{02}, u_{*2}, $u_2(z)$ and friction velocity corresponding to roughness z_{03}, u_{*3}, $u_3(z)$ friction velocity.

Obviously, the uniform roughness under the logarithmic wind profile is no longer applicable, according to the experimental observation summary, leeward wind profile roughness change can be described as:

$$u(z) \begin{cases} u' \dfrac{\ln(z/z_{01})}{\ln(0.30h/z_{01})}, z > 0.30h \\[2mm] u'' + [u' - u''] \dfrac{\ln(z/0.09)}{\ln(0.30h/0.09)}, 0.09h < z \leq 0.30h \\[2mm] u'' \dfrac{\ln(z/z_{02})}{\ln(0.09h/z_{02})}, z \leq 0.09h \end{cases} \tag{3}$$

The relationship between the friction speed under the influence of roughness change is:

$$\frac{u_{*n+1}}{u_{*n}} = \frac{\ln(h/z_{0n})}{\ln(h/z_{0n+1})} \tag{4}$$

Where z_{0n}, z_{0n+1}, respectively, on the direction of roughness and distance from the direction of the nearest roughness; u_{*n}, u_{*n+1} for the corresponding z_{0n}, z_{0n+1} friction speed.

A wind turbine location within the boundary layer height and wind turbine hub height. Equation (3) based on wind turbine tower measured wind speed at hub height roughness after position change on the fan hub height wind speed. Wind farms [6] are often obvious terrain changes, affected by the turbulent boundary layer airflow and stress were to occur disturbance. Effect of topography on the boundary layer to the theoretical calculations and experiments have brought great difficulties, taking into account the engineering application of computational time and accuracy requirements, this paper analyzes the changes in terrain analysis method based on potential flow theory influence. Terrain boundary layer wind of change is regarded with respect to disturbances in the boundary layer wind undisturbed on a small number of wind profilers disturbance, and disturbance of the terrain is divided into inner and outer layers, boundary layer turbulence disturbance stress significantly accelerated air flow is very obvious, and the outer layer is small perturbations stress can be ignored.

The outer boundary layer wind perturbation solving: the outer top of the boundary layer, the layer is fewer stress perturbations can be ignored, if small-scale problems without considering the Coriolis force, the outer flow field primarily inertial force and the pressure gradient force balance between the flow can be regarded as a non-viscous flow, irrotational motion of non-viscous flow of potential motion. Thus, the outer boundary layer flow field changes in terrain disturbance can be solved in accordance with the potential flow theory [7]. The terrain changes were not seen as a spoiler on a small field perturbation by potential flow theory are:

$$\overrightarrow{u'} = \nabla\chi \tag{5}$$

Where $\vec{u}'$ disturbance of the terrain changes was not spoiler field; $\chi(r, \phi, z)$ is the potential function under column coordinate; ∇ for the Hamiltonian operator.

In the location of wind turbines for the origin of coordinates, the outer boundary terrain perturbed flow field can be converted to a definite solution to solve the following problem

$$
\begin{cases}
\nabla^2 \varphi = 0 \\
\chi \mid_{r=R} = 0 \\
\chi \mid_{r=0} \text{ 有界} \\
\chi \mid_{r=L} = 0 \\
\dfrac{\partial_\chi}{\partial_z} \mid_{z=0} = \vec{u}_0 \bullet h(r, \varphi)
\end{cases}
$$

$$0 \le r \le R, 0 \le z \le L \tag{6}$$

Where $\vec{u}_0$ upper-level wind undisturbed wind velocity vector; $h(r, \phi)$ is a function of the height of the terrain; $R \ge 10\ km$ as the study area radius; terrain disturbance length in the vertical direction.

Laplace equation column coordinates using separation of variables to solve, and according to the superposition theorem of linear partial differential equations give $\chi(r, \varphi, z)$ general solution is:

$$\chi(r, \varphi, z) = \sum_{n=0}^{\infty} \sum_{j=1}^{\infty} J_n(\alpha_j \gamma)(A_{nj} \cos n\varphi + B_{nj} \sin n\varphi) \times \sinh \alpha_j (L - z) \tag{7}$$

Where: $J_n(\alpha_j \gamma)$ is n-order Bessel function.

By the boundary conditions, the Bessel function orthogonality and Eq. (5) becomes:

$$\vec{u}'_{WT} = \frac{1}{2R} \sum_{j=1}^{\infty} \sinh\left(1 - \frac{c_j^1}{R} z\right) c_j^1 \left[(A_{1j} + B_{1j})e_r + (B_{1j} - A_{1j})e_\varphi\right] \tag{8}$$

Where $\vec{u}'_{WT}$ the change of location for the terrain wind turbine outer flow field disturbance; c_j^1 is a first-order Bessel function of the first j zeros; e_r, e_φ, respectively radial and azimuthally direction unit vector; factor A_{1j}, B_{1j} is determined by the following formula:

$$A_{1j} + iB_{1j} = -\frac{2 \int_0^r \int_0^{2\pi} \vec{u}_0 \cdot \nabla h(r, \varphi) e^{i\varphi} r J_1\left(\frac{c_j^1}{R} r\right) d_\varphi d_r}{\pi R \cosh(1) c_j^1 \left[J_2\left(c_j^1\right)\right]^2} \tag{9}$$

Where $\nabla h(r, \varphi)$ contains the topography information.

According to Eqs. (7) and (8), for a given wind turbine location and topography surrounding topography information can be obtained on the location of wind turbines outer flow field disturbance.

The inner boundary layer wind perturbation solving: in flat terrain roughness and no significant changes, the pressure gradient in the boundary layer and viscous stress is the difference between the upper and lower layers (including molecular viscosity and turbulent viscosity) in equilibrium, so that the wind speed distribution achieves a steady state of equilibrium. If there are significant topography, the shear stress in a vicious, inertia and pressure gradient (i.e. positive stress) interaction equilibrium. Laminar Flow field changes as the result of nonlinear inertia force and frictional turbulence interaction. Therefore, the height of the flow field disturbance with a logarithmic wind profile changes, the inner flow field for the same height of the potential flow solutions correction value:

$$\Delta \vec{u}_j(z) = u_0(z) \frac{u_0^2(L_j)}{u_0^2\left(z_j'\right)} \nabla \chi_j u_{*n} \tag{10}$$

Where $u_0(z)$ is upwind undisturbed wind speed wind vector in the height z; L_j terrain disturbance in the vertical direction of the length scales; $z_j' = \max(z, l_j)$, $l_j << L_j$ inner boundary layer height is determined by the following formula:

$$l_j = 0.3 z_{oj} \left(\frac{L_j}{z_{0j}}\right)^{0.67} \tag{11}$$

Where z_{0j} is the corresponding l_j of the relative roughness, the wind direction is uniformly distributed roughness when $z_{0j} = z_0$. Otherwise:

$$z_{0j} = z_{0n} \left(\frac{z_{0n+1}}{z_{0n}}\right)^{w_n} \tag{12}$$

Where $w_n = \exp(-\frac{x_n}{D})$, $D = 5L_j$, x_n change for the n-thoroughness and fans distance.

Laminar Flow Field with highly disturbed by the number of law changes, so the correction value can be obtained in any of the inner height of the disturbance flow field with a high degree of outer disturbance potential flow solution. So far it has been a high degree of analytical solutions of different flow field of wind turbine locations disturbance, and upwind undisturbed wind vector obtained after superposition Winds perturbed by topography.

4   Modeling and Simulation Masta Law

Depending on the height of the wind speed measured masts, wind data, combined with the wind farm location, fans layout, and roughness changes, the use of wind profilers method and analysis method based on potential flow theory, physical model established by the wind speed wind towers Winds from the extrapolation per turbine hub height wind speed. Applications fan measured power curve modeling method to determine the measured power curve model for each fan, according to the wind tower wind farm each

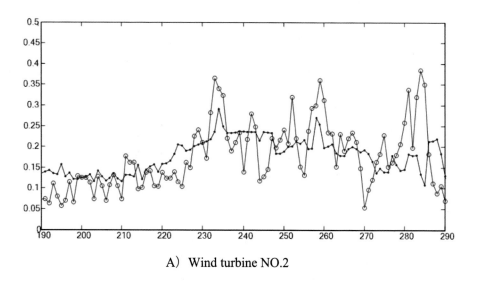

A）Wind turbine NO.2

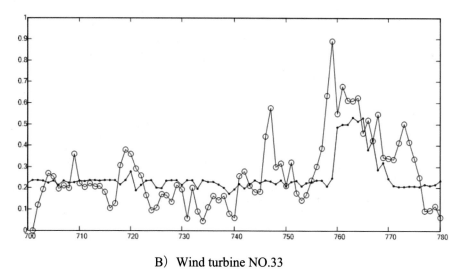

B）Wind turbine NO.33

Fig. 1. Comparison of the actual measured power curve of the wind turbine and the actual measured power curve by the maximum value method

unit measured data measured power curve model to calculate the wind farm power in the corresponding period of time, and then get abandoned wind power. Winds obtained from the physical model turbine hub height wind speed, the measured power input to the fan curve model method by the maximum output power curve obtained in the fan, and fan the measured power curve comparison shown in Fig. 1, to show clearly only part of the interception graphics [8, 9].

It can be seen from the Fig. 1, the output power curve of the abandoned wind power evaluation model based on the wind tower method is similar to the theoretical power curve. In graphics and data, it is also close to theoretical power.

5   Conclusions

In this paper, abandoned wind power assessment method based masts law. Firstly, wind speed physical model, which is used to predict the fan speed; then, using the maximum probability method to establish the measured power curve model wind turbine, the model is used to launch the fan should be sent theoretical power; Finally, according to wind tower the measured data of each unit measured wind farm power curve, calculate the theoretical generating capacity of wind farms in the corresponding period of time, then get abandoned wind power, through simulation concludes: masts law closer to theoretical power, is recommended.

References

1. Wu, Q. Zhou, M., Sun, L., et al.: Wind farm penetration limit calculation based on evolutionary programming algorithm. IEEE Innov. Smart Grid Technol. (2012)
2. Chang, T.-P., Ko, H.-H., Liu, F.-J., Chen, P.-H., Chang, Y.-P., Liang, Y.-H., Jang, H.-Y., Lin, T.-C., Chen, Y.-H.: Fractal dimension of wind speed time series. Appl. Energy (2011)
3. Azmy, A.M, Erlichi, I.: Impact of distributed generation on the stability of electrical power system. In: Proceedings of 2005 IEEE Power Engineering Society General Meeting (2005)
4. Billinton, R., Kumar, Y.: Transmission line reliability models including common mode and adverse weather effects. IEEE Trans. Power Apparatus Syst. (1981)
5. Campagnolo, F., Petrovi, V., Schreiber, J. et al.: Wind tunnel testing of a closed-loop wake deflection controller for wind farm power maximization. J. Phys. Conf. Ser. (2016)
6. James III, G.H., Carne, T.G., Lauffer, J.P.: The natural excitation technique (NExT) for modal parameter extraction from operating structures. Int. J. Anal. Exp. Modal Anal. (1995)
7. Ng, K.H., Spalding, D.B.: Some applications of a model of turbulence to boundary layers near walls. Phys. Fluids (1972)
8. Beltran, B., Ahmed-Ali, T, Benbouzid, M.E.H.: High-order sliding-mode control of variable-speed wind turbines. IEEE Trans. Ind. Electron. (2009)
9. Liang, T.T., Song, Y.D.: Robust adaptive individual pitch control of wind turbines. ASME J. Sol. Energy Eng. (2011)

Energy Consumption Regulation
for Substation Operation in Practice

Guangming Wang[(✉)], Xilan Wang, Yong Pei, Hui Jin, Hongren Yu,
Ju Zhang, Hong Gao, and Yan Zhao

Fushun Power Supply Branch,
State Grid Liaoning Electric Power Supply Co. Ltd., Fushun, China
405646232@qq.com

Abstract. This paper mainly introduces the operation process of the substation and introduces the reason and influence of power energy consumption. The energy function, classification, and application of substation equipment are briefly introduced. The monitoring application of power equipment in the substation is formed through the analysis of energy consumption data collection method, calculation basis, and energy consumption analysis method. The energy monitoring results of the substation are discussed. Energy saving is the main part of the design concept in substation operation. The energy saving substation provides good social, environmental and economic benefits for future engineering design. At present, the vacuum circuit breaker still has some limitations in the vacuum substation. If the circuit breaker fails, there will be undesirable consequences. This problem needs to be solved in time to avoid circuit breaker failure. The security of power consumption is hard to be guaranteed, which has a certain negative impact on the stability of the power supply. With the development of The Times, people's demand for power resources is increasing. In today's limited resources, how to meet the social power demand as far as possible, energy saving and consumption reduction are particularly important. The reduction of loss is equal to the transmission of more electricity under the same conditions, reducing the operating cost of power supply enterprises.

Keywords: Electric energy · Energy consumption · Energy-saving

1 Introduction

With the development of social economy, the power demand increases year by year, and the number of substations increases year by year. The company has a large number of 66 kV and 200 kV substations under its jurisdiction [1]. The substation has been unattended for the management, operation, and monitoring of the substation, which is completed by the control system, monitoring system and communication system [2]. Good temperature and humidity environment are needed to control and operate communication equipment in the substation to ensure the stable operation of equipment and continuous power supply [3]. To ensure the safe and stable operation of the main control room, protection room, SF6 [4] and other equipment of the substation. Equipment such as air conditioner, exhaust fan, new fan, and electric heating can

© Springer Nature Switzerland AG 2020
A. E. Hassanien et al. (Eds.): AMLTA 2019, AISC 921, pp. 809–816, 2020.
https://doi.org/10.1007/978-3-030-14118-9_79

reduce indoor temperature, maintain air circulation and improve indoor temperature in winter [5]. The operating rules of the substation are based on the indoor environment. These devices need a long-term operation and large power consumption [6]. In addition, the power consumption of the substation equipment needs to be run, which directly leads to the huge power consumption of the air conditioner of the power supply company in each substation [7]. Therefore, exhaust fan, fan, and electric heating equipment are the main reasons for the energy consumption of substation operation [8].

Power consumption equipment substation without interruption, resulting in a large number of lost power, not only to the power supply company, a large number of the resulting electricity loss, increase the substation operation cost, more to the environment caused great destruction. Substation impact energy consumption mainly related to environmental and economic effects of two aspects, (Refer to Fig. 1).

Fig. 1. The native gives the general shape of the transformer

Environmental impact: air conditioning, electric heating, and other equipment in the operation of substations emit large amounts of carbon dioxide and other waste gases, causing environmental pollution, which is not conducive to energy conservation and emission reduction. Economic impact: the long-term operation of energy consuming equipment in the substation takes up a large amount of power and increases the operating cost of the substation, causing losses to the power supply company.

Energy consumption of substation: active and reactive power losses are required in the operation of substation equipment. During the operation of the substation, the life and production of personnel on duty need to consume electric energy. Long-term use of lighting, air conditioning, wind turbines, and other building electrical appliances.

2 Energy Consumption Equipment in the Substation

The energy equipment of substation is the necessary equipment for the construction and operation of the existing substation. The air cooling equipment of the transformer is operated outdoors for a long time, which makes the air circulate, reduces the temperature of the main transformer and ensures the stable operation of the transformer. The air conditioner, exhaust fan new fan, electric heating equipment, and large energy consumption are all in the main control room, indoor substation protection room and SF6 room to maintain indoor temperature regulation.

The main function of energy consumption of substation equipment is to adjust the temperature. According to the function of the equipment, the energy consumption of the equipment is divided into heating equipment and cooling equipment. It is the most commonly used equipment for energy consumption of air conditioning. Reduce the main control room, switch the indoor temperature, and keep the indoor temperature constant. In summer, the indoor temperature rise drives the temperature control equipment of the substation. Meanwhile, the cooling temperature of the server needs to be controlled by the cooling equipment. When indoor temperature drops in winter, communication equipment, and control equipment of substation need to maintain a constant temperature, and low temperature will cause equipment shutdown. Therefore, electric heating and air conditioning equipment are adopted to improve indoor temperature (Table 1).

Table 1. Annual consumption statistics of various high power unit transformers. It's always under the illustration.

No	Name	YEAR(kw.h)	Pop
1	Main transformer	45.06	88.04%
2	Capacitance and reactance	5.40	10.54%
3	Office electricity	0.92	1.42%
Total		51.38	100.00%

3 Analysis of Energy Consumption Equipment in the Substation

It can be seen from the application and classification of energy consumption of substation equipment that the operating cycle and energy consumption of substation equipment are important factors that determine the level of the substation. Master operation cycle and operation energy consumption equipment, monitor the energy consumption of substation equipment reasonably and effectively, shorten the equipment operation time, reduce energy consumption, adjust the operation equipment and effectively reduce the energy consumption of substation on the premise of ensuring the stable operating environment of the substation.

3.1 The Operation Cycle of Energy Consumption Equipment

According to the use of energy equipment and the requirements of equipment operation environment, the operation cycle of various energy consumption equipment is in different operation modes. These require regular adjustments to energy consumption equipment. According to the operational requirements and seasonal changes, adjust the substation dispatch, remote control, operating environment communication equipment. The season is also a factor affecting the energy consumption of equipment operation, which refers to the influence of substation on the environmental state and reduces the movement of equipment energy.

Operation period of energy consumption equipment in substation including substation electric heating; 5 months (from November to March this year), substation operation air conditioning, refrigeration operation period of 4 months (June to September), and the heating period is 5 months (from November to March).

The operation mode of energy consumption equipment is defined using the following Equations:

$$Y_{ab} = Y_{cb} = Y \tag{1}$$

$$\triangle \dot{U}_a = \dot{I}_{ab} R_L \tag{2}$$

$$\triangle \dot{U}_b = -(\dot{I}_{ab} + \dot{I}_{cb}) R_L \tag{3}$$

$$\triangle \dot{U}_c = \dot{I}_{ab} R_L \tag{4}$$

$$\triangle \varepsilon_a = -\frac{\triangle \dot{U}_a}{\dot{U}_a} = -R_L Y(1.5 + j0.866) = -R_L Y_a \tag{5}$$

$$\triangle \varepsilon_c = -\frac{\triangle \dot{U}_c}{\dot{U}_c} = -R_L Y(1.5 - j0.866) = -R_L Y_c \tag{6}$$

Y is the admittance of the device, $\triangle \dot{U}_a$ is the voltage at A, $\triangle \dot{U}_b$ is the voltage at B, $\triangle \dot{U}_c$ is the voltage at C, R_L is the resistance value, I_{ab} is the current between A and B, $\dot{I}_{cb}$ is the current between C and B, $\Delta\varepsilon_a$ is the potential energy at point A, $\Delta\varepsilon_b$ is the potential energy at point A.

This substation adjusts the operating environment of the indoor substation according to a set of indoor environment indicators implemented in different geographical environments and seasons, which not only wastes a lot of energy but also affects the operation of the substation equipment. Existing indoor substation maintenance work process, operation, and maintenance personnel cannot automatically adjust the substation control room, protection room SF6, room environment. They cannot grasp the energy consumption data and energy efficiency of substations in real time and accurately and lack effective management mechanism. Indoor temperature control environment parameters cannot reasonable adjustment, depending on the operating and maintenance personnel, arrived at the scene of the substation regularly adjust the substation indoor air cooling, air conditioning, ventilation, the running status of electric equipment, waste a great deal of time and effort, result in large power loss, cannot adjust the substation equipment energy consumption will be the actual situation.

4 Energy Consumption Supervision Mode of the Transformer Substation

Energy-saving optimization control of substation operation and management of energy conservation and protection of substation chamber through the main control room, SF6 gas room, indoor air conditioning, exhaust fan, fan, electric heating equipment, energy consumption equipment real-time monitoring, data collection is complete, upload and monitoring, the data set is displayed in each substation equipment energy consumption operation time, including power, temperature, the numerical calculation.

The energy consumption of equipment of electricity, electricity, energy efficiency index of PUE, the formation of electricity usage, operation of air conditioning, air machine, eleEnergy consumption of power equipment, energy efficiency indexes of power and PUE, electricity consumption, air conditioning operation, fan operation, electric heating operation, temperature and humidity fluctuation, energy consumption data, reflect energy consumption and utilization rate of energy equipment of substation. The energy consumption of the staff working in the centralized operation team of the substation is governed by the substation, which is controlled through energy consumption monitoring, the real-time inquiry of the substation and energy equipment, and real-time control of the substation equipment. Based on the indoor substation operating environment parameters, underground substation location and seasonal changes, the management of energy consumption and energy efficiency analysis, the utilization of power equipment in the substation, operation curve, histogram display, and energy regulation are analyzed.

4.1 Energy Consumption Equipment Collection

Sensor technology based on Internet of things, temperature and humidity acquisition device, each switch state complete substation main control room, protection room, SF6 gas room, indoor air conditioner, fan, exhaust fan, electrical heating energy consumption data acquisition equipment, wireless cable uploaded to the system, the real-time data acquisition, data collection substation outdoor temperature humidity, such as the reference data, the optimization of wireless network to upload system. Energy consumption monitoring, energy consumption data collection and operation, and energy consumption data collection of substations are important databases for energy consumption analysis and energy efficiency management. Real-time and accurate energy consumption data of substations is a key research topic. Don't change your energy consumption. Power supply mode, avoid site construction, ensure the real-time and accurate collection of energy consumption data of substation equipment operation, all kinds of energy equipment. Complete the collection and centralized monitoring of energy consumption data of several substations.

4.2 Energy Consumption Data Analysis

Analysis of energy consumption and energy saving optimization of substation, substation equipment based on the energy consumption data, the seasonal and regional position parameters of each substation, according to the concentration of energy consumption data, using a diagram, chart, etc., according to operation curve of the air conditioning, ventilation operation curve, energy efficiency, energy consumption calculation of electric equipment, electric power, the value of PUE, clear substation transformer substation structure, evaluation of energy consumption and energy consumption between the transformer substation equipment energy consumption evaluation item by item. Make a comparative analysis of energy consumption, check the energy consumption data of substation, and compare with data comparison and energy consumption trend. According to the indoor environment, geographical location, season, energy consumption equipment operation data and real-time environmental data of the substation, energy saving optimization, air conditioning, fan, exhaust, operation time and electric heating switch state of the substation are formed. Check the best operation mode, power consumption trend and optimization plan of each energy consumption equipment, and view through the terminal push (Table 2).

Table 2. The annual consumption statistics of various low-power-consuming unit transformers

No	Name	YEAR(kw.h)	Pop
1	Main transformer	23.6	38.04%
2	Capacitance and reactance	45.8	60.54%
3	Office electricity	0.92	1.42%
Total		76.32	100.00%

Switch control and operation mode of energy equipment such as energy equipment, air conditioner, fan substation, air conditioner heating, etc. According to the data of the energy system of the substation and the energy saving optimization scheme, the remote control switch and temperature remote control terminal device of the remote air conditioner, fan, exhaust fan, electric heating equipment are completed by the control command execution and maintenance personnel.

5   The Effectiveness of Energy Consumption Monitoring in Substation Operation

Analyze energy consumption and energy efficiency management of substation, energy consumption data of substation equipment, trend, ultra-high power alarm, form energy-saving, and emission reduction plan, and optimize energy consumption of equipment in operation of substation equipment.

This analysis can be used for the power consumption rate of substation equipment, and the energy consumption trend can be understood through operation curve method, histogram method, energy efficiency evaluation of multidimensional substation, and operation and maintenance personnel. Combined with energy consumption of historical data, each substation forms optimization measures for energy saving equipment of substation, provides reasonable operation plan of energy equipment, equipment operation state and time consumption, and puts forward adjustment parameters for indoor environment parameter operation and underground substation location, seasonal variation, and other factors.

Substation air conditioning, exhaust machine, electric heating equipment, energy consumption of electricity, electricity, energy utilization ratio of energy consumption data, analysis of energy consumption trend of substation, the multiple dimensions of energy efficiency evaluation, based on the energy consumption data, environment, geographical location, seasonal factors, such as forming part of substation can adjust energy consumption optimization measures, remote control equipment and operation parameters, the energy optimization control for substation operation maintenance personnel. Get rid of manual methods and save manpower and material resources. Improve the cooling energy consumption life of equipment, reduce the operating cost of the substation, save a lot of electricity, improve the utilization rate of power consumption of substation and reduce emissions. We will implement energy management and energy-saving control of substations and improve their operation and management.

By the application of substation energy management and energy saving optimization, Focus on management of the Equipment operation condition, Monitor the Equipment operation condition of Air conditioning, exhaust fan, electric heating exactly. Energy consumption data of indoor energy consumption, power utilization ratio and other energy consumption showed in a Graphical way, manage about the Energy consumption of Substation, check the transformer substation energy consumption equipment operation state timely. Realize the energy consumption regulation and management of Substation to improve the level of ability of the Substation operation and maintenance management based on Energy consumption data and equipment environment data and reform the Energy saving optimization measures, remote adjustment of energy consumption equipment operation to get the best running environment. Remotely adjust the running conditions and parameters of energy equipment, air conditioning, and other energy consumption to improve the life of the equipment.

Provide energy consumption optimization and adjustment plan for substation equipment, reduce human error experience, adjust energy equipment operation remotely, it does not need to arrive at the operation site, operation and maintenance personnel work simply, save manpower and material resources. Applying substation operation and adjusting energy consumption, reducing power consumption of substation operation, saving power consumption, saving the operating cost of the substation, improving the power utilization rate of the substation, saving energy and reducing emission, bring considerable economic benefits to power companies.

6 Conclusions

The substation operation energy consumption control and management, we will strengthen the operation and management of the substation equipment energy consumption, energy consumption data acquisition equipment energy consumption, energy consumption, energy consumption trend analysis substation equipment operation regulation of energy consumption, reasonable adjustment of energy consumption, building energy consumption control mechanism of substation operation, reduce the energy consumption and has good energy saving effect. The operation of substation should be carried out in the whole life cycle, feasibility study, design, and construction and operation stage of the substation. The work of substation should be carried out according to different types of objectives, and at the same time, technical measures and management measures should be combined to reduce the demand for resources and energy, so as to better control the energy consumption of substation operation.

References

1. Jane, X., Yang, Y.: Substation secondary drawing review standard research. Natl. High-Tech Enterp. **3**(7), 80–83 (2016). (in Chinese)
2. Yang, A.: Research and implementation of an energy management system based on an energy consumption prediction model. South China University of Technology, Guangzhou **16**(8), 93–97 (2013). (in Chinese)
3. Fan, H.: Explore the analysis of energy consumption and energy saving in the process of the operation and maintenance of substation. J. Beijing Electric Power Univ. Nat. Sci. Ed. **12**(8), 27–29 (2012). (in Chinese)
4. Wang, D., Ji, L., Tan, T.: Intelligent substation secondary virtual circuit operation dimension. Design and implementation of the protection management system. Huadian Technol. **18**(8), 97–99 (2016). (in Chinese)
5. Xu, S., Liu, W., Shi, W.: Analysis of typical overvoltage of 500 kV substation based on EMTP Knowledge Research. Hydropower Energy Sci. **36**(04), 199–203 (2018)
6. Liu, X., Hong, T.: Comparison of energy efficiency between variable refrigerant flow systems and ground source heat pump systems. Energy Build. **42**(5), 127–133 (2010). (in Chinese)
7. Zeng, W., Li, Y., Qi, X.: Substation primary equipment overvoltage and its secondary circuit Research on influence and preventive measures. Electronic test **9**(Z1): 81–82+42 (2018)
8. Guo, R.: Plastic composite doors and aluminum windows and door broken bridge comparative analysis of the application. China's Foreign Trade **41**(12), 435–436 (2012). (in Chinese)

Implementation of Tower Grounding Resistance Measurement Based on Decreasing Overhead Grounding Line's Current Shunt Using High-Frequency Power Supply

Yujie Pei[✉], Jinglong Mu, Weijun Li, Jun Dong, Jianguo Xu, Chunmei Guan, Yi Qu, and Hongchuang Ma

Fushun Power Supply Branch,
State Grid Liaoning Electric Power Supply Co., Ltd., Fushun, China
Yujiepei.fssgcc@163.com

Abstract. This paper analyses the production of the current shunt and the influence factor. Then establish the tower overhead ground wire current shunt model. The current commonly used grounding resistance measurement method for shaking table method and clamp meter, shaking table method need to disconnect the ground lead line, the measuring error of clamp meter, based on the two methods proposed a new method of grounding measurement, continuous grounding the premise the high frequency power supply to achieve the measurement of grounding resistance. According to the model, this paper explained the relationship between the injecting current shunt coefficient and the overhead ground wire and voltage frequency when testing the grounding resistance. As a key to frequency characteristics of overhead ground wire, we offered a new method of measuring tower grounding resistance called high-frequency power method and got the best injection current frequency by tower split calculate. Compared with the shaking table method, the clamp table method is the most error, and the randomness of measured data is very strong. Followed by different frequency power measurement, the measuring result of the high-frequency power supply is close to that of the shake table method, and the error is less than 10%.

Keywords: Frequency · Current shunt · High-frequency power method

1 Introduction

The tower grounding resistance directly overhead transmission line lightning protection level and lightning trip rate, the measurement of grounding resistance of grounding device to check whether it is a necessary means for the ground wire and tower, and the tower is an important factor for shunt grounding resistance measuring device accuracy. The current commonly used grounding resistance measurement method for shaking table method and clamp meter, shaking table method need to disconnect the ground lead line, the measuring error of clamp meter, based on the two methods proposed a new method of grounding measurement, continuous grounding the premise the high frequency power supply to achieve the measurement of grounding resistance.

© Springer Nature Switzerland AG 2020
A. E. Hassanien et al. (Eds.): AMLTA 2019, AISC 921, pp. 817–825, 2020.
https://doi.org/10.1007/978-3-030-14118-9_80

At present, accurate, fast, simple and reliable method for measuring earthling resistance has become an urgent need for technical progress in lightning protection field. Among them, the three-pole method and the clamp gauge method recommended in the current standard specification are widely used in the measurement of the grounding resistance of the tower. Although they are simple and reliable in use, they are used in the actual engineering measurement. There are also certain problems. For example, the three-pole method, it mainly includes two methods, that is, 0.618 pole placement method and triangular pole layout method, of which, Although the 0.618 pole placement method is the most commonly used method in the measurement of the grounding resistance of the tower, it not only has a heavy workload but also has a low accuracy in the measurement process, at the same time, the measurement results are also subject to the voltage line. The effect of current line inductive voltage; Although the triple triangle method can shorten the distance of wiring and alleviate the difficulty of grounding resistance test in a complex environment to some extent, and the measurement error is relatively small, the current line is required by this method. Voltage line isobaric, the workload is also larger. At the same time, these two methods are greatly affected by the factors of topography, geology, and environment, especially in some fields where the topography is complex when these two methods are used to measure the grounding resistance. The electrode placement of the voltage measuring electrode is often due to the swamp or rock section and the location of the electrode goes underground according to the specified position, which makes both methods difficult to implement. Although the clamp meter method is simple and convenient to use, its measurement value is often larger than the actual grounding resistance value of the tower, and the stability and accuracy are not high, so the above three methods have some limitations.

2   The Tower and Overhead Ground Wire Model and Equivalent Parameters

The tower overhead ground wire model [2] as shown in Fig. 1, where Rl and Xl respectively represent the overhead ground line impedance and inductance, Rt and Xt represent the tower body impedance and inductance, Rg for grounding impedance. Shake table measurement method is adopted, if continuous grounding, the measuring circuit includes: injection current grounding wire by tower and overhead through the earth to form a loop; overhead ground wire and tower connected directly lead to the current through the overhead ground wire and tower adjacent to form a loop.

In the network of transmission lines in complex with circuit overhead line tower, adjacent overhead line tower, overhead ground wire to form multiple parallel networks on earth, resulting in the removal of grounding test of tower grounding resistance when the injection current shunt.

In multiple parallel network shunt carrier including tower grounding resistance, impedance, the tower is adjacent test overhead ground wire impedance and adjacent tower impedance test [3].

In the measurement process, the electrical current distribution principle of [2] as shown in Fig. 2, RX for grounding resistance grounding; RPZ voltage pole and tower

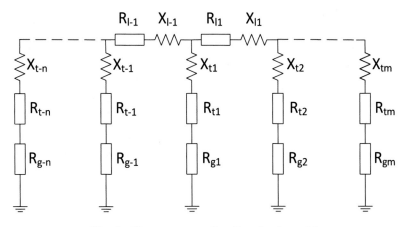

Fig. 1. The tower grounding line circuit model

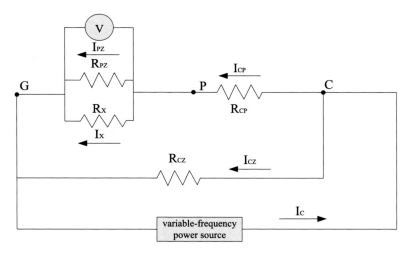

Fig. 2. Actual current distribution electrical schematic diagram

all around between the equivalent impedance; RCP grid current and voltage pole electrode impedance; RCZ electrode and the equivalent impedance of towers is all around. Variable frequency power supply output current to current shunt part of pole and tower all around between the impedance, part of a diversion to the current and voltage pole electrode impedance. The current into the voltage pole again shunts to around between impedance and tower grounding resistance. Can the grounding resistance of the shunt coefficient.

3   Diversion Tower Model

3.1   The Injection Current of Shunt Coefficient and the Adjacent Tower Base Model

In the same lines, the overhead ground wire and the effective impedance of tower grounding resistance can be obtained under the condition of knowing the shunt coefficient and tower base relationship. To find out the changes of the number of ground wire shunt impedance and tower base to determine ground wire shunt coefficient created with simplified shunt impedance model for simple lines. The measured tower as the center, in accordance with the current direction, the base tower is named L1L2...... L(m–1), Lm and R1, R2...... R(n–1), Rn. Based on the analysis method of phase component model proposed in [4–7], a simplified equivalent impedance of lightning line model, the impedance of all ground wire and tower span all tower grounding resistance values were set equal, base resistance is R0, lightning line impedance is Z0. The simplified model is shown in Fig. 3 [8].

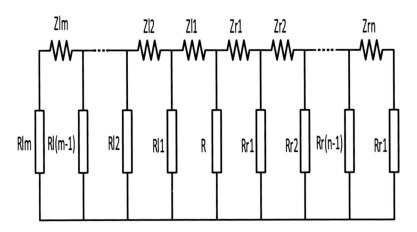

Fig. 3. Tower quantity model

According to Fig. 3, the (n–1) tower impedance is: $Zrn = Z0 + R0$; (n–2) tower impedance is $Zrr (n–1) = Z0 + R0//Zrn$. Followed by analogy, $Zr1 = Z0 + R0//Zrr2$, $Zl1 = Z0 + R0//Zll2$, the total impedance is $Zl1//Zr1$.

Based on the above formula, the shunt impedance of the lightning protection line can be calculated quickly by using the recursive method [9]. Analysis shows that the ground wire shunt impedance by m and n value of the tower base, shunt impedance and m value of n and a negative linear relationship; when m and n is greater than 15, with the increase of base tower number, shunt impedance decreases not obviously; the total number of base tower shunt impedance is the same, the difference was positively associated with shunt impedance m with n.

3.2 The Grounding Resistance of the Shunt Coefficient and Tower Space Model

The adjacent tower, with the line between the tower and the tower of shunt coefficient of the distance, the farther the distance, the smaller the shunt coefficient.

On the basis of the shunt coefficient and the adjacent tower base model, change model parameters, create ground resistance coefficient and split spacing tower model, simplified model of transmission line length setting unchanged, pole spacing increased to 2 times the original model, the resistance is R0, lightning line impedance Z20 = 2*Z0, tower base were m/2 and n/2, such as the simplified model shown in Fig. 4.

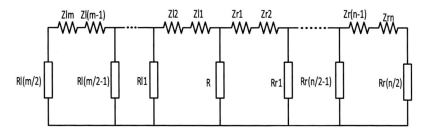

Fig. 4. Tower span model

According to the description, the n/2–1 impedance is Zrr (n/2) = 2 * Z0 + R0; then-2 impedance is Zrr (n/2–1) = 2 * Z0 + R0//Zrrn. Followed by analogy, Zr1 = 2 * Z0 + R0//Zrr2, Zl1 = 2 * Z0 + R0//Zll2, the total impedance is Zl1//Zr1.

Compared with the part 2 descriptions of Zrr (n–1) and the model of Zrr (n/2), the (Z0 + R0//(Z0 + R0)) (2 * Z0 + R0) is known, and using the recursive relationship, the Zrr (x–1) < Zrr (x/2) in part 2 described can be known.

3.3 Simulation Test Frequency and Tower Shunt

The whole tower loop can be equivalent to an LR lumped parameter circuit, the impedance formula of Z = R + jX = R + j. L. When the injected DC current or AC current, the inductive component is not obvious and can be neglected. Then the whole tower grounding impedance of the equivalent resistance of R, with the continuous improvement of the frequency, the inductive component is reflected gradually.

Compared with the ground wire inductive component of grounding impedance of inductive component is very small, for the simplification of the test model, assuming that the ground impedance is resistive, inductive component of all sources of tower lightning line.

On the basis of "industrial and civil electric device grounding design specification GBJ65-83" lightning conductor, often using the model for the LGJ-185 steel strand, the reactance value is about 12.8 uH. According to the calculation formula of inductance impedance Xl = 2, PI FL, 100 m measuring lightning line frequency signal when lightning line inductance values are about 0.4 if the measurement when the signal reaches the kHz lightning line inductance value will reach tens or even hundreds of ohms.

For the use of optical fiber composite overhead ground wire (OPGW), according to the structural parameters (diameter R1 = 2.48 mm, diameter R2 = 10.6 mm) provide by "GB/T 15972.10", the OPGW overhead ground wire inductance per unit length value can be calculated using the following Equation:

$$L = (0.08D \times d \times N)/(3D + 9W + 10H)$$

Where D is flux through the cable cross-section; d is permeability; N is the current through the optical; H-fiber cable length; W - radius cable.

Through the calculation, the OPEW overhead ground wire of the unit inductance value is. Its reactance value is even slightly larger than that of LGJ-185 steel strand.

According to the relationship between the inductance is proportional to frequency, can improve the measurement frequency to reduce the shunt, which uses high-frequency power supply method without removing the grounding measurement of tower grounding resistance.

4 Diversion Tower Frequency Calculation

The model of frequency and tower shunt coefficient on the above analysis is seen as a simplified model. In fact, there is still a very small perceptual component of grounding impedance. When the measuring frequency is too small, the inductive component of the lightning rod is small. The measurement frequency is too high, the measurement of the grounding resistance is higher. So, the measurement error can be minimized by determining the range of injection excitation current.

In order to solve the above problems, According to the tower overhead ground wire, the grounding grid structure, The integrated soil resistivity, and other factors, the theoretical analysis the tower range of frequency selection and then build the simplified circuit model. AC current is supplied by a power supply with different frequencies from 40 Hz to 10 kHz. And the simulation calculation and statistical measurement error should be demonstrated. The consistency between the simulation results and the theoretical calculation is verified. Continuing to open the ground deflectors to ensure the measurement error is as small as possible, and it must satisfy the conditions.

In the formula, the inductive component of ground impedance is l_0. The resistive component of the ground impedance is R_0; the inductive component of the lightning protection line is l_1. $2\pi f l\_0 \ll R\_0$ has indicated that the result of the measurement is too large because of the sensitivity of the grounding impedance. $R\_0 \ll 2\pi f l\_1$ has indicated that the measurement result is small due to the current shunt and obtained after conversion, because the actual measurement cannot guarantee the selected specific frequency can meet and that the value is 0. If and only if 0, it can satisfy the requirements of the maximum. In PI, it is the middle frequency.

Table 1. Frequency tradeoff under 10 Ω grounding resistance

L_0(uh)/l_1(mh)	10	20	30	40	50
2	11.26	7.96	6.50	5.63	5.04
4	7.69	5.63	4.60	3.98	3.56
6	6.50	4.60	3.75	3.25	2.91
8	5.63	3.98	3.25	2.81	2.52
10	5.04	3.56	2.91	2.52	2.25

According to "GB50169-2010 electrical equipment installation engineering construction and acceptance of earthed device specification", the grounding resistance value does not exceed 10. So, list the best frequency omega compromise in grounding resistance in the 10. Table 1 shows the frequency tradeoff under 10 Ω grounding resistance.

According to the relation between the frequency of compromise and grounding resistance is inversely proportional. It can be known that the measurement error of high-frequency power supply frequency is 12 kHz–2 kHz.

Considering the current injected into the earth due to the high-frequency characteristics, and it may be in contact with the earth to produce an electric spark. As well as the relationship between sampling points and computing speed in the FFT operation, the best injection frequency selection is 4 kHz.

5 Comparison of the Measured Field Data: Quality Management and Monitoring of Power Energy Saving Service Process

The subject comes from the research and application of them without removing the grounding measurement of tower grounding resistance based on Rogowski coil which cooperated project with Fushun power Supply Company of Liaoning Province. For the launch of the research, we measured the 500 kV power supply line to the Fushun electric power company which contains Qingdong line, Tiedong No. 1 line, Shunqian line. We have carried on the field test to the 220 kV power supply line which contain Beiping No. 1 line, Liaoyuan No. 1 line, Shuanghe No. 1 line, Qianli No. 1 line, Fuyang No. 2 line, Lishun No. 1 line, Pufu No. 1 line, Linbei No. 2 line, Linbei No. 1 line, Liaodong line, that line belongs to Fushun power supply company. The 66 kV power supply lines, that contain Shengping No. 2 line, Shengtai line, Shengma line, were carried Actual measurement as shown in Table 2.

Figure 5 shows the test environment. Compared with the shaking table method, the clamp table method is the most error, and the randomness of measured data is very strong. Followed by different frequency power measurement, the measuring result of the high-frequency power supply is close to that of the shake table method, and the error is less than 10%. The verification of the high-frequency power supply is almost no traffic which in the tower and the overhead lightning conductor of multiple loops.

Table 2. Field test data

Location	The voltage level	Shaking table method	Clamp table method	Differ-frequency (45 Hz)	High-pitched (4 kHz)
Qingdong line-226	500 kV	4.0	2.90	2.72	3.7
Tiedong No.1 line-180	500 kV	5.88	3.85	4.42	5.83
Shunqian line-2	220 kV	1.4	1.22	1.13	1.33
Liaoyuan No.1 line-36	220 kV	5.9	4.05	4.38	5.63
Linbei No.2 line-24	66 kV	6.4	5.14	5.32	6.55
Shengma line-3	66 kV	8.0	6.52	7.31	8.15

Fig. 5. Test environment

6   Conclusions

Based on the study of split free ground measurement of transmission line tower grounding resistance, it analyzes influence factors for the injected current shunt in the tower and established a practical impedance test mode. Meanwhile, it put forward the method of the high-frequency power supply. And through the detailed frequency analysis, it elaborates the optimal frequency characteristic and then it applied to the accurate measurement of the grounding wire to achieve without dismantling the tower grounding resistance.

References

1. Liu, Z., Xu, J.: Comparison of several kinds of the instrument for measuring grounding resistance of tower. High Volt. Eng. **27**(5), 76–77 (2001)
2. Ye, T.F.: Research and development for the measurement system of large grounding impedance based on the shunt function of overhead lightning conductor. North China Electri. Power Univ. Wuhan (2012)
3. Liu, W., Liu, Y., Liu, H.T.: Grounding resistance measurement error analysis for grounding device and overhead grounding line connection not disconnected. Insul. Surge Arresters **4**(2), 62–65 (2012)

4. Dawalibi, F.: Ground fault current distribution between soil and neutral conductors. IEEE Trans on Power Apparatus and Systems **99**(2), 452–461 (1980)
5. Niles, G.B.: Measurements and computations of fault current distribution on overhead transmission lines. IEEE Trans. Power Apparatus Syst. **103**(3), 553–560 (1984)
6. Zou, J.F.: Computation of fault current distribution and shielding coefficients of ground wires for overhead transmission lines. Power Syst. Technol. **24**(10), 27–30 (2000)
7. Zhao, B., Li, L., Cui, X.: Shunt currents computation of ground lines in power circuits. J. North China Electr. Power Univ. **29**(4), 5–7 (2002)
8. Zhang, Q.: New technology of surveying the pole lapping earth resistance based on the high-frequency parallel method and its equipment. High Volt. Eng. **33**(1), 194–195 (2007)
9. Li, D.J., Wang, W., Zhang, Y.: A new analysis approach on the coefficient of lightning shield line when the earth short circuit taking place in substation. Insul. Surge Arresters **8**(1), 48–51 (2014)

Wind Power Dispatching Method Based on High-Voltage and Large Capacity Electric Heat Storage

Weichun Ge[1], Lingwei Zhao[2], and Shunjiang Wang[1,2,3(✉)]

[1] State Grid Liaoning Electric Power Supply Co., Ltd., Shenyang, China
wangshunjiang@163.com
[2] University of Chinese Academy of Sciences, Beijing, China
[3] Shenyang Institute of Computing Technology, Chinese Academy of Sciences, Shenyang, China

Abstract. The limitations of electric restricted by heat can be overcome by the application of super large capacity thermal storage in the thermoelectric system so that the flexibility and regulation performance of the power system can be enhanced effectively. In this paper, high voltage and super large capacity heating supply can be brought out by introducing high voltage into high density and high heat capacity solid heat storage devices in which multiple storage units can be series-parallel connected flexibly. Besides, the insulation fault can be solved by insolation technology of water and electricity. Minimizing the wind power curtailment under the prerequisites of the safety of power grid and heating quality by coordinating and optimizing the running mode of electric heat storage equipment according to constraint conditions of electrothermal integrated dispatching models as well as by improving wind power receiving ability by means of applying high voltage and super large capacity electric heat storage technology in the generation side. Automatic Generation Control (AGC) would connect high voltage switch during the period of presupposed power grid light load, or wind power curtailment. AGC would cut off the high voltage switch as soon as the high-temperature heat storage units reached the given upper limit temperature or the period of power grid light load and the wind power curtailment was over. We can seek for the optimum wind power curtailment period and the wind power accommodation limit more frequent by AGC directive. Finally, combined with the actual power system, the rationality of electrothermal optimal scheduling mode can be tested.

Keywords: Wind power accommodation · Super large capacity · Electric thermal storage · Optimal scheduling

1 Introduction

The wind power industry in our country is developing very fast in recent years. In pace with the increase of installed capacity year by year, the problem of power supplies structure became more prominent. The uncertainty of wind power, anti-peak regulation, and other characteristics lead to large amounts of wind power curtailment. Wind power curtailment mostly happened during the heating period in the north, which mostly related

A. E. Hassanien et al. (Eds.): AMLTA 2019, AISC 921, pp. 826–836, 2020.
https://doi.org/10.1007/978-3-030-14118-9_81

to power supply characteristics, wind power output as well as the electric heat constraints of thermal power units during the heating period. The minimum output and speed regulation of thermal power units are restricted by the heating, especially the forced output of the conventional units is higher than the minimum technical output caused by the constraint of "electric restricted by heat", which decrease the wind power accommodation ability during the light system load. According to the surveys by relevant departments shows that the wind power curtailment rate in our country in 2016 is over 10%. Wind power curtailment exerts very serious influence on wind power development in our country, so dispatching should research how to make more wind power accommodation.

The power industry takes a series of measures as well as make a contrast of the specific wind power accommodation plans to enhance the wind power accommodation ability in our country [1–3], the key technology of which is heat storage technology. The super large heat storage technology of heat electricity dispatching can be used as the technology routine of wind power accommodation by researching the heat storage technology of the literature [4, 5], as well as take advantage of high voltage-large capacity heat storage devices and other modern technological equipment, finally the ability of wind power accommodation can be improved.

2 Working Principles of High Voltage and Super Large Capacity Heat Storage

2.1 The Introduction of Electric Heat Storage

High voltage electric heat storage devices are consists of heat storage units, heat preservation shell, heat exchanging unit, water system, heat control system and GIS system. Heating electric heat storage by powering the phase change heat storage materials of the super large capacity electric heat storage units by the high voltage power from the thermal power plants, leading out heat through heat conduction media. Wind temperature can be improved to 500–600 °C when fans transmit wind to the heat storage units according to circulating fan control heat transfer system. The high-temperature wind heat the circulating water to 90 °C by heat-exchange units, by which the conversion efficiency can be reached to over 95%. The heat produced by the electric heat storage units is imported into the heating system of the original power plants to heating the customers. The diagram of the electric heat storage is shown as Fig. 1.

Taking electric heat storage device with a total capacity of 260 MW in a thermal power plant as an example, it can be seen from the installation site of the electric heat storage device in Fig. 2 that the electric heat storage system is powered by the 66 kV side of the 220/66 kV step-down transformer in the plant. There are 4 electric heat storage units in the 66 kV side, whose capacities are 80 MW, 60 MW, 60 MW and 60 MW. The heat storage devices mainly power in the heating period. When need the thermal power units to make peak regulation, the heat storage parts would cater to the users' need to switch on or switch off the heat storage devices according to the dispatching instructions. Electric heat storage runs when light load in the winter night of the heating period, starting-up 4 heating units which use electric to store heat for 4 h which just can satisfy the daytime and heating demand.

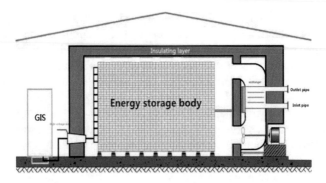

Fig. 1. Schematic diagram of electric heat storage

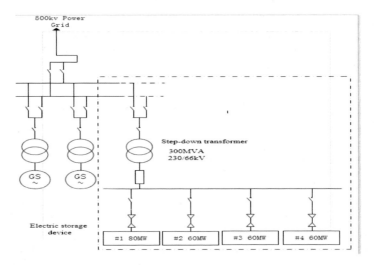

Fig. 2. Schematic diagram of the installation location of high voltage and a large capacity electric heat storage device Schematic diagram of electric heat storage

2.2   High Voltage and Large Capacity Implementation Technology

The high-temperature sintering synthesis of the inorganic salt materials reasonable ratio of sizing processing, made of the high-temperature regenerator, large capacity, and high-density heat storage material characteristics to achieve a breakthrough. Because the heat storage material has the characteristics of small volume, high density, high heat capacity and high-temperature resistance (above 1500°), the heat storage material is suitable for the occasions with strong heat storage capacity, heat release stability, and stable performance. According to the material characteristics of Table 1, it is found that the heat storage material used water or oil, the storage temperature is low under normal pressure, the stored heat is less, only 100–200° and the liquid material is not suitable for electric mixing, the safety is low. If the solid heat storage body was adopted, it could introduce high voltage directly into the heating body to realize the superpower heating and the security also could be improved. At the same time using the original water separation technology, there is no direct relationship between the body and the heat storage device of high temperature hot water output, the power supply heating circuit and regenerator is not integrated, but separated from each other, this separation will fully guarantee the safe operation of equipment in a variety of occasions, the solution to the problem of high voltage insulation.

Table 1. Material property comparison

Sequence number	Comparison programs	Resistance-type solid heat storage	Electrode type water heat storage
1	Heating method	Resistance	Electrode-water
2	Working voltage	10–66 kV	10–35 kV
3	Heat storage medium	Solid inorganic salts	Water
4	Heat storage mode	Heating and heat storage integration	Water pot/water tank
5	Working pressure	Non-pressure	Pressure-bearing
6	Heat storage temperature	500°	145°
7	The heat storage capacity of the 60 °C water outlet	300 kw·h/m^3	100 kw·h/m^3
8	The heat storage capacity of 100 °C water outlet	260 kw·h/m^3	53 kw·h/m^3
9	Electrothermal conversion efficiency	>99%	>99%
10	System comprehensive thermal efficiency	>95%	>92%
11	Safety assessment	Not need	Need to evaluate every year
12	Environmental assessment	Not need	The discharge of electrolyte needs to be evaluated
13	Operating consumables	No	Electrolyte material
14	Heating applications of various types of users	Extensive promotion	Not widely

2.3   Local Control System

Electricity for the high-voltage heating elements. High-voltage heating elements can be converted into heat at the same time being absorbed by the high-temperature regenerator. When the temperature of high-temperature re-generator reaches the set upper limit temperature or power grid trough is finished, the automatic control system will cut off the high-voltage switch and high voltage will stop supplying power. A heat output controller is used between the high-temperature regenerator and the high-temperature heat exchanger. The high-temperature heat exchanger converts the high-temperature heat energy stored in the high-temperature regenerator into hot water or steam output (Fig. 3).

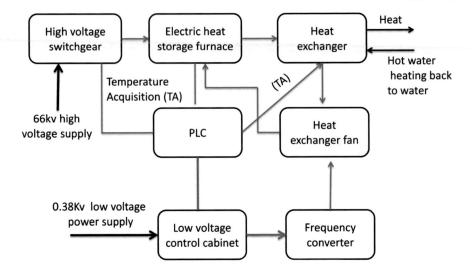

Fig. 3. Schematic diagram of high voltage heat storage control

3 Scheduling Model Contained Electric Heat Storage

3.1 Calculation Model of the Electric Heat Storage

Compared with the original power system, the active power dispatching model needs further improvement after the heat storage system is combined with the original heat storage system. In this paper, the model of heat storage device for active power dispatching in a power system is constructed by means of abstract physics model.

In Fig. 4, $P_{eh,i,t}$ is the power consumption of the I electric heat storage device in the thermoelectric model of the high voltage electric heat storage device, and the β_{eh} is the heat production efficiency of the electric heat storage device. When the electric heat storage is added, without considering the constraint conditions, the $P_{eh,i,t}$ electric power will flow to the electric heat storage device, and the corresponding $P_{eh,i,t} \beta_{eh}$ thermal power will be generated accordingly. In order to keep the heating load unchanged, the combined heat and power unit reduces the heat power of the heating load by $P_{eh,i,t} \beta_{eh}$. According to the characteristics of heat setting, the power output of the unit can be reduced by ΔP_{chp} because of the heat supply of the electric heat storage device. The power generation of the thermal power unit is reduced by $P_{eh,i,t} + \Delta P_{chp}$. In order to maintain the constant power load, wind power consumptive needs.

The heat exchange in ideal [6] is considered only in the literature. The relationship between the actual power and thermal system is complex, and the heat storage, return and storage heat release are constrained by the actual conditions. In addition, the electric heat storage, thermal storage device model active power dispatching of the power system is mainly affected by the three constraint conditions, respectively, heat storage capacity constraint power constraints and state constraints. The capacity constraint mainly sets the maximum heat storage capacity of the heat storage device modelHeat storage power constraint is mainly thermal storage device power and heat

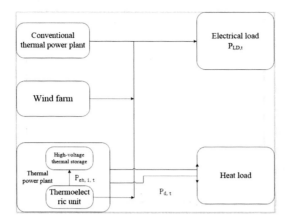

Fig. 4. Thermoelectric model of high voltage heat storage devices

storage power by exothermic heat transfer rate limit; state constraint mainly refers to the heat storage power and heat power electric heat storage device of heat storage change, at the same time to consider the power of abandoned wind frequency and heat storage insulation equipment efficiency, heat storage capacity storage device the heat should be able to guarantee the continuous heating of 1–2 days.

3.2 Optimal Scheduling Model

Enhance the absorptive capacity of wind power accommodation to the greatest extent, then construct the optimal scheduling model. According to the prediction of wind power, the optimal dispatching model is established by maximizing the objective of maximizing the profit of abandoned wind.

$$\min\left(\sum_{t=1}^{N}\sum_{i=1}^{N_W} P_{W,i,t}^{curt} + \varepsilon \sum_{t=1}^{N}\sum_{i=1}^{N_W} f_i\left(P_{G,i,t}\right)\right) \tag{1}$$

$P_{W,i,t}^{curt}$ presents for the abandoned air at time t of wind farm i;
$P_{G,\,i,\,t}$ presents for the output of the units
The coal consumption of thermal power units can be expressed as

$$f_i\left(P_{G,i,t}\right) = a_i\left(P_{G,i,t}^2\right) + b_i\left(P_{G,i,t}\right) + c_i \tag{2}$$

(a_i, b_i and c_i are determined by coal consumption coefficient of units and electrothermal properties).

It can be obtained from the full system active balance of Fig. 4 the formula:

$$\sum_{i=1}^{N_G} P_{G,i,t} + \sum_{i=1}^{N_W} \left(P_{W,i,t} - P_{W,i,t}^{curt} \right) + \sum_{i=1}^{N_C} P_{chp,i,t} = P_{LD,t} + \sum_{i=1}^{N_E} \left(P_{eh,i,t} + P_{d,t} \right)$$

$$(3)$$

(PW, i, t presents for the wind power forecasting of wind power farm at time i, Pd, t presents for the electric heating load of the thermoelectric unit).

In addition, the scheduling model is also constrained by heating constraints, unit constraints (such as operation interval and unit climbing speed, technical constraints), thermal storage constraints, network security and other constraints [7]. In order to 2 electric thermal storage device as an example, under normal operating conditions for the electric thermal storage heating heating equipment in 24 h not longer than 7 h, the regenerative device of temperature control within 550 °C, and will enter the electric energy output in the form of hot water in 24 h. On the premise of ensuring the reliable heating and the operation safety of the power grid, the optimal solution of the dispatching model is obtained according to the constraints.

When the power grid needs peak shaving, the provincial switch can be switched on through the automatic control system (AGC), and the high voltage switch can be put on the remote heat exchanger (which can be put into an arbitrary electric accumulator according to the actual situation, or all the input). At the same time, the electric heat storage device converts the electric energy into heat energy and is absorbed by the high-temperature heat storage body at the same time. When the temperature of the heat storage body reaches the preset upper limit temperature or the end of the peak regulation of the electric network, the high-voltage switch is cut off, and the electric heat storage device is out of operation. Although the optimal scheduling model can ensure the minimum wind curtailment in technology, due to the large wind power volatility and randomness, so when calculating the maximum limit of consumptive abandoned wind by optimal scheduling model, considering the frequency modulation effect of electric heat storage, we seek for the best abandoned wind time and wind power accommodation limit intensively by the AGC command in the actual peak dispatching operation process.

4   Scenario Verification

4.1   Simulation Example

In the simulation of this article analysis. As a case, the actual data is generated by a regional power grid, the basic data is shown in Fig. 5. According to the optimal scheduling model, it is verified that the high-voltage super-large-capacity electric heat storage can enhance the practical effect of absorbing the power of wind power.

For the moment, the influence of network constraints on the operation of the system is neglected. The changes in the active power balance are mainly evaluated, and a model of active optimization scheduling is constructed. By comparing the wind power output curve after increasing the electricity storage heat in Fig. 6, it is found that increasing the electricity storage heat energy greatly improves the wind power

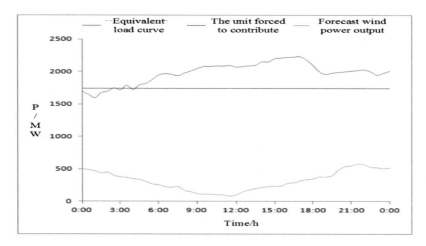

Fig. 5. Data curve of a regional power grid

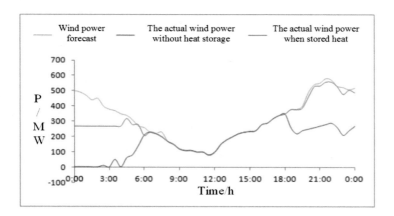

Fig. 6. Comparison of wind power output curve after increasing electric heat storage

receiving capacity. In accordance with the actual situation of the amount of abandonment of wind in the system, scientific and reasonable regulation of the operating conditions of the heat storage device to ensure that the cogeneration output and power supply load is scientific and reasonable, to maximize the acceptance of wind power.

Under the premise of following AGC conventional power grid security and stability strategy, adding real-time data constraints, to achieve more concentrated AGC unit output instructions. As shown in Fig. 7, it can be seen that the density of AGC adjustment is larger and the adjusted power generation curve is closer to the actual load curve of the power grid, thereby reducing the power grid stability problem of wind power consumption and improving wind power consumption abilities.

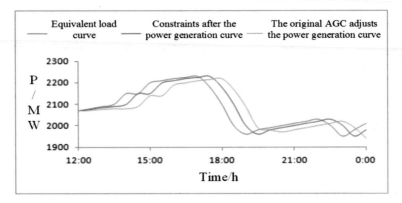

Fig. 7. Contrast diagram of AGC executive function

4.2 Example Analysis

Due to the characteristics such as flexible control, small size and constant output temperature, phase change thermal storage materials are used to configure thermal storage units with different temperatures according to the demand for heating or working conditions. At the same time, multiple energy storage units can be flexibly connected in series and parallel. This cascade can store the heat more than 100 MW to meet the needs of different heating capacity.

According to Figs. 8 and 9, we can find that at 01:41, the #1 electric storage device is on, 05:17 when the power grid peak end, exit the electric storage heaters. Power plant access to reduce electricity, increase wind power to absorb space.

During the actual operation, the temperature of the electric storage device needs to be monitored. Figure 10 shows the average temperature of the oven when the #1 electric storage device is put on the same day.

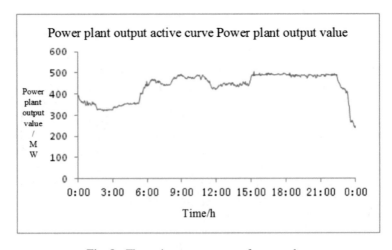

Fig. 8. The active power curve of power plant

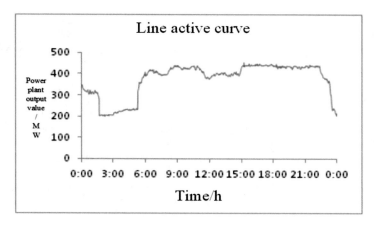

Fig. 9. The active power curve of the power plant line

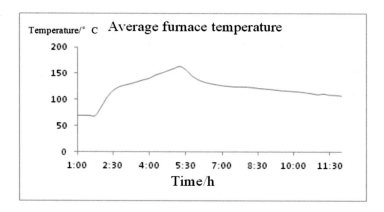

Fig. 10. Temperature curve of the electric heat storage device

5 Conclusion

In this paper, high-voltage electricity will be introduced into high-density and high-heat-capacity solid-state heat storage devices. The water and electricity separation technology will be used to solve the insulation problem and achieve ultra-high-power heating. Through the application of high-voltage large-capacity electric heat storage technology on the power supply side to break the "heat-power" coupling relationship, high-voltage high-capacity electric heat storage is configured on the power supply side with high flexibility and ease of dispatch. The use of peak-valley electricity through the solid electrical energy storage device to convert electrical energy into heat for urban heating, coal-fired thermal power units can be achieved without reducing the output of the case to achieve the depth of the power grid peak shaving, wind power to enhance capacity. High-voltage high-capacity electric heat storage device with flexible control of management, small size, and output temperature constant phase change thermal storage materials. According to the demand for heat or use conditions to configure the temperature of the heat storage unit. The multiple energy storage units can be flexible in series, parallel, with high flexibility and ease of scheduling features at the same time.

References

1. Wu, X., Wang, X.L., Li, J.: Joint dispatching model and solution in wind energy hybrid system. Chin. Soc. Electr. Eng. **33**(13), 10–17 (2013)
2. Yan, G.G., Liu, J., Cui, Y.: Economical evaluation of using wind energy to increase the scale of wind power dispatching network. Chin. Soc. Electr. Eng. **62**(2), 45–52 (2014)
3. Zheng, L., Hu, W., Lu, Q.Y.: Energy storage system to improve wind power access planning and operation of a comprehensive optimization model. Chin. Soc. Electr. Eng. **34**(16), 2533–2543 (2014)
4. Yuan, X.M., Cheng, S.J., Wen, J.Y.: Analysis on the application prospects of energy storage technology in solving the problems of large-scale wind power grid integration. Chin. Soc. Electr. Eng. **37**(1), 14–18 (2013)
5. Chen, L., Xu, F., Wang, X.: Implementation mode and effect analysis of thermal energy storage to improve wind power consumption. Chin. Soc. Electr. Eng. **37**(1), 46–54 (2013)
6. Lu, Q., Jiang, H., Chen, T.Y.: Wind power project consuming wind power based on electric boiler and its national economic evaluation. Autom. Electr. Power Syst. **38**(1), 6–12 (2014)
7. Long, H.Y., Ma, J.W., Zhang, J.B.: Research on grid keneng scheduling based on heating demand side management. Demand Side Manag. **13**(1), 24–27 (2011)
8. Gao, Q.S., Yang, J.: Review of fault diagnosis of power cables. Guizhou Electr. Power Technol. **5**(1), 54–58 (2016). (in Chinese)

Calculation of Energy Saving Based on Building Engineering

Bin Shao and Xingyu Liu[✉]

State Grid Liaoning Electric Power Supply Co. Ltd., Shenyang, China
xingyu.liuly@163.com

Abstract. The development of energy-saving buildings has updated building construction. The energy consumption of buildings is the focus of our research. By expounding the calculation process of heat transfer coefficient of the exterior wall, roof, floor, door, and window, energy saving calculation is discussed. Due to the diversity of building energy-saving materials, the excessive use of materials leads to huge waste of resources in order to meet the energy-saving standards. Therefore, how to make full use of the existing materials to meet the requirements of energy conservation standards. Building energy conservation is the implementation and establishment of the scientific concept of development. Actively promote construction. Energy conservation is not only conducive to alleviating and alleviating energy shortages. Air pollution, greenhouse gas emission reduction, protected areas. The ball's environment, it helps improve people's lives and work. We will create conditions to promote sustainable development of the national economy. To achieve the goal of energy conservation, on the one hand, we need to improve energy efficiency. People's awareness of energy conservation depends on the restrictions of relevant policies.

Keywords: Building energy conservation · Heat transfer coefficient · Correction factor · Exterior wall

1 Introduction

Building energy conservation is an important work to implement the scientific concept of development and build a conservation-oriented society [1, 2, 3]. Actively promoting building energy conservation can not only alleviate energy shortage, reduce air pollution, reduce greenhouse gas emissions and protect the earth's environment, but also improve people's living and working conditions and promote the sustainable development of the national economy. In order to achieve the purpose of building energy conservation, on the one hand, people who need to improve energy conservation awareness depend on the restrictions of relevant policies; On the other hand, we need to study and apply various building energy-saving technologies in the whole society. In recent years, building energy conservation has been widely applied in engineering [4]. This paper summarizes the calculation of building energy consumption through the heat calculation major of an office building and hopes that we can jointly study and discuss energy saving calculation [5].

© Springer Nature Switzerland AG 2020
A. E. Hassanien et al. (Eds.): AMLTA 2019, AISC 921, pp. 837–844, 2020.
https://doi.org/10.1007/978-3-030-14118-9_82

The experimental project is located in the cold area of Dalian weather belt [6]. The new urban road south, east, west, north to the construction land. South to the building red line 45.00 m, without shelter. Southeast, 9°. It is 57.63 m long and 17.13 m wide [7]. Plan comparison rules for the building. The structure is a frame structure, divided into four (five parts) [8]. The building area is 3660.12 m, the building volume is 13555.75 m, the building coefficient is 0.25, the south window wall area is 0.31, the east window wall area is 0.08, the west window wall area is 0.07, and the north window wall area is 0.25. The project completed the construction drawing design in January 2016.

2   Calculation of Heat Transfer Coefficient of the Enclosure Structure

The enclosure structure is the main heat dissipation part of the building. Improving the thermal insulation performance of buildings can fundamentally improve the thermal stability of buildings, reduce the energy consumption of heating and air conditioning, and realize building energy conservation.

2.1   Exterior Wall

China's traditional building structure is heavy, light protection, wall insulation performance is poor. The outer wall is the main heat dissipation part of the outer shell structure. In winter, heat is not easy to lose, while in summer, it can reduce solar radiation and indoor temperature to maintain stability, thus achieving the goal of energy conservation (Refer to Fig. 1).

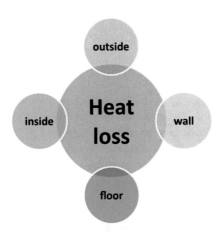

Fig. 1. Analysis of the proportion of heat loss in construction projects

The external wall of the project is made of new material, high heat insulation light aggregate concrete composite block (short composite block). The composite masonry is made of cement, and the slag and clay are coarse aggregates. The composite block is light in weight and high in strength. Good insulation performance, durability, simple construction, significant comprehensive benefits. The exterior wall is composed of 290 thick composite blocks and 90 thick composite blocks covering the frame column.

- **Wall heat transfer coefficient**

  Calculation formula

$$k = 1/R_0 = 1/R_1 + R + R_M \tag{1}$$

$$R = \sum R_i \tag{2}$$

$$R_i = \delta/\lambda_{ci} \tag{3}$$

$$\lambda_{ci} = \lambda_i \times \alpha \tag{4}$$

  Where

- R_0—Heat transfer resistance, the ability to characterize the impedance heat transfer of the maintenance structure, including the air boundary layer on both sides of the surface [W/(m^2.k)]
- R_i—Inner surface thermal resistance [(m^2.k/W)], Generally take $R_i = 0.11$[(m^2.k/W)]
- R_e—External surface thermal resistance [(m^2.k/W)], Generally take $R_e = 0.04$ [(m^2.k/W)]
- R—The thermal resistance of the wall structure is equal to the thermal resistance of the material layers that make up the wall [(m^2.k/W)].
- δ—The calculation of thermal conductivity of each material layer [W/(m^2.k)]
- λ_{ci}—The thermal conductivity of the material of each material is generally measured in the dry state of the laboratory [W/(m^2.k)]

- **External wall average heat transfer coefficient**

  The exterior surrounding bridge effect conditions, the calculation formula of the average heat transfer coefficient:

$$K_m = (K_p \cdot F_p + K_{B1} \cdot F_{B1} + K_{B2} \cdot F_{B2} + K_{B3} \cdot F_{B3})/(K_p + K_{B1} + K_{B2} + K_{B3}) \tag{5}$$

  Where

- K_m—The average heat transfer coefficient of the exterior wall [W/(m^2.k)];
- K_p—The average heat transfer coefficient of the exterior wall [W/(m^2.k)];
- K_{B1}, K_{B2}, K_{B3}—The heat transfer coefficient of the wall around the bridge site [W/(m^2.k)];
- F_p—The area of the main body part of the outer wall (m)
- F_{B1}, F_{B2}, F_{B3}—The exterior parts of the area around the bridge (m^2)

Calculation of the average heat transfer coefficient of the exterior wall: 7.2 m column, the 3.6 m layer used in this project.

$$F_0 = (7.2 - 0.45) \times (3.6 - 0.57 - 0.08) - 2.1 \times 2.05 \times 2 = 11.303\,(\mathrm{m}^2)$$

$$F_{B1} = 0.45 * 3.6 = 1.62\,(\mathrm{m}^2)$$

$$F_{B2} = 0.57 * (7.2 - 0.45) = 3.848\,(\mathrm{m}^2)$$

$$F_{B3} = 0.08 * (7.2 - 0.45) = 0.540\,(\mathrm{m}^2)$$

$$K_m = \left(K_p \times F_p + K_{B1} \times F_{B1} + K_{B2} \times F_{B2} + F_{B3} \times F_{B3}\right)/(F_p + F_{B1} + F_{B2} + F_{B3}$$
$$= 0.67 \left[\mathrm{W}\,/\,(\mathrm{m}^2.k)\right]$$

External wall thermal bridge heat transfer coefficient:

$$K_b = (K_{B1} \times F_{B1} + K_{B2} \times F_{B2} + F_{B3} \times F_{B3})/(F_{B1} + F_{B2} + F_{B3}) = 1.03 \cdot \left[\mathrm{W}/(\mathrm{m}^2.k)\right]$$

2.2 Roofing

The heat loss of the roof is not large in the total heat loss of the envelope, but the effect of roof insulation on the top layer is very significant. Because the roof is affected by solar radiation all the time, the indoor thermal environment is very poor in the summer, and the energy consumption of the air conditioning is high. The roof of this project is not a flat roof, and 80 thick polystyrene foam plastic plate is used. The effect picture is shown in Fig. 2. The method of calculating the heat transfer coefficient of the roof is the same as that of the outer wall.

Fig. 2. Roof rendering

2.3 Ground

The thermal design of the building ground should take different surface materials into consideration in terms of people's health, comfort, and heating. Considering the improvement of the heat preservation and moisture resistance of the ground floor, and use a plate and block insulation materials as a cushion to make the ground thermal resistance close to the ground thermal resistance of the building.

Calculation formula:

$$k = 1/R_0 = 1/R_1 + R + R_M \tag{6}$$

$$R = \sum R_i \tag{7}$$

$$R_i = \sum R_i \tag{8}$$

$$\lambda_{ci} = \lambda_i \times \alpha \tag{9}$$

Symbolic meaning in the same way

20 thick polystyrene foam plastic plates are used in the ground for insulation.

2.4 Doors and Windows

Exterior door and window is the weakest part of building heat loss. Besides reducing measures of window and wall area ratio, effective measures should also be taken to improve insulation performance of external doors and windows, improve airtightness of external windows and doors, and reduce infiltration of cold air.

Calculation formula:

$$K = \left(\sum A_Q K_G + \sum A_f K_f + \sum I_\psi \psi\right)/A_t \tag{10}$$

Where

- K—The heat transfer coefficient of the window [W/(m^2.k)]
- A_Q—Window glass area (m^2)
- A_f—The projection area of the window frame (m^2)
- I_ψ—Circumference of glass region (m)
- K_G—Heat transfer coefficient area in the central area of window glass [W/(m^2.k)]
- K_f—The heat transfer coefficient of the additional line between the window frame and the window glass [W/(m^2.k)]
- A_t—Total projection area of the whole window (m^2)

2.5 Problems to be Paid Attention to in Calculation of Heat Transfer Coefficient of the Enclosure Structure

The heat transfer coefficient of the exterior wall should be the average heat transfer coefficient. The thermal conductivity of the insulation material should be calculated by the coefficient of thermal conductivity, that is to multiply by the coefficient of thermal conductivity measured in the laboratory dry state by a correction factor greater than 1 Table 1 shows the ratio of construction systems A and B in the outer wall area, Table 2 shows the calculation of the heat transfer coefficient of roofing including Ceiling plastering, Concrete slab, Polystyrene board, Dry paving slag, and Leveling layer. While Table 3 shows the calculation of ground heat transfer coefficient for Rammed soil, Concrete cushion, Polystyrene board, Stone concrete, Binding layer, and Surface.

Table 1. The ratio of A and B in the outer wall area

Construction system	A	B
Brick and concrete structure	0.75	0.25
Frame structure	0.65	0.35
Frame shear structure	0.55	0.45
Shear wall structure	0.35	0.65

Table 2. Calculation of heat transfer coefficient of roof

Name	Practice	Thermal conductivity $[W/(m^2.k)]$	Correction factor	Thermal resistance 1 $[(m^2.k/W)]$	Thermal resistance 2 $[(m^2.k/W)]$	Heat transfer coefficient $[W/(m^2.k)]$
Roofing	Ceiling plastering	0.87	1.0	0.023	1.766	K = 0.52
	Concrete slab	1.74	1.0	0.057		
	Polystyrene board	0.041	1.2	1.626		
	Dry paving slag	0.420	1.5	0.048		
	Leveling layer	0.93	1.0	0.022		

Table 3. Calculation of ground heat transfer coefficient

Practice	Thickness (m)	Thermal conductivity $[W/(m^2.k)]$	Correction factor	Thermal resistance 1 $[(m^2.k/W)]$	Thermal resistance 2 $[(m^2.k/W)]$	Heat transfer coefficient $[W/(m^2.k)]$
Rammed soil	1.000	0.93	1.0	1.075	1.591	K = 0.59
Concrete cushion	0.060	1.28	1.0	0.047		
Polystyrene board	0.020	0.41	1.2	0.407		
Stone concrete	0.040	1.74	1.0	0.023		
Binding layer	0.030	0.93	1.0	0.032		
Surface	0.020	2.91	1.0	0.007		

3 Area Calculation

Area calculation is a simple but tedious work. Whether the area calculation is accurate or not has a great impact on the thermal results. This requires us to calculate carefully one by one. Therefore, the calculation of area is simple, but we must carefully calculate, calculate patiently, avoid calculation, leakage, miscalculation, etc.

4 Thermal Calculations and Epilogue

In the heat transfer coefficient and area calculation are finished, we should fill in the building thermal calculation sheet, statement, such calculations. Energy calculations cover, energy efficiency design, heat transfer coefficient calculation, thermal calculation, public building energy-saving design statement etc.

In order to popularize energy saving technology and design the thermal performance of building maintenance structure rationally, according to the national standard, standard and local standard, combined with my own practical experience, the author describes the thermal engineering calculation of the building specialty, without involving other majors. We should establish the concept of the life cycle of buildings, take all aspects of architectural planning, design, construction, maintenance, operation, and demolition into consideration, so as to maximize the benefits of energy saving technology and truly achieve the all directional energy-saving of buildings.

5 Conclusion

In order to promote energy-saving technology and rationally design the thermal performance of the building maintenance structure, the author elaborates on the thermal engineering calculation of the building according to national norms, standards, local standards and his own practical experience, and does not involve other majors. Building energy conservation is a systematic project. The application effect of single energy-saving technology is very limited. Only the overall system coordination and comprehensive consideration of various factors can achieve energy-saving goals. The concept of the whole life cycle of the building should be established, and all aspects of the planning, design, construction, maintenance operation and demolition of the building should be comprehensively considered, so that the energy-saving technology can maximize its benefits and truly realize the all-round energy conservation of the building.

References

1. Gabbar, H.A., Musharavati, F., Pokharel, S.: System approach for building energy conservation. Energy Procedia **62**, 666–675 (2014)
2. Song, D.: Architectural Design and Technology, vol. 4, no. 11, pp. 12–20. Tongji University Press, Shanghai (2010)

3. Xiao, S., Chen, L.: Energy-saving economy analysis of building of ultra-high-rise residential project – a case study of a super-high-rise residential building project in Hefei. Green Technol. **2**(12), 24–33 (2014)
4. Sun, M.: Comparison of AHP and ANP. Chin. Foreign Entrep. **3**(10), 103–110 (2014)
5. Xiao, J., Zhao, P., Jing, L.: An efficiency evaluation study of building energy-saving technology. Build. Technol. **1**(12), 55–70 (2012)
6. Yao, R., Li, B.: Energy policy and standard for the built environment in China. Renewable Energy **13**(30), 1973–1988 (2015)
7. Cabeza, L.: Experimental Study on the performance of insulation materials in mediterranean construction. Energy Build. **5**(42), 630–636 (2010)
8. Yael, P., Guide, C.: Climatic considerations in school building design in the hot-humid climate for reducing energy consumption. Appl. Ener. **2**(16), 421–427 (2009)

Heat Discharge of Molten Salt in Double Energy Storage Tank

Kaiyu Pang[✉], Yonghui Chen, Min Zhang, Jianxun Dong,
Xingye Zhou, Zhi Wang, and Bo Guo

Technical Services Department, Liaoning CPI Power Station Combustion
Engineering Research Center Co. Ltd., Shenyang, China
877203948@qq.com

Abstract. The heat storage performance of double - tank molten salt tank is better than a single - tank system. In order to obtain the heat storage mechanism of double pot molten salt, the spiral coil heat exchanger was placed in the annular gap composed of the cylindrical baffle and open end tank wall. The energy release characteristics of the double tank are analyzed in detail. The results show that the cylindrical baffle can adjust the molten salt flow field in the tank, the outlet temperature of heat exchanger, the heat transfer capacity and the transient heat transfer effect. The results of this study can be used as the basis of double tank design.

Keywords: Double tank heat storage system · Spiral coil heat exchanger · Heat discharge

1 Introduction

Energy shortage and environmental pollution have become the bottleneck of China's economic development. In particular, the severe haze in many regions has promoted the development of renewable energy in China. However, the actual volume of molten salt storage is never more than half of the volume of the heat storage tank, so the cost of the heat storage tank is too high. In order to reduce the cost of heat storage, people use double tank heat storage method [6]. Compared with the single tank heat storage system, the double tank heat storage system has better performance, which makes the heat storage system simple and reduces the cost of heat storage. In the process of heat storage and heat release of molten salt, the heat transfer process of the heat transfer fluid in molten salt in the tank is realized through a natural convection heat transfer process. The experimental and simulation results show that the cold plume generated by the spiral coil surface during the process of heat release will increase the natural convection of molten salt in the tank, leading to the increase of molten salt. In order to reduce the accumulation of heat storage medium, we use the partition water tank to reduce the heat released by the low-temperature molten salt of natural convection and the surrounding high-temperature molten salt wall for heat exchange and adjust the molten salt tank of natural convection to improve heat dissipation efficiency and heat transfer efficiency. Finally, the heat dissipation law, heat dissipation efficiency, and heat

© Springer Nature Switzerland AG 2020
A. E. Hassanien et al. (Eds.): AMLTA 2019, AISC 921, pp. 845–853, 2020.
https://doi.org/10.1007/978-3-030-14118-9_83

transfer and heat dissipation characteristics of the heat exchanger are analyzed for the proposed molten salt heat storage system of cylindrical baffles [2].

2 Experimental System and Method

2.1 Experiment System

Figure 1 shows the flow of molten salt in the tank during heat exchange between molten salt and the release medium in the tank under the action of a cylindrical partition. The arrow represents the direction of fluid flow, the solid line represents the hot fluid and the dashed line represents the cooled fluid.

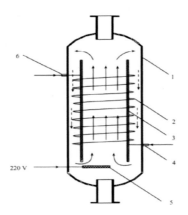

1. Molten salt tank	2. Helical coil heat exchanger
3. Cylindrical baffles	4. Heat exchanger population
5. Heating rod	6. Heat exchanger outlet

Fig. 1. Discharge principle of molten salt in a single tank with a cylindrical

A cylindrical partition with low thermal conductivity is arranged in a molten salt storage tank located at the bottom of the heat storage tank. An exothermic coil heat exchanger is arranged between the cylindrical partition and the wall of molten salt storage tank in a circular channel. In the heat storage process, the heating rod at the bottom heats the molten salt in the jar to the desired temperature. Cryogenic molten salt is confined to a circular channel. High temperature molten salt continues to flow from the upper to the annular channel, which can exothermic efficiency [4].

Figure 2 shows the profile of molten salt and the temperature measurement point. The diameter of molten salt tank is 600 mm and the height is 700 mm. The spiral center of the coil heat exchanger is 450 mm away from the spiral pitch of the coil heat exchanger, 25 mm in diameter in a coil, the bottom of the heat exchanger inlet groove

125 mm in diameter, and the bottom of an outlet 475 mm liner are vertically attached to the annular channel, and the salt bath wall is equidistant from the partition.

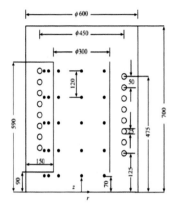

Fig. 2. Tank molten salt profile and temperature measurement point schematic diagram

In this paper, the design of molten salt trough is positive along the axis direction. Five layers of thermocouples are arranged at different heights in the tank. The temperature measurement points of the thermocouple at the bottom are 70 mm from the tank bottom and 120 mm vertically upward. The temperature measurement points are 70 mm, 190 mm, 310 mm, 430 mm and 550 mm respectively. Five temperature measuring points are arranged at the same height, the first and second columns are in the annular channel, and the third and fifth columns are outside the annular channel.

2.2 Data Analysis

(1) Heat taken away by the air during heat exchange Q (W):

$$Q = C_p q_m (T_{out} - T_{in}) \tag{1}$$

Where W is the specific heat capacity of the air at constant temperature (Tin + Tout)/2, qm is the mass flow of air into the spiral coil in kg/s, Tout is the Outlet temperature, Tin is the spiral coil heat exchanger inlet temperature.

(2) Sensible heat release rate f:

$$f = \frac{T_{s,0} - T_s}{T_{s,0} - T_r} \tag{2}$$

Sensible heat release rate f is the percentage of heat extracted from molten salt during heat transfer between molten salt and air. In Eq. (2), Ty is the initial temperature of molten salt tank, Ts is the average temperature of tank for a while, and the solidification temperature of molten salt of Tr.

(3) Instantaneous efficiency of the spiral coil heat exchanger:

$$\varepsilon = \frac{T_{out} - T_{in}}{T_s - T_{in}} \tag{3}$$

Instantaneous efficiency refers to the ratio between the actual heat transfer effect and the maximum heat transfer in the heat exchanger.

3 Results and Discussion

3.1 Regulation Effect of Cylindrical Partition on Molten Salt Flow Field

Figure 3 shows the temperature distribution of the molten salt outside the annular channel before and after the addition of the cylindrical partition. The temperature of the molten salt in each layer outside the annular channel takes the average temperature of the third to fifth columns.

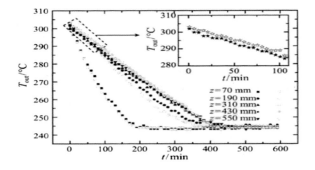

Fig. 3. Comparison of molten salt temperature outside annular channel with and without a baffle

Figure 4 shows the temperature distribution of the molten salt in the annular channel before and after the addition of the separator. The temperature of the molten salt in the annular channel takes the average values of the first and second columns. Consistent with Fig. 3, the molten salt in the process of heat release the same trend of the overall cooling, cooling rate in turn slow down from bottom to top. At 420 min, the molten salt enters into the phase transition and exothermic stage by sensible heat release.

The temperature of the molten salt at z = 70 mm and z = 190 mm is higher than the temperature of the molten salt at the same height when the baffle is not provided when the baffle is not used. The temperature of the molten salt at z = 310 mm, 430 mm and 550 mm is lower than that of the baffle When the molten salt temperature at the

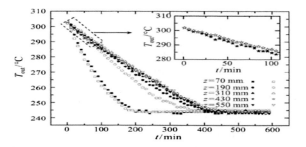

Fig. 4. Comparison of molten salt temperature inside annular channel with and without a baffle

same height. This result is consistent with the change of molten salt temperature outside the annular channel, which further shows that the cylindrical partition plays the role of regulating the molten salt flow field in the tank.

3.2 Comparison of Heat Transfer between Molten Salt and Air Before and After Adding Cylindrical Separator

Figure 5 shows the spiral coil air flow Qv = 28.5, 57, 85.5 m³/h under three operating conditions, plus the cylindrical partition before and after the heat exchange between molten salt and air.

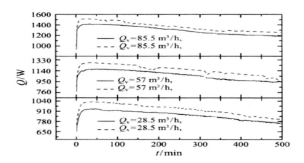

Fig. 5. Comparison of transient heat transfer Q, with and without a baffle

The initial heat release heat smaller, the reason is the beginning of heat exchange within 15 min, the heat exchanger around the molten salt temperature is high, almost molten salt tank temperature, so the tank temperature difference between the molten salt and the coil wall is very small, Not enough to produce the driving force of natural convection can only rely on the heat conduction coil wall heat away. With the extension of the heat release process, the temperature of the heat storage medium is gradually decreased, and the exit temperature of the heat extraction medium is gradually decreased, so the heat exchange capacity is also gradually reduced.

3.3   Comparison of Outlet Temperature of Spiral Coil Before and After Adding Cylindrical Separator

Figure 6 is the three working conditions, with or without cylindrical partition spiral coil outlet temperature comparison. At Q = 28.5 m³/h, the outlet temperature of the heat exchanger with cylindrical bulkheads is higher than the outlet temperature of the separator without heat exchangers and the temperature is maintained above 130 °C for longer. With a cylindrical bulkhead, the helical coil exit temperatures were above 344°F (344 °C), whereas the helical coil exit temperatures without the cylindrical partition fell below 130 °C after 291 min. Therefore, when the outlet temperature is higher than 130 °C, the time taken for the cylindrical partition to increase is about 18.2%.

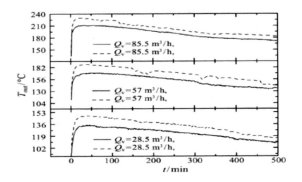

Fig. 6. Comparison of transient outlet air temperature with and without a baffle

Increase the flow of heat release medium, spiral coil outlet temperature increases accordingly. When the air flow Qv = 57 m³/h, the coil heat exchanger outlet temperature reaches about 1700 °C; when the air flow Qv = 85.5 m³/h, the coil heat exchanger outlet temperature further increases to about 200 °C. It can be seen that increasing the flow rate of the heat releasing medium in the coil can increase the outlet temperature of the coil heat exchanger.

3.4   Comparison of Efficiency of Front and Rear Coil Heat Exchanger with Cylindrical Baffle

The advantages of the cylindrical bulkhead are also reflected in the instantaneous efficiency of the heat exchanger, as shown in Fig. 7 for the instantaneous efficiency change of the heat exchanger with Qv = 28.5 m³/h at the helical coil. For both the presence or absence of a cylindrical baffle, the efficiency of the heat exchanger increases in a very short period of time and then fluctuates around one of the efficiencies, and then decreases as the heat release decreases.

Figure 8 shows the instantaneous efficiency of the coil heat exchanger with time under different conditions. It can be seen that with the increase of flow rate, the flow rate of the heat-releasing medium in the heat exchanger increases, the convection heat transfer coefficient increases, and the heat release efficiency of the heat exchanger

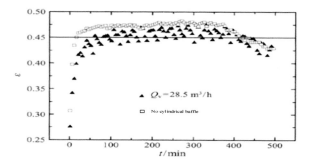

Fig. 7. Comparison of transient heat exchanger effectiveness with and without a baffle

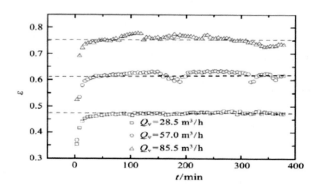

Fig. 8. Comparison of transient heat exchanger effectiveness in different flow rates with a baffle

increases accordingly. Increasing the rate of heat release medium can increase the coil Heat exchanger efficiency.

3.5 Comparison of Sensible Heat Release Rate and Heating Time Before and After Adding Cylindrical Separator

Figure 9 shows the energy release rate of the liquid during the sensible heat phase with time at Qv = 28.5 m³/h. After adding a cylindrical separator f value increased slightly, but the increase is not large. The main reason is that the value of f is related to the initial temperature of the molten salt, the temperature of the molten salt at a certain time t and the solidification temperature of the molten salt, and both the initial temperature of the molten salt and the solidification temperature of the molten salt are fixed values, a single-valued function of salt temperature. The temperature range of molten salt determines the change of energy release rate.

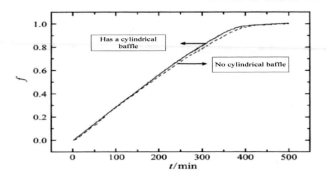

Fig. 9. Comparison of the cumulative fractional energy discharge, f, with and without a baffle

After 200 min of heat exchange, the value of f with a cylindrical baffle reached 56.07%, while the value of f without a cylindrical baffle was 54.11%. Continuation of heat transfer, when it reached 420 min, had a cylindrical baffle and no cylinder. The fins of the bulkhead reached about 100% and remained 100% for the following 80 min of heat exchange.

4 Concluding Remakes

- Cylindrical baffles serve to regulate the molten salt flow field, but the material, size, and location of the cylindrical baffles need to be optimally analyzed.
- With the deepening of the heat transfer process, the role of the cylindrical partition gradually reduced to reach a certain time when the heat has been basically ineffective.
- The cylindrical baffle can prevent the micro-convection mixed flow of heat and cold storage medium in the process of releasing heat, and increase the outlet temperature of the heating medium of the spiral coil, the heat exchange of the heat exchanger, the performance of the coil heat exchanger and the heat extraction efficiency. In addition, but also can effectively shorten the user to take the heat time.
- The heat transfer medium flow of coil heat exchanger is increased and the outlet temperature of heat transfer medium is increased. In addition, the instantaneous efficiency of heat exchanger and heat exchanger is also improved.

References

1. Mills, D.: Advances in solar thermal electricity technology. Sol. Energy **18**(76), 19–31 (2004)
2. Richard, T.: The global CSP market-its industry structure and decision mechanisms. Hanburg **49**(21), 45–61 (2003)
3. Herrmann, U., Kelly, B., Price, H.: Two-tank molten salt storage for parabolic trough solar power plants. Energy **94**(29), 883–887 (2004)

4. Liu, W., Davidson, H., Mantell, S.: Natural convection from a horizontal tube heat exchanger immersed in a tilted enclosure. ASME J. Sol. Energy **125**(1), 67–75 (2003)
5. Liu, W., Davidson, J., Kulacki, F.: Natural convection from a tube bundle in a thin inclined enclosure. ASME J. Sol. Energy **126**(2), 702–709 (2004)
6. Su, Y., Davidson, H.: Transient natural convection heat transfer correlations for tube bundles immersed in a thermal storage. ASME J. Sol. Energy **16**(129), 210–214 (2007)
7. Haltiwanger, J., Davidson, J.: Discharge of a thermal storage tank using an immersed heat exchanger with an annular baffle. Sol. Energy **41**(83), 193–201 (2009)
8. Su, Y.: Numerical Study of Natural Convection of Heat Exchangers Immersed in a Thermal Storage Vessel, vol. 35, no. 27, pp. 261–278. University of Minnesota, Minneapolis (2006)

Development of Insulation Parameter Monitoring System for Transmission Line Arrester

Fang Han[2(✉)], Dantian Zhong[1,3], Qiang Gao[1], Feng Yuan[1,3], and Yunhua Zhang[1,3]

[1] State Grid Liaoning Electric Power Search Institute, Shenyang 110006, China
[2] State Grid Shenyang Electric Power Supply Company, Shenyang 110006, China
1160668156@qq.com
[3] Liaoning Dongke Electric Power Technology Co., Ltd., Shenyang 110079, China

Abstract. Transmission line arresters play an important role in the safe operation of cable protection, and lightning now causes a large proportion of power accidents. To reduce line accidents caused by lightning strikes, at present, a large number of lightning arresters have been installed on the line, and maintenance person needs to carry out necessary regular maintenance of the line arresters. Because of the operation line of 66 kV and above voltage level, the tower is very high, the staff installs the line arrester, and it is very difficult for the maintenance staff to test the power failure of the line. This paper studies insulation parameter monitoring technology of transmission line arrester. This paper introduces the development of an effective measurement system for insulation parameters of lightning arrester, solar power supply system and remote communication module based on GPRS + Zigbee radio frequency technology. The measurement accuracy and stability of the system are verified by experiments.

Keywords: Arrester · Monitoring technology · Radiofrequency technology

1 Introduction

With the development of Chinese electric power industry, the length of the transmission line is increasing, line arrester is mainly to protect the safety of transmission lines, plays a key role to avoid tripping accident, when the line is overvoltage, arrester will act as to line protection and high power equipment safety. However, evaluation of operation state of lightning arrester is lack of effective test data, because of the difficulty of the test of 66 kV and above power transmission line is large, the existing technology is difficult to the live detection of transmission line lightning arrester, this means that important line arrester need regular monitoring and inspection, pressure test line lightning arrester need to introduce to installing 20 m above the height, close observation is difficult and dangerous, routine test requirements on the weather conditions are very high [1], so a research on the measurement of the running state of transmission line arresters is introduced in this paper, a set of insulation parameter measurement

A. E. Hassanien et al. (Eds.): AMLTA 2019, AISC 921, pp. 854–860, 2020.
https://doi.org/10.1007/978-3-030-14118-9_84

system of transmission line lightning arrester is developed, based on wireless remote communication technology, operating conditions to the maintenance of power operation arrester is facilitated, to avoid the power cut test and routine inspection by monitoring, the operating parameters of the lightning arrester are directly got from the background or handheld terminal [2].

2 Overall Scheme Design System

The full current and action times of the lightning arrester can be indicated by the mechanical counter at the bottom of the transmission line lightning arrester can indicate. However, due to the counter does not have the communication function and installation distance is high, it is difficult to observe the accurate measurement data. In this paper, a lightning arrester monitoring system based on electronic technology is designed, which is composed of a solar power supply system and a wireless communication module. the arrester current, resistance current, action number and action time parameter data is measured by Sensor, the sensor can be unified power supply by way of a cascade, and connected through RS485 interface to remote communication module, remote communication part by GPRS + Zigbee 2.4 GHz to remote data can be transmitted to the monitoring of the background, or transferred to hand-held devices of patrol officers, record test data is more convenient. The overall system design architecture is shown in Fig. 1.

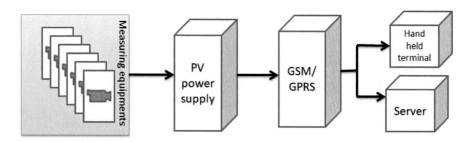

Fig. 1. An overall design system architecture

3 Implementation of the Overall Program

3.1 Acquisition Sensor Design

Hardware design: Study on the measurement of lightning arrester's current signal with small current precision mutual inductor, through the precision current transformer to measure current signal at first, secondly, the measured result is converted into the voltage signal which the processor can accept the range by using signal conditioning and conversion circuit, thirdly, using fast Fourier transform algorithm to calculate in processor, then the full current and resistive current of lightning arrester can be obtained in the end. As the leakage current will increase suddenly when lightning or over-voltage operation occurs to arrester, and the duration of large current is 4–10 uS, so the design adopts Transient Voltage Suppressor to limit voltage, then uses the surge change of the voltage to trigger a digital comparator, thus the measurement about the action number of arrester are finished.

The sensor also has the function of recording the real-time of action, which through the real-time clock chip inside the sensor, and when the MCU receives the action number and trigger command, then read the time immediately, and also the number and time of action are stored at the same time. The interface between the sensor and remote communication module adopts RS485 mode to increase the anti-jamming performance. The principle diagram of the sensor hardware design as shown in Fig. 2.

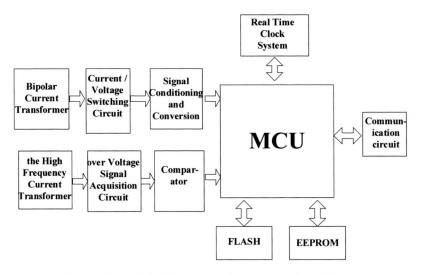

Fig. 2. The principle diagram of the sensor hardware design

Software algorithm of current sampling: Because of the nonlinear characteristics of lightning arrester, and its current is not a sine wave, which contains high harmonics. The third harmonic is very sensitive to temperature changes, and mainly behavior about the changes of resistive current in early aging period is the rising of the third harmonic component of resistive current [3]. By analyzing and comparing the monitoring methods of low voltage grade arrester, the third harmonic current in the leakage current can be measured by the harmonic analysis method, and the operation condition of the MOA is determined by the curve of the third harmonic current after aging and no aging. The software formula as follows:

Considering the condition of high order harmonic and noise in the signal, the signal X (t) can be expressed as:

$$X(t) = \sum_{k=0}^{+\infty} X_k \sin(n\omega t + \varphi_k) + N(t) \tag{1}$$

According to the orthogonality of trigonometric functions:

$$A_k = \frac{2}{T} \int_0^T X(t) \sin \omega t \, dt = X_k \cos \varphi_k$$
$$B_k = \frac{2}{T} \int_0^T X(t) \cos \omega t \, dt = X_k \sin \varphi_k$$

(2)

The above is for a continuous signal, for a discrete signal sampled can be expressed as a discrete form of integration:

$$A_k = \frac{2}{N} \sum_{n=0}^{N-1} X(n) \sin\left(\frac{2\pi}{N} kn\right)$$
$$B_k = \frac{2}{N} \sum_{n=0}^{N-1} X(n) \cos\left(\frac{2\pi}{N} kn\right)$$

(3)

Where N is the number of sampling points in a cycle.

Taking into account the higher harmonic content of the system, there are almost no even harmonics. By the above formula shows, the DFT transform only decompose the odd harmonics, and the part of odd harmonics which higher than 7 times are filtered out by a hardware circuit in the frequency domain. Thus the amplitude of the fundamental wave and the third harmonic wave can be calculated. After the algorithm is processed, the calculation method is in accordance with the IEC60099-5 standard, which ensures the accuracy of the resistive current extraction algorithm [4, 5].

3.2 Solar Power Supply System Design

The solar power supply system of this scheme mainly includes three parts: the solar cell array, the charge, and discharge controller and the storage battery, and the schematic diagram is shown in Fig. 3. Charge and discharge controller by setting the corresponding float voltage range and uniform charging voltage range, and combining the state of the battery capacity and voltage, to carry on float charging or uniform charging to the battery, also to supply the power to load. When the battery voltage is too high, it is required to terminate the battery charging, and when the battery voltage is too low, it is required to terminate the battery discharging to protect the battery. The controller is the key control component of the system. The storage battery is the core of system management, which is used to store the surplus electrical energy converted from the solar-cell array to ensure continuous and reliable power supply.

3.3 Communication Module Design

Communication model uses GPRS and Zigbee wireless communication radio frequency union, and the GPRS wireless remote communication can be transmitted through the mobile network to the remote monitoring system. The 2.4 GHz enhanced

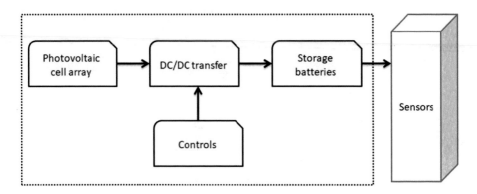

Fig. 3. Block diagram of the solar power system

power type radio frequency module to achieve data transmission between the sensor and hand-held devices, to facilitate local inspection personnel [6–8].

The wireless communication module adopts data transparent transmission mechanism, when the background monitoring system sends out the command of reading data, wireless module will transfer the command to MCU through internal UART serial bus after receiving it, and then MCU switches into RS485 communication mode through level translators and sends command to the sensor using the cable, similarly, the sensor will return the data to background monitoring system after receiving the command, which using communication module in a wireless way. The data transmission mechanism is also suitable for hand-held devices and sensors.

4   Overall Test Analysis

4.1   Sensor Accuracy Test

Due to the existing of harmonics and high voltage electromagnetic field will affect the measurement accuracy of the sensor, so the sensor not only adopts well-designed electromagnetic shield shell but also uses the method of combining hardware filter and software algorithm, apart from the above, the current real value and the third harmonic value can be extracted from the lightning arrester leakage current, and the change curve of third harmonic can be used as the reference value of the aging degree of the lightning arrester.

Sensor measurement accuracy according to the requirements of Q/GDW 537-2010 that Technical Specifications for Capacitive Equipment and MOA Line Insulation Monitoring Devices: the measurement error of full current, resistance current is $\leq \pm$ (standard reading $\times$ 2.5% + 5 uA). In the laboratory, the DANDICK DK-34B1 multi-function AC sampling transmitter calibration device is used to verify the accuracy of the sensor. Use the 66 kV arrester as an example, and its leakage current value in 0.2 mA–0.6 mA and resistive current does not exceed 20% of the total current, in accordance with the requirements of the above numerical range, measurement accuracy of the sensor tests were conducted. 0.00 mA–1 mA AC current is sent through the current calibration device, and third harmonic content is set in 0%–20%, sensor accuracy test data as shown in Table 1.

Table 1. Experimental data table of sensor measurement accuracy

Third harmonic content	Measurement parameters	Measurement data			
0%	Full current value	0.00	0.20	0.51	1.00
	Resistive current value	0.00	0.00	0.00	0.00
5%	Full current value	0.00	0.21	0.51	1.01
	Resistive current value	0.00	0.01	0.03	0.06
10%	Full current value	0.00	0.21	0.51	1.01
	Resistive current value	0.00	0.02	0.06	0.11
15%	Full current value	0.00	0.21	0.51	1.02
	Resistive current value	0.00	0.04	0.08	0.17
20%	Full current value	0.00	0.22	0.51	1.02
	Resistive current value	0.00	0.05	0.11	0.22

The test data show that the measurement accuracy of the sensor is less than 1%, which fully meets the requirements of the specification.

4.2 Reliability Test of Power Supply System

The solar power supply system has been tested for 3 months in winter, the voltage of the sensor is monitored and recorded in real time, and a total of 4800 load voltage data are collected. During this period the weather condition is changeable and the temperature is very low, which provides a good test condition.

The test data is used to plot the corresponding curve, the vertical axis represents the voltage of the load and the horizontal axis is the time, as shown in Fig. 4.

Although there are slight fluctuations about the voltage of the load as shown in Fig. 4, always in a reasonable range of load operating voltage, the experiment shows

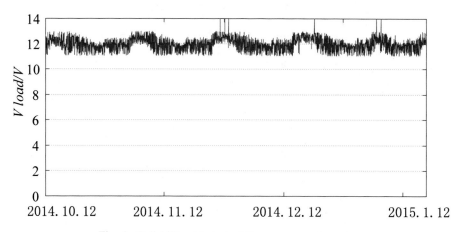

Fig. 4. Reliability data test of the power supply system

that the design of the solar power system can be a reliable power supply to the load. In winter −15 °C outdoor, the experiment is carried out to the solar power supply system in two situations, one condition is the battery panel is covered by snow and another situation is the snow on the battery panel melts after sun comes out, in this period of time the power supply is still normal, and under influence of the sun the snow on the battery panel can be eliminated itself, and the power supply is normal and reliable.

5   Conclusions

The research results of transmission line lightning arrester insulation parameter monitoring system can solve many problems caused by the power outage test of lightning arrester for cable terminal. Under the situation of the transmission line is charged, the test personnel can obtain accurate measurement data by monitoring the background system or hand-held devices.

High precision measurement of the sensor can be used in the situation of no voltage reference, directly calculating resistive current trends with software algorithms, and can accurately record the lightning arrester action and the time of action, to provide accurate data support for judging the lightning arrester running state. The use of a solar power system can solve the problem of outdoor power supply for an outdoor measuring device, and after testing under the extreme weather conditions, the reliability verification of power supply is well completed. The application of the whole monitoring system can reduce the work of a lot of routine power failure test, and it is useful to help the maintenance personnel to develop a better policy for replacement and extension of high-voltage cable terminal arrester.

References

1. Yin, X., Shao, T., Gao, Y., et al.: Current situation and development of MOA detection method. High Voltage Eng. **28**(6), 35–36 (2018)
2. Zhang, Z., Zang, D.: The research of on-line monitoring method about MOA. High Voltage Electr. **45**(5), 126–129 (2013)
3. Qian, J., Chang, Y.: The monitoring and research of non-uniform moisture and aging about the MOA. Electric Porcelain Arrester **31**(6), 42–46 (2017)
4. Flisowski, Z., Mazzetti, C., Wlodek, R.: New approach to the selection of effective measures for the lightning protection of structures containing sensitive equipment. J. Electrostat. **60** (2/4), 287–295 (2015)
5. Li, B., Shu, N., Li, Z., et al.: Development of high-precision standard device for ZnO arrester testing instrument. High Voltage Electr. **45**(1), 95–97 (2017)
6. Peng, B., Chen, J., Zhou, Z., et al.: Poor insulator detection system based on wireless transmission. High Voltage Electr. **43**(4), 310–312 (2018)
7. Zhou, N., Xuan, X., Zhao, Y., et al.: The research of arrester detector using wireless communication. Electr. Power Autom. Equip. **25**(11), 89–90 (2016)
8. Zakhidov, R., Kiseleva, E., Kiseleva, M., et al.: Experience and prospects of using solar-wind low-power energy complex in power supply systems of remote objects. Appl. Solar Energy **51**(2), 156–162 (2015)

Complex Networks and Intelligent Systems

Correlating Thermal Anomaly with Earthquake Occurrences Using Remote Sensing

Utpal Kanti Mukhopadhyay[1], Richa N. K. Sharma[2(✉)],
Shamama Anwar[2], and Atma Deep Dutta[2]

[1] Indian Institute of Technology Kharagpur, Kharagpur, West Bengal, India
[2] Birla Institute of Technology, Mesra, Ranchi, Jharkhand, India
richasharma.ranchi@gmail.com

Abstract. Earthquakes are the sudden tremors of the ground leaving behind damages to life and property, ranging from smaller to massive scale. Earthquake prediction has been in the limelight of the scientific community since very early times. Prediction of earthquakes has been found in texts of ancient civilizations based on planetary positions with respect to the earth. With the advent of real-time observations on various data sources many attempts are going in this direction. The present paper investigates and put forward some facts based on data obtained from satellites for an earthquake which occurred in Imphal, India, in 2016. It studies the thermal anomaly data that took place before the earthquake. MODIS Land Surface Temperature (LST) product was used wherein daily nighttime images of six years have been used for the study. Quality Assurance of the datasets was performed to identify the good quality pixels having maximum information. A change detection technique for satellite data analysis namely Robust Satellite Technique has been used and RETIRA index has been calculated. The study of this RETIRA index has been done for three years and it has been found that the RETIRA index is considerably high for the earthquake year. But it cannot be concluded that the high value of RETIRA index is a sure indicator for an earthquake and hence it leaves scope for future studies.

Keywords: Thermal anomaly · Robust Satellite Technique (RST) ·
Land Surface Temperature (LST) ·
MODIS (Moderate Resolution Imaging Spectro-radiometer) ·
Robust Estimator of TIR Anomalies (RETIRA) index

1 Introduction

Prediction of earthquakes, form in themselves, a specialized study. However, after their occurrences, there are ways and means to understand its cause as a lot of data can be reconstructed after its occurrence. Hence, by analyzing past earthquakes and their characteristics, perhaps a better understanding of the earthquake phenomenon can be made. The most widely acclaimed of all geological theories behind the reason of earthquake occurrence is tectonic movement, thereby stress develops in rocks beyond

© Springer Nature Switzerland AG 2020
A. E. Hassanien et al. (Eds.): AMLTA 2019, AISC 921, pp. 863–875, 2020.
https://doi.org/10.1007/978-3-030-14118-9_85

elastic limit and hence the manifestation of earthquakes. Researchers, since the last couple of decades, have been digging their minds to find the clues that nature leaves before an imminent earthquake. Some of the previously reported and tested precursory signals in this field include Ultra Low Frequency (ULF), magnetic field, Total Electron Content (TEC) changes in ionosphere, noises in communication signals across wide frequencies which can be observed in mobile network, TV and GPS signals etc., sudden bright light emission from the ground, Thermal/Infrared anomalies, gravity anomalies, abnormal cloud patterns, changes in animal behavior etc. [1]. But the main problem with these precursors is that they are not consistent, meaning they may occur prior to one earthquake but may not be observed in another. The uncertainty involved in the time and place of occurrence of earthquakes has lured scientist across the globe but decades of research have failed to reliably predict the time and magnitude of occurrence of earthquakes. Nevertheless, time to time, this topic of the earthquake of earthquake prediction has been rekindled by newer, more sophisticated ways to analyze the precursors before an earthquake and provide important clues for earthquake prediction. With the advent of remote sensing in the 1970s, our understanding of the earth and earth phenomenon has evolved.

Since the last decade, the fluctuations in Earth's thermally emitted radiation detected by thermal sensors onboard satellites has been regarded as an observable precursor prior to earthquakes. Although a large number of studies has been performed on this since the 80 s, more often than not these studies have been questioned and greeted with skepticism mainly due to insufficiency of validation dataset and scarce importance related to causes other than seismic event [2] and also due to prior methodologies employed to discern the existence of thermal anomalies. Recently, a statistical approach named Robust Satellite Technique (RST) has been extensively used as a suitable method for studying the anomalous behavior of TIR signals prior to and after earthquake events [3]. This present paper uses this approach to investigate the presence of thermal anomaly prior to Imphal Earthquake that occurred on 4th January 2016 (Indian Standard Time) by implementing RST technique and to understand the spatiotemporal correlational between the earthquake occurrence and the appearance of such transient anomaly in the space-time domain. To analyze the thermal behavior of the study area, 6 years of MODIS Land Surface Temperature data in terms of reference dataset has been used.

2   Materials and Methods

2.1   Study Area

The earthquake under investigation is the Imphal Earthquake that jolted Imphal, the capital of the state of Manipur in north-east India on January 4, 2016, with a moment magnitude of 6.7. This was the largest seismic event in the last six decades. Prior to this event, the region experienced an earthquake of magnitude 7.3 in 1957. The epicenter of the Imphal Earthquake of 2016 was located at 24.834°N and 93.656°E near Noney Village of Tamenglong district of Manipur. The focal depth of the earthquake was estimated to be 50 km. Eleven people lost their lives, 200 others were injured and various structures were damaged. The shake was widely felt in the eastern and north-eastern India and was even felt in Bangladesh. The entire north-east region of India comes under seismic zone V with regards to the seismic zone map of India. This means that the entire region is very much vulnerable to earthquakes. The study is a post-event analysis of this earthquake event (Fig. 1).

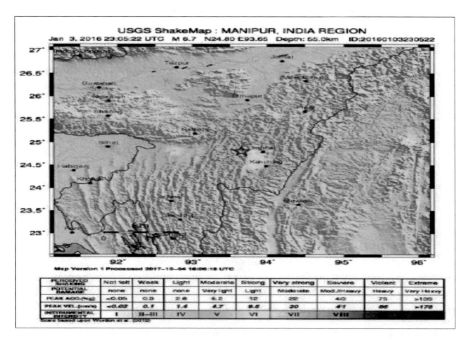

Fig. 1. ShakeMap of the study area providing a near-real-time map of ground motion and shaking intensity after the earthquake. (Image is taken from USGS website https://earthquake.usgs.gov/data/shakemap/)

2.2 Data Sets Used

The MODIS (Moderate Resolution Imaging Spectro-radiometer) LST, namely, MOD11A1 daily nighttime images of six years (2011–2016) was used to examine the ground thermal condition before and after the Imphal Earthquake. MODIS is an instrument aboard Terra and Aqua satellite. Terra orbits earth from the North Pole to the South Pole across the equator in the morning at 10:30 AM, whereas, Aqua orbits earth from the South Pole to the North Pole over the equator in the afternoon at 1:30 PM. MODIS consists of 36 spectral bands from 0.4 to 14 μm. MODIS also caters eight-day and monthly global gridded LST products. The daily LST product (Level 3) has a spatial resolution of 1 km. LST is gridded in the Sinusoidal Projection. A tile consists of 1200 × 1200 grids in 1200 rows and 1200 columns. The exact grid size is 0.928 km by 0.928 km at a spatial resolution of 1 km.

The daily MOD11 LST and Emissivity data are acquired by implementing the generalized split-window algorithm. The split window method corrects for atmospheric effects based on the differential absorption in adjacent infrared bands [4]. The product comprises of daytime/nighttime LSTs, quality assessment bit flags, satellite view angle, observation time and emissivity which are collectively termed as Subdatasets (SDS) [5]. All of the thermal data are downloaded from USGS Earth Explorer. Information on earthquake as regard to the location of the epicenter, focal depth, magnitude, time of event and casualty was acquired from The United States Geological Surveys (USGS) website and Indian Meteorological Department (IMD) website.

2.3   Data Pre-processing

In order to extract information from raw data, pre-processing of MOD11A1 night time products were essential. Three adjacent tiles of MODIS data cover the area of interest. Associated tiles were mosaicked and re-projected in ArcMap 10.3.1. Each MODIS HDF file contains Quality Science dataset layers which provide useful information regarding usability and usefulness of the data product. Information extracted from these quality sciences SDS are used for analysis. The LST Quality Science Dataset layers are binary encoded meaning each pixel has an integer value that must be converted to bit binary value for cloud masking and interpretation. MODIS LST image contains both good and bad quality pixels. The bad quality pixels are attributed due to cloud or sensor defects. In order to determine the pixel values in the Quality Control image to be retained as good quality pixels, a number of permutations and combinations were calculated. Different combinations and permutations of bit flags comprise an 8-bit binary number which has a decimal equivalent in Quality Control (QC) image within 0–255 range. Only those binary values which are acceptable are taken for calculations

Table 1. Table of defined bit flags for quality assurance scientific dataset QC night/day in MOD11A1 adapted from the Collection 6 MODIS land surface temperature products users guide

Bits	Bit fla name	Key
1 & 0	Mandatory QA flag	00: LST produced, good quality not necessary to examine more detailed QA 01: LST produced, other quality, recommended examination of more detailed QA 10: LST not produced due to a cloud effect 11: LST has not produced other than due to cloud effect
3 & 2	Data quality flag	00: good quality data 01: other quality data 10: TBD 11: TBD
5 & 4	Emissivity error flag	00: average emissivity error $<= 0.01$ 01: average emissivity error $<= 0.02$ 10: average emissivity error $<= 0.04$ 11: average emissivity error >0.04
7 & 6	LST error flag	00: average LST error $<= 1K$ 01: average LST error $<= 2K$ 10: average LST error $<= 3K$ 11: average LST error $>3K$

such as Mandatory QA flag 00 and 01, Data Quality flag 00 and 01, Emissivity error flag 00, 01, 10 and LST error flag 00 and 01. Here the Least Significant Bit (LSB) is 0. The bit flag values involved in analysis are as follows (Table 1):

Next, the QC layer is used to create a logical mask. Bits that are "No Data" in QC layer will be "No Data" in the destination if a pixel is not NA in "QC layer" and has an acceptable bit flag value like 00 or 17 v or 25, and then the value of the source is written in the destination.

The no data (NULL) value has to be set to zero. Unfortunately, the value zero also indicates the highest quality pixels. In order to preserve pixels with value zero, a pixel with a QC value of 0 and a valid LST value are of highest quality, while pixels with a QC value of 0 and an invalid LST value can be set to NULL.

The no data (NULL) value must be set to zero. Tragically, the value 0 also shows the most astounding quality pixels. Keeping in mind the end goal to save good quality pixels, pixels with a QC value equal to 0 and valid LST value are of good, while pixels with a QC value of 0 and an invalid LST value can be set to NULL (Fig. 2).

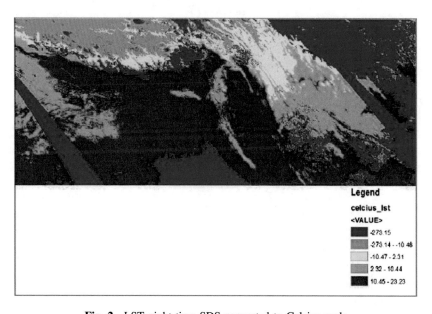

Fig. 2. LST night time SDS converted to Celsius scale

Information on how to convert the digital values of the satellite data can be obtained in the MODIS Land Surface Temperature product user guide [5]. The LST night time dataset is a 16-bit unsigned integer with a valid range of 7500–6553. It has a scale factor of 0.02 which means that the LST night time data is 50 times magnified. The scale factor is used for a linear DN value rescaling to temperatures in Kelvin scale. Temperatures in °C can be evaluated with the formula ((DN * 0.02–273.15). It must be remembered that the no data pixels are ignored in this conversion.

2.4   Robust Satellite Technique (RST)

Robust Satellite Technique is a change detection technique for satellite data analysis used for monitoring natural hazards and is invariant with the utilization of any satellite or sensor. Therefore, RST can be utilized on different satellite data and applied for different events such as earthquakes, volcanoes, floods, forest fires etc. [6]. Beside thermal anomaly detection prior to earthquakes, Robust Satellite Technique has been used for Ash Plume detection and Tracking [7], for oil spill detection and monitoring [8, 9], for monitoring sea water turbidity by RST [10], RST technique for Sahara dust detection and monitoring, RST for pipeline rupture detection [11], RST approach for volcanic hotspot detection and monitoring among others [12] (Fig. 3).

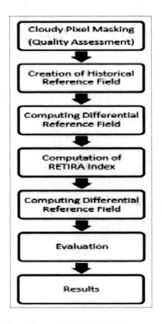

Fig. 3. Flowchart of RETIRA index computation

The methodology is premised on analyzing multi-temporal homogeneous satellite thermal IR images of several years which are co-located in time and space domain. Each pixel of the satellite image is processed in terms of expected and natural variability range of the TIR signal for each pixel in the satellite image. Anomalous TIR signal, therefore an anomalous pixel is identified as a deviation from its normal or expected value using a unitless index called Robust Estimator of TIR Anomalies (RETIRA) index [10]. Mathematically RETIRA index is as follows

$$RETIRA\ INDEX = \frac{\Delta LST(r,t) - \mu \Delta LST(r)}{\sigma \Delta LST(r)}$$

Where $\Delta LST(r, t)$ = the difference between the current LST value on the associated image $LST(r, t)$ at the location 'r' $\equiv (x, y)$, at the acquisition time 't' $(t = \tau)$ and its spatial average of the image 'T(t). $\mu\Delta LST(r)$ and $\sigma\Delta LST(r)$ are the time average and standard deviation of $\Delta LST(r, t)$ respectively, at the location (r) calculated on cloud-free pixels.

Time average and temporal mean are evaluated by processing several years of homogeneous cloud free reference datasets. It must be noted that only land pixels are taken into consideration leaving all sea pixels. The numerator part i.e. the difference between $\Delta LST(r, t)$ and $\mu\Delta LST(r)$ can be thought of as the signal to be investigated for possible relation with seismic activity while the standard deviation in the denominator $(\sigma\Delta LST(r))$ represents the noise due to natural and observational causes. That is to say, the computation of the RETIRA index is based on the comparison of the signal and its local variability which is noise. The index depicts the strength of the thermal anomaly. Temporal mean and the standard deviation is calculated using the cell statistics tool in ArcMap 10.3.1. The temporal mean is the mean LST calculated for every pixel over the reference time period. Standard deviation is the measure of the deviation of LST from the mean LST. For each day RETIRA index was calculated for two earthquake years and one non-earthquake year. For each day, a high value and a low value of RETIRA index were obtained. This paper takes under consideration on high values of RETIRA indices for analysis. A baseline value of 3.5 was considered above which RETIRA indices are abnormal.

3   Results and Discussions

On 26th December 2015, 2nd January 2016, 11th January 2016, 12th January 2016 RETIRA indices obtained on computation were abnormal. In fact, two days before the earthquake, the RETIRA Index was beyond the baseline. As the earthquake event passed, the values of RETIRA indices decreased. But a week after the earthquake event on 11th January and 12th January there was a surge in the values of RETIRA indices.

3.1   Analysis of Results of Night Time Retira Index Computation for Earthquake Year (2016)

In an attempt to identify any correlation between transient temporal thermal anomalies earthquake events, a graphical analysis was carried out for the main earthquake year. For each day under investigation, RETIRA indices were plotted. On 26th December 2015, the RETIRA index was abnormal. This was the first instance within the days under investigation when the RETIRA index was observed anomalous. 2 days prior to the main seismic event of magnitude 6.8 on Richter scale, the RETIRA index was again above the baseline value. After the earthquake gradual lowering of RETIRA indices were observed. Again 7 and 8 days after the earthquake, RETIRA indices were anomalous. The graph also depicts that the RETIRA indices increased gradually before it became anomalous on 11th and 12th January 2016. A fitting curve was implemented from the RETIRA indices as a function of time. To achieve a good degree of precision a polynomial of degree 5 was used to interpolate (Figs. 4 and 5).

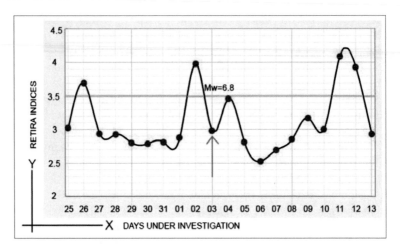

Fig. 4. Graphical analysis of the earthquake year (2015–2016)

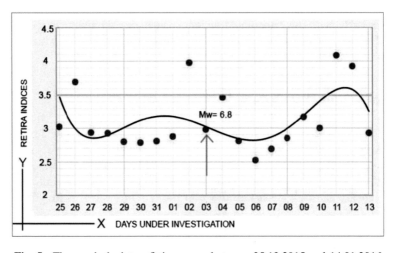

Fig. 5. The graph depicts a fitting curve between 25.12.2015 and 14.01.2016

3.2   Analysis of Graphical Result of Night Time Retira Index Computation for Secondary Earthquake Year (2014–2015)

With an aim to see if there exists similar kind of trend of RETIRA indices on some other earthquake year over the study area, another earthquake year was taken into consideration. Between the 25th December 2014 and 14th January 2015, two earthquakes, one on 28th December 2014 and the other on 6th January 2015, both of magnitude 4.8 were recorded in the north-eastern region of India, within the study area. Hence this period has been considered as another seismically perturbed year. Again, RETIRA indices were plotted as a function of time for each day under investigation. As it can be seen from the graph, the RETIRA indices were abnormal prior to the first seismic event of magnitude

4.8 on 28.12.2014. After the event, the value of RETIRA indices dropped. Before the second seismic event on 06.01.2015, the RETIRA indices once again gradually increased. A similar trend of RETIRA indices was seen for the main earthquake year where prior to the seismic event, the indices escalated. For this secondary earthquake year, a fitting curve was interpolated from the RETIRA indices as a function of time. Again, to achieve a good degree of precision, a 5th order polynomial was used to interpolate. But the curves were inconclusive of a strong relationship between the magnitude of the earthquake and the fluctuation of RETIRA indices. This is due to the fact that the main earthquake year saw a magnitude of 6.8 whereas the secondary earthquake year experienced two earthquakes of relatively smaller magnitudes i.e. 4.8, but if we closely look at both the graphs for the two earthquake years, the RETIRA indices had higher abnormal values for 2014–2015 (Figs. 6 and 7).

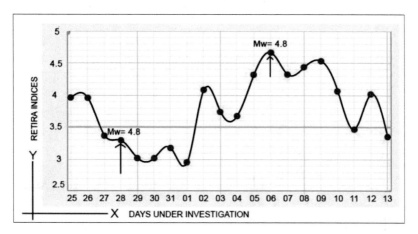

Fig. 6. Graph depicting RETIRA indices plotted against the days under investigation for secondary earthquake year.

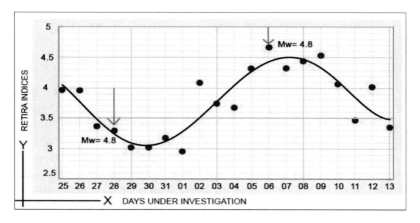

Fig. 7. Graph depicting a fitting curve of RETIRA indices plotted against the days under investigation for secondary earthquake year (2014–2015).

3.3   Analysis of Results of Night Time Retira Index Computation for Non-earthquake Year (2013–2014)

Similar graphical analysis was carried out for the non-earthquake year also where RETIRA indices were plotted for each day. The year 2013–2014 has been considered as a non-earthquake year because of the reason that no earthquake of magnitude 5.0 occurred within the time window under consideration i.e. 25th December 2013 to 14th January 2014. Only one seismic event of magnitude 4.2 occurred in the study area during this period. As the graph depicts, there is only one time on 8th January 2014, when the RETIRA indices touch the baseline that is exactly 3.5. Above this, the index would be abnormal. A fitting curve was interpolated from the RETIRA indices as a function of time. A polynomial of degree 5 was used to interpolate. The degree of order 5 was enforced to achieve a good degree of precision. But the fitting curve never crossed the baseline of 3.5. Analyzing the fitting curves reveals that the two earthquake years are similar and a non-earthquake year shows quite a different trend (Figs. 8 and 9).

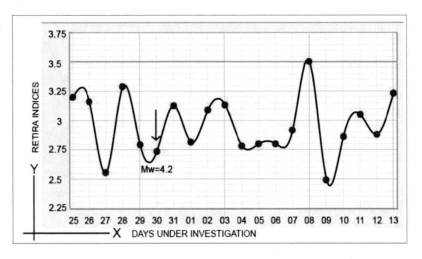

Fig. 8. Graphical analysis of the non-earthquake year (2013–2014)

3.4   Pixel Based Analysis of the Non-earthquake Year

Pixels labeled 'No data' hindered our understanding of the evolution of thermal anomaly. Cloud cover, among others, is the main reason for such 'no data' pixels. With this aim, a pixel-based analysis was carried out. The results shown in the table dictate that the number of pixels with abnormal RETIRA index surged from 3 (nine days prior to the earthquake) to 115 (two days prior to the earthquake). On the day of the earthquake, RETIRA index was normal but 7 and 8 days after the event the RETIRA indices were beyond normal value and number of anomalous pixels increased from 16 to 51 in 1 day. These anomalous pixel counts give us some idea about the evolution of thermal anomalies before and after the event (Table 2).

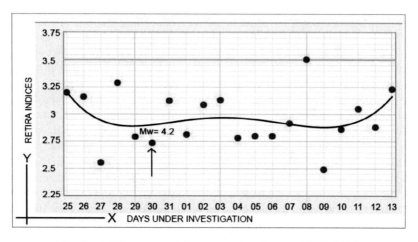

Fig. 9. A fitting curve of the non-earthquake year (2013–2014)

Table 2. Table depicting the number of anomalous pixels recorded on days before and after the earthquake event of 2016

Days under investigation with abnormal RETIRA index values	RETIRA index highest value	No. of pixels with RETIRA index >3.5
26.12.2015	3.69	3
02.01.2016	3.97	115
04.01.2016 (Mw = 6.8)	2.97	NIL
11.01.2016	4.08	16
12.01.2016	3.92	51

4   Conclusions and Future Scope of Study

Till date, a valid prediction specifying the date, time, place, and magnitude of the earthquake have not been standardized [13]. Study of earthquake precursors in many ways facilitate the understanding of their occurrences and thereby lead one step closer to predict earthquakes. Single precursor in pre-seismic monitoring research has its own limitations [14]. Devoting attention to multiple precursors is a promising choice which can raise the reliability and credibility of the relationship between thermal anomalies and imminent seismic events [3]. The region under study did show an anomalous behavior of LST prior to the main earthquake year 2015–2016. The RETIRA index showed thermal anomaly on 26th December 2015, 2nd January 2016, 11th January 2016 and 12th January 2016 for the main earthquake year, before and after the seismic event. In addition, the pixel-based analysis also reiterated the rise in temperature in the same year. For statistically validating the results of RETIRA index of the main earthquake year, a secondary earthquake year and anon-earthquake year were considered. Although, the present study showed that the RETIRA index has crossed over the baseline in both the earthquake years, yet the earthquake with relatively higher magnitude has relatively lower RETIRA index values whereas the earthquake with low magnitude has a higher RETIRA index above the baseline as in the case of secondary earthquake year (2014–2015). The present study is inconclusive in this aspect and leaves scope for future investigations. However, efforts in investigating precursory earthquake signals had been carried out with Outgoing Long Wavelength Radiation (OLR) [15] pertaining to this Imphal Earthquake of 2016. The authors' result showed that on 27th of December, 2015 an abnormal OLR flux of 303 W/m^2 was recorded against a mean of 279 W/m^2 [15]. Such precursor studies if combined and applied may pave way for further investigations in this direction to strengthen earthquake prediction.

Forecasting moderate and strong earthquakes in the future, prior to several months and days at the regional and global scale is highly demanded to mitigate the disaster. Up to this point, no single quantifiable geophysical variable and information investigation strategy represents the impressive potential for an adequately dependable earthquake forecast. Passive microwave remote sensing can be a good choice because of its ability to penetrate cloud masking observation of the surface in the cloud cover. Passive microwave remote sensing suffers from one main disadvantage, i.e. its spatial resolution (1–5 km) unlike thermal infrared imagery (nearly 1 km). The combination of TIR and microwave information proves to be an essential viewpoint for understanding land surface temperature changes before earthquakes. Such data combination can exploit the high accuracy and spatial resolution of satellite TIR data and the infiltrating cloud capacity of microwave data. Combined analysis of multiple precursors is a feasible option because it can reduce the detection of false thermal anomalies. Because utilizing only a single approach to identify the pre-earthquake anomaly is not statistically robust [16].

In the meantime, ground estimations, for example, air temperature data, geochemical observations, groundwater level changes along faults can provide a superior understanding and supplement the missing data attributed due to cloud cover. New machine learning algorithms can optimize this multi-method scheme. Big data analysis based on a long-term series of remote sensing data and global earthquake cases is an important means to establish the statistical relationship between thermal anomalies and earthquake events [14].

Acknowledgement. The authors are thankful to Dr. Virendra Singh Rathore, Assistant Professor, of the Department of Remote Sensing, Birla Institute of Technology, Mesra, Ranchi 835215, India, for his valuable suggestions.

References

1. Jeganathan, C., Gnanasekaran, G., Sengupta, T.: Analysing the spatiotemporal link between earthquake occurrences and orbital perturbations induced by planetary configuration. Int. J. Adv. Remote Sens. GIS Geogr. **3**(2), 123–146 (2015)
2. Pavlidou, E., van der Mejide, M., van der Werff, H.M.A., Ettma, J.: Time series analysis of remotely sensed TIR emissions: linking anomalies to physical processes. Presented at: AGU Fall Meeting 2013, 9–13 December 2013, San Francisco, United States of America, pp. 1–2 (2013)

3. Eleftheriou, A., Fizzolla, C., Genzano, N., Lacava, T., Lisi, M., Paciello, R., Pergolla, N., Vallianatos, F., Tramutoli, V.: Long-term RST analysis of anomalous TIR sequences in relation with earthquakes occurred in Greece in the period 2004–2013. Pure Appl. Geophys. **173**, 285–303 (2016)
4. Wan, Z., Dozier, J.: A generalized split-window algorithm for retrieving land surface temperature from space. IEEE Trans. Geosci. Remote Sens. **34**(4), 892–905 (1996)
5. Wan, Z.: New refinement and validation of the collection 6 MODIS land surface temperature/emissivity products. Remote Sens. Environ. **140**, 36–45 (2014)
6. Pergola, N., Aliano, C., Coveillo, I., Filzzola, C., Genzano, N., Lavaca, T., Lisi, M., Mazzeo, G., Tramutoli, T.: Using RST approach and EOS-MODIS radiances for monitoring seismically active regions: a study on the 6th April 2009 Abruzzo earthquake. Hazards Earth Syst. Sci. **10**, 239–249 (2010)
7. Marchese, F., Pergola, N., Telesca, L.: Investigating the temporal fluctuations in satellite AVHRR thermal signals measured in the volcanic area of Etna (Italy). Fluctuations Noise Lett. **6**, 305–316 (2006)
8. Grimaldi, C.S.L., Coviello, I., Lacava, T., Pergola N., Tramutoli, V.: A new RST based approach for continuous oil spill detection in TIR range: the case of the Deepwater Horizon platform in the Gulf of Mexico. In: Liu, Y., MacFadyen, A., Ji, Z.G., Weisberg, R.H. (eds.) Monitoring and Modeling the Deepwater Horizon oil spill, A Record-Breaking Enterprise, Geophysical Monograpph Series, American Geophysical Union (AGU), Washington DC, USA, pp. 19–31 (2011)
9. Casciello, D., Lacava, T., Pergola, N., Tramutoli, V.: Robust satellite technique (RST) for oil spill detection and monitoring. In: International Workshop on the Analysis of Multi-temporal Remote Sensing Images, 2007, MultiTemp 2007, 18th–20th July 2007 (2007)
10. Lacava, T., Ciancia, E., Coviello, I., Polito, C., Tramutoli, V.: A MODIS based robust satellite technique (RST) for timely detection of oil spilled areas. Phys. Chem. Earth **9**(2), 45–67 (2015)
11. Filizzola, C., Pergola, N., Pietrapertosa, C., Tramutoli, V.: Robust satellite techniques for seismically active areas monitoring: a sensitivity analysis on September 7, 1999 Athen's Earthquake. Phys. Chem. Earth **29**, 517–527 (2004)
12. Harris, A.J., Swabey, S.E.J., Higgins, J.: Automated threshold 463 of active lava using AVHRR data. Int. J. Remote Sens. **16**, 3681–3686 (1995)
13. Geller, R.S.: Predicting Earthquakes is impossible. Los Angeles Times, February 1997
14. Jiao, Z.H., Zhao, J., Shan, X.: Pre-seismic anomalies from optical satellite observations: a review. Nat. Hazards Earth Syst. Sci. **18**, 1013–1036 (2018)
15. Venkatanathan, N., Hareesh, V., Venkatesh, W.S.: Observation of earthquake precursors – a study on OLR scenario prior to the earthquakes of Indian and neighboring regions occurred in 2016. Int. J. Earth Sci. Eng. **9**(3), 264–268 (2016). ISSN 0974-5904
16. Bhardwaj, A., Singh, S., Sam, L., Bhardwaj, A., Martin-Torres, F.J., Singh, A., Kumar, R.: MODIS based estimates of strong snow surface temperature anomaly related to high altitude earthquakes of 2015. Remote Sens. Environ. **188**, 1–8 (2017)

IOT-Based Conceptual Framework for the Prevention of Acute Air Pollution Episodes for Reducing and Limiting Related Diseases in Egypt

Basmah El Haddad[✉] and Zainab Elsadi

Institute of National Planning, Cairo, Egypt
basmah.elhaddad@gmail.com, zainabelsadii@yahoo.com,
{basmah.elhaddad,zainab.elsadi}@inp.edu.eg

Abstract. Egypt is suffering from acute air pollution episodes as a result of Egyptian farmers' continuous burning of rice straw after harvest. The Egyptian government issued strict legislations and took measurements to prevent the burning processes and counter its pollution. Still the problem persists constituting a major public health issue. The paper introduces a proposed Internet of Things "IOT-Based conceptual Framework for the prevention of acute air pollution episodes caused by rice straw burning which negatively impacts the environment and public health. The proposed IOT-Based conceptual framework considers environmental and related healthcare dimensions and parameters. It argues that IOT technologies can efficiently solve this problem by preventing and controlling the fire counts and monitoring the black cloud movement to react or pro-act in dealing with health consequences. The paper starts explaining IOT definition, architecture, benefits and challenges. Then explains in details the acute air pollution episode and the black cloud, its reasons and consequences. It presents a proposed IOT-Based solution scenario and its corresponding phases. Finally, it contributes with a proposed general IOT-Based conceptual Framework to prevent and control the acute air pollution episode and limit its health-related problems in Egypt, concentrating on the proposed IOT layers and components used.

Keywords: Internet of Things "IOT" · IOT-Based framework ·
Environmental acute air pollution · Black Cloud · Health IOT network · Egypt

1 Introduction

Cairo, Egypt's capital suffers from acute air pollution episodes that are caused by various factors that include meteorological factors, the nature of topography of the city trapping pollutants in addition to the air pollution resulting from transport, commercial activities and the agricultural sector amongst others. Pollution from the agricultural sector are caused by the burning of rice straw after harvest by the Egyptian farmers. This leads to the Black Cloud phenomenon which is a thick layer of smog that remains in the atmosphere for more than three months in the autumn each year over cities of Egypt. Its first appearance was over the Nile Delta and Cairo 1997 and its impacts

© Springer Nature Switzerland AG 2020
A. E. Hassanien et al. (Eds.): AMLTA 2019, AISC 921, pp. 876–887, 2020.
https://doi.org/10.1007/978-3-030-14118-9_86

became visible two years later [2]. The black cloud moves along cities and governorates of Egypt in a certain path affected by the low air pressure, which causes huge health complications leading to the increase of respiratory and cardiovascular patients in particular as well as triggering allergies and autoimmune diseases. It even caused early deaths that have been estimated in 2010 to be 6000 premature mortality for Cairo and 35322 for Egypt. The number for premature mortality is expected to reach 8200 people by 2025 [3].

To combat air pollution caused by agriculture, the Egyptian Environmental Affairs Agency (EEAA) launched a fund mechanism to collect and recycle agriculture wastes. Additionally, it worked on identifying and preventing the rice straw burning process using 3 satellites to detect and allocate straw burning spots around the clock. In addition, EEAA has conducted capacity building for its officers to manage rice straw burning processes, as well as improving efficiency of air quality monitoring system by increasing the stations to be about 95 station all over Egypt. Lastly, EEAA provided the necessary tools and gears to collect, store and reuse the rice straw (477 rice straw storage centers). **Despite those efforts, the air pollution episodes still exist and results in emission of toxic fumes that accumulate in a black cloud**.

To solve this problem, the government needs significant manpower observing and monitoring extremely large areas of rice fields to prevent the burning processes. This constitutes a great financial and human burden. Therefore, Egypt's main target is to prevent the acute air pollution episode by detecting, monitoring, controlling and preventing/decreasing the open-field rice straw burning in the Nile Delta in an attempt to control and reduce the toxic effect of the black cloud and in order to reduce and limit related respiratory and cardiovascular health complications in Egypt. **Motivation**; thus the paper's goal is to support Egypt's main target to overcome this problem by adopting a conceptual framework using the most recent embedded technologies and tools like sensor technology, IOT "Internet of Things" and cloud computing to control and prevent the rice straw burning in order to reduce public health complications and mortality rates. It introduces an IOT-Based conceptual framework to prevent and control the acute air pollution episode aiming to reduce its negative impact on the environment and Egyptian public health. The first section of this paper consists of the introduction, problem definition and methodology, while the second section introduces the IOT technology, its definition, architecture, benefits and challenges. The third section introduces the acute air pollution episode, its reasons and consequences then it introduces a proposed IOT-Based Solution Scenario and the corresponding Phases. This section, also contributes with a proposed general IOT-Based conceptual Framework to prevent and control the acute air pollution episode caused by the rice straw burning processes and to limit the related diseases in Egypt, explaining in details the proposed IOT layers, entities, components and the data exchange and communication within the IOT relational framework. Finally the paper presents the limitations, conclusion and future work. **Methodology**; the paper follows a mixed multi approach as follows: descriptive analytical method depending on observing and evaluating existing status of the acute air pollution episode in Egypt and the IOT model building approach depending on IOT standards, layers and components.

2 Internet of Things "IOT"

2.1 IOT Definition

There is no one universally agreed upon definition of Internet of things 'IOT', nevertheless it's a key element of global digital transformation. It's a concept of how we interact with the physical world [4]. IOT is a huge network of internet-connected things; smart objects that are able to sense, communicate, collect, exchange and capture real-time data using embedded and smart technology. In IOT, things, devices or smart objects gather and share information in real-time directly with each other and the cloud, making it possible to collect, record and analyze new data streams faster and more accurately [5]. This enables objects to recognize themselves and others, talk to each other, connect with applications, deliver insights, obtain intelligent behaviors, drive transformation and make related decisions as they interact with their internal states as well as the external environment. With IOT anything's will be able to communicate to the internet from anywhere providing any services by any network enabling new types of services and applications as smart cities, smart homes, smart health, connected cars, environmental improvements, etc., which offers notifications, security, energy saving, automation, communication, computers and entertainment [6]. IOT objects with embedded electronics can transfer data over a network without human interaction; i.e. wearable devices, environmental/biomedical sensors, machinery in factories, components in a vehicle or devices in homes and buildings like cellphones, coffee makers etc. [7]. IOT vision has evolved due to a convergence of multiple technologies; wireless communication, real-time analytics, deep learning, commodity sensors, embedded systems …etc. and is very much related to Big Data, Cloud Computing, Machine learning and Artificial Intelligence. According to the US National Security and Telecommunication Advisory Committee (NSTAC) IOT is based on three shared common principles; "**Things/Devices:** within a network are instrumented so they can be addressed individually. **Platform**: Devices are interconnected through shared platform, such as a cloud service, **Intelligence**: Devices may perform functions adaptively, on their own or with other devices and applications based on programming and inputs from the physical world" [4].

2.2 IOT Architecture

The Architecture defines system components and how they work and relate to each other's and how they collect, exchange, analyze data and information. There are different IOT architectures according to various perspectives. The most common one defines IOT architecture as divided in three layers according to its certain functionality and components as follows [8–11]:

Perception/Device/Sensing Layer. Identifies, perceives and detects smart objects in the environment. It includes a set/group of internet-enabled/connected devices that are able to collect and exchange information through internet communication networks as sensors, actuators, cameras, radio frequency identification devices RFID's, global positioning system GPS, control gateway, smart appliances which collecting/sensing/

measuring the parameters of interest in certain areas (weather conditions, air pollution status, energy consumption, health indicators etc.).

Network/Data Communication Layer. It's a core layer responsible for routing and processing the data to relevant procedures. It forwards the data from the perception layer to the application layer and secures its exchange under certain constraints of devices' capabilities, network limitations and applications constraints. This layer makes use of wireless as well as wired communications infrastructure available using a combination of short-range networks communication technology as Bluetooth, ZigBee to transfer data to nearby gateways while Wi-Fi, 3G, 4G, Power Line Communication, satellites carry the information over longer distances.

Application/Service Layer. Is responsible for providing services to the users through various applications. It's the layer where the information is received and processed. It allows users to access various applications in different areas like transportation, healthcare, etc., that provides high-level intelligent solutions such as disaster monitoring, health monitoring, transposition, medical and ecological environment, and management relevant to all intelligent applications. This can be done using different APIs "Application Programming Interfaces" that serve as a bridge to connect useful information and data to the IOT. Thus making IOT useful by connecting various devices in a powerful network. APIs provide the interface between internet and devices because of its capability to collect, exchange and sense the data. There are many APIs as Natural Language Processing APIs, Video and Image Analytics APIs, Text analytics APIs as well as Machine Learning APIs that automates the data processing, monitoring and classifying according to learned priorities. It deals with sensors data to understand, analyze, predict and support the decision making processes. Researchers can adopt a four/five/six layer architecture consisting of fog/cloud/intelligent computing to provide management services and enhance security aspects.

2.3   IOT Benefits and Challenges

From one side, IOT addresses many benefits to/for individual, community and society levels supporting better decision making, increasing systems efficiency, introducing new services and environmental benefits. It provides major benefits of; tracking behavior for real-time marketing, sensor-driven decision analytics, process optimization, enhanced situational awareness and instantaneous control/response in complex autonomous systems. It assures accuracy, real-time analysis, availability and feasibility. From the other side there are common challenges regardless of the IOT-application used as security, reliability, large scale data, heterogeneity, legal and social aspects, policies, financial issues, lack of knowledge etc.

3   The Black Cloud Phenomenon as an Acute Air Pollution Problem in Cairo and the Proposed Solution

Cairo is one of the most air polluted megacities in the world; in 2017 it was the 9th most polluted city according to Reuters. "Egypt generally and Cairo specifically suffers from high ambient concentrations of atmospheric pollutants including particulates (PM), Carbon monoxide CM, Nitrogen oxides NOx, Ozone O3 and Sulphur dioxide SO2" [12]. This pollution can be attributed to natural and meteorological factors in addition to pollution caused by human activity. These episodes lead to chronic health complications, many short-term and chronic respiratory and cardiovascular diseases in children and asthma patients. World Health Organization "WHO", considered "Air pollution" as a major environmental risk to health, it stated that, "Air pollution is the silent killer of around 7 million people each year on the planet, almost all of them in poor countries in Asia and Africa. About a quarter of deaths from heart disease, stroke, lung cancer, chronic and acute respiratory diseases including asthma can be attributed to air pollution" [13]. The air pollution phenomenon over Cairo called "Black Cloud" is attributed to different reasons that gather every year during the autumn season including heavy traffic/industrial/residential/commercial/mixed emissions or biomass and rice straw burning which this paper concentrates on. These factors diverse from natural factors to human made ones as follows; **The nature of the low topography of Cairo** bounded by Giza and Mokatam the western and eastern highland traps pollutants over the city [1]. **Meteorological factors** and thermal inversions that concentrate pollutants in the air layers near the surface of the earth associated by low wind speed and elevated humidity. **Existing pollutants** caused by human activities (inhalable particulates matters less than 10 μ in diameter PM10, Gaseous pollutant; So2 and No2, Co caused by natural reasons, high traffic density and rice straw burning respectively). **This paper focuses on controlling, monitoring and dealing with the biomass and rice straw burning processes to prevent healthcare consequences using an integrated IOT-Based conceptual framework and its technologies.** Figure 1 shows how the black cloud occurs. The problem starts after the harvest season when farmers violate the Egyptian regulations and rules by burning illegally agriculture waste of the

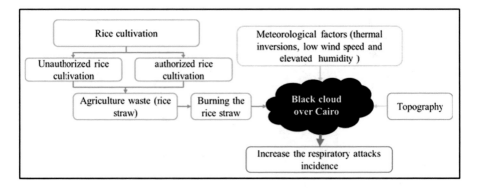

Fig. 1. Black cloud phenomenon, the air pollution problem in Cairo

authorized rice cultivation as well as the unauthorized fields. Burning processes combined with the mentioned factors cause the black cloud that brings pollution levels up to ten times the limits set by WHO. It can persist for days/weeks threatening people's health generally and asthma/respiratory/lungs patients specifically. It can have a small/large impact according to its intensity unless it escalates to get a crises when it hits districts/cities with least healthcare facilities.

3.1 The proposed IOT-Based Solution Scenario for Preventing Acute Air Pollution Episode and its Consequences

In an attempt to prevent the acute air pollution episode we concentrated on preventing the burning process from the early stages of rice cultivation. The proposed solution scenario aims to **prevent** the problem/crisis or **control** its consequences if it happens. It entails various tasks and activities like; **detecting** the rice fields, **detecting** the burning locations, **controlling** the burning processes, **evaluating** and **monitoring** the black cloud; **its movements; the weather condition, analyzing and predicting the related health problems and places, triggering appropriate alarms, proactively trying to manage and eliminate the episodes impact**. In order to realize this goal, the proposed IOT-Based Prevention Scenario consists of five main phases as follows:

Phase One. Allocate the Problem's Origin Sites. During this phase the main concern is to allocate the problem's origin sites; to allocate and find the rice fields. Thus be able to distinguish between the legally and illegally rice cultivated fields, as farmers keep on planting rice much more than they are permitted. In order to do this, we propose to use Remote Sensing-Based Mapping methods as follows [14].

"Optical Remote Sensing-Based Mapping Methods; Used for mapping rice areas worldwide. Through its images it discriminate land use/land cover and measures crop areas as one is able to view the Earth surface in the spectral range 0.4 to 2.5 µm. Satellite sensors are able to derive time-series of vegetation indices. Optical remote sensing sensors like NASA's Moderate Resolution Imaging Spectroradiometer MODIS (used in Egypt) can identify rice areas. *Microwave Remote Sensing-Based Mapping Methods:* Acquires images theoretically under any weather conditions; cloud cover, rain, etc. which are perfect for mapping rice areas. Satellite Microwave data retrieved are great to define rice areas" [14].

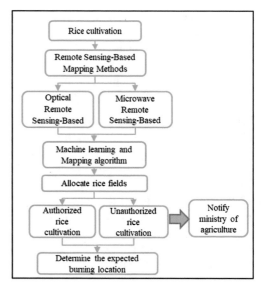

Fig. 2. Phase one; allocating rice fields

These methods consists of: maximum likelihood classifier; establishing relationship between classified image and GPS measured area; and estimation of the rice area under hill shades and non-visible area based on field survey [14]. Using Remote sensing-Based Mapping satellite sensors make it able to get different and various images and land maps that can be processed with mapping algorithms to finally detect the rice fields. At this point an alert to the ministry of agriculture should be autonomously sent through the IOT network or IOT Environmental cloud triggering an alarm of un-authorized rice cultivation fields to get a quick response. Using Machine Learning predictive analytics, the phase ends by predicting the fire locations as shown in Fig. 2.

Phase Two. Stop the Burning Processes. To prevent, monitor and stop the rice straw burning during phase two there are different methods. Two of them could be used in this case: Using fire sensors in expected fire location. This could be applied to the authorized rice fields that we already know and planned for. In this case as an example fire detectors using Arduino Uno interfaced with a temperature sensor for sensing heat, a smoke sensor for sensing smoke and a buzzer to give a direct alarm indication, can be used. One can also add an LCD display to Arduino boards and use IOT technology to smartly connect the monitor process to a certain Fire Security System application or webpage controlled by Arduino. Immediately officers will head to fire locations in order to stop the burning process. Nevertheless we don't advice using these fire detectors as they should be spread overall the whole fields in specific intervals, which will be very expensive due to large no. of fields over the Nile delta as well as the farmers themselves can easily destroy or stop the fire detection process [15]. Using Satellite Imaging;. To estimate the location and extent of the fire in the authorized as well as the unauthorized rice cultivation fields. We propose using MODIS, one of five scientific instruments onboard the satellite platform, Terra, part of NASA's Earth Observation system to produce fire count maps, which show the locations and intensity of fires and monitor atmospheric profiles. At the end of this phase burning process and fire should be allocated precisely, images and data sets will be communicated through the IOT network and platform to the EEAA data servers for further analysis using higher- resolution satellite imagery for example to get a complete picture of the situation. Then EEAA field officers will be alarmed according to their position by their mobile IOT APIs to go and stop the fire and prevent any escalation.

Phase Three. Monitor Air quality and the Black Cloud. It's supposed that we already stopped all known burning processes however this phase address accuracy concerns of the previous one that can result in missing some fire locations emitting a lot of smoke that must be monitored. "To trace the smoke path, one could use the multi-angle Imaging Spec-troradiometer (MISR) data from the NASA Atmospheric Science Data Center (ASDC). MISR Interactive Explorer and trajectory analysis models, can be used to combine the plume data with the fire data and atmospheric models, to learn where the smoke plumes come from, and where they will move" [1]. This information, the environmental IOT sensors data set of the 95 Egypt's "Air Quality Monitoring" stations and the weather stations meteorological data combined will be communicated/analyzed/exchanged predicting black cloud movement, pace and concentration.

Phase Four. Medical Care Facility Notification. Directly after knowing the smoke movement, the proposed system will directly alert specific hospitals and pharmacies through alarms from the IOT cloud and IOT applications informing expected harmful air pollution levels, to get ready, prepare medical protocols, request additional staff, prepare appropriate medicine in forehand before patients arrive or have troubles.

Phase Five. Monitor and Alert Asthma/Lung/Respiratory Patients. Phase five starts in parallel with phase four, when the problem occurs on a larger geographic area and will be difficult for health professionals to visit all patients at the same time giving personal assessment. Hence the proposed IOT solution concentrates more on asthma/lung/respiratory patients, who are registered in the environmental related patient's network. The IOT based environmental-healthcare network supports patients by giving appropriate trusted reliable health decision according to their health state and real on-time environmental parameters. It helps them through quick guidance and direct instructions to move to the nearest available and appropriate hospital or get proper medicine from certain pharmacies according to the situation and the problems range (GPS, Intelligent decision making). From one side, patients will be alarmed about the black cloud through their IOT mobile applications before it hits to get ready, then they will be advised and directed to certain hospitals/pharmacies providing the proper treatment. Alert could be also triggered at patient's families to follow up with their relatives. At the same time patient electronic health records and history will be autonomously gathered and collected from the environmental health related databases and servers at the IOT cloud to be ready if needed. From the other side, if the smoke already hits before patients got alerted, the IOT based health domain solution supposes that asthma patients can be provided with gateway devices and sensors; a smart phone can act as a gateway, while there are small biosensors attached to the patient's body or stuff acting like IOT devices, which can sense continuous and remotely human's body vital sign/parameters as blood pressure, rate of breathing, body temperature, etc. environmental sensors will measure air quality index, PM10, O3, CO, NOx, hydrocarbons etc. these datasets are continuously and automatically collected/processed or sent to relatives or emergency units to take right decisions [16].

3.2 The Proposed IOT-Based Conceptual Framework Components

Based on IOT definition, components and layers presented in Sect. 2, the five proposed phases of the IOT-Based Solution Scenario for preventing Acute Air Pollution Episode and its Consequences in Sect. 3, we present the proposed IOT-Based conceptual Framework components from the IOT perspective. It will use the Internet to merge various heterogeneous things, sensors and smart devices. It contains different layers and devices as follows;

Sensors. *Environmental sensors* at the air quality monitoring stations provide high quality readings for various environmental parameters as CO levels, humidity, hydrocarbons, particulate matter, chemical fumes, fragrances, dust etc. They send the collected datasets to the IOT cloud server to be processed via the IOT platform, tools and applications, doing analysis and creating alarms to be send respectively to smart IOT devices. • *Satellite Imaging*, Optical remote sensing satellite sensors such as

(MISR) and Moderate Resolution Imaging Spectroradiometer (MODIS), Microwave sensors to identify rice fields' locations, detect fire counts and trace the path of the smoke with help of machine learning and mapping algorithms. • *Body area network,* bio-medical sensors and embedded devises monitor human biomedical signals from the human body. They sense, collect and processes data and information everywhere and all the time. Wearable devices acquire the electrocardiogram (ECG) data …etc. while mobile health things deliver efficient healthcare services and new solutions. **Devices.** Different computers, smart phones, laptops, tablets…etc. **Networks.** Networks and connectivity depending on data rate versus data range; satellites, radio waves, mobile networks, WIFI, Zigbee, 4G and Bluetooth. **Applications.** IOT applications, software, smart technologies and programs. Table 1 introduces the proposed IOT - Based conceptual Framework components and layers during the solution phases.

Table 1. The proposed IOT - based conceptual framework components

	Phase 1	Phase 2	Phase 3	Phase 4	Phase 5
Sensors	Optical and Microwave Satellite imaging	(VIIRS) Suomi NPP satellite	MISR sensors and Environmental sensors in the air quality monitoring stations	Environmental sensors in the air quality monitoring stations	VOCs sensors, physiologic parameters sensors and Biomedical sensors
Devices	Computer	Computer, smart phone	Computer	Computer	a spirometer wristband
Network & connectivity	radio waves	radio waves, mobile network	radio waves, WiFi network, mobile network	WiFi network	WiFi network and Bluetooth
Data set	Rice field location	Fire location, fire counts	Aerosols concentration and location	Medical staff and medicine availability	Environmental ad Health parameters
Processing and interface	Remote sensing programme	Remote sensing programme, alarming application on computer and smart phone	Cloud computing, big data, predictive analytic, data mining	cloud computing, smart interface	cloud computing, smart application,
Human Sources	ministry of agriculture responsible	EEAA Officers extinguishing the fire	EEAA employee	Medical staff, pharmacy staff	Patients, relatives, emergency

3.3   The Proposed IOT- Based Conceptual Framework preventing Acute Air Pollution Episode and Its Consequences

After presenting the IOT-Based solution scenario, phases and components individually, now we present the whole proposed framework. The proposed solution argues that using IOT- Based framework will succeed in preventing/controlling the fire counts, monitoring/following the black cloud movement, and reacting and pro-acting to save Egyptians lives. It provides different IOT activities of monitoring, data classification, event triggering, real time alerting, decision supporting. The decisions can be made autonomously as a result of devices communicating with each other's sensing, collecting, exchanging real time information and performing intelligent processes. In addition, remote patient monitoring, handling of service deliverance, generating unprecedented amount of data, can be processed using cloud computing. Collected information can be directly sent to the cloud through the network layer for processing, cleansing, filtering, transformation as well as analyzing, inferencing and visualizing. Figure 3 presents the proposed IOT- Based conceptual framework for preventing the acute air pollution episodes and limiting its negative health impact.

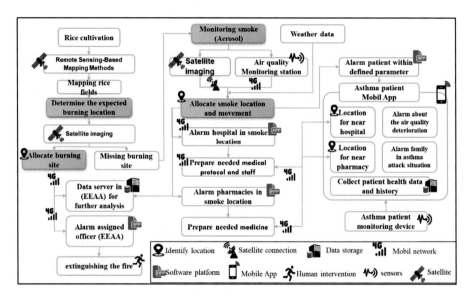

Fig. 3. Proposed IOT-Based conceptual framework to prevent acute air pollution & impact

3.4 Limitations and challenges of the Proposed IOT conceptual Framework

There are many challenges such as: **Security.** Connecting so many devices should force a strong security policy. **Privacy.** The internet connected devices exchanging real people and government private data raise the various privacy issues. **Integration and Interoperability.** It's complicated to exchange data and information between different devices with different protocols. Besides the regular challenges mentioned above facing many IOT frameworks generally, one can count specific limitations facing the proposed framework for the Egyptian context such as follows; **Financial** burden. The use of satellite imaging, sensors and devices is very expensive especially during the contemporary economic reform, which makes it hard to allocate financial resources for the proposed solution. **Lack of qualified** and skilled human resources that can deal and operate with such an advanced system.

4 Conclusion and Further Work

In this Paper we introduced and presented the acute air pollution problem and the black cloud phenomenon threating the environment and public health. We discussed its reasons concentrating on the rice straw burning processes. Accordingly, we proposed and analyzed an IOT-Based conceptual framework for the prevention of acute air pollution episodes and its negative impact on asthma and respiratory patients. The proposed solution argues that using IOT- Based framework will succeed in preventing/controlling the fire counts, monitoring/following the black cloud movement, and reacting and pro-acting to save Egyptians lives. The proposed IOT-Based conceptual framework considers environmental as well as related healthcare dimensions, parameters and networks. An IOT-Based Solution scenario with its five phases has been introduced and discussed thoroughly and the IOT technology used has been introduced in details. As a future work it is recommended to adapt this framework in an attempt to apply and realize it, to gain the IOT benefit of accessibility, reliability and real-time processing to solve such problems, further research addressing the general and specific limitations is needed.

References

1. Leitzell, K.: A black cloud over Cairo. Earthdata (2011). https://earthdata.nasa.gov/user-resources/sensing-our-planet/a-black-cloud-over-cairo
2. Ahram online: Toxic burning of rice straw in Egypt down 13–15% in 2017, 19 November 2017. http://english.ahram.org.eg/News/281781.aspx. Accessed 9 June 2018
3. Lelieveld, J.: The contribution of outdoor air pollution sources to premature mortality on a global scale. Nature **525**, 367–371 (2015). https://doi.org/10.1038/nature15371
4. Abendroth, B., et al.: Cybersecurity Policy for the Internet of Things. Microsoft (2017)
5. Niewolny, D.: How the Internet of Things is Revolutionizing Healthcare. White Paper, Document Number: IOTREVHEALCARWP REV 0 (2013)
6. Mohammeda, Z.K.A., Ahmed, E.S.A.: Internet of things applications, challenges and related future technologies. WSN **67**(2), 126–148 (2017). http://www.worldscientificnews.com/
7. Anbarasan, B.: The Internet of Things – The Thing to Watch. IJSRSET, Themed Section: Engineering and Technology, **3**(3) (2017). Print ISSN 2395-1990, Online ISSN 2394-4099
8. Alvarez, M., et al.: Smart CEI Moncloa: an IOT-based platform for people flow and environmental monitoring on a smart university campus. Sensors (2017). https://doi.org/10.3390/s17122856
9. Talari, S., et al.: A review of smart cities based on the internet of things concept. Energies **10**, 421 (2017). https://doi.org/10.3390/en10040421
10. Middha, K.: Internet of things (IOT), architecture, challenges, applications: a review. Int. J. Adv. Comput. Sci. **9**, 389–393 (2018). ISSN 0976-5697
11. Bilal, M.: A review of internet of things architecture, technologies and analysis smartphone-based attacks against 3D printers. Cornell University Library, arXiv.org arXiv:1708.04560 (2017)
12. El Askary, H.M., et al.: Analyzing Black Cloud Dynamics over Cairo, Nile Delta Region and Alexandria Using Aerosols and Water Vapor Data. IntechOpen (2011). https://doi.org/10.5772/16727, https://www.intechopen.com/

13. Miles, T.: These are the world's most polluted cities (2018). https://www.weforum.org/agenda/2018/05/these-are-the-worlds-most-polluted-cities
14. Mosleh, M., et al.: Application of remote sensors in mapping rice area and forecasting its production: a review. Sensors **15**, 769–791 (2015). https://doi.org/10.3390/s150100769
15. Sharma, A.K., et al.: IOT enabled forest fire detection and online monitoring system. Int. J. Curr. Trends Eng. Res. (IJCTER) (2017). https://www.ijcter.com/published-papers/volume-3/issue-5/iot-enabled-forest-fire-detection-and-online-monitoring-system/
16. Al-Hamadi, H., Chen, I.R.: Trust-based decision making for environmental health community of interest IOT systems. In: IEEE 12th International Conference on Wireless and Mobile Computing, Networking and Communications (WiMob) (2016). https://doi.org/10.1109/wimob.2016.7763201

Smart and Incremental Model to Build Clustered Trending Topics of Web Documents

Mona A. Abou-Of[1(✉)], Hassan M. Saad[2(✉)],
and Saad M. Darwish[2(✉)]

[1] Pharos University in Alexandria, Alexandria, Egypt
mona.abouof@pua.edu.eg
[2] Institute of Graduate Studies and Researches,
Alexandria University, Alexandria, Egypt
hasan.magdy@gmail.com, saad.darwish@alex-igsr.edu.eg

Abstract. The abstract Social media trends, which have become more popular nowadays, introduce a rich hub of a broad spectrum of topics. It is of great importance to track emerging related topics when major events occur. The source of such information would be available not only through social portals but also through news, articles and web portals. All this information is aggregated together, by the proposed news aggregator model, to be useful for retrieving the recent popular trends of a certain category or country. The proposed model addresses the identification of semantically related topics from user preferences and favorites that are added manually by the user. Their textual contexts are acquired from the news search and then a clustering technique is applied followed by tracking of trending topics in term space. By quantitative experiments on manually annotated trends, we compared the model with two other well-known algorithms, using three different online datasets. The presented results demonstrate that the model reliably achieves a better entropy and F-measure, and so outperforms the two other mentioned algorithms.

Keywords: News aggregator · WordNet · Detecting and tracking topics · Semantic similarity · Text mining · Summarization · NLP · Web mining · Incremental clustering · K-Means

1 Introduction

The increasing number of people contributing to the Internet, either deliberately or incidentally has created a huge set of data that gives us millions of potential insights into trending web topics with personal tastes and human behavior. Now, social features are presenting new media on many websites with a great idea called "Trends"; understanding user's interactions is a complex and far-reaching topic. The proposed model looks like a magazine which aggregates data from several data resources, such as prepared categories or user defined resources. All these data are collected and then filtered to be able to deliver the desired information to specific users. So, the main objective is to minimize user's wasted time on "scrolling" by producing trends under an appropriate category or country. As shown in Table 1, we demonstrate a dataset

© Springer Nature Switzerland AG 2020
A. E. Hassanien et al. (Eds.): AMLTA 2019, AISC 921, pp. 888–897, 2020.
https://doi.org/10.1007/978-3-030-14118-9_87

crawled in Oct 2017 under 3 regions and 2 categories; more than 400 data sources for each region covering about 8000 trending topics and 10K topic pairs labeled by a human expert.

Table 1. The trending topics among many regions in Oct 2017.

	KSA	Egypt	World
Sports	#Nassr Hilal Live	#World Cup Qualifier	#EPL watch Live
	#Bein Subscription	#Ahly in Africa	#Real Madrid Struggle
Politics	#Freezing Qatar	#Crush Extremists	#Citizenship Invest
	#Women in Stadiums	#Western Desert Attack	#Hamas cedes Gaza

In this paper, the contribution of the suggested technique has two-fold. The first one is a "Semantic Document Similarity" algorithm which is an enhanced measure of [1], which is used twice in the proposed model: at first in document summarization step then in the incremental clustering step. Document summarization is highly needed when we receive a long web document. The second one is an "Incremental K-Means" (IKMeans) algorithm, which is required to maintain high cohesiveness, expressed in terms of hot trending, while new documents are being added. The process allows new documents to be reassigned to clusters that have already been created. In the next section, we provide a related literature on the subject of trending topics. In Sect. 3, we discuss the ideas, the techniques, and an overview of the proposed model. Section 4 provides an analysis of "Semantic Document Similarity" algorithm. Then we present a novel approach in the second algorithm of incremental clustering schema for maintaining trending topics reports with high cohesiveness with daily news in Sect. 5. Finally, we provide the experimental results compared with other techniques [2] in Sect. 6, followed by the conclusion and the future work in Sect. 7.

2   Related Work

Trending of topics in social media provides an important source for current information about events around the world and can also provide us about user's interest [3]. One task in this area is the detection of topics of general interest the very moment they occur. Such trending topics [1] and temporal characteristics of Twitter topics were studied in [4], in addition to a significant trending topic detection method to summarize what happens in the real world introduced in [5]. Related studies in utilizing and studying trending are proposed in [6] as a two-level learning hierarchy of Fuzzy C-means (FCM) for sensing topics in Twitter. All studies [1–6] are focusing on one side (content only) but [7] tackles this challenge, by identifying coherent topic-dependent linking content generators with spreaders. A thorough understanding of social media discussions and the demographics involved in these discussions has become critical for many applications like business or political analysis [8, 9]. Clustering is usually the main step in topics aggregators. A broad survey of clustering algorithms has recently been given in [10]. In regards to these applications areas, incremental clustering thoughts have been gained from [11, 12]. In this paper, we combine these approaches together as described later to improve clustering performance within one day and in between multiple days capturing topic shift over time.

3   Proposed Approach

The architecture of the proposed model is divided into two phases as shown (in Fig. 1). In the first phase, the model starts by building the main hub repository using different data resources. Then a crawler and a summarizer are used to obtain the summarized documents. In this phase, user preferences are collected and data resources are added manually. Next, in the second phase, the model measures the similarity between documents to categorize the trending web topics with similar patterns by the traditional K-Means. This system tracks all daily news by an incremental building user profiles module which updates user profiles. The following sections describe the different components of the proposed model.

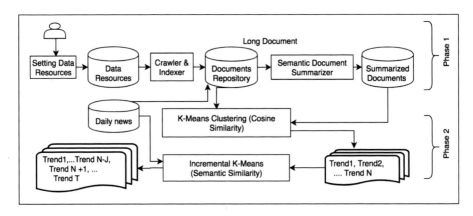

Fig. 1. The architecture of the smart trending web topics model

3.1 Set Data Resources by User. In this step, the user can follow web portals under countries, regions, and categories. In addition, he/she can follow hundreds of trends aggregated dynamically. All common resources are added in the system by default and contain all common pre-configured categories, attached to all sites containing news and all hot topics sources. Moreover, the user is allowed to augment his/her own favorite's sites by adding all URLs on the web portal or mobile applications.

3.2 Crawler and Indexer. In this step, all processing is based on linguistic features. The framework starts with an article, including its recursive visited URLs, as an input to the pre-processing stage to extract the corresponding suitable keywords. These keywords are then used in the next stages to extract the correspondence synonyms which are used in Word Sense Disambiguation (WSD) described in Sect. 4.2. The target here is to measure whether the news list is similar to the input tracked topics or not.

4   Semantic Document Similarity

Automatic text summarization has become an essential need to overcome the problem of long articles. It could be classified mainly as automatic extractive and abstractive. There are many proposals for measuring semantic similarity between two synsets. Here we used the similarity measure shown in (1), to get the normalized overlap of subjects in two sets S(t1) and S(t2), with an enhancement added to it:

$$sim(t_1, t_2) = \frac{1}{2} \cdot \left(\frac{\#(S(t_1) \cap (S(t_2))}{\#(S(t_1) + 1} + \frac{\#(S(t_1) \cap (S(t_2)))}{\#(S(t_2) + 1} \right) \tag{1}$$

Where #S(t) = number of node's subjects in the sentence t.

Given two sentences t1 and t2, the measurement in (1) determines how similar the meaning of two sentences is. The higher the score, the more similar meaning of the two sentences. This semantic similarity measure used in the summarizer and in the IKMeans phases. We enhanced this measure by preprocessing steps (steps 1–3) based on WordNet [13], and post-processing steps (steps 5 and 6 in Algorithm 1) that enhance the overall outcome in addition to the subject's normalized overlap. The proposed Algorithm 1 shows the steps for computing the semantic similarity between two sentences.

Algorithm 1. Semantic similarity between two sentences t1 and t2

---

1. Execute Wordnet Preprocessing: Tokenization -> Stemming -> Part of speech (POS).
2. Find the most appropriate senses of each subject (synset S) by Word Sense Disambiguation using Michael Lesk [14].
3. Optimize the synsets subject's normalized overlap.
4. Compute the similarity of each pair of synsets (S1, S2) of each subject in the two sentences based on the above formula (1).
5. Build a semantic similarity relative matrix $R_{m \times n}$ of each pair of word senses (synset), where m=|t1|, n=|t2|, and R[i, j] is the semantic similarity measure between the most appropriate senses of word at position i of sentence t1 and the most appropriate senses of word at position j of sentence t2. Thus, R [i,j] is also the weight of the edge connecting from i to j.
6. Combine the match results from the previous step to a single similarity value for the two sentences by matching average with Dice coefficient [15].

4.1 Step 1: NLP Preprocessing

4.1.1 Tokenization. Each sentence from the documents repository is partitioned into a list of words, and the stop words are removed. Stop words are frequently occurring, and defined as the insignificant words that appear in a database record, article, or a web page, etc.

4.1.2 Stemming Word. We use the Porter stemming algorithm [16] which is a process of removing the common morphological and inflexional endings of words.

4.1.3 Tagging Part of Speech. This task is to identify the correct part of speech of each word in the sentence. The algorithm takes a sentence and a specified tag set (either syntactic roles as subject, object, ... or functional roles as noun, verb, ...) as inputs. The output is a single best POS tag for each word. The rule-based Brill's tagger is used.

4.2 Step 2: Word Sense Disambiguation (WSD) is the process of finding out the most appropriate sense of a word in a given sentence. The Lesk algorithm [14] uses dictionary definitions (gloss) in WordNet database to disambiguate a polysemous word in a sentence context. The major objective of its idea is to count the number of words that are shared between the two glosses. The more overlapping the words, the more related the senses are.

4.3 Step 3: Normalization of Subjects Overlap. To score the overlap, we use a new scoring mechanism that differentiates between N-single words and N-consecutive word overlaps and effectively treats each gloss as a bag of words. It is based on Zipf's Law [17], which says that the length of words is inversely proportional to their usage. The shortest words are used more often, the longest ones are used less often. Measuring overlaps between two strings are reduced to solve the problem of finding the longest common substring with maximal consecutive. Each overlap which contains N consecutive words contributes N^2 to the score of the gloss sense combination.

4.4 Step 4: Similarity Computation. In this step, we do compute the similarity of the sentences based on the similarity of their subjects. As shown in formula (1), to get the normalized overlap of subjects in two synsets S1 and S2, with an enhancement added to it. sim(S1, S2) is how similar the meaning of two synsets is. The higher the score, the more similar meaning of the two sentences. So if S1 = S2 then sim(S1, S2) would be = 1 and if there are different then the value will be 0–1, this is basically the semantic similarity measure is applied in Algorithm 1.

4.5 Step 5: Building Semantic Similarity Relative Matrix. In this step, an $R_{m \times n}$ matrix is built, where each value represent a single semantic similarity measure between each pair of word senses and the total of each edge is the most appropriate senses of word at position i of sentence t1 and the most appropriate senses of word at position j of sentence t2. If a word does not exist in the dictionary, we use the edit distance similarity instead and output a lower associated weight, and if t1 and t2 are two sets of disjoint nodes, we compute the total matching weight of bipartite graph using the Hungarian [18] method.

4.6 Step 6: Dice Coefficient. The match results from the previous step are combined into a single similarity value for the two sentences. For each formula, we applied a Dice coefficient strategy to compute the overall score.

5   Incremental Documents Clustering Using K-Means

5.1 Document Representation. Each document is represented as a vector using the vector space model (VSM). This is an algebraic model for representing text document (or any object, in general) as vectors of identifiers. Documents and queries are represented as vectors.

5.2 TF-IDF. In this step, we present important words or terms in documents. VSM method measures the similarity value of a document to other document based on the weight values of terms which are gotten from the Term Frequency-Inverse Document Frequency (TF-IDF) method. TF-IDF and VSM methods are usually used for sorting and counting the similarity of documents.

5.3 Finding Normal Similarity Score. We have used cosine similarity to identify the similarity score of a document. The method Find Cosine Similarity takes two argument vecA and vecB as a parameter which is vector representation of document A and B and returns the similarity score which lies between 1 and 0, indicating that document A and B are completely similar and dissimilar respectively.

5.4 Incremental K-Means Algorithm Implementation

Algorithm 2. Incremental K-Means Algorithm Implementation

---

1. Define Threshold MainDist, Period T and Constant K
2. Initial Run (once per period T)
   2.1. Choose k cluster centers and initialize their centroids to predefined (threshold) points inside the dataset.
   2.2. Assign each sample of the dataset to its closest centroid using the cosine similarity measure.
   2.3. Update the centroids (prototypes) by computing the average of all the samples associated with this prototype.
   2.4. If the convergence criterion is not met return to step 2.2 The usual convergence criterion is where the decrease in the objective function is less than a threshold limit (or no centroids changes).
3. Incremental Run (for any arriving newcomer M)
   3.1. Calculate all semantic distance between M and all centers.
   3.2. Finding closest cluster center using semantic similarity measure (Algorithm 1).
   3.3. If this measure is within MainDist then assign M to this closest cluster, else build a new cluster with centroid = M.
4. Archive inactive clusters
5. Repeat Step 3 for any arriving newcomer within T period, otherwise repeat step2

5.4.1 Step 1. In this step, you have to define the three global constants before running the system. MainDist is a threshold value of the maximum allowed distance between newcomers M and all already existing centers, that means this M is belonging to an existing cluster else this M is a new center candidate. T is the interval time used in step 2, in the proposed model the trends are computed once a day. so each day we initialize new clusters centers sets. All trends will be discovered within all hot clusters. K is the initial number of centers in IKMeans, initialized once per period T.

5.4.2 Step 2: Normal K-Means Based on Cosine Similarity. In this step, we have used cosine similarity to identify the similarity score between summarized documents with initial user-defined cluster centers. To implement K-Means algorithm, in next iterations (T period), the centroids can be initialized based on the results previously obtained during the clustering process of the last day. After each document being assigned to its closest cluster center, we recalculate the mean of each cluster center which indicates the new position of cluster center (centroid).

5.4.3 Step 3: Incremental K-Means Based on Semantic Similarity. The initial step methodology is to do a full clustering first for all summarized documents and then do the incremental ones for the newcomer documents. The full clustering uses Cosine similarity for the whole scattered documents to build k clusters. This requires the usage of coarse-grained technique (cosine similarity) for time-consuming. In contrast, for the newcomer document, it requires a fine-grained technique (Semantic Similarity) to fit in one of the already made k clusters. In the proposed model, we are going to schedule the full clustering everyday morning, and then the incremental ones for all new items added within that day through the end of the day. The process then starts all over. So, we start the full clustering and then each incremental (in order) until the entire incremental clustering set is finished. To test this, the system would take a source directory and a directory to place the data hub into. The resulting clusters are a combination of the type of (Full and Incremental) depending on their arriving sequence. To detect the trends of the full and increment, we use all clusters as a pool of candidate trends. Each time an incremental clustering is done, the system updates the setting so the next time they use different centers. All newcomer documents either join the closest clusters or build new candidate clusters.

5.4.4 Step 4. In this steps we mainly detect all inactive clusters and then archive them, this step is highly needed specifically in the large pool of trends to enhance the overall system performance.

5.4.5 Step 5. When the entire incremental clustering set is finished which means that T is expired, restart the full clustering process.

6   Experiments

The clustering results of the proposed model have been evaluated by two well-known measures named Entropy and F-measure. Generally, higher values of F-measure and lower values of entropy are representative of better results in clustering. The results of the proposed IKMeans in comparison with K-Means and Gravitational Search Algorithm (GSA-KM) and K-Harmonic Means and Gravitational Search Algorithm (GSA-KHM) [2] have been reported on Reu_01, and Re0 datasets from Reuters-21587 [19] and Mini_Newsgroup from 20Newsgroups [20] sources in Figs. 2, 3, 4, 5, 6 and 7. The properties of these datasets are given in Table 2.

6.1 First Dataset [Reu_01] this dataset contains 1000 documents under 5 classes. It is represented under 5 countries (USA, Canada, Japan, UK, and Switzerland), based on low overlapping between classes. So IKMeans produces remarkable entropy and F-measures than the other two algorithms (GSA-KHM and GSA-KM) as shown in Figs. 2 and 3 respectively. The IKMeans entropy lies between 0.2 and 0.34 while the F-measure is between 0.74 and 0.85.

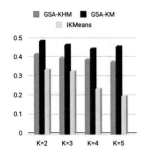

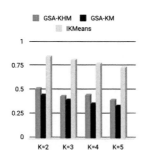

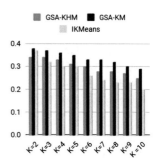

Fig. 2. The entropy comparison of GSA-KHM, GSA-KM, and IKMeans methods on Reu 01 dataset.

Fig. 3. The F-measure comparison of GSA-KHM, GSA-KM, and IKMeans clustering methods on Reu 01 dataset.

Fig. 4. The entropy comparison of GSA-KHM, GSA-KM, and IKMeans clustering methods on Re0 dataset.

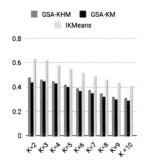

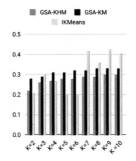

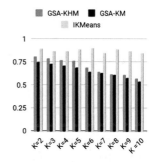

Fig. 5. The F-measure comparison of GSA-KHM, GSA-KM, and IKMeans methods on Re0 dataset.

Fig. 6. The entropy comparison of GSA-KHM, GSA-KM, and IKMeans clustering methods on Mini_Newsgroup.

Fig. 7. The F-measure comparison of between GSA-KHM, GSA-KM, and IKMeans clustering methods on Mini_Newsgroup dataset

Table 2. The properties of text document datasets.

Dataset	Dataset	# Documents	# Classes
Reu_01	Reuters-21587	1000	5
Re0	Reuters-21587	1500	10
Mini_Newsgroup	20Newsgroup	2000	20

6.2 Second Dataset [Re0] this dataset contains 1500 web documents under 10 classes. It is represented under 5 countries (USA, Canada, Japan, UK, and Switzerland) and 5 topics (cocoa, coffee, corn, plywood, and rice). In Fig. 4, the entropy comparison between GSA-KHM, GSA-KM, and IKMeans clustering methods on Re0 dataset shows that IKMeans produces the lowest values in most of the k values between 0.13 and 0.37. While in Fig. 5 the IKMeans presents the highest F-measure between 0.41 and 0.63.

6.3 Third Dataset [Newsgroup]. It contains 2000 documents under 20 classes: atheism, graphics, windows. misc, IBM hardware, Mac hardware, windows.x, for sale, autos, motorcycles, sport.baseball, rec.sport.hockey, crypt, electronics, sci.med, sci. space, religion Christian, talk politics guns, talk politics Mideast, politics.misc, talk. religion. In Fig. 6, the entropy measure lies between 0.21 to 0.43 while in Fig. 7 the IKMeans presents the highest F-measure between 0.85 to 0.90.

As shown by all experiments, presented in the mentioned figures, IKMeans produces very distinctive results better than GSA-KHM and GSA-KM in terms of entropy and F-measure for all used datasets. In some cases IKMeans's entropy is not good as in Mini_Groups dataset, this is because of high overlapping between classes for one item but in the same case, the F-Measure is very distinguishing.

7   Conclusions and Future Work

The presented approach towards trending topic aggregation is based on news context modeling and text mining. The usage of the semantic similarity algorithm in the incremental IKMeans clustering has refined the trend clustering. Such approach is therefore demonstrated to yield accurate clusters. The accuracy is achieved by average F-measure greater than 85% and Entropy less than 0.28 for the three different mentioned datasets. So it outperforms the two other algorithms (GSA-KHM and GSA-KM). The remaining key challenge for future work is the integration of other appropriate context sources in different languages with the right-to-left like Arabic style. Also, the yielded clusters might be improved using fuzzy algorithm since trending topics are not mutually exclusive by nature.

References

1. Fuchs, S., Borth, D., Ulges, A.: Trending topic aggregation by news-based context modeling. In: Proceedings of the 39th Annual German Conference, Advances in Artificial Intelligence, pp. 162–168. Springer, Cham (2016)
2. Mirhosseini, M.: A clustering approach using a combination of the gravitational search algorithm and k-harmonic means and its application in text document clustering. Turk. J. Electr. Eng. Comput. Sci. **25**, 1251–1262 (2016)
3. Sapul, M.S.C., Aung, T.H., Jiamthapthaksin, R.: Trending topic discovery of Twitter Tweets using clustering and topic modeling algorithms. In: Proceedings of 2017 14th International Joint Conference on Computer Science and Software Engineering, Thailand. IEEE (2017)

4. Zhang, Y., Ruan, X., Wang, H., He, S.: Twitter trends manipulation: a first look inside the security of Twitter trending. IEEE Trans. Inf. Forensics Secur. **12**, 144–156 (2016)
5. Georgiou, T., El Abbadi, A., Yan, X.: Privacy-preserving community-aware trending topic detection in online social media. In: Chap. 11 of DBSec 2017: Data and Applications Security and Privacy XXXI, pp. 205–224. Springer, Cham (2017)
6. Muliawati, T., Murfi, H.: Eigenspace-based fuzzy c-means for sensing trending topics in Twitter. In: AIP Conference Proceedings, Indonesia, vol. 1862, no. 1 (2017)
7. Recalde, L., Nettleton, D.F., Baeza-Yates, R.: Detection of trending topic communities: bridging content creators and distributors. In: Proceedings of the 28th ACM Conference on Hypertext and Social Media, Prague, Czech Republic, pp. 205–213. ACM (2017)
8. Georgiou, T., El Abbadi, A., Yan, X.: Extracting topics with focused communities for social content recommendation. In: Proceedings of the 20th ACM Conference on Computer-Supported Cooperative Work and Social Computing, USA. ACM (2017)
9. Morchid, M., Josselin, D., Portilla, Y., Dufour, R., Linarès, G.: A topic modeling based representation to detect tweet locations. In: The International Archives of the Photogrammetry, Remote Sensing and Spatial Information Sciences, France (2015)
10. Wang, J., Zelenyuk, A., Imre, D., Mueller, K.: Big data management with incremental k-means trees–GPU-accelerated construction and visualization. Inform. Open Access J. **4**, 24 (2017)
11. Islam, M.N., Seera, M., Loo, C.K.: A robust incremental clustering-based facial feature tracking. Appl. Soft Comput. **53**, 34–44 (2017)
12. Butnaru, A.M., Ionescu, R.T., Hristea, F.: ShotgunWSD: an unsupervised algorithm for global word sense disambiguation inspired by DNA sequencing. In: Proceedings of EACL 2017, Romania (2017)
13. Miller, G.: WordNet: a lexical database for English. Commun. ACM **38**, 39–41 (1995)
14. Corra, E., Lopes, A., Amancio, D.: Word sense disambiguation. Inf. Sci.–Inform. Comput. Sci. Intell. Syst. Appl.: Int. J. **442**(C), 103–113 (2018)
15. Shamir, R., Duchin, Y., Kim, J., Sapiro, G., Harel, N.: Continuous dice coefficient: a method for evaluating probabilistic segmentations. In: Proceedings of Radiotherapy and Oncology, Barcelona, Spain, vol. 127. Elsevier (2018)
16. Chan, G., Ong, K., Wong, T., Chow, L.: Intelligent context-based pattern matching approaches to enhance decision making. In: Proceedings of International Conference on Computational Science and Its Applications, vol. 10960, pp. 485–497. Springer, Cham (2018)
17. Lee, E.: Partisan intuition belies strong, institutional consensus and wide Zipf's law for voting blocs in US Supreme Court. J. Stat. Phy. **173**, 1722–1733 (2018)
18. Vu, D., Dao, N., Cho, S.: Downlink sum-rate optimization leveraging Hungarian method in fog radio access networks. In: Proceedings of International Conference on Information Networking (ICOIN). IEEE, Thailand (2018)
19. https://archive.ics.uci.edu/ml/datasets/reuters-21578+text+categorization+collection (2007)
20. http://qwone.com/~jason/20Newsgroups (2008)

Combining CMMI Specific Practices with Scrum Model to Address Shortcomings in Process Maturity

Sarah K. Amer[1]([⊠]), Nagwa Badr[2]([⊠]), and Ahmed Hamad[1]([⊠])

[1] The British University in Egypt, El Sherouk City, Egypt
{sara.amer, ahmed.hamad}@bue.edu.eg
[2] Ain Shams University, Abbasia, Cairo, Egypt
najwa_badr@cis.asu.edu.eg

Abstract. Software process improvement (SPI) is the approach to understand the software development process lifecycle and implement necessary changes to achieve a high-quality, maintainable product. The software industry is comprised of small and medium-sized enterprises that adopt agile models as their preferred model of development. The Capability Maturity Model Integration (CMMI) model for software process improvement and quality is used to address aspects of software quality and maturity that Scrum was not initially designed to consider. CMMI is used for its reliable practices and consideration of many aspects of software development in process, project and product. After consideration of literature, three process areas are selected to address necessary elements in Scrum used by software developers in offshoring destinations. The practices are selected based on their practicality in small, agile settings and the ability to be incorporated into Scrum activities without disrupting the models sprint cycles. This opens the way for smaller enterprises to create produce quality.

Keywords: Software Process Improvement ·
Capability Maturity Model Integration · Scrum · CMMI

1 Introduction

Today, quality is crucial to maintain in software more than ever before, hence Software Process Improvement – the systematic procedure for replacing ad-hoc with best practices and standards to ensure high quality, becomes necessary for long-term maintainability. With Software Process Improvement (SPI) comes Software Quality Assurance (SQA), the systematic set of actions to assure that the software development process and product conform to functional and managerial requirements. Offshoring software projects have also become commonplace in recent years due to its affordability, especially when offshored to countries with cheaper currencies. SPI models are many, such as and Capability Maturity Model Integration (CMMI) [1], the ISO 9000 series of standards [2], and Six Sigma [3]. More small and medium-sized enterprises (SMEs) are adopting quality standards by customizing tailoring them down to a more manageable model. SMEs comprise the majority of the global software industry, and

© Springer Nature Switzerland AG 2020
A. E. Hassanien et al. (Eds.): AMLTA 2019, AISC 921, pp. 898–907, 2020.
https://doi.org/10.1007/978-3-030-14118-9_88

are limited in resources and personnel [4–6]. Incorporating complex software process improvement models and quality assurance is extremely challenging and often unattainable despite their proven benefits in literature. This is all the more reason for the software engineering community to aid by introducing process improvement practices without compromising agility, for SMEs either work ad-hoc, or follow an agile methodology such as Scrum. The focus of most studies in literature is medium-or-large scale companies, hence SMEs find enormous challenge in finding customizable solutions to compete with the large conglomerates, taking advantage of their small size and agility (the latter not being a characteristic of large software enterprises) to deploy mature software. SMEs are targeted for offshoring contracts, and therefore a quality assurance standard is what can make one enterprise get the contract [7]. Also, agility for small developers cannot be compromised and there remains the fear for many developers that quality models hinder agility and creativity. In a highly competitive industry, delivering quality content the fastest is what can make an enterprise survive. It is important to consider the elements where agile methodologies like Scrum fall short, and borrow elements from a quality model to create an environment where smaller agile enterprises can increase the maturity of their products.

The rest of the paper is structured as follows: Sect. 2 briefly covers background information on software quality with specific focus on CMMI and Scrum. Section 3 explores relevant related works that focus on creating a bridge between CMMI and Scrum. Section 4 proposes three CMMI Specific Practices that are strong candidates for a customized merger model of CMMI and Scrum. Finally, Sect. 5 provides the conclusion and possible future work.

2 Background

2.1 Agile Methodology and Scrum

The Agile Alliance was formed in 2001, and published the Agile Manifesto that comprises of twelve principles that apply to any agile methodology. Projects done with an agile methodology follow a time-boxed iteration, and every iteration is a full development lifecycle consisting of planning, analysis, implementation, unit testing and acceptance testing. The client is involved until the product is shipped or deployed.

Scrum is a popular agile methodology, adopted in many SMEs around the world for its sprint culture. In the beginning, the Product Owner and Client develop user stories and finalize the Product Backlog. The Product Owner's experience allows them to detect development and technological constraints during those sessions during the validation of the Client's requirements. Then the Product Owner and Scrum master determine the priorities of the requirements, dependencies for the workflow and min-imize rework issues. The user stories are discussed with stakeholders and the difficulty and size of each story is estimated and grouped into proposed releases at the end of every sprint. The Scrum Master works with the development team very closely every sprint, which is typically between 2 and 4 weeks. Each sprint has a Planning Meeting to decide what the output of the sprint should be. A Sprint Backlog is developed at the end of each Sprint meeting.

During the sprint, a short daily meeting discusses current progress to solve technical issues and verify product components. Output for each sprint is a working product that the Product Owner, Client and other stakeholders use and test. Any changes in requirements are discussed during the Review Meeting, and updates are made to the Product Backlog. After the Review Meeting and before the next Planning Meeting, a Retrospective Meeting is held to analyze the problems and possible improvements to team performance and methods used. Feedback is incorporated into the following sprint.

2.2   Software Quality

Software quality exists to address the issue of problematic software solutions and billions of dollars in losses due to incomplete software, or software that does not fit the purpose for which it was made. Late discovery of defects is commonplace and can lead to a slip in process and product quality. The purpose of software quality is to conform to explicitly-stated functional and performance requirements in documentation [8]. It gives rise to systematically monitoring and evaluating software projects to ensure that predefined technical and project-related requirement and quality standards are being met.

Professionals and researchers over the past decades have worked to come up with possible solutions to measure quality. The measurement process can be divided into five main activities: Formulation, Collection, Analysis, Interpretation, and Feedback. Software quality assurance models use metrics that address different aspects of quality. There are three types: project, process and product metrics.

Software metrics are statistical predictions, and each number has 3 dimensions: error, bias and variance. If the software development team ignores them, over-confidence and over-optimism can take over and cause over-expensive software that does not follow requirements. It is unwise to use many metrics together; they are chosen depending on the software's purpose, company size and the stakeholders' priorities.

2.3   Software Process Improvement and CMMI

Software Process Improvement models do not cancel or avoid volatile business environments, but are designed to embrace change rather than respond lately to it. CMMI is a maturity model for the development of software products and services. However, it is not a development life cycle model, like Waterfall. CMMI was developed and is maintained by the Software Engineering Institute (SEI) of Carnegie Mellon University. SEI's principle behind the model's creation is that the quality of a software product is directly affected by the quality of the process used to develop it [10]. CMMI is adopted by thousands of enterprises worldwide, both in developed nations and the growing offshoring destinations in developing countries. The best way to implement CMMI is gradual institutionalization, as it becomes ingrained in the way various processes are performed within an enterprise.

The purpose of CMMI is to raise quality of software development processes rather than the final product, incorporate industrial best practices, and encourage speedy work and early arrival to the market [9, 10]. It embraces change and recognizes risk factors before their effects impair the overall development process.

CMMI Ver. 1.3 comprises of 22 process areas, each comprising of specific and generic goals and practices. The model follows a maturity scale, with Level 1 meaning no conscious implementation of quality, up to Level 5 meaning optimization of resources and renowned innovation and quality [11]. The maturity level and adherence to the model is formally appraised by an external certified party, normally every 3 years. CMMI is a living model due to it being constantly updated by its creators. During the writing of this paper, CMMI Ver. 2.0 was released.

CMMI is a popular choice framework by software organizations worldwide for its focus on process improvement instead of quality control, highly descriptive nature and comprehensive attention to detail, its basis on industry's best practices, and its reduction of costs long-term when properly implemented.

3   Related Work

3.1   CMMI as a Quality Model in Software Enterprises

Many enterprises all around the world have incorporated some kind of quality improvement methodology into their development processes due to low quality in software being a global concern. Countries like the United States, Canada, and Japan are prolific world leaders in software development and innovation. And yet, in Canada as a studied example, only 50% of the enterprises operate at a Level 3 maturity, and flawed expensive software is still an issue [6, 12].

Salman et al. [13] survey several enterprises and find that CMMI appraisal in such environments can be possible, challenging other literature that claim its impracticality. It could be assumed this notion exists because a developing country's economic situation and strength of software development education is slightly behind other countries like Canada and Japan. Another theory points to poor implementation of quality models, such as the study by Margarido et al. [14]. They propose a framework to evaluate the implementation quality of each practice by considering the performance results and level of compliance with standards. Enterprises properly implementing CMMI see an improvement in their performance, and eventually are able to quantitatively measure it.

Offshoring software projects happen across continents, but due to developing countries' cheaper currencies, large contractors target them. There is interest in researching the software industry and quality in those destinations to determine how software quality can be incorporated so as to produce quality software affordably even in the absence of enough resources to incorporate a full-on quality model like CMMI. There are many reasons software quality in developing countries need work, recurring ones being outdated education systems, lack of specialists, understaffing, limited R&D budgets, and management's non-committal to long-term planning of methodological quality. Al-Allaf et al. [15] survey a sample of software organizations in an offshoring destination in the Middle East, and unveil an issue common in many enterprises in the region – quality assurance frameworks in smaller enterprises are only partially adapted in engineering practices and technology. Otherwise, they are ignored in high-level organization issues, collection of metrics and control over the development process enterprise-wide. Larger enterprises were found to more likely implement CMMI successfully. Abdelaziz et al. [16] research CMMI in another offshoring destination, and common issues found include insufficient workforce – an epidemic in several developing nations. This forces employees to work on several unrelated tasks at once, decreasing focus and by extent the individual's productivity. Another is missed functions; rather than the organization implement all CMMI process areas that need to exist in order to improve the process quality, some 'cut corners' and ignore some crucial tasks. Other problems identified are resistance to change, scope creep, unmaintainable systems and a high number of defects. The authors provide advice based on their experience as professional CMMI appraisers on how to alleviate the damages caused by these issues.

3.2   CMMI with Scrum

Scrum is one of the most popular agile models in software enterprises [17]. Researchers have attempted to find alternatives for organizations to introduce manageable SPI frameworks that combine CMMI's comprehensiveness with agile programming's speedy iterations. Scrum is flexible, iterative and quick to client feedback. However, it falls back in quality assurance, process management, and engineering practices. Shortcomings of the model include: little consideration for process quality assurance, little support for building reusable artifacts, little support for building critical software, challenges to project integration, little consideration for risk management and project budget control, and no addressing of suppliers of subcontracting amongst others [19, 20].

Iqbal et al. study the trends of specific practices in the 2nd maturity level of CMMI and collect data to understand why smaller enterprises less likely adopt an SPI framework [5]. Factors included company size, infrastructure support, time constraints, and managerial disinterest. They study small and medium-sized enterprises (SMEs) that implement specific practices from Maturity Level 2, and their analysis indicates that specific practices associated with seven process areas of the second level of CMMI are already followed in many SMEs informally. However, they note that many small enterprises do not have the potential to fully incorporate CMMI, thus giving validity to the hypothesis that SPI is a large-scale expensive endeavor. Quality is what puts some organizations at a great advantage over competitors, and can mean their survival. The authors elaborate on differences in the implementation of practices from CMMI Level 2 between SMEs, and advise to incorporate practices of the Measurement and Analysis (MA) process area. This is due to larger enterprises having the advantage of quantitative measurement of their performance. The works of Chevers et al. [12, 21–23] discuss the impactful-ness of process areas of levels 2 and 3 on the quality of the process of development and quality through case studies in more than one software industry. The authors identify the most highly-ranked practices that can be used to develop a process improvement framework prototype. Amongst these areas were Risk Management (RSKM), Project Planning (PP) and Requirement Development (RD). The authors work on simplifying CMMI as a process improvement framework for small organizations by reducing the eighteen process areas of maturity levels 2 and 3 into ten deemed most influential. Suggested future work entails having a larger and more geographically-varied industry sample, and proposing a model that addresses small developers' limitations by being cost-effective and non-disruptive.

Tosun et al. [24] present their work, named software quality improvement project (SQIP) by selecting specific practices from the CMMI maturity levels 2 and 3. They use two projects to compare the results of the one with no quality processes (default) with the results of the one using SQIP. The case study is conducted in a small organization facing challenges in project management, defect rate and performance measurement. In the enterprise, there is no set process definition, and software is produced for the client by a group of skilled developers with little regard to quality assurance. To measure SQIP performance, the authors measure defect rates across both projects whilst keeping other factors constant. One such element of study is during Requirements Management, the practices of the implemented process area taking more than seven times as long to perform in the SQIP project. However, this considerably cut down coding, and testing time and effort later on. The coding time goes down from 31% to 20%, and testing from 65% to 33% after implementing SQIP. The coding time coding and testing times go down significantly after implementing SQIP. They define a set of best and worst practices to serve the organization's software quality and maintainability, and propose that future work finds solutions to automate metrics and bug data collection to be used by the enterprise for future forecasts.

Farid et al. in [17, 18] address small enterprises' need for structure and maturity in their processes without sacrificing agility. They provide an approach to map specific practices of CMMI to Scrum and then apply it to several companies. They focus on the project management category in particular by considering activities related to planning, monitoring and controlling the project. Afterwards they calculate the coverage of each specific practice using a set of functions. Each mapped practice is either completely satisfied, partially satisfied, or unsatisfied by Scrum's practices. They successfully show several CMMI practices satisfied without compromising agility or mandating extra personnel. Risk Management (RSKM) however remains largely unsatisfied by most Scrum practices, highlighting one element of quality assurance of process missing from the agile model. Their work does not address Supplier Agreement Management (SAM) at all due to there being no Scrum practices pertaining to offshoring aspects.

4 Proposed Specific Practices

The research and contributions in the previous section provide valuable insight into various efforts to bridge the gap between agile methodology and software maturity models. Several process areas from the CMMI model have been studied many times while others are not sufficiently explored, if at all. Notable examples are Measurement and Analysis (MA), Supplier Agreement Management (SAM) and Risk Management (RSKM). The latter two require to be addressed more frequently due to their relevance to offshoring enterprises. Some CMMI practices can neither be mapped to a standard Scrum model nor integrated into it to improve maturity and performance without compromising agility. However, other process area practices can be considered, such as those pertaining to the above mentioned. Process areas that have already been researched by others will not be covered in this section, such as Requirements Management (REQM). Related works indicates that process areas from maturity levels 2 and 3 are the most likely candidates to be added into Scrum without changing its iterative culture. However, they cannot be implemented in full, especially in smaller software enterprises. Each process area from the three mentioned above is broken down into its components of specific practices that are partially supported or unsupported by Scrum at any point in its development cycle. Practices that can be performed by the same development team, Scrum Master and Product Owner will be considered while the rest is eliminated. In Tables 1, 2 and 3, each process area is referred to by its designated abbreviation. Specific Goals are shortened to SG, and Specific Practices are shortened to SP. Specific Practices comprise Specific Goals in CMMI as of Ver. 1.3. The left column indicates the numeric value of each Specific Goal. The second column is the brief description of the specific goal, and the brief description and number of practice(s) selected from that goal to be incorporated into Scrum. The third column determines and elaborates on the level of support the practice has within Scrum, as it may be partially supported, or unsupported. The last column contains the proposed positions within the Scrum model to incorporate the selected practices.

Table 1. Measurement and Analysis (MA) specific goals and practices for Scrum

Goal	Specific practice	Level of support	Proposed positioning in Scrum
MA-SG1	Align measurement and analysis activities *SP1: Establish measurement objectives* *SP2: Specify measures to be collected* *SP3: Specify procedures for data analysis* *SP4: Specify procedures for data collection and storage*	No support No defined parameters Therefore industry-wide metrics used	Sprint planning Daily Scrum meeting
MA-SG2	Provide measurement results *SP1: Collect measurement data*	Partial support Only Burndown charts used No quantitative metric use in any part of lifecycle	Sprint Cycle/Iteration
MA-SG2	Provide measurement results *SP2: Analyze measurement data* *SP3: Store data and analysis results*	No support No defined parameters and methodology Therefore industry-wide metrics used Absence of historical data	Sprint Cycle/Iteration

Table 2. Supplier Agreement Management (SAM) specific goals and practices for Scrum.

Goal	Specific practice	Level of support	Proposed positioning in Scrum
SAM-SG1	Establish agreements with supplier *SP2: Select suppliers* *SP3: Establish agreements*	No support No offshoring/off-location support	Product owner meeting with the supplier Product backlog development
SAM-SG2	Satisfy agreements with supplier *SP3: Accept the acquired product* *SP4: Transition the product*	No support No offshoring support No support for integrating offshored components	Sprint planning Sprint Cycle/Iteration

Table 3. Risk Management (RSKM) specific goals and practices for Scrum.

Goal	Specific practice	Level of support	Proposed positioning in Scrum
RSKM-SG1	Prepare for risk management *SP1: Determine sources of risk and categorize them*	No support Absent active risk categorization and recognition	Product owner meeting with the Client/Product Backlog development Development team communication/Daily Scrum meeting Sprint planning
RSKM-SG2	Identify risks and analyze them *SP1: Identify risks* *SP2: Evaluate, categorize and set priorities to risks*	Partial support Little focus dedicated to active risk identification in any project	Development team communication/Daily Scrum meeting
RSKM-SG3	Mitigate risks *SP1: Develop plans for risk mitigation* *SP2: Implement plans for risk mitigation*	No support No mitigation plans	Development team communication/Daily Scrum meeting Sprint Cycle/Iteration Sprint retrospective

5   Conclusions and Future Work

Due to smaller enterprises with limited resources comprising the majority of the global software industry, it is necessary to find realistic solutions for them to improve the quality of their process, and by extension, the quality of their software. As such enterprises are commonly targeted for offshoring projects, it becomes a matter of sustainability and survival. This work offers beginning steps in the direction of incorporating specific practices from CMMI into Scrum that were not fully present or present at all beforehand. The three process areas in focus are Measurement and Analysis (MA), Risk Management (RSKM), and Supplier Agreement Management (SAM), with the last being unique to this work as no previous literature to our knowledge has attempted to incorporate it. If incorporating practices from Measurement and Analysis (MA) and Risk Management (RSKM) into Scrum operations requires no extra documentation or new personnel, then the model is considered a success. Future work entails creating a customized Scrum-based model that incorporates the three process area practices discussed in this work, and more from levels 2 and 3 of CMMI without changing Scrum's structure or mandating extra personnel.

References

1. Phillips, M.: CMMI® Version 1.3 and Beyond (2010)
2. Hoyle, D.: ISO 9000 Quality Systems Handbook-updated for the ISO 9001: 2015 Standard: Increasing the Quality of an Organization's Outputs. Routledge, Abingdon (2017)
3. Green Belt Body of Knowledge - International Association for Six Sigma Certification. https://www.iassc.org/body-of-knowledge/green-belt-body-of-knowledge/. Accessed 9 Aug 2018
4. Sánchez-Gordón, M.-L., O'Connor, R.V.: Understanding the gap between software process practices and actual practice in very small companies. Softw. Qual. J. 24(3), 549–570 (2016)
5. Iqbal, J., et al.: Software SMEs' unofficial readiness for CMMI®-based software process improvement. Softw. Qual. J. 24(4), 997–1023 (2016)
6. Pino, F.J., García, F., Piattini, M.: Software process improvement in small and medium software enterprises: a systematic review. Softw. Qual. J. 16(2), 237–261 (2008)
7. Beck, K., et al.: Agile manifesto. Softw. Dev. (2001)
8. Haddad, H.M., Ross, N.C., Meredith, D.E.: A framework for instituting software metrics in small software organizations. Int. J. Softw. Eng. 5(1), 69 (2012)
9. Chrissis, M.B., Konrad, M., Shrum, S.: CMMI for Development: Guidelines for Process Integration and Product Improvement. Pearson Education, London (2011)
10. Potter, N., Sakry, M.: Implementing SCRUM (agile) and CMMI together. Process. Group-Post Newsl. 16(2), 1–6 (2009)
11. Pino, F.J., García, F., Piattini, M.: Software process improvement in small and medium software enterprises: a systematic review. Softw. Qual. J. 16, 237–261 (2008)
12. Chevers, D.: The impact of CMMI Levels 2 and 3 practices on process maturity in Canadian software development firms. Int. J. Eng. Sci. Innov. Technol. (IJESIT), 3(4), 743–751 (2014)
13. Salman, Y.B., Karahoca, A., Cheng, H.-I.: Route to CMMI in Turkish Software Companies, pp. 679–682 (2007)
14. Margarido, I.L., Faria, J.P., Vidal, R.M., Vieira, M.: Towards a framework to evaluate and improve the quality of implementation of CMMI® practices. Lecture Notes in Computer Science (including subseries Lecture Notes in Artificial Intelligence and Lecture Notes in Bioinformatics), pp. 361–365 (2012)
15. Al-Allaf, O.N.A.: The usage of capability maturity model integration and web engineering practices in large web development enterprises: an empirical study in Jordan. J. Theor. Appl. Inf. Technol. 39, 150–166 (2012)

16. Abd El Aziz, A.: From CMMI ML2 to ML3 in One Year! Part3: New Practices. Egypt–SPIN Newsletter, Sponsored by SECC (2008). 4-11.Y
17. Farid, A.B., Elghany, A.S.A., Helmy, Y.M.: Implementing project management category process areas of CMMI version 1.3 using scrum practices, and assets. Int. J. Adv. Comput. Sci. Appl. **7**, 243–252 (2016)
18. Helmy, Y.M., Farid, A.B., Elghany, A.S.A.: Simplifying CMMI version 1.3 implementation by using agile practices: an empirical study. Int. J. Intell. Comput. Inf. Sci. (IJICIS) **14**(4), 1431–1446 (2015)
19. Turk, D., France, R., Rumpe, B.: Limitations of agile software processes. In: Third International Conference on Extreme Programming and Agile Processes in Software Engineering (XP 2002) (2002)
20. Palomino, M., Dávila, A., Melendez, K., Pessoa, M.: Agile practices adoption in CMMI organizations: a systematic literature review. In: Advances in Intelligent Systems and Computing (2017)
21. Chevers, D.A., Grant, G.: Software process improvement adoption and benefits in Canadian and English-speaking Caribbean software development firms. Electron. J. Inf. Syst. Dev. Ctries. **77**, 1–15 (2016)
22. Chevers, D.A., Moore, S.E., Duggan, E.W., Mills, A.M.: Identifying key software development practices in the English-speaking Caribbean using the nominal group technique. In: ACIS 2008 Proceedings of 19th Australasian Conference on Information Systems (2008)
23. Chevers, D., Chevers, D.: Software process improvement: awareness, use, and benefits in Canadian software development firms. Rev. Adm. Empres. **57**(2), 170–177 (2017)
24. Tosun, A., Bener, A., Turhan, B.: Implementation of a software quality improvement project in an SME: a before and after comparison. In: Conference Proceedings of the EUROMICRO (2009)

World Perception of the Latest Events in Egypt Based on Sentiment Analysis of the Guardian's Related Articles

Walid Gomaa[1,2(✉)] and Reda Elbasiony[3]

[1] Egypt Japan University of Science and Technology, Alexandria 21934, Egypt
walid.gomaa@ejust.edu.eg
[2] Faculty of Engineering, Alexandria University, Alexandria 11432, Egypt
[3] Faculty of Engineering, Tanta University, Tanta 31527, Egypt
reda@f-eng.tanta.edu.eg

Abstract. In order to infer how the world has perceived the unfolding of events in Egypt during the last eight years, we take the *Guardian newspaper* as a sample study to extract valuable information about the world viewpoints on the big events in Egypt during this period. We perform a sentiment analysis on all the articles in the 'World' section of the newspaper from the beginning of 2010 till the end of 2017 based on just the keyword 'Egypt'. We extracted *Unigram* tokens from each article and used them for making inference using three lexicons dictionaries: afinn, nrc, and bing. The results show that the general trend is slightly negative over all the selected period. Many conflicting feelings were prevalent during this period such as positive, negative, trust, fear, anger and anticipation. The results show also that years 2011 and 2013, where the world witnessed the two uprisings in Egypt, have witnessed the peaks in both positive and negative emotions.

Keywords: Sentiment analysis · Lexicon-based · Guardian · afinn · nrc · bing

1 Introduction

Understanding the world perception of the local events in any country plays an important role in political decision-making in this country. World newspapers are considered as the wealthiest resources for the world viewpoints about the big events in any country. However, extracting valuable information and knowledge from these resources is not an easy task. Crowdsourcing is one of the possible methods to perform this task, however, it is very costly in both time and money. Sentiment analysis via text mining has been proven to be more suitable and accurate method to extract the required information from the related articles [1,3].

Sentiment analysis is a method used to analyze written or spoken data to extract the peoples' opinions and attitudes [8]. Two main techniques are used to detect sentiments; machine-learning-based methods, lexicon-based methods

© Springer Nature Switzerland AG 2020
A. E. Hassanien et al. (Eds.): AMLTA 2019, AISC 921, pp. 908–917, 2020.
https://doi.org/10.1007/978-3-030-14118-9_89

and hybrid methods [7]. Machine-learning-based methods apply supervised and unsupervised machine learning algorithms on linguistic and syntactic features to classify the selected tokens into different opinions [2,6,12].

Lexicon-based methods depend on predefined sentiment lexicons, which are dictionaries of words or phrases and their sentiments as labels on the form of database tuples, to calculate the overall attitude or sentiment of the selected speech or document [4,5]. The sentiment label can be predefined using one of the following forms:

- A score that ranges from a negative value to a positive value representing sentiments starting from the most negative sentiment to the most positive one respectively, e.g. Afinn lexicon [11] which represent sentiments using real values range from -5 to $+5$.
- A predefined set of emotions, e.g. nrc lexicon [9] which classify sentiments into a set of eight emotions (anger, fear, anticipation, trust, surprise, sadness, joy, and disgust).
- A binary fixed categories, e.g bing lexicon [4] which classify sentiments into one of two categories; 'Positive' and 'Negative'.

In this paper, we perform a sort of text analysis, in particular, sentiment analysis, in order to infer how the world has perceived the unfolding of events in Egypt during the last eight years. We take as a sample study the Guardian newspaper. This choice is based on the following reasons: (1) it is one of the highly reputable newspapers in the world, (2) it provides easy accessibility to its articles through an API interface, (3) it is English based newspaper which can be readily analyzed using available sentiment dictionaries, and (4) it is very representative of the Anglo-Saxon perception of the events in Egypt. Using the API interface provided by the Guardian we have extracted all the articles in the 'World' section of the newspaper from the beginning of 2010 till the end of 2017. The extraction is based on just the keyword 'Egypt'. So we started one year before the political upheaval in order to study the preconditions that led to the consequent events in 2011 and thereafter.

2   Experiments and Results

We divided the experiments into two main parts; in the first part, we extracted the articles which mentioned the keyword 'Egypt' and shows the monthly numbers of articles. Then, in the second step, we performed the lexicon-based sentiment analysis on the extracted articles using three different lexicons (afinn, nrc, and bing).

2.1   Word Count

Figure 1 shows an (inverted) histogram for the number of articles that mentioned the keyword 'Egypt' from the beginning of 2010 till the end of 2017. The bins are monthly calculated, though the labels are bimonthly for clarity of visualization.

The histogram is overlaid with a red curve that indicates the general trend of frequency of the keyword 'Egypt' being in the international focus since the beginning of 2010. The grey envelope around the trend curve represents the 95% confidence level. It is apparent that there is a sharp increase since the beginning of 2010 that remains stable till the mid of 2014, after which there is a small decline of focus.

There are four sharp peaks (frequency is more than 100). Two of them correspond to the uprisings in Jan–Feb 2011 and the uprising in Jun–Jul 2014. The third peak is explained by the streak of events that happened towards the end of 2011 including Maspero demonstrations, clashes in Tahrir square, parliamentary elections, and the burning of the Institute d'Egypte. The last peak happened during November 2015 reflecting the drastic explosion of the Russian Metrojet Flight 9268. It is apparent that some of the major events in Egypt such as the Sinai mosque attack in November 2017 did not get much attention, implying that only events that have more of an international and/or regional consequences get more attention and international publicity.

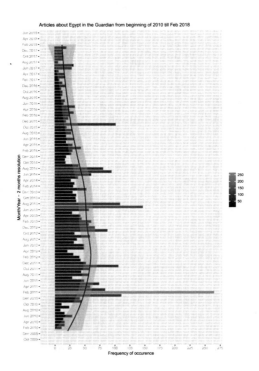

Fig. 1. Monthly histogram of the count of the keyword 'Egypt' over the period 2010–2017

2.2 Sentiment Analysis

The analysis is essentially done on a monthly basis for each year from the beginning 2010 till the end of 2017. *Unigram* tokens are extracted from each article and used for making inference using three lexicons dictionaries: afinn, nrc, and bing.

Sentiment Analysis Based on the afinn Lexicon. The afinn lexicon is a list of 2,476 pre-defined English words each assigned a score that ranges from -5, the most negative sentiment, to 5, the most positive sentiment. The list of unigram tokens, extracted from the articles, are inner joined with the afinn lexicon and the count for each score per month (fixing the year) is computed. Then, a weighted affin-sentiment average is computed for every month, where the relative frequencies are taken as the weights.

Figure 2 shows the average monthly sentiment score for every year 2010–2017. The general trend as can be seen is slightly negative over all these years. Most years exhibit cyclic behavior of increase and decrease in positive attitudes, which can be seen as a natural consequence of the ongoing of events that carry with it cyclic patterns of hope followed by depression and/or fear. However, it is apparent that these alterations are rather mild. However, for some years there is a rather strong trend. For example, the year 2010 (just prior to the first uprising), there was a steadily increase in positive feeling. This seems counter-intuitive given the later sudden unfolding of events at the beginning of 2011. So the question may arise whether the western media truly expresses the undercurrents in Egypt! Similar observation for the year 2016 which had witnessed a sharp increase of positive attitudes starting from May. We will below investigate more of the point clouds for these years.

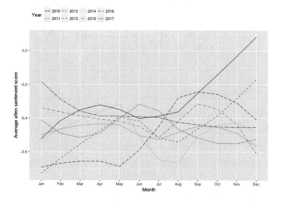

Fig. 2. Average monthly afinn sentiment score for every year spanning 2010–2017.

Figure 3 shows afinn-based sentiment analysis over the whole period 2010–2017 both at a monthly and yearly resolution. As can be seen from the trend curve (in red) there has been a steady decrease in positivity that reached its peak in the year 2013, then there has been a reverse direction since then, though slow. Note that there has been a positive peak around September–October 2015, though with the downing of the Russian air jet, this has been followed by a negative decline towards the end of 2015 and beginning of 2016. What can also be seen is that the overall trend is smooth, meaning that all sentiments, to a large extent, have been conservative; no sustained sharp optimism or pessimism.

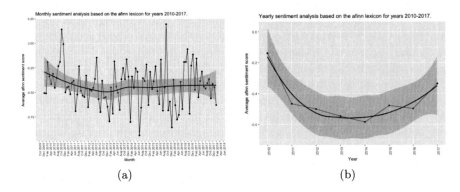

(a)   (b)

Fig. 3. Average monthly and yearly afinn sentiment analysis over the whole period 2010–2017. The red curve is the trend with 95% confidence envelope.

Sentiment Analysis Based on the nrc Lexicon. The nrc emotion lexicon is a list of words, the current collection has 13,901 words, and their associations with eight emotions (anger, fear, anticipation, trust, surprise, sadness, joy, and disgust) and two sentiments (negative and positive). The annotations were manually done through Amazon's Mechanical Turk [9,10].

Figure 4 shows nrc-based emotion histogram for every year in the range 2010–2017. It is evident that strong emotions are not prevalent; these include: surprise, sadness, joy, disgust. The most prevalent emotions are: positive, negative, trust, and fear. These are followed by: anger and anticipation. These results are in harmony with the afinn-based analysis in that, the general mood is overall conservative with mild transitions between positive and negative sentiments. It can also be seen that the year 2013 has witnessed the peaks in both positive and negative emotions, which can be described as post- and pre-uprising in June-July that year respectively. Along side these are peaks of trust and fear in the same year as also markings of the post-and pre-uprising in June-July. The same four emotions also pop up in the respective post- and pre-uprising in January 2011, though at a mild scale compared to that of year 2013. It is of interest that all emotions exhibit the least degree in the two years 2010 and 2017 which can be considered the years that predates and postdates the political upheavals in between.

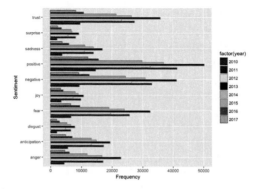

Fig. 4. Yearly sentiment analysis based on the nrc 10 emotions.

Figure. 5a shows the same analysis done aggregately over the whole period 2010–2017. It is evident that the 'positive' sentiment is dominant followed by 'negative' indicating the constant fluctuations as events are unfolded in constant alternations between good and bad. This periodicity is also manifested in the next domination of the opposing emotions of 'trust' and 'fear'. Followed again by the rather opposing emotions of 'anger' and 'anticipation'. This indicates in general the great deal of uncertainty in the last few years, though, this uncertainty has been met wisely by the people reflecting a deep historical confidence in the unfolding of the future.

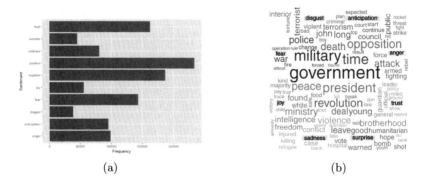

(a)   (b)

Fig. 5. (a) Sentiment analysis based on the nrc 10 emotions over the whole period 2010–2017 and (b) A word cloud, based on nrc lexicon, over the whole period 2010–2017.

A *word cloud* is a visualization of all words inside a text corpus in which the size of each word is proportional to its relative frequency in the text and colors of words are indicative of their associative emotion. Figure 5b shows the word cloud, based on the Guardian articles, over the whole period 2010–2017. It is based on the nrc lexicon, where the 'positive' and 'negative' sentiments are excluded for visualization feasibility. Emotions are shown in black font highlighted in grey. The major words include 'government', 'military' and are associated with the emotion of 'fear'; 'revolution' is associated with 'surprise' as the unfolding of the uprising in Jan 2011 was not expected. 'brotherhood' is associated with 'trust', as it is here understood in literal meaning rather than associated with the particular political faction.

Figures 6 and 7 show the word clouds around the period of the major uprisings in Jan–Feb 2011 and Jun–Jul 2013. In the former we illustrate two periods, the first includes Jan–Feb 2011 and the second adds the month of March. Notice that the major words include 'government', 'opposition', and 'revolution'; followed to a lesser extent by 'president', 'military', 'police', and 'corruption'. Notice that 'brotherhood' is a minor word, and is wrongly associated with the 'trust' sentiment as its semantical interpretation in the nrc lexicon (fraternity), as a unigram, does not have any political association. Including March in the word cloud does not change much; this can be attributed to the relaxation of the major events in addition to less coverage from the international media as major events are smoothed out. On the other hand the word clouds in Fig. 7 reflect a different reality consistent with the state of affairs at that time. We see in this case 'brotherhood' as the main word as the Muslim Brotherhood were in power at that time while absent, even in participation in the first uprising in Jan–Mar 2011. Subsequent main words include 'military', 'coup', and 'revolution' which might indicate some confusion in the Guardian's, and maybe in the western media in general, perception of the events in Egypt during Jun–Jul 2013. Other expected main words include 'opposition', 'government', 'violence', and 'president'.

(a)   (b)

Fig. 6. Sentiment word cloud based on the nrc lexicon. (a) during Jan–Feb 2011, (b) during Jan–Mar 2011.

A *radar chart* is a graphical method for displaying the data in a two-dimensional chart of three or more variables (sentiments in our case) represented on corresponding axes all starting from the same point. Figure 8a shows the radar chart for the years 2011, 2013, 2017, and over the whole period 2010–2017.

(a)   (b)

Fig. 7. Emotional word cloud based on the nrc lexicon. (a) during Jun–Jul 2013, (b) during Jun–Aug 2013.

It is constructed based on the nrc lexicon using relative frequencies for each corresponding period. As shown it is almost the same for the studied periods, and the main emotions are both 'trust' and 'fear', followed by 'anger' and 'anticipation'.

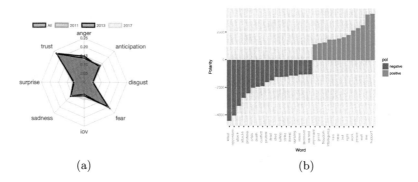

(a)   (b)

Fig. 8. (a) Radar chart for the years 2011, 2013, 2017, and over the whole period 2010–2017. Sentiments are based on the nrc lexicon. (b) Major sensational words along with their polarities ($|polarity| \geq 1000$) over the whole period 2010–2017.

Sentiment Analysis Based on the bing Lexicon. The bing lexicon categorizes words in a binary fashion into 'positive' and 'negative' categories/sentiments. This list is compiled over many years by Liu and collaborators starting from their first paper [4]. In this experiment the frequency of each sentiment is counted, then a polarity score is computed $p = n_{pos} - n_{neg}$, where n_{pos} and n_{neg} are the frequencies of positive and negative sentiment respectively. Figure 9 shows the polarity score both on a yearly basis and over the whole period 2010–2017. Generally, there is a tendency and attraction towards the neutrality position ≈ 0, which resembles the case above with the nrc lexicon where counterparts come together in almost equality, for example, the 'positive' and 'negative' sentiments are together the most dominant. For half the years the polarity is almost constant. The second half, including 2011, 2012, 2013, 2014, 2015, witnessed sharp changes. The years 2011 and 2013 have witnessed the two major uprisings, where eventually towards the end of these years the polarity leaned upwards towards neutrality and even positivity in 2013. The mid of 2012 witnessed the presidential decree of calling into session the dissolved parliament and the corresponding rejection to that decision by Egypt's Supreme Constitutional Court, followed by the resignation of the Minister of Defense and Chief of Staff; a lot of protests and demonstrations; then in November that year came the presidential declaration that in effect immunized any presidential action from any legal challenge which caused fury in the streets. So all such uncertainty, instability, and confusion induced sharp negative sentiment towards the end of 2012. From Fig. 9(b), it is evident that the most negative polarization occurred during the year 2013 after which recovery occurred until it reached a state similar to that of 2010.

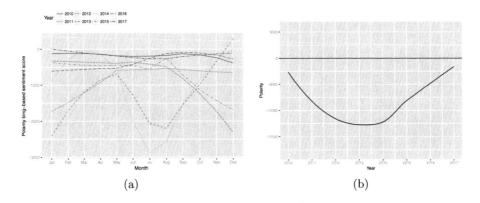

Fig. 9. Polarity score based on the bing lexicon.

Figure 8b depicts the major sensational words along with their polarities. Only words with significant polarities are depicted $|polarity| \geq 1000$. It is evident that there are more negative words and more intense of them. The most prominent negative words include variants of violence words in addition to those of protesting, crisis, and attacking. The positive words include 'support', 'peace', 'right', 'freedom', etc. The aggravation, intense, and frequency of the negative words over the positive ones indicate the nature of the stream of events during the last seven years as being rather more violent and desperate.

3   Conclusions

During the recent seven years, many critical events have occured in Egypt; two main uprisings in years 2011 and 2013, two presidential elections, and some other serious accidents. The world newspapers reflected the world anxiety about these events and their influence on the middle east region. We studied the Egypt's related articles in the Guardian newspaper in this period using lexicon-based sentiment analysis using 3 different lexicons. Based on our study, we found that the general emotion was slightly negative during this period. Many contradicted feelings were prevalent such as (positive, negative, trust, and fear) especially in the two years of uprisings which may reflected the political polarization in Egypt during this period.

References

1. Balahur, A., Steinberger, R., Kabadjov, M., Zavarella, V., Van Der Goot, E., Halkia, M., Pouliquen, B., Belyaeva, J.: Sentiment analysis in the news. arXiv preprint arXiv:1309.6202 (2013)
2. Boiy, E., Moens, M.F.: A machine learning approach to sentiment analysis in multilingual web texts. Inf. Retr. **12**(5), 526–558 (2009)
3. Godbole, N., Srinivasaiah, M., Skiena, S.: Large-scale sentiment analysis for news and blogs. Icwsm **7**(21), 219–222 (2007)
4. Hu, M., Liu, B.: Mining and summarizing customer reviews. In: Proceedings of the Tenth ACM SIGKDD International Conference on Knowledge Discovery and Data Mining. KDD 2004, pp. 168–177. ACM, New York (2004). http://doi.acm.org/10.1145/1014052.1014073
5. Kim, S.M., Hovy, E.: Determining the sentiment of opinions. In: Proceedings of the 20th International Conference on Computational Linguistics, p. 1367. Association for Computational Linguistics (2004)
6. Maas, A.L., Daly, R.E., Pham, P.T., Huang, D., Ng, A.Y., Potts, C.: Learning word vectors for sentiment analysis. In: Proceedings of the 49th Annual Meeting of the Association for Computational Linguistics: Human Language Technologies, vol. 1, pp. 142–150. Association for Computational Linguistics (2011)
7. Maynard, D., Funk, A.: Automatic detection of political opinions in tweets. In: Extended Semantic Web Conference, pp. 88–99. Springer, Heidelberg (2011)
8. Medhat, W., Hassan, A., Korashy, H.: Sentiment analysis algorithms and applications: a survey. Ain Shams Eng. J. **5**(4), 1093–1113 (2014)
9. Mohammad, S., Turney, P.D.: Crowdsourcing a word-emotion association lexicon. Comput. Intell. **29**, 436–465 (2013)
10. Mohammad, S.M., Turney, P.D.: Emotions evoked by common words and phrases: using mechanical turk to create an emotion lexicon. In: Proceedings of the NAACL HLT 2010 Workshop on Computational Approaches to Analysis and Generation of Emotion in Text. CAAGET 2010 (2010)
11. Nielsen, F.Å.: A new ANEW: evaluation of a word list for sentiment analysis in microblogs. arXiv preprint arXiv:1103.2903 (2011)
12. Pang, B., Lee, L., Vaithyanathan, S.: Thumbs up?: sentiment classification using machine learning techniques. In: Proceedings of the ACL-02 Conference on Empirical Methods in Natural Language Processing, vol. 10, pp. 79–86. Association for Computational Linguistics (2002)

Parallel Computation for Sparse Network Component Analysis

Dina Elsayad[(⊠)], Safwat Hamad, Howida A. Shedeed,
and M. F. Tolba

Faculty of Computer and Information Sciences,
Ain Shams University, Cairo, Egypt
{dina.elsayad, shamad, Dr_Howida}@cis.asu.edu.eg,
fahmytolba@gmail.com

Abstract. The Gene regulatory network analysis is one of the gene expression data analysis tasks. Gene regulatory network goal is determining the topological order of genes interactions. Moreover, the regulatory network is a vital for understanding genes influence on each other. However, the main challenge confronting gene regulatory network algorithms is the massive data size. Where, the algorithm runtime is relative to the data size. This paper presents a Parallel computation for Sparse Network Component Analysis (PSparseNCA) with application on gene regulatory network. PSparseNCA is a parallel version of SparseNCA. PSparseNCA enhanced the computation of SparseNCA using a distributed computing model. Where, the workload is distributed among P processing nodes, PSparseNCA is more efficient than SparseNCA. It achieved a better performance and its speedup reached 12.33. In addition, PsparseNCA complexity is $O(NM^2/P)$ instead of $O(NM^2)$ for SparseNCA.

Keywords: Gene data · Bioinformatics · Component analysis ·
High performance · Parallel · Regulatory network

1 Introduction

Gene data analysis is one of the Bioinformatics research areas that aims to understand the gene functionality [1]. Bioinformatics [2] is interdisciplinary research filed that combines computer science, biology, information engineering, mathematics and statistics to manage, process, interpret and analyze biological data. Bioinformatics primary goal is increasing the understanding of biological processes [3]. The main research areas in Bioinformatics are: Computational evolutionary biology [4], Gene expression analysis [1], Sequence analysis [5], Genome annotation [6], Protein expression analysis [7], Protein-protein docking [8], Analysis of regulation [9], Analysis of mutations in cancer [10], Predictions of protein structure [11], Comparative Genomics [12], High throughput image analysis [13], Modeling biological systems [14] and Microarrays [15].

One of the vital research fields in bioinformatics is Microarrays. It is a multiplex lab-on-a-chip that use high throughput screening methods for assaying massive amounts of biological material [16]. Microarrays has different types, each depends on this biological material. One of the microarrays type is DNA microarrays [17].

© Springer Nature Switzerland AG 2020
A. E. Hassanien et al. (Eds.): AMLTA 2019, AISC 921, pp. 918–927, 2020.
https://doi.org/10.1007/978-3-030-14118-9_90

The DNA microarray can be defined as a high throughput experimental technique, which measure expression levels of a large numbers of genes simultaneously. The gene expression level is controlled by the amount of gene mRNA. Where, more mRNA indicates more gene activity. Furthermore, the gene activity controls the gene transcription. Where, the gene will be transcribed if and only if the gene is active [18]. The DNA microarrays is a useful technique for gene expression analysis, identification of disease genes and development of new diagnostic tool. The gene data analysis is a big challenge because of the massive data size and the analysis process that involves many computational tasks [19].

Gene data analysis process starts with subset extraction where the differentially expressed genes extracted. Those expressed genes called discriminator genes. Furthermore, the input of the other analysis tasks such as clustering, search and network analysis can be the differentially expressed genes [20]. The gene regulatory network [20] goal is to analyze the genes interaction topological organization. The recent gene regulatory network analysis techniques can be classified into four main categories: reverse engineering techniques [21–26], the regression techniques [27–31], mutual information based techniques [32–37] and component analysis techniques [38–48]. In fact, almost all of these techniques are computationally intensive and time consuming. Therefore, parallel techniques are required to enhance the techniques performance and speedup.

The rest of the paper is organized as follows; the needed background and the related work are provided in the next section. Section 3 provides the proposed algorithm PSparseNCA (Parallel Computation for Sparse Network Component Analysis). Section 4 discusses the implementation and results. Finally, Sect. 5 provides the conclusions and future work.

2 Background and Related Work

The gene regulatory network (GRN) [20] is one of computational tasks for the gene data analysis. The focus of gene regulatory network is to study the topological order of gene interaction. In addition, it is concerned with investigating how the genes influence each other. Figure 1 illustrates the regulatory network of transcription factor family GLI in human. As shown in Fig. 1 the gene regulatory network components is a set of nodes and sets of edges. Where, nodes represent genes or transcription factor and edges represent regulatory interactions between genes. Transcription factor (TF) [49]; also known as sequence specific DNA binding factor, is a protein that controls the genetic information transcription from DNA to mRNA. This control is done through binding to a specific DNA sequence. As shown in Fig. 1 the node has different shapes based on its type. Where the ellipses represent TFs and boxes represent genes. On the other hand, hexagons represent clustered genes. Moreover, in case of known protein-DNA binding, the connection is represented by red line.

As stated before, the gene regulatory network techniques can be classified into four categories (Fig. 2): regression based techniques, reverse engineering techniques, mutual information based techniques and component analysis techniques. Furthermore, the component analysis technique is categorized into a number of sub-categories.

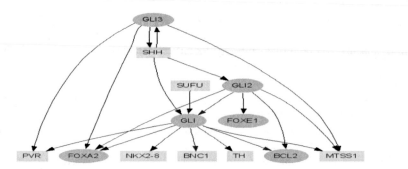

Fig. 1. Gene regulatory network of TF family GLI in human (Image source: http://rulai.cshl. edu/TRED/GRN/Gli.htm)

The component analysis technique sub-categories are: Principal Component Analysis (PCA) [38], Singular value decomposition [39], Independent component analysis [40–42] and Network component analysis (NCA).

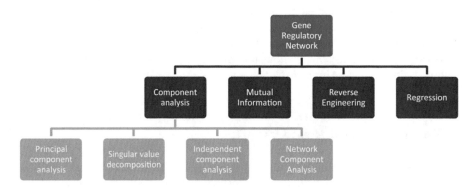

Fig. 2. Gene regulatory network techniques

The singular value decomposition (SVD) [39] objective is to discover the underlying patterns in the gene expression data. SVD is based on the idea that a small number of the characteristic modes can capture the complicated gene expression profiles. These characteristic modes capture the temporal change pattern of gene expression data. As preprocessing step before applying the SVD, the rows and columns must have a zero mean. This is applicable by subtracting the mean values of the data and performing iterative normalization procedures for the rows and columns [50]. The singular values are the square root the Eigen-values of $[A^T A]$ where [A] is the gene expression data matrix.

The principal component analysis technique (PCA) [38] is a dimensional reduction technique that determines the key variables. These key variables illuminate the distinction in the data and improve the data investigation and perception. In other words, when PCA is applied on a data set [A] of M observation in N variables; in the case of the gene expression data, the observation are the genes and the variables are the samples or experiment conditions, it finds new x variables which reduce the dimensionality of [A]. These new x variables; called the principle components, must be mutually uncorrelated and orthogonal. In addition, it must cover the variance in the original data as much as possible. These principle components is a linear combination of the original variable.

Another gene regulatory network technique is the independent component analysis (ICA) [40–42]. The ICA is an unsupervised analysis technique that based on the usage of some hidden variables [Y] to model the gene data matrix [A] as expressed in (1). Where [X] is the new independent component variables. Moreover, the statistical dependences; computed by mutual information (MI), between the columns of matrix [X] must be minimized. The limitation of ICA is data sensitivity particularly the regulatory variables whose influence of the genes follows the Super Gaussian distribution [42].

To avoid limitations of Independent Component Analysis (ICA) technique, Chang et al. presented the Network Component Analysis (NCA) [44–48]. As expressed in (2) the model of NCA considers the noise of data. Where, the gene expression data is denoted by matrix [Y] of size (M × K); M is number of genes, K represent number of experiment (samples). The connectivity matrix is denoted by matrix [A] of size (M × N); N denotes number of TFs. In addition, the transcription factor (TF) activity is presented by matrix [S] of size (N × K). Furthermore, the Γ matrix denotes the Gaussian noise, which affects the model measurement certainty. SparseNCA (Sparse Network Component Analysis) [48] is the recent attempt to solve the NCA model especially in case of incomplete prior information about the connectivity matrix [A].

$$A = XY \tag{1}$$

$$Y = AS + \Gamma \tag{2}$$

SparseNCA (Fig. 3) is a two-steps algorithm that uses subspace-based method to estimate the connectivity matrix [A] while imposing a sparsity constraint to incorporate the influence of incomplete prior information. This prior information determines indices of zeros in connectivity matrix [A]. Either this information may be missing because of an error in measurement process or existence of unknown additional zeros [51]. After estimation of [A] the TFA matrix [S] is estimated using least squares approach. At each iteration the [A] is estimated by (3) and weight vector is estimated by (7). As the other GRN techniques, SparseNCA is time-consuming algorithm because of the massive size of gene data. So, parallel algorithms are needed to enhance the speedup and computational runtime. The following section provides a parallel version of SparseNCA called PSparseNCA.

$$a_m(j+1) = \frac{(Q_{11} + \lambda W_m(j))^{-1} 1}{1^T (Q_{11} + \lambda W_m(j))^{-1} 1} \tag{3}$$

$$Q_{11} = H_m^T Q H_m \tag{4}$$

$$H_m = \begin{bmatrix} I_{(N-L_m)} \\ 0_{L_m X (N-L_m)} \end{bmatrix} \tag{5}$$

$$Q = U_0^\wedge \, U_0^{\wedge T} \quad \text{Where } U_0^\wedge \text{ is the M-N columns of } U^\wedge of \text{ SVD}(Y) \tag{6}$$

$$w_{i(j+1)} = \left[\left(a_{mi}^2 (j+1) + \in (j+1) \right) \right]^{-1} \tag{7}$$

Input	[Y] : Gene Data matrix
Step 1:	Initialize A(0) = I
Step 2:	For n = 1,2,....,N do
Step 2.1:	For j=1,2,..... do
Step 2.1.1:	Update a_n(j) using (3)
Step 2.1.2:	Update weight vector W using (7)
Step 3:	Form updated matrix A = [a^T_1 a^T_2 ... a^T_N]^T
Step 4:	Update S = (A^T A)$^{-1}$ A^T Y

Fig. 3. SparseNCA algorithm

3   PSparseNCA

The gene regulatory network plays a vital role in understanding the genes interaction and its influence on each other. Unfortunately, most of the recent techniques is time consuming because the huge volume of gene data. Therefore, parallel algorithms should be used to enhance time complexity and speedup. PSparseNCA (Parallel computation for Sparse Network Component Analysis) algorithm is a parallel version of SparseNCA algorithm (shown in Figs. 4 and 5).

There are two data parallelism model, functionality parallelism and data parallelism. In functionality parallelism, the algorithm is divided into independent functions. Where, each independent function is executed on different processing node simultaneously. One the other hand, in data parallelism model; the data is distributed among the processing nodes. Where, each processing node process chunk of the data at the same time. The PSparseNCA use the data parallelism technique.

From hardware architecture perspective, there are two parallelism type, shared memory model and distributed model. In distributed model, each processing node has its private memory and nodes communication is performed through sending/receiving messages. While in shared memory model, the processing nodes share common memory space. Moreover, to take full advantage of the available computing resources, the two models can be combined together. This model called the hybrid model. In hybrid model, each processing node has its own memory as the distributed model. In addition, each node share the same memory as in the shared memory model.

The proposed algorithm PSparseNCA implement the distributed model. Where the data is distributed among P processing nodes; each node process an N/P data item. After that, the master node gathers the results back from the P processing nodes.

Input	[Y] : Gene Data matrix
Step 1:	Initialize A(0) = I
Step 2:	Estimate [A] and weight vector W in parallel
Step 2.1:	For j=1,2,..... do
Step 2.1.1:	Send chunk of the data for each processing node
Step 2.1.2:	Collect data from the processing nodes
Step 3:	Form updated matrix A = [a_1^T a_2^T ... a_N^T]^T
Step 4:	Update S = $(A^T A)^{-1} A^T Y$

Fig. 4. PSparseNCA: master node

Input	[A] : Connectivity sub matrix
	W : weight sub vector
Step 1:	Update $a_n(j)$ using (3)
Step 2:	Update weight vector W using (7)
Step 3:	Send connectivity sub matrix [X] and weight sub vector W to the master node

Fig. 5. PSparseNCA: worker node

4 Implementation and Results

This section demonstrates some experiments to evaluate the performance of the proposed algorithm PSparseNCA against the original algorithm SparseNCA. The PSparseNCA is implemented using C++ with MPI. While, SparseNCA is implemented using C++. PsparseNCA complexity is $O(NM^2/P)$ instead of $O(NM^2)$ for SparseNCA; where P is number of processing nodes. The experiments were conducted on the Bibliotheca Alexandrina High Performance Computing cluster, where each processing node is Intel ® Xeon® CPU E5-2680 v3 @ 2.50 GHz. For comparison, three gene datasets are used. Table 1 shows the size (number of genes x number of samples) of each dataset.

Table 2 indicates the overall execution time (in seconds) of PSparseNCA and SparseNCA. Where, PSparseNCA was running using 40 processing nodes for the first two datasets and 80 processing nodes for the third dataset; the number of processing nodes is limited by maximum number of TFs in dataset. The table shows that PSparseNCA is more efficient than SparseNCA providing better performance and reaching a speed up of 12.33 for the third dataset. Furthermore, the computational speedup of PSparseNCA using different numbers of processing nodes is depicted in Fig. 6 using the third dataset. As shown in Fig. 6, the speedup is increased in case of 16 processing node duo to the balanced work load and communications cost between nodes. After that it decreased duo to communication cost between nodes. The speedup peak value 12.33 occurred at 80 processing nodes.

Table 1. Gene datasets used for comparison

Dataset No.	1	2	3
Size	1247 × 69	921 × 40	1240 × 88

Table 2. Algorithm overall execution time (in seconds)

Dataset No.	1	2	3
SparseNCA	974.58	393.79	32698.56
PSparseNCA	84.94	40.68	2651.08
Speedup	11.47	9.68	12.33

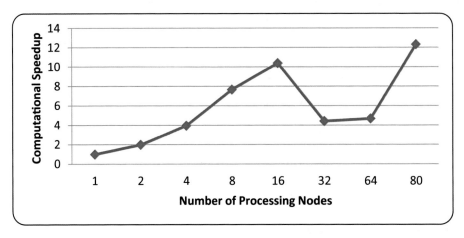

Fig. 6. Algorithm computational speedup of dataset No. 3

5   Conclusion and Future Work

One of gene data analysis tasks is gene regulatory network. The gene regulatory network goal understands the gene interaction topological order and how the gene influences each other. One of the gene regulatory techniques is SparseNCA that construct reliable GRN even in case of incomplete prior information. As well as the other GRN techniques, SparseNCA is computational intensive and time-consuming algorithm. Therefore, PSparseNCA is introduced to enhance speedup and time complexity of SparseNCA. PSparseNCA uses the data parallelism technique on distributed memory model to enhance the speedup of SparseNCA. Where the gene data is distributed among P processing nodes. Furthermore, each processing node has its own memory and communicating with each other by sending/receiving messages. PSparseNCA successfully reduced the runtime and the speedup of SparseNCA algorithm. Using number of gene dataset, the experimental result showed that PSparseNCA outperformed SparseNCA algorithm runtime. The measured speedup factor reached 12.33. For future work, we aim to enhance the performance of another algorithm of gene regulatory network algorithms and compare between different gene regulatory network techniques categories.

References

1. Velculescu, V.E., Zhang, L., Vogelstein, B., Kinzler, K.W.: Serial analysis of gene expression. Science **270**(5235), 484–487 (1995)
2. Isea, R.: The present-day meaning of the word bioinformatics. Glob. J. Adv. Res. **2**, 70–73 (2015)
3. Nair, A.: Computational biology & bioinformatics - a gentle overview. Commun. Comput. Soc. India **30**(1), 7–12 (2007)
4. Cosmides, L., Tooby, J.: From Function to Structure: The Role of Evolutionary Biology and Computational Theories in Cognitive Neuroscience. The MIT Press (1995)
5. Durbin, R.: Biological Sequence Analysis: Probabilistic Models of Proteins and Nucleic Acids. Cambridge university press (1998)
6. Kelley, L.A., MacCallum, R.M., Sternberg, M.J.: Enhanced genome annotation using structural profiles in the program 3D-PSSM. J. Mol. Biol. **299**(2), 501–522 (2000)
7. Ghaemmaghami, S., Huh, W.-K., Bower, K., Howson, R.W., Belle, A., Dephoure, N., O'Shea, E.K., Weissman, J.S.: Global analysis of protein expression in yeast. Nature **425** (6959), 737–741 (2003)
8. Dominguez, C., Boelens, R., Bonvin, A.M.: HADDOCK: a protein-protein docking approach based on biochemical or biophysical information. J. Am. Chem. Soc. **125**(7), 1731–1737 (2003)
9. Janssen, P.J., Jones, W.A., Jones, D.T., Woods, D.R.: Molecular analysis and regulation of the glnA gene of the gram-positive anaerobe Clostridium acetobutylicum. J. Bacteriol. **170** (1), 400–408 (1988)
10. Berrozpe, G., Schaeffer, J., Peinado, M.A., Real, F.X., Perucho, M.: Comparative analysis of mutations in the p53 and K-ras genes in pancreatic cancer. Int. J. Cancer **58**(2), 185–191 (1994)
11. Shortle, D.: Prediction of protein structure. Curr. Biol. **10**(2), 49–51 (2000)
12. Rubin, G.M., Yandell, M.D., Wortman, J.R., Gabor, G.L., Nelson, C.R., Hariharan, I.K., Fortini, M.E., Li, P.W., Apweiler, R., Fleischmann, W.: Comparative genomics of the eukaryotes. Science **287**(5461), 2204–2215 (2000)
13. Dowsey, A.W.: High-throughput image analysis for proteomics, Citeseer (2005)
14. Haefner, J.W.: Modeling Biological Systems: Principles and Applications. Springer Science (2005)
15. Churchill, G.A.: Fundamentals of experimental design for cDNA microarrays. Nat. Genet. **32**(1), 490–495 (2002)
16. Culf, A., Cuperlovic-Culf, M., Ouellette, R.: Carbohydrate microarrays: survey of fabrication techniques. OMICS: J. Integr. Biol. **10**(3), 289–310 (2006)
17. Gasch, A., Spellman, P., Kao, C., Carmel-Harel, O., Eisen, M., Storz, G., Botstein, D., Brown, P.: Genomic expression programs in the response of yeast cells to environmental changes. Mol. Biol. Cell **11**(12), 4241–4257 (2000)
18. Schena, M., Shalon, D., Davis, R., Brown, P.: Quantitative monitoring of gene expression patterns with a complementary DNA microarray, in Science, Washington (1995)
19. Yang, Y., Choi, J., Choi, K., Pierce, M., Gannon, D., Kim, S.: BioVLAB-microarray: microarray data analysis in virtual environment. In: IEEE Fourth International Conference on eScience (2008)
20. Aluru, S.: Handbook of Computational Molecular Biology. CRC Press (2006)
21. Jostins, L., Jaeger, J.: Reverse engineering a gene network using an asynchronous parallel evolution strategy. BMC Syst. Biol. **4**(1), 17–33 (2010)

22. Klinger, B., Bluthgen, N.: Reverse engineering gene regulatory networks by modular response analysis-a benchmark. Essays Biochem. **62**(4), 535–547 (2018)
23. Perkins, M., Daniels, K.: Visualizing dynamic gene interactions to reverse engineer gene regulatory networks using topological data analysis. In: 2017 21st International Conference on Information Visualisation (IV) (2017)
24. Liu, Z.-P.: Reverse engineering of genome-wide gene regulatory networks from gene expression data. Curr. Genomics **16**(1), 3–22 (2015)
25. de Souza, M.C., Higa, C.H.A.: Reverse engineering of gene regulatory networks combining dynamic bayesian networks and prior biological knowledge. In: International Conference on Computational Science and Its Applications (2018)
26. Villaverde, A.F., Banga, J.R.: Reverse engineering and identification in systems biology: strategies, perspectives and challenges, vol. 11, no. 91 (2014)
27. Pirgazi, J., Khanteymoori, A.R.: A robust gene regulatory network inference method base on Kalman filter and linear regression. PLoS ONE **13**(7), e0200094 (2018)
28. Lam, K.Y., Westrick, Z.M., Muller, C.L., Christiaen, L., Bonneau, R.: Fused regression for multi-source gene regulatory network inference. PLoS Comput. Biol. **12**(12), e1005157 (2016)
29. Omranian, N., Eloundou-Mbebi, J.M.O., Mueller-Roeber, B., Nikoloski, Z.: Gene regulatory network inference using fused LASSO on multiple data sets. Scientific Reports **6**, 20533 (2016)
30. Guerrier, S., Mili, N., Molinari, R., Orso, S., Avella-Medina, M., Ma, Y.: A predictive based regression algorithm for gene network selection. Front. Genet. **7**, 97 (2016)
31. Gregoretti, F., Belcastro, V., Di Bernardo, D., Oliva, G.: A parallel implementation of the network identification by multiple regression (NIR) algorithm to reverse-engineer regulatory gene networks. PLoS ONE **5**(4), e10179–e10183 (2010)
32. Sales, G., Romualdi, C.: Parmigene—a parallel R package for mutual information estimation and gene network reconstruction. Bioinformatics **27**(13), 1876–1877 (2011)
33. Shi, H., Schmidt, B., Liu, W., Muller-Wittig, W.: Parallel mutual information estimation for inferring gene regulatory networks on GPUs. BMC Res. Notes **4**(1), 189–194 (2011)
34. Zhang, X., Zhao, X.-M., He, K., Lu, L., Cao, Y., Liu, J., Hao, J.-K., Liu, Z.-P., Chen, L.: Inferring gene regulatory networks from gene expression data by path consistency algorithm based on conditional mutual information. Bioinformatics **28**(1), 98–104 (2011)
35. Meyer, P.E., Lafitte, F., Bontempi, G.: Minet: AR/Bioconductor package for inferring large transcriptional networks using mutual information. BMC Bioinform. **9**(1), 461 (2008)
36. Lachmann, A., Giorgi, F.M., Lopez, G., Califano, A.: ARACNe-AP: gene network reverse engineering through adaptive partitioning inference of mutual information. Bioinformatics **32**(14), 2233–2235 (2016)
37. Barman, S., Kwon, Y.-K.: A novel mutual information-based Boolean network inference method from time-series gene expression data. PloS one **12**(2) (2017)
38. Raychaudhuri, S., Stuart, J.M., Altman, R.B.: Principal components analysis to sum-marize microarray experiments: application to sporulation time series. In: Pacific Symposium on Biocomputing, NIH Public Access, pp. 455–466 (2000)
39. Holter, N.S., Mitra, M., Maritan, A., Cieplak, M., Banavar, J.R., Fedoroff, N.V.: Fundamental patterns underlying gene expression profiles: simplicity from complexity. Proc. Nat. Acad. Sci. **97**(15), 8409–8414 (2000)
40. Hyvarinen, A., Karhunen, J., Oja, E.: Independent Component Analysis. John Wiley & Sons (2001)
41. Aapo, H.: Fast and robust fixed-point algorithms for independent component analysis. IEEE Trans. Neural Netw. **10**(3), 626–634 (1999)

42. Liebermeister, W.: Linear modes of gene expression determined by independent component analysis. Bioinformatics **18**(1), 51–60 (2002)
43. Liao, J.C., Boscolo, R., Yang, Y.-L., Tran, L.M., Sabatti, C., Roychowdhury, V.P.: Network component analysis: reconstruction of regulatory signals in biological systems. In: Proceedings of the National Academy of Sciences (2003)
44. Chang, C., Ding, Z., Hung, Y.S., Fung, P.C.W.: Fast network component analysis (FastNCA) for gene regulatory network reconstruction from microarray data. Bioinformatics **24**(11), 1349–1358 (2008)
45. Jayavelu, N.D., Aasgaard, L.S., Bar, N.: Iterative sub-network component analysis enables reconstruction of large scale genetic networks. BMC Bioinform. **16**(1), 366 (2015)
46. Shi, Q., Zhang, C., Guo, W., Zeng, T., Lu, L., Jiang, Z., Wang, Z., Liu, J., Chen, L.: Local network component analysis for quantifying transcription factor activities. Methods **124**, 25–35 (2017)
47. Noor, A., Ahmad, A., Serpedin, E., Nounou, M., Nounou, H.: ROBNCA: robust network component analysis for recovering transcription factor activities. Bioinformatics **29**(19), 2410 (2013)
48. Noor, A., Ahmad, A., Serpedin, E.: SparseNCA: sparse network component analysis for recovering transcription factor activities with incomplete prior information. IEEE/ACM Trans. Comput. Biol. Bioinform. **15**(2), 387–395 (2018)
49. Latchman, D.S.: Transcription factors: an overview. Int. J. Biochem. Cell Biol. **29**(12), 1305–1312 (1997)
50. Eisen, M.B., Spellman, P.T., Brown, P.O., Botstein, D.: Cluster analysis and display of genome-wide expression patterns. Proc. Nat. Acad. Sci. **95**(25), 14863–14868 (1998)
51. Zhu, X., Gerstein, M., Snyder, M.: Getting connected: analysis and principles of biological networks. Genes Develop. **21**(9), 1010–1024 (2007)

Influence Ranking Model for Social Networks Users

Nouran Ayman$^{(\boxtimes)}$, Tarek F. Gharib, Mohamed Hamdy, and Yasmine Afify

Information Systems Department, Faculty of Computer and Information Sciences,
Ain Shams University, Cairo 11566, Egypt
{nouran_ayman,tfgharib,m.hamdy,yasmine.afify}@cis.asu.edu.eg

Abstract. Microblogging is the most important feature for Social Networks (SN) nowadays. It allows users to interact together by sharing and posting contents. The concept of spreading content between users raises an important question: Who are the users responsible for this content? In other words, the detection of content spreaders becomes one of the most important analytic issues. The common belief is that the best content spreaders are the best connected users (the most central users within network). Specifically, k-shell decomposition methodology defines the most efficient content spreaders as those located within the core of the network. In this paper, influence ranking model (IRM) is presented to rank SN users based on their contribution in spreading a specific content. The proposed model is inspired by the pruning process of the powerful k-shell decomposition methodology. IRM has been evaluated in realistic experiments using the famous datasets of Advogato trust network and Bitcoin Alpha trust weighted signed network. The proposed model was assessed in terms of distinction of nodes ranking and dissemination capability.

Results have shown that IRM has promising results in SN users ranking.

Keywords: Influential users · Microblogging · Social Networks · K-shell

1 Introduction

Social Network (SN) became one of the most noticeable traits for this era. Users around the world exploited it as a platform where they can share their ideas, find other users with the same interests and advertise their products [1].

As a result of the extensive growth of SN and microblogging services, SN became a very fruitful environment for many applications such as viral marketing and advertisement, monitoring people's opinions and traits, social psychology analysis and communities' discovery. Such applications depend mainly on users' interactions and users' effect on each other. Such effect appears as a change in users behavior or opinion by spreading content and interacting with other users content that matches their interest. So, organizations -such as marketing organizations- tend to discover influential microblog users who can greatly

© Springer Nature Switzerland AG 2020
A. E. Hassanien et al. (Eds.): AMLTA 2019, AISC 921, pp. 928–937, 2020.
https://doi.org/10.1007/978-3-030-14118-9_91

promote their products so that the advertising investment can reap its highest benefits. This problem is known as "influence maximization" [2,3]. SN users with the highest ability to spread opinions across SN resulting in a huge change in their behavior and attitude can be defined as influential users. Moreover, influential users have the highest contributive role in content diffusion in SN.

The analysis of SN structure and different users' behavior helps in ranking users. This kind of ranking can be utilized to determine the roles played by users in content diffusion [2,4].

K-shell methodology is considered one of the robust network decomposition methodologies. It identifies influential users based on their centric position in SN. This centric position emphasis their role in spreading content through network [5,6]. Therefore, this work is motivated to investigate the suitability of applying k-shell decomposition methodology to SN to achieve users ranking goal.

This paper aims to introduce a new model that coincides the nature of SN as well as its ability of rank users based on their role to disseminate content. The driving factor behind is to exploit users' interactions, content spread.

The remainder of this paper is structured as follows. Section 2 presents the surveyed papers. Section 3 presents some important preliminaries for the proposed model. Section 4 presents a detailed explanation for proposed model. Section 5 presents the experiments and Sect. 6 includes conclusion and future work.

2   Related Work

SNs are usually represented as a graph, where nodes correspond to SN users and edges correspond to relationships between users or their interactions. In this section, an overview is presented on approaches related to our work.

The Community Scale-Sensitive Maxdegree (CSSM) algorithm was introduced. It aimed to put a new definition for the centrality metric of nodes in a graph. It assumed that the node centrality depends on the centrality of all its neighboring nodes besides the node degree. Despite the appealing results of CSSM, but misleading cases can take place easily as a result of spammers and fake friends [7]. In order to tackle this problem, a new model was proposed that compared between different properties of node, K-Shell and Community centrality model (KSC) that took in consideration both the internal and external properties of the node. Experiments using Susceptible-Infected-Recovered (SIR) model showed that the best internal property used was k-shell property. In spite of the proved fact that k-shell property has the best influence indication, k-shell depends on unweighted graphs, which means all the graph edges are treated equally [8].

To solve this challenge, a new approach was proposed in with some improvements to traditional k-shell to adapt to weighted graphs. It used the sum of degree of the 2 vertices of the edge as the weight for the edge. The basic idea of this weighting method is that an edge becomes more important when its end nodes have greater degree. Accordingly, the importance of each node is calculated using both node degree and weights of incident edges [5].

In [9], a new approach was proposed to detect influential users using a directed graph of users' interactions. First fake friends and spammers were detected then cluster value and link strength metrics were used to detect actual influential users.

In [1], the ACQR framework introduced user activeness, reputation and posts' quality features beside centrality of user in network.

In [10], Xinrank approach integrated the active level metric with Tunrank algorithm to identify influential users. In order to discover users' interest, Naïve Bayes classifier was used to classify users to predefined interests using weight distribution.

Some adjustments were introduced to PageRank in [11] to discover influential users in twitter using retweet count as a weight for edges. In [12], new algorithm was introduced PR4MB (PageRank for Micro-Blog) that integrates PageRank with activity, quality and credibility of user where user influence depend on the previously mentioned metrics and out-degree of graph node.

Based on the conducted survey, it is observed that many approaches are developed to detect influential users in micro-blog networks ranging from: 1. Graphical approaches 2. Non-graphical approaches 3. Hybrid approaches that make use of both graphical and user features models.

First, graphical approaches depend only on the network topology, which ignores the main advantage of micro-blogging service which is user interactions and content dissemination among users. Second, non-graphical approaches depending on both user and content features. Third, hybrid approaches tend to combine graphical features with user and content features. Such combination matches the nature of online social network users and their interactions. Thus, hybrid approaches make a good use of users' interactions and behaviors to enhance the detection of influential users.

Finally, k-shell decomposition methodology is a robust graphical based approach. It targets network decomposition instead of users ranking. Accordingly, it needs some adjustments to match the dynamic nature of SNs and the users ranking process.

3   Dynamic Nature of Social Networks and Users Ranking

The microblogging feature of SN causes numerous interactions between SN users. These interactions vary based on the users interest in the disseminated content. In our opinion, users ranking approaches must take in consideration the dynamic nature of SN and the interest of users in the disseminated content.

As shown in Fig. 1(a), SN users have long time relationships between each other. This kind of relationships do not vary with time (i.e. the followship relationship in Twitter and the friendship in Facebook). Long time rigid relationships between users neither reflect the real degree of influence of each user, nor reflect the role of users in disseminating content across SN. Accordingly, rigid relationships are very misleading in SN users ranking.

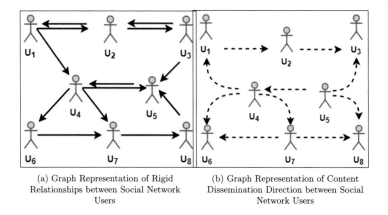

(a) Graph Representation of Rigid Relationships between Social Network Users

(b) Graph Representation of Content Dissemination Direction between Social Network Users

Fig. 1. Rigid relationship vs content dissemination direction between Social Network users

In Fig. 1(b), this kind of dynamic relationship is short-term relationship. It varies with the dynamic interest of users in the disseminated content. Moreover, short-term relationship demonstrates the real role of users in propagating content in SN. Accordingly, they are very beneficial in ranking users.

By comparing the graph representation of in Fig. 1, it is observable that there is a radical change in the graph topology of both representations. Consequently, it is more favorable to use the short-term relationship in users ranking process as it reflects the interactions between users to propagate a specific content in SN.

As shown in Fig. 1(b), the out-degree edges of each user highlight the real contribution of users in disseminating specific content in SN. Accordingly, it should be used in users ranking with respect to a specific content.

3.1   K-Shell Decomposition Methodology

K-shell methodology is a robust method that aims to decompose network into a number of shells. It starts with peripheral users until reaching the core users in the network. it depends on a users pruning process where all users with degree $k = 1$ are removed. This process continues until all the remaining users have a degree $k >= 2$. All the pruned users are assigned an integer value $k_s = 1$. This pruning process is repeated until all users are assigned a k_s value [6].

K-shell pruning process can be used to distinguish between both core shells and peripheral shells. Moreover, these shells can include more than one user. Accordingly, this process is used as a users decomposition process not as a users ranking process.

3.2   Mathematical Model

The following mathematical model demonstrates SN representation:

Consider SN as a directed weighted graph $G(N, E)$ where N is set of nodes representing the users in SN and E is set of directed weighted edges representing the interactivities between users in SN.

- $N = \{n_1, n_2, n_3, \ldots, n_k\}$ where k is the number of users in SN.
- $E = \{e_1, e_2, e_3, \ldots, e_m\}$ where m is the number of interactivities between SN users.

$\forall$ edge $e \in E$:

- e is a directed edge between two users (nodes) in SN. The direction of edge e shows the flow to content/interactivity from source to target of edge e.
- e is a weighted edge, where $weight(e)$ indicates the amount of interactivities propagated from source ($source(e)$) to target ($target(e)$) of edge e.

$\forall$ node $n \in N$:

- $OUT(n)$ is set of out-degree edges from n.
- The influence weight (IW) of node n can be demonstrated using two factors, the number of out-degree edges of node and the weight of these edges. Accordingly α is used as a favorable parameter in Eq. 1

$$IW(n) = \alpha \ |OUT(n)| + (1 - \alpha) \sum_{e \in OUT(n)} weight(e) \qquad (1)$$

- The pruning of node n from graph leads to a new set of nodes consequently a new graph structure $G/\{n\}$. This set contains all graph nodes except pruned node n.

4   Proposed Influence Ranking Model

The proposed Influence Ranking Model (IRM) aims to adapt the K-shell decomposition methodology to match the nature of SN with massive user interactions. Uniquely, the IRM introduces usage of weighted and directed graph with K-shell in SN analysis. IRM is inspired by the pruning process of k-shell decomposition methodology. It implements a pruning phase for nodes ranking instead of nodes decomposition into number of shells. IRM consists of three phases (As shown in Fig. 2).

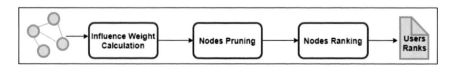

Fig. 2. Proposed IRM pipeline

4.1   Influence Weight Calculation Phase

The influence weight for each node in SN is calculated, which is an indicator of the contributive role played by node to spread content. Accordingly, the out-degree edges of each node are considered as well as their weights using Eq. 1.

4.2   Node Pruning Phase

Each node is pruned and the total graph weight is calculated for all the other nodes using Eq. 2. Node pruning takes place by eliminating only the out-edges of node. These edges reflect the influence of node in SN.

$$T(G/n) = \sum_{i \in G\{n\}} \alpha \; |OUT(i)| + (1-\alpha) \sum_{e \in OUT(i)} weight(e) \times IW(target(e)) \quad (2)$$

4.3   Node Ranking Phase

Nodes are sorted ascendingly based on correspondent graph weight.
The following pseudo code illustrates how IRM ranks SN users.

Algorithm 1. Influence Ranking Model for SN Users Ranking

Input   Directed Network G (N, E) , Edge Weight Table (EWTable) , α
Output Node Influence Weight Table (NIWTable) ,
            Node Pruning Weight Table (NPWTable)

```
1:  for each node n ∈ 𝒩 do
2:      for each edge e ∈ Out − degree(n) do
3:          EdgesWeights += EWTable(e)
4:      end for
5:      NIWTable[n] = α(|Out-degree(n)|)+(1-α)(EdgesWeights)
6:  end for
7:  for each node n ∈ 𝒩 do
8:      for each node x ∈ 𝒩 do
9:          if n != x then
10:             for each edge e ∈ Out − degree(x) do
11:                 EdgesWeights += EWTable(e) × NIWTable[target(e)]
12:             end for
13:             GraphWt += α(EdgesWeights)+(1-α)(|Out-degree(x)|)
14:         end if
15:     end for
16:     NPWTable[n] = GraphWt
17: end for
18: Sort NPWTable ASC
19: return NPWTable, NIWTable
```

Based on the pseudo code, the complexity of IRM is divided into 3 parts:

- First, the complexity of influence weight calculation phase is $O(N)$.
- Second, the complexity of node pruning phase is $O(N^2)$.
- Third, the complexity of ranking nodes phase is $O(NLog(N))$

Although the complexity of IRM is high, parallel execution of IRM is implemented to guarantee scalability.

5   Evaluation, Results and Discussions

With the purpose of evaluating the proposed model, multiple experiments are carried out to assess the performance from different perspectives. IRM model aims to rank nodes based on a pruning process inspired by the powerful k-shell decomposition methodology. Accordingly, the distinction of nodes ranking is evaluated using a new evaluation metric. Moreover, an experiment is conducted to evaluate the effectiveness of ranked node in content dissemination.

5.1   Datasets

In order to evaluate the proposed model, two real datasets are used. First dataset is the Advogato trust network. It consists of 6,541 nodes and 51,127 edges with network density equals 0.0011957 $edges/nodes^2$ [13]. The second dataset is Bitcoin Alpha trust weighted signed network. It consists of 3,783 nodes and 24,186 edges with network density equals 1.169022 $\times 10^{-3} edges/nodes^2$ [14].

5.2   Evaluation Metrics

In order to evaluate the proposed model, two evaluation metrics are used. First, the new Average Shell Load (ASL) metric is introduced to evaluate the uniqueness of nodes ranking in specific number of shells. It compares the number of nodes ranks against number of shells (As shown in Eq. 3). ASL optimal value is 1, which means that each node is ranked in a separate shell.

$$ASL \ (Top \ K \ Shells) \ = \ \frac{\sum_{j=1}^{K} \#nodes \ in \ shell \ j}{K} \tag{3}$$

Furthermore, the effectiveness of ranking is evaluated using the Independent Cascade (IC) model. IC illustrates the concept of information diffusion starting by specific nodes (seed nodes).

5.3   Benchmark Approach

Weighted k-shell decomposition methodology (WKs) is used as a benchmark approach. WKs applies the concept of edge weighting with k-shell decomposition methodology using the same pruning process [5].

5.4   Experimental Results

Experiments are conducted on a PC with intel core i5 processor and 6 GB RAM. Microsoft Visual Studio is used with C# programming language and SQL Server is used for dataset storage. Also, a parallel execution is used to guarantee scalability of IRM model.

First, in order to evaluate the rank uniqueness, the number of nodes in top K shells is evaluated for both IRM and WKs using $\alpha = 0.5$. As shown in Fig. 3, there is a crucial difference between number of nodes in each shell for both IRM and WKs. Moreover, it is observed that in Advogato trust network the number of nodes in top 5 shells reaches up to 11 nodes in case of IRM while it reaches 474 nodes in case of WKs. Furthermore, in Bitcoin Alpha trust weighted signed network, number of nodes is 26 in case of WKs. While in IRM all the top 5 shells contain only 1 node per shell. Accordingly, ASL evaluation metric is applied on all the shells obtained by IRM and WKs respectively to assess the distinction of ranking.

As shown in Table 1, the number of nodes in all shells is evaluated. It is shown that ASL for IRM is on average 1.19 for both networks. On the other hand in case of WKs, the ASL is on average 284.84. Accordingly, The proposed IRM significantly superpasses WKs in uniquely ranking SN nodes.

Table 1. Average Shell Load (ASL)

Dataset	#Shells		ASL	
	IRM	WKS	IRM	WKS
Advogato trust network	3849	12	1.05	140.1
Bitcoin Alpha trust network	3594	27	1.33	429.58

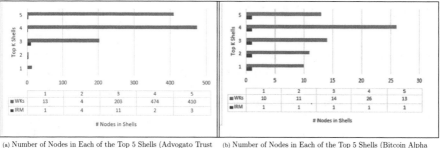

(a) Number of Nodes in Each of the Top 5 Shells (Advogato Trust Network)   (b) Number of Nodes in Each of the Top 5 Shells (Bitcoin Alpha Trust Network)

Fig. 3. Number of nodes in each of the Top 5 shells of IRM vs WKs

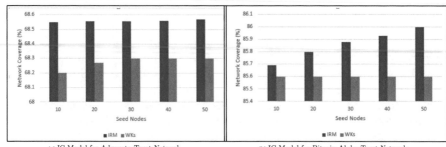

(a) IC Model for Advogato Trust Network   (b) IC Model for Bitcoin Alpha Trust Network

Fig. 4. Network coverage of IRM vs WKs

After the evaluation of ranks uniqueness, the assessment of ranking effectiveness in terms of content dissemination capability is required. Consequently, IC model is used, which assesses the dissemination capability using network coverage. Based on Fig. 4, the network coverage of ranked nodes using IRM on both Advogato trust network and Bitcoin Alpha trust network is 68.59% and 86% respectively. Consequently, this experiment proves the capability of ranked nodes in propagating content through network. Therefore, the IRM can be efficiently employed in applications where minimizing the number of seeds is crucial.

6   Conclusion

Social Network is considered a beneficial platform for individuals to express their opinion, interact with each other and disseminate content. The analysis of these interactivities is very crucial to reveal critical knowledge about SN users. One of the most vital knowledge is the role of users in SN and the ability to disseminate content to others. This role can be used in determination of the user rank in SN, which can be used in many applications such as viral marketing, monitoring public opinion and event prediction. Therefore, the influence rank model (IRM) is proposed. IRM aims to rank SN users based on their interactivities with respect to a specific disseminated content. It applies a pruning process inspired by that of robust k-shell methodology.

IRM is evaluated from different perspectives. First, the distinction of users ranking is assessed using the introduced evaluation metric ASL. It was observed that ASL for IRM in both networks is very close to optimal value 1. This experiment proves the distinction ranking capability of IRM compared to WKs. Moreover, the dissemination capability of the ranked nodes is evaluated for both IRM and WKs using IC model. It was found that the network coverage of IRM is on average 77.29% while network coverage is on average 76.9% in case of WKs. Accordingly, network coverage of IRM superpasses that of WKs. All aspects of evaluation reflect the significant capability of IRM. This proves that IRM can be effectively employed in many applications that require SN users ranking.

References

1. Chai, W., et al.: ACQR: a novel framework to identify and predict influential users in micro-blogging. In: PACIS (2013)
2. Sadri, A.M., Hasan, S., Ukkusuri, S.V., Lopez, J.E.S.: Analysis of social interaction network properties and growth on Twitter. Soc. Netw. Anal. Min. **8**(1), 56 (2018)
3. Tabassum, S., Pereira, F.S.F., Fernandes, S., Gama, J.: Social network analysis: an overview. Wiley Interdisciplinary Reviews: Data Min. Knowl. Disc. **8**(5), e1256 (2018)
4. Chen, W., Teng, S.-H.: Interplay between social influence and network centrality: a comparative study on Shapley centrality and single-node-influence centrality. In: 26th International Conference on World Wide Web. WWW Conferences Steering Committee (2017)
5. Wei, B., Liu, J., Wei, D., Gao, C., Deng, Y.: Weighted k-shell decomposition for complex networks based on potential edge weights. Physica A **420**, 277–283 (2015)
6. Kitsak, M., et al.: Identification of influential spreaders in complex networks. Nat. Phys. **6**(11), 888 (2010)
7. Hao, F., Chen, M., Zhu, C., Guizani, M.: Discovering influential users in micro-blog marketing with influence maximization mechanism. In: GLOBECOM. IEEE (2012)
8. Hu, Q., Gao, Y., Ma, P., Yin, Y., Zhang, Y., Xing, C.: A new approach to identify influential spreaders in complex networks. In: International Conference on Web-Age Information Management. Springer (2013)
9. Rad, A.A., Benyoucef, M.: Towards detecting influential users in social networks. In: International Conference on E-technologies. Springer (2011)
10. Wu, X., Wang, J.: Micro-blog in China: identify influential users and automatically classify posts on Sina micro-blog. J. Ambient Intell. Humanized Comput. **5**(1), 51–63 (2014)
11. Nafis, M.T., et al.: To find influential's in Twitter based on information propagation. Int. J. Comput. Appl. (2015)
12. Mao, G.-J., Zhang, J.: A PageRank-based mining algorithm for user influences on micro-blogs. In: PACIS (2016)
13. Massa, P., Salvetti, M., Tomasoni, D.: Bowling alone and trust decline in social network sites. In: DASC 2009. IEEE (2009)
14. Kumar, S., Spezzano, F., Subrahmanian, V.S., Faloutsos, C.: Edge weight prediction in weighted signed networks. In: ICDM. IEEE (2016)

Turning Caregivers into Informed Agents as a Strategy to Disseminate Scientific Information About Cancer

Ali Ruiz Coronel[1(✉)] and Fernando Ramirez Alatriste[2]

[1] National Council on Science and Technology Assigned to the National Cancer Institute, 14080 Mexico City, Mexico
aliruiz@conacyt.mx
[2] Department of Complexity Science, Autonomous University of Mexico, 03100 Mexico City, Mexico
fernando.ramirez@uacm.edu.mx

Abstract. This paper reports on a case study at the National Cancer Institute (INCan), Mexico. Applied research in complex networks was conducted in order to tackle the problem of how to spread scientific information on cancer among economically and educationally disadvantaged social groups. The proposed model is based on a generalization of Axelrod's Cultural Dissemination Model, with a bounded opinion formation, public authority model and a target culture. Informed agents were introduced in a social network in order to influence the culture in the network towards the scientific target culture, we analyze this dynamics within four different network topologies. We found that turning caregivers into informed agents is an actionable strategy highly dependent on the network's topology. KEYWORDS: informed agents, information dissemination, applied complex networks, cancer.

Keywords: Agent based model · Complex network · Culture dissemination

1 Introduction

The National Institute of Cancer (INCan) was founded in 1946 with the explicit aim of attending the needy cancer patients who had no access to any other health service provider, and so has been. According to the socioeconomic data gathered by the INCan in 2016, from 5167 patients, 2701 reported a usual total family income of one and one and a half minimum wages, this is 88.36 Mexican pesos, 4.6 U.S. dollars, more less. Yet, this amount was even less at that moment, as many of them were unable to work due to the cancer symptoms. The low literacy level of the patients, is exhibited in the fact that 68.7 of those who declared being employed, performed unqualified informal jobs that demand physical effort. Actually, 9.75 of the patients in 2016, could neither read nor write and 33.55 of them, had only attended primary school. The growing body

© Springer Nature Switzerland AG 2020
A. E. Hassanien et al. (Eds.): AMLTA 2019, AISC 921, pp. 938–944, 2020.
https://doi.org/10.1007/978-3-030-14118-9_92

of research exploring the relationship between literacy, poverty, health status and health outcomes, has led to the formation of a field of study that brings together research and practice from diverse fields referred to as *health literacy* [1,2].

After a review of 3,015 titles and abstracts and 684 full articles on health literacy, De Walt et al. [3] concluded that there is enough scientific evidence to sustain that patients with low literacy, have poorer health outcomes. Although poverty, literacy and health are related in multiple and complex ways, it is a fact that in order to make decisions, people rely on the health information that is available to them and that they can understand. Unfortunately, many times the information on cancer that is available to low literate people and that is presented to them in a more understandable way, is inaccurate, insufficient or just patently false [4]. This sort of information, has the characteristics of a rumor: it spreads through low degree users, contains negation and speculation, and lacks verification [5]. Mis-information on cancer can be tremendously dangerous to patients. Through qualitative research conducted at the INCan we sadly endorse that mis-information often leads to a delay on cancer diagnosis and treatment, and detachment or premature cessation of the medical treatment, reducing the patient's possibilities of recovery and increasing the Institute's burden. Thus, the communication of scientific information on cancer between health professionals, patients, caregivers and other potential future stakeholders is vital, yet, the most adequate means to doing so remain a challenge [6,7].

2   Turning Caregivers into Informed Agents

At the INCan, each patient is obliged to be accompanied by a "caregiver". In 2016, 5167 oncological patients were treated and 211100 medical consultations were offered. So, caregivers compose a large population that has to come to the hospital periodically. As the patient relies on his (her) caregiver, the latter is charged with enforcing an active role. Caregivers are exposed to previously unknown health information that they must understand and process rapidly. This information is transmitted by health care professionals, whose language is specialized and whose sociocultural background is quite different from theirs, so they have to develop skills to sustain a fruitful communication. Under these stressing circumstances they have to do a lot of decision-making knowing that any decision may compromise the patient's health and even his (her) life expectancy. Also, they have to master the formal and informal rules to access the services provided in the hospital and other attached facilities like hostels and soup kitchens. This development of personal knowledge and capability fostered by emotional motivation, is a personal asset recognized by other members of the network, so they approach the caregivers seeking for trustable information on cancer.

In a landmark publication on medical anthropology, Kleinman [8] differentiated between *disease, sickness* and *illness*. For him, disease is a strictly biomedical disorder. Sickness is a concept that comprises all the sociocultural aspects related to the disease. Illness is the comprehensive experience of symptoms and problems as felt by the patient and family and the personal meanings developed to cope with the disease. At the INCan, health-care practitioners understand and develop information on cancer as a disease. But because of the huge sociocultural chasm, they know little of their patients' experience of cancer as sickness and as illness. Caregivers on the other hand, owe first hand information on cancer as sickness and illness and are typically longing to get information of it as a disease. If they were provided with it, they would become informed agents as they would bridge information of cancer as a disease, as sickness and as illness and would be able to transfer it to the rest of the agents in their network.

At the INCan, while navigating the health system, caregivers are exposed to previously unknown health information which they must assimilate and use to the benefit of the patient. This development of their health literacy is acknowledged by other members of their network. Information is known to spread more easily within communities or groups whose members are in close physical proximity and linked with social ties that depict similar social status [9]. Face to face communications, increase the strength of persuasiveness and supportiveness and allow information to transit via decentralized horizontal dynamics [10]. This is a reason why health-care professionals are less influential than caregivers in spreading information among economically and educationally disadvantaged people -they are physically an socially too far apart. Caregivers, on the other hand, belong to the same social network, share their socio-cultural background and are easily reachable. We argue that if they were provided with scientific information, they could translate it into understandable and meaningful information and spread it widely among the INCan's potential future stakeholders.

3   The Model

In the literature of agent-based models, an *informed agent* is one that has more information than the rest of the agents in the network [11–13]. In our model, they also have *public authority*, which is the capability to influence their neighbors in opinion exchange because their expertise and knowledge are publicly acknowledged [14]. In this case, the public authority k_i of agent i is taken to be its degree.

Our model is an extension of Robert Axelrod's Model of Dissemination of Culture [15]. Instead of a discrete culture, we propose a continuous culture and two different types of agents interacting in a social network. The culture of the agent i is a vector $\in \mathbb{R}^F$

$$x_i(t) = [x_{i,1}, \ldots, x_{i,F}], \tag{1}$$

where F is the number of attributes of the culture $x_{i,s} \in (-1, 1)$, the negative values correspond to the myths ant the positive correspond to the scientific knowledge.

Informed agents influence their neighbors and induce them to change their opinion towards a preset target culture x^*. In order to do so, the culture $x_i^{inf}(t)$

of an informed agent i must be close to the culture of his neighborhood $N_i(t)$, so that each attribute s of the culture $x_{i,s}^{inf}$ of an informed agent i is proportional to the corresponding attribute s of the culture of his neighbor

$$\overline{x}_{i,s}(t) = |N_i|^{-1} \sum_{j \in N_i} x_{j,s}(t) \tag{2}$$

with a bias toward a target culture x^*

$$x_i^{inf}(t) = \overline{x}_i(t) + \mu(x^* - \overline{x}_i(t)) \tag{3}$$

where μ is the opinion convergence parameter.

At each time step, the interaction of the regular agent i with one random neighbor j takes place with probability $P_{i,j}(t)$, which is defined as the fraction of the overlapping features $O_{i,j}(t)$ that comply with the bounded confidence limit h at time t

$$|x_{i,s}(t) - x_{j,s}(t)| < h, \tag{4}$$

i.e.

$$P_{i,j}(t) = \frac{O_{i,j}(t)}{F} \tag{5}$$

when two agents interact, they update their cultures [12]

$$x_i(t+1) = x_i(t) + \mu \frac{S_j}{S_i + S_j}(x_j(t) - x_i(t)) \tag{6}$$

$$x_j(t+1) = x_j(t) + \mu \frac{S_i}{S_i + S_j}(x_i(t) - x_j(t)) \tag{7}$$

where

$$S_i(t) = (k_i)^\alpha \tag{8}$$

Because the public authority k_i is the degree of the agent i, its distribution depends on the topology of the network. We analyzed four different network topologies: scale-free, random, small-world and spatially clustered networks.

4 Results

We carried out a sensitivity analysis of the model, here we only report the simulations of the model in the software Netlogo [16] for 100 agents. The initial conditions of the model were 10% of informed agents and 100% of regular agents with negative culture, the number of features per culture $F = 5$, the bounded confidence $h = 0.15$, the exponent in the social power (8) $\alpha = 1$, and the opinion convergence parameter $\mu = 0.1$.

Figure 1 shows the different network topologies that were analyzed. The blue scale agents have positive culture, the red scale agents have negative cultures, and the green ones are the informed agents, the size of the agent is proportional to its public authority (8). (a) shows the small world network with rewiring probability

a)   b)

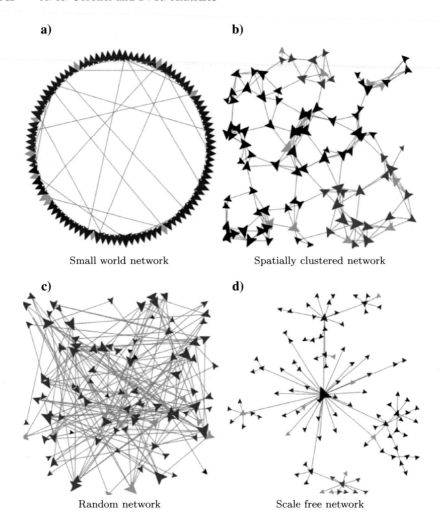

Small world network   Spatially clustered network

c)   d)

Random network   Scale free network

Fig. 1. In all the four topologies, blue scale agents have a positive culture, red scale agents have a negative culture, green agents are the informed agents, the size of the agent is proportional to its public authority

of $p^{sw} = 0.1$; (b) shows the spatially clustered network with mean degree $<k> = 5$; (c) shows the random network with connection probability $p^{rnd} = 0.03$ and (d) shows a scale-free network generated with the preferential attachment algorithm. Figure 2 shows the density of regular agents with positive culture for the four network topologies as a function of time. We found that the density and the time to reach a stable state depends of the topology of the network. Our results are preliminary, yet the model yields that for being most efficient, this strategy has to be part of a wider one that assures a constant, close and long lasting relationship between the health professionals and the caregivers to keep them being informed

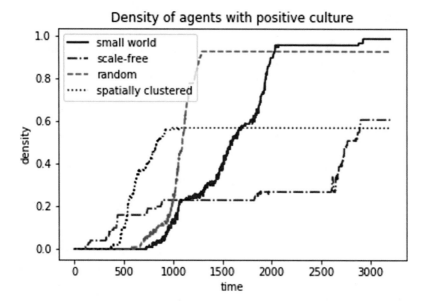

Fig. 2. Density of regular agents with positive culture versus time, we show the evolution for four different network topologies

agents. Real-world applications of networked systems require a tremendous effort of an interdisciplinary team to characterize, model and analyze the networks in order to better understand the impact of their structure and dynamics in the real-world system.

5   Conclusion

The model supports the contention that turning caregivers into informed agents is a feasible strategy for the communication and spread of useful scientific information on cancer within economically and educationally disadvantaged social groups while also providing scientific guidance on how to achieve this goal. Real-world applications of networked systems may contribute to a more effective policy implementation by integrating the knowledge of interdisciplinary teams of scientists; by providing means to measure the impact of a policy; and to constantly adjust its implementation in order to obtain better results. More generally, it emphasizes that positive change in public health requires the engagement and participation of non-medical communities.

Acknowledgements. ARC acknowledges support from the National Council on Science and Technology, through the Program Catedras Conacyt 479. FRA thanks the Centre for Complexity Science (C3), UNAM for hosting him as visiting researcher.

References

1. Berkman, N.D., Davis, T.C., McCormack, L.: Health literacy: what is it? J. Health Commun. **15**, 9–19 (2010)
2. Nutbeam, D.: The evolving concept of health literacy. Soc. Sci. Med. **67**, 2072–2078 (2008)
3. De Walt, D.A., Berkman, N.D., Sheridan, S., Lohr, K.N., Pignone, M.P.: Literacy and health outcomes. J. Gen. Intern. Med. **19**, 1228–1239 (2004)
4. Black, P., Penson, D.: Prostate cancer on the internet: information or misinformation? J. Urol. **175**, 1836–1842 (2006)
5. Kwon, S., Cha, M.: modeling bursty temporal pattern of rumors. In: Proceedings of the Eighth International Conference, pp. 650–651. Association for the Advancement of Artificial Intelligence (2014)
6. Sheridan, S., Halpern, D., Viera, A., Berkman, N., Donahue, K., Crotty, K.: Interventions for individuals with low health literacy: a systematic review. J. Health Commun. **16**, 30–54 (2011)
7. Easton, P., Entwistle, V., Williams, B.: Health in the "hidden population" of people with low literacy. A systematic review of the literature. Public Health **10**, 459 (2010)
8. Kleinman, A.: The Illness Narratives: Suffering, Healing, and the Human Condition. Basic Books, New York (1988)
9. Herriman, N.: The great rumor mill: gossip, mass media, and the ninja fear. J. Asian Stud. **69**(3), 723–748 (2010)
10. Kang, C., Kraus, S., Molinaro, C., Spezzano, F., Subrahmanian, V.S.: Diffusion centrality: a paradigm to maximize spread in social networks. In: Advances in Social Networks Analysis and Mining, Conference Paper, pp. 558–568 (2012)
11. Afshar, M., Asadpour, M.: Opinion formation by informed agents. J. Artif. Soc. Soc. Simul. **13**(4), 5 (2010)
12. Askarisichani, O., Jalili, M.: Influence maximization of informed agents in social networks. Appl. Math. Comput. **254**, 229–239 (2015)
13. Fan, K., Pedrycz, W.: Opinion evolution influenced by informed agents. Physica A **06**, 01–11 (2016)
14. Chen, X., Xiong, X., Zhang, M., Li, W.: Public authority control strategy for opinion evolution in social networks. Chaos **26**, 083105 (2016)
15. Axelrod, R.: The dissemination of culture: a model with local convergence and global polarization. J. Conflict Resolut. **41**(2), 203–226 (1997)
16. Wilensky, U.: Netlogo 6.04 center for connected learning and computer-based modeling. Northwestern University, Evanston. http://ccl.northwestern.edu/netlogo

Comparative Analysis of Unmixing Algorithms Using Synthetic Hyperspectral Data

Menna M. Elkholy[1]([✉]), Marwa Mostafa[2]([✉]) [iD],
Hala M. Ebeid[1]([✉]) [iD], and Mohamed F. Tolba[1]([✉]) [iD]

[1] Faculty of Computer and Information Sciences,
Ain Shams University, Cairo, Egypt
{menna.elkholy,halam,fahmytolba}@cis.asu.edu.eg
[2] Data Reception, Analysis and Receiving Station Affairs,
National Authority for Remote Sensing and Space Science, Cairo, Egypt
marwa@narss.sci.eg

Abstract. Hyperspectral imaging (HS) records hundreds of continuous bands for each pixel in an image. Due to coarse spatial resolution of HS, and multiple scattering, the spectral measured by the hyperspectral cameras (HSCs) are mixtures of spectra of materials in each pixel. Thus, at each pixel, a spectral unmixing process is required to utilize an accurate estimation of the number of endmembers, their signatures and their abundances fraction. In this paper, we present a large-scale comparison of endmember extraction algorithms. The algorithms explored Vertex Component Analysis algorithm (VCA), Minimum Volume Simplex Analysis (MVSA), N-FINDR, Alternating Volume Maximization (AVMAX), Pixel Purity Index (PPI), Simplex Identification Via Split Augmented Lagrange (SISAL). Three categories of experiments were carried out; accuracy assessment, robustness to the noise, and execution time. The performance of algorithms was evaluated using two different metrics (MSE and SAD). We use simulated hyperspectral dataset sampled from USGS library - The experimental results show that MVSA and SISAL demonstrate robust performance to the changes in the size of the scene. PPI had the least performance compared with other algorithms. AVMAX and VCA have almost identical performance.

Keywords: Spectral unmixing · Hyperspectral images ·
Endmember extraction · Simulated dataset

1 Introduction

Hyperspectral imaging is widely used in remote sensing applications such as agriculture, biotechnology, geoscience, object identification, environmental monitoring, and material identification [1, 2]. Hyperspectral imagery is a 3D cube of hundreds of contiguous bands, providing information about different materials on the ground. Due to the tradeoff between spatial and spectral sensors resolutions in the hyperspectral instrument, hyperspectral images suffer from limited spatial resolution compared with multi spectral and panchromatic images. Hyperspectral unmixing is considered as a critical and challenging task to overcome this limitation.

© Springer Nature Switzerland AG 2020
A. E. Hassanien et al. (Eds.): AMLTA 2019, AISC 921, pp. 945–955, 2020.
https://doi.org/10.1007/978-3-030-14118-9_93

Hyperspectral unmixing also called spectral mixture analysis. It aims to separate the pixel spectra in the hyperspectral image into a collection of spectral signatures (endmembers) and a set of their fractional (abundances map).

Depending on the mixing gauge at each pixel, the mixture of constituent materials is either linear or nonlinear [2]. Linear spectral unmixing assumes that the reflectance measured within each pixel is a linear combination of the reflectance of each endmember, weighted by its abundance fraction. In contrast, Nonlinear spectral unmixing holds when the microscopically pure components are intimately mixed inside the pixel and it requires a detailed prior knowledge about the materials. In practice, nonlinear mixtures oftentimes happen in real scenarios, but linear models are the widely used technique to approximate these complex mixtures.

Hyperspectral unmixing is a complex task, which consists of three interdependent steps: dimensionality reduction, endmember determination, and inversion. The first step, dimensionality reduction, is considered an optional step and used by some algorithms to reduce computational load. Principle Component Analysis (PCA), Independent Component analysis (ICA) and Minimum Noise Fraction (MNF) Algorithms are commonly used in this step. The second step is the endmember determination, which mainly estimates the set of distinct spectra (endmembers) in the mixed pixels in the scene. This step is considered the primary and the most critical step in unmixing. Finally, inversion is applied to generate abundance planes in order to estimate the fractional abundances for each mixed pixel from its spectrum and the endmember spectra.

In this paper, we introduce a large-scale comparison between Vertex Component Analysis algorithm (VCA) [11], Minimum Volume Simplex Analysis (MVSA) [3], modified MVSA [4], N-FINDR [12], Alternating Volume Maximization (AVMAX) [13], Pixel Purity Index (PPI) [14], and Simplex Identification Via Split Augmented Lagrange (SISAL) [6]. This paper studies the behavior of the aforementioned endmember extraction method to define the optimal embedding technique used with complex hyperspectral data. Different experiments were conducted using different simulated hyperspectral to assess the performance of the seven algorithms in terms of accuracy assessment, robustness to the noise, and execution time.

The remainder of this paper is organized as follows: Sect. 2 reviews the related work on hyperspectral unmixing. Section 3 describes the methodologies used to assess the unmixing algorithms. Section 4 explores the simulated data set generation method. The results are discussed in Sect. 5. Finally, Sect. 6 concludes the paper.

2 Related Work

This section reviews the recent and the famous related work for hyperspectral linear unmixing methods. Linear unmixing techniques classified into statistical, geometrical, and sparse based unmixing. The statistical category is the methods that assumes that spectral unmixing algorithm adopted the Bayesian paradigm to quantify the total behavior of hyperspectral data, used usually with highly mixed data. Nascimento, and Bioucas-Dias [9] used dependent component analysis (DECA) to unmixing. They proposed a method to overcome the limitations of unmixing methods based on Independent Component Analysis (ICA) and on geometrical based approaches.

Geometrical approach works under the assumption that the distribution of data samples is a simplex set whose vertices correspond to the endmembers. The geometrical approaches had two sub-categories: Pure Pixel (PP) and Minimum Volume (MV). Pure pixel-based algorithms assume that the scene contains at least one pure pixel per endmember. Some popular algorithms implemented using this assumption are the simplex growing algorithm (SGA) [5] that iteratively grows a simplex by finding the vertices corresponding to the maximum volume.

If the pure pixel assumption is not fulfilled, the unmixing process is a rather challenging task and it is a more realistic scenario. Some popular algorithms implemented using this assumption are the minimum-volume enclosing simplex (MVES), and the minimum volume simplex analysis (MVSA) algorithms. In [7], a simple heuristic method for approximately solving the optimization problem based on the idea of alternating projected sub-gradient descent was introduced and the experimental results demonstrated superior performance compared with classical methods.

Sparse regression based unmixing methods assume that the observed spectra be expressed as linear combinations of known spectral signatures which can be identified in spectral libraries, which makes them semi-supervised in nature.

Following the success of deep learning methods in computer vision, natural language processing and speech recognition applications, in the last few years, deep learning methods have been successfully empowered in remote sensing applications. In [10], they proposed a novel two-staged neural network auto-encoder that is specialized to extract endmembers with their pixel abundances from hyperspectral data. The main advantage of this method that it is scalable for large-scale data.

3 Methods

In this section, we highlight the mathematical formulation of linear unmixing and summarize the endmember extraction algorithms considered in this work: Vertex Component Analysis algorithm (VCA), Minimum Volume Simplex Analysis (MVSA) [3, 4], N-FINDR, Alternating Volume Maximization (AVMAX), Pixel Purity Index (PPI), and Simplex Identification Via Split Augmented Lagrange (SISAL).

To define unmixing problem in mathematical terms, we assume the linear mixing scenario. The observed spectra are given by

$$y_n = \sum_{j=1}^{M} m_j \alpha_{nj} + n_n, \tag{1}$$

$$s.t. \ \alpha_{nj} \geq 0, \sum_{j=1}^{M} \alpha_{nj} = 1 \tag{2}$$

where α_{nj} is the abundance fraction for the jth endmember at the nth pixel, n is the number of pixels, M is the number of endmembers, n_n is additive noise, and the endmember set $\{m_j : j = 1, \ldots, M\}$. Owing to physical constraints (2), abundance fractions are non-negative and their sum is equal to one.

For each considered method, there are myriads of variations proposed in the literature. Our strategy is to consider the basic version of each method. The rationale is that most users will more likely consider the basic form. The reason for generally selecting these considered seven methods is that they are some of the most commonly used. Below is a short description of the models considered. For more details, we refer readers to the references given for each approach. Vertex Component Analysis algorithm (VCA) [11] iteratively projects data onto a direction perpendicular to the subspace spanned by the endmembers already determined. The new endmember signature corresponds to the extreme of the projection.

Minimum Volume Simplex Analysis (MVSA) fits a minimum volume simplex to the hyperspectral data by constraining the abundance fractions to belong to the probability simplex. The resulting optimization problem (3), which is computationally very complex, is solved in modified MVSA [4] with the interior point algorithm, whereas in MVSA [3], a sequential quadratic programming (SQP) approach was used, and it is the main contribution and difference between the two versions. The modified MVSA optimization greatly reduces the computational complexity of the algorithm and allows for its practical utilization with reasonably large and complex hyperspectral datasets.

The optimization problem simplifies to:

$$Q_b = \arg \max \log |\det(Q)| \tag{3}$$

$$s.t : QY \geq 0, 1_P^T Q = q_p$$

Simplex Identification Via Split Augmented Lagrange (SISAL) [6] reformulates the optimization problem (3) with respect to the inverse of the matrix of estimated endmembers. This optimization problem was solved by a sequence of variable splitting augmented Lagrangian optimizations. The optimization problem simplifies to:

$$Q_b = \arg \min - \log |\det(Q)| \tag{4}$$

Simplex Identification Via Split Augmented (SISAL) [6] aims to find the minimum volume simplex containing the hyperspectral vectors.

N-FINDR [12] finds the set of pixels defining the largest volume whose vertices are specified by purest pixels. NFINDR uses exhaustive search to find endmembers. For an arbitrary set of p data sample vectors $e_1, e_2, \ldots, e_p$ form a p-vertex simplex and define its volume by $V(e_1, e_2, \ldots, e_p)$

$$\left\{ e_1^*, e_2^*, \ldots, e_p^* \right\} = \arg \{ max_{\{e_1, e_2, \ldots, e_p\}} V(e_1, e_2, \ldots, e_p) \} \tag{5}$$

The set of $\left\{ e_1^*, e_2^*, \ldots, e_p^* \right\}$ is the desired set of endmembers needed to be found.

Alternating Volume Maximization (AVMAX) [13] is inspired by N-FINDR in terms of the algorithmic structures. Both algorithms attempt to maximize the simplex volume by some forms of one-at-a-time pixel search. Among the various N-FINDR algorithms, AVMAX is notably similar to the SC-N-FINDR algorithm [11]. The problem in (5) may be handled in a convenient manner by the idea of cofactor expansion and alternating optimization.

The Pixel Purity Index (PPI) [14] generates the maximum number of most pure pixels which were similar to endmembers but does not provide identified endmembers. PPI continues by generating a large number of skewers through the number of wavelength bands and dimensional data. For each skewer, every data point is projected onto the skewer. The data points which correspond to extreme (or near extreme) in the direction of skewer are identified and placed on the list of endmembers.

4 Synthetic Hyperspectral Image

This section presents the methodology to generate a complex simulated hyperspectral data. The spectral signatures used are from the United States Geological Survey (USGS) spectral library released in 2007 [15]. Twenty random spectral signatures are selected, and their signatures are shown in Fig. 1. Second, we use the Dirichlet distribution method [16] to generate the abundance maps, which has the following steps:

1. Divide an image of N pixels into number of endmember (P) regions and randomly fill up each region with a type of ground material
2. To generate the mixed pixels, we use the Dirichlet distribution to replace the pixels whose largest abundance is larger than 0.8 with a mixture of two endmembers with equal abundances

After endmembers $\hat{M}$ and abundances $\hat{A}$ have been generated, and based on them, a hyperspectral image is created as $\hat{Y} = \hat{M}\hat{A}$. To simulate a more real hyperspectral

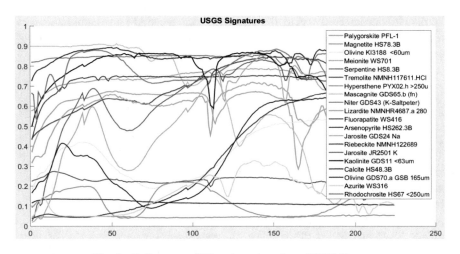

Fig. 1. Reflectance of 20 endmembers from USGS library

image that contains complex noises, we add the zero-mean Gaussian (or some other) noise to the generated image data using Signal-to-Noise Ratio (SNR).

$$SNR = 10 \log_{10} \left\{ \frac{\mathbb{E}[y^t y]}{\mathbb{E}[n^t n]} \right\}. \tag{6}$$

5   Experimental Results

In this section, we conducted four experiment sets to evaluate the performance of the endmember extraction algorithms (Vertex Component Analysis algorithm (VCA), Minimum Volume Simplex Analysis (MVSA), N-FINDR, Alternating Volume Maximization (AVMAX), Pixel Purity Index (PPI), Simplex Identification Via Split Augmented Lagrange (SISAL)) on simulated data. We use spectra obtained from USGS library [15]. The abundance fractions are calculated according to Dirichlet distribution [16].

5.1   Quality Metrics

To evaluate the performance of the aforementioned algorithms, the estimated spectral signatures are compared with true ones. We use two metrics to evaluate the present methods. The first matrix is the Spectral Angle Mapper (SAM) [17] and is defined as:

$$\alpha = \cos^{-1} \left(\frac{\sum_{i=1}^{n} t_i r_i}{\sqrt{\sum_{i=1}^{n} t_i^2} \sqrt{\sum_{i=1}^{n} r_i^2}} \right) \tag{7}$$

Where α is the spectral angle between a true spectrum and a reference spectrum, n is the number of bands, t: true spectrum and r: reference spectrum. We use this method according to its insensitivity to illumination since the SAM uses only the vector direction and not the vector length. The second matrix considered in our experiments is mean square error (MSE), computed by the following formula

$$MSE = \frac{1}{n} \sum_{i=1}^{n} \left(Y_i - \hat{Y}_i \right)^2 \tag{8}$$

5.2   Results

In the first experiment set, we used a hyperspectral simulated scene with size 2000×2000 pixels (N = 2000), and four endmembers (P = 4). We evaluate the robustness to noise of the aforementioned algorithms using different values for SNR. Figure 2 shows that the computed SAM and MSE obtained using different SNR. In terms of SAM and MSE, PPI displays the lowest performance in all cases nearly, whereas MVSA [3] and modified MVSA [4] algorithms have almost identical performances. MVSA [3], modified MVSA [4] and SISAL [6] demonstrated the best accuracy in cases that SNR >= 50. On the other hand, SISAL algorithm suppresses modified MVSA [4], in case of SNR < 50. In case of SNR <= 10, the MVSA [3] algorithm displays the lowest recorded result compared with other algorithms w. N-FINDER displays the best result in the case of SNR <= 10; and hardly affected by change of SNR. VCA slightly changes when SNR is low.

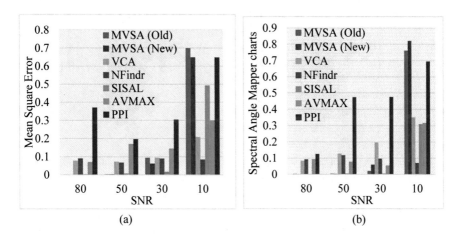

(a)   (b)

Fig. 2. Results of valuate the robustness to noise of the aforementioned algorithms using different SNR values, using (a) Mean Square Error and (b) Spectral Angle Mapper charts.

In the Second experiment set, we used a simulated hyperspectral scene with four endmembers (P = 4) and signal to noise ratio (SNR = 50) to assess the endmember extraction algorithms by varying the scene size N. The number of pixels N ranges from 2000 to 500,000. Results obtained by using the s even comparative algorithms for the considered scene using different number of pixels are shown in Fig. 3. In terms of SAM and MSE, PPI algorithm attains the lowest results in all cases. Modified MVSA [4] and SISAL have the best performance in all cases and their error percentages marginally grow when the number of pixels is large. AVMAX algorithm had just the same accuracy in all cases which indicate that the number of pixels has no effect on the algorithm. N-FINDR and VCA have satisfying accuracy results even in worst case. For illustrative purposes, Fig. 4 shows the first endmember sampled from MVSA modified version, VCA and ground truth from USGS library with N = 20,000.

In the third experiment set, we used a simulated hyperspectral scene with size of 2000 × 2000 pixel (N = 2,000) and added SNR = 50 to utilize the effect of varying the number of endmembers. The number of endmembers P ranges from 4 up to 50. As shown in Fig. 5, MVSA [3], Modified MVSA [4] and SISAL demonstrated better results when the number of endmembers P <= 10. While when the number of end-members P > 10, the percentage of the results significant increase. In case of end-members (P = 50), the MVSA [3], Modified MVSA [4] and SISAL showed the worst results. Using VCA, N-FINDR and PPI, the result percentage slightly raised with the growth of the number of endmembers. In case of endmembers (P = 20), almost all algorithms have the same accuracy. The MVSA [3] grows by large percentage with the

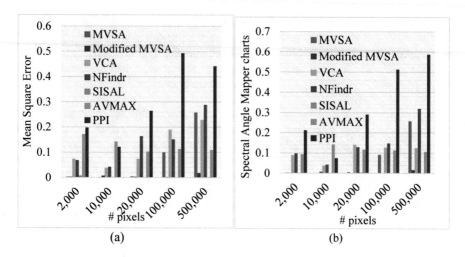

Fig. 3. Study varying the scene size N on the endmember extraction algorithms, using (a) Mean Square Error and (b) Spectral Angle Mapper charts.

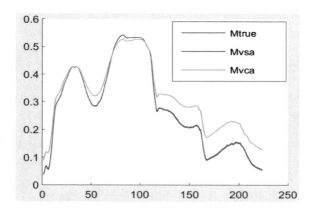

Fig. 4. Using scene size with N = 20,000: first endmember sampled using modified MVSA, VCA and ground truth

growth of endmember. PPI has less accuracy compared with other algorithms. VCA and AVMAX grow by intermediate percentage with growth of endmembers.

In the fourth experiment set, we used a simulated hyperspectral scene that has different pixels size to assess the computation time of each algorithm, as shown in Table 1. The simulated hyperspectral scene has endmembers (P = 4) and SNR = 50. Overall, the change in image size affects the computation time of each algorithm. AVMAX and VCA were slightly affected unlike the MVSA [3]. The computation time increased significantly with the increase in image size. The modified MVSA [4] is affected much faster compared with its older version. Both the modified MVSA and SISAL have almost equal processing time, and they both are better than N-FINDR and PPI.

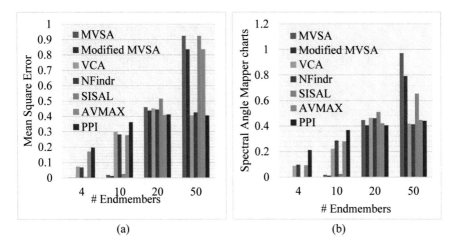

Fig. 5. Results of studying the effect of varying the number of endmembers using simulated hyperspectral scene with 2000 × 2000 pixels and added SNR = 50, using (a) Mean Square Error and (b) Spectral Angle Mapper charts

Table 1. Study of the effect of change of scene size on the run time of each algorithm

Methods	Image size				
	2000	10,000	20,000	100,000	500,000
MVSA	1.92	16.90	41.76	767.69	25422.50
Modified MVSA	0.45	0.98	6.96	11.11	30.79
VCA	0.07	0.14	0.19	0.74	3.52
NFINDR	0.37	1.44	2.95	13.38	69.30
SISAL	0.32	0.80	1.30	8.07	45.55
AVMAX	0.14	0.18	0.23	0.61	2.52
PPI	0.12	1.02	2.23	12.99	62.83

6 Conclusion

Hyperspectral unmixing (HU) is a critical preprocessing step for hyperspectral applications. In this paper, we extensively compare seven algorithms (MVSA, Modified MVSA, VCA, AVMAX, NFINDR, PPI and SISAL). Three experiments sets were carried out and evaluated the aforementioned algorithms in terms of accuracy assessment, robustness to the noise, and execution time. Two different metrics (MSE and SAD) were utilized to measure the performance of the algorithm. Several experiments with the simulated data led to the conclusion that Modified MVSA and SISAL are hardly affected by the increase of the number of pixels but rather affected by the expanding the numbers of the endmember and value of SNR. MVSA is the slowest algorithm in all experiments, unlike Modified MVSA. MVSA is affected by the change of the number of pixels, number of endmember and SNR value. N-FINDR had the best accuracy when SNR has a small value. AVMAX and VCA had almost identical performance, which is better than NFINDR in case of the growth of the number of pixels and number of endmembers. PPI slightly was affected by the change in the number of endmembers but sharply affected by the change in the number of pixels and SNR values. Overall, PPI is considered the worst algorithm in comparison of the other algorithms. In future work, we will test all these algorithms in real datasets and track the changes in accuracy and computational complexity of each algorithm with each dataset having different parameters.

References

1. Bioucas-Dias, J.M., Plaza, A., Dobigeon, N., Parente, M., Gader, P.: Hyperspectral unmixing overview: geometrical, statistical, and sparse regression-based approaches. IEEE J. Sel. Top. Appl. Earth Obs. Remote Sens. 5(2), 354–379 (2012)
2. Yu, J., Chen, D., Lin, Y.: Comparison of linear and nonlinear spectral unmixing approaches: a case study with multispectral TM imagery. Int. J. Remote Sens. 38(3), 773–795 (2016)
3. Li, J., Bioucas-Dias, J.M.: Minimum volume simplex analysis: a fast algorithm to unmix hyperspectral data. In: IEEE International Geoscience and Remote Sensing Symposium, vol. 3, p. III-250 (2008)
4. Li, J., Agathos, A., Zaharie, D., Bioucas Dias, J.M., Plaza, A.: Minimum volume simplex analysis: a fast algorithm for linear hyperspectral unmixing. IEEE Trans. Geosci. Remote Sens. 53(9), 5067–5082 (2015)
5. Chang, C., Wu, C., Liu, W., Ouyang, Y.: A new growing method for simplex-based endmember extraction algorithm. IEEE Trans. Geosci. Remote Sens. 44(10), 2804–2819 (2006)
6. Bioucas-Dias, J.M.: A variable splitting augmented Lagrangian approach to linear spectral unmixing. In: IEEE GRSS Workshop on Hyperspectral Image and Signal Processing-WHISPERS, pp. 1–4 (2009)
7. Wang, F., Chi, C., Chan, T., Wang, Y.: Nonnegative least-correlated component analysis for separation of dependent sources by volume maximization. IEEE Trans. Pattern Anal. Mach. Intell. 32(5), 875–888 (2010)
8. Zymnis, A., Kim, S., Skaf, J.: Hyperspectral image unmixing via alternating projected sub gradients. In: Record of the Forty-First IEEE Asilomar Conference on In Signals, Systems and Computers, pp. 1164–1168 (2007)
9. Nascimento, J.M.P., Bioucas-Dias, J.M.: Hyperspectral unmixing algorithm via dependent component analysis. In: IEEE International in Geoscience and Remote Sensing Symposium, pp. 4033–4036 (2007)
10. Ozkan, S., Kaya, B., Bozdagi Akar, G.: EndNet: sparse autoencoder network for endmember extraction and hyperspectral unmixing. IEEE Trans. Geosci. Remote Sens. 8(99), 1–5 (2018)
11. Nascimento, J., Dias, J.: Vertex component analysis: a fast algorithm to unmix hyperspectral data. IEEE Trans. Geosci. Remote Sens. 43(4), 898–910 (2005)
12. Winter, M.E.: N-FINDR: an algorithm for fast autonomous spectral end-member determination in hyperspectral data. In: SPIE Conference on Imaging Spectrometry, pp. 266–275 (1999)

13. Ambikapathi, A., Chan, T., Ma, W., Chi, C.: A robust alternating volume maximization algorithm for endmember extraction in hyperspectral images. In: 2nd Workshop on Hyperspectral Image and Signal Processing: Evolution in Remote Sensing (2010)
14. Wu, h., Chaudhry, F., Liu, W.: Pixel purity index-based algorithms for endmember extraction from hyperspectral imagery. In: Recent Advances in Hyperspectral Signal and Image Processing, pp. 30–62 (2006)
15. Clark, R.N., Swayze, G.A., Gallagher, A.J., Calvin, W.M.: The U.S. Geological Survey, Digital Spectral Library: Version 1 (0.2 to 3.0um) (1993)
16. Minka, T.P.: Estimating a Dirichlet distribution. Technical report (2000)
17. Yang, C., Everitt, J., Bradford, J.: Yield estimation from hyperspectral imagery using Spectral Angle Mapper (SAM). Trans. ASABE 51(2), 729–737 (2008)

Author Index

© Springer Nature Switzerland AG 2020
A. E. Hassanien et al. (Eds.): AMLTA 2019, AISC 921, pp. 957–960, 2020.
https://doi.org/10.1007/978-3-030-14118-9

Printed in the United States
By Bookmasters